D0385112

Incompressible Flow

Incompressible Flow

RONALD L. PANTON

Mechanical Engineering Department
University of Texas
Austin

A Wiley-Interscience Publication

JOHN WILEY & SONS

New York • Chichester • Brisbane • Toronto • Singapore

Library of Congress Cataloging in Publication Data:

Panton, Ronald L. (Ronald Lee), 1933–
Incompressible flow.
"A Wiley-Interscience publication."
Bibliography: p.
Includes index.
1. Fluid dynamics. I. Title.

TA357.P29 1984 620.1′064 83-23342
ISBN 0-471-89765-5

Printed in the United States of America

10 9 8 7 6 5 4 3 2 1

Preface

This book is written as a textbook for students beginning a serious study of fluid dynamics, or for students in other fields who want to know the main ideas and results in this discipline. A reader who judges the scope of the book by its title will be somewhat surprised at the contents. The contents not only treat incompressible flows themselves, but also give the student an understanding of how incompressible flows are related to the general compressible case. For example, one cannot appreciate how energy interactions occur in incompressible flows without first understanding the most general interaction mechanisms. I subscribe to the philosophy that advanced students should study the structure of a subject as well as its techniques and results. The beginning chapters are devoted to building the concepts and physics for a general, compressible, viscous fluid flow. These chapters taken by themselves constitute the fundamentals that one might study in any course concerning fluid dynamics. Beginning with Chapter 6 our study is restricted to fluids that obey Newton's viscosity law. Only when we arrive at Chapter 10 do we find a detailed discussion of the assumptions that underlie the subject of incompressible flow. Thus, roughly half the book is fundamentals, and the rest is incompressible flow.

Applied mathematicians have contributed greatly to the study of fluid mechanics, and there is a tendency to make a text into a sampler of known mathematical solutions. A conscious effort was made in writing the book to strike an even balance among physics, mathematics, and practical engineering information. The student is assumed to have had calculus and differential equations; the text then takes on the task of introducing tensor analysis in index notation, as well as various special methods of solving differential equations that have been developed in fluid mechanics. This includes an introduction to several computer methods and the method of asymptotic expansions.

The book places heavy emphasis on dimensional analysis, both as a subject in itself and as an instrument in any analysis of flow problems. The advanced worker knows many shortcuts in this area, but the student needs to study the foundations and details in order to be convinced that these shortcuts are valid. Vorticity, vortex lines, and the dynamics of vorticity also receive an expanded treatment, which is designed to bring the serious student more information than is customary in a textbook. It is apparent that advanced workers in fluid mechanics must be able to interpret flow patterns in terms of vorticity as well as in the traditional terms of forces and energy.

The study of how changes in the Reynolds number influence flow patterns occupies a large part of the book. Separate chapters describe flows at low, moderate, and high Reynolds numbers. Because of their practical importance, the complementary subjects of inviscid flows and boundary-layer flows are treated extensively. Introductory chapters on stability and turbulence are also given. These last two subjects are so large as to constitute separate fields. Nevertheless, a beginning student should have an overview of the rudiments and principles.

The book is not meant to be read from front to back. The coverage is rather broad so that the instructor may select those chapters and sections that suit his or her goals. For example, I can imagine that many people, considering the level and background of their students, will skip Chapter 2 on thermodynamics or Chapter 3 on tensor index notation. I placed these chapters at the beginning, rather than in an appendix, with the thought that the student would be likely to review these subjects even if they were not formally assigned as a part of the course. Students who want more information about any chapter will find a supplemental reading list at the back of the book.

A chapter usually begins with an elementary approach suitable for the beginning student. Subsections that are marked by an asterisk contain more advanced material, which either gives a deeper insight or a broader viewpoint. These sections should be read only by the more advanced student who already has the fundamentals of the subject well in hand. Likewise, the problems at the end of each chapter are classified into three types: (A) problems that give computational practice and directly reinforce the text material, (B) problems that require a thoughtful and more creative application of the material, and finally (C) more difficult problems that extend the text or give new results not previously covered.

Several photographs illustrating fluid flow patterns have been included. Some illustrate a simplified flow pattern or single physical phenomenon. Others were chosen precisely because they show a very complicated flow that contrasts with the simplified analysis of the text. The intent is to emphasize the nonuniqueness and complexity possible in fluid motions. In most cases only the major point about a photograph is explained. The reader will find a complete discussion in the original references.

Writing this book has been a long project. I would like to express my appreciation for the encouragement that I have received during this time from my family, students, colleagues, former teachers, and several anonymous reviewers. The people associated with John Wiley & Sons should also be mentioned: At every stage their professional attitude has contributed to the quality of this book.

RONALD L. PANTON

Austin, Texas
January 1984

Contents

Incompressible Flow

1 Continuum Mechanics

The science of fluid dynamics describes the motions of liquids and gases and their interaction with solid bodies. There are many ways to further subdivide fluid dynamics into special subjects. The plan of this book is to make the division into compressible and incompressible flows. Compressible flows are those where changes in the fluid density are important. The major specialty concerned with compressible flows is called gas dynamics. It deals with high-speed flows where density changes are large and wave phenomena frequently occur. Incompressible flows, of either gases or liquids, are flows where density changes in the fluid are not an important part of the physics. The study of incompressible flow includes such subjects as hydraulics, hydrodynamics, aerodynamics, and boundary-layer theory. It also contains the background information for such special subjects as hydrology, lubrication theory, stratified flows, turbulence, rotating flows, and biological fluid mechanics. Incompressible flow not only occupies the central position in fluid dynamics, but is also fundamental to the practical subjects of heat and mass transfer.

Figure 1-1 shows a photograph of a ship's propeller being tested in a water tunnel. The propeller is rotating, and the water flow is from left to right. The most prominent feature of this photograph is the line of vapor which leaves the tip of each blade and spirals downstream. The vapor is not itself important, but it marks a region of very low pressure in the core of a vortex which leaves the tip of each blade. This vortex would exist even if the pressure were not low enough to form water vapor. The convergence of the vapor lines into a smaller spiral indicates that the flow is faster behind the propeller than in front of the propeller.

A photograph of an airplane in level flight is shown in Fig. 1-2. A smoke device has been attached to the wing tip so that the core of the vortex formed there is made visible. The vortex trails nearly straight back behind the aircraft. From the sense of the vortex we may surmise that the wing is pushing air down on the inside while air rises outside the tip.

There are obviously some differences in these two situations. The wing moves in a straight path, whereas the ship's propeller blades are rotating. The propeller operates in water, a nearly incompressible liquid, whereas the wing operates in air, a very compressible gas. The densities of these two fluids differ by a factor of 800 : 1. In spite of these obvious differences, these two flows are governed by the same laws, and their fluid dynamics are very similar. The purpose of the wing is to lift the airplane, while the purpose of the propeller is to produce thrust on the boat. The density of the air as well as that of the

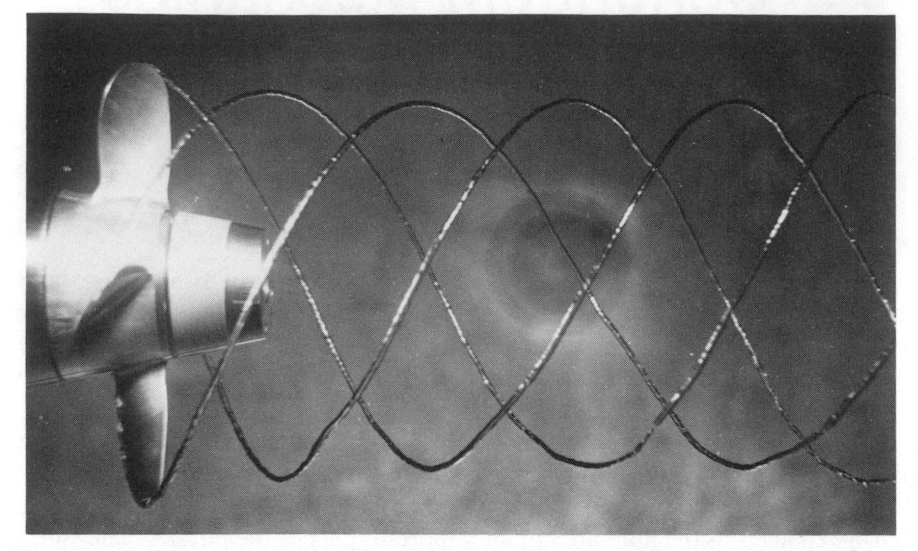

Figure 1-1 Water-tunnel test of a ship's propeller. Cavitation vapor marks the tip vortex. Photograph taken at the Garfield Thomas Water Tunnel, Applied Research Laboratory, Pennsylvania State University was supplied by B.R. Parkin.

water is nearly constant throughout the flow. Both flows have a vortex trailing away from the tip of the surface. This and many other qualitative aspects of these flows are the same. Both are incompressible flows.

In this book we shall learn many characteristics and details of incompressible flows. Equally as important, we shall learn when a flow may be considered

Figure 1-2 Aircraft wing tip vortices. Smoke is introduced at the wing tip to mark the vortex cores. Photograph by W.L. Oberkampf.

as incompressible, and in exactly what ways the physics of a general flow simplifies for the incompressible case. This chapter is the first step in this direction.

1.1 THE CONTINUUM ASSUMPTION

Fluid mechanics, solid mechanics, electrodynamics, and thermodynamics are all examples of physical sciences in which the world is viewed as a continuum. The continuum assumption simply means that physical properties are imagined to be distributed throughout space. Every point in space has finite values for properties such as velocity, temperature, stress, electric field strength, and so on. From one point to the next the properties may change value, and there may even be surfaces where some properties jump discontinously. For example, the interface between a solid and a fluid is imagined to be a surface where the density jumps from one value to another. On the other hand, the continuum assumption does not allow properties to become infinite or to jump discontinuously at a single isolated point.

Sciences that postulate the existence of a continuum are essentially macroscopic sciences and deal, roughly speaking, with events that may be observed with the unaided eye. Events in the microscopic world of molecules, nuclei, and elementary particles are not governed by continuum laws, nor described in terms of continuum ideas. However, there is a connection between the two points of view. Continuum properties may be interpreted as the averages of events involving a great number of microscopic particles. The construction of such interpretation falls into the disciplines of statistical thermodynamics (statistical mechanics) and kinetic theory. We shall discuss, from time to time, some of the simpler microscopic models that are used for continuum events. This aids in a deeper understanding of continuum properties, but in no way does it make the ideas "more true". The fundamental assumptions of continuum mechanics stand by themselves without reference to the microscopic world.

The continuum concept developed slowly over the course of many years. Leonhard Euler (a Swiss mathematician, 1707–1783) is generally credited with giving a firm foundation to the ideas. Previously scientists had not distinguished clearly between the idea of a point mass and that of a continuum. Sir Isaac Newton (1643–1727), in his major contributions, actually used a primitive form of the point mass as an underlying assumption (he did at times, however, also employ a continuum approach). What we now call Newton's mechanics or classical mechanics refers to the motion of point masses. In the several centuries following Newton, problems concerning the vibration of strings, the stresses in beams, and the flow of fluids were attacked. In these problems it was necessary to generalize and distinguish point-mass properties from continuum ones. The continuum assumption is on a higher level of abstraction and cannot be mathematically derived from the point-mass concept. On the other hand, by introducing notions such as the center of mass and moments of inertia, we can derive the laws governing a macroscopic point mass from the continuum laws. Hence, the continuum laws include, as a special case, the laws for a point mass.

1.2 FUNDAMENTAL CONCEPTS, DEFINITIONS, AND LAWS

It is hard to give a precise description of a *fundamental concept* like mass, energy, or force. They are hazy ideas. We can describe their characteristics, tell how they act, express their relation to other ideas, but when it comes to saying what they are, we must resort to vague generalities. This is not really a disadvantage, because once we work with a fundamental concept for a while and become familiar with its role in physical processes, we have learned the essence of the idea. This is actually all that is required.

Definitions, on the contrary, are very precise. For example, pressure may be precisely defined after we have the ideas of force and area at hand. Once we have made a definition of a certain physical quantity, we may explore its characteristics and deduce its exact relation to other physical quantities. There is no question how pressure is related to force, but there is a certain haziness about what a force is.

The situation is analogous to the task of writing a dictionary. How can we write out the meaning of the first word? By the very nature of a dictionary we must use other words in defining the first word. The dilemma is that those words have not yet been defined. The second word is not much easier than the first. However, after the meanings of a few key words are established, the task becomes much simpler. Word definitions can then be formulated exactly, and subtle distinctions between ideas may be made. As we use the language and see a word in different contexts, we gain a greater appreciation of its essence. At this stage, the problem of which words were the very first to be defined is no longer important. The important thing is the role the word plays in our language and the subtle differences between it and similar words.

Stretching the analogy between a continuum and a dictionary a little bit further, we can draw a correspondence between the molecules of a continuum and the letters of a word. The idea conveyed by the word is essentially independent of our choice of the language and letters to form the word. In the same way, the continuum concepts are essentially independent of the microscopic particles. The microscopic particles are necessary but unimportant.

The mathematical rules by which we predict and explain phenomena in continuum mechanics are called *laws*. Some *restricted laws* apply only to special situations. The equation of state for a perfect gas or Hooke's law of elasticity are examples of this type of law. Laws that apply to all substances we shall distinguish by calling them *basic laws*. There are many forms for the basic laws of continuum mechanics, but in the last analysis they may all be related to four laws: the three independent conservation principles for mass, momentum, and energy plus a fundamental equation of thermodynamics. These suffice when the continuum contains a "simple substance" and gravitational, electrical, magnetic, and chemical effects are excluded. In fluid mechanics, however, we frequently want to include the gravity force. In such cases another basic law for this force should be added to the list. Problems dealing with electrical, magnetic, and chemical effects would require correspondingly more basic laws.

Newton's second law is familiar to all students from their earliest course in physics:

$$F = Ma = \frac{d^2x}{dt^2}$$

This law relates the ideas of force, mass, and acceleration. It should not be considered as a definition of force. It is our responsibility to identify and formulate all the different types of forces. In this law we usually consider distance, time, mass, and force to be fundamental concepts, and acceleration to be a defined quantity. Newton's law tells us that these quantities cannot take on independent values but must always maintain a certain relationship.

Which concepts are taken to be fundamental and which are defined is a matter of tradition and convenience. For example, we usually take length and time as fundamental and consider velocity to be defined by the time derivative of the position. On the other hand, we might take velocity and time as fundamental concepts and then consider distance to be defined by the integral

$$x = \int_0^t v \, dt$$

This would be unusual and awkward; however, it is conceptually as valid as defining velocity from the ideas of distance and time.

In this book we shall not emphasize the philosophical aspects and the logical construction of continuum mechanics. This task belongs to a branch of mathematics called *rational mechanics*. Our efforts will fall short of its standards of rigor. Our purpose is to understand the physics and to quantify (if possible) practical situations in fluid mechanics. We do not intend to sacrifice accuracy, but we cannot afford the luxury of a highly philosophical approach.

1.3 SPACE AND TIME

The natural independent variables of continuum mechanics are three-dimensional space and time. We assume all the concepts and results of Euclidean geometry: length, area, parallel lines, and so on. Euclidean space is the setting for the progress of events as time proceeds independently. With these assumptions about the nature of time and space we have ruled out relativistic effects and thereby limited the scope of our subject.

In order to measure space and other physical quantities it is necessary to introduce a coordinate system. This then brings up the question of how a quantity such as the energy might depend on the coordinate system in which it is calculated. One of the major facts of physics is the existence of special coordinate systems called *inertial frames*. The laws of physics have exactly the same form when quantities are measured from an inertial coordinate system. The magnitude of the momentum or the magnitude of the energy will be different when measured in different coordinates, however, the physical laws deal only with changes in these quantities. Furthermore, the laws have a structure such that the same change will be observed from any inertial system. All inertial coordinate systems are related by *Galilean transformations*, in which one coordinate system is in uniform translational motion with respect to the other. Furthermore, any coordinate system that is in uniform translational motion with respect to an inertial system is also an inertial system. We sometimes say that a coordinate system that is fixed with respect to the "distant stars" is an inertial coordinate system. Of course, we cannot be too precise about this concept, or we run into relativity. The laboratory is not an inertial coordinate system because of the earth's rotation and acceleration. Nevertheless, many events occur in such a short time that earth rotation may be neglected and laboratory coordinates may be taken as an inertial system.

As mentioned above, all the facts of Euclidean geometry are assumed to apply to space, while time is a parameterlike independent variable which proceeds in a forward direction. At any instant in time we may define a *control volume*, or control region, as any closed region which we may specify in any manner we choose. It is our invention. The boundary is called a *control surface*, and we prescribe its motion in any manner we choose. The purpose of a control region is to focus our attention upon physical events at the boundary and within the region.

It will be useful to define three types of regions which depend upon how the surface of the region moves with time (Fig. 1-3). A *fixed region* (FR) is one where the control surface does not move at all but is fixed in space. We might imagine a fixed region as enclosing a compressor as shown in Fig. 1-3. The region surface cuts through the inlet and outlet pipes, and fluid flows across these surfaces into or out of the region. At another place the control surface must cut through the shaft which drives the compressor. Here we imagine that the control surface is stationary even though the material which composes the shaft is moving tangentially to the surface. When we use a fixed region, we must allow material to either cross the surface or slide along it.

The second type of region is called a *material region* (MR), because the surface moves with the local velocity of the material. Consider a bubble of gas that is rising through a liquid. As the bubble rises, it expands in size and the gas inside exhibits a circulatory motion. A material region that just encloses

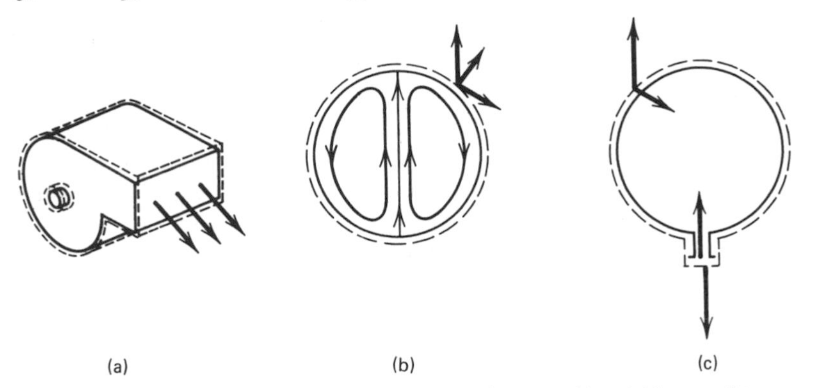

Figure 1-3 Control regions: (a) fixed region around a centrifugal blower, (b) material region around a bubble rising in a liquid, (c) arbitrary region around a balloon with escaping air.

the gas has a local velocity composed of three parts: the rising velocity of the bubble, the expansion velocity of the bubble, and the gas velocity at the interface due to the internal circulation (a sliding velocity tangent to the surface). If we omit the velocity of the internal circulation, the region will no longer strictly fit the definition of a material region. The surface will still always enclose the same material, but the surface will not have the local material velocity.

Any control region that does not fall into the first two categories is called an *arbitrary region* (AR). An example of an arbitrary region is given by a toy balloon which has been turned loose to move freely through the air. Choose the surface of the region to coincide with the balloon everywhere except at the mouth, where air is escaping. At this point the surface cuts across the plane of the exit and the air crosses the surface of the region. Such a region is very useful for an analysis; however, it must be classed as an arbitrary region.

In the examples above, the regions have been of finite size and have obviously been chosen in order to perform an engineering analysis. Control regions are also very useful for conceptual and theoretical purposes. When they are used for these purposes, one often considers a sequence of regions that become smaller and smaller. An example of this type of reasoning will occur in the next section.

1.4 DENSITY, VELOCITY, AND INTERNAL ENERGY

Density is the mass per unit volume of a substance and is one of our fundamental concepts. We consider that the continuum has a density at every point in space. The following thought experiment is a popular way to illustrate the concept. Consider a specific point in space, and choose a fixed control region that encloses the point. Imagine that we freeze the molecules and then count the number of them within the region. With this information we form the ratio of the mass of the material to the volume of the region; that is, the average density of the control region. Let L be a measure of the size of the control region: L might be the distance across the central point to a certain position on the control surface. Now, the experiment is repeated with a smaller but geometrically similar control region. Each time the results are plotted as in Fig. 1-4. A logarithmic scale for L is used, because L ranges over many orders of magnitude. When L is very large, say a mile, the measurement represents an average that might have little to do with the local fluid density. As L becomes small the experiment produces a consistent number for M/V even as L ranges over several orders of magnitude. This number is the density at the point P. Finally, the control region becomes so small that L approaches the distance between molecules. With only a few molecules within the volume, the ratio M/V jumps as the control region shrinks past a molecule. To continue the process produces even more scatter in M/V.

If we start the process again with a different-shaped control region, we find a different curve for very large values of L, but as the length becomes a millimeter or so, the same plateau in M/V may occur. If so, then it will be valid to take a continuum viewpoint and define a density at the point P. Mathematically, the definition is expressed by

$$\rho = \lim_{L \to 0} \frac{\Sigma m_i}{V} \tag{1.4.1}$$

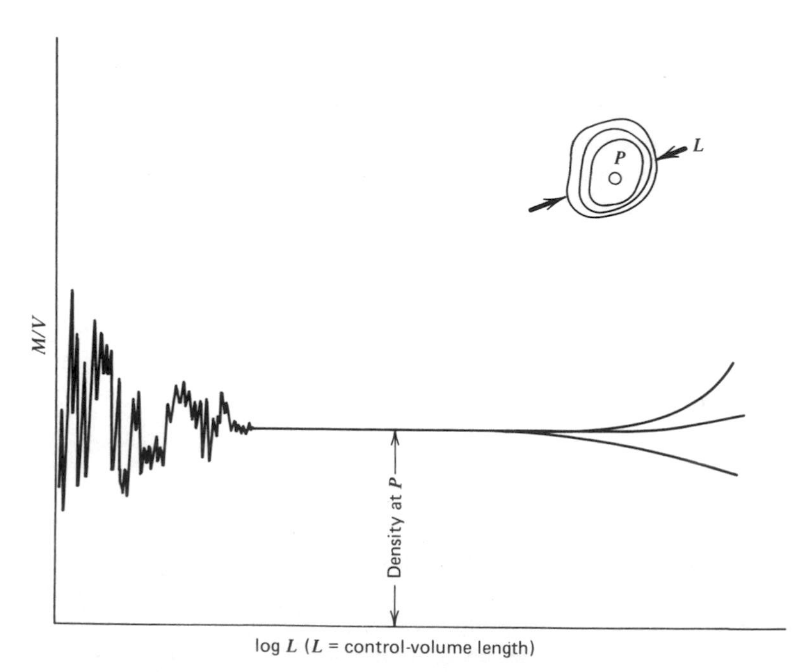

Figure 1-4 Thought experiment to define density.

where the summation occurs over all particles within the region. The limit process $L \to 0$ is understood to go toward zero but never to reach a molecular scale.

In a flow where the number of molecules changes rapidly over a distance comparable to intermolecular distances, the continuum assumption will be suspect. To illustrate this consider the problem of computing the internal structure of a shock wave. The thickness of a shock wave is only a few times the mean free path (the average distance a molecule travels before colliding with another molecule). Over this distance the density may increase by a factor of two. Can the density profile be computed using continuum assumptions? This problem is a borderline case, and it turns out that the continuum calculation gives reasonable answers. In ordinary engineering situations density gradients occur over distances of the order of centimeters and the continuum assumption is unquestionably valid.

We can gain a better insight into the continuum assumption by reviewing some of the molecular properties of air. Air at atmospheric conditions contains 3×10^{19} molecules in one cubic centimeter. Numbers like this are hard to comprehend. How long would it take to count the molecules in one cubic millimeter of air? Suppose a superfast electronic counter can count at the rate of one million molecules per second. A simple calculation shows that for a cubic millimeter of air we would have to let the counter run for

$$3 \times 10^{10} \text{ sec} = 8.3 \times 10^{6} \text{ hr} = 3.5 \times 10^{5} \text{ days} = 1000 \text{ yr}.$$

A cubic millimeter was chosen for this example because the time to count for a cubic centimeter would also be hard to comprehend.

A few other facts about air at standard conditions are worth noting. The mean free path is about 8×10^{-8} m ≈ 0.1 μm, and this is about 25 times the distance between molecules (3×10^{-9} m). In other words, a molecule passes about 25 molecules before it collides with another molecule. The number of molecules in a cube that is one mean free path on each side is 15,000: still a large number. It can be predicted by kinetic theory that the density of this volume will fluctuate in time by only 0.8% (rms). If we reduce the side of our volume to 0.1 mean free path, we now have only 15 molecules and the density fluctution will be 25%. These numbers show that the mean free path also offers a convenient dividing line between the continuum and microscopic worlds. Another interesting facts about simple gases (at standard conditions) is that the distance between molecules is about 10 times the size of a simple molecule. (The nucleus of an atom is about 1/100,000 of the size of the atom.)

In liquids, the size required for the continuum hypothesis to be valid is somewhat smaller than for gases; however, the mean free path concept is not valid for liquids. The distances between molecules and the sizes of the molecules are roughly the same in liquids, so smaller volume is required for a reasonable formulation of the density.

Velocity is another fundamental continuum concept that is based upon the volume limiting process. There are actually two ways to define fluid velocity: the *molar-averaged velocity* and the *mass-averaged velocity*. They may have different values if the fluid is a chemical mixture. The mass-averaged velocity is formed by the vector sum of all particle velocities with the mass used as a weighting factor:

$$\mathbf{v} = \lim_{L \to 0} \frac{\Sigma m_i \mathbf{v}_i}{\Sigma m_i} \tag{1.4.2}$$

The mass-averaged velocity is natural for problems of fluid flow where the momentum equation is important. The product $\rho\mathbf{v}$ gives the momentum per unit volume averaged over all particles. If the fluid is a chemical mixture, the average motion of one chemical species may not be in the direction of $\mathbf{v}$. We define the molar-averaged velocity of chemical species k by summing only over molecules of that species:

$$\mathbf{V}^{(k)} = \lim_{L \to 0} \frac{\Sigma \mathbf{v}_i^{(k)}}{n^{(k)}}$$

In this expression $n^{(k)}$ is the number of molecules of species k within the volume. The molar-averaged velocity of the entire mixture is the vector sum over all molecules divided by the total number of molecules,

$$\mathbf{V} = \lim_{L \to 0} \frac{\Sigma \mathbf{v}_i}{n}$$

If the fluid has a uniform chemical composition, the two velocities are equal, $\mathbf{V} = \mathbf{v}$. In situations where there is mass diffusion or chemical reactions, it is sometimes more convenient to employ a molar-averaged velocity. Since we shall deal only with fluids of uniform composition in this book, the mass-averaged velocity will always be used.

The term *fluid particle* has at least two meanings in common usage. The first is a point concept. Here we envision a point which floats along, moving with the local fluid velocity at each place in space at that particular time. A line traced through the flow field by this method is called a *particle path*. We say that the point that traces the path is a fluid particle, or *material point*. For some purposes—for instance, to talk about the expansion of the fluid—it is necessary to consider a small chunk of fluid. This second meaning for the term fluid particle is made precise by considering a small material region (MR) and allowing the size of the region to tend to zero. Which of the two meanings is intended is usually obvious from the context. Note that because of molecular diffusion, a fluid particle does not always consist of the same molecules. As a particle moves through the flow it gains and loses molecules because of random molecular motions.

The third fundamental concept which we want to cover in this section is *internal energy*. The particle velocity defined above is the average velocity of the molecules, which we observe from our macroscopic world. As far as the macroscopic world is concerned, the kinetic energy of this bulk motion is

$$\text{bulk-motion kinetic energy per unit mass} = \tfrac{1}{2}\mathbf{v} \cdot \mathbf{v} \qquad 1.4.3$$

However, this will not completely account for all the energy of the molecular translational motions. The true total energy sums the molecular velocities,

$$\text{kinetic energy of translation} = \lim_{L \to 0} \frac{\Sigma m_i \tfrac{1}{2}\mathbf{v}_i \cdot \mathbf{v}_i}{\Sigma m_i} \qquad 1.4.4$$

The missing energy, which is hidden from direct macroscopic observation, is the thermodynamic internal energy due to random translational motion. We can formulate an expression for the internal energy by introducing the random molecular velocities (denoted by a prime). To do this we subtract from each molecular velocity $\mathbf{v}_i$ the average fluid velocity $\mathbf{v}$:

$$\mathbf{v}' = \mathbf{v}_i - \mathbf{v}$$

In terms of $\mathbf{v}_i'$ the translational internal energy is expressed by

$$\text{internal energy from random translational velocities} = \frac{\Sigma m_i \tfrac{1}{2}\mathbf{v}_i' \cdot \mathbf{v}_i'}{\Sigma m_i} \qquad 1.4.5$$

Thus the total molecular kinetic energy is split into two parts: a macroscopic part, which is observable as bulk motion, and a microscopic part, which is called the internal energy. There are many other forms of microscopic energy that are hidden from our continuum world: molecular vibration, rotation, potential energies of molecular configurations, potentials of molecules close to each other, and so on. All of these forms of microscopic energy must be accounted for in the thermodynamic internal energy.

1.5 THE INTERFACE BETWEEN PHASES

The interface between two phases offers some special difficulties in continuum mechanics. The most obvious problem is that the thickness of the interface is small compared to intermolecular distances.

Consider for a moment a gas in contact with a liquid. In the liquid the molecules are closely packed and exert strong attractive forces on each other. For a molecule that is deep within the liquid these forces come from all directions. As we approach the surface the situation changes, as the neighboring liquid molecules are only on one side. The other side is occupied by the gas. Gas molecules are constantly bombarding the surface, becoming mingled with liquid, and sometimes being absorbed. If we idealize the interface as a surface with zero thickness, we must in general assign it physical properties. It is a two-dimensional world. Each physical property then has a two-dimensional analogue: corresponding to density, for example, we have the mass per unit area (the adsorbed mass). Energy per unit volume has a surface analogue in energy per unit area. This includes not only the energy associated with the motions of interface molecules, but also the energy of the special configuration of molecules at the interface.

The two-dimensional interface world is much more complicated than our world. The geometry is non-Euclidean. Conservation laws are complicated because mass, momentum, and energy may change through interactions with the three-dimensional world. Deviations from theory are usual, because a few foreign molecules contaminating the surface can have a great influence. We shall not go into the thermodynamics and fluid mechanics of interfaces; the interested reader should consult Scriven (1960) [or a synopsis given in Aris (1962)].

Interfacial physics and chemistry are subjects in themselves. In order to make progress in our main interest, fluid mechanics, we shall have to assume a very simple model of the interface. In a great many practical applications this model will suffice. We assume that an interface is a surface of zero thickness which contains no mass, momentum, or energy. Across the interface the density is allowed to jump discontinuously. On the other hand, the temperature and the tangential velocity are assumed to be continuous. This assumption is justified because the molecules from both sides are constantly colliding and equilibrating within the surface layer. These ideas are illustrated in Fig. 1-5, where a gas flows over a liquid. Molecules leaving the surface and moving back into either fluid have the same tangential velocity. In other words, the velocity of fluid within the interface has only one value. This assumption is called the "no-slip condition". It is not an obvious fact. Indeed, it was once the subject of a long debate [see Goldstein (1965, p. 676) for a brief history]. The debate concerned surface tension and the fact that some liquids are attracted to certain solids while others are not. It turns out that wetability is not important, and that the no-slip condition applies in general to all substances.

The velocity perpendicular to the interface is discontinuous whenever mass is transferred across the surface. This situation is illustrated by considering a vaporizing liquid. There is a continuous flow of vapor away from the surface with a mass flux $\rho v|_{\text{vap}}$. This must be balanced by an equal flux into the surface from the liquid side of $\rho v|_{\text{liq}}$. Since the two densities are quite different, the velocities must also be different. The discontinuity in normal velocity and the continuity of tangential velocity apply even if the surface itself is in motion.

Several macroscopic fluid phenomena require giving the surface a property called surface tension. It is the surface analogue of pressure in the three-dimen-

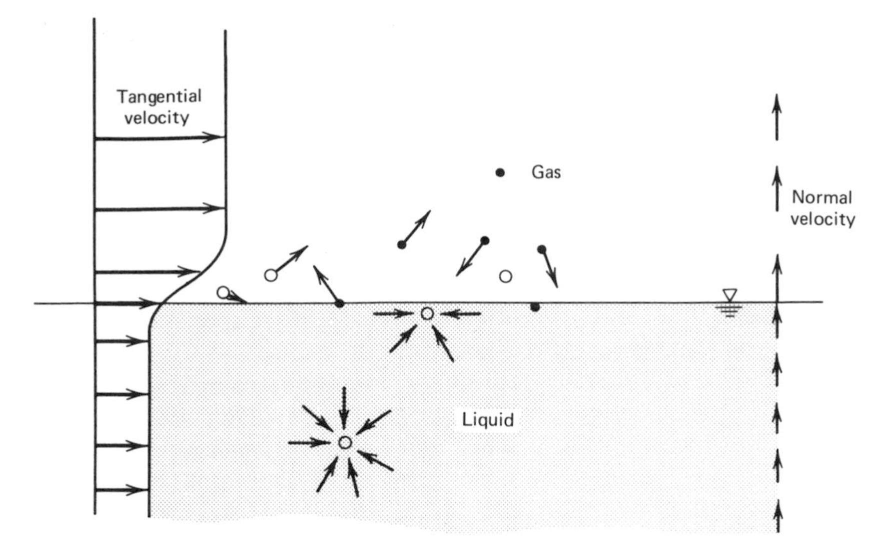

Figure 1-5 Liquid–gas interface. The tangential velocity is continuous, but the normal velocity may have a discontinuity.

sional world. From point to point within the surface, the tension may vary. Its exact value depends upon the flow on both sides, the curvature of the surface, and the thermodynamic state (temperature and composition). The following simplified formula relates the surface tension to the flow properties on either side of the interface:

$$[p - \tau_{nn}]_a - [p - \tau_{nn}]_b = \sigma\left[\frac{1}{R_1} + \frac{1}{R_2}\right]$$

In this equation p is the pressure, τ_{nn} the normal viscous force, σ the surface tension, and R_1 and R_2 the principal radii of curvature. Another formula, which is probably more familiar, gives the surface tension when there is no flow and the surface is spherical. It is

$$p_a - p_b = \frac{2\sigma}{R}$$

This formula shows that the pressure within a droplet or a bubble is larger than that of the bulk fluid.

Even though surface tension is conceptually a variable like pressure, we are usually forced to assume it is constant at some value measured in a static experiment. We shall not review the static fluid phenomena where surface tension is important. Batchelor (1967) contains a good survey of these problems.

1.6 CONCLUSION

In this first chapter we have attempted to define the scope and nature of fluid mechanics. The three fundamental continuum concepts of density, velocity, and energy were introduced. We shall introduce many more concepts as they are needed in later chapters. In all of our work we shall limit ourselves to exclude magnetic, electrical, chemical, and interfacial effects. The fluids in the problems which we will study will always be assumed to be homogeneous, simple, compressible substances. Even with all of these restrictions there will be plenty of material to cover.

Perhaps the most fundamental restriction in our subject is the continuum assumption. The characteristic size of the flow must be a continuum scale length. There is a famous physical phenomenon called Brownian motion which illustrates this restriction very nicely. The botanist Robert Brown, while observing life forms in a water droplet by means of a microscope, noticed that some pollen particles in the water had a jittery motion. The motion was actually like a random vibration where the velocity was abruptly changing direction at a high frequency. It gave the particles a fuzzy appearance. The pollen particles were a few micrometers in size, maybe 100 times the intermolecular spacing in water. Later, the reason for this random meandering of the particles was correctly ascribed to unequal and fluctuating molecular forces. The particle was not large enough so that molecular bombardment on one side was always exactly counterbalanced on the other side.

A calculation of the motion was finally made by Einstein and by Smoluchowski. They used an ad hoc mixture of molecular and continuum ideas. The random driving force was taken from molecular concepts, and a continuum viscous retarding force was assumed. Situations of this type, in the gray area between continuum mechanics and kinetic theory, have grown into what is now called colloidal science. It marks a boundary of continuum fluid mechanics where body sizes become comparable with molecular sizes (see Fig. 1-6).

Another boundary for the continuum assumption occurs for finite-size bodies in gas flow. As the density is reduced and vacuum conditions approached, either at high altitudes or in vacuum systems, the distance between molecules may become several centimeters. Now the body size may be comparable to the mean free path. Consider a sphere shooting through a rarefied gas. Molecules that collide with the front of the sphere are sent forward several sphere diameters before they interact with other molecules and influence the gas motion. Behind the sphere there is a partial vacuum swept out by its motion. Several diameters back, the random molecular velocities fill this region in once more. This flow field is much different than the one we would find if the mean free path were very small compared to the diameter. The extension of fluid mechanics into this region is called rarefied-gas dynamics.

These illustrations show two ways in which the continuum assumption may fail: the characteristic length in the flow (the body diameter) may be so small

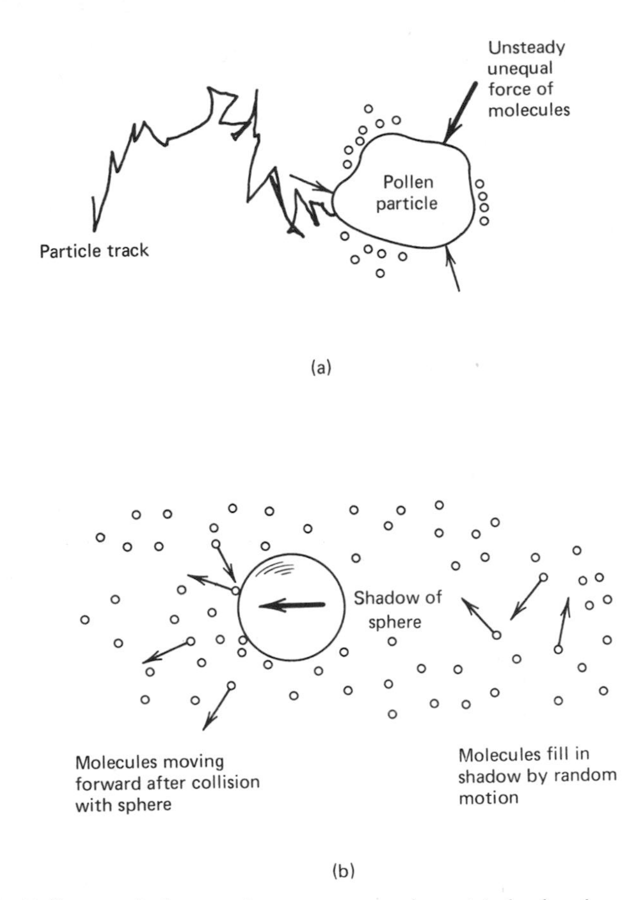

Figure 1-6 Failures of the continuum assumption: (a) body size compares with molecular dimensions (very small particle in a liquid); (b) body size compares with distance between molecules (sphere moving through a rarefied gas).

that it compares with the molecular dimensions, or the mean free path of the fluid may be comparable with the characteristic length of the body.

PROBLEMS

1.1 (B) Consider an unsteady one-dimensional flow where the density and velocity depend upon x and t. A Galilean transformation into a new set of variables x', t' is given by the equations $x = x' + Vt'$, $t = t'$, where V is a constant velocity. For the moment let $f = f(x, t)$ stand for a function which we wish to express in the x', t' coordinate system. By careful use of the chain rules of calculus, find expressions for $\partial f/\partial t'$ and $\partial f/\partial x'$. Next, consider the "substantial derivatives" of ρ and v, which are

$$\frac{\partial \rho'}{\partial t'} + v'\frac{\partial \rho'}{\partial x'}, \qquad \frac{\partial v'}{\partial t'} + v'\frac{\partial v'}{\partial x'}$$

Show that the substantial derivatives above have exactly the same mathematical form when transformed into the x, t coordinate system (note that $\rho' = \rho$ and $v' = v - V$).

1.2 (A) A droplet of liquid is moving through a gas. It evaporates uniformly, does not deform, and has no internal circulation. A control region coinciding with the liquid is what type of region?

1.3 (A) A droplet of liquid is moving through a gas. It does not evaporate or deform, but it does have an internal (and surface) circulation. Describe the velocity of a material region whose surface encloses the droplet.

1.4 (B) A material region was defined as one where the surface velocity $\mathbf{w}$ is everywhere equal to the fluid velocity $\mathbf{v}$. Such a region always contains the same fluid. Can you define $\mathbf{w}$ in a less restrictive way and still have a region always containing the same material?

1.5 (B) Prove that the average of the random molecular velocities $\mathbf{v}_i'$ (see 1.4.5) is zero, that is,

$$\lim_{L \to 0} \sum m_i \mathbf{v}_i' = 0$$

1.6 (B) Prove that the total kinetic energy (per unit mass) of translational molecular motion may be split into two parts as follows:

$$\sum m_i \tfrac{1}{2}\mathbf{v}_i \cdot \mathbf{v}_i = \sum m_i \tfrac{1}{2}\mathbf{v}_i' \cdot \mathbf{v}_i' + \tfrac{1}{2}\mathbf{v} \cdot \mathbf{v} \sum m_i$$

1.7 (C) Using A to denote Avogadro's number, the number of moles of species k is $n^{(k)}/A$, and the mole fractions are $\chi_k = N^{(k)}/N$. Prove that the molar-averaged velocities of each species and the molar-averaged velocity of the mixture are related by

$$\mathbf{V} = \sum_k \chi_k \mathbf{V}^{(k)}$$

1.8 (C) We have found that the continuum assumption cannot be applied to events with a characteristic length which is of the order of molecular scales. Is there a characteristic time interval for which the continuum assumption is not valid?

1.9 (B) The momentum $\mathbf{p}_i$ of a molecule is equal to the product $m_i \mathbf{v}_i$. From the definitions of ρ and $\mathbf{v}$, show that the product $\rho\mathbf{v}$ is the total momentum of all molecules per unit volume.

2 Thermodynamics

Students may well wonder why a chapter entitled "Thermodynamics" should appear in a book on incompressible flow. This chapter would in fact be unnecessary if we restricted our study entirely to the laws and phenomena that occur in incompressible flow. We might then simply discuss the events and display the appropriate equations and results. However, we have chosen to obtain a broader understanding by investigating how incompressible flows arise as a special class of more general flows. Indeed, it turns out that incompressible flows include more than those special cases where the fluid itself is incompressible. This broader objective requires that the thermodynamics of the fluid be considered.

There are two ways to study thermodynamics: the classical approach and the axiomatic or postulational approach. We shall take the second approach. However, before we do so, it will be helpful to contrast the two.

In the early development of classical thermodynamics workers handicapped themselves by adopting the attitude that all fundamental concepts should be subject to direct measurement. Classical theory starts directly from experimental results, using work and heat as fundamental concepts. Since energy can only be measured by using the energy conservation law, it was necessary to define internal energy through the use of this law and the concept of a process. The law expressing the conservation of energy became known as the first law of thermodynamics. Energy interactions between a system and its surroundings are one major aspect of classical thermodynamics. The relations between the state properties of a substance (thermodynamic properties) is the second major aspect of the subject.

For a given substance, such as oxygen for example, there is a certain relation between the pressure, the temperature, and the internal energy. The key to property relationships is entropy concept. In classical thermodynamics entropy is defined using the first law and the concept of a reversible process. This, however, is not enough. One of the characteristics of entropy (the fact that irreversible processes in an isolated system must increase the entropy) must be stated as an additional postulate called the second law of thermodynamics. This classical approach is still the standard in most introductory courses.

Usually in the historical development of a field of science we progress from experimentation and the formulation of special laws to a higher degree of abstraction where only a few principles are required to cover all special cases.

This abstract development of principles aids us in relating seemingly distinct problems and helps us solidify the essential characteristics of our physical concepts. At the beginning, the field of mechanics had many special laws for the motion of planets and for bodies falling under the influence of gravity. These laws were proposed as the result of observations and experiments. As the science of mechanics developed, we realized the central importance of Newton's second law and that previous results could be found as special cases.

In the axiomatic approach to thermodynamics, we propose abstract postulates and proceed to investigate their meaning and implications. The truth or validity of our concepts is then tested, after the fact, by our ability to predict various phenomena and the results of experiment.

Another characteristic of axiomatic thermodynamics is that energy transfers and state-property relationships may be studied separately. In fact, the axiomatic approach gives an emphasis to thermodynamic state properties. The major postulate (fundamental assumption) is equivalent to the second law of thermodynamics. It allows us to determine the characteristics of entropy, pressure, temperature, and the chemical potential. We are then able to develop the thermodynamics of state properties without explicitly using the first law for energy conservation. In approaching the subject by this path, we emphasize that the thermodynamics of state properties is distinct from a first-law analysis of energy transfers. (The first law will be included in Chapter 5 with all the other dynamical laws.)

The postulational formulation of thermodynamics has not been a satisfactory approach for beginners, because they have great difficulty in trying to attach a physical meaning to entropy. Entropy cannot be directly measured, and we cannot experience it with our physical senses. In fact, some scientists contend that it might be well not to refer to entropy as an entity but always call it the "entropy function". In this way its mathematical properties are emphasized, and the beginner is not disappointed by a lack of physical interpretations. Since most readers are already familiar with classical thermodynamics, it should not be too burdensome to present a simple postulational formulation of the subject. The presentation is similar to the approaches of H. B. Callen (1960), A. Sommerfeld (1956), and E. A. Guggenheim (1949).

2.1 SYSTEMS, PROPERTIES, AND PROCESSES

A *simple system* is a special kind of control region. The matter contained within the region is homogeneous, isotropic, chemically inert, and not moving. The system is not subject to gravitational, electric, magnetic, or interfacial effects. It may receive work only through the normal pressure force. Thus, shear forces are taken to be zero. Special properties are given to the walls of the system, whereby we allow interchanges of material, heat, or volume between the system and its adjacent surroundings. The walls may be either real or imaginary; that is, they may be actual solid surfaces containing a fluid, or they may be imaginary surfaces which cut through the fluid at any location. Different types of walls are given special names, which we shall discuss a little later.

Certain properties of the system, called *thermodynamic properties* (internal energy, entropy, temperature, etc.), are related to each other, and a change in one property may cause changes in the others. When we speak of the *state* of the system we mean that all the thermodynamic properties have definite, unique values. A *process* is any mechanism by which the state of the system is changed. Some processes that we imagine may be very difficult to achieve experimentally; however, this has no bearing upon our results. Most of the time we imagine a *reversible process*, which by definition consists of a sequence of equilibrium thermodynamic states.

Actual processes are never truly reversible. The fact that a wall is moved or becomes hotter than the material inside the system causes the pressure or the temperature of the system to be nonuniform. This, in turn, means the process must be irreversible. A process where either the pressure, temperature, or composition of the system is nonuniform is an *irreversible process*. The effect of a moving wall is transferred to the fluid by compression or expansion waves, which travel very rapidly and tend to keep the system at a uniform pressure. In most practical cases, the moving of a wall can be considered as a reversible process. Conduction heat transfer, on the other hand, is a relatively slow process. Because of this, a reversible heat addition process can only be approached by slowly increasing the wall temperature.

At the opposite extreme, we model very rapid processes as being *adiabatic*. For such processes the heat transfer is so slow that most of the material in the system will not see any heat-transfer effects during the process.

Thermodynamic properties of the system may be classed into two categories: extensive and intensive. In order to define these terms, consider a system with uniform properties. Since we get to choose the system boundaries, we can redraw the boundary so that the size of a new system is a fraction λ of the old system, that is, $V_2 = \lambda V_1$. Any property of the system, X, that is reduced in proportion to the size of the system is called an *extensive property*. Extensive properties obey the relation

$$X_2 = \lambda X_1 \tag{2.1.1}$$

Energy, mole number, and the volume itself are examples of extensive properties. Any property y that is unchanged is called an *intensive property*. Intensive properties obey the relation

$$y_2 = y_1 \tag{2.1.2}$$

Pressure and temperature are examples of intensive properties.

Another way to define extensive and intensive is to consider X as a function of V:

$$X = f(V)$$

Now we let $V \to \lambda V$ and $X \to \lambda^n X$, where n is some undetermined power. The equation above now reads

$$\lambda^n X = f(\lambda V)$$

and may be written as

$$\lambda^n f(V) = f(\lambda V)$$

If $n = 1$ the property is extensive, and in mathematical terminology f is said to be a homogeneous function of degree one. If $n = 0$ the property is intensive, and f is a homogeneous function of degree zero. We shall follow the standard practice of using capital letters for extensive properties and lowercase letters for intensive properties (with the exception of temperature).

2.2 THE INDEPENDENT VARIABLES

It is a fact of experience that fixing three independent properties will determine the thermodynamic state of a simple system containing a single chemical species. Some caution must be exercised, because just any choice of three variables may not give an independent set. For example, p, T, and ρ are not independent, but p, T, and V are independent. If one is only interested in the intensive state of the system, then only two independent intensive properties need be chosen. Of the several choices of independent variables that may be made, we choose the internal energy E, the volume V, and the number of moles of substance, N.

The amount of matter in the system can be measured by any of several equivalent variables. Systems containing a single chemical species are readily specified by N, the number of moles. When a system consists of a mixture of chemically inert species, we must in principle allow for changes in composition. There will be an additional independent variable required for each species present. Usually N_i, the number of moles of species i, is used. As an alternative to using the number of moles, we could use the mass of each species. Recall that the mass and the number of moles are related by $M = N/\mathcal{M}_i$, where $\mathcal{M}_i$ is the mass of a mole of molecules. A mixture such as air, where the composition is uniform and does not change during a process, can be modeled as a pure substance. This is done by using an average molecular weight and average values of other thermodynamic properties.

Liquids and gases require that the volume V be used as a thermodynamic variable. The shape of the system is unimportant to the thermodynamic state. Certain aspects of the thermodynamics of solids are also independent of the shape, but others are not. We can do reversible work on a solid and change its shape but not its volume. This work changes the state of strain of the solid, and additional variables describing the shape of the solid are required for a complete thermodynamic formulation. In all of our work, solids will be considered as rigid and a detailed thermodynamic formulation will be unnecessary. Then they fit the definition of a simple system.

2.3 TEMPERATURE AND ENTROPY

Many of the concepts in thermodynamics—for example, mass, pressure, energy, and volume—are familiar from mechanics. The two completely new concepts in thermodynamics are temperature and entropy. They are not only complementary in the theory but are also complementary in our intuitive understanding. Of all the thermodynamic properties, we have the best intuitive understanding of temperature and the worst intuitive understanding of entropy.

We are all aware that when two bodies of different temperature are brought into contact they exchange internal energy on the microscopic level by heat transfer. During this energy exchange we can detect no macroscopic motions or forces. This is the essential character of heat transfer. Heat transfer is called conduction if it occurs locally by the interaction of the molecules, and radiation if the energy is transferred by electromagnetic fields.

Let us consider the molecular mechanism of conduction in a solid or liquid. The molecules are closely packed, and they behave somewhat like oscillators which vibrate about a mean position. The restoring force of the oscillator is the molecular repulsion force of the neighboring molecules. A large part of the internal energy of a liquid or solid is attributed to these random oscillatory motions. Now if one part of the substance has more energetic oscillators than another part, there is a tendency for the energy to redistribute itself so that all oscillators have the same energy. This is the microscopic energy transport process of heat conduction. Temperature is the macroscopic property which measures the possibility of heat transfer. The temperature of a solid or liquid is proportional to the amount of energy in oscillating motions of the molecules.

The oscillator-to-oscillator mechanism is the dominant mode of transfer in nonmetals. This is not true in the case of metals. Energy transport within metals also occurs by free electrons, which can travel rapidly for great distances before they collide with an oscillator or another electron. That is why metals have a higher thermal conductivity than nonmetals.

Next, we will consider the microscopic interpretation of temperature for substances in the gas phase. Gases have kinetic energy in translational molecular velocities. Consider two nearby positions in the gas, where the average kinetic energy of the molecules is slightly different. A molecule leaving the high-energy region enters the low-energy region, where it collides, and after a time it becomes indistinguishable from other molecules in the low-energy region. The result of this process is that the kinetic energy of the low region is increased. Likewise, some molecules from the low region, because of their random motion, find themselves migrating into the high-energy region. They collide with high-energy molecules and cause a net reduction in the kinetic energy of the high region. Again we have a microscopic process for transporting internal energy. The temperature of the gas is a measure of the translational kinetic energy of the molecules. For a perfect gas the precise formula is

$$\tfrac{3}{2}kT = \tfrac{1}{2}m\overline{U^2}$$

In this formula k is Boltzmann's constant, m the molecular mass, and $\overline{U^2}$ the mean squared random translational velocity of the molecules.

We can summarize as follows: There are microscopic mechanisms whereby internal energy, which is hidden at the molecular level, can be transported. The transport mechanism involves motions which equilibrate high- and low-energy particles. The tendency to transport energy in this way depends upon the energy level itself, and temperature is defined to be proportional to the energy in the transferring mode. Internal energy, on the other hand, is the total energy in all microscopic modes of motion (translation, rotational, vibrational, potential, etc.).

So far the microscopic interpretations of continuum properties have all been familiar terms from geometry and mechanics—concepts we feel comfortable with. Entropy, our next subject, does not have such satisfying interpretations. On the other hand, it is a fundamental concept, so we cannot expect to say what it is, but only describe what it does. Here are a few "what it does" statements. Entropy is something that is constant in a reversible process where there is no heat transfer. Entropy measures irreversibility in that irreversible effects always cause the entropy to increase. For a reversible process, the change in entropy is the heat flux divided by the temperature: $dS = dQ/T$. Entropy is the dependent variable of the fundamental thermodynamic equation of a substance. All of these statements tell us about the continuum nature of entropy. As with the previous properties, we might inquire into its microscopic foundation.

The microscopic interpretation of entropy is not very simple. We can give a brief idea, but a course in statistical mechanics is really required to appreciate the necessary concepts. The Austrian physicist Boltzmann related the entropy and the *thermodynamic probability* $\mathscr{W}$ by the famous equation

$$S = k \ln \mathscr{W}$$

In order to explain thermodynamic probability, consider a system in a fixed thermodynamic state. The system has certain values of the energy E and volume V, and a certain number of particles, N. Microscopically there are many different arrangements (specific particle positions and velocities) of the N particles that will possess the same total energy. The thermodynamic probability is the number of different microscopic arrangements that will produce the given macroscopic thermodynamic state. The logarithmic scale between probability and entropy is required because thermodynamic probabilities are multiplicative in cases where entropies are additive. Thus, if two systems are considered as a composite system, the entropy is the sum $S = S_1 + S_2$, and the thermodynamic probability is the product $\mathscr{W} = \mathscr{W}_1 \mathscr{W}_2$. The logarithm is the only mathematical function between S and $\mathscr{W}$ that will give this characteristic.

2.4 FUNDAMENTAL EQUATION OF A SUBSTANCE

Choosing values of E, V, and N fixes the thermodynamic state and determines the values of all the other thermodynamic properties. In particular, there is a relation for the entropy,

$$S = S(E, V, N) \tag{2.4.1}$$

The major characteristics of entropy can be found only after we state the second law. For the moment, however, the major point to make is that $S = S(E, V, N)$ contains all the thermodynamic information about a substance. In this sense it is a *fundamental equation* for the system. If this single function is known, then all thermodynamic properties may be found. This is not true for just any thermodynamic function. For example, if we knew the function $S = (p, T, N)$, we would not have complete thermodynamic information about the system.

For a given set of values of independent properties, a fundamental equation allows all other thermodynamic properties to be found. Assume that we know the function $S = S(E, V, N)$. We can also find all the partial derivatives of this function. The following list of equations shows how all thermodynamic properties may be determined for arbitrary values of E, V, and N:

Entropy $$S = S(E, V, N)$$

Temperature $$T = \left(\frac{\partial S}{\partial E}\bigg|_{V, N} \right)^{-1}$$

Pressure $$p = T \frac{\partial S}{\partial V}\bigg|_{E, N}$$

Chemical potential $$\mu = T \frac{\partial S}{\partial N}\bigg|_{E, V}$$

Enthalpy $$H \equiv E - pV$$

Gibbs free energy $$G \equiv H - TS$$

Helmholtz free energy $$F \equiv E - TS$$

Compressibility factor $$Z \equiv \frac{pV}{NR_0T}$$

Specific heat at constant volume $$C_V = T\left.\frac{\partial S}{\partial T}\right|_{V,N}$$

Specific heat at constant pressure $$C_p = T\left.\frac{\partial S}{\partial T}\right|_{p,N}$$

Joule-Thomson coefficient $$j = \left.\frac{\partial T}{\partial p}\right|_{H,N}$$

Thus, one function contains all the thermodynamic information. $S(E, V, N)$ is not the only form for a fundamental equation. We shall study alternate forms in Section 2.11.

There are some quasithermodynamic properties that are not related to the fundamental equation: transport coefficients such as viscosity, thermal conductivity, and mass diffusivity; electrical properties; optical properties; and so on. They are functions of the thermodynamic state, but they are not involved in classical thermodynamic theory in any way.

2.5 DEFINITION OF TEMPERATURE, PRESSURE, AND CHEMICAL POTENTIAL

A straightforward application of calculus to 2.4.1 gives the equation for the differential of S,

$$dS = \left.\frac{\partial S}{\partial E}\right|_{V,N} dE + \left.\frac{\partial S}{\partial V}\right|_{E,N} dV + \left.\frac{\partial S}{\partial N}\right|_{E,V} dN. \qquad 2.5.1$$

So many different independent variables may occur in thermodynamics that it is standard practice to list the other independent variables as subscripts on any partial derivative. In this way it is apparent that $(\partial S/\partial E)_{V,N}$ means that S is considered to be a function of E, V, and N.

We now define the *temperature* as the following derivative of $S(E, V, N)$:

$$\frac{1}{T} \equiv \left.\frac{\partial S}{\partial E}\right|_{V,N} \qquad 2.5.2$$

Since a derivative of $S(E, V, N)$ is also a function of E, V, N, the definition 2.5.2 implies that $T = T(E, V, N)$. This relation is called an *equation of state* of the substance. A substance has two other equations of state, $p = p(E, V, N)$

and $\mu = \mu(E, V, N)$, which follow from the definition of pressure,

$$\frac{p}{T} \equiv \left.\frac{\partial S}{\partial V}\right|_{E,N}, \qquad \text{or} \quad p = T\left.\frac{\partial S}{\partial V}\right|_{E,N} \tag{2.5.3}$$

and from the definition of the *chemical potential*,

$$\frac{\mu}{T} \equiv \left.\frac{\partial S}{\partial N}\right|_{E,V}, \qquad \text{or} \quad \mu = T\left.\frac{\partial S}{\partial N}\right|_{E,V} \tag{2.5.4}$$

We shall investigate the physical characteristics of the newly defined properties shortly.

Let us continue the main thread of the development by substituting 2.5.2, 2.5.3, and 2.5.4 into 2.5.1. The result is

$$dS = \frac{1}{T}dE + \frac{p}{T}dU + \frac{\mu}{T}dN \tag{2.5.5}$$

We might even go so far as to call this equation the second law of thermodynamics. Later on (Section 5.12) we shall see how this equation can be substituted into the energy equation to produce the more common statements of the second law. A more usual form of 2.5.5 is

$$T\,dS = dE + p\,dV + \mu\,dN \tag{2.5.6}$$

In either form it is the *fundamental differential equation of thermodynamics*. It plays the central role in *thermodynamics* that Newton's second law plays in mechanics. Different substances will have different fundamental equations $S = S(E, V, N)$, but each and every fundamental equation satisfies the fundamental differential equation (2.5.5).

It is most common to find thermodynamic data in the form of equations of state. A little reflection will show that a knowledge of the three equations of state is equivalent to a knowledge of the fundamental equation. Substitution of the state equations into 2.5.5 and integration produces the fundamental equation to within a constant of integration. Later on we shall see that by another method, even the integration constant is determined in principle.

2.6 COMPOSITE SYSTEMS AND INTERNAL CONSTRAINTS

Let us return now to the discussion of special walls for the system. First of all, a wall is either fixed or movable. The familiar piston in a cylinder is an example of a movable wall. Not only does a movable wall cause the volume to change, but because of the work it does, the energy of the system may also change.

A wall that allows heat to pass freely in either direction is called a *diathermal* wall. Such a wall allows the energy of the system to change while the volume and mole numbers are held constant. The opposite extreme, a wall that does not allow heat to cross, is termed an *adiabatic* wall.

A final characteristic of walls is that of being permeable. A *permeable* wall is a device by which we introduce more material into the system. In its simplest form we might think of it as a wall with microscopic holes connected to another system in the same intensive state as the system under study. Now, by a slight increase or decrease in the pressure of the auxiliary system, we may add or subtract material from the system. When material comes into the system, both the number of moles and the energy increase. We may not be able to say exactly how much energy is received (this would depend upon the detailed structure of the permeable wall), but there must be some. Since there was no macroscopically observed motion we must class this mechanism of energy transport as a heat-transfer process. In light of the fact that a permeable wall allows both mass and heat transfer simultaneously, we cannot have a permeable, adiabatic wall. That would be contradiction. A wall that is permeable must also be diathermal.

To be completely general we must imagine that there are walls that allow only one chemical species to pass. In the construction of thermodynamic theory it is necessary to allow a system to receive or give up any one of its chemical constituents independently.

The next topic of this section is that of a composite system. A *composite system*, as shown in Fig. 2-1, is two simple systems with an internal wall

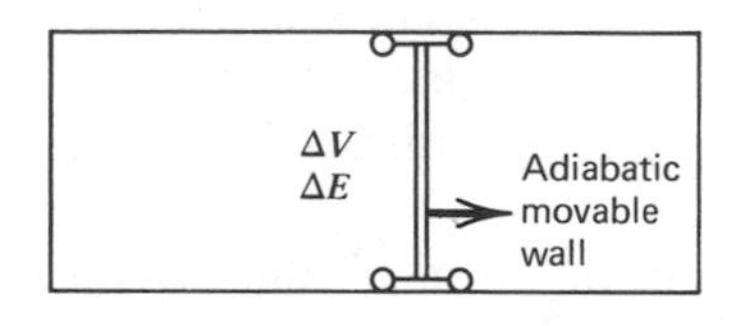

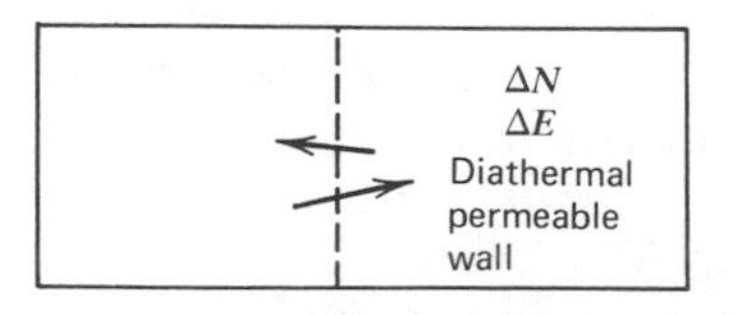

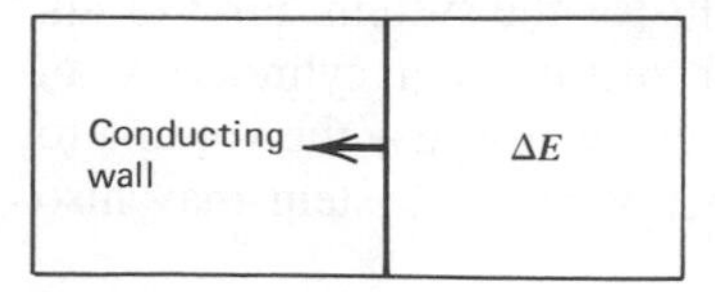

Figure 2-1 Composite systems with different internal constraints.

between them. The outer walls are rigid, solid, and adiabatic; this extra restriction will always apply to composite systems used in this book. Such a system is sometimes said to be *isolated*. This means that the total volume, total energy, and total number of molecules in the composite system are constant. Any restrictive characteristics of the internal wall are known as *internal constraints*.

Different types of composite systems are constructed by giving special characteristics to the internal wall. For example, if the internal wall is a movable, solid, diathermal wall, then in principle volume and energy, but not mass, can be interchanged between the two systems that constitute the composite system. All the while, the total volume and total energy of the composite system are constant. It is possible to expand the definition of composite system to include several simple systems; however, we shall have no need for composite systems composed of more than two systems.

2.7 THE BASIC POSTULATE OF THERMODYNAMICS

The basic postulate of thermodynamics is a method to answer the following problem. Consider a composite system with definite equilibrium states defined by E_A, V_A, N_A in one system and by E_B, V_B, N_B in the other. We call this composite system the *primary system*. The internal wall has specified constraints; for example, the wall might be fixed, adiabatic, and solid. Next, we allow one or more of these constraints to change. The adiabatic constraint becomes diathermal and allows energy to cross, the fixed wall is unfastened so that it is free to move, or the solid wall becomes permeable so that material and energy may cross. After one or more constraints are removed, the systems interact with each other, generally through irreversible processes, and finally come to a new equilibrium state. The central question is: what are the new values of $E_A, V_A, N_A, E_B, V_B, N_B$ in the new equilibrium state? The basic postulate of thermodynamics gives a way to answer this question.

Let us illustrate this problem with a specific example. Consider a primary system with a fixed, adiabatic, solid wall. The initial state of the system is completely defined by specifying the energies E_A and E_B, the volumes V_A and V_B, and the mole numbers N_A and N_B. Now we let the constraining wall change to a diathermal wall so that energy can be freely exchanged between systems A and B. After a time the energy exchange stops and the two systems are in equilibrium once again. Since the composite system is isolated, the total energy $E = E_A + E_B$ is constant during this process. In this particular example the volumes and mole numbers of the two systems do not change during the process so the initial and final values are the same. Furthermore, once the final equilibrium state is achieved, we may change the constraining wall back to the adiabatic condition, the original constraint condition, and have no effect on the final thermodynamic state.

To find the final state of the composite system we take a devious route. Let us construct in our imaginations a system that will be in the same state as the final state of the primary system. We begin by taking volumes V_A and V_B and mole numbers N_A and N_B to be the same as those of the original state of the primary system. Since the volumes and mole numbers did not change when the wall became diathermal, we know that these are the correct values for the final state. The problem hinges upon the distribution of the energy between systems A and B. We know the total amount of energy but we don't know how it is to be split. We make an arbitrary choice, say 40% to system A and 60% to system B. This defines a specific thermodynamic state, since we now have fixed the independent properties for each subsystem. There is one final detail we need to consider. The energy distribution we have chosen is probably not in equilibrium when a diathermal wall is used between the systems. Thus, in our imaginary system we use a fixed, solid, adiabatic wall. As we noted above, the final state of the primary system is unaffected if the diathermal wall is changed back to an adiabatic wall.

If we now compare our imaginary system having a 40–60% energy distribution with the final state of the primary system, we shall more than likely find that they are in different states. The chance that we have guessed the proper distribution of energy is very remote. Since it didn't take too much effort to

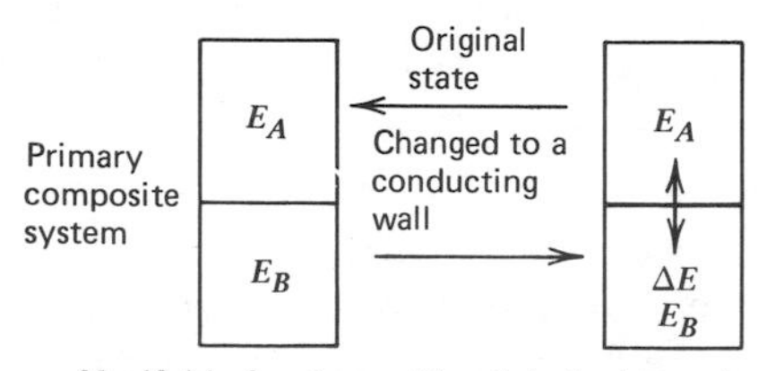

Manifold of systems with adiabatic, impervious, fixed walls

E_{A1}	E_{A2}	E_{A3}	E_{A4}	E_{A5}	E_{A6}	E_{A7}
E_{B1}	E_{B2}	E_{B3}	E_{B4}	E_{B5}	E_{B6}	E_{B7}

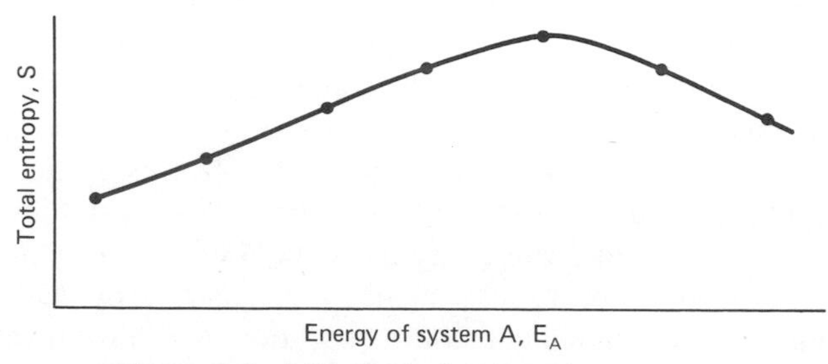

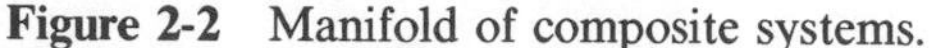

Figure 2-2 Manifold of composite systems.

construct the first imaginary system, we can make another try. This time we put in a different energy distribution, say 39–61%. As we continue this process, we produce what is called a *manifold of constrained systems*. The complete manifold has an infinite number of members, each with a slightly different distribution of energy between systems A and B. All members have the same volume, the same number of molecules, and the same constraining walls as the primary system in its original state. Figure 2-2 depicts this situation.

The dependent thermodynamic variables of each imaginary system have different values. When we give a different distribution of the energies E_A and E_B, even though the total is constant, we find that the temperatures T_A and T_B, the pressures p_A and p_B, and all other dependent thermodynamic variables change (we are taking E, V, N as independent). Also the dependent extensive properties for the composite system change. $H = H_A + H_B$, $G = G_A + G_B$, and $S = S_A + S_B$ take on different values as we move from one member of the manifold of constrained states to another member.

Because the manifold has composite systems with all possible distributions of energy, one of the imaginary systems is in exactly the same state as the final state of the primary system. The question is which one. The basic postulate of thermodynamics says that if we find the member of the manifold that has the largest value of the total entropy, it is in the same state as the primary system. This postulate is the prescription to find the final state.

The ideas developed in the specific example above can be stated in general terms. First we review the problem: a primary composite system is in a definite equilibrium state and has an internal wall of known constraining characteristics. One or more of the constraints are relaxed so that the two subsystems may interact. The independent variables for each subsystem are E, V, N, and depending upon which constraints are relaxed, some independent variables remain fixed and some take on new values. After the interaction is completed, the subsystems are once again in equilibrium. The question is: what is the new state of the primary system?

We solve the problem with the aid of a manifold (or ensemble) of constrained states. The manifold is an imaginary collection of an infinite number of composite systems. The members of the manifold are constructed by the following rules: (a) Each member has the same internal constraining walls as the original primary systems. (b) The total values of E, V, and N are the same for the primary system and for all members of the manifold. (c) Those independent variables that were fixed variables in the primary composite system have those same values for all subsystems in the manifold. (d) Values of the changing independent variables are selected for each member of the manifold so that each member has different values. All possible values of the changing independent variables are represented in the manifold by some member.

The *basic postulate of thermodynamics* is as follows: the final equilibrium state after one or more internal constraints have been relaxed in a primary composite system is the state of the member of the manifold of constrained systems which has the maximum entropy. Thus, we need only find the system in the manifold which has the largest entropy. The values of the independent variables for this system are the same as corresponding variables of the primary system in its final state.

This postulate is in essence the second law stated in terms of a characteristic property of entropy. It will also allow us to establish the physical characteristics of temperature, pressure, and chemical potential. We previously defined these properties in Section 2.5 as certain derivatives of the entropy function. At this stage these definitions are hollow shells; we actually need to explore the characteristics of temperature, pressure, and chemical potential and confirm that the abstract definitions 2.5.2, 2.5.3, and 2.5.4 have the proper physical behavior.

2.8 CHARACTERISTICS OF THE TEMPERATURE

In this section we investigate some of the consequences of defining temperature by 2.5.2:

$$\frac{1}{T} = \left.\frac{\partial S}{\partial E}\right|_{V,N} \tag{2.5.2}$$

To do this we consider a primary composite system with a fixed, solid, and adiabatic wall. Then we relax the adiabatic constraint by changing the wall to a diathermal wall. In this way we are allowing heat transfer between the systems. We shall show, with the aid of the basic postulate of thermodynamics, that the final equilibrium state is one where the temperatures of the two systems are equal. Our first step is to mentally construct a manifold of constrained composite systems. The members of the manifold differ from each other in that they have different energies E_{A1}, E_{A2}, E_{A3}. However, we require that the total energy of the composite system be the same for all members; for instance

$$E = E_{A2} + E_{B2} \tag{2.8.1}$$

Imagine that the members of the sequence are arranged in order of increasing E. Now we compute the total entropy of any one of the composite systems,

$$S = S_A + S_B \tag{2.8.2}$$

The change in entropy between the adjacent members of the manifold is found using 2.5.5 and noting that dV and dN for both system A and system B are zero:

$$dS = dS_A + dS_B$$

$$dS = \frac{1}{T_A} dE_A + \frac{1}{T_B} dE_B$$

Because of the constraint 2.8.1 on the total energy, $dE_A = -dE_B$, so we obtain

$$dS = \left(\frac{1}{T_A} - \frac{1}{T_B}\right) dE_A \qquad 2.8.3$$

This equation gives the entropy change of the composite system in terms of the energy of system A.

The basic postulate of thermodynamics states that the member of the manifold that has the maximum entropy is in the same thermodynamic state as the primary system. Hence we find this member by setting

$$\frac{dS}{dE_A} = 0 \qquad 2.8.4$$

Thus from 2.8.3 we have

$$0 = \frac{1}{T_A} - \frac{1}{T_B} \qquad 2.8.5$$

This means that $T_A = T_B$ for the equilibrium state of the primary system. Furthermore, we can show that our definition of temperature requires energy to go from the high temperature to the low temperature. Consider a composite system that is slightly displaced from equilibrium. Because S is a maximum as the system returns to equilibrium, dS must be positive. If $T_A < T_B$, 2.8.3 shows that dE_A becomes positive as the system receives energy. Thus, the definition of temperature, 2.5.2, has the proper characteristics. Temperature indicates the potential for heat transfer from regions of high temperature to regions of low temperature. When the temperatures of two systems are equal, then the heat transfer is zero.

2.9 CHARACTERISTICS OF THE PRESSURE

The characteristics of the pressure are found by considering a composite system with a solid, fixed, adiabatic wall. The original state of the system is known. In this problem we release two constraints: the internal wall becomes diathermal and movable. Thus, both energy and volume may be exchanged between the systems as the final state is approached.

As our procedure requires, we now imagine a manifold of constrained composite systems with solid, fixed, adiabatic walls. Each member has the same numbers of molecules, N_A and N_B, as the primary systems. Also, each member of the manifold has the same total volume and total energy as the primary system:

$$V = V_A + V_B \qquad 2.9.1$$

$$E = E_A + E_B \qquad 2.9.2$$

The manifold is now doubly infinite. For any fixed values of V_A and V_B, all possible values of E_A and E_B are represented by distinct members of the manifold.

The change in entropy between members of the manifold may be found from 2.5.5:

$$dS = dS_A + dS_B$$

$$= \frac{1}{T_A} dE_A + \frac{p_A}{T_A} dV_A + \frac{1}{T_B} dE_B + \frac{p_B}{T_B} dV_B \qquad 2.9.3$$

This is simplified by differentiating the conditions 2.9.1 and 2.9.2 and substituting:

$$dS = \left(\frac{1}{T_A} - \frac{1}{T_B}\right) dE_A + \left(\frac{p_A}{T_A} - \frac{p_B}{T_B}\right) dV_A \qquad 2.9.4$$

The entropy of any member of the manifold is a function of two variables, E_A and V_A. When a maximum of S occurs we have

$$\frac{\partial S}{\partial E_A} = 0 \quad \text{and} \quad \frac{\partial S}{\partial V_A} = 0 \qquad 2.9.5$$

These derivatives are the coefficients of dE_A and dV_A in 2.9.4. Hence for a maximum of S we must have thermal equilibrium

$$T_A = T_B \qquad 2.9.6$$

and mechanical equilibrium

$$p_A = p_B \qquad 2.9.7$$

The effect of pressure equilibrium may be attributed completely to the movability of the internal wall, because we have previously found that temperature is associated with the diathermal characteristics of the wall.

The reader might have expected that we would investigate the characteristics of pressure by changing the wall constraint from fixed to movable while retaining the solid and adiabatic constraints. The problem stated in this way does not yield a unique solution. From a physical point of view releasing the wall sets up an oscillation as it overshoots and then returns in the opposite direction. The oscillation would continue indefinitely if it were not for the fact that the compression and expansion waves generated by the wall are damped by viscosity. Thus, energy from the pistonlike motion is deposited within a system by viscous dissipation. This means that the final distribution of energy between the systems depends on the details of an irreversible process. If we changed the viscosity of one system, a different final state would ensue (cf. Callen, 1960, Appendix C). It is only by changing the wall to a diathermal wall that energy can be redistributed between the systems so that a definite final state is attained.

2.10 EULER'S EQUATION FOR HOMOGENEOUS FUNCTIONS

Euler derived a relationship for mathematical functions which are homogeneous and of degree one. Since entropy is a function of this type we may derive the relationship with that in mind. The result actually proves that the three equations of state contain all the thermodynamic information and are completely equivalent to the fundamental equation 2.4.1. We begin by noting that S, E, V, and N are all extensive properties. Suppose that two systems are in the same intensive state, but one is λ times larger than the other. The entropy of the second system is λ times the entropy of the first:

$$S(\lambda E, \lambda V, \lambda N) = \lambda S(E, V, N) \qquad 2.10.1$$

Differentiation of this expression with respect to λ yields

$$\frac{\partial S(\lambda E, \lambda V, \lambda N)}{\partial(\lambda E)} \frac{d(\lambda E)}{d\lambda} + \frac{\partial S(\lambda E, \lambda V, \lambda N)}{\partial(\lambda V)} \frac{d(\lambda V)}{d\lambda}$$

$$+ \frac{\partial S(\lambda E, \lambda V, \lambda N)}{\partial(\lambda N)} \frac{d(\lambda N)}{\partial\lambda} = S(E, V, N)$$

which may be written as

$$\frac{\partial S(\lambda E, \lambda V, \lambda N)}{\partial(\lambda E)} E + \frac{\partial S(\lambda E, \lambda V, \lambda N)}{\partial(\lambda V)} V$$

$$+ \frac{\partial S(\lambda E, \lambda V, \lambda N)}{\partial(\lambda N)} N = S(E, V, N)$$

Now note that

$$\frac{\partial S(\lambda E, \lambda V, \lambda N)}{\partial(\lambda E)} = \frac{\partial S}{\partial E}(E, V, N)$$

Similar statements are true for the other partial derivatives; therefore the relation above reduces to

$$\left.\frac{\partial S}{\partial E}\right|_{V,N} E + \left.\frac{\partial S}{\partial V}\right|_{E,N} V + \left.\frac{\partial S}{\partial N}\right|_{E,V} N = S \qquad 2.10.2$$

Substituting 2.5.2, 2.5.3, and 2.5.4 produces Euler's equation as applied to thermodynamics:

$$S = \frac{1}{T}E + \frac{p}{T}V + \frac{\mu}{T}N \qquad 2.10.3$$

Substitution into 2.10.3 of known equations of state for T, p, and μ produces the fundamental equation without any integration constants. Thus, we have shown complete equivalence between the fundamental equation 2.4.1 the three equations of state 2.5.2, 2.5.3 and 2.5.4.

2.11 THE GIBBS–DUHEM EQUATION

The Gibbs–Duhem equation shows that the three equations of state are not completely independent. The derivation is very simple. Differentiate Euler's equation (2.10.3) to arrive at

$$dS = \frac{1}{T}dE + \frac{p}{T}dV + \frac{\mu}{T}dN + E\,d\left(\frac{1}{T}\right) + V\,d\left(\frac{p}{T}\right) + N\,d\left(\frac{\mu}{T}\right) \qquad 2.11.1$$

Subtracting the fundamental differential equation (2.5.5) gives the final result,

$$0 = E\,d\left(\frac{1}{T}\right) + V\,d\left(\frac{p}{T}\right) + N\,d\left(\frac{\mu}{T}\right) \qquad 2.11.2$$

Application of the Gibbs–Duhem equation can be illustrated by assuming that two equations of state,

$$\frac{1}{T}(E, V, N) \quad \text{and} \quad \frac{p}{T}(E, V, N)$$

are given and we want to find the third equation,

$$\frac{\mu}{T}(E, V, N)$$

From 2.11.2 we may formulate the answer as the integral

$$\frac{\mu}{T} = \int d\left(\frac{\mu}{T}\right) + \text{const} = -\int \frac{E}{N}d\left(\frac{1}{T}\right) - \int \frac{V}{N}d\left(\frac{p}{T}\right) + C \qquad 2.11.3$$

The conclusion is that knowledge of just two equations of state determines all thermodynamic information with the exception of a constant. For practical applications this is all that is required (in fact we cannot determine the absolute energy anyway).

2.12 ALTERNATE FORMS OF THE FUNDAMENTAL EQUATION

The fundamental equation of a substance was proposed in the form $S = S(E, V, N)$ as containing all the thermodynamic information. Obviously we do not lose any information if we invert the equation and use energy as a dependent variable (it is a fact that S is a monotonic function of E):

$$E = E(S, V, N) \tag{2.12.1}$$

The fundamental differential equation corresponding to 2.12.1 is

$$dE = T\,dS - p\,dV - \mu\,dN \tag{2.12.2}$$

Other forms of the fundamental equation are also possible. Consider the definition of enthalpy,

$$H = E + pV$$

Taking the derivative of the equation above gives

$$dH = dE + p\,dV + V\,dp \tag{2.12.3}$$

If we substitute 2.12.2 into 2.12.3 we arrive at

$$dH = T\,dS + V\,dp - \mu\,dN \tag{2.12.4}$$

This is the fundamental differential equation for a function

$$H = H(S, p, N) \tag{2.12.5}$$

There is a formal method, the Legendre transformation, of changing a differential equation such as 2.12.2 into new variables in such a way that no information is lost. We shall accept the fact that the function $H(S, p, N)$ contains all thermodynamic information [but $H(E, V, N)$ would not]. It is not by accident that the properties H, G, and F have been defined in a certain way. They are all capable of containing all thermodynamic information when expressed in certain sets of independent variables. Sometimes these functions are called thermodynamic potential functions or canonical equations in order to distinguish them from a thermodynamic equation which would contain only a partial description of the substance.

2.13 INTENSIVE FORMS

So far in our development we have dealt with systems, and the thermodynamic properties have been considered to be properties of the system. When the size of the system was increased, all extensive properties increased. This simple dependence of thermodynamic properties upon system size allows us to formulate thermodynamics on an intensive basis. There are several choices: a unit-volume basis, a unit-mole basis, a unit-mass basis are all used. Since fluid mechanics deals with Newton's second law, it is most useful for us to employ a unit-mass basis. A lowercase letter will be used for any extensive property that is put on a unit-mass basis:

$$x \equiv \frac{X}{\mathcal{M}N} \tag{2.13.1}$$

where $\mathcal{M}$ is the mass of one mole of substance.

Consider again the fundamental equation for systems of different sizes (2.10.1):

$$\lambda S(E, V, N) = S(\lambda E, \lambda V, \lambda N)$$

Let us view this as a strictly mathematical equation. Since λ is an arbitrary number, let us choose $\lambda = 1/\mathcal{M}N$; then the entropy per unit mass is

$$s \equiv \frac{S}{\mathcal{M}N} = S\left(\frac{E}{\mathcal{M}N}, \frac{V}{\mathcal{M}N}, \frac{1}{\mathcal{M}}\right)$$

Because we want to reserve the lowercase v for velocities, replace the specific volume by the density:

$$s = S\left(e, \frac{1}{\rho}, \frac{1}{\mathcal{M}}\right) \tag{2.13.2}$$

$$= s(e, \rho) \tag{2.13.3}$$

Equation 2.13.2 says that simply replacing the extensive symbols S, E, V by their intensive counterparts and replacing N by $1/\mathcal{M}$ gives an equation which has only intensive variables. From another point of view, it means that N must occur in the fundamental equation in such a way that it may be divided through and grouped with E and V. Another aspect of 2.13.2 which needs more explanation is the appearance of the molar mass $\mathcal{M}$, numerically equal to the molecular mass ratio W (molecular weight). The fundamental equation $S(E, V, N)$ contains a universal dimensional constant R_0 (the gas constant), which we do not list as a variable. In the intensive form 2.13.2, $\mathcal{M}$ appears explicitly, but it has been dropped in the next line. What actually happens is that $\mathcal{M}$ in 2.13.3 has been absorbed into the dimensional constant by defining a specific gas constant $R \equiv R_0/\mathcal{M}$. Even though this "constant" depends upon the molecular weight (molecular mass ratio) of the substance, it is not customary to list R as a variable. (If we had formulated thermodynamics using the mass M instead of the number of moles, N—for example, if we had taken $S(E, V, M)$ as our fundamental equation—then we would have arrived at 2.13.3 directly.)

The fundamental differential equation for the intensive function $s(e, \rho)$ is

$$ds = \frac{1}{T}de - \frac{p}{\rho^2 T}d\rho \qquad 2.13.4$$

There are two intensive equations of state corresponding to 2.5.2 and 2.5.3:

$$\frac{1}{T} = \left.\frac{\partial s}{\partial e}\right|_{\rho}, \qquad \text{that is,} \quad T = T(e, \rho) \qquad 2.13.5$$

and

$$\frac{p}{\rho^2 T} = -\left.\frac{\partial s}{\partial \rho}\right|_{e}, \qquad \text{that is,} \quad p = p(e, \rho) \qquad 2.13.6$$

The third equation of state cannot be derived directly from $s(e, \rho)$, but is found either by integration of the Gibbs–Duhem equation (2.11.3) or by substitution into the intensive form of Euler's equation (2.10.3):

$$\frac{\mu}{\mathcal{M}} = Ts - e - \frac{p}{\rho} = -(h - Ts) = -g \qquad 2.13.7$$

Dividing the chemical potential by $\mathcal{M}$ changes it from a mole basis to a unit-mass basis. The combination $h - Ts$ is defined as the Gibbs free energy. For a pure substance the Gibbs free energy is the negative of the chemical potential.

2.14 THE PERFECT GAS

All gases behave as perfect gases when the pressure is low compared to the critical pressure. The fundamental equation for a perfect gas with constant specific heats (the Sackur–Tetrode equation) is as follows:

$$S = S_0 \frac{N}{N_0} + NR_0 \ln\left[\left(\frac{E}{E_0}\right)^{C}\left(\frac{V}{V_0}\right)\left(\frac{N}{N_0}\right)^{-C-1}\right] \qquad 2.14.1$$

where the exponent is $C = C_V/R_0$, and S_0, E_0, V_0, N_0 correspond to a reference state. The same equation in intensive form is

$$s = s_0 + c_v \ln\frac{e}{e_0} - R\ln\frac{\rho}{\rho_0} \qquad 2.14.2$$

This equation is on a unit-mass basis with $c_v = C_V/\mathcal{M}$ and $R = R_0/\mathcal{M}$. These equations are approximations which are valid at high temperatures. Intensive equations of state are found using the definitions 2.13.5 and 2.13.6. They are

$$\frac{1}{T} = \left.\frac{\partial s}{\partial e}\right|_{\rho} = c_v \frac{1}{e}, \qquad \text{or} \quad e = c_v T \qquad 2.14.3$$

and

$$-\frac{p}{\rho^2 T} = \left.\frac{\partial s}{\partial T}\right|_{e} = -\frac{R_0}{\mathcal{M}\rho}, \qquad \text{or} \quad p = \rho R T \qquad 2.14.4$$

Note that the energy of a perfect gas depends only upon the temperature.

The validity of the perfect-gas law is most easily assessed by using the compressibility function. The *compressibility factor* is defined as

$$Z \equiv \frac{p}{\rho R T} = Z(P_r, T_r) \qquad 2.14.5$$

where P_r and T_r are the *reduced pressure* and *reduced temperature*, the values nondimensionalized by the temperature and pressure at the critical point p_c, T_c:

$$P_r \equiv \frac{p}{p_c}, \qquad T_r \equiv \frac{T}{T_c} \qquad 2.14.6$$

Figure 2-3 is a plot of $Z(P_r, T_r)$ constructed from experimental data on 26 different gases with simple molecules. These gases all fit the chart within 2.5%. (If the molecular structure of a gas is very complex the Z function has the same character with a slightly different shape.) From this chart one can see that the perfect-gas law ($Z = 1$) is valid not only at low pressures, but also for higher pressures as long as the temperature is high.

The assumption that the specific heat of a gas is constant is reasonable for modest temperature ranges. From 2.14.3 we interpret C_V as a proportionality constant between internal energy and temperature. In gases composed of monatomic molecules the only form of internal energy is the kinetic energy of random translational motion. Because this is the same motion that the temperature measures, we expect a constant value of c_v. Kinetic theory, in fact, predicts that $C_V/R_0 = \frac{3}{2}$. In more complex molecules internal energy also resides in molecular rotation and possibly in vibrations between atoms. A rule of physics, the equipartition of energy, says roughly that energy will be equally distributed between all available distinct modes of motion. A key word in the rule is "available". Some modes have their first quantum energy level so high that they are not excited at ordinary temperatures.

For instance, a diatomic molecule has three translational modes but only two rotational modes; rotation about the axis connecting the atoms is not available at room temperature. A diatomic molecule has a value of $C_v/R_0 = \frac{5}{2}$: more energy is needed to raise the temperature. At extremely high tempera-

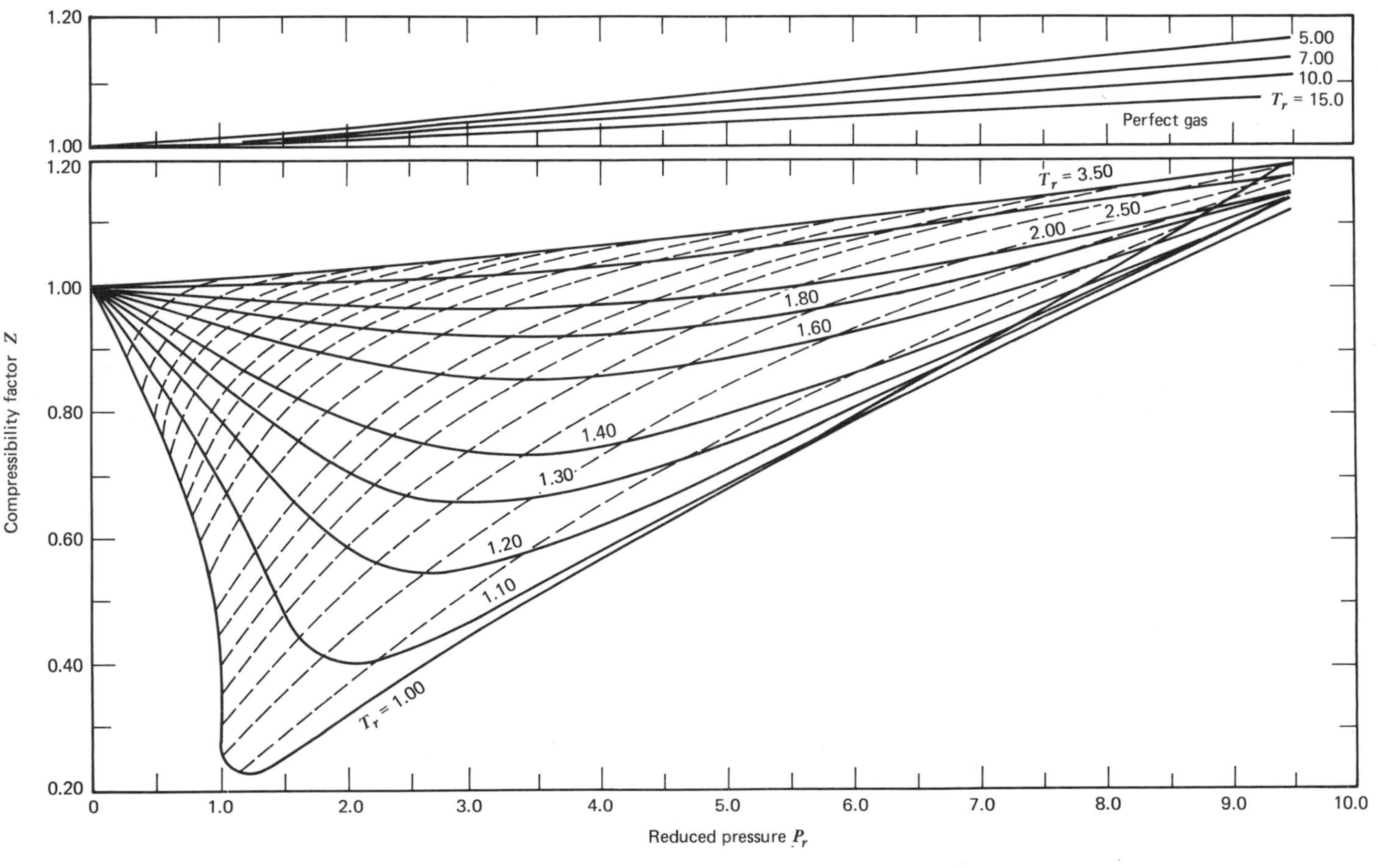

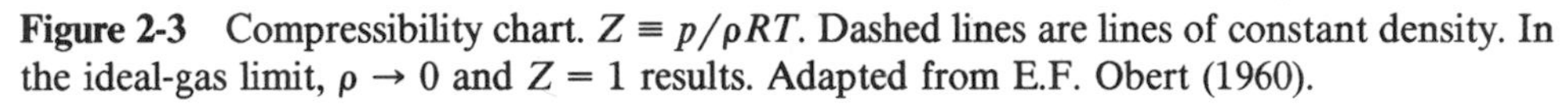

Figure 2-3 Compressibility chart. $Z \equiv p/\rho RT$. Dashed lines are lines of constant density. In the ideal-gas limit, $\rho \to 0$ and $Z = 1$ results. Adapted from E.F. Obert (1960).

tures a vibration mode begins to appear. This mode is not yet fully excited when disassociation of the atoms begins to occur.

2.15 MAXWELL RELATIONS

There are an amazing number of different ways to express thermodynamic information: $s = s(e, \rho)$, $e = e(s, \rho)$, $T = T(\rho, p)$, $T = T(e, \rho)$, and so on. These functions are not all completely independent of each other. Likewise, the partial derivatives of the various functions are not completely independent. Certain theoretical relationships between two partial derivatives of thermodynamic functions are given by Maxwell relations. A Maxwell relation is most easily derived if we turn the fundamental equation around and consider e as a function of s and ρ:

$$e = e(s, \rho) \tag{2.15.1}$$

The differential of this function is

$$de = \left.\frac{\partial e}{\partial \rho}\right|_s d\rho + \left.\frac{\partial e}{\partial s}\right|_\rho ds \tag{2.15.2}$$

Now consider the fundamental differential equation 2.13.4 arranged as follows:

$$de = \frac{p}{\rho^2} d\rho + T\,ds \tag{2.15.3}$$

Comparing the two equations above, we may equate the coefficients of $d\rho$ and ds. This produces intensive equations of state equivalent to 2.5.2 and 2.5.3. They are

$$\left.\frac{\partial e}{\partial \rho}\right|_s = p\rho^{-2}, \qquad \left(\frac{\partial e}{\partial s}\right)_\rho = T \tag{2.15.4}$$

A Maxwell relation is derived by noting that the mixed partial derivative of $e = e(s, \rho)$ can be formed in either order; that is

$$\frac{\partial^2 e}{\partial s\,\partial \rho} = \frac{\partial^2 e}{\partial \rho\,\partial s} \tag{2.15.5}$$

Therefore, if we substitute 2.15.4 into 2.15.5, we find

$$\rho^{-2} \left.\frac{\partial p}{\partial s}\right|_\rho = \left.\frac{\partial T}{\partial \rho}\right|_s \tag{2.15.6}$$

This relation is known as the Maxwell relation from $e(s, \rho)$.

There are many Maxwell relations, because we can express the fundamental differential equation in a variety of forms. Let us consider another example and at the same time obtain a relation for use in a later section. Consider the Helmholtz free energy, which is defined by

$$f \equiv e - Ts \tag{2.15.7}$$

Differentiation gives

$$df = de - T\,ds - s\,dT \tag{2.15.8}$$

Next, we substitute the fundamental differential equation 2.13.4 to obtain a fundamental differential equation for f. The result is

$$df = -s\,dT + p\rho^{-2}\,d\rho \tag{2.15.9}$$

From this equation we see that f as a function of T and ρ will be a fundamental thermodynamic function. The Maxwell relation from f is computed by noting that the coefficients of dT and $d\rho$ in 2.15.9 are the first derivatives of f, that is,

$$\left.\frac{\partial f}{\partial T}\right|_{\rho} = -s, \qquad \left.\frac{\partial f}{\partial \rho}\right|_{T} = \frac{p}{\rho^2} \tag{2.15.10}$$

Differentiating the first equation in 2.15.10 with respect to ρ, the second with respect to T, and then equating gives the Maxwell relation

$$\left.\frac{\partial s}{\partial \rho}\right|_{T} = -\rho^{-2}\left.\frac{\partial p}{\partial T}\right|_{\rho} \tag{2.15.11}$$

Other Maxwell relations are derived by considering the functions $h(s, p)$ and $g(T, p)$.

2.16 WORKING EQUATIONS

Many different choices of independent and dependent variables may be made in thermodynamics, each being advantageous for one type of problem or another. The most common way to specify thermodynamics information is by two equations of state of the form $e = e(T, \rho)$ and $p = p(T, \rho)$. From the theoretical standpoint these equations are connected to the equations of state 2.13.5 and 2.13.6 in the following way. The first is readily found by solving 2.13.5, $T = T(e, \rho)$, for e. This is always possible because T is always a monotonically increasing function of e. The second form is found by substituting the relation $e = e(T, \rho)$ into 2.13.6: $p = p(e, \rho) = p(e = e(T, \rho), \rho)$.

Knowledge of the thermodynamic properties of a substance has now been reduced to finding two equations of state. This requires two types of experiments: a p–ρ–T measurement and an energy measurement. The p–ρ–T measurements can be made with great accuracy, but the energy measurement (adding heat to a system and measuring its change in temperature) is very imprecise. The modern method of finding the energy equation of state is to use spectroscopy to find the specific heat

$$c_v = \left(\frac{\partial e}{\partial T}\right)_\rho \tag{2.16.1}$$

If c_v is known as a function of temperature at one value of the density, it is possible to determine the energy equation of state $e = e(T, \rho)$. We shall now derive the equation that will show how this is done. This equation will also be useful in later sections because it will allow us to replace the energy variable in equations with the more practical variable temperature.

If $e = e(T, \rho)$, then the derivative is

$$de = \left.\frac{\partial e}{\partial T}\right|_\rho dT + \left.\frac{\partial e}{\partial \rho}\right|_T d\rho \tag{2.16.2}$$

The left-hand partial derivative is, by the definition 2.16.1, the specific heat. The right-hand one is changed into a useful form as follows. The two functions

$$e = e(s, \rho) \quad \text{and} \quad s = s(T, \rho) \tag{2.16.3}$$

can be combined to get

$$e = e(s(T, \rho), \rho) \tag{2.16.4}$$

By the chain rules of differentiation we have

$$\left.\frac{\partial e}{\partial \rho}\right|_T = \left.\frac{\partial e}{\partial s}\right|_\rho \left.\frac{\partial s}{\partial \rho}\right|_T + \left.\frac{\partial e}{\partial \rho}\right|_s \tag{2.16.5}$$

Substitution of both equations of 2.15.4 and the Maxwell relation 2.15.11 into 2.16.5 gives

$$\left.\frac{\partial e}{\partial \rho}\right|_T = -\frac{T}{\rho^2}\left.\frac{\partial p}{\partial T}\right|_\rho + \frac{p}{\rho^2} \tag{2.16.6}$$

This derivative depends only upon the p–ρ–T equation of state. Thus, the final form for 2.16.2 is

$$de = c_v\, dT + \left[\frac{p}{\rho^2} - \frac{T}{\rho^2}\left.\frac{\partial p}{\partial T}\right|_\rho\right] d\rho \tag{2.16.7}$$

The energy is found by integrating the first term to the required temperature at constant density employing the experimental value of c_v, then substituting the p–ρ–T data into the bracketed term and integrating (at constant temperature) to the required density. Note that 2.16.7 is completely general and is valid for variable $c_v = c_v(\rho, T)$.

A working differential equation for the entropy is found by substituting 2.16.7 into 2.13.4. This produces $s = s(\rho, T)$.

Later on in our theoretical work we shall want to know how sensitive the density is to changes in pressure and temperature. The equation used for this purpose is

$$\frac{d\rho}{\rho} = \frac{1}{\rho} \left.\frac{\partial \rho}{\partial T}\right|_p dT + \frac{1}{\rho} \left.\frac{\partial \rho}{\partial p}\right|_T dp \qquad 2.16.8$$

The coefficients in this equation are known as the *bulk expansion coefficient* (or thermal expansion coefficient)

$$\beta(p, T) \equiv -\frac{1}{\rho} \left.\frac{\partial \rho}{\partial T}\right|_p \qquad 2.16.9$$

and the *isothermal compressibility coefficient*

$$\alpha(p, T) \equiv \frac{1}{\rho} \left.\frac{\partial \rho}{\partial p}\right|_T \qquad 2.16.10$$

Both of these functions may be found from the p–ρ–T equation of state. Substituting these definitions into 2.14.1 yields

$$\frac{d\rho}{\rho} = \alpha\, dp - \beta\, dT \qquad 2.16.11$$

This is essentially a differential form of the p–ρ–T equation of state.

2.17 THE SPEED OF SOUND

The speed of sound in a substance is another thermodynamic property we will need in later chapters. To illustrate how the speed of sound is important in fluid mechanics we consider the following problem. Suppose a certain steady flow is established. Now, for some reason the body is moved, or a portion of the flow passage changes its position. The flow field must adjust itself and reach a new steady state corresponding to the new geometry. The first adjustment is accomplished by waves that travel from the moving surface and change the flow field. Several reflections may be required before the new steady state is reached. Also, the speed of the waves can depend upon their strength. The lowest wave speed is that for an infinitesimal motion of the wall. It is the speed of sound.

The equations of acoustics are a special case of the general equations of fluid mechanics. When these acoustic equations are formulated it turns out that the speed of wave propagation is a thermodynamic property. The speed of sound, a, is given by

$$a = \left(\left. \frac{\partial p}{\partial \rho} \right|_s \right)^{1/2} \qquad 2.17.1$$

The appearance of s as the property held constant is the result of the fact that acoustic waves are reversible adiabatic processes.

We can find a more convenient equation for the speed of sound where only derivatives of the p–ρ–T equation of state occur. It is easiest to approach this equation by an indirect path. The properties of partial derivatives result in the following equation:

$$1 = \left. \frac{\partial T}{\partial e} \right|_\rho \left. \frac{\partial e}{\partial s} \right|_\rho \left. \frac{\partial s}{\partial T} \right|_\rho \qquad 2.17.2$$

Substituting 2.13.5 and 2.16.1 into 2.17.2 gives

$$c_v = T \left. \frac{\partial s}{\partial T} \right|_\rho \qquad 2.17.3$$

Next, we recall another partial-derivative relation,

$$\left. \frac{\partial s}{\partial T} \right|_\rho \left. \frac{\partial T}{\partial \rho} \right|_s \left. \frac{\partial \rho}{\partial s} \right|_T = -1 \qquad 2.17.4$$

and substitute this into 2.17.3 to obtain the intermediate result

$$c_v = -T \left. \frac{\partial \rho}{\partial T} \right|_s \left. \frac{\partial s}{\partial \rho} \right|_T \qquad 2.17.5$$

A set of similar equations can be formulated to derive the following equation for c_p:

$$c_p = -T \left. \frac{\partial p}{\partial T} \right|_s \left. \frac{\partial s}{\partial p} \right|_T \qquad 2.17.6$$

Dividing 2.17.6 by 2.17.5 and simplifying with relations similar to 2.17.2 produces

$$\gamma \equiv \frac{c_p}{c_v} = \frac{\left. \frac{\partial p}{\partial T} \right|_s \left. \frac{\partial s}{\partial p} \right|_T}{\left. \frac{\partial \rho}{\partial T} \right|_s \left. \frac{\partial s}{\partial \rho} \right|_T} = \left. \frac{\partial p}{\partial \rho} \right|_s \left. \frac{\partial \rho}{\partial p} \right|_T \qquad 2.17.7$$

The final equation is found from the definitions of a (2.17.1) and α, (2.16.10) and the use of 2.17.7. The speed of sound is given by

$$a^2 = \frac{\gamma}{\rho\alpha} \tag{2.17.8}$$

Through the formulas above we see that the speed of sound may be related to the isothermal compressibility of the fluid.

2.18 INCOMPRESSIBLE SUBSTANCE

A completely incompressible fluid has zero for both the compressibility coefficient α and the thermal expansion coefficient β. This leads to an infinite speed of sound. Obviously, if we were considering an acoustic problem in a liquid, this would not be a very good approximation.

Frequently, however, we make the approximation that the density of a liquid (or solid) is constant and not a thermodynamic variable. The thermodynamic theory in this approximation has one less independent variable than that for a simple compressible substance. The fundamental equation for such a substance is

$$s - s_0 = c_v \ln \frac{e}{e_0} \tag{2.18.1}$$

The corresponding fundamental differential equation (from 2.13.4) is

$$ds = \frac{1}{T} de \tag{2.18.2}$$

The thermal equation of state is found as

$$\frac{1}{T} = \left.\frac{\partial s}{\partial e}\right|_\rho = \frac{c_v}{e}$$

or

$$e = c_v T \tag{2.18.3}$$

The specific heats of liquids and solids are approximately constant.

Recall that the pressure equation of state is usually found from the definition (2.13.6),

$$\frac{p}{T} = -\rho^2 \left.\frac{\partial s}{\partial \rho}\right|_e \tag{2.18.4}$$

For an incompressible substance we cannot find the pressure because s is not a function of ρ. To be more precise, we cannot define a "thermodynamic" pressure. There is, of course, a pressure in an incompressible fluid (normal stress in a solid); however, it does not influence the thermodynamic state. Here the pressure is simply the force per unit area, and we say that it is only a *mechanical pressure* variable. This leads to conceptual difficulties in some previously defined variables which involve pressure. For instance, the enthalpy $h = e + p/\rho$ is no longer a purely thermodynamic variable.

A change in enthalpy has a thermodynamic part and a mechanical part:

$$\Delta h = \Delta e + \frac{1}{\rho}\Delta p$$

$$= c_v \Delta T + \frac{1}{\rho}\Delta p$$

As a practical matter the mechanical part is usually much smaller than the thermodynamic part. A conclusion that follows immediately is that an incompressible substance has only one specific heat:

$$c_p \equiv \left.\frac{\partial h}{\partial T}\right|_p = \frac{\partial}{\partial T}[e + p(\rho)] = \frac{de}{dT} = c_v \tag{2.18.5}$$

The ratio of specific heats, $\gamma = c_p/c_v$, is thus unity.

2.19 CONCLUSION

Thermodynamics, as we have developed it, deals with systems containing a uniform fluid at rest. We need to apply thermodynamics to a moving continuum where each point in space has different properties.

Recall that the size of a system can be effectively eliminated by making use of intensive variables. Since we may choose a control volume in any manner whatsoever, we may consider any point in the continuum and define the internal energy and density exactly as we did in Chapter 1. These two independent quantities, together with the equations of state of the substance, determine all the intensive thermodynamic properties at that point. Hence, we have no difficulty in putting thermodynamics on a local basis.

The extension of these ideas to a moving continuum is very simple, but it requires an additional assumption. The assumption is that the bulk motion of the fluid does not affect the thermodynamic state. From the microscopic view this assumption means that if we subtract from each particle the average motion of all the molecules, we will still have the same pattern of random molecular motion that would exist without the bulk motion. In other words, the bulk motion does not affect the statistical averages of the molecular properties. The thermodynamic properties, in effect, are determined by an observer floating with the local fluid velocity.

PROBLEMS

2.1 (A) Which of the following are simple thermodynamic systems? (a) A cavity that contains a liquid where the top surface is a moving belt. (b) A container with a liquid in one-half in equilibrium with its vapor in the other half. (c) A steel rod being pulled in a tensile test machine.

2.2 (A) Is the ratio of two extensive properties a thermodynamic property? Is the ratio S/E intensive or extensive?

2.3 (A) Find the three equations of state for a system with the fundamental equation given below. Substitute the state equations into Euler's equation.

$$S = R_0 N \left(\frac{E}{E_0} \right)^{1/2} \left(\frac{V}{V_0} \right) \left(\frac{N}{N_0} \right)^{-3/2}$$

2.4 (A) If the internal energy E is considered as the dependent variable in the fundamental differential equation, what are the proper definitions for T, p, and μ?

2.5 (B) Consider a composite system containing a monatomic perfect gas (see 2.14). System A contains twice the number of molecules that system B contains. The wall between the systems is fixed, adiabatic, and solid. Plot a graph of the entropy of a manifold of composite systems where $E_A + E_B = E$. Use E_A/E as the independent variable and construct a nondimensional dependent variable as

$$\frac{S - S_{\text{reference}}}{\text{scaling constant}}$$

You need not explicitly state what the reference constant is, but you must define the scaling constant exactly. Where is the maximum of the curve? Compute the temperatures T_A and T_B. Show that they are equal when S is a maximum.

2.6 (A) Consider an insulated frictionless piston–cylinder system containing a gas. (a) A weight is suddenly added to the piston, so that it must seek a new equilibrium value. The equations of motion for the pressure show that the piston will oscillate. Why will the piston stop oscillating? (b) Extremely small weights are added one at a time over a long period of time. The final total weight is the same as in part (a). Will the final thermodynamic state be the same as in part (a)? Discuss the work done in each instance. How is this problem relevant to the situation in a composite system with a fixed, adiabatic, impervious wall, which is changed to a movable, adiabatic, impervious wall?

2.7 (B) Using the appropriate manifold of constrained systems, derive the characteristics of the chemical potential μ as defined by 2.5.4.

2.8 (B) A Mollier chart is a plot of $h(p, s)$ in the form of h as a function of s for lines of constant pressure. A chart for a certain substance has two lines given by the following equations:

at 30 psia $\qquad h = 1050s - 650 \text{ Btu/lb}_m$

at 35 psia $\qquad h = 1050s - 630 \text{ Btu/lb}_m$

At the point $s = 1.8 \text{ Btu/lb}_m\,°\text{R}$ and $p = 30$, find the temperature, density, and internal energy.

2.9 (B) The fundamental differential equation of thermostatics may be rearranged so that different variables are used as dependent and independent. Suppose that the Gibbs function G is to be taken as the dependent variable. What independent variables must be used if the result is to be a fundamental equation and contain all the thermodynamic information about the system?

2.10 (A) Find the equation of state for the chemical potential of a monatomic gas (2.14) by two methods: differentiation of the fundamental equation, and integration of the Gibbs–Duhem equation.

2.11 (A) Find the Maxwell relation associated with the enthalpy function.

2.12 (A) Find the Maxwell relation associated with the Gibbs free-energy function.

2.13 (B) Verify the following equation is valid for the enthalpy function $h(T, p)$:

$$dh = c_p\, dT + \left[\frac{1}{\rho} + \frac{T}{\rho^2} \left. \frac{\partial \rho}{\partial T} \right|_p \right] dp$$

(This problem uses results from 2.12, 2.13, and 2.17.)

2.14 (C) Verify the following relation:

$$c_p - c_v = \frac{T}{\rho} \frac{\beta^2}{\alpha}$$

Begin by considering $s = s(T, p)$ as a composite function formed from $s = s(T, \rho)$ and $\rho = \rho(T, p)$.

2.15 (A) Using the expression given in Problem 2.14, show that a perfect gas gives

$$c_p - c_v = R$$

2.16 (B) Prove that the definitions of c_v and c_p given in section 2.4 are equivalent to the customary definitions,

$$C_V = \left.\frac{\partial E}{\partial T}\right|_V, \qquad C_p = \left.\frac{\partial H}{\partial T}\right|_p$$

(Hint: Take note of 2.17.2).

2.17 (A) Consider an incompressible fluid. What is the expression for Δh for arbitrary changes ΔT and Δp?

2.18 (A) Find α and β for a perfect gas.

2.19 (A) Derive 2.17.6.

2.20 (A) Find the formula for the speed of sound of a perfect gas.

2.21 (A) Find a differential equation for the entropy $s(\rho, T)$ which could be integrated for known functions p–ρ–T and c_v $(T, \rho = \text{const})$.

2.22 (A) Derive the entropy equation for a perfect gas with a constant specific heat using the result of Problem 2.21.

2.23 (B) A standard form of the Gibbs–Duhem equation is

$$-S\,dT + V\,dp + N\,d\mu = 0$$

Derive this equation from 2.12.2.

3 Vector Calculus and Index Notation

Mathematics is the language we use to quantify our physical ideas. The development of mathematics and the development of science have taken place simultaneously, with a great deal of interaction. In some instances scientific needs have inspired mathematical progress, while in others, originally abstract mathematical results have found later applications to science. Nevertheless, the best pedagogical viewpoint is to separate the subjects and clearly distinguish between physical and mathematical assumptions.

The purpose of this chapter is to introduce vector and tensor calculus. There are two ways we can approach the subject. One approach uses "symbolic" or "Gibbs" notation, and the other uses "index" or "Cartesian" notation. When Gibbs notation is employed we are essentially looking at vector calculus as a separate mathematical subject. Scalars, vectors, and tensors are viewed as different types of things. That is to say, a vector is a single entity with special mathematical properties. The plus sign between two vectors has a different meaning than a plus sign between two scalars. We must make new definitions for vector and tensor addition, multiplication, integration, and so on. To study vector calculus in the Gibbs notation requires us to define many new operations and investigate which are allowed and which are disallowed. For example, if **a**, **b**, and **c** are vectors, we all know the following fact:

$$(\mathbf{a} \cdot \mathbf{b})\mathbf{c} \neq \mathbf{a}(\mathbf{b} \cdot \mathbf{c})$$

The associative law is not true for multiplication.

There is a certain economy of effort in using this symbolic notation and many experienced workers prefer to use it. These workers, of course, know what is legal and what is illegal. Another advantage of the notation is a philosophical one. The symbols make no specific reference to a coordinate system. This, however, is of little consequence to us, as we are not usually satisfied with an abstract result, but must write out a component equation in order to find a numerical result.

The other approach to vector analysis uses index notation. This notation always deals with scalar variables. Whenever we write an equation we use the scalar component of a vector or the scalar component of a tensor. In this way we don't have to worry about legal and illegal operations. All our previous knowledge of algebra and calculus is immediately applicable. It is true that we will need a few new symbols and rules, but they will be essentially shorthand conventions. Index notation is frequently thought of as being restricted to a Cartesian coordinate system. This is its most straightforward interpretation, but it is not actually restricted to Cartesian systems. One can actually use index notation for any orthogonal coordinate system. (Another type of index notation with subscripts and superscripts applies to nonorthogonal systems as well.) To do so requires that we generalize the meaning of the symbols slightly. Since most vector-calculus results are tabulated in the common coordinate systems, we can usually find them (see Appendix B) and can use them without delving into their derivation.

Workers in fluid mechanics must have a knowledge of both symbolic and index notations. The two notations are used with equal frequency in the literature. Our plan will be to learn how to convert expressions from one notation into the other. In this way we can perform all algebra and calculus operations in index notation, and then as the final step convert the equation to Gibbs notation. Similarly, when we encounter an equation in symbolic notation, we shall be able to write out its equivalent in index notation. By this means we do not have to learn all the special rules for vector products and differential operations. As more experience is gained, the student may choose for himself the notation which he prefers to use.

It is not necessary that the student master all the rules of index notation at this time. The next two chapters, Kinematics of Local Fluid Motion and Basic Laws, offer enough practice to reinforce the material adequately. The last two-thirds of the book have been written assuming the student has a working knowledge of both symbolic tensor notation and index notation.

3.1 ROTATION OF AXES

The measurement of certain types of physical quantities such as position, velocity, or stress requires that a coordinate system be introduced. The value of the x-direction velocity depends upon how we set up the x-direction coordinate. The key to classifying quantities as scalars, vectors, or tensors is how their components change if the coordinates are rotated to point in new directions.

A scalar, such as density or temperature, is unchanged by such a rotation. It has the same value in either coordinate system. This is the defining characteristic of a scalar.

In order to characterize vectors and tensors, we first investigate how the coordinates of a point in space change if we rotate the coordinate axes. Consider the directed line segment from the origin of a rectangular coordinate system to the point P having coordinates x, y, z (Fig. 3-1). The projection of OP upon the x axis is the number x. Similarly for the y and z axes. Now let the coordinate axes be rotated to new directions. We denote the new coordinate system as x', y', z'. The new coordinate system is still orthogonal and right-

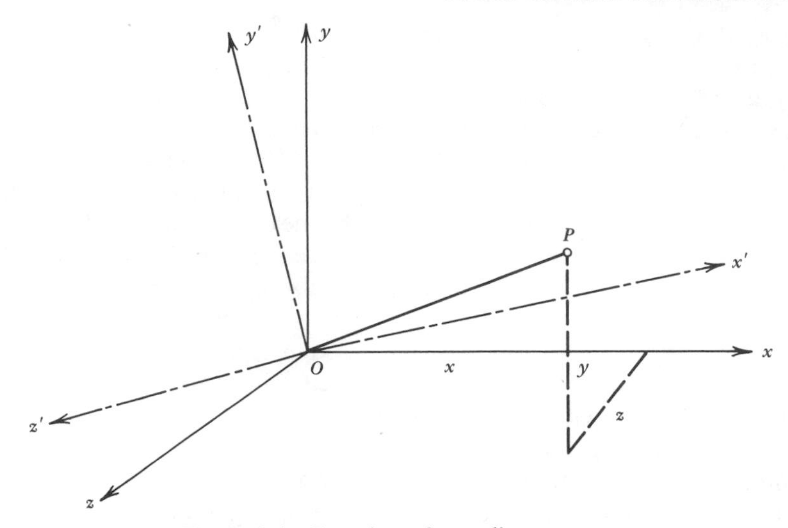

Figure 3-1 Rotation of coordinate axes.

handed. The projection of OP onto the x' axis is the number x'. We want to find a relation between x' and the old values x, y, and z. For simplicity we consider a two-dimensional problem first. Let α be the angle between the x' and x axes, and let β be the angle between the x' and y axes. A two-dimensional picture is given in Fig. 3-2. The segment Ox' may be expressed as the sum of OB and Bx'. The first segment is $OB = x \cos \alpha$. The second segment, Bx', is equal to xC, which in turn is the same as OA. We see that OA is readily computed as $y \cos \beta$. Hence, for this figure, the value of x' is

$$x' = x \cos \alpha + y \cos \beta$$

In three dimensions the geometry is almost too complicated to diagram. The formula that results is simply the previous one plus another term. If we use the notation $\cos(x, x')$ to denote the cosine of the angle from the x to the x' axis, we have

$$x' = x \cos(x, x') + y \cos(y, x') + z \cos(z, x') \qquad 3.1.1a$$

Similar arguments give the new y' and z' components of the point P. The equations are

$$y' = x \cos(x, y') + y \cos(y, y') + z \cos(z, y') \qquad 3.1.1b$$

$$z' = x \cos(x, z') + y \cos(y, z') + z \cos(z, z') \qquad 3.1.1c$$

These rotation equations have similar structure, and we can simplify the matter by introducing some index notation rules.

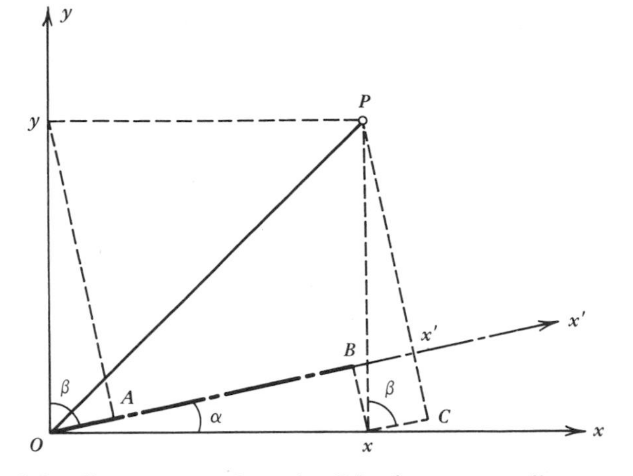

Figure 3-2 Components of a point P in the new coordinate system.

Instead of using x, y, z we write x_1, x_2, x_3. We also shorten the cosine notation by definitions such as

$$
\begin{aligned}
c_{13'} &\equiv \cos(x_1, x_{3'}) = \cos(x, z') \\
c_{3'1} &\equiv \cos(x_{3'}, x_1)
\end{aligned}
\qquad 3.1.2
$$

In the new notation 3.1.1 turns into

$$
\begin{aligned}
x_{1'} &= x_1 c_{11'} + x_2 c_{21'} + x_3 c_{31'} \\
x_{2'} &= x_1 c_{12'} + x_2 c_{22'} + x_3 c_{32'} \\
x_{3'} &= x_1 c_{13'} + x_2 c_{23'} + x_3 c_{33'}
\end{aligned}
\qquad 3.1.3
$$

Next, we introduce the idea of a *free index* and represent the three equations as one typical equation where the index j may take on the value 1, 2, or 3; that is,

$$x_{j'} = x_1 c_{1j'} + x_2 c_{2j'} + x_3 c_{3j'} \qquad \text{for} \quad j' = 1', 2', \text{or } 3' \qquad 3.1.4$$

There are a few rules for using a free index. First, the same free index must appear once and only once in each term of an additive expression. The free index implies three distinct equations as it takes on the values 1, 2, or 3. Since we always deal with vectors with three components, we can agree to omit the words "for $j = 1, 2,$ or 3". Lastly, we may change a free index to another letter if we change it in every term in the expression. Hence the expression

$$x'_k = x_1 c'_{1k} + x_2 c'_{2k} + x_3 c'_{3k}$$

is completely equivalent to 3.1.4.

We can introduce another simplification into 3.1.4 by using the conventional summation sign,

$$x'_j = \sum_{i=1}^{3} x_i c'_{ij} \qquad 3.1.5$$

It turns out that a summation will always have the *dummy index* (the summation index) appearing twice in the term to be summed. Moreover, the index will always range from 1 to 3. Thus, we adopt the convention that whenever an index letter is repeated in a term, the summation sign will be omitted. For example, the expression $a_i b_i$ stands for $a_1 b_1 + a_2 b_2 + a_3 b_3$. Equation 3.1.5 is now expressed as

$$x'_j = c'_{ij} x_i \qquad 3.1.6$$

Index notation deals with scalar numbers, and therefore the order in which we write multiplications is immaterial; in symbolic notation this is not true. It is permissible to change the dummy index to another letter if we avoid any letter used as a free index. An equivalent form of 3.1.6 is

$$x'_k = c'_{jk} x_j \qquad 3.1.7$$

Remember that in this compact notation, because it appears twice, j is a dummy index implying the summation of three terms, and k, since it appears only once, is a free index, which means that there are three equations of this type with indices $k = 1, 2,$ or 3.

The inverse relation for 3.1.6 is found by noting that the role of the primed and unprimed coordinates may be interchanged. Considering the first set of coordinates as primed and the second as unprimed, 3.1.6 gives

$$x_j = c'_{ij} x'_i \qquad 3.1.8$$

Another variation is obtained by observing that the cosine is the same whether we measure the angle from the primed axes to the unprimed axes or vise versa:

$$\cos\left(x'_i, x_j\right) = c'_{ij} = c_{ji}' = \cos\left(x_j, x'_i\right) \qquad 3.1.9$$

(This is not the same as taking the transpose. The transpose of c'_{ij} is c'_{ji}.)

Employing the relation above, 3.1.8 becomes

$$x_j = c_{ji'} x_{i'} \tag{3.1.10}$$

Hence, either 3.1.10 or 3.1.8 can be considered as the inverse relation for 3.1.6.

Next we introduce the *Kronecker delta* or "substitution" tensor. In order to motivate the need for this tensor, we derive the law of cosines. Substitute 3.1.10 into 3.1.7 to arrive at

$$x_{k'} = c_{jk'} c_{ji'} x_{i'} \tag{3.1.11}$$

In order to investigate this equation further, the $k = 1$ component is written out as

$$\begin{aligned} x_{1'} &= c_{j1'} c_{ji'} x_{i'} \\ &= c_{j1'} c_{j1'} x_{1'} + c_{j1'} c_{j2'} x_{2'} + c_{j1'} c_{j3'} x_{3'} \end{aligned} \tag{3.1.12}$$

Note that $x_{1'}$, $x_{2'}$, and $x_{3'}$ are independent; for example, assuming a fixed rotation, we may let the pointin space change so that $x_{1'}$ changes while $x_{2'}$ and $x_{3'}$ remain the same. Since $x_{1'}$, $x_{2'}$, and $x_{3'}$ are independent, the coefficients in 3.1.12 must be

$$c_{j1'} c_{j1'} = 1 \tag{3.1.13}$$

$$c_{j1'} c_{j2'} = 0, \qquad c_{j1'} c_{j3'} = 0 \tag{3.1.14}$$

In general, when k is equal to i in 3.1.11 the cosine term is one, and when k is not the same as i the cosine term is zero. The *Kronecker delta* is defined to select out terms in exactly this manner. It is defined by the relations

$$\begin{aligned} \delta_{ki} &\equiv 1 \qquad \text{if} \quad k = i \\ \delta_{ki} &\equiv 0 \qquad \text{if} \quad k \neq i \end{aligned} \tag{3.1.15}$$

The law of cosines (3.1.13 and 3.1.14) is expressed in index notation as

$$c_{jk'} c_{ji'} = \delta_{ki} \tag{3.1.16}$$

A second form of the expression is found by using 3.1.9:

$$c_{k'j} c_{i'j} = \delta_{ki}$$

(Note that the role of the primed and unprimed coordinates can be interchanged in these equations. This gives $\delta_{ki} = c_{kj'} c_{ij'}$.)

One further point is to be made by substituting 3.1.16 into 3.1.11. This yields

$$x'_k = \delta_{ki} x'_i \qquad 3.1.17$$

The name "substitution tensor" comes from the fact that whenever δ_{ki} occurs in a term, it selects out only those components with $k = i$; hence we can substitute k for i in that term and eliminate the delta. For example, 3.1.17 is really

$$x'_k = x'_k \qquad 3.1.18$$

Equation 3.1.17 therefore reduces to an identity.

3.2 DEFINITION OF VECTORS

The position vector is the archetype of a vector. It has a magnitude and a direction which may be measured with respect to a chosen coordinate system. An alternative description of the position vector is to give its three components. Thus, a vector is something that has three scalar components. This is of course, an incomplete description. Not every set of three scalars is a vector. The essential extra property of a vector is found if we rotate the coordinate axes used to measure the components. A mathematical definition of a vector is as follows: three scalar quantities v_i $(i = 1, 2, 3)$ are the scalar components of a *vector* **v** if they transform according to

$$v'_j = c_{ij'} v_i \qquad 3.2.1$$

under a rotation of coordinate axes. (Without exception, all velocities, forces, and other vectors are measured in coordinate systems with the same orientation as the space coordinates; only the scale units are different. This means the rotation is given by 3.1.6.)

A special symbol for a vector in symbolic or Gibbs' notation is required. Boldface type, an arrow, an underline, or an overbar are common means of denoting vectors. The symbol v_j used in our notation does not, strictly speaking, represent the vector itself, but only a typical scalar component of the vector. This is important to remember when questions arise about proper mathematical operations. Most of the time, however, we shall not call v_j "a typical scalar component of the vector" but simply refer to it as "the vector v sub j". This terminology is mathematically imprecise, but on the other hand is brief and suggestive.

An alternative definition of a vector is as follows: a vector associates a scalar with any chosen direction in space by an expression that is linear in the direction cosines of the chosen direction. If we choose any direction α in space, then a vector associates a scalar with this direction (the component of v in the direction α) by a relation containing the direction cosines of α to the first power. The linear equation can be found from our previous definition. Let the direction α coincide with the $x_{1'}$ axis of a new coordinate system. Equation 3.2.1 for this axis reads

$$v_{1'} = c_{i1'} v_i = c_{11'} v_1 + c_{21'} v_2 + c_{31'} v_3 \tag{3.2.2}$$

or, denoting $v_{1'}$ by $v^{(\alpha)}$,

$$v^{(\alpha)} = \cos(x_1, \alpha)\, v_1 + \cos(x_2, \alpha)\, v_2 + \cos(x_3, \alpha)\, v_3 \tag{3.2.3}$$

This is the alternative defining equation for a vector. The maximum value of $v^{(\alpha)}$ is the magnitude of $\mathbf{v}$, and the direction that gives the maximum value is called the direction of the vector. Note that if α assumes the direction of a coordinate axis, then $v^{(\alpha)}$ is the number v_1, v_2, or v_3 as the case may be.

To conclude this section, note that the inverse relation for 3.2.1 is found by interchanging the roles of the primed and unprimed coordinates. It is

$$v_j = c_{i'j} v_{i'} \tag{3.2.4}$$

or equivalently

$$v_i = c_{j'i} v_{j'} \tag{3.2.5}$$

The last equation is obtained by changing indices: $j \to i$ and $i' \to j'$.

3.3 ALGEBRA WITH VECTORS

Index notation allows us to use the standard algebra of scalars, since we do not employ the vector itself but only a typical scalar component. For example, if a is a scalar, the ith component of $a\mathbf{v}$ is av_i. Multiplication of 3.2.1 by a proves that this combination is indeed a vector:

$$\left(av_{j'}\right) = c_{ij'}\left(av_i\right) \tag{3.3.1}$$

The addition of two vectors is defined as the addition of the separate components. In index notation we write

$$w_i = u_i + v_i \tag{3.3.2}$$

If u_i is ($u_1 = 3$, $u_2 = 4$, $u_3 = 5$) and v_i is ($v_1 = 1$, $v_2 = 2$, $v_3 = 3$) then

$w_1 = 3 + 1 = 4$, $w_2 = 4 + 2 = 6$, and $w_3 = 5 + 3 = 8$. The equivalent vector addition in symbolic notation is

$$\mathbf{w} = \mathbf{u} + \mathbf{v} \qquad 3.3.3$$

The plus sign in symbolic notation does not have the same meaning as the plus sign in ordinary algebra. It is a special symbol for the addition of vector entities.

We turn next to the scalar or inner product of two vectors. The scalar product b is defined as the sum $u_i v_i$. In symbolic notation it is denoted as the dot product $\mathbf{u} \cdot \mathbf{v}$. We have

$$b = u_i v_i = u_1 v_1 + u_2 v_2 + u_3 v_3 \qquad 3.3.4$$

Substitution of 3.2.5 into 3.3.4 shows how b is affected by rotating the axes

$$\begin{aligned} b = u_i v_i &= c'_{ji} u'_j c'_{ki} v'_k \\ &= c'_{ji} c'_{ki} u'_j v'_k \\ &= \delta''_{jk} u'_j v'_k \\ &= u'_j v'_j \end{aligned}$$

Thus the product $u_i v_i$ is a scalar, since its value does not change when the axes are rotated.

A geometric interpretation of the dot product is very useful. In order to make this interpretation we consider a primed and an unprimed coordinate system. We choose the unprimed coordinate system so that $u_1 = u$ and $u_2 = u_3 = 0$. The primed system is chosen so that $v'_1 = v$ and $v'_2 = v'_3 = 0$. The inner product is then formed from 3.3.4 with 3.2.4 substituted for v_i. This is

$$\begin{aligned} b = u_i v_i &= u_i c'_{ki} v'_k \\ &= u_1 c'_{11} v'_1 \\ &= uv \cos(x_1, x'_1) \\ &= uv \cos(\alpha, \beta) \end{aligned} \qquad 3.3.5$$

In the equation above, u and v are the magnitudes, and α and β are the directions, of u_i and v_i respectively. Equation 3.3.5 is the familiar formula for the dot product as the product of the magnitudes of the vectors and the cosine of the angle between the vectors.

An often used consequence of 3.3.5 is if two nonzero vectors are perpendicular, their inner product is zero. Another important result is that the inner product of a vector with itself is the square of the magnitude:

$$v_i v_i = v^2 \qquad 3.3.6$$

It is tempting to write $v_i v_i$ as v_i^2, but the single i does not indicate the summation rule should be applied.

An especially important interpretation of the scalar product happens when one of the vectors is a unit vector. Let a unit vector $\boldsymbol{\alpha}$ have a magnitude of one and an arbitrary direction α. Equation 3.3.5 shows that $\boldsymbol{\alpha} \cdot \mathbf{v}$ is

$$\alpha_i v_i = v \cos(\alpha, \beta) \qquad 3.3.7$$

This is the component v_i in the direction α. Hence, if we wish to project v_i in any direction, we form the scalar product of v_i and the unit vector in that direction.

It is easily shown that the components of a unit vector are the direction cosines of its direction. If a_i is a unit vector in the x_1 direction, then the projection of α_i onto a_i is α_1

$$\begin{aligned} a_i \alpha_i &= \alpha \cos(\alpha, x_1) \\ \alpha_1 &= \cos(\alpha, x_1) \end{aligned} \qquad 3.3.8$$

This fact is also easily established by geometry.

It would be natural to discuss the vector or cross product at this point. However, it will be easier to interpret this idea if we wait until some tensor concepts are introduced.

3.4 DEFINITION OF TENSORS

A tensor is defined by a generalization of the vector definition 3.2.1. A (rank-2) tensor is defined as nine scalar components that change under a rotation of axes according to the formula

$$T_{i'j'} = c_{ki'} c_{lj'} T_{kl} \qquad 3.4.1$$

A double sum on the dummy indices k and l is indicated on the right-hand side, while the free indices i and j imply nine equations of this type as the indices range over the values 1, 2, 3 independently. Symbolic notation requires a special symbol to separate tensors from vectors and scalars. The notation is not uniform; we choose to use boldface sans serif type (**T**).

As with a vector, there is a useful alternative definition of tensor: a *tensor* associates a vector with each direction in space by an expression that is linear in the direction cosines of the chosen direction. To show the equivalence of this definition with 3.4.1, we let the chosen direction α be the $i' = 1'$ direction and write out 3.4.1 for $i' = 1'$:

$$T_{1'j'} = c_{k1'} c_{lj'} T_{kl} \qquad 3.4.2$$

Expanding the sum on k produces

$$T_{1'j'} = c_{11'} c_{lj'} T_{1l} + c_{21'} c_{lj'} T_{2l} + c_{31'} c_{lj'} T_{3l} \qquad 3.4.3$$

The equation is clarified somewhat if we change the notation by using superscripts to show the direction with which a particular vector is associated; that is, we let

$$T_{j'}^{(\alpha)} \equiv T_{1'j'}, \quad T_l^{(1)} \equiv T_{1l}, \quad T_l^{(2)} \equiv T_{2l}, \quad T_l^{(3)} \equiv T_{3l} \qquad 3.4.4$$

Inserting these definitions into 3.4.3 and multiplying the equation by $c_{ij'}$ yields

$$c_{ij'} T_{j'}^{(\alpha)} = c_{11'} c_{ij'} c_{lj'} T_l^{(1)} + c_{21'} c_{ij'} c_{lj'} T_l^{(2)} + c_{31'} c_{ij'} c_{lj'} T_l^{(3)} \qquad 3.4.5$$

The left-hand side is the vector associated with the α direction, whose components are measured in the primed coordinate system and transformed to the unprimed system. By recalling the formula 3.1.16 for the product of two cosines, 3.4.5 may be written as

$$T_i^{(\alpha)} = c_{11'} \delta_{il} T_l^{(1)} + c_{12'} \delta_{il} T_l^{(2)} + c_{13'} \delta_{il} T_l^{(3)} \qquad 3.4.6$$

Changing subscripts to eliminate the substitution tensor gives

$$T_i^{(\alpha)} = c_{11'} T_i^{(1)} + c_{12'} T_i^{(2)} + c_{13'} T_i^{(3)} \qquad 3.4.7$$

This is the desired definition; a tensor associates a vector $T_i^{(\alpha)}$ with any direction in space by an equation that is linear in the direction cosines of the α direction, Note that $T_i^{(1)}$ is the vector *associated* with the x_1 direction. It does not usually lie along the x_1 direction. Likewise $T_i^{(2)}$ is associated with the x_2 direction but does not necessarily lie along that direction.

The formula 3.4.7 may be written in a slightly different form by using the unit vector α_i, which you will recall is aligned with the $x_{1'}$ axis. From 3.3.8 we know that the components of α_i are the direction cosines. Hence,

$$T_i^{(\alpha)} = \alpha_1 T_i^{(1)} + \alpha_2 T_i^{(2)} + \alpha_3 T_i^{(3)}$$

Now we revert to the previous notation using 3.4.4:

$$T_i^{(\alpha)} = \alpha_j T_{ji} \qquad 3.4.8$$

The vector associated with the α direction is the product of the unit vector in that direction with the tensor.

Frequently the components of a tensor are displayed as a matrix. The customary arrangement is

$$\begin{bmatrix} T_{11} & T_{12} & T_{13} \\ T_{21} & T_{22} & T_{23} \\ T_{31} & T_{32} & T_{33} \end{bmatrix} \quad 3.4.9$$

In this form the components of $T_i^{(1)}$ are the first row, those of $T_i^{(2)}$ the second row, and those of $T_i^{(3)}$ the last row.

Definitions analogous to 3.4.1 are used to define tensors of higher rank. Such tensors are seldom needed in fluid mechanics. An exception to this is a selector tensor called the *alternating unit tensor*. This tensor is defined to be 1, 0, or -1 according to

$$\varepsilon_{ijk} = \begin{cases} 1 & \text{if} \quad ijk = 123, 231, \text{or } 312 \\ 0 & \text{if} \quad \text{any two indices are alike} \\ -1 & \text{if} \quad ijk = 321, 213, \textit{or } 132 \end{cases} \quad 3.4.10$$

The appearance of ε_{ijk} in an index-notation equation is equivalent to a cross in symbolic notation. By the definition 3.4.10 the indices may be rearranged according to the following rules: Moving an index from front to back or from back to front is permitted:

$$\varepsilon_{ijk} = \varepsilon_{jki} = \varepsilon_{kij} \quad 3.4.11$$

Interchanging two adjacent indices causes a change in sign:

$$\begin{aligned} \varepsilon_{ijk} &= -\varepsilon_{jik} \\ \varepsilon_{ijk} &= -\varepsilon_{ikj} \end{aligned} \quad 3.4.12$$

It also follows from the definitions that the equation below is valid:

$$\varepsilon_{ijk}\varepsilon_{ilm} = \delta_{jl}\delta_{km} - \delta_{jm}\delta_{kl} \quad 3.4.13$$

An easy way to remember 3.4.13 is to write down the four δ's with a minus sign between them. The two free indices of the first ε are distributed to the first position in each δ, keeping the same order:

$$\varepsilon_{ijk}\varepsilon_{i-\,-} = \delta_{j-}\delta_{k-} = \delta_{j-}\delta_{k-}$$

The second positions on each δ are filled by the free indices of the second ε using the same order in the first group and reversing the order for the second group.

3.5 SYMMETRIC AND ANTISYMMETRIC TENSORS

The *transpose* of a tensor is the tensor obtained by interchanging two indices; the transpose of T_{ij} is T_{ji}. To be specific, if we let the symbol $(T^t)_{ij}$ be the *ij* component of the transpose of the tensor T_{ij}, then $(T^t)_{ij} = T_{ji}$; for example, $(T^t)_{12} = T_{21}$. In the matrix representation the transpose interchanges the components about the diagonal:

$$T_{ij} = \begin{bmatrix} T_{11} & T_{12} & T_{13} \\ T_{21} & T_{22} & T_{23} \\ T_{31} & T_{32} & T_{33} \end{bmatrix}, \qquad T_{ji} = \begin{bmatrix} T_{11} & T_{21} & T_{31} \\ T_{12} & T_{22} & T_{32} \\ T_{13} & T_{23} & T_{33} \end{bmatrix} \qquad 3.5.1$$

The symbolic notation for the transpose is $\mathbf{T}^t$. A tensor is said to be *symmetric* if it is equal to its transpose. Q_{ij} is symmetric if

$$Q_{ij} = Q_{ji} \qquad 3.5.2$$

A tensor R_{ij} is called *antisymmetric* if it is equal to the negative of its transpose,

$$R_{ij} = -R_{ji} \qquad 3.5.3$$

The first tensor below is symmetric, while the second is antisymmetric:

$$Q_{ij} = \begin{bmatrix} 3 & 4 & 1 \\ 4 & 5 & -2 \\ 1 & -2 & 2 \end{bmatrix}, \qquad R_{ij} = \begin{bmatrix} 0 & 3 & 1 \\ -3 & 0 & -5 \\ -1 & 5 & 0 \end{bmatrix}$$

A symmetric tensor has only six independent entries, since 3.5.2 relates the off-diagonal elements. An antisymmetric tensor has only three independent entries. The off-diagonal elements are related by 3.5.3, and the diagonal elements can satisfy this equation only if they are zero.

An arbitrary tensor T_{ij} may be decomposed into the sum of a symmetric and an antisymmetric tensor. To show this we start with T_{ij} and add and subtract one-half its transpose:

$$T_{ij} = \tfrac{1}{2}T_{ij} + \tfrac{1}{2}T_{ji} + \tfrac{1}{2}T_{ij} - \tfrac{1}{2}T_{ji} \qquad 3.5.4$$

Now we use bracketing around the indices to denote the following combinations:

$$T_{(ij)} \equiv \tfrac{1}{2}(T_{ij} + T_{ji}) \qquad 3.5.5$$

$$T_{[ij]} \equiv \tfrac{1}{2}(T_{ij} - T_{ji}) \qquad 3.5.6$$

With these definitions we express 3.5.4 as

$$T_{ij} = T_{(ij)} + T_{[ij]} \tag{3.5.7}$$

We illustrate equation 3.5.7 with a specific example,

$$\begin{bmatrix} 1 & 2 & 3 \\ 4 & 0 & 5 \\ 2 & 1 & 3 \end{bmatrix} = \begin{bmatrix} 1 & \dfrac{2+4}{2} & \dfrac{3+2}{2} \\ \dfrac{4+2}{2} & 0 & \dfrac{5+1}{2} \\ \dfrac{2+3}{2} & \dfrac{1+5}{2} & 3 \end{bmatrix} + \begin{bmatrix} 0 & \dfrac{2-4}{2} & \dfrac{3-2}{2} \\ \dfrac{4-2}{2} & 0 & \dfrac{5-1}{2} \\ \dfrac{2-3}{2} & \dfrac{1-5}{2} & 0 \end{bmatrix}$$

$$= \begin{bmatrix} 1 & 3 & \frac{5}{2} \\ 3 & 0 & 3 \\ \frac{5}{2} & 3 & 3 \end{bmatrix} + \begin{bmatrix} 0 & -1 & \frac{1}{2} \\ 1 & 0 & 2 \\ -\frac{1}{2} & -2 & 0 \end{bmatrix}$$

One can readily show that $T_{(ij)} = T_{(ji)}$ and it is therefore symmetric, and that $T_{[ij]} = -T_{[ji]}$ and it is therefore antisymmetric. Hence 3.5.7 is the decomposition of an arbitrary tensor into symmetric and antisymmetric tensors. $T_{(ij)}$ is the *symmetric part* and $T_{[ij]}$ is the *antisymmetric part*.

We define the *inner product of two tensors* as the double summation

$$a = T_{ij}S_{ji} \tag{3.5.8}$$

Symbolically this is **T:S** [note that another product, $T_{ij}S_{ij}$, is possible and is denoted by **T:(S)**^{t}].

There is a very important and useful fact about products of the form $T_{ij}S_{ji}$. If one of these tensors is symmetric and the other is antisymmetric, then this product is zero. Simply writing out the terms and using the definitions 3.5.2 and 3.5.3 will establish this fact.

The *dual vector* d_i of a tensor T_{jk} is defined by the inner product

$$d_i = \varepsilon_{ijk}T_{jk} \tag{3.5.9}$$

(It may be proved that this product is indeed a vector.) Breaking T_{jk} into symmetric and antisymmetric parts gives

$$d_i = \varepsilon_{ijk}T_{(jk)} + \varepsilon_{ijk}T_{[jk]} \tag{3.5.10}$$

Now ε_{ijk} by its definition is antisymmetric with respect to any two indices. Therefore, the first term is zero because it is the inner product of a symmetric and an antisymmetric tensor. Equation 3.5.10 becomes a statement that the dual vector depends only on the antisymmetric part of a tensor:

$$d_i = \varepsilon_{ijk} T_{[jk]} \qquad 3.5.11$$

The inverse of 3.5.9 is found by multiplying both sides by ε_{ilm}, that is

$$\varepsilon_{ilm} d_i = \varepsilon_{ilm} \varepsilon_{ijk} T_{jk}$$

Employing 3.4.13 gives

$$\begin{aligned} \varepsilon_{ilm} d_i &= \left(\delta_{lj}\delta_{mk} - \delta_{lk}\delta_{mj}\right) T_{jk} \\ &= T_{lm} - T_{ml} \\ &= 2T_{[lm]} \end{aligned}$$

or

$$T_{[lm]} = \tfrac{1}{2}\varepsilon_{ilm} d_i \qquad 3.5.12$$

The three independent components of an antisymmetric tensor are equivalent, in the information they give, to the three components of the dual vector. With this result, the decomposition 3.5.7 can also be expressed as

$$T_{ij} = T_{(ij)} + \tfrac{1}{2}\varepsilon_{ijk} d_k \qquad 3.5.13$$

An arbitrary tensor may be expressed by its symmetric part plus its dual vector.

3.6 ALGEBRA WITH TENSORS

Considering tensors, vectors, and scalars together produces a great many different ways to multiply the components together. We have already covered the tensor multiplications that result in scalars: $S_{ij}T_{ji}$ and $S_{ij}T_{ij}$. The summation conventions make the meaning of these operations apparent. Many other types are possible. For example, $S_{ij}T_{jk} = R_{ik}$ ($\mathbf{S} \cdot \mathbf{T} = \mathbf{R}$) is the *tensor product* of two tensors. A *vector product* of a tensor and a vector is defined as

$$u_j = v_i T_{ij} = T_{ij} v_i \qquad 3.6.1$$

The symbolic notation for this expression is $\mathbf{v} \cdot \mathbf{T}$ where, in contrast to the index notation above, the order of symbols is important. The symbolic formula $\mathbf{T} \cdot \mathbf{v}$ stands for a different vector which, in index notation is given by

$$w_i = T_{ij} v_j = v_j T_{ij} \qquad 3.6.2$$

The *dyadic* or tensor product T_{ij} of two vectors u_i and v_i is defined to be

$$T_{ij} = u_i v_j = v_j u_i$$

$$\mathbf{T} = \mathbf{uv} \qquad 3.6.3$$

Again the order **uv** is important in symbolic notation but is immaterial in index notation. The opposite order **vu** in symbolic notation is the transpose of 3.6.3. The index notation for **vu** is described as follows. Let

$$\mathbf{Q} = (\mathbf{T})^t = \mathbf{vu}$$

Then

$$Q_{ij} = T_{ji} = u_j v_i = v_i u_j \qquad 3.6.4$$

Hence $u_j v_i$ is the transpose of $u_i v_j$.

If we have a tensor expression for T_{ij} and wish to write out the T_{11} component, we simply substitute $i = 1$ and $j = 1$. As an example, consider the equation below (no physical interpretation is implied):

$$T_{ij} = v_k w_i S_{kj} + a\delta_{ij} + \varepsilon_{ijk}\omega_k \qquad 3.6.5$$

The component T_{11} is

$$\begin{aligned} T_{11} &= v_k w_1 S_{k1} + a\delta_{11} + \varepsilon_{11k}\omega_k \\ &= v_k w_1 S_{k1} + a \\ &= v_1 w_1 S_{11} + v_2 w_1 S_{21} + v_3 S_{31} + a \end{aligned}$$

In simplifying this expression we make use of the properties of δ_{ij} and ε_{ijk}. A second example is given by finding the T_{12} component of 3.6.5. It is

$$\begin{aligned} T_{12} &= v_k w_1 S_{k2} + a\delta_{12} + \varepsilon_{12k}\omega_k \\ &= v_1 w_1 S_{12} + v_2 w_1 S_{22} + v_3 w_1 S_{32} + \omega_3 \end{aligned}$$

In a similar manner we can find any component of 3.6.5.

The process called *contraction* on i and j selects out the diagonal components of T_{ij} and adds them together. In index notation, contraction is accomplished by changing i and j to the same symbol. In the example 3.6.5, contraction of T_{ij} produces

$$\begin{aligned} T_{ii} &= T_{11} + T_{22} + T_{33} \\ &= v_k w_i S_{ki} + a\delta_{jj} + \varepsilon_{iik}\omega_k \\ &= v_k w_i S_{ki} + 3a \end{aligned}$$

In symbolic notation T_{ii} is denoted by tr(**T**) and is known as the *trace* of T. Notice that the contraction of δ_{ij} is 3 and the contraction of ε_{ijk} on any two indices is 0.

3.7 VECTOR CROSS PRODUCT

The *vector product* of two vectors ($\mathbf{u} \times \mathbf{v}$ in symbolic notation) is defined as

$$w_i = \varepsilon_{ijk} u_j v_k \tag{3.7.1}$$

The components are found by expanding the summations:

$$w_1 = \varepsilon_{123} u_2 v_3 + \varepsilon_{132} u_3 v_2 = u_2 v_3 - u_3 v_2$$

$$w_2 = \varepsilon_{231} u_3 v_1 + \varepsilon_{213} u_1 v_3 = u_3 v_1 - u_1 v_3$$

$$w_3 = \varepsilon_{312} u_1 v_2 + \varepsilon_{321} u_2 v_1 = u_1 v_2 - u_2 v_1$$

In symbolic notation the order of writing **u** and **v** is important, since $\mathbf{u} \times \mathbf{v} = -\mathbf{v} \times \mathbf{u}$. In 3.7.1 we are dealing with scalar components and the expression below is equally valid:

$$w_i = v_k \varepsilon_{ijk} u_j \tag{3.7.2}$$

An important point about the notation is that the first index on ε_{ijk} must be the component of the vector **w**, the second index must be the same as that of the first vector of the product $\mathbf{u} \times \mathbf{v}$, and the last index must be associated with the last vector of the product $\mathbf{u} \times \mathbf{v}$ (of course ε_{ijk} may be replaced by any of its equivalent forms as given in 3.4.11). As an aid in translating between notations it is usually best to keep the order of **u** and **v** the same as that used in symbolic notation.

There are actually two types of vectors: true vectors and pseudovectors. A true vector passes the rotation-of-axes test (3.2.1) even if we switch from a right-handed system to a left-handed one (that is, a rotation plus a reflection of one axis). A pseudovector changes sign when such a switch is made. We bring this up at the present time because the cross product defined by 3.7.1 is really a pseudovector defined for a right-handed system. (The direction of angular momentum, for example, depends upon the handedness of the coordinate system.) Having brought the matter up for the sake of accuracy, we can agree to use only a right-handed system and simply dismiss this distinction and call everything a vector.

The cross product produces a vector that is perpendicular to the plane of the two vectors and directed in the sense of right-handed rotation of the first vector onto the second. We may prove that $\mathbf{w} = \mathbf{u} \times \mathbf{v}$ is perpendicular to **v** by showing that the dot product $\mathbf{v} \cdot \mathbf{w}$ is zero. The product is

$$v_i w_i = v_i \varepsilon_{ijk} u_j v_k = \varepsilon_{ijk} v_i v_k u_j = 0 \tag{3.7.3}$$

The dyadic $v_i v_k$ is symmetric and ε_{ijk} is antisymmetric, therefore the product must be zero. A similar argument shows that $\mathbf{u}\cdot\mathbf{w}$ is zero and proves that $\mathbf{w}$ is perpendicular to the plane formed by $\mathbf{u}$ and $\mathbf{v}$. The same type of argument shows immediately that the cross product of a vector with itself, $\mathbf{u} \times \mathbf{u}$, must always be zero.

The magnitude of the vector product is equal to the product of the magnitude of the vectors and the sine of the angle between the vectors.

$$w = uv \sin(\alpha, \beta) \tag{3.7.4}$$

The proof is left as an exercise.

There are several complicated formulas in symbolic notation where various combinations of vector and scalar products occur. For example

$$(\mathbf{a} \times \mathbf{b})\cdot(\mathbf{c} \times \mathbf{d}) = (\mathbf{a}\cdot\mathbf{c})(\mathbf{b}\cdot\mathbf{d}) - (\mathbf{a}\cdot\mathbf{d})(\mathbf{b}\cdot\mathbf{c}) \tag{3.7.5}$$

We illustrate the use of Cartesian algebra by proving this equation. First, we write out the left-hand side in index notation and collect terms:

$$\text{LHS} = \varepsilon_{ijk} a_j b_k \varepsilon_{ipq} c_p d_q = \varepsilon_{ijk} \varepsilon_{ipq} a_j b_k c_p d_q$$

Using 3.4.13,

$$\begin{aligned} \text{LHS} &= (\delta_{jp}\delta_{kq} - \delta_{jq}\delta_{kp}) a_j b_k c_p d_q \\ &= \delta_{jp} a_j c_p \delta_{kq} b_k d_q - \delta_{jq} a_j d_q \delta_{kp} b_k c_p \end{aligned}$$

The properties of the substitution tensor $\boldsymbol{\delta}$ allows us to set $p = j$ and $q = k$ in the first term and $q = j$ and $p = k$ in the second term. This produces

$$\text{LHS} = a_j c_j b_k d_k - a_j d_j b_k c_k$$

This is the index notation form of the right-hand side of 3.7.5. Hence, we have proved the validity of 3.7.5.

3.8 DERIVATIVE OPERATIONS

We next consider the calculus operation in index notation. Let a scalar ϕ, a vector component v_i, or a tensor component T_{ij} be a function of position x_i in space. The notation $\phi(x_i)$ means $\phi(x_1, x_2, x_3)$, and the notation $v_i(x_i)$ stands for the three functions $v_1(x_1, x_2, x_3)$, $v_2(x_1, x_2, x_3)$, and $v_3(x_1, x_2, x_3)$. When x_i is enclosed within parenthesis to indicate a function, the index notation rules do not apply to the independent variable: $\phi(x_i)$ is obviously not a vector, and $v_i(x_i)$ is a vector and not a dot product. We say that $\phi(x_i)$ is a scalar field and that $v_i(x_i)$ is a vector field.

There are several derivative operations that may be formed with tensor functions. To begin, let the differential in space be represented by dx_i or equivalently by a unit vector α_i, to indicate direction, times ds to indicate magnitude. That is,

$$dx_i = \alpha_i \, ds \tag{3.8.1}$$

Using the summation convention, the calculus expression for the dependent differential is

$$d\phi = \frac{\partial \phi}{\partial x_i} dx_i = \alpha_i \frac{\partial \phi}{\partial x_i} ds \tag{3.8.2}$$

This notation can be made shorter by simplifying the partial derivative symbol as follows:

$$\frac{\partial(\ )}{\partial x_i} = \partial_i(\ ) \tag{3.8.3}$$

Then 3.8.2 is written

$$d\phi = \partial_i \phi \, dx_i \tag{3.8.4}$$

As an example, let us consider the function $\phi = 3x_1 x_2 + 4\exp(x_3)$. The derivatives $\partial_1 \phi = 3x_2$, $\partial_2 \phi = 3x_1$, $\partial_3 \phi = 4\exp(x_3)$ can be thought of as a vector field with components $3x_2, 3x_1, 4\exp(x_3)$. Equation 3.8.4 says that $d\phi$ is the inner product of $\partial_i \phi$ with the vector dx_i:

$$d\phi = 3x_2 \, dx_1 + 3x_1 \, dx_2 + 4\exp(x_3) \, dx_3$$

The vector $\partial_i \phi$ exists at every point in space. It points in the direction for which $d\phi$ is a maximum and the magnitude is the amount $d\phi/ds$ for that direction. Symbolic notation uses the symbol ∇ for differentiation, and the equivalent form of 3.8.4 is

$$d\phi = \nabla\phi \cdot d\mathbf{x} = \boldsymbol{\alpha} \cdot \nabla\phi \, ds \tag{3.8.5}$$

It is easy to demonstrate that the three quantities $\partial_i \phi$ constitute a vector. We rewrite 3.8.5 in index notation and insert 3.3.8 for the unit vectors:

$$\frac{d\phi}{ds} = \partial_1 \phi \cos(x_1, \alpha) + \partial_2 \phi \cos(x_2, \alpha) + \partial_3 \phi \cos(x_3, \alpha)$$

Since this has the form of 3.2.3, $\partial_i \phi$ must be a vector.

Formulas for vector and tensor functions are found by letting ϕ be the typical scalar component v_i or T_{ij}. The equations are

$$dv_i = \partial_j v_i \, dx_j, \qquad \text{or} \quad d\mathbf{v} = \boldsymbol{\alpha} \cdot \nabla \mathbf{v} \, ds$$
$$= d\mathbf{x} \cdot \nabla \mathbf{v}$$
$$= d\mathbf{x} \cdot \operatorname{grad} \mathbf{v} \qquad 3.8.6$$

$$dT_{ij} = \partial_k T_{ij} \, dx_k, \qquad \text{or} \quad d\mathbf{T} = \boldsymbol{\alpha} \cdot \nabla \mathbf{T} \, ds$$
$$= d\mathbf{x} \cdot \nabla \mathbf{T}$$
$$= d\mathbf{x} \cdot \operatorname{grad} \mathbf{T} \qquad 3.8.7$$

The partial derivatives in the expressions above are called *gradients*, and alternative symbolic notations are grad ϕ, grad $\mathbf{v}$, and grad $\mathbf{T}$. Notice that the gradient always raises the rank of a tensor by one: the gradient of a scalar is a vector, the gradient of a vector is a tensor, and the gradient of a tensor is a third-rank tensor. Let $v_1 = 4x_1x_2^2$, $v_2 = 3x_2x_3$, and $v_3 = x_1 \exp(x_2)$. The nine partial derivatives that may be formed constitute the components of a tensor function,

$$\partial_i v_j = \begin{bmatrix} \partial_1 v_1 & \partial_1 v_2 & \partial_1 v_3 \\ \partial_2 v_1 & \partial_2 v_2 & \partial_2 v_3 \\ \partial_3 v_1 & \partial_3 v_2 & \partial_3 v_3 \end{bmatrix} = \begin{bmatrix} 4x_2^2 & 0 & \exp(x_2) \\ 8x_1x_2 & 3x_3 & x_1 \exp(x_2) \\ 0 & 3x_2 & 0 \end{bmatrix}$$

If the contraction process is performed on a gradient, the result is called a *divergence*. Expressions for the *divergence* of a vector are

$$\partial_i v_i = \partial_1 v_1 + \partial_2 v_2 + \partial_3 v_3$$
$$\nabla \cdot \mathbf{v}, \qquad \operatorname{div} \mathbf{v} \qquad 3.8.8$$

and those for a tensor are

$$\partial_i T_{ij}, \qquad \nabla \cdot \mathbf{T}, \qquad \operatorname{div} \mathbf{T} \qquad 3.8.9$$

Of course there is no divergence defined for a scalar function. Notice that $\nabla \cdot \mathbf{v}$ is a scalar and $\nabla \cdot \mathbf{T}$ is a vector. The divergence decreases the rank by one.

In addition to the derivatives discussed above, we can select terms from $\partial_i v_j$ and form a vector function called the *curl* (actually the dual vector of the tensor $\partial_i v_j$). The notations are

$$s_i = \varepsilon_{ijk} \partial_j v_k, \qquad \mathbf{s} = \nabla \times \mathbf{v}, \qquad \mathbf{s} = \operatorname{curl} \mathbf{v} \qquad 3.8.10$$

By using the properties of the tensor ε_{ijk} we find that the components of 3.8.10 are

$$s_1 = \varepsilon_{123}\partial_2 v_3 + \varepsilon_{132}\partial_3 v_2 = \partial_2 v_3 - \partial_3 v_2$$

$$s_2 = \varepsilon_{231}\partial_3 v_1 + \varepsilon_{213}\partial_1 v_3 = \partial_3 v_1 - \partial_1 v_3$$

$$s_3 = \varepsilon_{312}\partial_1 v_2 + \varepsilon_{321}\partial_2 v_1 = \partial_1 v_2 - \partial_2 v_1$$

The curl is not necessarily perpendicular to the vector v_i (an exception of great importance is plane or axisymmetric flow where v_i is the velocity).

Second derivatives frequently occur in physical expressions. For example, the divergence of the gradient of a scalar function is

$$\partial_i \partial_i \phi, \qquad \nabla \cdot (\nabla \phi), \qquad \text{div}(\text{grad } \phi) \tag{3.8.11}$$

This particular expression is also called the Laplacian of ϕ and is usually written in symbolic notation as $\nabla^2\phi$:

$$\partial_i \partial_i \phi = \partial_1 \partial_1 \phi + \partial_2 \partial_2 \phi + \partial_3 \partial_3 \phi = \nabla^2 \phi \tag{3.3.12}$$

The Laplacian and all other derivative operations are tabulated in several coordinate systems in Appendix B.

We can also treat a vector function the same way. The divergence of a vector gradient is

$$\partial_i \partial_i v_j, \qquad \nabla \cdot (\nabla \mathbf{u}), \qquad \text{div}(\text{grad } \mathbf{v}) \tag{3.8.13}$$

The result of these differential operations is a vector. It is common to use the ∇^2 symbol with this derivative operation also, that is,

$$\nabla \cdot (\nabla \mathbf{v}) = \nabla^2 \mathbf{v} \tag{3.8.14}$$

Some people prefer to reserve the symbol ∇^2 for use as the Laplacian of a scalar, $\nabla^2\phi$. The symbol $\nabla^2\mathbf{v}$ causes no problems in rectangular coordinates, as $\nabla^2\mathbf{v}$ has three components of the type $(\nabla^2\mathbf{v})_1 = \nabla^2 v_1$, where the right-hand side is the Laplacian of v_1. The difficulty comes in nonrectangular coordinate systems. Confusion can arise because

$$(\nabla^2\mathbf{v})_{\text{component } i} \neq \nabla^2(\mathbf{v}_{\text{component } i})$$

For examples, check the Tables in Appendix B concerned with cylindrical or spherical coordinate systems. As long as one is aware that the components of the Laplacian of a vector are not equal to the Laplacians of its components (except in rectangular coordinates), there should be no problem with using $\nabla^2\mathbf{v}$. Be aware that others will prefer a special symbol such as that introduced by Moon and Spencer (1971) ✡$v \equiv \nabla \cdot (\nabla \mathbf{v})$.

3.9 INTEGRAL FORMULAS OF GAUSS AND STOKES

The fundamental theorem of integral calculus is the formula relating the integral and the derivative of the integrand. For the integrand $f = d\phi/dx$,

$$\int_{x=a}^{x=b} f\,dx = \int_{x=a}^{x=b}\left(\frac{d\phi}{dx}\right) dx = \phi(b) - \phi(a) \qquad 3.9.1$$

The equivalent theorem for a volume integral is called Gauss's theorem. We write Gauss's theorem for an arbitrary tensor function $T_{jk\cdots}(x_i)$:

$$\int_R \partial_i(T_{jk\cdots})\,dV = \int_S n_i T_{jk\cdots}\,dS \qquad 3.9.2$$

$T_{jk\cdots}$ may be a scalar, vector, or tensor function of any rank. In 3.9.1 ϕ is evaluated at the end points of the line. The analogy with 3.9.2 is that $T_{jk\cdots}$ is evaluated on the surface S bounding the region R. When T_{jk} is evaluated in the surface integral, it must be multiplied by the local outward unit normal n_i as depicted in Fig. 3-3 (the symbol n_i is reserved for a unit vector pointing outward from an area element dS). It is also important to note that in the volume integral of 3.9.2, the ∂_i must operate on the entire function. An integrand of the form $w_i \partial_i v_j$ is not of the proper type for 3.9.2 to apply. As an example we take $T_{jk\cdots}$ as a scalar function ϕ. Then we have

$$\int_R \partial_i \phi\,dV = \int_S n_i \phi\,dS \qquad 3.9.3$$

As another example, let $T_{jk\cdots}$ be a vector function v_i; then

$$\int_R \partial_i v_i\,dV = \int n_i v_i\,dS \qquad 3.9.4$$

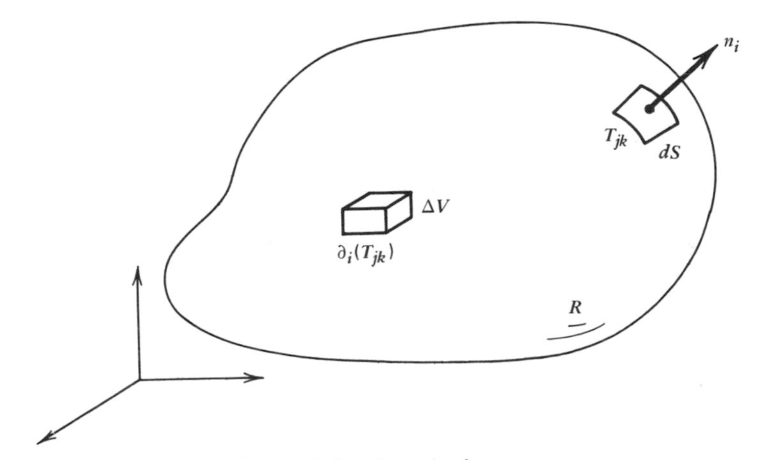

Figure 3-3 Gauss's theorem.

Recall that our index-notation rules are in effect, so 3.9.3 stands for three equations ($i = 1, 2, 3$), while 3.9.4 is one equation where each integrand contains three terms.

A useful fact can be derived from 3.9.3 by letting ϕ be the constant value one. Since $\partial_i 1 = 0$, 3.9.3 becomes a proof that the integral of any component of the outward normal around a closed surface is zero. That is,

$$0 = \int_S n_i \, dS \qquad 3.9.5$$

From this equation we can also deduce the geometric interpretation of $n_i\,dS$. Consider Fig. 3-4, where dS is the area of the right hand surface which is oriented in the direction n_i. The projection of dS onto the x_1 plane is $dS]_1$ with an outward normal $(-1, 0, 0)$. The element dS, its projection $dS]_1$, and their connecting side-wall surface enclose a volume; hence we may apply 3.9.5 with $i = 1$. The $i = 1$ component of 3.9.5 does not contain a contribution from the side walls, because $n_1 = 0$ there. Thus,

$$0 = \int_S n_1 \, dS = -1\,dS]_1 + n_1\,dS$$

$$n_1\,dS = dS]_1, \qquad \text{or} \quad \cos(\alpha, x_1)\,dS = dS]_1 \qquad 3.9.6$$

The conclusion is that $n_1\,dS$ gives us the projection of a surface onto a plane normal to the x_1 direction.

Next, we discuss another special integral formula called Stokes's theorem. Consider the surface in Fig. 3-5, and choose one side to be the outside. The curve bounding the surface is L, and a unit tangent t_i on L is directed according to the right-hand-screw convention (as we proceed along L in the direction of t_i, the interior is on the left). Finally, we let $\nabla \times \mathbf{v}$ be evaluated on

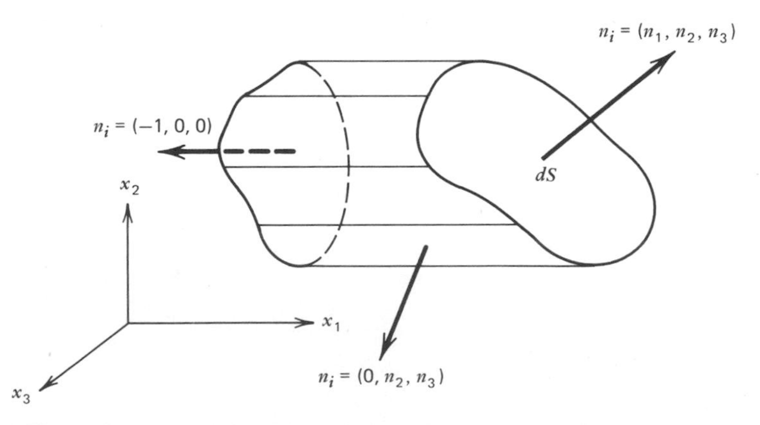

Figure 3-4 Projection of an element dS onto a coordinate plane.

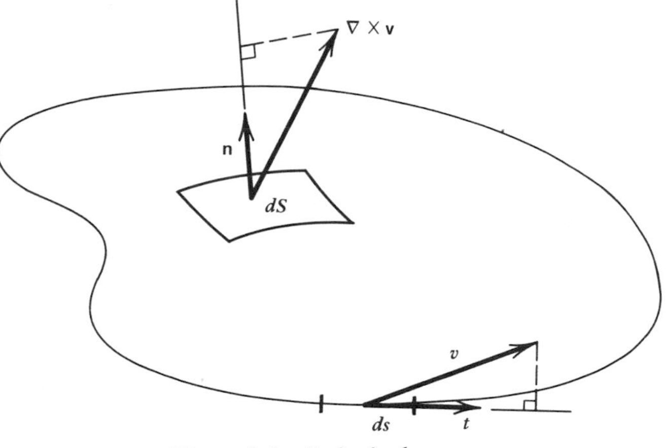

Figure 3-5 Stokes's theorem.

dS, and then take its component along the outward normal. Stokes's theorem says

$$\int \mathbf{n} \cdot \nabla \times \mathbf{v}\, dS = \oint \mathbf{t} \cdot \mathbf{v}\, ds$$

or in index notation

$$\int n_i \varepsilon_{ijk} \partial_j v_k \, dS = \oint t_i v_i \, ds \tag{3.9.7}$$

When v_i is the velocity, the line integral is the *circulation*. The quantity $\nabla \times \mathbf{v}$ is the *vorticity*, which will be discussed in the next chapter.

3.10 LEIBNITZ'S THEOREM

Integrals that involve a parameter often occur in fluid mechanics. In most cases time plays the role of a parameter and the integrals are of the form

$$I(t) = \int_{R(t)} T_{ij\cdots}(x_i, t)\, dV \tag{3.10.1}$$

Here $T_{ij\cdots}$ stands for any scalar, vector, or tensor function of interest. Not only does time change the integrand, but the region of integration $R(t)$ may be moving. We let $\mathbf{w}$ be the velocity of the surface of R. In addition to translating, the surface may be expanding or contracting. The velocity w_i is any prescribed function of position on the surface. The theorem of Leibnitz allows us to find

dI/dt in a convenient manner. The theorem is

$$\frac{d}{dt}\int_{R(t)} T_{ij\cdots}(x_i, t)\, dV = \int_R \frac{\partial T_{ij}}{\partial t}\, dV + \int_S n_k w_k T_{ij}\, dS \qquad 3.10.2$$

A short notation for the derivative with respect to time will be $\partial_0 T_{ij}$. Equation 3.10.2 states that we may move the derivative with respect to time inside the integral if we add a surface integral to compensate for the motion of the boundary. The surface integral tells how fast T_{ij} is coming into R because of the surface velocity w_i. If the boundary does not move, then $w_i = 0$ and the theorem merely says it is permissible to interchange the order of differentiation and integration.

As a specific example take T_{ij} as the constant scalar function one ($T_{ij} = 1$). The integral on the left of 3.10.2 is the volume of the region. Since $\partial_0 1 = 0$, 3.10.2 becomes

$$\frac{dV_R}{dt} = \frac{d}{dt}\int_R dV = \int_S n_k w_k\, dS \qquad 3.10.3$$

The rate of change of the volume of a region is the integral of the normal component of the surface velocity over the region.

The one-dimensional version of Leibnitz's theorem is also very useful:

$$\frac{d}{dt}\int_{x=a(t)}^{x=b(t)} f(x, t)\, dx = \int_a^b \frac{\partial f}{\partial t}\, dx + \frac{db}{dt} f(x = b, t) - \frac{da}{dt} f(x = a, t)$$

3.10.4

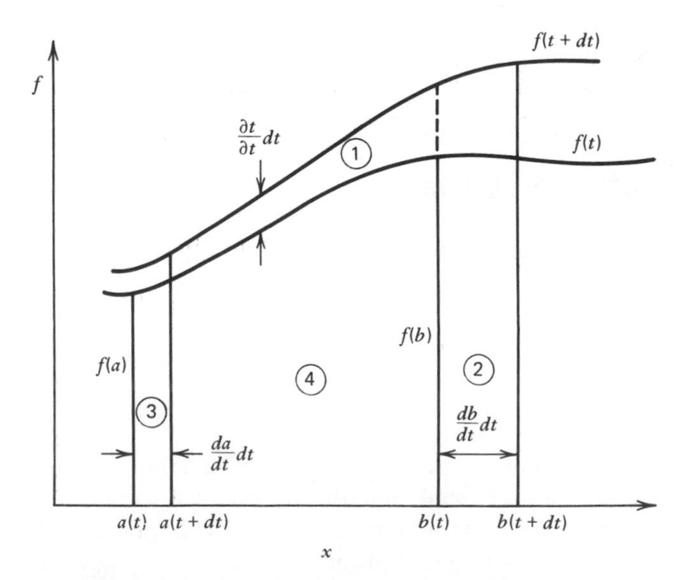

Figure 3-6 Leibnitz's theorem in one dimension.

In this form the left-hand side is an integral where the integrand and the limits of integration are a function of the parameter t. The rate of change of this integral with respect to t is equal to the sum of three terms. The first term is the contribution due to the increase $\partial f/\partial t$ between a and b. The second term is the contribution because the right-hand limit is moving. The integral changes because f at $x = b$ is brought into the integral with the velocity db/dt. The third term is similarly the result of the motion of the left-hand limit, da/dt. Figure 3-6 depicts the terms in this equation (after the equation has been multiplied by a time increment dt).

3.11 CONCLUSION

A discussion of the basic principles and the derivation of conservation laws can be done satisfactorily in rectangular coordinates. Index-notation expressions have a direct interpretation in these systems. However, the most general coordinate systems are curvilinear systems where the axes are not orthogonal. Symbolic notation such as $\nabla \cdot \mathbf{T}$ is more general in the sense that it has meaning in a nonorthogonal system, whereas, in the present sense, $\partial_i T_{ij}$ does not. Fortunately, only a very few areas in fluid mechanics have need for nonorthogonal coordinates.

Algebra operations such as $\mathbf{v} \cdot \mathbf{T}$ or $\mathbf{v} \times \mathbf{w}$ are identical in orthogonal coordinates and rectangular coordinates. That is to say, we may erect a local rectangular coordinate system at any point in the field, whose axes coincide with those of the orthogonal system, so that the local tensor components can be identified with rectangular components.

Derivative operations involve the values of components not only at the local point, but at neighboring positions also. The distance vector in an orthogonal system includes scale factors, and the coordinate directions are constantly changing. The two expressions below offer an example of a component of $\nabla \mathbf{v}$ in rectangular and spherical coordinates:

$$\nabla v|_{xy} = \frac{\partial v_y}{\partial x} \qquad \nabla v|_{\theta r} = \frac{1}{r}\frac{\partial v_r}{\partial \theta} - \frac{v_\theta}{r}$$

There is a systematic method by which index-notation expressions can be interpreted in orthogonal systems. The symbol $\partial_\theta v_r$ can be generalized to essentially stand for $\nabla \mathbf{v}|_{\theta r}$; that is, when a symbol such as $\partial_\theta v_r$ is used, v_r is no longer the physical component of $\mathbf{v}$ in the r direction, and $\partial_\theta(\ )$ is no longer simply $\partial(\ )/\partial\theta$. Procedures to interpret and generalize index notation use Christoffel symbols. The reader should consult the mathematical references if he wishes further information on this subject.

Differential expression in the major coordinate systems are used so frequently and are so complicated that they are tabulated in many places. Appendix B gives an extensive list for cylindrical and spherical coordinates.

One final item should be mentioned. Recall that the nine numbers T_{ij} will change if the measuring coordinate system is rotated. If T_{ij} is symmetric an especially important result occurs. There is one particular set of axes, called the *principal axes*, where all the off diagonal elements of T_{ij} become zero. The three nonzero diagonal elements are known as the *principal values* or *eigenvalues*.

PROBLEMS

3.1 (A) The point P is at $x_1 = 5$, $x_2 = 4$, $x_3 = 0$. What will be the coordinate of P in a coordinate system that is rotated 20° counterclockwise (x_1 axis toward x_2 axis) about the x_3 axis?

3.2 (A) Which of the following expressions are allowed in index notation (a, b, c, d, and e are arbitrary quantities)?

$$a = b_i c_{ij} d_j \qquad a_i = b_i + c_{ij} d_{ji} e_i$$

$$a = b_i c_i + d_j \qquad a_l = \varepsilon_{ijk} b_j c_k$$

$$a_c = \delta_{ij} b_i + c_i \qquad a_{ij} = b_{ji}$$

$$a_k = b_i c_{ki} \qquad a_{ij} = b_i c_j + e_{jk}$$

$$a_k = b_k c + d_i e_{ik} \qquad a_{kl} = b_i c_{ki} d_l + e_{ki}$$

3.3 (A) Consider the three vectors: $\mathbf{u} = (3, 2, -7)$, $\mathbf{v} = (4, 1, 2)$, and $\mathbf{w} = (6, 4, -5)$. (a) Are $\mathbf{u}$ and $\mathbf{v}$ perpendicular? (b) What are the magnitudes of $\mathbf{v}$ and $\mathbf{w}$? (c) What is the angle between $\mathbf{v}$ and $\mathbf{w}$? (d) What are the components of a unit vector in the direction of $\mathbf{w}$? (e) What is the projection of $\mathbf{u}$ in the direction of $\mathbf{w}$?

3.4 (C) Do the nine numbers c'_{ij} of 3.1.6 constitute the components of a tensor? In the text a tensor component measured in the new (rotated) coordinate system was denoted by primes attached to the subscripts. Would it be equally acceptable to associate the prime with the basic symbol, that is v'_i or T'_{ij}? What conceptual problem arises if this type of notation is applied to the direction cosines? Why is this conceptual problem actually inconsequential?

3.5 (C) Why is it that the definition of the Kronecker delta makes no reference to a coordinate system? Is δ_{ki} a tensor?

3.6 (B) Prove that the following equations are true by using index notation:

$$(\mathbf{a} \times \mathbf{b}) \cdot \mathbf{c} = \mathbf{a} \cdot (\mathbf{b} \times \mathbf{c}) = (\mathbf{c} \times \mathbf{a}) \cdot \mathbf{b}$$

$$\mathbf{t} \times (\mathbf{u} \times \mathbf{v}) = \mathbf{u}(\mathbf{t} \cdot \mathbf{v}) - \mathbf{v}(\mathbf{t} \cdot \mathbf{u})$$

$$\mathbf{u} \times \mathbf{v} = -\mathbf{v} \times \mathbf{u}$$

3.7 (A) Consider the tensor T_{ij} defined below. Compute $T_{(ij)}$ and $T_{[ij]}$, find the dual vector for this tensor, and verify equations 3.5.7 and 3.5.13:

$$T_{ij} = \begin{bmatrix} 6 & 3 & 1 \\ 4 & 0 & 5 \\ 1 & 3 & 2 \end{bmatrix}$$

3.8 (A) Prove that the product $S_{ij}T_{ji}$ is zero if S_{ij} is symmetric and T_{ji} is antisymmetric.

3.9 (B) Consider the vector $\mathbf{w} = \mathbf{n} \times (\mathbf{v} \times \mathbf{n})$ where $\mathbf{v}$ is arbitrary and $\mathbf{n}$ is a unit vector. In which direction does $\mathbf{w}$ point, and what is its magnitude?

3.10 (B) Prove 3.7.4 by using primed and unprimed coordinates as was done in the proof of 3.3.5.

3.11 (B) Let the vector b_j be given by the function $b_j = x_j$. What is a simple relation for the vector gradient $\partial_i b_j$?

3.12 (B) Write the following formulas in Gibbs notation using the symbol ∇. Convert the expressions to Cartesian notation and prove that the equations are correct.

$$\operatorname{div}(\phi \mathbf{v}) = \phi \operatorname{div} \mathbf{v} + \mathbf{v} \cdot \operatorname{grad} \phi$$

$$\operatorname{div}(\mathbf{u} \times \mathbf{v}) = \mathbf{v} \cdot \operatorname{curl} \mathbf{u} - \mathbf{u} \cdot \operatorname{curl} \mathbf{v}$$

$$\operatorname{curl}(\mathbf{u} \times \mathbf{v}) = \mathbf{v} \cdot \operatorname{grad} \mathbf{u} - \mathbf{u} \cdot \operatorname{grad} \mathbf{v} + \mathbf{u} \operatorname{div} \mathbf{v} - \mathbf{v} \operatorname{div} \mathbf{u}$$

3.13 (B) Is the operator $\partial_i \partial_j(\ )$ symmetric or antisymmetric? Prove the following: $\operatorname{curl} \operatorname{grad} \phi = 0$; $\operatorname{div} \operatorname{curl} \mathbf{v} = 0$.

3.14 (B) Verify the form of Gauss's theorem (3.9.2) for the special case that $T_{jk\cdots}$ is the scalar function $\phi = x_1$ and the region of integration is $x_1 = (0, L_1)$, $x_2 = (0, L_2)$, $x_3 = (0, L_3)$.

3.15 (C) Derive the one-dimensional Leibnitz formula 3.10.4 by considering a suitable function and region in the 3-dimensional formula 3.10.2.

3.16 (C) The viscous stress law for a Newtonian fluid is $T_{ij} = -\frac{2}{3}\mu\delta_{ij}\partial_k v_k + 2\mu\partial_{(i}v_{j)}$. Prove that $\partial_i T_{ij} = \mu \partial_i \partial_i v_j$ in an incompressible flow where $\partial_i v_i = 0$.

3.17 (C) Verify that $v_j \partial_j v_i = \partial_i(\frac{1}{2}v^2) - \varepsilon_{ijk} v_j \omega_k$; where $\omega_k = \varepsilon_{klm} \partial_l v_m$ is the vorticity.

4 Kinematics of Local Fluid Motion

The characteristic that distinguishes between solids and fluids is how they respond to shear stresses. A solid responds with an angular strain: two lines originally at right angles are distorted to another angle. The strain continues until the displacement is sufficient to generate internal forces that balance the imposed shear force. Hooke's law of elastic solids states that the stress is proportional to the deformation. Fluids, on the other hand, cannot withstand an imposed shear force. They continue to deform as long as the stress is applied. Thus, in a fluid, we must relate the shear stress not to finite deformations, but to rates of deformation. One of the reasons for studying kinematics is to find the exact mathematical expression for the rate of deformation. This is really part of the larger problem of breaking the motion of two neighboring fluid particles into elementary parts.

We will examine fluid movements in a small neighborhood and classify them into the elementary motions of translation, solid-body rotation, and deformations. Deformation can be further classified into two types: extension and shear. All of these motions are continual in a fluid and are dealt with on a rate basis. The translational rate is, of course, just the local particle velocity. The rotation rate and deformation rate are major concepts to be formulated and interpreted in this chapter.

Prior to discussing elementary motions, we devote the first three sections to a review of the two different methods of describing fluid flows. These methods differ essentially in the choice of independent variables. The dependent quantities are the same for both descriptions. Once a viewpoint for describing flows and the kinematics of local motion are established, we shall be in a position to take up the dynamic equations in the next chapter.

4.1 LAGRANGIAN VIEWPOINT

The *Lagrangian viewpoint* of fluid mechanics is a natural extension of particle mechanics. We focus attention on material particles as they move through the flow. Each particle in the flow is labeled, or identified, by its original position x_i^0. The temperature in Lagrangian variables is given by

$$T = T\left(x_i^0, \hat{t}\right) \tag{4.1.1}$$

The independent variables in the Lagrangian viewpoint are the initial position x_i^0 and the time $\hat{t}$. Let us use r_i for the position of a material point, or fluid particle. Initially the fluid particle is at the position x_i^0, and the particle path through space is given in the form

$$r_i = r_i(x_i^0, \hat{t}) \tag{4.1.2}$$

These equations give the paths of the particles with time $\hat{t}$ as a parameter.

The *velocity* and *acceleration* of a particle are defined by

$$v_i = \frac{\partial r_i}{\partial \hat{t}} \quad \text{and} \quad a_i = \frac{\partial^2 r_i}{\partial \hat{t}^2} \tag{4.1.3}$$

In the Lagrangian description these quantities are functions of the particle identification tag x_i^0 and the time $\hat{t}$ as shown in Fig. 4-1.

We will illustrate these ideas with a problem known as the ideal stagnation-point flow. Assume that a two-dimensional blunt body is placed in a steady stream flowing from the top to the bottom of the page. When the flow goes around the body there must be a streamline (a formal definition of streamline will be given shortly) which divides the flow so that part of it proceeds to the right of the body and the remainder flows to the left. This streamline is called the stagnation streamline. There is a small neighborhood where the stagnation streamline intersects the body, and in this region the surface may be treated as flat. In certain cases we may neglect viscosity and allow the fluid to slip along the wall (a correction for viscosity will be discussed later in the book). Fig. 4-2 shows the resulting flow pattern. It is known that the particle positions for this

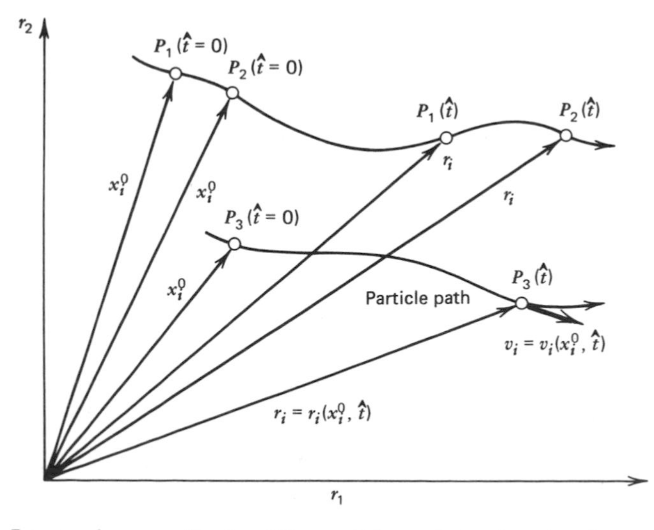

Figure 4-1 Lagrangian coordinates. The particle path is given by the position vector $r_i = r_i(x_i^0, \hat{t})$.

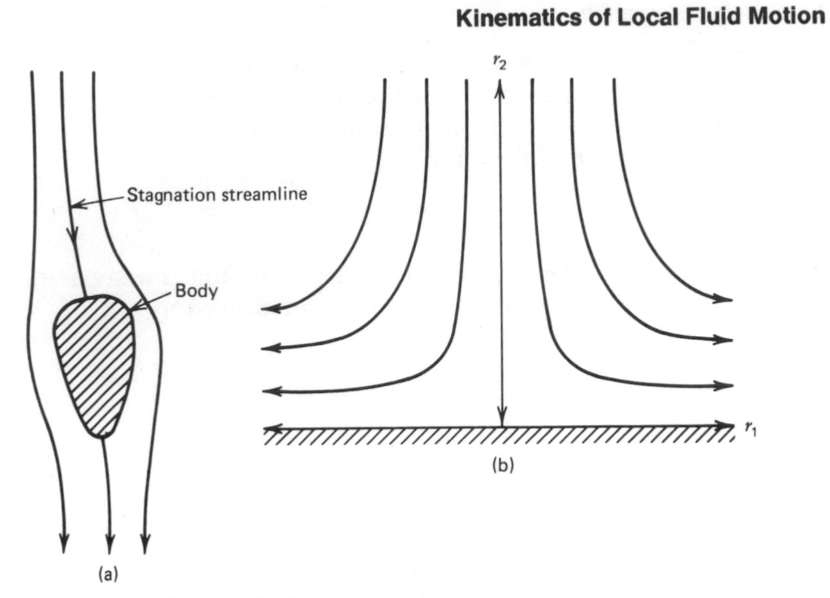

Figure 4-2 Stagnation-point flow pattern.

problem are given by

$$r_1 = x_1^0 \exp(c\hat{t})$$
$$r_2 = x_2^0 \exp(-c\hat{t}) \qquad 4.1.4$$
$$r_3 = x_3^0$$

The constant c is determined by the size and shape of the body and the free-stream velocity. The corresponding velocities are

$$v_1 = \frac{\partial r_1}{\partial \hat{t}} = c x_1^0 \exp(c\hat{t})$$
$$v_2 = \frac{\partial r_2}{\partial \hat{t}} = -c x_2^0 \exp(-c\hat{t}) \qquad 4.1.5$$
$$v_3 = \frac{\partial r_3}{\partial \hat{t}} = 0$$

It is easy to find an explicit particle-path equation in this case. Multiplying the first two equations of 4.1.4 yields the equation for a hyperbola,

$$r_2 = \frac{x_1^0 x_2^0}{r_1} \qquad 4.1.6$$
$$r_3 = x_3^0$$

Next, let us consider the meaning of the equations

$$dr_i = \frac{\partial r_i}{\partial x_j^0} dx_j^0 \qquad 4.1.7$$

dr_i is the vector distance between two particles which were initially separated by dx_j^0. Notice that in writing 4.1.7 we have suppressed the role of $\hat{t}$ as an independent variable. We are making a physical distinction between space and time. The quantity r_i has a vectorlike behavior in space, and we give time the role of what might be called an "independent parameter variable". It makes physical sense to differentiate 4.1.7 with respect to time:

$$\frac{\partial(dr_i)}{\partial \hat{t}} = \frac{\partial}{\partial \hat{t}}\left(\frac{\partial r_i}{\partial x_j^0}\right) dx_j^0$$

$$= \frac{\partial}{\partial x_j^0}\left(\frac{\partial r_i}{\partial \hat{t}}\right) dx_j^0$$

$$= \frac{\partial v_i}{\partial x_j^0} dx_j^0 = dv_i \qquad 4.1.8$$

Thus the rate of change of the distance between two particles is simply the difference in their velocities.

Returning now to the stagnation point example, we compute the distance between two particles as

$$dr_1 = \frac{\partial r_1}{\partial x_1^0} dx_1^0 + \frac{\partial r_1}{\partial x_2^0} dx_2^0 = \exp(c\hat{t})\, dx_1^0$$

$$dr_2 = \exp(-c\hat{t})\, dx_2^0 \qquad 4.1.9$$

Fig. 4-3a shows two particles which are initially displaced from each other by dx_1^0 with $dx_2^0 = 0$. According to 4.1.9, at a later time these particles will still have zero displacement dr_2 and an exponentially increasing displacement dr_1. A similar analysis (Fig. 4-3b) can be made for two particles with initial displacement $dx_1^0 = 0$, dx_2^0. It shows that for later times the particles have no dr_1 displacement and an exponentially decreasing dr_2 displacement.

The Lagrangian analysis of fluid motion is usually quite difficult and is seldom attempted. Furthermore, if we employ velocity as our major dependent quantity, instead of the particle position vector, we can usually find out all we want to know about a flow pattern. The Eulerian viewpoint is much more useful because physical laws written in terms of it do not contain the position vector r_i, and the velocity appears as the major variable.

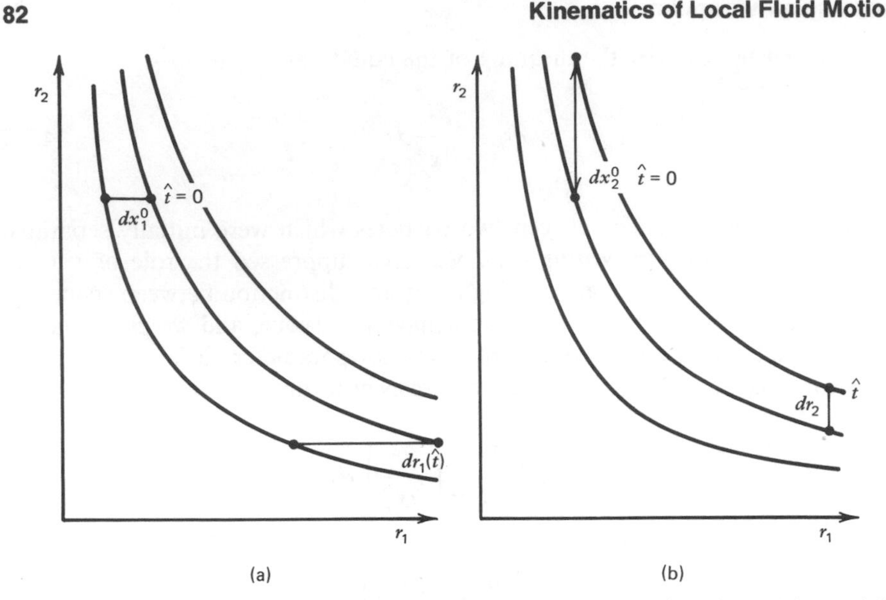

Figure 4-3 Relative motion of two particles in the stagnation flow: (a) particles initially separated by dx_1^0, (b) particles initially separated by dx_2^0.

4.2 EULERIAN VIEWPOINT

The *Eulerian viewpoint* has us watch a fixed point in space x_i as time t proceeds. All flow properties such as r_i, v_i are considered as functions of x_i and t. The temperature of the fluid is given by $T(x_i, t)$. At fixed time $T(x_i, t)$ tells how the temperature changes in space; at a fixed point $T(x_i, t)$ gives the local temperature history. The particle position vector in Eulerian variables is simply

$$r_i = x_i \tag{4.2.1}$$

The position vector in Eulerian variables has as components the local coordinates of the particle. (Note that the position vector, a variable, and the particle path equation, a function, are distant concepts.)

Substituting $r_i = x_i$ into 4.1.2 and noting the obvious equivalence between the time variables, we have the transformation between Lagrangian and Eulerian variables as

$$\begin{aligned} x_i &= r_i\left(x_i^0, \hat{t}\right) \\ t &= \hat{t} \end{aligned} \tag{4.2.2}$$

These relations connect the Eulerian variables x_i, t and the Lagrangian variables x_i^0, $\hat{t}$.

Particle-path equations in Lagrangian variables are obtained by substituting $\hat{t} = t$ in 4.2.2 and relegating x_i^0 to the role of a parameter:

$$x_i = r_i\left(x_i^0, t\right) \tag{4.2.3}$$

We retain r_i as the function symbol in 4.2.3 to denote that this relation is a particle-path equation (to be precise, we used r_i with two meanings in 4.1.2: on the left-hand side it is the position vector, a dependent variable, while on the right-hand side it is the particle-path function).

Streamlines in a flow are defined as lines that, at any instant, are tangent to the velocity vectors. If dx_i is a differential along a streamline, the tangency condition is expressed by the three equations

$$\frac{dx_1}{v_1} = \frac{dx_2}{v_2} = \frac{dx_3}{v_3} \qquad 4.2.4$$

The form of 4.2.4 in vector calculus is

$$\varepsilon_{ijk} v_j \, dx_k = 0, \qquad \text{or} \quad \mathbf{v} \times d\mathbf{x} = 0 \qquad 4.2.5$$

The cross product of two nonzero vectors is zero only if they are parallel. A unique direction for the streamline is determined at all points in space where the velocity is not zero. If the velocity becomes zero at a point, or along a line, it is possible for two or more streamlines to exist at that point. This is what happens at the stagnation point, where the streamline splits and moves around the body on each side. The term stagnation point comes from the fact that the velocity at this point must be zero.

Again consider the ideal stagnation-point flow. The Eulerian-Lagrangian transformation is given by

$$\begin{aligned} x_1 &= x_1^0 \exp(c\hat{t}) \\ x_2 &= x_2^0 \exp(-c\hat{t}) \\ t &= \hat{t} \end{aligned} \qquad 4.2.6$$

Equations for the particle paths are found by substituting $\hat{t} = t$ into the first two equations above. In these equations x_1^0 and x_2^0 are parameters. The term *velocity field* means the Eulerian function $v_i(x_i, t)$. The velocity field is found by substituting 4.2.6 into 4.1.5. The result is

$$\begin{aligned} v_1 &= c x_1 \\ v_2 &= -c x_2 \end{aligned} \qquad 4.2.7$$

A flow where the velocity field is independent of time is called a *steady flow*. Streamlines are obtained by substituting 4.2.7 into 4.2.4:

$$\frac{dx_2}{dx_1} = \frac{v_2}{v_1} = -\frac{x_2}{x_1} \qquad 4.2.8$$

Integration produces

$$x_2 = \frac{A}{x_1} \tag{4.2.9}$$

where A is an arbitrary constant. This is the same equation as the particle-path equation.

It is a general result that path lines and streamlines are identical in a steady flow. This is of course not true in an unsteady flow. Another aspect of steady flow is that it depends upon the coordinate system. A flow may be steady in one coordinate system and unsteady in another. A body moving with a uniform velocity through a stationary fluid produces an unsteady flow with respect to a stationary coordinate system: the flow around a boat is unsteady to an observer on the shore. However, the same flow is steady when observed from a coordinate system attached to the body: an observer on the boat itself finds that the flow is steady.

4.3 SUBSTANTIAL DERIVATIVE

When we adopt the Eulerian viewpoint, our attention is focused upon specific points in space at various times. We lose the ability to easily track the history of a particle. In many instances we are required to express the time rate of change of a particle property in the Eulerian variables (x_i, t). The *substantial* (or "material") *derivative* is an expression that allows us to formulate, in Eulerian variables, a time derivative evaluated as we follow a material particle.

Let F be a property of the flow under consideration. F may be expressed in Lagrangian variables by the function $F_L(x_i^0, \hat{t})$, or in Eulerian variables by the function $F_E(x, t)$, that is,

$$\begin{aligned} F &= F_L\left(x_i^0, \hat{t}\right) \\ F &= F_E\left(x_i, t\right) \end{aligned} \tag{4.3.1}$$

Equating these functions makes sense only if we substitute in the transformation 4.2.2:

$$F = F_L\left(x_i^0, \hat{t}\right) = F_E\left(x_i = r_i\left(x_i^0, \hat{t}\right), t = \hat{t}\right) \tag{4.3.2}$$

Now the rate of change of F as we follow a particle is found from the chain rules of calculus,

$$\frac{\partial F_L}{\partial \hat{t}} = \frac{\partial F_E}{\partial x_i}\frac{\partial r_i}{\partial \hat{t}} + \frac{\partial F_E}{\partial t}\frac{\partial t}{\partial \hat{t}} \tag{4.3.3}$$

But since $v_i = \partial r_i/\partial \hat{t}$, we have

$$\frac{\partial F_L}{\partial \hat{t}} = \frac{\partial F_E}{\partial t} + v_i \frac{\partial F_E}{\partial x_i} \tag{4.3.4}$$

We might now substitute 4.2.2 into the right-hand side of 4.3.4 so that $\partial F_L/\partial \hat{t}$ would appear as a function of x_i^0 and $\hat{t}$. Actually, however, what we are interested in is the physical interpretation of 4.3.4. We keep the right-hand side in Eulerian variables and note that this particular combination has the physical interpretation of the time derivative following a particle. This substantial derivative occurs so frequently in fluid mechanics that it is given a special symbol:

$$\frac{\partial(\ )}{\partial \hat{t}} = \frac{D(\ )}{Dt} \equiv \frac{\partial(\ )}{\partial t} + v_i \partial_i(\ ) \tag{4.3.5}$$

or in symbolic notation

$$\frac{D(\ )}{Dt} \equiv \frac{\partial(\ )}{\partial t} + (\mathbf{v} \cdot \nabla)(\ )$$

The first term on the right-hand side is called the *local rate of change* because it vanishes unless F is changing with time at a fixed local point. The second term is called the *convective change* in F. It vanishes unless there are spatial gradients in F, that is, F has a different value in the neighborhood. This different value is convected (or advected) into the point by the flow velocity v_i.

We illustrate the substantial derivative with two short examples. First let us take F to be the position vector r_j. Equation 4.2.1 says that in Eulerian variables $r_j = x_j$; hence

$$\begin{aligned} \frac{Dr_j}{Dt} &= \frac{\partial r_j}{\partial t} + v_i \partial_i r_j \\ &= 0 + v_i \partial_i x_j \\ &= v_i \delta_{ij} \\ \frac{Dr_j}{Dt} &= v_j \end{aligned} \tag{4.3.6}$$

This equation is consistent with our previous definition of velocity and is the Eulerian counterpart to 4.1.3.

The second example is the acceleration of a fluid particle. It is given by

$$\frac{Dv_j}{Dt} = \frac{\partial v_j}{\partial t} + v_i \partial_i v_j \tag{4.3.7}$$

The terms on the right-hand side are called the local and convective accelerations respectively. Let us compute the accelerations of the particles in the stagnation-point flow. Recall that $v_1 = cx_1$ and $v_2 = -cx_2$. The accelerations are

$$\frac{Dv_1}{Dt} = \partial_0 v_1 + v_1 \partial_1 v_1 + v_2 \partial_2 v_1$$

$$= cx_1 \partial_1 (cx_1) = c^2 x_1$$

and

$$\frac{Dv_2}{Dt} = \partial_0 v_2 + v_1 \partial_1 v_2 + v_2 \partial_2 v_2$$

$$= c^2 x_2$$

The acceleration of a material particle is a function of the position in the flow in the Eulerian viewpoint. In this example the flow is steady and only the convective acceleration term is nonzero.

4.4 DECOMPOSITION OF MOTION

We are now ready to begin the decomposition of the local fluid motion into elementary parts. We consider a primary material point called P and a neighboring point called P' as shown in Fig. 4-4. The vector position of P' relative to P is dr_i, which can also be represented by a unit vector α_i and a distance ds. After an infinitesimal time, P and P' will have moved to new positions. P will move according to the local velocity v_i, while P' will move with the velocity $v_i + dv_i$. The particle P is considered as the main particle, and after its translation velocity is subtracted, the motion of P' is then described as if we were observing it from the main particle. The statements made below are valid only locally in the limit as the distance between P and P' becomes small.

The motion of P and P' may be decomposed into three distinct components: a translation, a solid-body-like rotation, and a deformation. The translational motion is simply the velocity of P itself. All the other motions taken together are dv_i, the velocity of P' with respect to P. The velocity increment is given by the calculus expression

$$dv_j = \partial_i v_j \, dr_i$$

Recall that the velocity gradient may be decomposed into symmetric and antisymmetric parts. Thus

$$dv_j = \partial_{(i} v_{j)} \, dr_i + \partial_{[i} v_{j]} \, dr_i \tag{4.4.1}$$

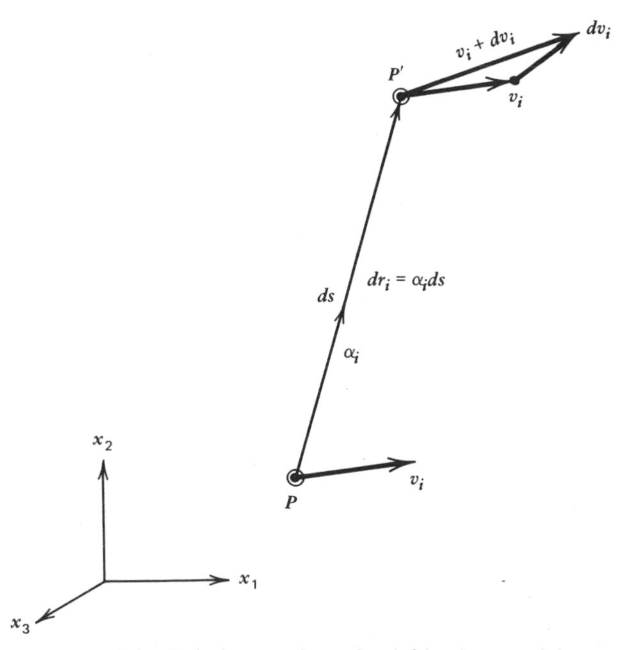

Figure 4-4 Relative motion of neighboring particles.

It turns out that the symmetric part indicates the straining motions of P' with respect to P (strain and deformation are equivalent terms). We denote it by

$$dv_j^{(s)} \equiv \partial_{(i}v_{j)}\, dr_i \tag{4.4.2}$$

The antisymmetric part of 4.4.1 turns out to be associated with the rotational motion of P' around P. Thus, we let

$$dv_j^{(r)} \equiv \partial_{[i}v_{j]}\, dr_i \tag{4.4.3}$$

This decomposition is uniquely determined for every point in the flow. In the present section we only define, describe, and interpret these elemental motions. A detailed derivation and proof is contained in the next two sections.

The motion of P' about P that is like a solid-body rotation must have the form of the rotation equation $\mathbf{V} = \mathbf{\Omega} \times \mathbf{R}$. To arrive at this form, recall that the antisymmetric part of a tensor may be replaced by its dual vector (3.5.9 and 3.5.12). Let ω_i be the dual vector defined by

$$\omega_i = \varepsilon_{ijk}\partial_j v_k, \qquad \boldsymbol{\omega} = \nabla \times \mathbf{v} \tag{4.4.4}$$

Since $\partial_{[i}v_{j]} = \frac{1}{2}\varepsilon_{ijk}\omega_k$, the rotational component of the motion is given by writing 4.4.3 as

$$dv_j^{(r)} = \tfrac{1}{2}\varepsilon_{ijk}\omega_k \, dr_i$$

$$= \varepsilon_{jki}\left(\tfrac{1}{2}\omega_k\right) dr_i \qquad 4.4.5$$

This is in the rotation form $\mathbf{V} = \boldsymbol{\Omega} \times \mathbf{R}$. The vector $\boldsymbol{\omega}$ corresponding to an angular velocity in 4.4.5 is called the *vorticity*. Each point in the flow has a vorticity. The physical interpretation of 4.4.5 is that vorticity is twice the angular velocity of the solid-body rotation of P' about P.

Let us compute the vorticity of the stagnation-point flow. The velocity components for that flow are

$$v_1 = cx_1, \qquad v_2 = -cx_2, \qquad v_3 = 0$$

The vorticity is

$$\omega_1 = \varepsilon_{1jk}\partial_j v_k = \varepsilon_{123}\partial_2 v_3 + \varepsilon_{132}\partial_3 v_2 = 0$$

$$\omega_2 = \varepsilon_{2jk}\partial_j v_k = \varepsilon_{231}\partial_3 v_1 + \varepsilon_{213}\partial_1 v_3 = 0$$

$$\omega_3 = \varepsilon_{3jk}\partial_j v_k = \varepsilon_{312}\partial_1 v_2 + \varepsilon_{321}\partial_2 v_1 = 0$$

The particles in this flow do not have any solid-body rotation. Flows with $\omega_i = 0$ are called *irrotational.*

As a second example, consider viscous flow through a slot of width $2h$. The velocity is given by

$$v_1 = v_0\left[1 - \left(\frac{x_2}{h}\right)^2\right], \qquad v_2 = 0$$

In this flow the only nonzero component of the vorticity is perpendicular to the plane of the flow. It is

$$\omega_3 = \varepsilon_{321}\partial_2 v_1 + \varepsilon_{312}\partial_1 v_2 = \frac{2v_0}{h}\left(\frac{x_2}{h}\right)$$

The vorticity is a maximum at either wall and is zero on the centerline.

From these examples it is obvious that vorticity does not have anything to do with curvature of the streamlines. In the first example the streamlines were curved but the vorticity was zero, while in the second example the streamlines were straight and the vorticity was finite.

Vorticity plays an important role in fluid mechanics. Sometimes the mathematical solution of flow problems is formulated using vorticity as a major dependent variable. We shall return to the study of vorticity in Chapter

13. While we are on the subject, however, there is one important distinction to be made. The words "vorticity" and "vortex" are used with very different meanings in fluid mechanics. Vorticity is a local property of the flow field, while the word vortex is used to describe any type of swirling flow pattern. As a matter of fact, the vorticity is zero in an ideal vortex.

Next, we will take up the straining motions. Straining or deformation is important because it is related to the stresses in the fluid. The total straining velocity is given by 4.4.2. It is directly proportional to the symmetric part of the velocity-gradient tensor, $\partial_{(i}v_{j)}$, which is called the *strain-rate tensor* or the *rate-of-deformation tensor* (other common notations are $\partial_{(i}v_{j)} = S_{ij} = \epsilon_{ij} = \dot{\gamma}_{ij}$ or def $\mathbf{v}$). Our next major task is to find the physical interpretation for each component of the strain-rate tensor.

The extensional component of the deformation is, by definition, the component in the direction of dr_j. Since the dot product $\alpha_j \, dv_j^{(s)}$ is the magnitude of the projection of $dv_j^{(s)}$ onto α_j (recall that α_j is a unit vector), we may express the extension velocity as

$$dv_k^{(es)} = \alpha_k \alpha_j \, dv_j^{(s)}$$

Inserting 4.4.2 into this equation gives

$$dv_k^{(es)} = \alpha_k \alpha_i \alpha_j \partial_{(i}v_{j)} \, ds \tag{4.4.6}$$

We can gain an insight into the physical meaning of the strain rate components $\partial_{(i}v_{j)}$ if we consider some special cases. Consider two particles P and P' that are separated only in the x_1 direction; that is $\alpha_1 = 1$, $\alpha_2 = 0$, $\alpha_3 = 0$. The extension velocity is found by evaluating 4.4.6. For these two particles the result is

$$dv_1^{(es)} = \partial_{(1}v_{1)} \, ds$$

From this we interpret $\partial_{(1}v_{1)}$ as the extension rate between two particles separated in the x_1 direction (per unit separation distance). A similar argument shows that $\partial_{(2}v_{2)}$ is the extension rate for two particles originally separated by one unit in the x_2 direction. In general, we may state that the diagonal entries of the strain-rate tensor are equal to the extension rates for particles separated in the coordinate directions.

The shearing strain is the component of $dv_i^{(s)}$ perpendicular to the direction dr_i. It may be most easily found by subtracting 4.4.6 from 4.4.2:

$$\begin{aligned} dv_j^{(ss)} &= dv_j^{(s)} - dv_j^{(es)} \\ dv_j^{(ss)} &= \left[\alpha_i \partial_{(i}v_{j)} - \alpha_j \alpha_i \alpha_k \partial_{(i}v_{k)}\right] ds \end{aligned} \tag{4.4.7}$$

Although this is a complicated-looking expression, we may simplify it greatly by considering some specified cases. First, take P and P' separated along the x_1 direction ($\alpha_1 = 1$, $\alpha_2 = 0$, $\alpha_3 = 0$). The shearing velocity of P' must be in the x_2 and x_3 directions. From 4.4.7 we find

$$dv_2^{(ss)} = \partial_{(1}v_{2)}\, ds$$

$$dv_3^{(ss)} = \partial_{(1}v_{3)}\, ds \qquad 4.4.8$$

$$dv_1^{(ss)} = 0$$

A physical interpretation of the off-diagonal component $\partial_{(1}v_{2)}$ is that it gives the shearing velocity in the x_2 direction of a particle P' that is originally separated from P in only the x_1 direction (per unit separation distance). Likewise, $\partial_{(1}v_{3)}$ indicates the shearing velocity in the x_3 direction of the same particles. In general, an off-diagonal element of $\partial_{(i}v_{j)}$ is proportional to the shearing velocity in the j direction for a particle P' that is separated in the i direction from a particle P. (We do not use the term "shear strain rate" for $dv_i^{(ss)}$, as that term is reserved for the rate of closure of the angle between two perpendicular lines.)

4.5 ELEMENTARY MOTIONS IN A LINEAR SHEAR FLOW

We consider a linear shear flow with an arbitrary constant c:

$$v_1 = cx_2 \qquad 4.5.1$$

This example has elementary motions that are typical of any fluid flow. We shall compute all the different motions considering a variety of points as illustrated in Fig. 4-5. First we display the strain-rate tensor and the vorticity. From 4.5.1 we find that

$$\partial_{(1}v_{1)} = \partial_{(2}v_{2)} = 0$$

$$\partial_{(1}v_{2)} = \tfrac{1}{2}[\partial_1 v_2 + \partial_2 v_1] = \frac{c}{2} \qquad 4.5.2$$

$$\partial_{(2}v_{1)} = \partial_{(1}v_{2)} = \frac{c}{2}$$

and

$$\omega_3 = -\partial_2 v_1 = -c \qquad 4.5.3$$

The vorticity of this flow has only one nonzero component.

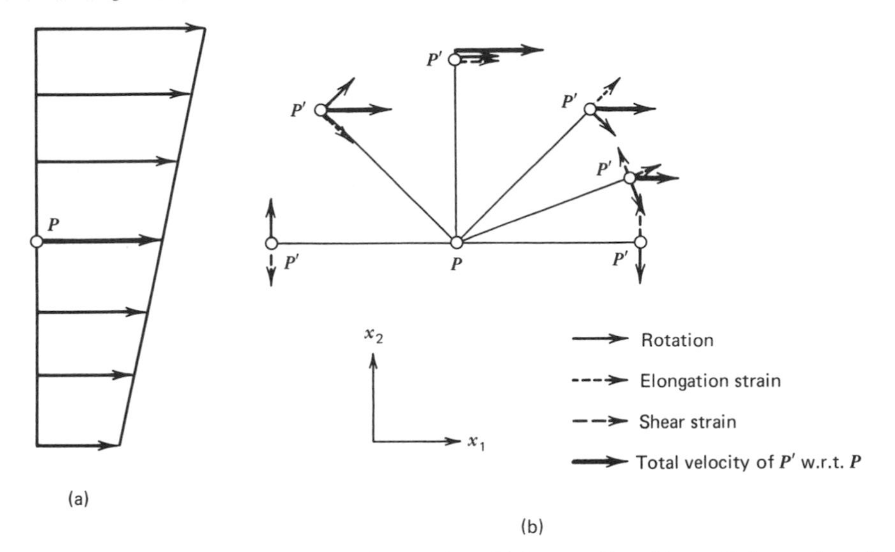

Figure 4-5 Kinematics of shear flow: (a) velocity profile, (b) particle motion for several choices of the particle position.

Now we are in a position to calculate the elementary motions of P' with respect to P. The velocity due to solid-body rotation is found from 4.4.5. The two components of the rotation velocity are

$$dv_1^{(r)} = \frac{c}{2}\alpha_2\, ds$$
$$dv_2^{(r)} = -\frac{c}{2}\alpha_1\, ds \qquad 4.5.4$$

Figure 4-5 shows a central point P and several choices for the second point P'. We have taken the distance between the points, ds, to be unity. This is not an important assumption, as all the various velocity components are proportional to ds. The rotational velocity $dv_i^{(r)}$ has the same magnitude for all choices of P'.

Next, compute the straining (deforming) motions. They are given by 4.4.2; for this flow the components are

$$dv_1^{(s)} = \partial_{(2}v_{1)}\alpha_2\, ds = \frac{c}{2}\alpha_2\, ds$$
$$dv_2^{(s)} = \partial_{(1}v_{2)}\alpha_1\, ds = \frac{c}{2}\alpha_1\, ds \qquad 4.5.5$$

These components are not plotted, because we want to further subdivide the straining motion into elongation and shearing components. The elongation velocities are found from 4.4.6; for our linear shear flow this equation reduces to

$$dv_j^{(es)} = \alpha_j \alpha_1 \alpha_2 c \, ds \qquad 4.5.6$$

The components of this equation are obtained by inserting $j = 1$ and $j = 2$. The elongational velocities are given on the figure by dotted arrows. In this particular example the elongation strain rate becomes zero for particles that are separated along the coordinate axes, because either $\alpha_1 = 0$ or $\alpha_2 = 0$ for such configurations.

Shearing motions of P' with respect to P can be found by subtracting the elongational from the total deformation rate:

$$dv_k^{(ss)} = dv_k^{(s)} - dv_k^{(es)}$$

The components are computed from 4.5.5 and 4.5.6 above. The results are

$$\begin{aligned} dv_1^{(ss)} &= \left(\tfrac{1}{2} - \alpha_1^2\right)\alpha_2 c \, ds \\ dv_2^{(ss)} &= \left(\tfrac{1}{2} - \alpha_2^2\right)\alpha_1 c \, ds \end{aligned} \qquad 4.5.7$$

These equations reveal that the maximum shearing deformation occurs on the coordinate axis $\alpha_1 = 0$ or $\alpha_2 = 0$. We also can see that the shearing deformation will be zero when both $\alpha_1 = \pm 1/\sqrt{2}$ and $\alpha_2 = \pm 1/\sqrt{2}$. This occurs at points that form a set of axes (the principal axes) rotated 45° from the original axes. If we expressed the strain-rate tensor $\partial_{(i}v_{j)}$ with components measured in a coordinate system rotated 45° from the flow direction, the off-diagonal elements would be zero.

In Fig. 4-5 the shearing strains are shown as a dashed arrow, while the total velocity of P' with respect to P is given by a heavy line arrow. The point at 22.5° is the only point pictured that has all the different types of elementary motions.

*4.6 PROOF OF VORTICITY CHARACTERISTICS

The *vorticity vector* at every point in the flow was defined by

$$\boldsymbol{\omega} = \nabla \times \mathbf{v} \qquad 4.6.1$$

We shall prove that ω_i is twice the angular velocity of the solid-body rotation motion of P' with respect to P.

Let the material line from P to P' be dr_j, and recall that 4.1.8 gives the motion of P' with respect to P,

$$\frac{\partial (dr_j)}{\partial \hat{t}} = dv_j \qquad 4.6.2$$

We may reexpress this by changing the left-hand side according to 4.3.5 and expanding the dv_j in Eulerian variables as follows:

$$\frac{D(dr_j)}{Dt} = dv_j = \partial_i v_j \, dr_i \qquad 4.6.3$$

This equation governs the changes in both the direction and the length of a material line element. The differential dx_i is changed to dr_i (4.2.1) so that we can remind ourselves that we are dealing with material particles. Now, the velocity gradient $\partial_i v_j$ may be separated into parts as was done in 4.4.1. Furthermore, we can eliminate $\partial_{[i}v_{j]}$ in favor of the vorticity:

$$dv_j = \partial_{(i}v_{j)} \, dr_i + \tfrac{1}{2}\varepsilon_{jki}\omega_k \, dr_i \qquad 4.6.4$$

We have now separated the motion into two components and isolated the effect of the vorticity. The first component is the straining motion, which we have denoted by $dv_j^{(s)}$ in 4.4.2. The second component obeys an equation exactly like a solid-body rotation. This component was denoted by $dv_j^{(r)}$ in 4.4.5.

Referring to 4.6.4, we argue that if the components $dv_j^{(s)}$ are zero ($\partial_{(i}v_{j)} = 0$), the motion of P' is like a solid-body rotation about P. The axis of rotation is along ω_i, and the angular velocity is $\omega_i/2$.

Next, we take up the converse problem: If the motion in the neighborhood of P is a solid-body rotation, does the second term in 4.6.4 give this motion? We can prove this is true if we can show that a solid-body rotation implies that $\partial_{(i}v_{j)} = 0$.

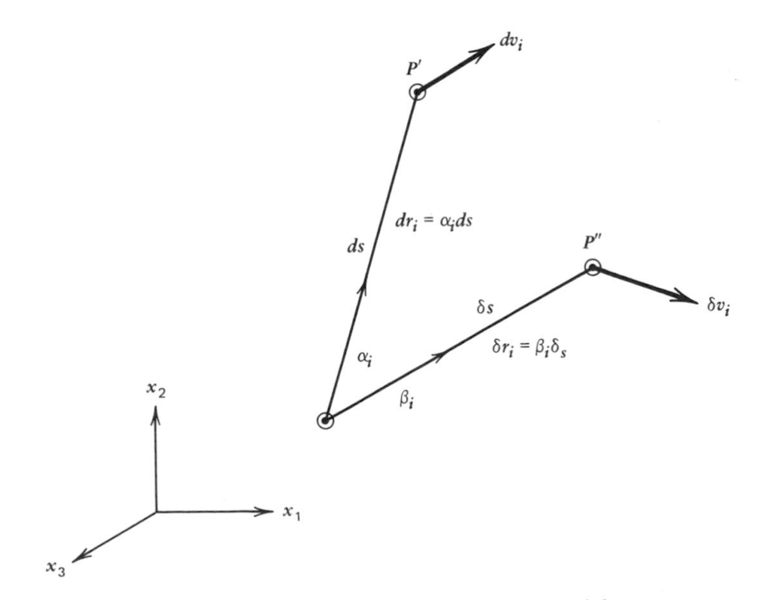

Figure 4-6 Relative motion of three particles.

Consider another point P'' (Fig. 4-6) in the neighborhood of P and a distance $\delta r_j = \beta_j \delta s$ away. We shall use d for increments associated with P', and δ for those associated with P''. Form the inner product

$$dr_i \, \delta r_i = ds \, \delta s \cos\theta \tag{4.6.5}$$

If the motion is a solid-body rotation, neither ds, δs, nor θ will change with time. Hence, a solid-body rotation implies that

$$\frac{D(dr_i \, \delta r_i)}{Dt} = 0 \tag{4.6.6}$$

In order to explore the consequences of 4.6.6, we expand the left-hand side as

$$\begin{aligned} \frac{D(dr_i \, \delta r_i)}{Dt} &= \delta r_i \frac{D(dr_i)}{Dt} + dr_i \frac{D(\delta r_i)}{Dt} \\ &= \delta r_i \, dv_i + dr_i \, \delta v_i \end{aligned}$$

Inserting $dv_i = \partial_j v_i \, dr_j$ and a similar equation for δv_i produces

$$\begin{aligned} \frac{D(dr_i \, \delta r_i)}{Dt} &= \partial_j v_i \, dr_j \, \delta r_i + \partial_j v_i \, \delta r_j \, dr_i \\ &= \left[\partial_j v_i + \partial_i v_j\right] dr_j \, \delta r_i \\ &= 2\partial_{(j} v_{i)} \, dr_j \, \delta r_i \end{aligned} \tag{4.6.7}$$

$$\frac{D(dr_i \, \delta r_i)}{Dt} = 2\partial_{(j} v_{i)} \, \alpha_j \beta_i \, ds \, \delta s$$

For a solid-body rotation the left-hand side is zero; hence $\partial_{(j} v_{i)}$ must be zero, and the velocity in 4.6.4 consists entirely of the vorticity component. Therefore, we have proved that solid-body rotation and nonzero vorticity are equivalent.

*4.7 RATE-OF-STRAIN CHARACTERISTICS

All of the deformation (straining) motions are the result of the symmetric strain-rate tensor $\partial_{(i} v_{j)}$. Consider again the three material points P, P', and P'' and the inner product $dr_i \, \delta r_i$. Another expression for the time rate of change of this product is

$$\begin{aligned} \frac{D(dr_i \, \delta r_i)}{Dt} &= \frac{D}{Dt}(ds \, \delta s \cos\theta) \\ &= -ds \, \delta s \sin\theta \frac{D\theta}{Dt} + \cos\theta \left[ds \frac{D(\delta s)}{Dt} + \delta s \frac{D(ds)}{Dt} \right] \end{aligned} \tag{4.7.1}$$

We shall equate 4.7.1 and 4.6.7 and investigate several special cases.

First take P' and P'' to be the same point. This means that $\alpha_i = \beta_i$, $ds = \delta s$, and $\theta = 0$. Equations 4.7.1 and 4.6.7 yield the relation

$$\frac{1}{ds}\frac{D(ds)}{Dt} = \alpha_i \alpha_j \partial_{(j}v_{i)} \qquad 4.7.2$$

The left-hand side of 4.7.2 is called the *extensional strain rate*. It gives the extension between P and P' for an arbitrary choice of direction α_i. We may take P and P' to be separated only along the x_1 axis by choosing $\alpha_i = (1, 0, 0)$:

$$\frac{1}{ds}\frac{D(ds)}{Dt} = \partial_{(1}v_{1)} \qquad 4.7.3$$

The $\partial_{(1}v_{1)}$ component of the strain-rate tensor gives the rate of extension of two material particles separated in the x_1 direction as shown in Fig. 4-7a. Similar arguments show that the $\partial_{(2}v_{2)}$ and $\partial_{(3}v_{3)}$ components are extension rates for particles originally separated in the x_2 and x_3 directions respectively.

The angular, or shearing, deformation will be discussed next. Consider another special choice such that PP' and PP'' form a right angle, $\theta = \pi/2$. Equations 4.7.1 and 4.6.7 now reduce to

$$\left.\frac{D\theta}{Dt}\right|_{\theta=\frac{\pi}{2}} = -2\alpha_i \beta_j \partial_{(i}v_{j)} \qquad 4.7.4$$

This formula gives the shearing deformation rate between any two material lines that are originally at right angles (see Fig. 4-7b). In analogy with solid mechanics, this is called the *shear strain rate*. If we specialize to lines originally directed along the x_1 and x_2 directions [$\alpha_i = (1, 0, 0)$, $\beta_i = (0, 1, 0)$] we obtain

$$\left.\frac{D\theta}{Dt}\right|_{x_1 - x_2} = -2\partial_{(1}v_{2)} \qquad 4.7.5$$

The shearing strain rate between material lines along the x_1 and x_2 directions is measured by the $\partial_{(1}v_{2)}$ off-diagonal component of the strain-rate tensor. Note that the shearing deformation does not depend upon which axis is chosen first. This is reflected in the fact that $\partial_{(i}v_{j)}$ is symmetric and thus $\partial_{(1}v_{2)} = \partial_{(2}v_{1)}$.

Extensions of these arguments show that the off-diagonal elements of the strain-rate tensor have a physical interpretation as the shearing strain rates between lines coinciding with coordinate directions.

In Section 4.5 we interpreted $\partial_{(1}v_{2)}$ as measuring the shearing velocity $dv_2^{(ss)}$ when the point P' is separated in the x_1 direction from P. Now consider another point P'' separated from P along the x_2 direction. This point would have a shearing velocity $dv_1^{(ss)}$. The angle formed by $P'PP''$ is $\pi/2$, and both of

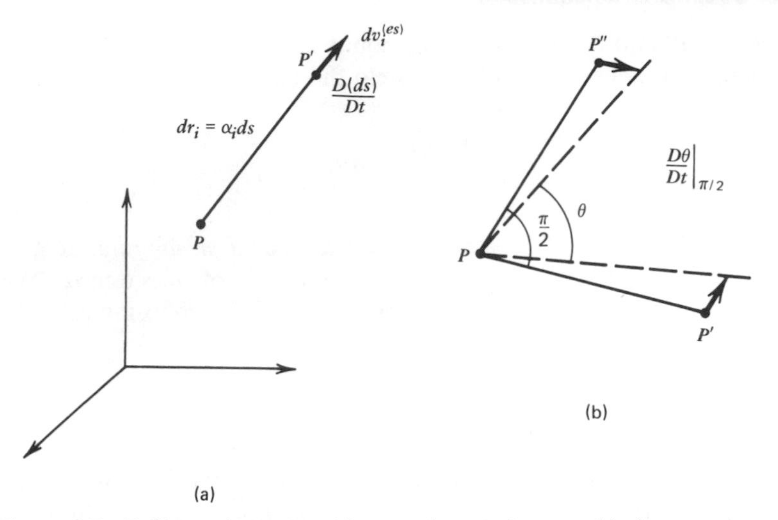

Figure 4-7 Deformation rates: (a) extension strain rate, (b) shear strain rate.

these velocities tend to close the angle. Hence we may write

$$\left.\frac{D\theta}{Dt}\right|_{x_1-x_2} = -\left.\frac{dv_2^{(ss)}}{ds}\right|_{P'P} - \left.\frac{dv_1^{(ss)}}{ds}\right|_{PP''} = -2\partial_{(1}v_{2)} \qquad 4.7.6$$

This equation illustrates the connection between 4.4.7 and 4.7.5.

4.8 RATE OF EXPANSION

As a material particle (i.e., a small piece of fluid) moves through the fluid, its size and shape may change. It is important to know when the volume of a fluid particle is changing. For instance, if a particle is expanding or contracting, it is doing work on the remaining fluid or vice versa. The volume of a material region is given by the integral

$$V_{\mathrm{MR}} = \int_{R(t)} dV \qquad 4.8.1$$

The surface velocity of the region R is equal to the local fluid velocity. Differentiating 4.8.1 with respect to time and applying Leibnitz's theorem (3.10.2) with $w_i = v_i$, we find

$$\frac{DV_{\mathrm{MR}}}{Dt} = \frac{d}{dt}\int_R dV = \int_S n_i v_i \, dS \qquad 4.8.2$$

Next, the surface integral is converted into a volume integral by Gauss's theorem (3.9.4),

$$\frac{DV_{\mathrm{MR}}}{Dt} = \int_R \partial_i v_i \, dV \tag{4.8.3}$$

Now the mean-value theorem for integrals is used to arrive at

$$\frac{DV_{\mathrm{MR}}}{Dt} = (\partial_i v_i)^* V_{\mathrm{MR}} \tag{4.8.4}$$

The asterisk indicates that the integrand is evaluated at the appropriate point. This point must be within R. When we divide by V_{MR} and allow the volume to approach zero about a specific point, the term $(\partial_i v_i)^*$ will be evaluated at the point in question. Hence

$$\lim_{V_{\mathrm{MR}} \to 0} \frac{1}{V_{\mathrm{MR}}} \frac{DV_{\mathrm{MR}}}{Dt} = \partial_i v_i \tag{4.8.5}$$

Equation 4.8.5 gives us a physical interpretation for $\nabla \cdot \mathbf{v}$ as the *rate of expansion* of a material region (or, if you like, a fluid particle). The rate of expansion is also known as the dilation rate.

We might also note that the sum $\partial_i v_i = \partial_1 v_1 + \partial_2 v_2 + \partial_3 v_3$ is equal to the trace of the strain-rate tensor $\partial_{(i} v_{j)}$, that is, the sum of the extension rates in the three coordinate directions. The major subject of this book is incompressible flow, where the rate of expansion is nearly zero ($\nabla \cdot \mathbf{v} = 0$). This condition requires that the sum of the extension rates in the deformation tensor be zero. If extension is occurring along one coordinate axis, a compensatory contraction must occur along another set of axes.

*4.9 STREAMLINE COORDINATES

The streamlines of a flow may be used as the basis of a local orthogonal coordinate system. If the flow is smooth enough, the coordinate system can even be global. In general, however, the streamlines will contain knots or other complicated patterns that restrict the coordinates to a local definition. In this section we discuss the coordinate definitions and give the velocity and vorticity components in streamline coordinates.

Let r_i be the position vector to any point on a certain streamline as shown in Fig. 4-8. From some arbitrary origin, s will denote the distance along the streamline. We can consider $r_i(s)$ as describing the streamline completely. The unit vector tangent to the streamline is

$$t_i \equiv \frac{dr_i}{ds} \tag{4.9.1}$$

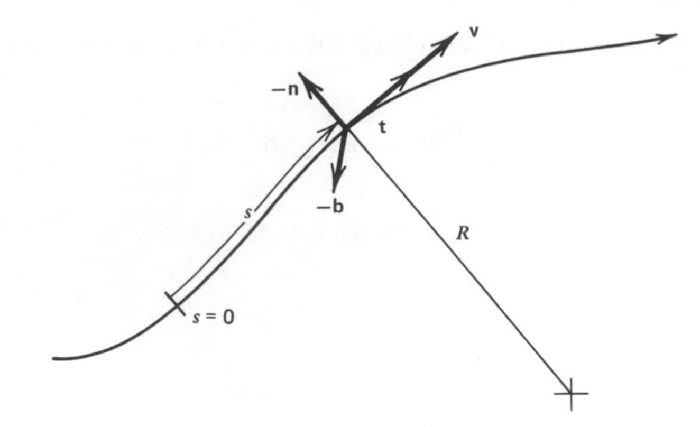

Figure 4-8 Streamline coordinates.

Since t_i has unit magnitude, it can change only in direction. This change must be perpendicular to t_i itself. Therefore, the principal normal direction is defined by

$$n_i \equiv R(s)\frac{dt_i}{ds} \qquad 4.9.2$$

where $R(s)$ is a scale factor to insure that n_i is of unit length. $R(s)$ is called the radius of curvature ($k = 1/R$ is the curvature). A local orthogonal coordinate system is completed by defining the binormal direction to be perpendicular to t_i and n_i:

$$\mathbf{b} \equiv \mathbf{t} \times \mathbf{n} \qquad 4.9.3$$

The vectors t_i, n_i, and b_i are unit vectors in an orthogonal streamline coordinate system.

The velocity v_i is in the t_i direction and is simply given by $v_t = v$, $v_n = 0$, $v_b = 0$. In a local region near the streamline we may consider v as a function of **t**, **n**, and **b**. The derivative dv/dt is the change of v along the streamline, while dv/dn and dv/db are changes in the normal and binormal directions respectively.

It is also possible to express the vorticity in the streamline coordinates (see Truesdell, 1954). The components are

$$\omega_t = (\mathbf{t} \cdot \nabla \times \mathbf{t})v \qquad 4.9.4$$

$$\omega_n = \frac{dv}{db} \qquad 4.9.5$$

$$\omega_b = \frac{v}{R} - \frac{dv}{dn} \qquad 4.9.6$$

When ω_t is zero, the velocity and the vorticity are perpendicular to each other.

This component depends on a geometric property of the streamlines, $\mathbf{t} \cdot \nabla \times \mathbf{t}$, multiplying the magnitude of the velocity. The normal vorticity component ω_n is directly related to the velocity gradient in the binormal direction. The last component ω_b has contributions from the streamline curvature R and from the local velocity gradient dv/dn.

The component ω_b is the only nonzero component in two-dimensional flows. It is typically the largest component in any flow. The two terms in 4.9.6 express a streamline-curvature effect v/R and a local velocity-gradient effect $-dv/dn$. Vorticity can exist because of either effect. For example, a solid-body rotation is given by $v = R\Omega$. This results in $dv/dn = -\Omega$ and hence $\omega_b = 2\Omega$. This corresponds to the interpretation of ω as twice the solid-body rotation. Another example is furnished by a solid wall with arbitrary curvature. The velocity on the wall must be zero; hence $\omega_t = 0$, $\omega_n = 0$ (b_i is tangent to the wall), and

$$\omega_b = -\left.\frac{dv}{dn}\right|_{\text{wall}} \qquad 4.9.7$$

Further discussion of wall vorticity and wall streamlines is given in Section 13.2.

4.10 CONCLUSION

The motion in a small neighborhood of fluid has been investigated by considering the motion of two particles separated by a small amount. Instantaneously the relative motion of these material particles is determined by the velocity-gradient tensor $\partial_i v_j$ in an Eulerian description of the flow field. Further separation of the motion was made by decomposing $\partial_i v_j$ into its symmetric and antisymmetric parts. The symmetric part $\partial_{(i} v_{j)}$ is termed the strain-rate tensor because it produces deformation motions consisting of extension and shearing strains (the sum of the extensional strains in the coordinate directions $\partial_{(i} v_{i)} = \partial_i v_i$ indicates the volumetric expansion rate). The antisymmetric part $\partial_{[i} v_{j]}$ causes a solid-body-like rotational motion. The three independent entries of $\partial_{[i} v_{j]}$ may be expressed in the form of a vorticity vector ω_i ($\boldsymbol{\omega} \equiv \nabla \times \mathbf{v}$). Thus, vorticity becomes another local property of the flow field in the same way that linear momentum v_i, kinetic energy $\frac{1}{2} v_i v_i$, and angular momentum $\mathbf{r} \times \mathbf{v}$ are local flow properties. Frequently in the remainder of the book we shall interpret flows in terms of vorticity and the physical events that establish certain vorticity patterns.

PROBLEMS

4.1 (B) How long will it take a particle traveling on an ideal stagnation streamline to reach the stagnation point?

4.2 (A) In a table of vector differential operators, look up the expressions for $\nabla \times \mathbf{v}$ in a cylindrical coordinate system. (a) Compute the vorticity for the flow in a round tube where the velocity profile is

$$v_z = v_0\left[1 - \left(\frac{r}{R}\right)^2\right]$$

(b) Compute the vorticity for an ideal vortex where the velocity is

$$v_\theta = \frac{\Gamma}{2\pi r}, \qquad \Gamma \text{ constant}$$

(c) Compute the vorticity in the vortex flow given by

$$v_\theta = \frac{\Gamma}{2\pi r}\left[1 - \exp\left(-\frac{r^2}{4\nu t}\right)\right]$$

Sketch all velocity and vorticity profiles.

4.3 (A) Find the elemental motions $dv_j^{(r)}$, $dv_j^{(es)}$, and $dv_j^{(ss)}$ for the viscous flow in a slot. The velocity profile is

$$v_1 = v_0\left[1 - \left(\frac{x_2}{h}\right)^2\right] \qquad v_2 = v_3 = 0$$

Take the primary point P to be at $x_2/h = \frac{1}{2}$, and express the incremental motions as a fraction of v_0. Assume the point P' takes on several positions at 0, 30, 45, and 90° from the x_1 axis. In all cases the distance ds is $0.1h$. Diagram the results in a vector sketch including the total motion dv_j.

4.4 (A) The surface temperature of a lake changes from one location to another as $T(x_1, x_2)$. If you attach a thermometer to a boat and take a path through the lake given by $x_i = b_i(\hat{t})$, find an expression for the rate of change of the thermometer temperature in terms of the lake temperature.

4.5 (A) The energy is a thermodynamic function $e(\rho, T)$. The density and temperature are in turn functions of space and time, $T(x_i, t)$, $\rho(x_i, t)$. By using results of Section 2.16, show that

$$\frac{De}{Dt} = c_v\frac{DT}{Dt} + \frac{1}{\rho^2}\left[p - T\left.\frac{\partial p}{\partial T}\right|_\rho\right]\frac{D\rho}{Dt}$$

4.6 (A) Consider a two-dimensional flow with velocity components $v_1 = cx_1$, $v_2 = -cx_2$. Find expressions for the vorticity and the strain-rate tensor.

4.7 (A) Consider a point at $x_2 = h/2$ in Problem 3. Find the rate of closure of the angle between two material lines in the x_1 and x_2 directions. Find the rate of closure of the angles between an x_1 line and lines at 45° from it.

4.8 (B) Compute the circulation Γ (3.9.7) around a circuit including the origin for the velocity profiles of Problem 2(b) and (c).

4.9 (A) Find the rate of expansion for the stagnation point flow: $v_1 = cx_1$, $v_2 = -cx_2$.

4.10 (A) Consider the two-dimensional flow given in cylindrical coordinates by $v_r = Q/(2\pi r)$, $v_z = v_\theta = 0$. Compute the components of the strain-rate tensor for this flow.

4.11 (B) Show that an alternative expression for $dv_j^{(ss)}$ is $d\mathbf{v}^{(ss)} = \boldsymbol{\alpha} \times (\boldsymbol{\alpha} \times d\mathbf{v}^{(s)})$. Prove that this is equivalent to $d\mathbf{v}_j^{(ss)} = d\mathbf{v}_j^{(s)} - d\mathbf{v}_j^{(es)}$.

5 Basic Laws

The laws we shall formulate in this chapter are of such a fundamental nature that they cannot be proven in the mathematical sense. They are the starting point. They are also basic in the sense that they apply to all substances, solids as well as fluids. The truth of the basic laws has been established by a sort of scientific evolution. Of all the propositions that have been put forward, these laws have survived the test of time. Results predicted from them correspond to our experience.

Through the years the laws and the concepts involved in the laws have undergone subtle changes. As different viewpoints in physics have developed, the concepts have been generalized, adapted, and reinterpreted. In fluid mechanics today, we have a mature subject where the basic concepts and laws are well developed. This does not mean that we completely understand the multitude of phenomena that occur in fluid mechanics. Even though the basic equations have been written, it often happens that their solution can be obtained if at all, only by employing simplifying assumptions.

A derivation of a basic law is really the mathematical formulation of the relationships between several physical concepts. As we formulate the laws, we want to pay particular attention to the nature of the concepts such as surface force, work, heat flux, and so forth. The final form of the laws will be differential equations that are valid at every point in the continuum.

There are three major independent dynamical laws in continuum mechanics: the continuity equation, the momentum equation, and the energy equation. These laws will be formulated in the first part of the chapter. There are several additional laws that may be derived from the momentum equation. The first of these governs kinetic energy, the second angular momentum, and the third vorticity. The first two are treated in this chapter; the law governing vorticity will be introduced in Chapter 13. The last law we will study in this chapter is the second law of thermodynamics. A final section gives the integral or global form of the laws.

5.1 CONTINUITY EQUATION

The equation derived in this section has been called the *continuity equation* to emphasize that the continuum assumptions (the assumption that density and velocity may be defined at every point in space) are a prerequisite. The physical principle underlying the equation is the conservation of mass, and it is sometimes simply called the mass conservation law. The continuum assumption is, of course, a foundation for all the basic laws. The mass conservation principle may be stated in terms of a material region as follows: "The amount of matter in a material region is constant." An equivalent statement is: *The time rate of change of the mass of a material region is zero*. The mass of the material region is computed by integrating the density over the region. Thus in mathematical terms we have

$$\frac{dM_{\mathrm{MR}}}{dt} = \frac{d}{dt}\int_{\mathrm{MR}} \rho \, dV = 0 \qquad 5.1.1$$

The bounding surface of the material region is moving with the local fluid velocity v_i (Fig. 5-1). We use Leibnitz's theorem (3.10.2) with $w_i = v_i$ to move the time differentiation inside the integral:

$$\int_{\mathrm{MR}} \partial_0 \rho \, dV + \int_{\mathrm{MR}} n_i v_i \rho \, dS = 0 \qquad 5.1.2$$

Next, the theorem of Gauss changes the surface integral into a volume integral:

$$\int_{\mathrm{MR}} [\partial_0 \rho + \partial_i(\rho v_i)] \, dV = 0 \qquad 5.1.3$$

(Note that this equation applies at every instant, and thus the restriction to a material region is no longer necessary. We could, in principle, choose a different region for each instant in time. The material region has already played out its role in that the surface velocity in 5.1.2 is the fluid velocity v_i.) Since the specific choice of the integration region is arbitrary, the only way

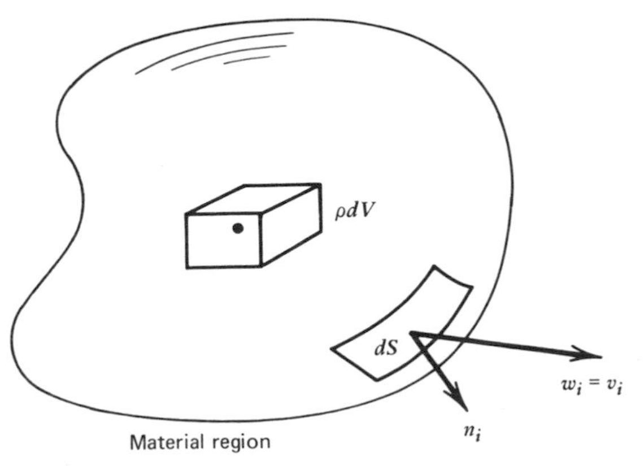

Figure 5-1 Continuity equation for a material region. Surface velocity is equal to fluid velocity.

5.1.3 can be true is if the integrand is zero. It cannot happen that the integrand is positive in one part of the region and negative in another so that they always cancel out. If there were a part of space where the integrand was positive (say), we could immediately choose that place as the region of integration and violate Eq. 5.1.3. Thus, the integrand is identically zero everywhere. This is the differential form of the continuity equation:

$$\partial_0 \rho + \partial_i(\rho v_i) = 0 \qquad 5.1.4$$

or, in symbolic notation,

$$\frac{\partial \rho}{\partial t} + \nabla \cdot (\rho \mathbf{v}) = 0$$

The special form of 5.1.4 for steady flow is $\partial_i(\rho v_i) = 0$, and for incompressible flow it is $\partial_i v_i = 0$.

In order to get a better physical understanding of the terms in the continuity equation, let us evaluate the equation at a point P in the center of a fixed differential volume element $\Delta x \, \Delta y \Delta z$. If we multiply 5.1.4 by $\Delta x \, \Delta y \Delta z$, the first term will be

$$\frac{\partial}{\partial t}(\rho \Delta x \, \Delta y \Delta z) \qquad 5.1.5$$

The physical interpretation of ρ is the mass per unit volume, and $\Delta x \, \Delta y \Delta z$ is the volume of the fixed element. So their product is the mass of the fixed element. We then interpret the original term $\partial_0 \rho$ as the rate of change of mass per unit volume at a fixed point in space (Fig. 5-2).

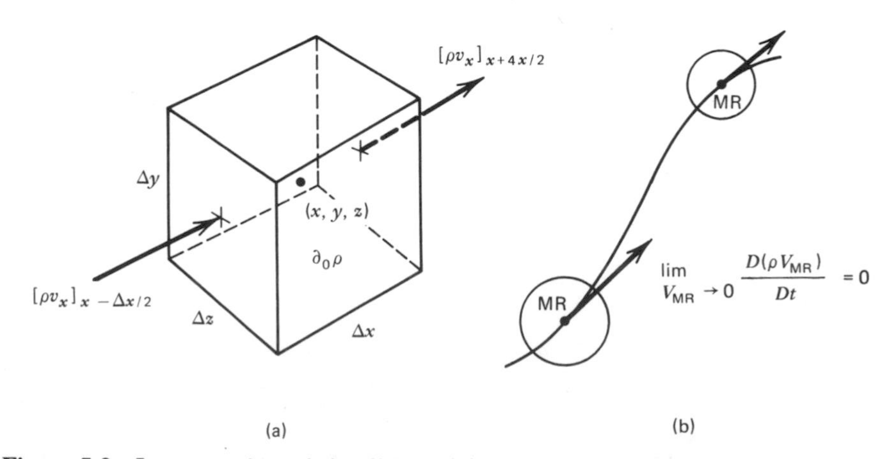

Figure 5-2 Interpretation of the differential continuity equation: (a) for an element fixed in space, (b) for a fluid particle.

The next term in 5.1.4 is actually three terms. Writing them out produces

$$\partial_i(\rho v_i) = \partial_x(\rho v_x) + \partial_y(\rho v_y) + \partial_z(\rho v_z)$$

We take the first term as typical, multiply it by $\Delta x \, \Delta y \Delta z$, and group the symbols as follows;

$$\frac{\partial}{\partial x}(\rho v_x \, \Delta y \Delta z) \, \Delta x \qquad 5.1.6$$

The product $v_x \, \Delta y \Delta z$ is the volume flow rate across a $\Delta y \Delta z$ surface. Multiplying by ρ gives the mass flow rate $\rho v_x \, \Delta y \Delta z$. The net mass flow rate out of the element through the two surfaces $\Delta y \Delta z$ is given by $\partial_x(\rho v_x \, \Delta y \Delta z) \, \Delta x$. This is readily seen by computing the mass flow rate at each surface and subtracting. Consider the mass flow through the surfaces at $x + \Delta x/2$ and $x - \Delta x/2$. The net difference is

$$[\rho v_x]_{x+\Delta x/2} \, \Delta y \Delta z - [\rho v_x]_{x-\Delta x/2} \, \Delta y \Delta z$$

Expand these terms in a Taylor series about x, and keep only the terms that will not drop out if we divide by $\Delta x \, \Delta y \Delta z$ and go to the limit $\Delta x \to 0$:

$$\left\{ \left[\rho v_x + \frac{\partial}{\partial x}(\rho v_x) \left(x + \frac{\Delta x}{2} - x \right) + \cdots \right] - \left[\rho v_x + \frac{\partial}{\partial x}(\rho v_x) \left(x - \frac{\Delta x}{2} - x \right) + \cdots \right] \right\} \Delta y \Delta z$$

or

$$\frac{\partial}{\partial x}(\rho v_x) \, \Delta x \, \Delta y \Delta z$$

Thus 5.1.6 has the physical interpretation as the net x direction mass flow rate out of the element. The other two terms, $\partial_y(\rho v_y)$ and $\partial_z(\rho v_z)$, are the net mass flow rates out of the element in the y and z directions respectively. All these terms are called convective terms and represent the net rate of mass efflux out of the element. From the point of view of a fixed point in space, the continuity equation 5.1.4 is a balance between the rate of accumulation of mass and the net outflow:

$$\underbrace{\partial_0 \rho}_{\substack{\text{rate of accumulation} \\ \text{of mass per unit} \\ \text{volume at } P}} + \underbrace{\partial_i(\rho v_i)}_{\substack{\text{net flow rate of} \\ \text{mass out of } P \\ \text{per unit volume}}} = 0 \qquad 5.1.4$$

The continuity equation may also be viewed from the standpoint of a material particle moving through the flow. Differentiate the second term in

5.1.4 and identify the substantial derivative (4.3.5),

$$\partial_0 \rho + v_i \partial_i \rho = -\rho \partial_i v_i$$

$$\underset{\substack{\text{rate of change of}\\\text{the density of a}\\\text{fluid particle}}}{\frac{D\rho}{Dt}} = -\underset{\substack{\text{mass}\\\text{per}\\\text{unit}\\\text{volume}}}{\rho} \underset{\substack{\text{particle}\\\text{volume}\\\text{expansion}\\\text{rate}}}{\partial_i v_i} \qquad 5.1.7$$

We can give a physical interpretation to 5.1.7 by recalling from 4.8.5 that $\partial_i v_i$ is the rate of increase of the volume of a material particle. Substitution of 4.7.5 into 5.1.7 gives

$$\underset{\substack{\text{fractional rate of}\\\text{change of the density}\\\text{of a material particle}}}{\frac{1}{\rho}\frac{D\rho}{Dt}} = -\underset{\substack{\text{fractional rate of}\\\text{change of the volume}\\\text{of a material particle}}}{\frac{1}{V_{MR}}\frac{dV_{MR}}{dt}} \quad \text{as} \quad V_{MR} \to 0 \qquad 5.1.8$$

The change in density of a material particle is entirely the result of changes in its volume. Equation 5.1.8 implies that the mass of a material particle ρV_{MR} is a constant (Fig. 5-2b).

5.2 MOMENTUM EQUATION

The momentum equation for a continuum is the analogue of Newton's second law for a point mass. It is not possible to derive the momentum equation from Newton's second law because the concepts of a point mass and of a continuum are distinctly different. The momentum principle is: *The time rate of change of the linear momentum of a material region is equal to the sum of the forces on the region.* Two types of forces may be imagined: body forces, which act upon the bulk of the material in the region, and surface forces, which act on the boundary surface. We let F_i stand for a body force per unit mass, and R_i stand for a surface force per unit area. The net force on the region consists of the two integrals,

$$\text{net force on material region} = \int_{MR} \rho F_i \, dV + \int_{MR} R_i \, dS$$

Next we must compute the momentum within the region. We usually think of the velocity v_i as the rate of change of position, but v_i can also play the role of the "i-direction momentum per unit mass." The product ρv_i is the i-direction momentum per unit volume. Therefore, $\rho v_i \, dV$ is the i-direction momentum within the element dV. The rate of change of momentum of the region is

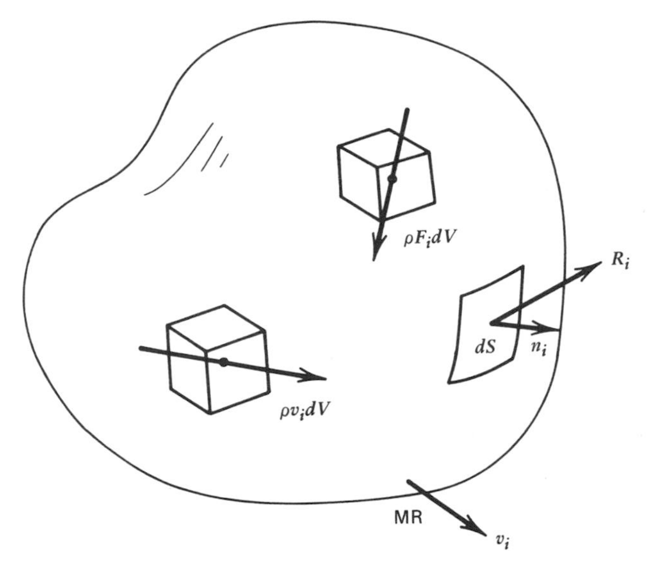

Figure 5-3 Momentum equation for a material region.

computed as

$$\text{rate of change of momentum of a material region} = \frac{d}{dt}\int \rho v_i \, dV$$

The momentum principle is then given by the equation (see Fig. 5-3)

$$\frac{d}{dt}\int_{\text{MR}} \rho v_i \, dV = \int_{\text{MR}} \rho F_i \, dV + \int_{\text{MR}} R_i \, dS \qquad 5.2.1$$

The left-hand side can be changed by using Leibnitz's and Gauss's theorems as we did for the continuity equation. The result is

$$\int \left[\partial_0(\rho v_i) + \partial_j(\rho v_j v_i) \right] dV = \int \rho F_i \, dV + \int R_i \, dS \qquad 5.2.2$$

This is not the final form of the equation; however, it is as far as we can proceed without knowing some details about the surface force R_i.

5.3 SURFACE FORCES

When we construct an imaginary closed surface, it divides the fluid into its inside and its outside portions. The direct action of the outside fluid upon the inside fluid is replaced by the concept of a surface force. Essentially, we imagine the outside fluid has vanished and been replaced by forces to produce the actual effect on the inside fluid. Surface forces are really momentum and force effects at the microscopic level.

The surface-force concept is instantaneous. We construct the imaginary surface and evaluate the force at an instant in time. The bulk motion of the fluid plays no direct role in this process. Neither does the prescription of the surface motion as a function of time. We can apply it the surface-force concept to a material surface, which follows the fluid, or we can apply it to a fixed surface, which allows fluid to cross. If a material region and fixed region coincide at a given time, then the surface forces on them are identical.

The surface force per unit area is taken to be a function of the position P in space and also the orientation of the surface that passes through the point P. We may let the orientation of the surface be given by the outward unit normal vector n_i. Then our proposition is that the surface force per unit area is the function $R_i(n_i;\ x_i)$. The purpose of the next section is to show that the dependence of R_i upon n_i may be given by introducing a *stress tensor* T_{ij} that obeys the equation

$$R_j = n_i T_{ij} \qquad 5.3.1$$

The stress tensor depends on the position of P, but not on the orientation of the plane. Although we speak of fluids, the results are applicable to solids as well.

Equation 5.3.1 is all that is needed to complete the derivation of the momentum equation, and the reader may want to skip the next section upon first reading.

*5.4 STRESS-TENSOR DERIVATION

First we shall investigate how $R_i(n_i)$ changes if we change the direction of n_i by 180°. That is, we shall prove that the force due to the outside fluid on the inside is exactly equal and opposite to the force due to the inside fluid on the outside. Consider a small volume centered at point P (Fig. 5-4). The two ends are parallel with area ΔS and located a distance Δl apart. The normal vector for side 1 is n_i^{I}, while that for side 2 is $n_i^{\mathrm{II}} = -n_i^{\mathrm{I}}$. The perimeter of ΔS is denoted by s, and the normal vector at any point on the side is n_i^{III}. We write the momentum equation 5.2.2 for this region using the mean-value theorem for the integrals,

$$\left[\partial_0(\rho v_i) + \partial_j(\rho v_j v_i)\right]^* \Delta S\,\Delta l = [\rho F_i]^* \Delta S\,\Delta l + R_i^*\left(n_i^{\mathrm{I}}\right)\Delta S + R_i^*\left(n_i^{\mathrm{II}}\right)\Delta S + R_i^*\left(n_i^{\mathrm{III}}\right) s\,\Delta l \qquad 5.4.1$$

Here $R_i^*(n_i^{\mathrm{I}})$ stands for the stress on the surface with normal n_i^{I}. The asterisks indicate a mean value somewhere within the region of integration. As the region is shrunk to zero thickness ($\Delta l \to 0$), the body force and inertia terms

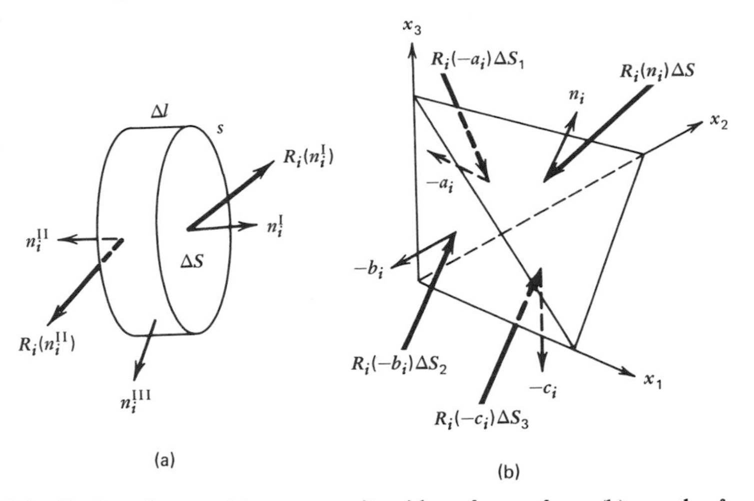

Figure 5-4 Surface forces: (a) on opposite sides of a surface, (b) on the faces of a tetrahedron.

drop out, leaving the surface forces in exact balance, that is,

$$0 = R_i^*\left(n_i^{I}\right)\Delta S + R_i^*\left(n_i^{II}\right)\Delta S \qquad 5.4.2$$

Now letting $\Delta S \to 0$ requires that the mean values on both faces take on the value of R_i at the point P. Letting $n_i = n_i^{I} = -n_i^{II}$ yields

$$R_i(n_i) = -R_i(-n_i) \qquad 5.4.3$$

This proves that the force due to the inside fluid on the outside fluid is equal and opposite to the force due to the outside fluid on the inside fluid.

We proceed to see how the stress on a plane of arbitrary direction is related to the stress on planes in the coordinate directions. Figure 5-4b shows a tetrahedron with three surfaces ΔS_1, ΔS_2, and ΔS_3 parallel with the coordinate planes. The triangular face ΔS has the unit normal vector n_i while the unit normals in the coordinate directions are a_i, b_i, and c_i. Let ΔL be some typical dimension of the tetrahedron. We will let $\Delta L \to 0$ while keeping the same direction n_i for the triangular face. The volume of the tetrahedron is proportional to $(\Delta L)^3$ and all the surfaces are proportional to $(\Delta L)^2$. If we write the momentum equation 5.2.2 for the region and again estimate the integrands by mean values, the equation will take the form

$$[\text{inertia terms}]^*(\Delta L)^3 = [\text{body-force term}]^*(\Delta L)^3 + [\text{surface-force terms}]^*(\Delta L)^2 \qquad 5.4.4$$

Dividing by $(\Delta L)^2$ and letting $\Delta L \to 0$ shows that the surface forces are in exact balance:

$$0 = [\text{surface-force terms}]^* \qquad 5.4.5$$

Writing out the surface forces explicitly produces the equation

$$0 = R_i^*(n_i)\,\Delta S + R_i^*(-a_i)\,\Delta S_1 + R_i^*(-b_i)\,\Delta S_2 + R_i^*(-c_i)\,\Delta S_3 \qquad 5.4.6$$

The next step is to relate the side areas to the top area ΔS. In Chapter 3 (3.9.6) we proved that the i component of the unit normal vector times the area equals the projection of the area on an i plane. Applying this to the tetrahedron gives

$$\begin{aligned} n_1\,\Delta S &= \Delta S_1 \\ n_2\,\Delta S &= \Delta S_2 \\ n_3\,\Delta S &= \Delta S_3 \end{aligned} \qquad 5.4.7$$

Substituting 5.4.7 into 5.4.6, using the fact that $R_i(n_i) = -R_i(-n_i)$, and letting $\Delta S \to 0$ yields

$$R_i(n_i) = n_1 R_i(a_i) + n_2 R_i(b_i) + n_3 R_i(c_i) \qquad 5.4.8$$

This equation has exactly the same form as 3.4.7 which defines a tensor. We can see this if we change the notation by letting the stresses associated with the x_i plane be $R_i(a_i) = T_{1i}$, $R_i(b_i) = T_{2i}$, and $R_i(c_i) = T_{3i}$. Equation 5.4.8 can now be written as

$$\begin{aligned} R_i(n_i) &= n_1 T_{1i} + n_2 T_{2i} + n_3 T_{3i} \\ R_i &= n_j T_{ji} \end{aligned} \qquad 5.4.9$$

The surface force per unit area R_i depends upon the orientation of the plane through P by the relation above. The components of the stress tensor are functions of the position in space.

5.5 INTERPRETATION OF THE STRESS-TENSOR COMPONENTS

A component of the stress tensor T_{ij} is equal to the stress in direction j on a plane with a normal in direction i. This is illustrated in Fig. 5-5. Consider an x_2–x_3 plane through P with unit normal $n_i = (1, 0, 0)$. From

$$R_j = n_i T_{ij} \qquad 5.3.1$$

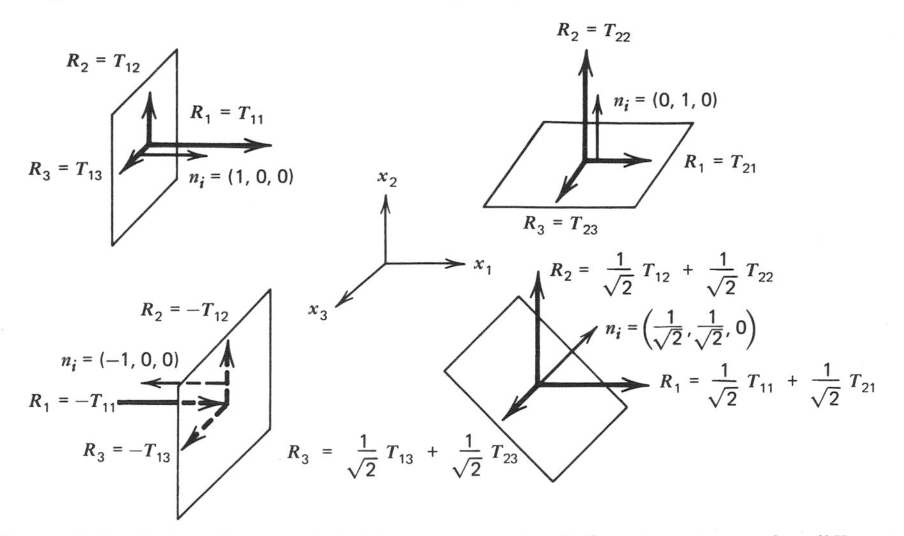

Figure 5-5 Surface forces related to components of the stress tensor for different orientations of the surface.

of Section 5.3, written in the form

$$R_j = 1T_{1j} + 0T_{2j} + 0T_{3j}$$

we find

$$R_1 = T_{11}, \qquad R_2 = T_{12}, \qquad R_3 = T_{13} \qquad 5.5.1$$

Similarly, for a plane with normal $n_i = (0, 1, 0)$ the stress components are

$$R_1 = T_{21}, \qquad R_2 = T_{22}, \qquad R_3 = T_{23} \qquad 5.5.2$$

The components of T_{ij} are sometimes loosely called "stresses", but they are not actually the stresses. However, under certain conditions they are equal to the stresses. The stress R_i is related to the stress tensor T_{ij} by $R_j = n_i T_{ij}$. For example, consider again an x_2–x_3 plane, but this time take the normal as downward, $n_i = (-1, 0, 0)$. This reverses the designations of the inside and outside of the plane. Now from $R_j = n_i T_{ij}$ we have

$$R_1 = -T_{11}, \qquad R_2 = -T_{12}, \qquad R_3 = -T_{13} \qquad 5.5.3$$

For this plane the stresses are equal to the negative of the stress-tensor components.

One final example will close this section. Let us find the stress on a plane with normal $(1/\sqrt{2}, 1/\sqrt{2}, 0)$. Evaluating the components of $R_j = n_i T_{ij}$, we find

$$R_1 = \frac{1}{\sqrt{2}} T_{11} + \frac{1}{\sqrt{2}} T_{21}$$

$$R_2 = \frac{1}{\sqrt{2}} T_{12} + \frac{1}{\sqrt{2}} T_{22} \qquad 5.5.4$$

$$R_3 = \frac{1}{\sqrt{2}} T_{13} + \frac{1}{\sqrt{2}} T_{23}$$

These components are shown in Fig. 5-5.

5.6 PRESSURE AND VISCOUS STRESS TENSOR

The pressure that was used in the thermodynamical equation 2.5.3 is a function of the thermodynamic state. We shall use the subscript t to denote thermodynamic pressure:

$$p_t = f(e, \rho) \qquad 5.6.1$$

When a fluid is not moving, we expect the normal stress to be the pressure. Will this still be true when the fluid moves? If it is not true, how do the pressure and the normal stress differ?

The thermodynamic pressure has a different conceptual origin than that of the surface forces that we discussed in the previous section. Surface forces in $R_i = n_j T_{ji}$ are mechanical force concepts. Because of this ambiguity it is necessary to relate the normal surface stress and the thermodynamic pressure.

In order to see the question in a different light, we break the stress tensor into two parts by subtracting out the thermodynamic pressure. In essence we are defining the *viscous stress tensor* τ_{ij} by the equation

$$T_{ij} = -p_t \delta_{ij} + \tau_{ij} \qquad 5.6.2$$

When a substance is not moving, we know that the normal stress is the same as the thermodynamic pressure. This requirement implies that the viscous stress τ_{ij} must vanish when there is no motion.

In general the normal stress is the sum of the pressure and a normal viscous stress. For planes with normals in the coordinate directions the normal stresses are

$$R_1 = T_{11} = -p_t + \tau_{11}$$

$$R_2 = T_{22} = -p_t + \tau_{22} \qquad 5.6.3$$

$$R_3 = T_{33} = -p_t + \tau_{33}$$

Normal viscous stresses are frequently small compared to the pressure, so it is common to neglect τ_{11} in comparison with p_t in many engineering calculations. However, there is an easily imagined experiment that illustrates a situation where the normal viscous stress is important. When we pour a very viscous liquid such as honey from a jar, the column of fluid obviously does not accelerate as fast as the local acceleration of gravity. A falling ball accelerates much more rapidly than the honey falling from the jar. The force that retards the honey column is an imbalance in the normal viscous force. If one were to cut the column instantaneously, the two parts would separate because the normal viscous tension stress would no longer retard the lower part.

The normal stress, unlike the pressure, can have different values for different directions of the vector n_i. We can average the normal surface force and call this average the mechanical pressure p_m. This gives

$$p_m \equiv -\tfrac{1}{3}(T_{11} + T_{22} + T_{33}) = -\tfrac{1}{3}T_{ii} \qquad 5.6.4$$

An incompressible fluid (a thermodynamic term) does not have a thermodynamic pressure, but it does have a mechanical pressure. When we are dealing with an incompressible fluid, the pressure variable is always interpreted as the mechanical pressure.

For a compressible fluid, what is the difference between the mechanical pressure and the thermodynamic pressure? As a first approximation, people have proposed that the difference between the pressures is a linear function of the rate of expansion (4.8.5). If the rate of expansion is zero, the fluid is behaving as if it were incompressible. So in this sense the assumption is consistent. Mathematically the assumption is

$$p_m - p_t = k\,\partial_i v_i = -\frac{k}{\rho}\frac{D\rho}{Dt} \qquad 5.6.5$$

The coefficient k is called the bulk viscosity. The second relation is obtained using the continuity equation (5.1.7). For common fluids it is nearly always assumed that k is zero, thus $p_m = p_t$ and there is no need to distinguish between mechanical and thermodynamic pressure. This is called Stokes's assumption.

Stokes's assumption implies that the average normal viscous stress is zero. To show this we take the trace of 5.6.2 and divide by three:

$$\tfrac{1}{3}T_{ii} = -p_t + \tfrac{1}{3}\tau_{ii}$$

By Stokes's assumption

$$p_t = p_m = -\tfrac{1}{3}T_{ii}$$

Combining the relations above, we find the desired result:

$$0 = \tau_{ii} = \tau_{11} + \tau_{22} + \tau_{33}$$

This is an equivalent statement of Stokes's assumption. More discussion of this assumption will be given in the next chapter.

5.7 DIFFERENTIAL MOMENTUM EQUATION

The derivation of the differential momentum equation, which was started in Section 5.2, can now be completed. Substitution of the surface-stress equation 5.3.1 into 5.2.2 and application of Gauss's theorem to the surface force yields

$$\int \left[\partial_0(\rho v_i) + \partial_j(\rho v_j v_i) - \rho F_i - \partial_j T_{ji} \right] dV = 0 \qquad 5.7.1$$

Since the region of integration is arbitrary, the integrand must be zero everywhere. Hence,

$$\partial_0(\rho v_i) + \partial_j(\rho v_j v_i) = \rho F_i + \partial_j T_{ji} \qquad 5.7.2$$

We can now introduce the pressure and viscous stress tensor by substituting 5.6.2 into 5.7.2. The result is

$$\partial_0(\rho v_i) + \partial_j(\rho v_j v_i) = \rho F_i - \partial_i p + \partial_j \tau_{ji}$$

In symbolic notation the equation is

$$\frac{\partial}{\partial t}(\rho \mathbf{v}) + \nabla \cdot (\rho \mathbf{v}\mathbf{v}) = -\nabla p + \nabla \cdot \boldsymbol{\tau} + \rho \mathbf{F} \qquad 5.7.3$$

This equation and the continuity equation are two of the most important equations of fluid mechanics.

In order to get a better idea of the physical role of each symbol in the momentum equation, we again consider an elementary cube $\Delta x \, \Delta y \Delta z$ located at the fixed point P (Fig. 5-6). We shall rederive the momentum equation by counting up the forces and momentum fluxes for this cube. This type of derivation is frequently done in elementary books. It is not as general as the derivation given above, but it has the advantage of displaying the physical meaning of the terms from the point of view of a fixed position in space.

First, consider that the i direction momentum within $\Delta x \, \Delta y \Delta z$ is given by

$$(\rho v_i)^* \Delta x \, \Delta y \Delta z$$

where the asterisk indicates an average value that exists somewhere within the

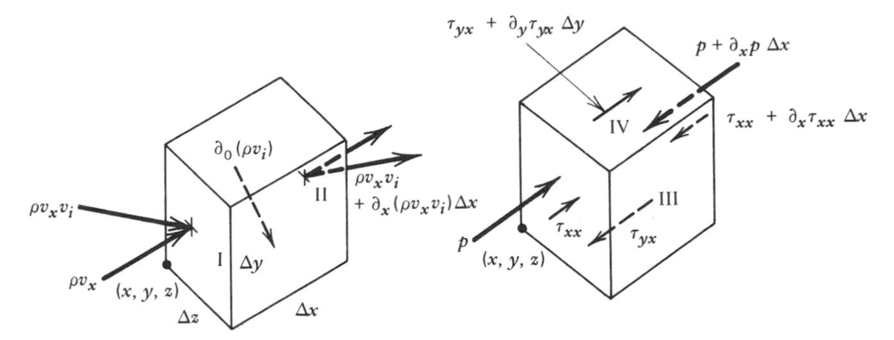

Figure 5-6 Interpretation of the differential momentum equation for a fixed element.

region. In this term v_i has the physical interpretation of the i-direction momentum per unit volume. The rate of change of i momentum within the fixed region of space $\Delta x\,\Delta y\,\Delta z$ is

$$\Delta x\,\Delta y\,\Delta z\;\partial_0(\rho v_i)^* \qquad 5.7.4$$

We see that if 5.7.4 is divided by $\Delta x\,\Delta y\,\Delta z$ and then the element is shrunk to zero size, the average value must occur at the point P. This results in the first term of 5.7.2; $\partial_0(\rho v_i)$ is the rate of increase of i momentum per unit volume at the point P.

Next, we consider the fact that i momentum is carried into and out of the fixed region by fluid flow across the surfaces. Across face I of Fig. 5-6 there is a mass flow of $\rho v_x\,dy\,dz$ carrying with it the i momentum v_i per unit mass. Hence, across this face the i momentum going into the region is

$$\int_{\mathrm{I}} \rho v_x v_i\,dy\,dz = (\rho v_x v_i)^*\,\Delta y\,\Delta z \qquad 5.7.5$$

The asterisk indicates an average value that occurs somewhere on face I. On face II, which is on the opposite side of the cube, a positive flow ρv_x carries i momentum out of the region. The rate momentum leaves is the integral of $\rho v_x v_i$ over II. Instead of evaluating this on face II, we can expand $[\rho v_x v_i]_{\mathrm{II}}$ using a Taylor series from face I to face II, that is,

$$\begin{aligned}\int_{\mathrm{II}} \rho v_x v_i\,dy\,dz &= \int_{\mathrm{II}} \left\{[\rho v_x v_i]_{\mathrm{I}} + \frac{\partial}{\partial x}[(\rho v_x v_i)]_{\mathrm{I}}\,\Delta x + \cdots\right\} dy\,dz \\ &= \int_{\mathrm{I}} \rho v_x v_i\,dy\,dz + \frac{\partial}{\partial x}\int_{\mathrm{I}} \rho v_x v_i\,dy\,dz\,\Delta x + \cdots \\ &= (\rho v_x v_i)_{\mathrm{I}}^*\,\Delta y\,\Delta z + \frac{\partial}{\partial x}(\rho v_x v_i)^*\,\Delta x\,\Delta y\,\Delta z + O(\Delta x^2\,\Delta y\,\Delta z)\end{aligned}$$

5.7.6

Again the asterisk stands for mean values. The net i momentum crossing the surface of the fixed region by fluid flow across faces I and II is found when 5.7.6 is subtracted from 5.7.5. The result is

$$-\frac{\partial}{\partial x}(\rho v_x v_i)^* \Delta x \, \Delta y \Delta z + O(\Delta x^2 \, \Delta y \Delta z)$$

Thus the term $-\partial_x(\rho v_x v_i)$ is the net convection of i momentum into a fixed unit volume at P by x-direction flow.

Fluid also flows across the Δx–Δy faces with a flow rate ρv_y. This leads to a net convection of i momentum

$$-\frac{\partial}{\partial y}(\rho v_y v_i)^* \Delta x \, \Delta y \Delta z + O(\Delta y^2 \, \Delta x \, \Delta z)$$

Similarly, across the two Δx–Δy surfaces, the net i-momentum gain by flow in the z direction is

$$-\frac{\partial}{\partial z}(\rho v_z v_i)^* \Delta z \, \Delta y \Delta x + O(\Delta z^2 \, \Delta x \, \Delta y)$$

When the three terms above are divided by $\Delta x \, \Delta y \Delta z$ and the limit Δx, Δy, $\Delta z \to 0$ is taken, the higher-order terms go to zero, leaving

$$-\partial_j(\rho v_j v_i) \qquad 5.7.7$$

as the net rate of increase of i momentum because of a fluid flow into a fixed unit volume.

Next we turn our attention to the forces. Forces which act on the cube are the body and surface forces. The body force is

$$\int \rho F_i \, dx \, dy \, dz = (\rho F_i)^* \Delta x \, \Delta y \Delta z \qquad 5.7.8$$

where F_i represents the force per unit mass. This is the most convenient terminology, since the weight of an object, $W_i = M g_i$, is given on a unit-mass basis by $F_i = g_i$.

Surface forces on the element consist of pressure and viscous forces. On face I the pressure force is

$$\int_{\mathrm{I}} p \, dy \, dz = p_{\mathrm{I}}^* \, \Delta y \Delta z \qquad 5.7.9$$

Counteracting this force is a force on face II,

$$\int_{\mathrm{II}} p \, dy \, dz = \int_{\mathrm{II}} \left[p_{\mathrm{I}} + \frac{\partial p}{\partial x} \Delta x + \cdots \right] dy \, dz$$

$$= \left[p_{\mathrm{I}}^* + \frac{\partial p_{\mathrm{I}}^*}{\partial x} \Delta x \right] \Delta y \Delta z + O(\Delta x^2 \, \Delta y \Delta z) \qquad 5.7.10$$

The net force is found by subtracting 5.7.10 from 5.7.9. This leaves the unbalanced force as

$$-\frac{\partial p_{\rm I}^*}{\partial x}\Delta x\,\Delta y\Delta z + O(\Delta x^2\,\Delta y\Delta z) \qquad 5.7.11$$

Division by $\Delta x\,\Delta y\Delta z$ and letting Δx, Δy, $\Delta z \to 0$ produces $-\partial p/\partial x$. This is the x-direction pressure force on the unit element. Repeating the reasoning for pairs of faces in the Δy and Δz directions leads to a net i-direction pressure force equal to the gradient

$$-\partial_i p \qquad 5.7.12$$

The normal viscous force τ_{xx} acts on faces I and II in exactly the same way that the pressure does. Hence, corresponding to 5.7.11 we have a net x-direction force of

$$\frac{\partial \tau_{xx}}{\partial x}\Delta x\,\Delta y\Delta z + O(\Delta x^2\,\Delta y\Delta z) \qquad 5.7.13$$

The sign is positive for a tension force.

Also acting in the x direction are shear forces on the side surfaces. On face III the force is

$$-\int_{\rm III}\tau_{yx}\,dx\,dy = -\tau_{yx}^*\,\Delta x\,\Delta z + \cdots \qquad 5.7.14$$

The companion force on face IV is

$$\int_{\rm IV}\tau_{yx}\,dx\,dz = \int_{\rm IV}\left\{[\tau_{yx}]_{\rm III} + \left[\frac{\partial \tau_{yx}}{\partial y}\right]_{\rm III}\Delta y + \cdots\right\}dx\,dz$$

$$= \tau_{yx}^*\,\Delta x\,\Delta y + \frac{\partial \tau_{yx}^*}{\partial y}\Delta y\Delta x\,\Delta z + \cdots$$

Combining this with 5.7.14 gives the net x-direction force for faces III and IV as

$$\frac{\partial \tau_{yx}^*}{\partial y}\Delta x\,\Delta y\Delta z + O(\Delta y^2\,\Delta z\,\Delta x)$$

The remaining two faces, V and VI, produce the x-direction force

$$\frac{\partial \tau_{zx}^*}{\partial z}\Delta z\,\Delta x\,\Delta y + O(\Delta z^2\,\Delta x\,\Delta y)$$

The x-direction viscous forces for the element, after the limit has been taken, add up to

$$\partial_j(\tau_{jx}) \tag{5.7.15}$$

Similar expressions exist for the y- and z-direction viscous forces. In general, the i-direction viscous forces per unit volume at the point P are

$$\partial_j \tau_{ji} \tag{5.7.16}$$

We put all of the preceding effects together to form the momentum equation,

$$\partial_0(\rho v_i) = -\partial_j(\rho v_j v_i) + \rho F_i - \partial_i p + \partial_j \tau_{ji} \tag{5.7.17}$$

$\partial_0(\rho v_i)$	$-\partial_j(\rho v_j v_i)$	$+\rho F_i$	$-\partial_i p$	$+\partial_j \tau_{ji}$
rate of i momentum increase at the fixed point P	net rate i momentum is carried into P by fluid flow ρv_j	i body force at P	net i pressure force at P	net i viscous force at P

All terms are on a unit-volume basis because we divided by $\Delta x \Delta y \Delta z$. In other words, $-\partial_i p$ is the net i-direction pressure force per unit volume at the point P.

We may also interpret the momentum equation from the viewpoint of a material particle moving in the flow. To do this we split the convective terms of 5.7.17 into parts and group the terms as follows:

$$\rho \partial_0 v_i + \rho v_j \partial_j v_i + v_i\left[\partial_0 \rho + \partial_j(\rho v_j)\right] = \text{forces} \tag{5.7.18}$$

The term in brackets is zero by the continuity equation, and the first two terms are by definition the substantial derivative. Thus the final form for the particle viewpoint is

$$\rho \partial_0 v_i + \rho v_j \partial_j v_i = -\partial_i p + \partial_j \tau_{ji} + \rho F_i$$

$$\rho \frac{Dv_i}{Dt} = -\partial_i p + \partial_j \tau_{ji} + \rho F_i \tag{5.7.19}$$

This equation states that the mass per unit volume (ρ) times the acceleration of a material particle (Dv_i/Dt) is equal to the net force acting on the particle. This is "Newton's law" for a continuum particle. Note that the forces are the same whether we interpret the inertia terms from the viewpoint of a fixed point in space (5.7.17) or from the viewpoint of a moving material particle (5.7.19). Forces act instantaneously without regard to the past or future position of the volume or surface used for their evaluation. They have no memory or ability to anticipate the motion of the surface. Our choice for the surface motion can play no essential role in force concepts.

In Newton's mechanics, particles are of finite size with a fixed mass M. Their linear momentum is governed by an equation stating that mass times acceleration equals net force:

$$M\frac{dv_i}{dt} = \sum \mathscr{F}_i \qquad 5.7.20$$

Sometimes students are introduced to this equation as stating that the rate of change of momentum is equal to the net forces:

$$\frac{d}{dt}(Mv_i) = \sum \mathscr{F}_i \qquad 5.7.21$$

This is proper, since the mass is a constant. The reason 5.7.21 is used is that it leads one more naturally to treat momentum as a fundamental property on the microscopic scale. Also, when relativity is considered it is easier to modify 5.7.21 to the proper form. What is not proper on the continuum scale is to say that 5.7.21 is the more basic form and that 5.7.20 results as a special case for finite particles of constant mass. In Section 5.15, where global forms of the equations are applied to several problems, we shall study the motion of a uniformly evaporating drop. The results of that problem show that 5.7.20 governs the motion of the droplet even when the mass is changing. Thus, 5.7.21 does not describe continuum particles with variable mass.

*5.8 ANGULAR-MOMENTUM EQUATION AND SYMMETRY OF T_{ij}

On the microscopic scale the angular-momentum equation is an independent law. In continuum mechanics we shall find that the linear-momentum equation may be used to derive the angular-momentum equation and they are thus not independent. There is one "if" in this statement. If the angular momentum of the microscopic particles is randomly oriented, then the vector sum for a large number of particles will be zero. On the other hand, if we imagine that the microscopic particles have their axes of rotation aligned in a special direction, then the summation will give a net angular momentum on the continuum level. If this were the case, we would need to postulate a surface couple in addition to the surface force. Fortunately, in common fluids and flow situations, the microscopic angular momentum is randomly oriented and the couple does not exist. When this is true, we shall be able to show that the stress tensor T_{ij} is symmetric.

First we derive the angular-momentum equation including a term for microscopic angular momentum. Then we shall show that the existence of a net microscopic angular momentum implies that the stress tensor is antisymmetric. Conversely, if the net microscopic angular momentum vanishes, then the stress tensor is symmetric and the angular-momentum equation is identical with the moment of the linear-momentum equation.

The angular-momentum law is stated as follows: *The rate of change of the angular momentum of a material region is equal to the sum of the torques*. Let the vector r_i be the position of a surface or volume element with respect to the origin. The relevant terms are formulated as follows:

Angular momentum of material in element dV about origin	$\mathbf{r} \times (\rho \mathbf{v})\, dV$	or	$\rho \varepsilon_{ijk} r_j v_k\, dV$
Torque of body force about origin	$\mathbf{r} \times \rho \mathbf{F}\, dV$	or	$\rho \varepsilon_{ijk} r_j F_k\, dV$
Torque of surface force about origin	$\mathbf{r} \times \mathbf{R}\, dS$	or	$\varepsilon_{ijk} r_j n_p T_{pk}\, dS$
Net microscopic angular momentum	$\mathbf{n} \cdot \mathbf{\Omega}\, dS$	or	$n_k \Omega_{ki}\, dS$

In the last term we have postulated an angular-momentum tensor Ω_{ki}, giving the transport of i-direction angular momentum across a k-direction plane by microscopic processes. The angular-momentum law then takes the following form:

$$\frac{d}{dt}\int_{\mathrm{MR}} \rho \varepsilon_{ijk} r_j v_k\, dV = \int_{\mathrm{MR}} \rho \varepsilon_{ijk} r_j F_k\, dV + \int_{\mathrm{MR}} \left[\varepsilon_{ijk} r_j n_p T_{pk} + n_k \Omega_{ki}\right] dS \tag{5.8.1}$$

The left side is converted using Leibnitz's and Gauss's theorems, while the last integral on the right-hand side is changed to a volume integral by Gauss's theorem. Since the region of integration is arbitrary, the integrands must be equal:

$$\varepsilon_{ijk}\partial_0(\rho r_j v_k) + \varepsilon_{ijk}\partial_p(v_p \rho r_j v_k) = \varepsilon_{ijk}\partial_p(r_j T_{pk}) + \partial_k \Omega_{ki} + \varepsilon_{ijk}\rho r_j F_k$$

The equation for the position vector $r_i = x_i$ is independent of time, and its space derivative is $\partial_i r_j = \delta_{ij}$. Using these facts, the equation above may be rearranged to

$$\varepsilon_{ijk} r_j\left[\partial_0(\rho v_k) + \partial_p(\rho v_p v_k) - \rho F_k - \partial_p T_{pk}\right] = \varepsilon_{ijk} T_{jk} + \partial_k \Omega_{ki}$$

The term in brackets is the linear-momentum equation and is equal to zero. The term $\varepsilon_{ijk} T_{jk}$ depends only upon the antisymmetric part of T_{jk} (see 3.5.11). Thus we have

$$-\varepsilon_{ijk} T_{[jk]} = \partial_k \Omega_{ki}$$

If we assume that the net microscopic flux of angular momentum, Ω_{ki}, is zero, then $T_{[jk]}$ is zero and we have proved that T_{ij} is symmetric. Furthermore, the angular-momentum equation is then exactly equal to $\mathbf{r} \times$ the linear-momentum equation. In continuum mechanics the two principles are not independent.

5.9 ENERGY EQUATION

The first law of thermodynamics states that the increase in energy of a material region is the result of work and heat transfers to the region. We shall discuss each of these concepts separately before combining them as required by the first law.

The energy of a material region contains contributions from all motions of the matter contained in it. The internal energy consists of microscopic motions such as random translation motion, molecular vibrations, molecular rotation, and any other microscopic energy modes. The sum of all these energies is the absolute thermodynamic internal energy e. The second form of energy is the kinetic energy of the bulk motion. For a unit mass this is $\frac{1}{2}v_i v_i = \frac{1}{2}v^2$. With these two forms of energy the total energy of the material region is (Fig. 5-7)

$$\text{total energy of material element } dV = \rho\left[e + \tfrac{1}{2}v^2\right] dV \qquad 5.9.1$$

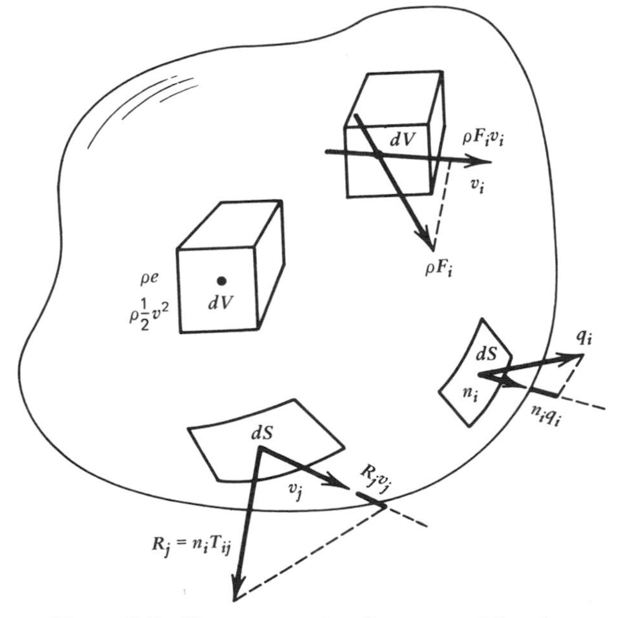

Figure 5-7 Energy equation for a material region.

A third type of energy is sometimes associated with the material. This is the potential energy that arises from a force field. A conservative force field will allow a representation by a potential. Gravity is the only body force we shall deal with in this book, and it may be represented by $F_i = \partial_i \hat{\phi}$. We have a choice of either considering that the potential energy $\hat{\phi}$ is associated with the gravity field, or computing the work done by the force. We shall take the latter route, and therefore not treat potential energy as such.

Work is the energy change when a force causes material to move. Before we formulate this concept for a continuum, let us review how work is formulated in classical particle mechanics. If we take Newton's law for a solid particle, $M\,dv_i/dt = \mathcal{F}_i$, and multiply both sides by v_i, we arrive at

$$M\frac{d}{dt}\left(\tfrac{1}{2}v_i v_i\right) = v_i \mathcal{F}_i$$

The work rate is the projection of the force along the instantaneous direction of the velocity. Only this component of the force increases the kinetic energy of the particle. The component of force perpendicular to v_i causes the trajectory to curve, but it does not increase the kinetic energy. Hence, it does no work.

The rate of work in a continuum is simply the component of the force in the direction of motion times the velocity of the matter. For the gravity force this is

$$\text{work rate of } F_i \text{ on element } dV = \rho v_i F_i \, dV \qquad 5.9.2$$

Likewise the surface force (both pressure and viscous) at dS produces

$$\text{work rate of } R_i \text{ at element } dS = v_i R_i \, dS = n_j T_{ji} v_i \, dS \qquad 5.9.3$$

Note that the past or future motion of the surface dS has nothing to do with the work. The important thing is the velocity of the material and not the velocity of the surface.

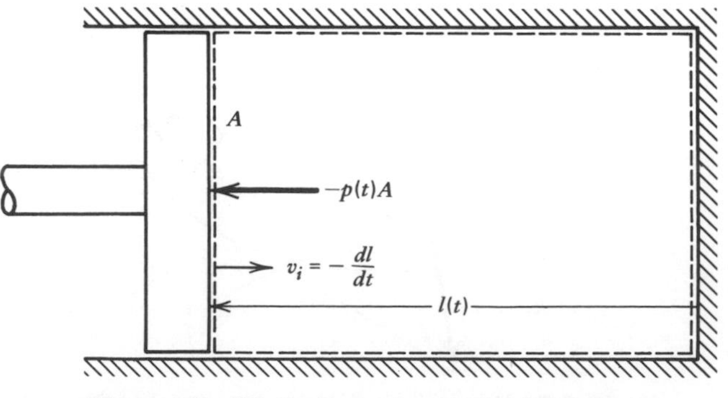

Figure 5-8 Work done at a moving boundary.

As an example let us compute the work done by a piston moving into a cylinder as shown in Fig. 5-8. The position of the piston is $l(t)$, and the area A. The piston force on the region is $p(t)A$. The matter at the side walls does not move, so the work is done only at the piston face. At the piston face, the fluid velocity is the same as the piston velocity. The work rate is then

$$\dot{W} = Ap(t)\frac{dl}{dt}$$

From time 1 to time 2 the work is

$${}_1W_2 = \int_{t_1}^{t_2} \dot{W}\,dt = \int_{t_1}^{t_2} pA\frac{dl}{dt}\,dt = \int_{l_1}^{l_2} pA\,dl = \int_{V_1}^{V_2} p\,dV$$

We are now obligated to consider p as a function of V.

Heat transfer is the second way energy is transferred into a region. The heat flux is the sum of all microscopic modes of energy transfer. Conduction is the most common type of microscopic energy transfer that we shall encounter. Other modes of energy transfer are radiation and the transport of energy by diffusion of different chemical species. All of these modes may be represented by a heat flux vector q (with dimensions of energy/area · time) which gives the magnitude and direction of the flux. The component of q perpendicular to a surface element dS is $n_i q_i$; thus

$$\text{rate of heat loss from } dS = n_i q_i\, dS \qquad 5.9.4$$

It is a matter of convention to define $\mathbf{q}$ as positive for a heat flux from the inside of the surface to the outside.

We are now ready to formulate the energy equation. *The rate of change of energy within a material region is equal to the rate that energy is received by heat and work transfers*. The mathematical statement is

$$\frac{d}{dt}\int_{\text{MR}} \rho\left(e + \tfrac{1}{2}v^2\right) dV = -\int_{\text{MR}} n_i q_i\, dS + \int_{\text{MR}} n_i T_{ij} v_j\, dS + \int_{\text{MR}} \rho F_i v_i\, dV \qquad 5.9.5$$

At this point the mathematical processes are familiar. Using Leibnitz's and Gauss's theorems on the left-hand side and converting the surface integrals on the right-hand side with Gauss's theorem, we get an equation containing only volume integrals. The integrand of this equation is the energy equation. It is

$$\partial_t\left[\rho\left(e + \tfrac{1}{2}v^2\right)\right] + \partial_i\left[\rho v_i\left(e + \tfrac{1}{2}v^2\right)\right] = -\partial_i q_i + \partial_i\left(T_{ij} v_j\right) + \rho v_i F_i \qquad 5.9.6$$

In symbolic notation the energy equation is

$$\underbrace{\frac{\partial}{\partial t}\left[\rho\left(e+\tfrac{1}{2}v^2\right)\right]}_{\substack{\text{rate of increase}\\ \text{of energy per}\\ \text{unit volume}}} + \underbrace{\nabla\cdot\left[\rho\mathbf{v}\left(e+\tfrac{1}{2}v^2\right)\right]}_{\substack{\text{convection of}\\ \text{energy into a}\\ \text{point by flow}}} = \underbrace{-\nabla\cdot\mathbf{q}}_{\substack{\text{net}\\ \text{heat}\\ \text{flow}}} + \underbrace{\nabla\cdot(\mathbf{T}\cdot\mathbf{v})}_{\substack{\text{work of}\\ \text{surface}\\ \text{forces}}} + \underbrace{\rho\mathbf{v}\cdot\mathbf{F}}_{\substack{\text{work of}\\ \text{body}\\ \text{forces}}} \qquad 5.9.7$$

This is the differential equation governing the total energy at any point in the continuum. All of the terms are on a unit-volume basis (i.e., energy change rate per unit volume). The energy equation is not usually used in this form, but is split into two equations: the mechanical energy equation and the thermal energy equation. We shall discuss these equations in the next section.

5.10 MECHANICAL- AND THERMAL-ENERGY EQUATIONS

The equation that governs kinetic energy is not an independent law, but is derived from the momentum equation. The dot product of v_i with the momentum equation (and some algebraic manipulation) will yield the mechanical (kinetic) energy equation, which has the form

$$\partial_0\left(\rho\tfrac{1}{2}v^2\right) + \partial_i\left(\rho v_i \tfrac{1}{2}v^2\right) = -v_i\partial_i p + v_i\partial_j\tau_{ji} + \rho v_i F_i$$

$$\frac{\partial}{\partial t}\left(\rho\tfrac{1}{2}v^2\right) + \nabla\cdot\left(\rho\mathbf{v}\tfrac{1}{2}v^2\right) = -\mathbf{v}\cdot\nabla p + \mathbf{v}\cdot(\nabla\cdot\boldsymbol{\tau}) + \rho\mathbf{v}\cdot\mathbf{F} \qquad 5.10.1$$

Note that all of the work of the body force goes to accelerate the fluid and increase its kinetic energy.

The thermal-energy equation is obtained by subtracting the mechanical-energy equation from the total-energy equation (5.9.7). The result is

$$\partial_0(\rho e) + \partial_i(\rho v_i e) = -p\,\partial_i v_i + \tau_{ji}\partial_j v_i - \partial_i q_i \qquad 5.10.2$$

When this equation is written in symbolic notation, (it is customary to employ the fact that τ_{ij} is symmetric so that $\tau_{ji}\partial_j v_i = \tau_{ij}\partial_j v_i$),

$$\partial_0(\rho e) + \nabla\cdot(\rho\mathbf{v}e) = -p\nabla\cdot\mathbf{v} + \boldsymbol{\tau}:\nabla\mathbf{v} - \nabla\cdot\mathbf{q}$$

Note that all of the heat flux goes to increase the internal energy. As with the continuity and momentum equations, the thermal-energy equation can be put into a form containing the substantial derivative.

The surface work terms in 5.9.6, 5.10.1 and 5.10.2 are very interesting. The total work of surface forces may be split into two parts: pressure work and viscous work. Then these terms may be split again as diagrammed below:

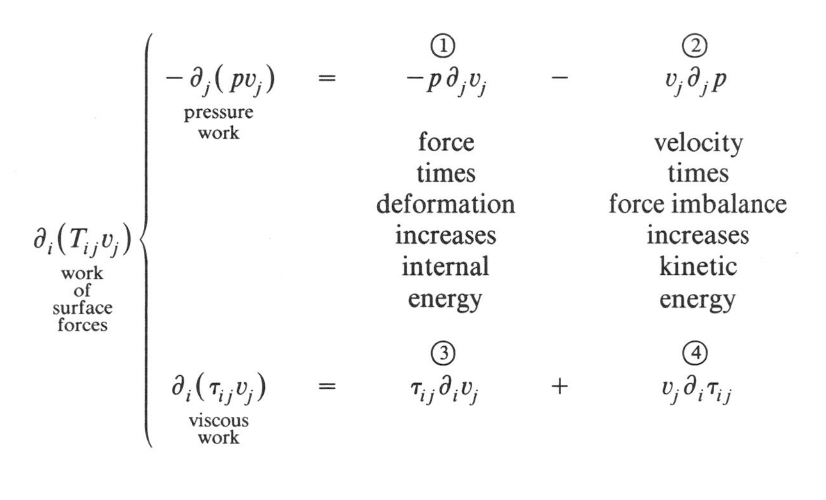

Terms ② and ④ are the velocity times gradients of forces. The gradients indicate an imbalance in the forces which directly accelerates the fluid and increases its kinetic energy. Thus, these terms appear in the mechanical-energy equation. Terms ① and ③ are forces multiplied by fluid deformations. They occur in the thermal-energy equation. Recall that $\partial_i v_i$ has the physical interpretation of the volumetric rate of expansion or contraction. Thus, term ① is the heating or cooling of the fluid by compression or expansion, depending upon the sign of $\partial_i v_i$. This is a reversible process. Term ③ is called *viscous dissipation*. It is responsible for heat generation in bearings and aerodynamic heating of spacecraft as they reenter the earth's atmosphere. Since τ_{ij} is symmetric, the product $\tau_{ij}\partial_i v_j$ is equal to $\tau_{ij}\partial_{(i}v_{j)}$ (see Section 3.5.8). This helps the physical understanding as we recognize $\partial_{(i}v_{j)}$ as the rate-of-deformation or strain-rate tensor. Viscous dissipation is always positive and produces internal energy (all viscosity laws have a form such that when $\partial_{(i}v_{j)}$ changes sign, so does τ_{ij}). This is an irreversible process, as we shall see when we study the entropy equation. In summary, surface forces have two effects: forces times deformations change the internal energy, while the velocity times an unbalanced force accelerates the fluid to change its kinetic energy.

Sometimes the work term in the kinetic energy equation is replaced by $v_j\partial_i\tau_{ij} = \partial_i(\tau_{ij}v_j) - \tau_{ij}\partial_i v_j$. This might be done for mathematical reasons. The difficulty with this form is that one is tempted to imagine a two-step process where all of the shear work accelerates the fluid, then subsequently kinetic energy of motion is changed into thermal energy by viscous dissipation. In actuality only the unbalanced forces $\partial_i\tau_{ij}$ accelerate the fluid.

5.11 ENERGY EQUATION WITH TEMPERATURE AS DEPENDENT VARIABLE

In the most useful form of the thermal-energy equation, temperature replaces internal energy as the major variable. Let us consider e as a function of ρ and T. In turn ρ and T are functions of space and time. Then by using chain rules we can write the substantial derivative of e as

$$\frac{De}{Dt} = \left.\frac{\partial e}{\partial T}\right|_{\rho} \frac{DT}{Dt} + \left.\frac{\partial e}{\partial \rho}\right|_{T} \frac{D\rho}{Dt} \qquad 5.11.1$$

The partial derivatives of e are known from thermodynamic theory (2.16.1 and 2.16.6). Substituting these equations and the continuity equation 5.1.7 into 5.11.1 yields

$$\rho \frac{De}{Dt} = \rho c_v \frac{DT}{Dt} + \left[-p + T \left.\frac{\partial p}{\partial T}\right|_{\rho} \right] \partial_i v_i \qquad 5.11.2$$

Next the thermal energy equation 5.10.2 may be put in a form where the substantial derivative of e appears on the left-hand side. Combining 5.11.2 and 5.10.2 gives the final form

$$\rho c_v \frac{DT}{Dt} = -T \left.\frac{\partial p}{\partial T}\right|_{\rho} \partial_i v_i - \partial_i q_i + \tau_{ij} \partial_j v_i \qquad 5.11.3$$

Thermodynamic equation-of-state information enters this equation through $c_v(\rho, T)$ and $\partial p/\partial T|_{\rho}$. The specific heat has not been assumed to be constant.

There are many forms the energy equation may take. Problem 5.9 gives what is probably the most useful form where c_p replaces c_v. Bird, Stewart, and Lightfoot (1960) give a convenient list of many forms for the energy equation.

*5.12 SECOND LAW OF THERMODYNAMICS

The fundamental differental equation of thermodynamics (2.13.4) and the energy equation can be combined to form an equation governing entropy. The thermodynamic equation is written in substantial derivative form,

$$T \frac{Ds}{Dt} = \frac{De}{Dt} - \frac{p}{\rho^2} \frac{D\rho}{Dt} \qquad 5.12.1$$

The last term changed by using the continuity equation 5.1.7:

$$\rho T \frac{Ds}{Dt} = \rho \frac{De}{Dt} + p \partial_i v_i \qquad 5.12.2$$

Substituting the thermal-energy equation 5.10.2 into 5.12.2 yields

$$\rho\frac{Ds}{Dt} = -\frac{1}{T}\partial_i q_i + \frac{1}{T}\tau_{ij}\partial_j v_i \qquad 5.12.3$$

To facilitate the physical interpretation we rewrite 5.12.3 as

$$\rho\frac{Ds}{Dt} = -\partial_i\left(\frac{q_i}{T}\right) + \frac{1}{T^2}q_i\partial_i T + \frac{1}{T}\tau_{ij}\partial_j v_i \qquad 5.12.4$$

The first term on the right-hand side is the entropy change of a material particle as a reversible effect of heat transfer. The sign of this term changes with the sign of the heat flux. The second and third terms are always positive. They represent irreversible increases in entropy because of heat transfer and viscous dissipation. The third irreversible process (in a nonreacting fluid) is due to diffusion. It does not appear, because we have not allowed the fluid to be a chemical mixture of varying composition. The entropy equation shows that the flow of a fluid without viscosity and heat conduction must be isentropic. Thermodynamically, irreversible processes are the direct result of heat conduction and viscous forces within the bulk of the fluid.

5.13 INTEGRAL FORM OF THE CONTINUITY EQUATION

Frequently we are interested in applying the basic laws to a finite region. Such equations are called global equations, overall equations, or simply integral forms of the equations. We have already postulated the integral forms for the special case of a material region. Here we extend the continuity law so that it applies to a region with arbitrary motion. The motion of the region is specified by w_i, the arbitrarily chosen velocity of its surface.

The starting point of the derivation is Leibnitz's rule for differentiating an integral that has limits that depend upon time (3.10.2):

$$\frac{d}{dt}\int_{AR} f\,dV = \int_{AR}\frac{\partial f}{\partial t}\,dV + \int_{AR} n_i w_i f\,dS$$

Let us choose $f = \rho$, and substitute the continuity equation 5.1.4 for $\partial_0\rho$ in the volume integral on the right-hand side. This yields

$$\frac{d}{dt}\int_{AR}\rho\,dV = -\int_{AR}\partial_i(\rho v_i)\,dV + \int_{AR} n_i w_i \rho\,dS \qquad 5.13.1$$

Application of Gauss's theorem and collecting terms gives the mass conservation law for a region with arbitrary motion:

$$\frac{d}{dt}\int_{AR}\rho\,dV = -\int_{AR}\rho(v_i - w_i)n_i\,dS \qquad 5.13.2$$

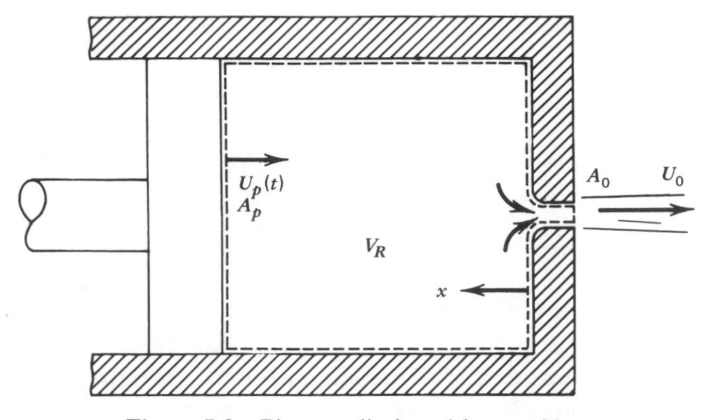

Figure 5-9 Piston-cylinder with an orifice.

The rate of change of mass within the region is equal to the integral of the mass flow relative to the moving boundary. The special cases of a material region or a fixed region are obtained by choosing $w_i = v_i$ or $w_i = 0$ respectively.

Consider as an example, a control region containing the fluid within the cylinder shown in Fig. 5-9. The piston moves with speed $U_p(t)$, causing an incompressible fluid to shoot out from the orifice with velocity U_0. Let $M = \rho V_R$ be the mass of fluid within the region. Equation 5.13.2 for this situation is

$$\frac{dM}{dt} = -\int \rho(v_i - w_i) n_i \, dS$$

The surface velocity w_i is zero everywhere except at the piston, where it is U_p. However, the fluid velocity at the piston is also U_p, so $v_i - w_i$ is zero at the piston. There is no flow across the piston face. The only contribution of the surface integral comes from the orifice; hence

$$\frac{dM}{dt} = -\rho U_0 A_0$$

or, since the density is constant,

$$\frac{dV_R}{dt} = -U_0 A_0$$

Furthermore, the volume of the region changes according to the geometric relation 3.10.3:

$$\frac{dV_R}{dt} = \int_S n_i w_i \, dS = -U_p A_p$$

Combining the two equations above shows that

$$U_p A_p = U_0 A_0$$

This result holds (because the fluid is incompressible) even though the flow is unsteady.

The exact history of the continuity principle is not known. Early Romans, although they tried to tax users according to the amount of water they received, did not really understand the continuity law and the relationship between velocity, area, and flow rate. The first known accurate quantitative statements of the continuity principle are those of Leonardo da Vinci (see Rouse and Ince 1957, Truesdell 1968). da Vinci was a keen observer of fluid motions and made many statements that showed his understanding of continuity. For example, he wrote, "By so much as you will increase the river in breadth, by so much you will diminish the speed of its course." He may even have been aware of the unsteady effects: "If the water is not added to or taken away from the river, it will pass with equal quantities in every degree of its length...". Subsequent to da Vinci, the principle was probably rediscovered by many others.

5.14 INTEGRAL FORM OF THE MOMENTUM EQUATION

The Leibnitz theorem (3.10.2) with $f = \rho v_i$ is

$$\frac{d}{dt}\int_{AR} \rho v_i \, dV = \int_{AR} \partial_0(\rho v_i) \, dV + \int_{AR} n_j w_j \rho v_i \, dS$$

We solve the momentum equation 5.7.2 for $\partial_0(\rho v_i)$, substitute in the equation above, and convert as many volume integrals as possible into surface integrals by using Gauss's theorem. The final result is the momentum principle for a region with arbitrary motion:

$$\frac{d}{dt}\int_{AR} \rho v_i \, dV = -\int_{AR} \left[\rho n_j (v_j - w_j) v_i + n_i p - n_j \tau_{ji}\right] dS + \int_{AR} \rho F_i \, dV \tag{5.14.1}$$

The rate of change of momentum within the region is equal to the rate that momentum is convected across the surface by the relative mass flow, plus the sum of the forces. In this equation the body force cannot be converted into a surface integral unless it is conservative and has a representation as a potential. It is interesting that the motion of the region does not affect the forces, nor does it influence the momentum instantaneously within the region. The motion of the surface has its only effect (other than the integration limit, of course) in the convection of matter in or out of the region. Moreover, any motion along the surface that we might assign to the surface velocity w_i (i.e., a sliding motion perpendicular to n_i) is irrelevant. Only the normal component $n_i w_i$ appears in the equation. The special cases of a material region ($w_i = v_i$) and a fixed region, ($w_i = 0$) are readily found from 5.14.1.

We apply 5.14.1 to the piston-orifice problem shown in Fig. 5-9. The fluid is again assumed to be incompressible, and in addition we neglect viscous and gravity forces. The integral of x momentum over the region is split up into an integral across the area and an integral along the axis from the piston to the orifice. Equation 5.14.1 becomes

$$\frac{d}{dt}\int_{x=p}^{x=0}\left[\int \rho v_x \, dA\right] dx = -\rho U_0^2 A + F^{(p)} \tag{5.14.2}$$

where the net pressure force is

$$F^{(p)} \equiv -\int n_x p \, dS = p_p A_p - \int_{x=0} n_x p \, dS \tag{5.14.3}$$

At $x = 0$ the pressure integral is over the end wall and the orifice. A continuity analysis shows that the average value of U_x over any station x is U_p; hence we may express 5.14.2 as

$$\frac{d}{dt}\left(\rho U_p V_R\right) = -\rho U_0^2 A_0 + F^{(p)}$$

In Section 5.13 we found that $U_p A_p = U_0 A_0$ and that $dV_R/dt = -U_p A_p$. Using these facts, we may put the momentum equation in the form

$$\rho U_p A_p \left(U_0 - U_p\right) + \rho V_R \frac{dU_p}{dt} = F^{(p)} \tag{5.14.4}$$

The first term accounts for the change in the momentum of the fluid between the piston and orifice. The net pressure force between the piston and the end wall must supply this momentum. The second term occurs only when the piston is accelerating. The pressure at the piston face must increase in order to accelerate the mass of fluid ρV_R within the cylinder. For known geometry and piston motion 5.14.4 gives the net pressure difference. In order to find the pressure at the piston face, some information about the nature of the flow must be known or assumed (see problem 12.15).

*5.15 MOMENTUM EQUATION FOR A DEFORMABLE PARTICLE OF VARIABLE MASS

An arbitrary control region is shown in Fig. 5-10. We let r_i stand for the position vector and define the mass M, the center of mass R_i, the velocity $\dot{R}_i$, of the center of mass, and the momentum P_i of the region by integrals over the

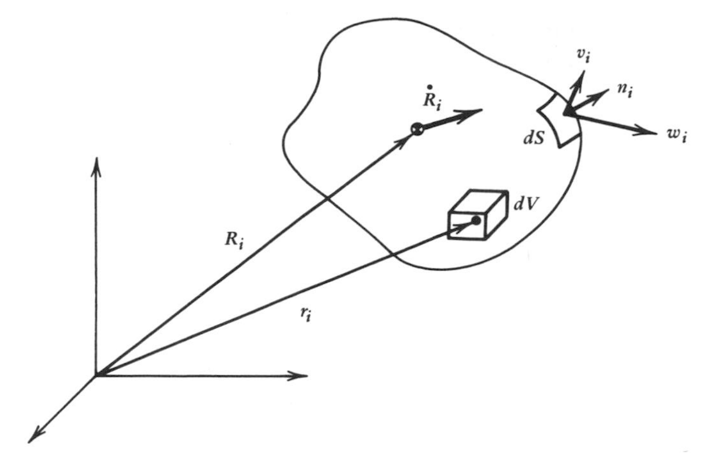

Figure 5-10 Finite deformable particle of variable mass.

region,

$$M = \int \rho \, dV, \qquad R_i = \frac{1}{M} \int \rho r_i \, dV$$

$$\dot{R}_i = \frac{dR_i}{dt}, \qquad P_i = \int \rho v_i \, dV \qquad 5.15.1$$

Later on we shall need the continuity equation (5.13.2) for the region. It is

$$\frac{dM}{dt} = -\int \rho n_j (v_j - w_j) \, dS \qquad 5.15.2$$

Another preliminary step is to find a relation between the momentum P_i of the region and the mass times the velocity of the center of mass, $M\dot{R}_i$. Consider Leibnitz's theorem applied to the definition of MR_i,

$$\frac{d}{dt}(MR_i) = \frac{d}{dt} \int \rho r_i \, dV = \int \partial_0 (\rho r_i) \, dV + \int n_j w_j \rho r_i \, dS \qquad 5.15.3$$

Now by using the product rule for differentiation, we have the identity (note that $\partial_{ri} = \partial_j x_i = \delta_{ji}$)

$$\begin{aligned} \partial_j (r_i \rho v_j) &= r_i \partial_j (\rho v_j) + \rho v_j \partial_j r_i \\ &= r_i \partial_j (\rho v_j) + \rho v_i \end{aligned}$$

Moreover, since the position vector $r_i = x_i$ is independent of time, $\partial_0 (\rho r_i) =$

$r_i \partial_0 \rho$. By the continuity equation this becomes $\partial_0(\rho r_i) = -r_i \partial_j(\rho v_j)$. Combining this with the result above shows that

$$\partial_0(\rho r_i) = -\partial_j(r_i \rho v_j) + \rho v_i$$

Substituting this expression into 5.15.3 produces

$$M\frac{dR_i}{dt} + R_i\frac{dM}{dt} = \int\left[-\partial_j(r_i\rho v_j) + \rho v_i\right] dV + \int n_j w_j \rho r_i \, dS$$

The first volume integral is converted to a surface integral, and the continuity equation 5.15.2 is used to obtain the desired relation,

$$M\dot{R}_i = P_i - \int \rho(r_i - R_i) n_j (v_j - w_j) \, dS \qquad 5.15.4$$

The mass times the velocity of the center of mass of a region is not necessarily equal to the momentum of the region if the region gains or loses mass.

The effect of the integral term in 5.15.4 is depicted in Fig. 5-11a. Consider a region that surrounds a liquid every particle of which is moving with a constant velocity $v_i = V_i$. The momentum is $P_i = MV_i$. At the back of the region small particles of liquid are being stripped away and leave the control region. Consequently, the center of mass of the region must move forward; thus $\dot{R}_i > V_i$, and $M\dot{R}_i > MV_i = P_i$. The integral term in 5.15.4 accounts for the fact that the center of mass of the region may move due to an asymmetric loss of mass from the region.

The momentum equation will be considered next. Introducing the definition of P_i into 5.14.1 gives

$$\frac{dP_i}{dt} = -\int \rho n_j (v_j - w_j) v_i \, dS + F_i^{(p)} + F_i^{(\nu)} + F_i^{(b)} \qquad 5.15.5$$

For convenience the following forces have been defined,

$$F_i^{(p)} = -\int n_i p \, dS, \qquad F_i^{(\nu)} = \int n_j \tau_{ji} \, dS, \qquad F_i^{(b)} = \int \rho F_i \, dV$$

In 5.15.5 the momentum P_i is eliminated in favor of R_i by substituting 5.15.4. The continuity equation 5.15.2 is employed to yield the final form

$$M\ddot{R}_i = F_i^{(p)} + F_i^{(\nu)} + F_i^{(b)} - \int \rho n_j (v_j - w_j)(v_i - \dot{R}_i) \, dS$$

$$+ \frac{d}{dt}\int \rho n_j (v_j - w_j)(r_i - R_i) \, dS \qquad 5.15.6$$

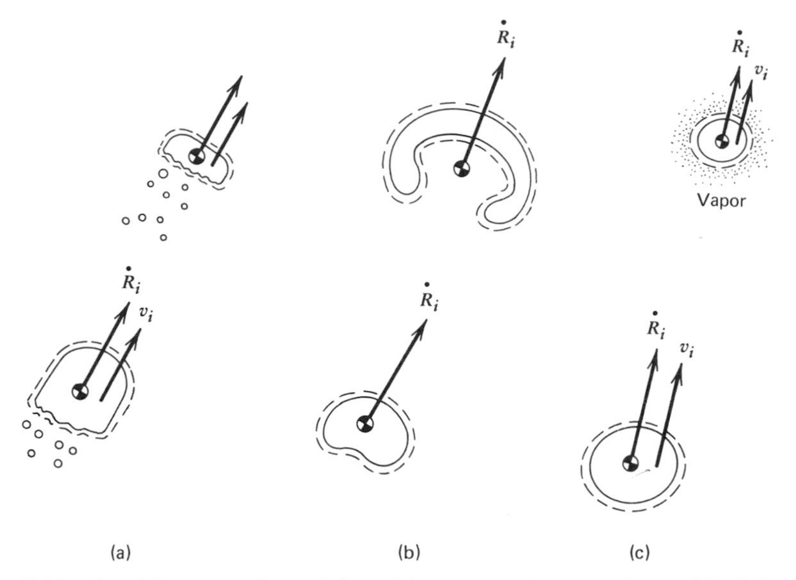

Figure 5-11 Special cases of particles: (a) asymmetric mass loss, (b) deforming particle, (c) vaporizing particle.

This is the momentum equation for a finite-size deformable particle of variable mass.

The first integral accounts for momentum $v_i - \dot{R}_i$ that leaves the region with the mass flux $\rho n_j(v_j - w_j)\, dS$. Physically this is a jet or rocket effect. The second integral accounts for the movement of the center of mass due to an asymmetric mass loss.

Several special cases will help interpret the momentum equation. In Fig. 5-11b a droplet of liquid moving in a gas is envisioned. Allow the droplet to deform but not to vaporize. Since there is no mass leaving the region, 5.15.6 is

$$M\ddot{R}_i = F_i^{(p)} + F_i^{(\nu)} + F_i^{(b)} \qquad 5.15.7$$

The mass times the acceleration of the center of mass equals the forces. Also, from 5.15.5 we see that $M\dot{R}_i = P_i$. If the droplet breaks into several parts, the control volume following the various parts, 5.15.7 is still true. The center of mass of the parts obeys 5.15.7.

Next, we consider a droplet that has an unusual shape and is also vaporizing. Furthermore, we assume that the liquid velocity at the surface is exactly equal to the velocity of the center of mass (Fig. 5-11b). For this example the first integral in 5.15.5 vanishes, but we are left with the second integral, which describes the movement of the center of mass because of shape changes.

Finally, we consider a spherical droplet (Fig. 5-11c) with the assumption that the vaporization is uniform and the fluid velocity is again uniform; hence,

$v_i = \dot{R}_i$. In 5.15.6 the second integral becomes [let the uniform vaporization velocity be $\dot{m}/A = \rho n_j(v_j - w_j)$]

$$\frac{d}{dt}\left[\frac{\dot{m}}{A}\int (r_i - R_i)\, dS\right]$$

For every surface element of dS and a positive value of $r_i - R_i$ there is a symmetrically positioned element with a negative value. Therefore, the integral over the surface is zero. The final equation for a vaporizing droplet is exactly the same as 5.15.7 for a nonvaporizing droplet:

$$M\frac{d^2R_i}{dt^2} = F_i^{(p)} + F_i^{(\nu)} + F_i^{(b)} \qquad 5.15.8$$

If vaporization is not uniform, then the droplet is propelled by the jet effect and the full equation 5.15.6 must be used.

The proper form of the momentum equation for a deforming particle (control region) of variable mass is a recent advance (see for instance Thorpe 1962). It was only clarified after problems of rockets and space vehicles became important.

*5.16 OTHER PHYSICAL LAWS IN INTEGRAL FORM

Any local differential law may be cast into an integral form for an arbitrary region. The procedure is to take f as the quantity of interest in Leibnitz's theorem, substitute the differential law for $\partial_0 f$, and convert all volume integrals of the form $\partial_i(\ )\, dV$ into surface integrals.

The energy equation is derived in a similar fashion. The result is

$$\frac{d}{dt}\int_{AR}\rho\left(e + \tfrac{1}{2}v^2\right) dV = -\int_{AR}\rho n_i(v_i - w_i)\left(e + \tfrac{1}{2}v^2\right) dS - \int_{AR} n_i q_i\, dS$$

$$-\int_{AR}\left[n_i v_i p - n_i \tau_{ij} v_j\right] dS + \int_{AR}\rho v_i F_i\, dV \qquad 5.16.1a$$

If the body-force potential $F_i = -\partial_i\hat{\phi}$ is used, then the work integral is eliminated and this effect appears as $e + \frac{1}{2}v^2 + \hat{\phi}$ in the first two integrals.

The work done by pressure forces at the surface of the control volume, $pn_i v_i\, dS$, may be interpreted in several ways. If the control surface is fixed in space, then $n_i v_i$ is the volume flux of fluid into the region. This type of work is known as "flow work" in thermodynamics. If the control surface is impervious but is moving, then $n_i v_i$ represents, because of the no-slip condition, the velocity of the boundary in the normal direction. This is the work done by a moving boundary.

These two effects may also be identified for the case where fluid flows across a boundary that has an arbitrary velocity w_i. Consider the following mathematical identity:

$$\int_{AR} p n_i v_i \, dS = \int_{AR} \frac{p}{\rho} \rho n_i (v_i - w_i) \, dS + \int_{AR} p n_i w_i \, dS \qquad 5.16.1b$$

The work of pressure forces, which is the basic concept, may be split into a flow-work part and a boundary-work part. This nomenclature is common in engineering thermodynamics. It is acceptable as long as we understand that it is just an arbitrary division of the work done by pressure forces.

The kinetic-energy equation 5.10.1 can also be cast into an integral form. In doing so, it is customary to use the identities

$$v_i \partial_i p = \partial_i (v_i p) - p \partial_i v_i \quad \text{and} \quad v_i \partial_j \tau_{ij} = \partial_j (\tau_{ji} v_i) - \tau_{ji} \partial_j v_i .$$

Thus we are replacing the pressure work that accelerates the fluid with the total work minus the compression work, and the accelerating shear work is replaced by the total shear work minus the viscous dissipation. The final equation reads

$$\begin{aligned} \frac{d}{dt} \int_{AR} \rho \tfrac{1}{2} v^2 \, dV = {} & - \int_{AR} \rho n_i (v_i - w_i) \tfrac{1}{2} v^2 \, dS + \int_{AR} \rho F_i v_i \, dV \\ & - \int_{AR} [n_i v_i p - n_i \tau_{ij} v_j] \, dS \\ & + \int_{AR} \left[p \partial_i v_i - \tau_{ij} \partial_i v_j \right] dV \end{aligned} \qquad 5.16.2$$

The next to the last volume integral is zero in incompressible flow, since $\partial_i v_i = 0$. This equation is especially useful in incompressible flow. We defer giving examples until Sections 7.1 and 19.9–19.12.

5.17 CONCLUSION

The basic laws we have formulated govern all continuum processes of all substances. They do not form a complete set of equations; there are more unknowns than equations even after we specify the thermodynamic equations of state. The two outstanding tasks are to relate the heat flux to the temperature field (a conduction law) and to relate the stress tensor to the fluid deformations (a viscosity law). These tasks will be taken up in the next chapter.

PROBLEMS

5.1 (B) The incompressible flow around a circular cylinder of radius r_0 is given in cylindrical coordinates as

$$v_r = -U\cos\theta\left[1 - \left(\frac{r_0}{r}\right)^2\right]$$

$$v_\theta = U\sin\theta\left[1 + \left(\frac{r_0}{r}\right)^2\right]$$

$$p = \rho U^2\left[1 - \left(\frac{r_0}{r}\right)^2\sin^2\theta + \left(\frac{r_0}{r}\right)^4\right]$$

Consider the following fixed surfaces one unit in length: S_{I} with $r = R_0$ as $-\pi/2 \le \theta \le \pi/2$, S_{II} with $\theta = \pi/2$ as $r_0 \le r \le R_0$, S_{III} with $r = r_0$ as $-\pi/2 \le 0 \le \pi/2$, and S_{IV} with $\theta = -\pi/2$ as $r_0 \le r \le R_0$. These surfaces form a fixed region. Compute the following quantities and explain their physical significance: (a) $\int_{\text{FR}} \rho v_x \, dV$, (b) $\int_{S_{\text{I}}} \rho n_i v_i \, dS$, (c) $-\int_{S_{\text{II}}} n_x p \, dS$, (d) $\int_{S_{\text{II}}} \rho n_i v_i v_x \, dS$, (e) $-\int_{S_{\text{III}}} n_x p \, dS$, (f) $-\int_{S_{\text{IV}}} n_r pr \, d\theta$.

5.2 (B) Look up the continuity equation in cylindrical and in spherical coordinates. Write out the equations for the special case that the density is constant. Consider a flow that is purely radial [i.e., the only velocity is $v_r(r)$], and find the velocity as a function of r for each case. Sketch a graph of the velocity. Compute the vorticity for each of these flows.

5.3 (A) The velocity profile in a two-dimensional flow is $v_x = v_0[1 - (y/h)^2]$. The stress tensor T_{ij} is $T_{xx} = T_{yy} = T_{zz} = -5$, and $T_{xy} = T_{yx} = -2\mu v_0(y/h^2)$. All other components are zero. Find the stress normal and tangential to a plane located at $y/h = \frac{1}{2}$ with its normal at a 30° angle to the flow direction.

5.4 (B) Prove that for any continuous fluid property f,

$$\frac{d}{dt}\int_{\text{MR}} \rho f \, dV = \int \rho \frac{Df}{Dt} dV$$

A physical law states that the rate of change of ρf for a material region comes about by a volume effect Q per unit volume and a surface effect $n_i P_i$ per unit area. Show that the law has the differential form

$$\rho \frac{Df}{Dt} = Q + \nabla \cdot \mathbf{P}$$

Generalize this result for f as a second-order tensor component.

5.5 (B) Prove the Rayleigh transport theorem:

$$\frac{d}{dt}\int_{\mathrm{MR}} \rho f \, dV = \frac{d}{dt}\int_{\mathrm{FR}} \rho f \, dV + \int_{\mathrm{FR}} \rho n_i v_i f \, dS$$

5.6 (B) Consider the incompressible flow in a two-dimensional contraction section. The velocity along the centerline is given by

$$v_1 = v_0\left[1 - \frac{a}{h}\left(\frac{x}{L}\right)\right]^{-1}$$

Find an equation for the velocity v_2 that will be valid in the region near the centerline of the flow. Find an equation for streamlines in this same neighborhood.

5.7 (B) Verify that the mechanical-energy 5.10.1 is a combination of the momentum equation and the continuity equation and therefore is not a separate physical law.

5.8 (B) Introduce the gravity potential $F_i = -\partial_i \phi$ into 5.9.6 and show that the equation may be rewritten as

$$\partial_0(\rho e_t) + \partial_i(\rho v_i e_t) = -\partial_i q_i + \partial_i(T_{ij} v_j)$$

where

$$e_t \equiv e + \tfrac{1}{2}v^2 + \hat{\phi}$$

5.9 (B) Derive the following equation starting from the thermal energy equation 5.10.2 and the definition of enthalpy, $h \equiv e + p/\rho$:

$$\rho c_p \frac{DT}{Dt} = -\nabla \cdot \mathbf{q} + \boldsymbol{\tau} : \nabla \mathbf{v} + \beta T \frac{Dp}{Dt}$$

5.10 (A) Derive the global form of the energy equations 5.16.1 and 5.16.2.

5.11 (B) Consider a rocket whose outer case has a nearly uniform velocity W_i. Let $\hat{u}_i$ stand for velocities measured with respect to the rocket case; that is, $\hat{u}_i = v_i - W_i$, where v_i is the absolute velocity. At the rocket exhaust plane $\hat{u}_i$ is uniform at the value U_i. Derive an equation for $M\, dW_i/dt$ by starting from 5.14.1. Eliminate all references to v_i in favor of $\hat{u}_i$ and W_i. Give the physical interpretation of each term.

5.12 (B) A very viscous liquid in laminar flow comes downward out of a long round tube. After the fluid exits the tube, viscous forces smooth the parabolic exit velocity profile to a uniform value. This happens in a short distance from the exit, so that gravity forces are negligible. Apply the momentum equation to find the area of the jet when the uniform flow is first established.

5.13 (C) Let the flow properties ρ^{I} and v_i^{I} be steady and uniform in a small region of space dS in cross section. The properties change smoothly and continuously through a distance ε to new values ρ^{II} and v_i^{II}. Consider a fixed region dS in area and ε long that encloses the place where the properties change. Let $n_i = n_i^{\mathrm{I}} = -n_i^{\mathrm{II}}$ be a unit normal in the flow direction. To obtain the "jump conditions" for the plane discontinuity in the flow properties ρ and v_i, we postulate that mass cannot be created within the region and let $\varepsilon \to 0$. Show that the global continuity equation yields the jump condition

$$\rho^{\mathrm{I}} n_i v_i^{\mathrm{I}} = \rho^{\mathrm{II}} n_i v_i^{\mathrm{II}}$$

Write the momentum equation for the direction normal to the discontinuity, and reduce it to another jump condition. Do the same for the tangential momentum equation. Use the fact that all fluids are viscous to simplify this result. Finally, find the jump conditions imposed by the energy equation at a plane discontinuity.

5.14 (A) A very long tube 3 cm in diameter carries water at an average velocity of 5 m/sec. A short nozzle attached to the end accelerates the flow with a 5:1 area reduction. Find the force between the pipe and the nozzle when the exit pressure is atmospheric (some prior knowledge of fluid mechanics relations is needed in the following problems).

5.15 (B) Do Problem 14 when the flow has 1-cm/sec average velocity in the tube.

5.16 (A) Do Problem 14 when the nozzle turns the flow by 120°.

5.17 (C) A plunger, a very loose fitting piston, with cross section A_p is centered in a cylinder with cross section A_c. Oil fills the cavity. Let the plunger have an arbitrary downward motion $V_p(t)$. Find an approximate expression for the force on the plunger in terms of V_p and the geometry.

5.18 (B) Evaluate the kinetic energy equation 5.10.1 for the Poiseuille flow in a slot. Sketch the distribution of the following quantities across the slot: $\partial_i(v_i p)$, $v_i \partial_i p$, $p \partial_i v_i$, $\partial_i(v_j \tau_{ij})$, $v_j \partial_i \tau_{ij}$, and $\tau_{ij} \partial_i v_j$. What is the physical interpretation of each of the quantities above? How are the integrals of these quantities over a volume of length L related?

6 Newtonian Fluids and the Navier–Stokes Equations

In this chapter we study the equations that relate the stress to the deformation and those that relate the heat flux to the temperature. Such relations are called *constitutive equations*. A given constitutive formula may be good for a large group of fluids, but in general one formula cannot describe all fluids. The simplest relations are linear equations: the stress is proportional to the rate of strain (Newton's viscosity law), or the heat flux is proportional to the temperature gradient (Fourier's law). At ordinary pressures and temperatures all gases obey these relations, as do many simple liquids. Liquids made up of complex molecules, liquid mixtures, and slurries of fine particles in a liquid (including blood) do not obey linear relations and are said to be non-Newtonian.

The idea of a linear relation between stress and rate of strain was first put forward by Newton, and for this reason the viscosity law bears his name. Much later, George S. Stokes (English mathematician, 1819–1903) and C. L. M. H. Navier (French engineer, 1785–1836) produced the exact equations that govern the flow of Newtonian fluids. These equations, or the appropriately generalized ones for compressible flow, are called the Navier–Stokes equations.

6.1 NEWTON'S VISCOSITY LAW

In Chapter 5 we proposed a surface stress R_i to describe the net intermolecular forces and microscopic momentum transport from one side of an imaginary surface to the other. The problem was decomposed into two effects by introducing a stress tensor T_{ij} such that $R_j = n_i T_{ij}$. Flow effects are contained in T_{ij} and surface orientation effects are contained in n_i. From point to point, the local flow situation changes and causes the stress tensor to vary. We now formulate an expression for T_{ij} by assuming that it is a function $T_{ij}(\rho, e, \partial_k v_l)$ of the local thermodynamic state and the local velocity gradients. We cannot include the velocity by itself in this expression, because then a Galilean transformation would change the stress. This would certainly not conform to physical reality. The simplest form for $T_{ij}(\rho, e, \partial_k v_l)$ is a linear function of the velocity gradients with coefficients that depend upon the thermodynamic state,

$$T_{ij} = A_{ij} + B_{ijkl} \partial_k v_l \qquad 6.1.1$$

The linearity assumption prohibits terms involving $\partial_k v_j \partial_j v_l$, the square of the velocity gradient.

There are two geometric properties that further restrict 6.1.1. The first is that most fluids are isotropic, having no preferred directions. Secondly, because we assumed no moment acts on the surface (Section 5.8), the stress tensor is symmetric. If these restrictions are imposed, it may be shown (Jeffreys 1963, Prager 1961, Yih 1969, Batchelor 1967, Aris 1962) that 6.1.1 must have the mathematical form

$$T_{ij} = A_1 \delta_{ij} + A_2 \partial_k v_k \delta_{ij} + A_3 \partial_{(i} v_{j)} \qquad 6.1.2$$

The coefficients A_1, A_2, A_3 are in principle functions of the thermodynamic state. We can fix A_1 by arguing that when there is no motion, the equation must reduce to give the thermodynamic pressure; therefore, $A_1 = -p$. The common symbols for A_2 and A_3 are λ and 2μ. They are called the second and first viscosity coefficients respectively. With this notation 6.1.2 becomes

$$T_{ij} = -p\delta_{ij} + \lambda \partial_k v_k \delta_{ij} + 2\mu \partial_{(i} v_{j)} \qquad 6.1.3$$

To continue the development, recall the definition of the mechanical pressure as the average normal stress,

$$-p_m \equiv \tfrac{1}{3} T_{ii} \qquad 6.1.4$$

The difference between the thermodynamic pressure (given by the equation of state $p(\rho, e)$ when the local values of ρ, e are inserted) and the mechanical pressure is computed from 6.1.3 by contracting on i and dividing by 3. The result may be rearranged to give

$$p - p_m = \left(\lambda + \tfrac{2}{3}\mu\right) \partial_k v_k = -\left(\lambda + \tfrac{2}{3}\mu\right) \frac{1}{\rho} \frac{D\rho}{Dt} \qquad 6.1.5$$

The last equality is obtained using the continuity equation.

Recall that an incompressible fluid has only a mechanical pressure. Equation 6.1.5 shows that the viscosity law will be consistent for an incompressible fluid ($D\rho/Dt = -\rho \partial_i v_i = 0$), as the right-hand side is zero and the symbol p will take on the meaning of the mechanical pressure.

Since the second term in 6.1.3 drops out in incompressible flow, the second viscosity λ can play no role in such flows. In order to discuss λ we must turn our attention to compressible flows. The question is, should there be any difference between the thermodynamic and mechanical pressures if the fluid is undergoing an expansion or compression? The assumption that the two pressures are equal is known as Stokes's assumption, and it means that

$$\lambda = -\tfrac{2}{3}\mu \qquad 6.1.6$$

This assumption is supported by kinetic theory when the fluid is a monatomic gas. Sherman (1955) measured the structure of shock waves in a rarefied gas flow where the thickness can become large enough to measure. He found that the continuum equations with Stokes's assumption produced the proper results for a monatomic gas. For air, a mixture of diatomic gases, Sherman found a slight deviation from 6.1.6, which varied with the strength of the shock wave. There have also been experiments in liquids, such as measurements of the absorption of sound waves, that show that Stokes's assumption is not quite true. Fortunately, all of these experiments are rather extreme cases of rapid compressions, which are outside the range of engineering situations. Stokes's assumption is reasonably accurate for all engineering situations for both gases and simple liquids, at least for those liquids that are Newtonian in the first place. Thus, deviations from it is not important in practice. Stokes's assumption is commonly taken as another characteristic of Newtonian fluids.

The final form of the stress relation is usually written as

$$T_{ij} = -p\delta_{ij} + \tau_{ij} \tag{6.1.7}$$

where τ_{ij} is the *deviatoric stress tensor* defined by

$$\tau_{ij} = -\tfrac{2}{3}\mu\,\partial_k v_k \delta_{ij} + 2\mu\,\partial_{(i}v_{j)} \tag{6.1.8}$$

This is in essence Newton's viscosity law. We also loosely refer to τ_{ij} as the shear stress tensor, although it also contains normal viscous components that add or subtract from the pressure. A very common term for τ_{ij} is the *viscous stress tensor*.

The first term in 6.1.8 contributes only to the normal stresses. A typical normal viscous stress

$$\tau_{11} = 2\mu\left[\underset{\substack{\text{average}\\ \text{rate of}\\ \text{extension}}}{-\tfrac{1}{3}\partial_k v_k} + \underset{\substack{x_1 \text{ direction}\\ \text{rate of}\\ \text{extension}}}{\partial_1 v_1}\right] \tag{6.1.9}$$

In this viscous stress, the extensional deformation $\partial_1 v_1$ is compared with the average extension rate for all three directions. If the extension rate is exactly equal to the average extension rate, the normal viscous stress τ_{11} is zero. Only extension rates greater or less than the average produce a normal viscous stress. Note that as a consequence of Stokes's assumption, the average normal viscous stress is always zero.

Now we turn our attention to the shearing stresses. A typical off-diagonal stress is

$$\tau_{21} = 2\mu\,\partial_{(2}v_{1)} = \mu[\partial_2 v_1 + \partial_1 v_2] \tag{6.1.10}$$

The strain-rate tensor is the only contributor to the shear stresses. Recall that the off-diagonal elements of the strain-rate tensor are the angular or shearing strains. Thus, 6.1.10 is a statement that the shear stress is proportional to the shearing strain rate.

We should remind ourselves of the theoretical status of Newton's viscosity law. It is not a fundamental law, but merely a reasonable approximation for the behavior of many fluids. As applied to gases it has some theoretical support because the kinetic theory of dilute monatomic gases produces Newton's viscosity law (it also predicts the dependence of the viscosity on the temperature). The law is valid for simple liquids, but it fails for complex liquids. In any case, the important point is that if a fluid is Newtonian, it must have certain characteristics. Newton's viscosity law implies that a fluid has the following properties:

1. Stress is a linear function of strain rate.
2. The coefficients in the expression for the stress are functions of the thermodynamic state.
3. When the fluid is stationary, the stress is the thermodynamic pressure.
4. The fluid is isotropic.
5. The stress tensor is symmetric.
6. The mechanical and thermodynamic pressures are equal. The average normal viscous stress is zero. Stokes's assumption applies: $\lambda = -\frac{2}{3}\mu$.

Fluids that fail to be Newtonian usually do not satisfy the first property. They have nonlinear and sometimes time-dependent (elastic) relationships to the strain rate.

6.2 MOLECULAR MODEL OF VISCOUS EFFECTS

The molecules of an ideal gas are so far apart that the intermolecular forces are very, very small. Molecules spend most of their time in free flight between brief collisions in which their direction and speed are abruptly changed. If we imagine a plane separating the gas into two regions, the molecules do not attract or repel each other across this plane (contrary to the situation in liquids). The primary source of "shear stress" is the microscopic transport of momentum by random molecular motions. Molecules migrating across the plane carry with them the momentum of the bulk velocity from their region of origin. A simple kinetic-theory model of this mechanism gives a good insight into the process, and at the same time produces an equation for the viscosity. We now take up a discussion of this model.

Recall the shear stress equation 6.1.10,

$$\tau_{21} = \mu[\partial_2 v_1 + \partial_1 v_2] \tag{6.1.10}$$

We specialize this equation by considering a flow $v_1(x_2)$ as show in Fig. 6-1. An x_2 plane, one unit square in area, separates the fluid into inside and outside parts. It is located at an arbitrary position x_2. The shear stress on the plane is

$$\tau_{21} = \mu \frac{dv_1}{dx_2} \tag{6.2.1}$$

In order to derive this equation we must use four facts from kinetic theory. First, molecules that cross the plane begin their free flight, on the average, a distance $\frac{2}{3}$ of a mean free path (l) away from the plane. Secondly, the mean free path is related to the molecular diameter d and the number density n by

$$l = \frac{1}{\sqrt{2}\,\pi d^2 n} \tag{6.2.2}$$

Thirdly, the flux of molecules across the plane from one side to the other is

$$\text{rate that molecules cross a unit area} = \tfrac{1}{4} n\bar{v}. \tag{6.2.3}$$

where $\bar{v}$ is the average random molecular speed (without regard for direction). Lastly, the average molecular speed is related to the temperature by

$$\bar{v} = \sqrt{\frac{8kT}{\pi m}} \tag{6.2.4}$$

where k is Boltzmann's constant and m is the molecular mass.

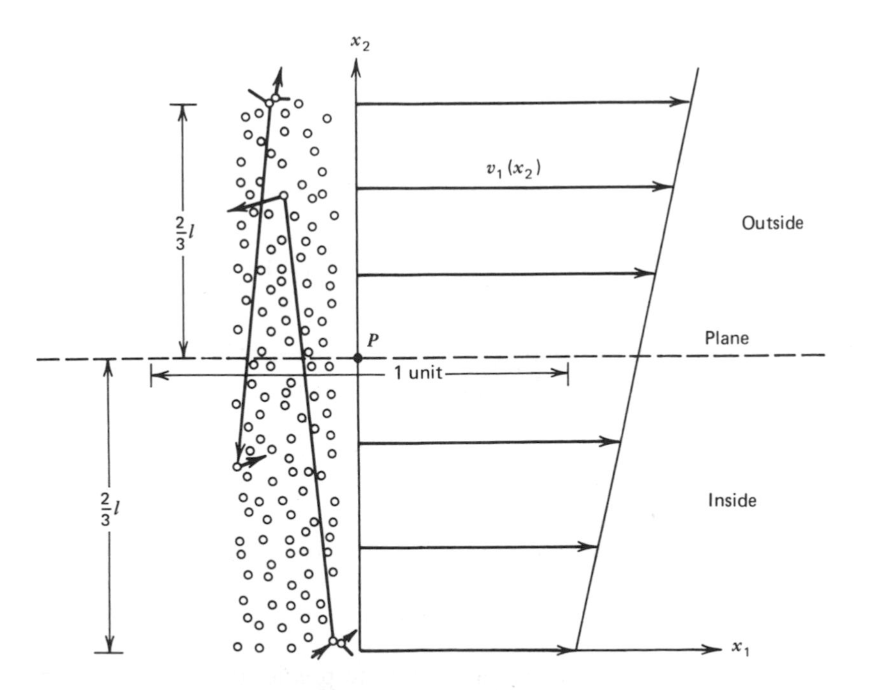

Figure 6-1 Molecular model of the viscosity of a gas.

We formulate the shear stress by postulating that the force is equal to the rate momentum crosses the plane,

$$\text{shear stress} = \frac{x_1 \text{ force}}{\text{unit area}}$$

$$= \frac{\text{rate of change}}{\text{unit area}} \text{ of } [x_1 \text{ momentum of inside fluid}]$$

$$\tau_{21} = \text{flux of momentum received}$$

$$- \text{flux of momentum lost}$$

$$\tau_{21} = \text{flux of molecules received from outside}$$

$$\times [x_1 \text{ momentum of outside fluid}]$$

$$- \text{flux of molecules lost from inside}$$

$$\times [x_1 \text{ momentum of inside fluid}]$$

The flux of molecules is given by 6.2.3 for both outgoing and incoming molecules. Molecules arriving from the outside carry an average x_1 momentum associated with the position $x_2 + \frac{2}{3}l$. The momentum of one of these particles is

$$[mv_1]_{x_2+\frac{2}{3}l} = m\left[v_1 + \frac{dv_1}{dx_2}(\tfrac{2}{3}l) + \cdots\right]_{x_2} \qquad 6.2.5$$

Molecules leaving the inside fluid cause a momentum loss of

$$[mv_1]_{x_2-\frac{2}{3}l} = m\left[v_1 + \frac{dv_1}{dx_2}(-\tfrac{2}{3}l) + \cdots\right]_{x_2}$$

The shear stress is then computed as the difference of the expressions above multiplied by the molecular flux 6.2.3. The stress is

$$\tau_{21} = \tfrac{1}{4}n\bar{v}m\tfrac{4}{3}l\frac{dv_1}{dx_2} \qquad 6.2.6$$

This equation has the same form as 6.2.1, and we can identify the viscosity as

$$\mu = \tfrac{1}{3}n\bar{v}ml \qquad 6.2.7$$

Substituting 6.2.2 and 6.2.4 gives a formula to predict the viscosity in terms of the molecular properties and the temperature,

$$\mu = \frac{2}{3d^2}\sqrt{\frac{mkT}{\pi^3}} \tag{6.2.8}$$

The gas viscosity increases for heavier molecules and decreases for larger molecules. In agreement with experiments, there is no effect of pressure. The viscosity increases as the square root of the absolute temperature according to this simple theory. Actually, the temperature effect is somewhat stronger. A more sophisticated kinetic-theory model, which includes the intermolecular forces, gives a much better prediction of the temperature dependence. Detailed procedures for calculating the viscosity of gases are contained in Bird, Stewart, and Lightfoot (1960).

The viscosity of liquids is a much more difficult task to model on the molecular level. The molecules are closely packed, and the intermolecular forces are very important. Experiments again show that there is little influence of the pressure upon the viscosity. The temperature influence in liquids is opposite to that of gases: increasing temperature causes a decrease in viscosity.

Since liquids are much like solids in that the molecules are closely packed, it will aid our thinking to review the stress–strain mechanism in solid materials. In a solid the stresses are directly proportional to the deformation as described by Hooke's law. We can more easily imagine the molecular situation if we consider a crystal with a definite lattice structure. If we imagine a plane slicing through the crystal, the sum of the intermolecular forces across this plane must balance the imposed shear. As a shear force is applied, the atoms move—that is, the average position of a vibrating atom changes—and the lattice is distorted. Because of the new directions and distances between atoms, the intermolecular forces are now different. The sum of forces on the plane now has a tangential component, which is the shear stress.

Note that during the deformation process, work is done on the crystal by the shear stress. Once the deformation stops, there is no longer work because there is no motion. When the stress is withdrawn, energy is retrieved as the crystal does work on the agent that supplies the force. In the strained state, the crystal has an extra internal energy associated with the deformed configuration. This is essentially a spring effect, as the energy is reversibly recovered when the deforming force is removed.

Liquids have intermolecular distances in the same range as solids; however, the molecules are not fixed in one position, and the configuration constantly changes. When a shear stress is applied to a liquid, the deformation continues as long as the force is applied. A velocity gradient must occur simultaneously with the shear stress, so there is a relative movement between molecules. The straining between molecules causes them to separate and brings them into new force fields of other molecules. The analogy with a deformed solid is that, on the average, the configuration of molecules in a fluid subject to shear is such that there is a net intermolecular shear stress across our imaginary plane. (In kinetic theory this is sometimes referred to as a momentum transfer by "collisions" occuring at the plane. Closely packed molecules are always within the force field of their neighbors and hence are always in a "collision" state.) The configuration of molecules in a liquid at rest is such that only a net normal force is transmitted across the plane. When a velocity gradient occurs, the net force is no longer normal but has a tangential component.

Because the process of forming new configurations and breaking up old configurations is ongoing, it requires a continuing input of work. In contrast to the solid, a liquid cannot store energy in a strained configuration. All the work done by a constant shear force is irreversible and eventually becomes random thermal motion of the molecules. This is, of course, the process of viscous dissipation, which was introduced in Section 5.10.

The shear force arises from two molecular mechanisms. The first is the net force field of the closely packed molecules of a liquid. A velocity gradient in the liquid gives rise to a "strained" molecular configuration, which in turn rotates the net force vector so that a shear stress exists. The second mechanism is momentum transport by random motion at the microscopic level, due to the mobility of fluid molecules. This kinetic contribution to the viscosity of liquids is small compared to the "average strained configuration" contribution previously discussed.

Even though the microscopic mechanisms in liquids and gases are quite different, the same continuum viscosity law governs both situations. The major effect is found in the viscosity coefficient itself: it has opposite temperature dependences for liquids and for gases.

6.3 NON-NEWTONIAN LIQUIDS

Many industrially important chemicals and products do not obey Newton's viscosity law. A good example of non-Newtonian behavior is a class of materials called high polymers. These materials may consist of from 100 to over 10,000 monomer units chained together. The resulting molecular weight can be over a million. Thus they are sometimes referred to as macromolecules. Although they form a chain in terms of chemical bonding, the monomers coil up in a random fashion to produce a ball-like molecule that is up to 100 times the diameter of a simple molecule such as oxygen. Of course, this molecule does not have a force field that extends out three or four molecular diameters as simple molecules do. In fact, a macromolecule is a somewhat spongy thing and changes its shape, especially when subjected to a shearing strain. Furthermore, a high-polymer material should not be viewed as a band of uniformly large giants. The method of producing the chemicals results in a size variation of several orders of magnitude. The molecular weight quoted is actually the average of a statistical distribution of molecular weights. For all the above reasons, these substances show unusual viscous behavior. Even a very dilute solution of macromolecules in ordinarily Newtonian liquids such as water produces some non-Newtonian effects.

Another class of non-Newtonian liquids consists of mixtures, slurries, and suspensions. The particles in these fluids range from below the continuum length scale to several orders of magnitude above it. Examples of these fluids are clay suspended in water, toothpaste, blood, paper pulp suspended in water, oil-well drilling fluid, and so on. A mixture with continuum size particles is actually a "two-phase flow" problem. There is no well-defined particle size above which mixtures should cease to be considered as uniform fluids and be treated as a two-phase systems.

Fluids may be non-Newtonian in several ways. The most common departures from Newtonian behavior are: (a) the stress is a nonlinear function of the strain, (b) additional normal viscous stresses are produced by shearing, and (c) the fluid is elastic as well as viscous. Some fluids show only one of these effects, and others all of them.

Figure 6-2 is a sketch of stress versus rate of strain for a simple shearing motion. Newtonian fluids produce a straight line on this graph, and the slope is the viscosity. A fluid is said to be "shear-thinning" if the apparent viscosity decreases with increasing strain rate. This behavior is characteristic of polymers. The most extreme case is a substance that has infinite viscosity, acting like a solid, until a certain level of stress is exceeded. Then the material becomes fluid with a Newtonian characteristic. A model of this behavior, called the Bingham plastic, gives the shear stress as

$$\tau = \pm\tau_0 + \mu\frac{dv}{dy} \qquad \text{if} \quad |\tau| > \tau_0 \tag{6.3.1}$$

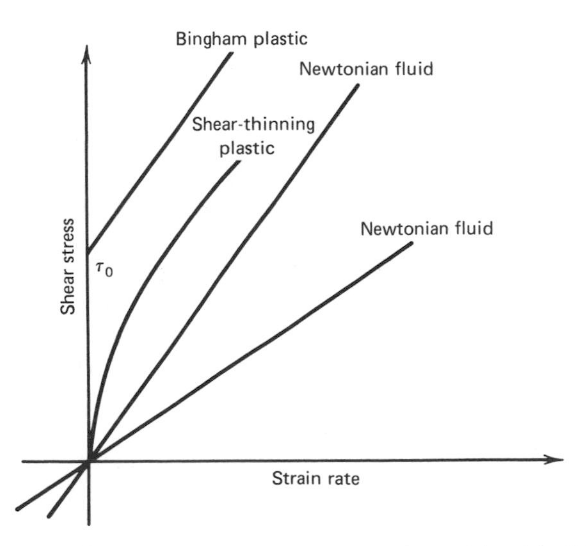

Figure 6-2 Shear stress as a function of strain rate for fluids with different viscous characteristics.

When a Bingham plastic flows in a tube, there is a core of fluid in the middle that moves as a solid plug. In this region the shear is less than τ_0, while next to the wall the stress is larger than τ_0 and the material flows as a fluid. Very concentrated slurries, those with a high volume of particle material, display this characteristic.

Fluids that exhibit a normal stress upon shearing give rise to many unusual and interesting flow phenomena. In most normal-stress fluids a shear flow also results in a viscous tension stress along the streamline. Such fluids display the *Weisenberg effect*: they climb a rotating cylinder immersed in a container of the fluid. In ordinary Newtonian fluids the free surface is depressed near the rotating shaft. A higher pressure is required on the outside to balance the centrifugal force associated with the circular streamlines. A hydrostatic pressure gradient is established within the fluid when the free surface sinks near the rotating cylinder. On the other hand, a fluid that produces a normal stress upon shearing climbs the cylinder, because the extra viscous tension along the streamlines tends to pull them toward the center, and this effect is more than enough to cancel the centrifugal force. A pressure gradient from inside to outside is also needed to keep the streamlines in equilibrium. The liquid climbs the center rod to supply this pressure gradient.

Another normal-stress effect of commercial importance is called die swell. In the process of extruding plastics or of manufacturing threads, the fluid is forced through a die in a continuous manner. Outside the die, in the free air, the fluid may expand to a larger diameter than the die. While it is in the die, the normal tension stress is developed along the streamlines. As the fluid emerges from the die, only atmospheric pressure is imposed and the wall shear is released. The material contracts along streamlines and hence swells. A Newtonian fluid does not show this behavior (except—for another reason—at very low Reynolds number). A jet of Newtonian fluid with straight streamlines contracts slightly as it exits from an orifice.

Many fluids that produce normal-stress effects are also viscoelastic. Since the transmit time across the die is short, the fluid remembers its state in the reservoir. Because it was forced to elongate in going through the die, it contracts once it leaves the die. Elastic behavior may greatly accentuate the die-swell phenomenon.

Elastic behavior in fluids is extremely complicated. Most of us have seen materials that will rebound if a stress is applied rapidly enough. James Clerk Maxwell (Scottish physicist, 1831–1879) made the first model equation for these substances by the following argument. Consider Hooke's law where the stress is proportional to the strain Θ,

$$\tau = \eta\Theta \qquad 6.3.2$$

Differentiate this law with respect to time to produce

$$\frac{1}{\eta}\frac{\partial \tau}{\partial t} = \frac{\partial \Theta}{\partial t} \qquad 6.3.3$$

We now have the equation in the same dimensions as a viscosity law, and we identify the strain rate (in a simple shear flow) as $\partial\Theta/\partial t \approx \partial u/\partial y$. Maxwell's viscoelastic equation is a sum of both types of behavior; that is, the rate of deformation is proportional to the stress for fluidlike behavior plus the rate of change of stress as the elastic contribution:

$$\tau + \frac{\mu}{\eta}\frac{\partial\tau}{\partial t} = \mu\frac{\partial u}{\partial y} \qquad 6.3.4$$

This is a linear viscoelastic law. It is the basis for generalizations of many types.

Unfortunately, most viscoelastic fluids, including the high polymers, are nonlinear. A nonlinear law is in fact obtained if we correct a logical error we made in deriving 6.3.4. We should have considered Hooke's law to apply as we follow a material particle. To do this, the differentiation with respect to time should be replaced with the substantial derivative in 6.3.3 and the corresponding term of 6.3.4. This will result in a nonlinear expression because the convective term in the substantial derivative is nonlinear. A more sophisticated and mathematically more complex model is produced by replacing the substantial derivative with a derivative that not only follows the particle velocity, but also rotates with the solid-body rotation. It is not our purpose to delve into the details of shear-thinning, viscoelastic constitutive equations; the interested reader may consult Bird, Armstrong and Hassager (1977).

The last topic we wish to discuss in this section is Thom's phenomenon, which is also called the drag-reduction phenomenon. It illustrates how even a very dilute mixture of macromolecules in water can have a dramatic effect on fluid flows. Solutions with only a few hundred parts per million concentration of polymers still maintain a Newtonian viscosity characteristic of the bulk fluid. In high-velocity flows these solutions become turbulent just as the pure fluid would. The startling difference is that the turbulence has a different structure. High frequency turbulence components are suppressed by the addition of the polymer, and in particular, the character of the viscous sublayer is markedly changed. The net result is a smaller velocity gradient next to the wall and hence a smaller shear stress. In boundary-layer flows this leads to a drag reduction and in pipe flows a smaller pressure drop for the same flow rate. There are several possible practical applications of this effect. One application that has been prohibited is in sailboat racing. It is illegal to coat the bottom of a sailboat with a polymer which would dissolve during the race.

*6.4 THE NO-SLIP CONDITION

The flow conditions at a contact surface between a solid and a fluid are important, as they enter into the mathematical formulation of flow problems. There are actually two conditions: one on the normal velocity and one on the tangential velocity. We consider a body surrounded by a flowing fluid, or a channel that confines a fluid flowing through it. A kinematic condition we impose is that the particle paths cannot go into the solid. Mathematically, the requirement is that the fluid velocity perpendicular to the wall vanish. If n_i is a local unit normal to the surface, the condition is expressed as

$$[n_i v_i]_{\text{wall}} = 0 \tag{6.4.1}$$

If the solid is moving with a local velocity V_i, then the equivalent condition is

$$[n_i(V_i - v_i)]_{\text{wall}} = 0 \tag{6.4.2}$$

In the case of steady flow the particle paths and streamlines are equivalent, and hence the condition 6.4.1 may be interpreted as a statement that the solid walls are loci of streamlines.

The kinetic restriction makes no statement about the velocity component that is tangent to the wall. Viscosity is responsible for the tangential condition, whose proper form was discussed throughout the nineteenth century. Today we accept the no-slip condition as an experimental fact. The condition is

$$[v_i]_{\text{wall}} = V_i \tag{6.4.3}$$

(Equation 6.4.3 includes the previous conditions 6.4.2 as well as the no-slip condition.)

Goldstein (1965) reviews the history of the no-slip condition. The condition itself is quite old; Daniel Bernoulli thought that it was necessary in order to account for the discrepancies between measured results and results calculated for (ideal) flows where viscosity was ignored. When one considers the various effects of surface tension, it is natural to suspect that the velocity of the fluid next to a wall might be influenced by the same things that influence surface tension: the chemical nature of the fluid or solid, the curvature of the surface, and so on. Coulomb provided some early evidence that this was not so. He experimented with a flat metal disk oscillating in water like a clock pendulum. Placing grease or grease together with powdered stone on the disk did not change the fluid resistance to the motion.

During the development period for this condition there were three alternate viewpoints. The first was that there was no slip at all, irrespective of the material, in accordance with the views of Bernoulli and Coulomb. The second was that a layer of stagnant fluid existed near the wall. Various things were supposed to determine the thickness of the layer: wall curvature, temperature, wall material, and fluid composition; it was supposed to be zero if the fluid did not wet the wall. At the outer edge of the stagnant layer the fluid was allowed to slip. The third viewpoint, due to Navier, was in effect that the slip velocity v_0

should be proportional to the stress:

$$[av_0]_{\text{wall}} = \mu \left. \frac{\partial v_1}{\partial x_2} \right|_{\text{wall}} \tag{6.4.4}$$

Navier proposed that an adjustment of the constant μ/a could reproduce the same effects as the assumption of a slipping stagnant layer. As it turns out, 6.4.4 is close to the truth, but the coefficient μ/a is always so small that v_0 is effectively zero. Thus, what was originally proposed to modify and explain the second viewpoint turns out to be a fairly accurate model that in practical cases reduces to the first viewpoint—the no-slip condition.

We can obtain a better idea of why a little slip seems to be required by doing a crude kinetic-theory calculation. Consider a gas bounded by a solid wall as shown in Fig. 6-3. In the neighborhood of the wall, the velocity may be approximated by a slip value v_0 and a uniform gradient. In this small region, of thickness say 10 mean free paths, the shear stress is almost constant at the value

$$\tau_{21} = \mu \left. \frac{dv_1}{dx_2} \right|_0 \tag{6.4.5}$$

Now we want to redo our previous molecular calculation of Section 6.2 for the special case that the x_2 plane is at the interface between the gas and the solid. The momentum carried from the fluid to the wall is again given by 6.2.5:

$$[mv_1]_{x_2+\frac{2}{3}l} = mv_0 + m\tfrac{2}{3}l \left. \frac{dv_1}{dx_2} \right|_0 + \cdots$$

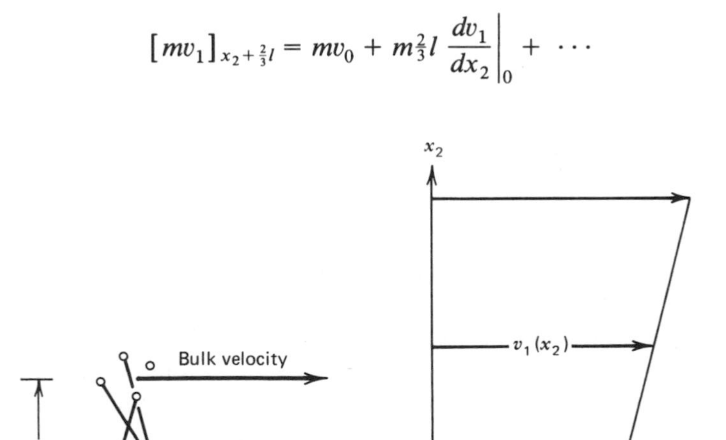

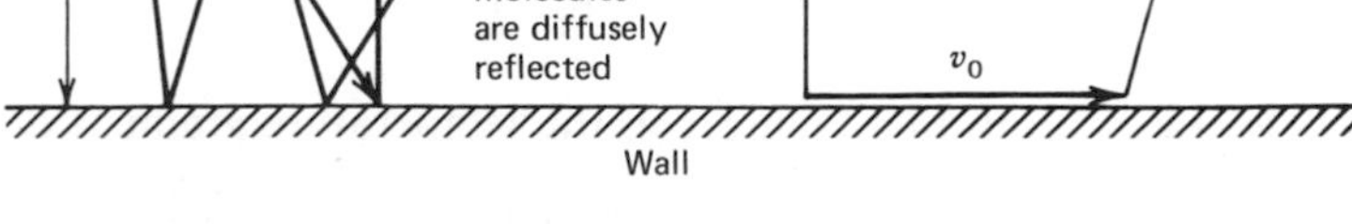

Figure 6-3 Kinetic model of slip flow at a solid wall.

The thing that is different about a wall is that the returning gas molecules have interacted or collided with a dense collection of solid molecules that have no bulk velocity. A fairly good assumption is that the molecules carry no x_1 momentum when they return from the wall (diffuse reflection). Using this assumption, the expression for the wall shear stress is

$$\tau_{21} = \tfrac{1}{4}n\bar{v}m\left[v_0 + \tfrac{2}{3}l \left.\frac{dv_1}{dx_2}\right|_0 + \cdots \right]$$

Substituting 6.4.5 for τ_{21} and using 6.2.7, we arrive at

$$v_0 = \tfrac{2}{3}l\frac{dv_1}{dx_2} \qquad 6.4.6$$

To interpret this equation, let U and L be a characteristic velocity scale and a characteristic length scale of the continuum flow. Then $d(v_1/U)/d(x_2/L)$ is of order one, and we can use 6.4.6 to find out how the slip velocity v_0 compares to U:

$$\frac{v_0}{U} = \frac{2}{3}\frac{l}{L}\frac{d(v_1/U)}{d(x_2/L)} \qquad 6.4.7$$

The slip velocity becomes zero as the mean free path becomes small compared to the continuum length L. For practical purposes there is no slip at the wall.

The conditions on the temperature at the interface between a solid and a fluid are analogous to those on the tangential velocity. Kinetic theory predicts a temperature jump that is again of the order of l/L, that is, zero for most practical purposes.

Our faith in the no-slip condition is backed up by experiments. Detailed measurements on many flows show agreement with predictions where the no-slip condition has been assumed as part of the analysis.

6.5 FOURIER'S HEAT-CONDUCTION LAW

The formulation of a heat-conduction law is a simpler task than that of the viscosity law. Following the same line of reasoning as for viscosity, we propose that the heat flux is a function of the thermodynamic state and the temperature gradient,

$$q_i = f(\rho, e, \partial_j T) \qquad 6.5.1$$

The most general relation that is linear in the temperature gradient is

$$q_i = A_i + B_{ij}\partial_j T \qquad 6.5.2$$

The coefficients A_i and B_{ij} are, in principle, functions of the thermodynamic state. Now, we require that the heat flux vanish when the temperature gradient vanishes; therefore A_i is zero. This leaves only B_{ij} as a tensor conductivity, and 6.5.1 reduces to

$$q_i = B_{ij}\partial_j T \qquad 6.5.3$$

This equation is frequently used to describe conduction in anisotropic solids, which exhibit a preferred direction for heat conduction. If the material is isotropic, we can assume $B_{ij} = -k\delta_{ij}$ (since δ_{ij} is the only isotropic second-order tensor). The final equation becomes

$$q_i = -k\partial_i T \qquad 6.5.4$$

The minus sign is dictated by the fact that heat flux is defined to be positive when energy is received.

The molecular interpretation of heat conduction was discussed in a qualitative way in the chapter on thermodynamics. The only thing which we add to that discussion here is a sketch of the kinetic theory of conduction in gases. The development is handled in the same way as for viscosity. Referring again to Fig. 6-1, we are now seeking to compute the internal energy that crosses the plane. The flux of molecules across the plane is the same as before:

$$\text{rate molecules cross a unit area} = \tfrac{1}{4}mv \qquad 6.5.5$$

These molecules originate at $x_2 \pm \frac{2}{3}l$, on the average, and they carry with them the internal energy energy of that location. In Fig. 6-1 we assume there is a temperature gradient $T(x)$ as well as the velocity gradient that is actually shown. The energy of each particle is evaluated at $x_2 + \frac{2}{3}l$ for particles coming from above:

$$\begin{aligned}\text{energy of a particle above the plane} &= mc_v[T]_{x_2+\frac{2}{3}l} \\ &= mc_v\left[T + \frac{\partial T}{\partial x_2}(-\tfrac{2}{3}l) + \cdots\right]_{x_2}\end{aligned} \qquad 6.5.6$$

and at $x_2 - 2l/3$ for particles coming from below:

$$\text{energy of a particle below the plane} = mc_v\left[T + \frac{\partial T}{\partial x_2}(\tfrac{2}{3}l) + \cdots\right]_{x_2} \qquad 6.5.7$$

The net transport of internal energy by molecular mechanisms is the heat flux

$$-q = \tfrac{1}{4}n\bar{v}mc_v\tfrac{4}{3}l\frac{\partial T}{\partial x_2} \qquad 6.5.8$$

Comparing 6.5.8 with the conduction law 6.5.4 shows that the gas conductivity is

$$k = \tfrac{1}{3} n \bar{v} m l c_v \qquad 6.5.9$$

Substituting 6.2.2 and 6.2.4 gives a formula for the thermal conductivity:

$$k = \frac{2c_v}{3d^2}\sqrt{\frac{mkT}{\pi^3}} = c_v \mu \qquad 6.5.10$$

This equation is independent of pressure, a fact verified by experiments, and has a square-root temperature variation. Experiments show the true temperature variation is somewhat stronger, just as was the case for viscosity. A more refined calculation, where account is taken of the intermolecular forces, gives a stronger temperature dependence that is more in line with experiments (especially for monatomic gases). Notice that any energy carried across the plane in internal molecular modes is accounted for in c_v, which we leave as constant in the equation.

The ratio of viscosity to thermal conductivity is an important quantity, as it compares the rate of diffusion of momentum with that of energy. This ratio is called the Prandtl number,

$$\text{Pr} \equiv \frac{\mu c_p}{k} = \frac{\mu c_v}{k}\frac{c_p}{c_v} \qquad 6.5.11$$

The theory above shows that the Prandtl number should be constant at the value $\text{Pr} = c_p/c_v = \gamma$. The prediction of a constant Pr is correct, but the value is about a factor of 2 too high. The more sophisticated kinetic theory gives $\text{Pr} = \frac{2}{3}\, c_p/c_v$, which is roughly correct.

6.6 NAVIER–STOKES EQUATIONS

The continuity equation, the momentum equations with Newton's viscosity law, and an energy equation with Fourier's conduction law are commonly referred to as the Navier–Stokes equations. Alternatively, in incompressible flow, we also use the term to apply to the continuity equation and the momentum equations for a Newtonian fluid where the density, viscosity, and thermal conductivity are constant. In flows governed by these equations, one may find the velocity without using the energy equations.

The Navier–Stokes equations in rectangular coordinates are easily found by substituting 6.1.8 into 5.7.19. Appendix C contains these equations and the corresponding forms in cylindrical and spherical coordinates.

Mathematically the Navier–Stokes equations are a set of partial differential equations with v_i and p (v_i, T, and p for the compressible case) as dependent variables. The highest terms, which come from the viscous forces, are linear and of second order. Since several first-order convective terms are nonlinear, the set is called quasilinear. Any set of second-order equations may be classified mathematically as either parabolic, elliptic, or hyperbolic. The type of boundary conditions required and the general character of the solutions are determined by the mathematical category. Interestingly enough, the Navier–Stokes equations, under various assumptions and simplifications, exhibit all three types of behavior.

6.7 CONCLUSION

A fluid-flow problem for a general fluid is governed by several equations. First, there are the basic relations for continuity, three momentum equations, and an energy relation. Second, there are the constitutive equations for the surface stresses and the heat flux. These equations are not basic, but they do apply to groups of substances. Various transport coefficients are introduced in the constitutive relations. They are quasithermodynamic properties that depend on the composition of the fluid and its thermodynamic state. Third, the thermodynamics of the fluid must be specified. This may be done through the fundamental equation $s(\rho, e)$ for the substance or, more commonly, through two equations of state $p(\rho, T)$ and $e(\rho, T)$. All of these equations are required to give a well-posed problem for a general flow situation.

PROBLEMS

6.1 (A) Stokes flow (low Reynolds number) over a sphere has velocity components

$$v_r = U\cos\theta\left[1 + \frac{1}{2}\left(\frac{r_0}{r}\right)^3 - \frac{3}{2}\left(\frac{r_0}{r}\right)\right]$$

$$v_\theta = U\sin\theta\left[-1 + \frac{1}{4}\left(\frac{r_0}{r}\right)^3 + \frac{3}{4}\left(\frac{r_0}{r}\right)\right]$$

Compute all components of the viscous stress tensor in r, θ, z coordinates.

6.2 (A) In Problem 1 find the maximum τ_{rr} and compare it with the dynamic pressure $\frac{1}{2}\rho U^2$.

6.3 (A) An ideal "inviscid" flow over a cylinder has the velocity components given in Problem 1 of Chapter 5. Compute all components of the viscous stress tensor.

6.4 (B) Prove that a Newtonian fluid with constant viscosity in incompressible flow obeys the relation $\nabla \cdot \boldsymbol{\tau} = \mu \nabla^2 \mathbf{v}$.

6.5 (C) Evaluate the Navier–Stokes equations for the velocity profiles of the Hill's spherical vortex given in Section 13.6. Integrate to find the pressure.

6.6 (A) Two long trains carrying coal are traveling in the same direction side by side on separate tracks. One train is moving at 40 ft/sec and the other at 50 ft/sec. In each coal car is a man shoveling coal and pitching it across to the neighboring train. The rate of coal transfer is 4 tons/min for each 100 ft of train length. This rate is the same for both trains. Find the extra force on each train per unit length caused by this mechanism.

7 Some Incompressible Flow Patterns

In the previous chapters we have dealt with the basic physics and the general ideas that apply to flow fields. We have found that these laws are local in nature and that they apply to each point in space. What makes one flow situation different from another is the boundary conditions. Boundary conditions include the location and motion of walls, imposed pressure differences, prescribed velocities, assumptions of symmetry, and so forth. The formulation of boundary conditions follows a few rules, for example the no-slip condition, but in general we must use physical intuition and make reasonable assumptions. The purpose of this chapter is to present some examples of simple flows and the arguments used in their analysis.

In this chapter we assume that all the flows are incompressible. Later in the book we shall make a detailed analysis of what it means for a flow to be incompressible. For our present purposes we can simply assume that an incompressible flow has a constant density, viscosity, specific heat capacity, and thermal conductivity. With these assumptions the velocity field can be found using the continuity and momentum equations without regard for the energy equation and equations of state. Thus the mechanical and thermal aspects of the flow can be separated.

7.1 PRESSURE-DRIVEN FLOW IN A SLOT

Consider two reservoirs in which the fluid surfaces are at two different elevations as shown in Fig. 7-1. A tube connects the reservoirs so that water may flow between them. We may assume that the reservoirs are so large that the flow into or out of the reservoir causes only a very slow rise or fall in the surface elevation. In other words, for purposes of the analysis, the fluid in the reservoirs is at a constant height. Since the hydrostatic pressures near the entrance and exit of the tube are different, we expect a flow to develop and attain a steady state.

The entrance to the tube is well rounded, and the fluid enters smoothly from the reservoir, having an almost constant velocity across the tube. The acceleration of the fluid from nearly zero velocity in the reservoir to the average value is accomplished by pressure forces, the pressure p_1 at the tube entrance being somewhat smaller than the hydrostatic pressure p_0 at the same level in the reservoir. Since friction is not important in the acceleration region, the exact

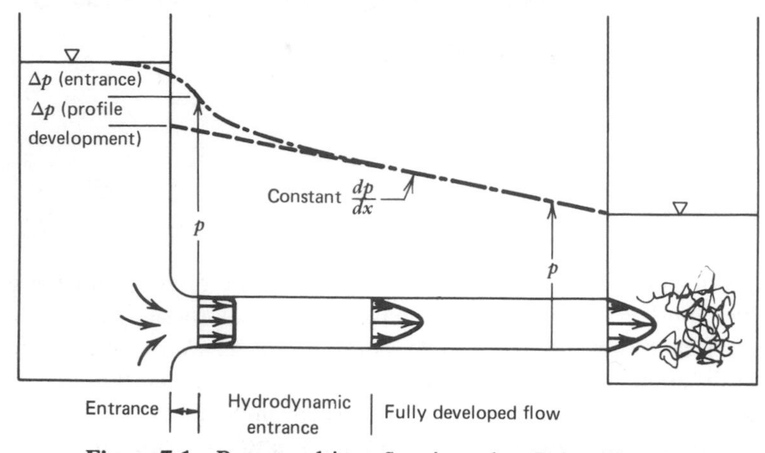

Figure 7-1 Pressure driven flow in a slot (Poiseuille flow).

value of the pressure can be computed using Bernoulli's equation, applied along the central streamline (readers who are not familiar with this equation will find it in Section 17.4):

$$p_0 - p_1 = \tfrac{1}{2}\rho V_1^2$$

Although the bulk of the flow satisfies the no-friction assumption, flow along the streamlines very close to the wall must be retarded by friction. In fact, the no-slip condition requires zero velocity on the wall itself. Thus, the entrance profile has a small portion near the wall where the velocity drops from V_1 to zero because of viscosity.

While the subject is at hand, we might note another aspect of the flat entrance profile that is inaccurate. It turns out that no matter how carefully the entrance is shaped, the profile is not completely flat, but contains a slightly lower velocity than V_1 on the centerline and bulges of higher velocity further out. This effect may be explained as follows. As the flow enters the tube, the streamlines are curved. In order to force the particles to follow a curved path, a normal pressure force must exist. The pressure on the outside of the streamline, toward the centerline, is higher than the pressure on the inside of the streamlines. Through Bernoulli's equation we know that a low pressure implies a high velocity and vice versa. Even though the wall becomes flat, there is still some curvature of the interior streamlines at this point. The streamlines will be very nearly parallel only after we go downstream in the flat-wall section for about one or two slot widths. This effect is very slight for a well-rounded entrance and can be ignored for most engineering applications.

The entrance profile, nearly flat with steep dropoffs next to the wall, undergoes further change as the flow proceeds down the tube. The viscous shear stress is at first confined to particles near the wall, but gradually it affects particles further and further from the wall. Each cross section of the tube must have the same mass flow rate, so when particles are slowed down near the wall, other particles in the center must be accelerated. Pressure forces are responsible for accelerating the center particles, so the pressure must continue to decrease in the flow direction. Finally, when a balance between the pressure forces and the shear forces is attained, the profile no longer changes as we go to new positions down the tube; the profile is *fully developed*. The region where the flow profile is developing is called the *entrance region* or, more precisely, the hydrodynamic entrance region. The hydrodynamic entrance is usually long: 50 to 100 tube widths is not uncommon in engineering situations. The entrance region becomes short only when the flow is very slow (in the sense that the Reynolds number approaches zero).

In the fully developed region it is possible to quantify the analysis with very little effort. To make things even simpler, we assume that the tube is a two-dimensional slot. Taking an x, y coordinate system as shown in Fig. 7-2, we assume that nothing changes with z and that v_z is zero. We also assume that v_x is not a function of x, since the profile is fully developed. When these assumptions are inserted into the continuity equation

$$\frac{\partial v_x}{\partial x} + \frac{\partial v_y}{\partial y} + \frac{\partial v_z}{\partial z} = 0$$

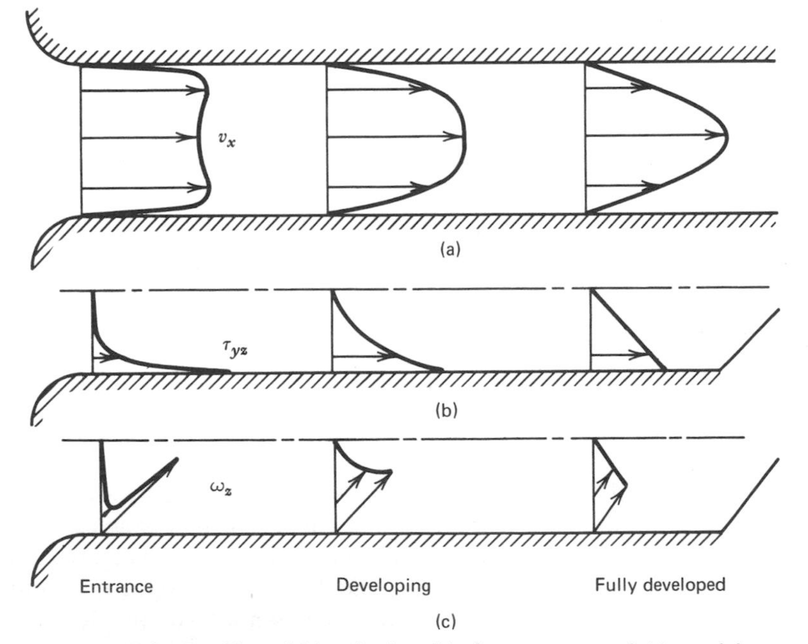

Figure 7-2 Profiles of (a) velocity, (b) shear stress, and (c) vorticity.

it reduces to

$$\frac{\partial v_y}{\partial y} = 0 \qquad 7.1.1$$

Upon partial integration we find $v_y = v_y(x \text{ only})$. The no-slip condition at the walls requires $v_y = 0$ for all x; hence v_y must be zero everywhere.

Turning now to the y-direction momentum equation, we have

$$\rho v_x \frac{\partial v_y}{\partial x} + \rho v_y \frac{\partial v_y}{\partial y} = -\frac{\partial p}{\partial y} - \rho g + \mu \frac{\partial^2 v_y}{\partial x^2} + \mu \frac{\partial^2 v_y}{\partial y^2}$$

Since v_y is zero, this simplifies to

$$\frac{\partial p}{\partial y} = -\rho g \qquad 7.1.2$$

which means that pressure changes in the y direction are solely the result of hydrostatic effects. Partial integration yields

$$p = -\rho g y + P(x) \qquad 7.1.3$$

The function $P(x)$ is the pressure along the bottom wall where $y = 0$. In anticipation of the x-momentum equation, we note that the pressure gradient in the x direction is at most a function of x. This is found from 7.1.3 since

$$\frac{\partial p}{\partial x} = \frac{dP}{dx} \qquad 7.1.4$$

We employ this fact as we write the x-direction momentum equation,

$$\rho v_x \frac{\partial v_x}{\partial x} + \rho v_y \frac{\partial v_x}{\partial y} = -\frac{dP}{dx} + \mu \frac{\partial^2 v_x}{\partial x^2} + \mu \frac{\partial^2 v_x}{\partial y^2}$$

Simplifying produces

$$0 = -\frac{dP}{dx} + \mu \frac{\partial^2 v_x}{\partial y^2} \qquad 7.1.5$$

This relation states that at every point in the flow, the net pressure force is exactly balanced by the net shear stress. As a result of this balance, particles are not accelerated and continue down the tube with an unchanging velocity. Another thing to notice about 7.1.5 is that the density has dropped out and will not influence the velocity profile. When the tube is not level, the density does play a role in determining the force due to gravity.

The solution of 7.1.5 is begun by noting that the pressure term is a function of x while the viscous term is a function of y. Thus, the equation is of the form $0 = g(x) + f(y)$. Since x can change independently of y, we arrive at the familiar conclusion that f and g are constant. Physically this means that in the fully developed region the pressure must decrease linearly with distance. The actual magnitude of the pressure drop is determined by the reservoirs, and we solve the problem assuming that dP/dx is a known parameter. Equation 7.1.5, together with the no-slip boundary conditions at the walls, constitutes the mathematical problem for v_x. The problem statement is

$$\frac{d^2 v_x}{dy^2} = \frac{1}{\mu}\frac{dP}{dx} = \text{const} \tag{7.1.6}$$

$$v_x(y=0) = v_y(y=h) = 0 \tag{7.1.7}$$

Integrating twice gives

$$v_x = \frac{1}{2\mu}\frac{dP}{dx}y^2 + C_1 y + C_2$$

Applying the boundary conditions determines the velocity profile as a parabola in y,

$$v_x = -\frac{h^2}{2\mu}\frac{dP}{dx}\left[\left(\frac{y}{h}\right) - \left(\frac{y}{h}\right)^2\right] \tag{7.1.8}$$

When the pressure gradient is negative, the flow is in the positive direction.

The velocity profile is a key result in any fluid-flow analysis because many other flow properties are found from it by simple relations. We begin a review of these properties by computing the flow rate across a section that is one unit deep in the z direction:

$$Q = \int_{A_x} n_i v_i \, dA = \int_0^h v_x \, dy$$

$$Q = -\frac{h^3}{12\mu}\frac{dP}{dx} = \frac{h^3}{12\mu}\frac{\Delta P}{L} \tag{7.1.9}$$

In the relation above, the (positive) pressure difference between two points (at the same elevation) a distance L apart has replaced the pressure gradient.

The flow-rate equation corresponding to 7.1.9 for round tubes was first given independently by G. Hagen (1839) (German engineer, 1797–1884) and J. Poiseuille (1840) (French physician, 1799–1869). They formulated the equation after careful experiments done with water in tubes of a variety of sizes and lengths. A good discussion of these researchers is given by Prandtl and Tietjens (1934); a copy of the data from this source is also given by Schlichting (1950). At the time of the experiments, the entrance-length effect was not completely understood, and when short tubes were used it caused a deviation from 7.1.9. Poiseuille could not explain this deviation, but Hagen came very close when he attributed it to an extra pressure drop required to accelerate the fluid. It is interesting that Poiseuille's motive for performing the experiments was to learn more about human blood flow.

Equation 7.1.9 is the basis for two types of measuring devices. The first is an apparatus to measure viscosity. Typically a reservoir is used to impose pressure on a vertical capillary tube, which is open to the atmosphere. The time that it takes a given quantity of fluid to flow through the apparatus is a direct indication of the viscosity (in fact, viscosities are sometimes quoted in terms of "Stabolt seconds" or "Redwood seconds" instead of in dimensionally correct units). In some early designs of these viscometers, the capillaries were too short to make the entrance length negligible. The linear relationship between viscosity and time was then lost. Of course, this design deficiency was corrected, and today these devices offer an inexpensive and accurate method of measuring viscosities.

The second apparatus is for measuring flow rates. Any parallel, laminar flow has a flow-rate equation that is linear in the pressure drop. Hence, if one is sure the flow through a tube, or a bank of tubes, is laminar and fully developed, then the pressure drop is directly proportional to the flow. Unfortunately, most industrially important flows are so fast that they are turbulent and the linear relation is not valid.

Another form of 7.1.9 contains the average velocity (defined by the relation $Q = v_{ave} h$),

$$v_{ave} = \frac{h^2}{12\mu} \frac{\Delta P}{L} \qquad 7.1.10$$

The maximum velocity is found from 7.1.8 by evaluating at $y/h = \frac{1}{2}$:

$$v_{max} = \frac{h^2}{8\mu} \frac{\Delta P}{L} \qquad 7.1.11$$

Comparing these expressions shows that

$$v_{ave} = \tfrac{2}{3} v_{max} \qquad 7.1.12$$

If flow in a circular tube is investigated, one finds that an even smaller fraction of the cross-section contains the high-speed flow, as the average velocity is only one-half the maximum velocity.

A more convenient form for the velocity profile is found by substituting 7.1.11 into 7.1.8:

$$\frac{v_x}{v_{\max}} = 4\left[\left(\frac{y}{h}\right) - \left(\frac{y}{h}\right)^2\right] \qquad 7.1.13$$

From this we compute the shear stress,

$$\tau_{yx} = \mu \frac{dv_x}{dy}$$

$$\frac{\tau_{yx}}{\mu v_{\max}/h} = 4\left[1 - 2\left(\frac{y}{h}\right)\right] \qquad 7.1.14$$

and the vorticity profile,

$$\omega_z = -\frac{dv_x}{dy}$$

$$\frac{\omega_z}{v_{\max}/h} = -4\left[1 - 2\left(\frac{y}{h}\right)\right] \qquad 7.1.15$$

In all parallel-flow problems there is a proportionality between vorticity and shear stress. Figure 7-2 gives the profiles of velocity, shear stress, and vorticity at several sections along the tube. Note that the core of the entrance profile has no vorticity or shear stress.

There are a few more aspects of the flow that can be brought out if we apply the integral kinetic energy equation (5.16.2) to the flow. For a steady, incompressible flow through a fixed region, the equation is

$$0 = -\int_{FR} \rho n_i v_i \tfrac{1}{2} v^2 \, dS + \int_{FR} \rho g_i v_i \, dV$$

$$- \int_{FR} n_i v_i p \, dS + \int_{FR} n_i \tau_{ij} v_j \, dS - \int_{FR} \Phi \, dV \qquad 7.1.16$$

where we have used the symbol Φ for the viscous dissipation $\tau_{ij}\partial_i v_j$. The fixed region of length L with pressure drop ΔP is shown in Fig. 7-3. As depicted in the figure, the first term of 7.1.16 drops out because the fully developed region has the same flux of kinetic energy at each end. The second term is also zero because the gravity force is perpendicular to the velocity. The fourth term is likewise zero: the normal stresses are zero on the end surfaces, and the velocity is zero at the wall (the shear forces on the walls do no work). Therefore 7.1.16 becomes

$$0 = \Delta P \, v_{\text{ave}} A - \int_{FR} \Phi \, dV \qquad 7.1.17$$

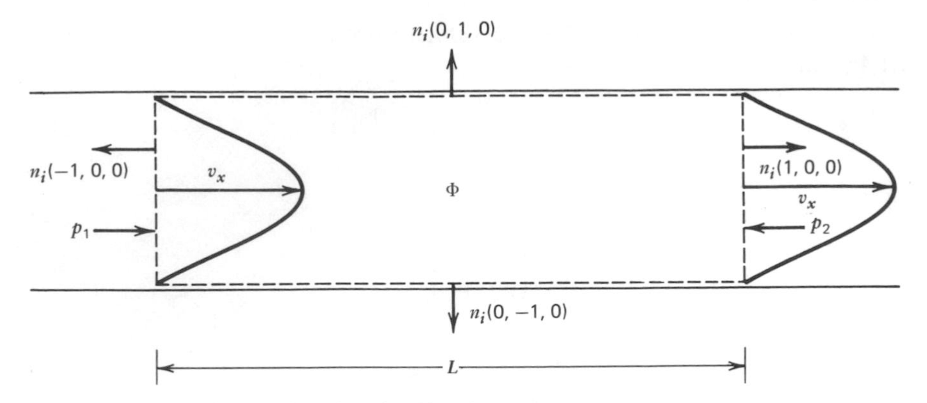

Figure 7-3 Fixed control region for kinetic-energy analysis of fully developed flow.

Again we use the symbol $\Delta P \equiv P_1 - P_2$ for the pressure difference at the same elevation. The work rate of pressure forces is exactly balanced by the rate that energy is dissipated by viscous action. The *head loss* is the viscous dissipation rate in a region per unit mass flow through the region:

$$h_L = \frac{1}{\rho v_{\text{ave}} A} \int \Phi \, dV \tag{7.1.18}$$

Head loss has the dimensions of energy per unit mass. In terms of it, 7.1.17 reads

$$\frac{\Delta P}{\rho} = h_L \tag{7.1.19}$$

This formula is valid for any fully developed tube flow. Hydraulic engineers prefer to define a head loss with the dimensions of a length; they would divide 7.1.18 by the acceleration of gravity g as part of the definition.

The head loss, as defined in 7.1.18, is very general and quite useful in the analysis of flow-system problems. Each component of the flow system has its head loss. A nondimensional head-loss coefficient is formed by taking the kinetic energy per unit mass of the entering (or exiting) fluid as a reference:

$$K \equiv \frac{h_L}{\frac{1}{2} v_{\text{ref}}^2}, \qquad \text{or} \quad h_L = K \tfrac{1}{2} v_{\text{ref}}^2 \tag{7.1.20}$$

Notice that K may be much greater than one. For example, a valve that is slightly open has a high-speed jet through the small open area. The energy of this jet is completely dissipated and constitutes the head loss. This energy may be 100 times the kinetic energy of the flow into the valve body, which is the reference kinetic energy.

The head loss for straight tubes with a fully developed profile increases linearly with the length of the tube. This fact leads to the introduction of the *friction factor* f so that

$$K = \frac{L}{h} f \tag{7.1.21}$$

The physical interpretation of f is the fraction of reference kinetic energy that is dissipated in a length of tube equal to the width h. This may be seen by combining 7.1.20 and 7.1.21 into the form

$$f = \frac{h_L}{(L/h)\frac{1}{2}v_{\text{ref}}^2} \tag{7.1.22}$$

Inserting 7.1.19 yields

$$f = \frac{\Delta P}{(L/h)\frac{1}{2}\rho v_{\text{ref}}^2} \tag{7.1.23}$$

This equation is a form of Darcy's law, named after Henri Darcy, (French engineer, 1803–1858), who experimentally determined f for turbulent pipe flow. In our case of laminar flow we can use 7.1.10 together with 7.1.23 to find

$$f = \frac{24}{\text{Re}}$$

where the Reynolds number is

$$\text{Re} = \frac{v_{\text{ave}} h}{\nu}$$

We have presented the kinetic-energy analysis of tube flow so that the reader could see how the concepts of head loss and friction factor arise. This method is actually better suited to turbulent flow, where f is frequently constant, than to laminar flow. In fact, 7.1.23 is not a good form for the pressure-drop equation in laminar flow because it gives the false impression that the density is important. The preferred equation is 7.1.9.

We mentioned previously that the hydrodynamic entrance region suffers an extra pressure drop. The reason is found if we apply the kinetic-energy equations between the entrance, where the velocity is uniform at v_{ave}, and a downstream station where the parabolic profile exists. The term for convection of kinetic energy contains the velocity cubed, and since the parabolic profile has a maximum equal to $\frac{3}{2}$ times the average, there is significantly more kinetic energy in the fully developed profile than in the entrance profile. This increase in energy requires an extra pressure drop to accelerate the flow. The viscous dissipation in the entrance region is concentrated near the wall, but if the entrance is smooth, it is not significantly higher than the fully developed value. As a matter of fact, in a round tube the convection of kinetic energy by the fully developed profile is twice the convection of kinetic energy by the entrance profile.

We turn now to a discussion of the events at the exit of the tube into the second reservoir. A sharp exit as shown in Fig. 7-4 will always cause flow separation, and a jet of fluid will issue into the reservoir. It is common to assume that the streamlines at the exit plane are still exactly parallel to the walls. If this is so, the y-momentum equation again simplifies to 7.1.2, the hydrostatic pressure equation. Furthermore, if we consider the y momentum near the wall of the reservoir and neglect any motion of the reservoir fluid, we again have the hydrostatic pressure equation. Our picture of the jet exit is then one where the pressure at the exit plane is the hydrostatic pressure that normally occurs in the reservoir in the absence of motion. Likewise, the pressure at the entrance is determined by the height of the fluid in the entrance reservoir. This pressure drop causes the flow to be established at a rate just sufficient to satisfy the pressure drop predicted by 7.1.9 (for very long tubes we can neglect the extra pressure drop of the hydrodynamic entrance region).

A jet issuing into a reservoir is a very unstable flow where small perturbations grow very rapidly. Turbulence develops in the entrance region, at first being confined to the edge between the reservoir fluid and the jet fluid. This is

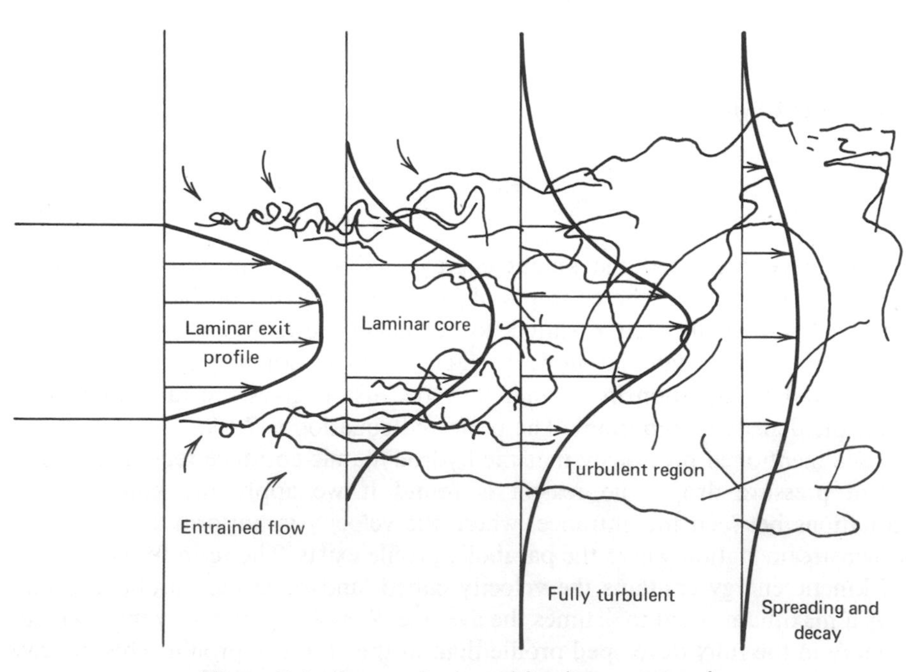

Figure 7-4 Decay of a jet exiting into a reservoir.

a region of high shear. As we proceed further away from the exit, the turbulence grows and eventually the jet is completely turbulent. At the same time the turbulence grows; it entrains fluid from the reservoir, which in turn induces a slight flow of reservoir fluid toward the jet. In computing the pressure in the jet exit we neglected this flow. Once the jet becomes completely turbulent, the centerline velocity begins to decay, accompanied by a spreading of the jet. The decay continues until all the directed kinetic energy of the jet is transformed into the random kinetic energy of turbulent eddies.

The ultimate fate of turbulent eddies is to be destroyed by the action of viscosity. Viscous dissipation finally claims all the turbulent energy and transforms it into random molecular motion. With a sensitive thermometer we would find a slight increase in the fluid temperature.

The picture we have drawn of the flow through a tube corresponds to reality if the Reynolds number of the flow is moderate to high. At Reynolds numbers ($Re = v_{ave}d/\nu$) above 1400–1600 the flow becomes turbulent, and therefore our assumption of steady flow is invalid. This brings up a point that is well to remember about any analysis of fluid flow phenomena. A solution to the Navier–Stokes equations for a given geometry may not be unique.

7.2 PLANE COUETTE FLOW

The simple shear flow named after M. F. A. Couette (1858–1943) is often used to introduce the concept of viscosity. We imagine two concentric circular cylinders with the gap between them filled with fluid (Fig. 7-5). One of the cylinders, say the inner one, is rotated while the other is stationary. We can make the analysis simpler if we consider that the gap width is very small

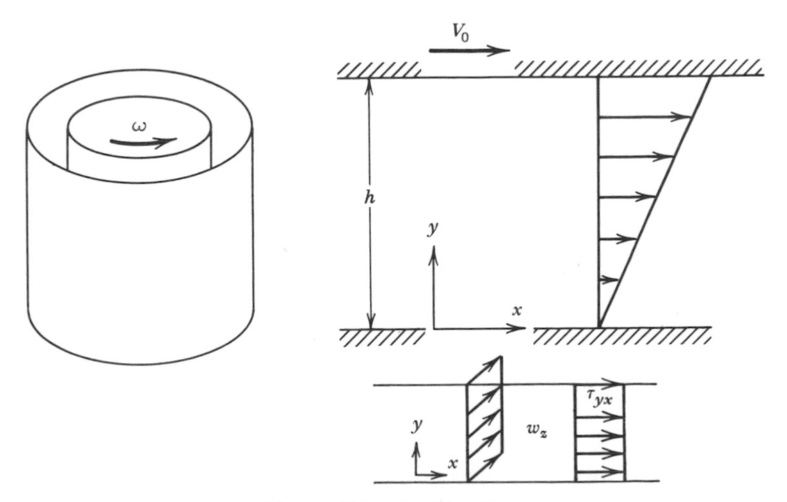

Figure 7-5 Couette flow.

compared to the inner radius. This allows us to model the flow as the flow in a plane, two-dimensional slot with one moving wall and one stationary wall.

The analysis to find the velocity profile is very similar to that for Poiseuille flow in the previous section. We assume a unidirectional flow $v_x(y)$ that is independent of x. In the x-direction momentum equation we make the additional assumption that there is no pressure gradient in the x direction. This yields

$$0 = \mu \frac{d^2 v_x}{dy^2} \tag{7.2.1}$$

which means that the shear stress on each side of a particle is exactly balanced: the shear stress is a constant across the gap. Notice that the viscosity can be divided out of 7.2.1, implying that neither viscosity nor density affects the profile. The boundary conditions for the problem are the no-slip condition applied to each wall,

$$\begin{aligned} v_x(y = 0) &= 0 \\ v_x(y = h) &= V_0 \end{aligned} \tag{7.2.2}$$

The solution to this problem is a linear profile

$$v_x = \frac{V_0 y}{h} \tag{7.2.3}$$

Calculating the shear stress from this profile gives a constant value, as we anticipated above:

$$\tau_{yx} = \mu \frac{dv_x}{dy} = \mu \frac{V_0}{h} \tag{7.2.4}$$

The vorticity is also a constant:

$$\omega_z = -\frac{dv_x}{dy} = -\frac{V_0}{h} \tag{7.2.5}$$

These quantities are plotted in Fig. 7-5. The corresponding analysis for flow in a cylindrical tube is also easily accomplished.

7.3 PRESSURE-DRIVEN FLOW IN A SLOT WITH A MOVING WALL

In problems we analyzed in the two previous sections, the velocity was governed by linear differential equations and linear boundary conditions. Here we study a composite flow. Let the Poiseuille flow velocity be $v_x^{(1)}$ and the

Couette flow velocity be $v_x^{(2)}$; then the algebraic sum represents the flow in a slot with a moving wall and an imposed pressure gradient, both acting simultaneously:

$$v_x = v_x^{(1)} + v_x^{(2)}$$

$$v_x = -\frac{h^2}{2\mu}\frac{dP}{dx}\left[\left(\frac{y}{h}\right) + \left(\frac{y}{h}\right)^2\right] + V_0\left(\frac{y}{h}\right)$$

or in a slightly different form

$$\frac{v_x}{V_0} = (1 + \mathbb{P})\left(\frac{y}{h}\right) + \mathbb{P}\left(\frac{y}{h}\right)^2$$

In this equation $\mathbb{P}$ is a nondimensional parameter which indicates the relative effects of the pressure gradient and the wall motion. $\mathbb{P}$ is given by

$$\mathbb{P} \equiv -\frac{h^2}{V_0\mu}\frac{dP}{dx} = 4\frac{v_{\max}^{(1)}}{V_0}$$

Velocity profiles are shown in Fig. 7-6. We shall find that these velocity profiles are good local approximations to the flow in slider bearings.

For the sake of completeness let us list the differential equation and boundary conditions that govern the complete problem. The momentum equation for the complete flow,

$$\frac{1}{\mu}\frac{dp}{dx} = \frac{d^2v_x}{dy^2}$$

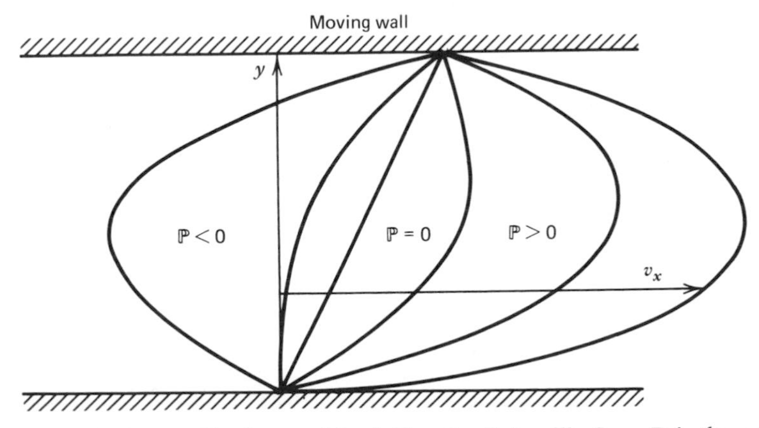

Figure 7-6 Velocity profiles for combined Couette–Poiseuille flow. $\mathbb{P}$ is the pressure-gradient parameter.

is the sum of

$$\frac{1}{\mu}\frac{dp}{dx} = \frac{d^2v_x^{(1)}}{dy^2} \quad \text{and} \quad 0 = \frac{d^2v_x^{(2)}}{dy^2}$$

The boundary conditions are

$$v_x(0) = v_x^{(1)}(0) + v_x^{(2)}(0) = 0$$

and

$$v_x(h) = V_0$$

$$v_x^{(1)}(h) + v_x^{(2)}(h) = 0 + V_0$$

Both the boundary conditions and the differential equations are linear, as is required for the superposition of solutions.

7.4 DOUBLE FALLING FILM ON A WALL

The double falling film is a problem designed to illustrate the boundary conditions between two immiscible liquids and the boundary condition between gases and liquids. Consider the flow situation depicted in Fig. 7-7, where a solid smooth plane is inclined at an angle θ to the vertical. Two immiscible liquid films flow down the plane under the influence of gravity. The actual thickness of each film is controlled by the method by which the flow is established. We do not concern ourselves with how the flow is established or how long it takes to reach a steady profile independent of x. These problems are avoided by assuming the film thicknesses h_a and h_b have known fixed values.

We take a rectangular coordinate system aligned with the flow and having the x axis on the plate. As with all problems in this chapter, we assume that there is only one nonzero velocity component, which is a function of y alone. This assumption was shown previously to satisfy the continuity equation identically. The y-direction momentum equation is once again the hydrostatic balance

$$0 = -\frac{\partial p}{\partial y} - \rho g \sin\theta \qquad 7.4.1$$

Here ρ is either ρ_a or ρ_b as needed. A partial integration of this equation gives

$$p = -\rho g \sin\theta\, y + f(x) \qquad 7.4.2$$

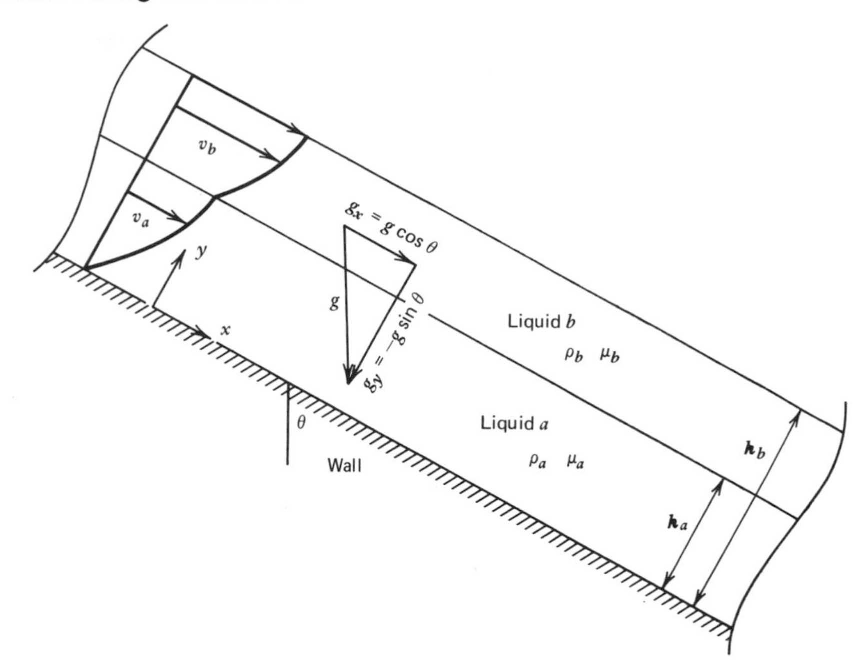

Figure 7-7 Films of two immiscible fluids falling down an inclined wall.

The arbitrary function $f(x)$ is evaluated using the fact that atmospheric pressure p_0 exists on the top of liquid b; hence

$$p = \rho_b g(h_b - y)\sin\theta + p_0, \qquad h_a \le y \le h_b \tag{7.4.3}$$

The pressure at the interface we denote as $p(h_a) = p_1$. Thus

$$p_1 = \rho_b g(h_b - h_a)\sin\theta + p_0 \tag{7.4.4}$$

Finally the pressure in liquid a is found from 7.4.2 and the condition above:

$$p = \rho_a g(h_a - y)\sin\theta + p_1, \qquad 0 \le y \le h_a \tag{7.4.5}$$

The pressure on the plate surface is in general slightly higher than ambient. If the plate is horizontal ($\theta = \pi/2$), the entire weight of the fluid is imposed, while if the plate is vertical, there is no pressure change through the liquid and the plate is at atmospheric pressure.

The flow is driven by the component of gravity along the plate. In this direction the momentum equation simplifies to

$$0 = \mu\frac{d^2 v_x}{dy^2} + \rho g\cos\theta \tag{7.4.6}$$

We drop the subscript x from the velocity in the equations which follow. This should cause no confusion, since there is only one nonzero component. It will allow us to adopt v_a and v_b as symbols for the velocity in liquids a and b respectively.

Next we turn to the boundary conditions. The no-slip condition applies at the wall and also at the interface; hence

$$v_a(0) = 0 \tag{7.4.7}$$

$$v_a(h_a) = v_b(h_b) \tag{7.4.8}$$

The next assumption is that the shear stress is the same on each side of the interface:

$$n_i^a\left[\tau_{ij}^a - \tau_{ij}^b\right] = 0$$

$$\tau_{yx}^a = \tau_{yx}^b \tag{7.4.9}$$

$$\mu_a \frac{dv_a}{dy} = \mu_b \frac{dv_b}{dy} \qquad \text{at} \quad y = h_a$$

The remaining boundary condition is not so obvious. Because of the no-slip condition, the air above liquid b must be moving at the same velocity as the liquid, and the shear stress must be continuous across the interface. However, instead of applying the correct boundary condition, which would require us to solve for the motion of the air, we assume that the air exerts only a negligible shear stress on the liquid:

$$0 \approx \tau_{yx}^{\text{air}} = \tau_{yx}^b = \mu_b \frac{dv_b}{dy} \qquad \text{at} \quad y = h_b \tag{7.4.10}$$

Since the viscosity μ_b is not zero, the velocity gradient in the liquid must vanish at the interface.

Integration of 7.4.6 produces the velocity profiles (note that $\mu/\rho = \nu$) for both liquids films:

$$v_a = -\frac{g}{2\nu_a} y^2 \cos\theta + C_1 y + C_2 \tag{7.4.11}$$

$$v_b = -\frac{g}{2\nu_b} y^2 \cos\theta + C_3 y + C_4 \tag{7.4.12}$$

Applying the no-slip condition, 7.4.7 gives $C_2 = 0$, while at $y = h_b$ the vanishing shear 7.4.10 shows that

$$C_3 = \frac{g}{\nu_b} h_b \cos\theta = \frac{h_b \nu_a}{h_a \nu_b} \frac{g}{\nu_a} h_a \cos\theta \tag{7.4.13}$$

The interface stress condition 7.4.9 yields

$$C_1 = \frac{gh_a}{\nu_a}\cos\theta\left[\frac{\rho_b}{\rho_a}\left(\frac{h_b}{h_a}-1\right)+1\right] \qquad 7.4.14$$

and the interface velocity condition yields

$$C_4 = \frac{gh_a^2\cos\theta}{\nu_a}\left\{\frac{1}{2}+\frac{1}{2}\frac{\nu_a}{\nu_b}-\frac{\rho_b}{\rho_a}+\left(\frac{\rho_b}{\rho_a}-\frac{\nu_a}{\nu_b}\right)\left(\frac{h_b}{h_a}\right)\right\} \qquad 7.4.15$$

The velocity profiles are

$$v_a = \frac{gh_a^2\cos\theta}{\nu_a}\left\{\left[1+\frac{\rho_b}{\rho_a}\left(\frac{h_b}{h_a}-1\right)\right]\left(\frac{y}{h_a}\right)-\frac{1}{2}\left(\frac{y}{h_a}\right)^2\right\}$$

and

$$v_b = \frac{gh_a^2\cos\theta}{\nu_a}\left\{\frac{1}{2}+\frac{1}{2}\frac{\nu_a}{\nu_b}-\frac{\rho_b}{\rho_a}+\left(\frac{\rho_b}{\rho_a}-\frac{\nu_a}{\nu_b}\right)\left(\frac{h_b}{h_a}\right)\right.$$

$$\left.+\frac{\nu_a h_b}{\nu_b h_a}\left(\frac{y}{h_a}\right)-\frac{1}{2}\frac{\nu_a}{\nu_b}\left(\frac{y}{h_a}\right)^2\right\}$$

An example of these velocity profiles is given in Fig. 7-7 for the case of a very viscous fluid on top of a less viscous fluid. We also note that the case of a single falling film is retrieved from the v_a equation by setting $h_b/h_a = 1$.

7.5 THE RAYLEIGH PROBLEM

In this section we consider the impulsive motion of a flat plate in its own plane. The plate is infinite and coincides with the x axis as shown in Fig. 7-8. For times greater than zero the plate has a constant velocity V_0. The no-slip condition requires that the fluid next to the wall also move at velocity V_0. At first, the particles near the wall are accelerated by an imbalance of the shear forces. As time proceeds this effect is felt further and further from the plate, inducing more and more fluid to move along with the plate.

The mathematical solution to this problem is begun by assuming that the velocity is only in one direction and is a function of y and t only:

$$v_x = v_x(y, t) \qquad 7.5.1$$

Equation 7.5.1 satisfies the continuity equation identically. Substituting this assumption into the y-direction momentum equation shows that the pressure is

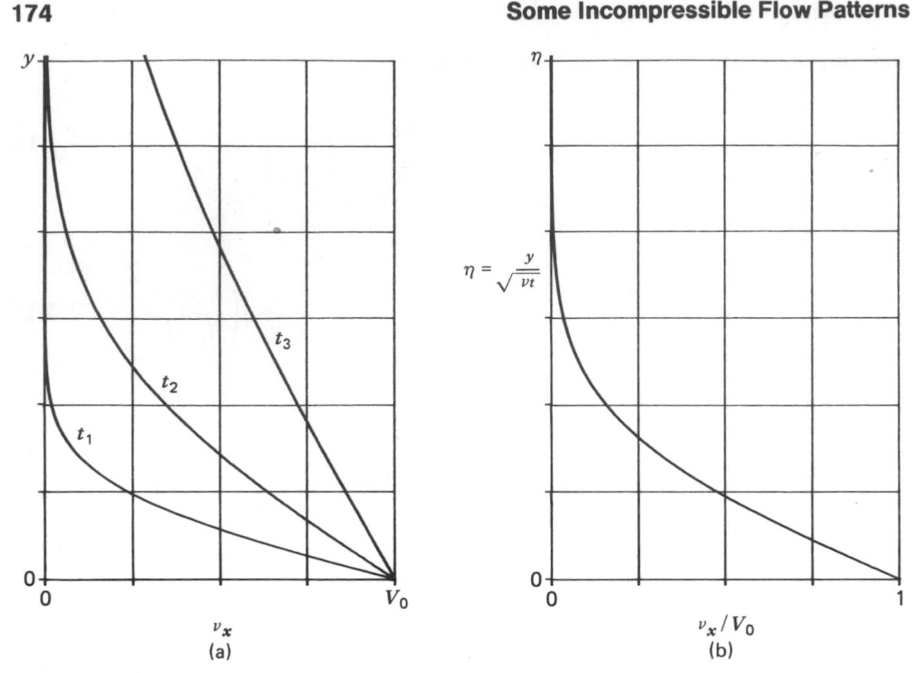

Figure 7-8 Impulsive motion of a flat plate in its own plane (the Rayleigh problem): (a) velocity profile at various times, (b) velocity profile in similarity variables.

governed by the hydrostatic equation

$$0 = -\frac{\partial p}{\partial y} - \rho g$$

Integration gives 7.5.2

$$p = -\rho g y + p_0$$

where we take the pressure to be uniform on the plate at a value p_0. Although 7.5.2 shows an infinite negative pressure at $y \to \infty$, we realize that the finite extent cf any apparatus would prohibit this result. The important point is that p is not a function of x.

The flow is governed by the x-momentum equation, which simplifies to the form

$$\rho \frac{\partial v_x}{\partial t} = \mu \frac{\partial^2 v_x}{\partial y^2} \tag{7.5.3}$$

Acceleration of a fluid particle is the result of an imbalance in the shear forces acting on the particle. We also note that ρ and μ do not enter the problem independently but only in the combination μ/ρ, which is by definition the kinematic viscosity,

$$\nu \equiv \frac{\mu}{\rho}$$

We will find that ν is a much more important parameter in fluid mechanics than the absolute viscosity μ (except in low-Reynolds-number flows).

Mathematically 7.5.3 is also called the *heat* or *diffusion equation*, as it is important to mass diffusion and heat conduction. It is a parabolic differential equation. The proper conditions to prescribe for parabolic equations are an initial condition for all space,

$$v_x(y, t = 0) = 0 \qquad 7.5.4$$

and boundary conditions at two positions in space for all time,

$$v_x(y = 0, t) = V_0 \qquad 7.5.5$$

$$v_x(y \to \infty, t) = 0 \qquad 7.5.6$$

Equation 7.5.5 is the no-slip condition at the plate surface.

The mathematical solution of 7.5.3 is of interest in itself, since it is an example of a similarity solution. A *similarity solution* is one where the number of independent variables in a partial differential equation is reduced by one; in this case two independent variables reduce to one. The similarity variable for this problem (a partial differential equation *problem* means the equations and boundary conditions considered together) is

$$\eta = \frac{y}{2\sqrt{\nu t}} \qquad 7.5.7$$

Since many different values of y and t give the same η, and the answer depends only upon η, there are many points y and t that have a "similar" answer (in this case exactly the same answer).

In order to transform 7.5.3 we note that

$$\frac{\partial(\ )}{\partial t} = \frac{d(\ )}{d\eta}\frac{\partial \eta}{\partial t} = -\frac{y t^{-3/2}}{4\sqrt{\nu}}\frac{d(\ )}{d\eta} = -\frac{1}{2}\frac{\eta}{t}\frac{d(\ )}{d\eta}$$

$$\frac{\partial(\ )}{\partial y} = \frac{d(\ )}{d\eta}\frac{\partial \eta}{\partial y} = \frac{1}{2\sqrt{\nu t}}\frac{d(\ )}{d\eta}$$

$$\frac{\partial^2(\ )}{\partial y^2} = \frac{\partial}{\partial y}\left(\frac{\partial(\ )}{\partial y}\right) = \frac{1}{2\sqrt{\nu t}}\frac{d}{d\eta}\left(\frac{1}{2\sqrt{\nu t}}\frac{d(\ )}{d\eta}\right) = \frac{1}{4\nu t}\frac{d^2(\ )}{d\eta^2}$$

For the dependent variable we introduce the symbol f and note that f is assumed to be a function of η alone:

$$\frac{v_x}{V_0} \equiv f(\eta) \qquad 7.5.8$$

Substituting the relations above into 7.5.3 produces an ordinary differential equation as promised,

$$f'' + 2\eta f' = 0 \tag{7.5.9}$$

If the choice of a trial similarity variable does not produce an ordinary differential equation, then the trial is unsuccessful. As a matter of fact Bluman and Cole (1969) have given (in principle) all the similarity variables for the diffusion equation.

Even if a similarity variable is found for a particular differential equation, it may not work. In order to be applicable, the similarity variable must make the boundary conditions on the original problem transform so that the new problem makes sense. The boundary conditions for the present problem transform as follows:

$$\begin{aligned} v_x(y, t = 0) = 0 \quad &\Rightarrow \quad f(\eta \to \infty) = 0 \\ v_x(y = 0, t) = V_0 \quad &\Rightarrow \quad f(0) = 1 \\ v_x(y \to \infty, t) = 0 \quad &\Rightarrow \quad f(\eta \to \infty) = 0 \end{aligned} \tag{7.5.10}$$

When the original problem, consisting of a partial differential equation and three boundary conditions, changes into one consisting of a second-order ordinary differential equation, we can allow only two boundary conditions. Accordingly, our similarity variable is successful, since it has collapsed two boundary conditions of 7.5.10 to the same thing. The three original conditions become only two conditions in the transformed variables.

Equation 7.5.9 may be integrated once, considering f' as the dependent variable. This yields

$$f' = C_1 \exp(-\eta^2) \tag{7.5.11}$$

Integrating again,

$$f = C_1 \int_0^{\eta} \exp(-\xi^2)\, d\xi + C_2$$

The integral above looks simple but is in fact not an elementary function. It is a "higher" function called the error function. The error function is defined as

$$\operatorname{erf}(\eta) \equiv \frac{2}{\sqrt{\pi}} \int_0^{\eta} \exp(-\xi^2)\, d\xi \tag{7.5.12}$$

One can see from the definition that erf(0) = 0, and the factor in front of the integral has been chosen so that erf(∞) = 1. The final answer, which satisfies the boundary conditions, is

$$f(\eta) = 1 - \operatorname{erf}(\eta)$$

$$\frac{v_x}{V_0} = 1 - \operatorname{erf}\left(\frac{y}{2\sqrt{\nu t}}\right) \qquad 7.5.13$$

This answer is graphed in Fig. 7-8. Before discussing it, we also compute the shear stress and the vorticity,

$$\omega_z = -\frac{\partial v_x}{\partial y} = -\frac{V_0}{2\sqrt{\nu t}}\frac{df}{d\eta}$$

$$\omega_z = \frac{V_0}{\sqrt{\pi \nu t}} \exp(-\eta^2) \qquad 7.5.14a$$

and

$$\tau_{yx} = \mu\frac{\partial v_x}{\partial y} = -\mu\omega_z$$

$$\tau_{yx} = -\frac{\mu V_0}{\sqrt{\pi \nu t}} \exp(-\eta^2) \qquad 7.5.14b$$

(Note that the shear stress is a function of the similarity variable only if it is scaled by the square root of the time; that is, $\tau_{yx}\sqrt{\pi \nu t}/\mu V_0$ is a nondimensional shear stress that is a function of η alone.) We are most interested in the shear stress on the plate itself. It is

$$\tau_{yx}(0) = \frac{\mu V_0}{\sqrt{\pi \nu t}} \qquad 7.5.15$$

The stress is infinite at the instant the plate starts moving and decreases as $1/\sqrt{t}$.

The velocity profile shows that the influence of the plate extends to infinity immediately after the plate starts moving. At large distances the error function vanishes exponentially [actually $\operatorname{erf} \eta \sim \eta^{-1}\exp(-\eta^2)$ as $\eta \to \infty$], but there is still a minute viscous influence throughout the flow. We can rationalize the influence at infinity by considering the molecular model of gas viscosity. Molecules that collide with the plate absorb some extra momentum before returning to the fluid. Although for the most part the molecules collide with other molecules several times before getting very far from the plate, in principle there is the possibility of molecules traveling to infinity without a collision. Nevertheless, the most important influence of viscosity is confined to a region near the plate.

Let us consider the place where the velocity has dropped to 1% of the plate value (v_{0x}/V_0 is 0.01). Figure 7-8 shows that η is about 1.8 at this position, which we denote by $y = \delta$. Then

$$\eta\left(\frac{v_x}{V_0} = 0.01\right) = 1.8 = \frac{\delta}{2\sqrt{\nu t}}$$

and

$$\delta = 3.6\sqrt{\nu t} \qquad 7.5.16$$

The diffusion of viscous effects is a basic phenomena in fluid mechanics and we frequently need to estimate how far diffusion has progressed. The quantity δ, called the viscous-diffusion distance, is useful for this purpose. The major effects of viscosity are contained between the wall and $y = \delta$. Notice that diffusion slows down as time goes on, that it depends upon the kinematic (not the absolute) viscosity, and that it is independent of the plate velocity. In terms of viscous diffusion, air is more viscous than water by a factor of about 10. We amplify this remark by computing the diffusion length after one minute for air,

$$\delta = 10.8 \text{ cm} \qquad (\nu = 0.150 \text{ cm}^2/\text{sec})$$

and for water,

$$\delta = 2.8 \text{ cm} \qquad (\nu = 0.010 \text{ cm}^2/\text{sec})$$

In general, viscous diffusion is a slow process; in most flow fields a particle travels a great distance in a minute. (Only $\frac{1}{3}$ sec is required for a particle to go from the nose of a Boeing 747 to the tail when the flight speed is 500 mph.)

The mathematical solution to this problem was first given by Stokes (1851, Note B). We now call it the Rayleigh problem because Rayleigh (1911) used the results in a creative way to derive a skin-friction law. The problem concerned the skin friction for laminar flow over a flat plate of length L. Lanchester had given a skin-friction law in his *Aerodynamics* (1907), but Rayleigh sought a physical derivation. He argued that we should watch the plate move through a stationary fluid and imagine that we are looking at one point in space as the plate moves by. When the leading edge of the plate passes our vantage point, it is similar to the initial instant of the infinite plate motion. Further back along the plate, the shear stress decreases because the particles have been in "contact" with the plate for a longer time. Rayleigh proposed that the flow at any position on the finite plate is the same as that on an implusively started infinite plate after a time t equal to the time since the leading edge passed the vantage point. The key to Rayleigh's argument was the idea of replacing t by x/V_0, where x is the distance from the leading edge. If we make this substitution in 7.5.15, the local shear stress on the finite plate becomes

$$\tau = \frac{\mu V_0}{\sqrt{\pi\nu}}\sqrt{\frac{V_0}{x}} \qquad 7.5.17$$

The total drag force is found by integrating over the length of the plate. The resulting formula is not very accurate, but it has the proper trends with all the parameters. Perhaps more important than the drag formula itself was the argument Rayleigh used to obtain it. The same argument is frequently employed in order to estimate the proper trends for phenomena involving viscous diffusion. For many purposes it does not matter if the magnitude is exactly right.

In closing this section we take note that the Rayleigh problem also applies to the situation of a stationary plate when the bulk fluid is started impulsively with a uniform velocity. The two problems are related by a Galilean transformation.

7.6 CONCLUSION

This chapter has illustrated the analytical approach to fluid-flow problems. An essential ingredient at the outset of any analysis is an assumption about how the flow varies in space and time. Unless such assumptions are made, the Navier–Stokes equations are too complicated to solve.

We have no guarantee that our solution will actually occur in reality, nor can we expect a unique answer to a general problem. The Navier–Stokes equations are known to produce several solutions for exactly the same boundary conditions.

The most important example in this chapter is probably Rayleigh's flat-plate problem. The impulsive motion of a plate in an infinite fluid is the simplest example of viscous diffusion. The resulting estimate for the depth of penetration of viscous diffusion as a function of time is often taken as a basis for thinking about diffusion in more complicated problems. A second aspect of this problem is the way Rayleigh used it to find a drag formula for a wing moving through a still fluid. The steady flow in a wing-fixed coordinate system is a transient flow in a ground-fixed system. The initial-value problem for the unsteady flow has the same character as the steady flow when one identifies x and Ut (though this analogy is only approximate). A third important aspect of this problem is that it introduces the idea of similarity. Since similarity reduces the number of independent variables, it is very often used in field problems with two or more variables.

PROBLEMS

7.1 (A) Consider the annulus formed between a rod of radius r_0 and a tube of radius r_1. Find the velocity profile for Couette flow where the inner rod is rotated with speed Ω.

7.2 (A) For the same geometry as in Problem 1, find the velocity profile if the rod is pulled in the axial direction at a speed $v_z = V_0$.

7.3 (A) For the same geometry as in Problem 1, find the velocity profile if a pressure gradient $\Delta p/L$ is applied in the direction of the rod axis.

7.4 (A) Show that the linear sum of velocity profiles in Problems 1, 2, and 3 represents the flow in an annulus with an imposed pressure gradient and a rotating, translating rod.

7.5 (B) A vertical pipe of radius r_0 has a film of liquid flowing downward on the outside. Find the velocity profile with the flow rate Q as a given constant.

7.6 (A) Let the pipe in Problem 5 turn with a speed Ω. Find the velocity profile for this situation.

7.7 (C) Consider an arbitrary region with an incompressible flow. Derive an "engineering Bernoulli equation" from 5.16.2 by introducing the gravity potential $F_i = \partial_i \phi$ and the concepts of flow work and boundary work (see Section 5.16). Rewrite the equation so that $\frac{1}{2}v^2 + \hat{\phi} + p/\rho$ is the quantity convected across the surface. Which term represents the head loss? Which term represents the work of a rotating shaft crossing the control region? Which term represents the work of a translating shaft? What form does the equation take for a fixed region? Let the flow in the fixed region be steady where the flow crosses the boundary, but periodic (average plus an oscillatory component) within the region. What is the time-averaged form of the equation?

7.8 (B) Consider a wire with an infinitely small radius, accelerated impulsively to a speed $v_z = V_0$ along its axis. Find the similarity variable and the differential equation that governs the velocity profile in the fluid surrounding the wire.

7.9 (B) Consider the Rayleigh problem, but allow the plate velocity to be an arbitrary function of time, $U(t)$. By differentiation show that the shear stress $\tau = \mu \, du/dy$ obeys the same diffusion equation that the velocity does. Suppose that the plate is moved in such a way as to produce a constant surface shear stress. What are the velocity profile and the surface velocity for this motion?

8 Dimensional Analysis

Up to this point our attention has been focused on describing physical concepts and formulating the laws that govern them. The existence of measurement methods—procedures to assign a number to a variable—was taken for granted. In this chapter we shall study the measurement and dimensional nature of physical variables. The major fact used in dimensional analysis is that no natural or fundamental units of measure exist for the physical variables. There are, of course, many universal constants in physics—the charge of an electron, Planck's constant, the gravitational constant between attracting masses, the speed of light, and so on. The point is that these constants are not relevant to all physical processes. The charge on an electron is not a fundamental unit to measure the current in an electric motor. The speed of light is not a fundamental unit to measure the speed of a water wave. Lacking any universally relevant measuring units in the physical world, we are obliged to construct our own scales. Our measuring scales are arbitrary inventions and as such can play no essential role in physical processes. If we change the size of the length unit, all variables involving length must increase or decrease in an appropriate way. By considering the dimensional aspect of a problem alone, one can simplify the problem and find out important information. This can be done even if the problem is too complex to allow us to analyze it in detail.

Dimensional analysis is a confusing and mystifying subject to many beginning students, and even among some practitioners it has an intimidating reputation. This attitude exists in spite of the fact that dimensional analysis is of great importance to the theoretician as well as to the experimentalist. The more difficult and complicated the problem, the greater value dimensional analysis has in simplifying and understanding it. To be proficient in fluid mechanics one must appreciate dimensional analysis and understand its limitations as well as its powers.

Some difficulties in understanding dimensional analysis arise because of different viewpoints that may be taken toward the subject; however, most of the confusion arises because an experienced worker will implicitly use extra information in applying dimensional analysis to a problem. The student, quite naturally, has difficulty in distinguishing when extra information is being used and when the principles of dimensional analysis are being applied. Of course, there is nothing wrong with introducing supplementary information into a problem. The important thing is that it be done in a straightforward and obvious way so that the critical assumption is explicitly known.

Lest the reader become discouraged at this point, let it be stated that dimensional analysis is not full of mystery and confusion. The subject itself is well founded, and with a little work one can master its principles.

8.1 MEASUREMENT AND DIMENSIONS

There are two major classes of quantities: things that are counted and things that are measured. A quantity that is counted—for instance, a number of molecules—is dimensionless. A quantity that is measured typically has a dimension associated with it.

The most elementary form of measurement is simply a comparison of the object we want to measure with a defined scale. A length of interest is compared with a meter scale, for example. The scale defines a unit, and the prescribed measuring method must include a procedure to extrapolate and interpolate. Sometimes even the simplest comparison requires an external device. A balance is a mechanism to compare two masses; a clock is a device that allows us to compare two times.

The key element of any measurement is the definition of the unit. We cannot begin to measure length, for example, without first defining a foot, a cubit, a light year, or some other length unit.

All the length concepts that have been defined have a common property called the *length dimension*. We can give the term length dimension a symbol, L, and with a little imagination L can be considered as a number. To do this we need to imagine that there is a basic, primitive, absolute, fundamental, universal length unit that is important to all physical processes. The length dimension L is a number that gives the size of a man-made length unit in terms of the basic absolute length unit. This viewpoint requires imagination because, in fact, nature does not have any such universal length unit and therefore we cannot assign such a value to L. This point need not cause concern, as L is used only for abstract purposes. We are never required to find its numerical value. The important characteristic of L is that it measures a unit of length. Furthermore, since L is a real number, we can perform algebraic operations on it without any conceptual difficulties.

If a man-made unit has an absolute size, then any length quantity also has absolute size. For a length that is l man-made units long, the absolute length is

$$\underset{\text{absolute size}}{\hat{l}} = \underset{\text{number of units}}{l} \times \underset{\text{absolute size of the measuring unit}}{L} \tag{8.1.1}$$

For example, an absolute length might be 5.5 meter units times the absolute size of the meter unit. The purpose of the formula above is to introduce a distinction between the magnitude of a variable, as symbolized by l, and its dimension (or physical nature), as symbolized by L.

In a similar manner a mass quantity would be written as

$$\hat{m} = m \times M$$

or a time quantity

$$\hat{t} = t \times T$$

In general, the circumflex means an absolute size, a lowercase letter means the size in man-made units, and the corresponding capital letter means the size of the unit in terms of a universal scale, that is, the dimension. Exceptions to this in the remainder of the chapter will be obvious and will not cause any confusion.

A alternate point of view, which some may prefer, is to consider two measuring units of different sizes. The size of the variable in the new unit is $\hat{l}$, the size in the old unit is l, and L is the ratio of the new measuring unit to the old measuring unit. Equation 8.1.1 then has the interpretation

$$\hat{l} = l \times L \qquad 8.1.1$$

$\hat{l}$	=	l	×	L
size of variable in terms of new measuring scale		size of variable in terms of old measuring scale		ratio of size of new unit to old unit

Note the L has the same value no matter what third scale of units is used to measure the old and the new units. This viewpoint avoids the artifice of a fictitious universal length unit.

Physical equations are relations between the l-type variables. When we substitute numbers into an equation they are l-type numbers. However, physical equations must be valid in any system of measurement units we choose (this is one of the fundamental hypotheses of dimensional analysis). Thus, the equations must also be valid if the absolute ($\hat{l}$-type) variables are substituted into them. We shall need these facts in proving the pi theorem.

In the early history of dimensional analysis the word dimension and the symbols M, L, and so on (which, incidentally, were introduced by Maxwell in 1871), had a more abstract and vague meaning than we have defined. The term "dimension" was used to refer to some ultimate physical nature of a concept. As time went on, it was found that physical concepts do not have any ultimate dimensions. For our purposes we take the restricted viewpoint that a dimension is a number as defined by 8.1.1 or a similar equation. We shall always use the word "dimension" to mean the absolute size of a measuring unit (or a scale-change ratio if you prefer).

Not all variables need have their own measuring units. Velocity, defined as the rate of change of a length, has a magnitude which changes if we change either the length or the time unit. The absolute size of a velocity $\hat{v}$ is related to the size v in a certain system of units by the equation

$$\hat{v} = vLT^{-1}$$

As another example, Newton's law $\mathscr{F} = m\, dv/dt$ implies that any force (measured in an M, L, T system) obeys the relation

$$\hat{\mathscr{F}} = \mathscr{F} M^1 L^1 T^{-2}$$

The dimensions that we choose as a basis to measure other quantities are called *primary dimensions*. A *secondary dimension* is expressible as a product of power of the primary dimensions, such as $V = LT^{-1}$ or $F = MLT^{-2}$.

In a later section it is shown that for any physical quantity x the absolute size of $\hat{x}$ is given by

$$\hat{x} = xM^a L^b T^c \qquad 8.1.2$$

The meaning of the symbols is analogous with 8.1.1: M is the size of the mass scale, L is the size of the length scale, and T is the size of the time scale. The exponents a, b, and c are always fractions, but theoretically they could be irrational numbers. (The fact that only fractions occur as exponents is a result of the structure of physics. In constructing physical theory we have not defined, at least not yet, a physical concept that is a combination of M, L, and T raised to an irrational power.) In writing 8.1.2 we have implied that M, L, and T are sufficient to express any other physical variable. A more general statement would be given by Bridgman's equation,

$$\hat{x} = xP_1^d P_2^e P_3^f \cdots \qquad 8.1.3$$

The symbol P stands for a primary dimension. The difference between 8.1.2 and 8.1.3 emphasizes that we must choose which dimensions are primary and that M, L, and T are not the only possible choices.

One of the key questions in dimensional analysis is the minimum number of primary dimensions that are required in 8.1.3. The answer is three. Three primary dimensions are sufficient to express the dimensions of all physical variables. This statememt applies to mechanical, electrical, and thermodynamic variables all taken at the same time. As examples consider the following quantities and their dimension in the M, L, T system:

$$\text{temperature} = ML^2/T^2$$

$$\text{electric charge (statcoulombs)} = M^{1/2}L^{3/2}/T$$

$$\text{thermal conductivity} = 1/LT$$

Here we interpret temperature as an energy per mole, and thus the perfect-gas law is $T = pV/N$. As a practical matter, however, all standard unit systems employ temperature as a primary dimension. Take note that the specification of three primary dimensions is only a sufficient number. It is possible and practical to take four or even five primary dimensions. When this is done, something must be inserted into the variable list to account for the redundancy. This topic will be discussed in more detail in Section 8.5.

8.2 VARIABLES AND FUNCTIONS

Relationships between physical variables are expressed by mathematical functions. A mathematical function, as you know, relates one dependent variable to one or more independent variables. Physical equations also contain "constants", which come from boundary conditions, appear in the governing laws, or arise from thermodynamic data. For example, consider the incompressible flow over a sphere (Fig. 8-1). Let the pressure p at any point $\mathbf{x}$ in the flow be the dependent variable. Mathematically,

$$p = f(\mathbf{x}, U, \rho, \mu, d, p_\infty) \tag{8.2.1}$$

In this equation, U is the free-stream velocity, ρ the density, μ the viscosity, d the diameter, and p_∞ the free-stream pressure.

The object of dimensional analysis is to group several variables together to form a new variable that is nondimensional. In our present example, the ratio

$$\Pi = \frac{p}{\rho U^2} \tag{8.2.2}$$

is a nondimensional pressure. Since Buckingham's (1867–1940, American Physicist) statement of the "pi-theorem", it has been tradition to use Π (the mathematical symbol for product) as a symbol for a nondimensional variable. The exponents of the primary dimensions of a *nondimensional variable* are all zero; that is, Bridgman's equation for a nondimensional variable is simply

$$\hat{\Pi} = \Pi M^0 L^0 T^0$$

$$\hat{\Pi} = \Pi \tag{8.2.3}$$

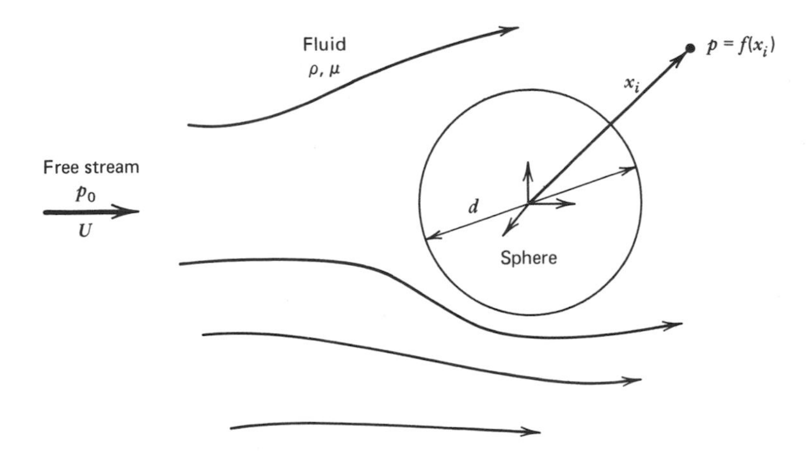

Figure 8-1 Flow over a sphere.

The value of a nondimensional variable is completely independent of the measuring units. In this sense it is a universal or natural variable, since it does not involve any man-invented scales. The combination ρU^2 is regarded as a natural measuring scale for the pressure p. A natural scale may change from one problem to another; thus ρU^2 is not always the natural scale for pressure.

Dividing 8.2.1 by ρU^2 yields

$$\Pi \equiv \frac{p}{\rho U^2} = \frac{1}{\rho U^2} f(\mathbf{x}, U, \rho, \mu, d, p_\infty) \qquad 8.2.4$$

The pi theorem, to be stated formally in the next section, proves that it is possible to reorganize 8.2.4 and group variables together into a new function that contains only nondimensional variables:

$$\frac{p}{\rho U^2} = F\left(\frac{\mathbf{x}}{d}, \frac{\rho d U}{\mu}, \frac{p_\infty}{\rho U^2}\right) \qquad 8.2.5$$

This equation contains only four variables, compared to the seven required in the original dimensional form 8.2.1. This is one of the most useful aspects of dimensional analysis. When a function is expressed in nondimensional variables, the number of variables is less that when the same relation is expressed in dimensional variables.

The importance of reducing the number of variables is often illustrated by the following analogy. Consider how a function would be represented graphically. A function of two variables can be represented by a single line graphed on a single page. A function of three variables requires several lines, one for each value of the third variable, but it still fits on one page. Functions of four variables require a book of graphs, and five variables a shelf full. For six or seven variables one would need a library of books to contain the required information. Considered in this light, reducing the number of variables from seven to four is a great simplification. To do the experiments or to make the calculations for a single book of graphs is a reasonable task, but to fill the library shown in Fig. 8-2 is an undertaking to which you could dedicate a lifetime.

Let us consider in more detail the types of variables that occur in a physical function. The dimensions of measured items must obey Bridgman's equation. For example, a pressure measured in decibels is not a proper form for changing the scale unit. Dimensional analysis cannot be applied to an equation where decibels are used.

Everything that is measured must be included in the variable list, since the value of each quantity depends upon the choice of measurement units. Distinctions between independent variables, boundary values, parameters, physical constants, universal constants, and dimensional constants are made according to the physical role that the variable plays in the function. In dimensional

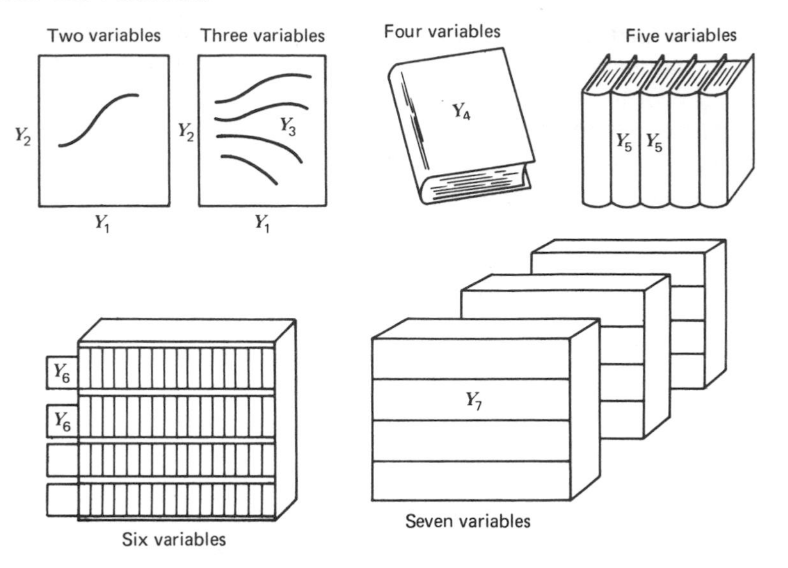

Figure 8-2 Records of information expands geometrically as the number of variables increases.

analysis these distinctions are unimportant, and all such items must be included in the list of variables. If a process involves the gravitational attraction force F_{12} between two masses M_1 and M_2 a distance r apart, the gravitational constant k ($F_{12} = kM_1M_2/r^2$) must be included in the variable list. The numerical value of k depends upon the chosen units of measurement.

The pi theorem deals with functions that describe physical processes. Furthermore, the functions must be in a proper form where all variables are shown. Frequently, practicing engineers use equations where specific units are assumed. For example, the speed of sound in air is given by

$$a = 49.1\sqrt{T} \qquad 8.2.6$$

where a is in ft/sec and T is in °R. The fact that 49.1 is a dimensional constant is not noted explicitly. The proper form of this equation is

$$a = (\gamma g_c RT)^{1/2} \qquad 8.2.7$$

Unhappily, there are some equations in use today (for pipe friction and channel friction) that are dimensionally inconsistent. These equations are mere correlations of experimental data and must be used only for the fluids employed in the experiments.

A popular example of an improper physical function is the sum of the equations for the position and for the velocity of a falling body, $s = \frac{1}{2}gt^2$ and $v = gt$. This gives

$$v = gt + s - \tfrac{1}{2}gt^2$$

which implies that $v = f(g, s, t)$. This is not true. We cannot substitute independent numbers for g, t, and s to retrieve values of v. A mathematician would not accept this as a function in the first place.

For purely physical and chemical processes, we know the governing laws and physical constants even though the situation may be much too complicated to solve. The task of dimensional analysis in these problems is to find the appropriate nondimensional variables, and thus the simplest answer. Dimensional analysis has also been applied to physical–biological processes. For example, how does the speed of a swimming fish depend upon the size and other physical properties of the fish? How does the oxygen consumption of a four-legged animal depend upon its size (and other characteristics) and the running speed? In formulating problems such as these, one must consider that a physical–biological constant—say the strength of muscle per unit area—may be a necessary item in the variable list. Some problems may need more than one constant—say the response time as well as the strength. The results of dimensional analysis of these types of problems need to be carefully tested to verify that the assumed variables and dimensional constants are indeed proper.

Some biological–social problems are clearly outside the realm of dimensional analysis—for example, how much study time is needed to make a grade of 98 on a German exam, or how much TV time is required to get 50% of the vote. Other questions may or may not be subject to dimensional analysis—for example how the spread of flu through the population depends upon the population density, the climate, and the number of vaccinations, or how the growth of a cancer depends upon the life span of the animal, the intensity of the exposure to a carcinogenic, and the size of the animal. The existence of dimensional constants in many of these processes must be speculated.

8.3 THE PI THEOREM

The pi theorem tells how many nondimensional variables are required for a given set of dimensional variables. It is based on two assumptions: (1) all variables obey Bridgman's equation 8.1.2, and (2) proper functions expressing a physical result are valid irrespective of the scale units for the primary dimensions. Changing the size of the units changes the size of the answer in a consistent manner.

Before stating the pi theorem we need to introduce the concept of the dimensional matrix. The dimensional matrix is formed by listing the exponents (a, b, and c in 8.1.2) of the primary dimensions of each variable. The sphere problem has the dimensional matrix

	p	p_∞	x	μ	d	ρ	U
M	1	1	0	1	0	1	0
L	−1	−1	1	−1	1	−3	1
T	−2	−2	0	−1	0	0	−1

The purpose of the matrix is to check for linear independence of the dimensions of the variables in terms of the chosen primary dimensions. This is done by finding the rank of the matrix.

To do so one must check the determinant of all possible square submatrices, beginning with the largest, until one is found that is nonzero. *The* rank *of the matrix is the size of the largest square submatrix that has a nonzero determinant.* The rank of the matrix above is 3, since the determinant of the last three columns is nonzero. If the determinants of all possible 3×3 submatrices are zero, one then proceeds to check all 2×2 submatrices until one with a nonzero determinant is found. The rank of the dimensional matrix tells how many fewer variables will occur when a function is expressed in nondimensional variables.

We now give a statement of the *pi theorem*: Given a proper physical function of dimensional variables that are n in number,

$$x_1 = f(x_2, x_3, x_4, \ldots, x_n) \qquad 8.3.1$$

Furthermore, all the variables in this function obey Bridgman's equation

$$\hat{x}_i = x_i P_1^a P_2^b P_3^c \qquad 8.3.2$$

where the P's are primary dimensions and a, b, and c are exponents. Under these assumptions it is possible to organize the original variables into nondimensional forms—that is, find some values for α's such that

$$\Pi = x_1^{\alpha_1} x_2^{\alpha_2} \cdots x_n^{\alpha_n} \qquad 8.3.3$$

Moreover, when the original function is expressed in nondimensional variables, it is simpler in that it contains only $m = n - r$ variables:

$$\Pi_1 = f(\Pi_2, \Pi_3, \ldots, \Pi_m) \qquad 8.3.4$$

where r is the rank of the dimensional matrix.

The name pi theorem comes from Buckingham's (1914) discussion of dimensional analysis. In this paper he explained the repeating-variable method of forming the Π groups and stated how many nondimensional variables would occur. Buckingham felt that the principle of dimensional homogeneity, which had been initiated by Fourier (1822), was the cornerstone of the method (Euler's writings also touch on the subject of measurement and dimensions). Buckingham's paper stimulated the subject, and his method was widely adopted. However, he was not the first to publish a pi theorem. Independently, the French engineer A. Vashy (1892) and the Russian physicist D. Riabouchinsky (1911) published statements that were equivalent to the pi theorem. Even earlier, Rayleigh (1879), although he did not give a pi theorem, formalized an indicial method, or power-product method, of finding nondimensional relationships. This method still finds favor among many workers.

The pi theorem was not adequately proved in the early papers, and some confusion remained. It concerned the number of pi variables required, $m = n - r$. The first statements of the theorem did not use the rank r of the dimensional matrix. Instead, they said the number of pi's was equal to the number of dimensional variables minus the number of "necessary" primary dimensions. The word necessary was vague, because a method for determining what was necessary was not provided. Bridgman (1922) pointed out that for some problems using the *MLT* system gives a different answer than using the *FLT* system. The variables of Bridgman's example could be expressed by two dimensions in one primary system, while three were required in the other. Van Driest (1946) gave the solution by pointing out that the number of pi variables will be different if the dimensions of the variables are not independent in terms of the chosen primary dimensions. By using the rank of the dimensional matrix in our statement of the pi theorem, we have automatically tested for independence. Further information on the history of dimensional analysis is in Macagno (1971).

There are several methods of finding a set of pi variables for a given problem. As one gains more and more experience, short cuts are found, and finally a trial-and-error method becomes the quickest. We first explain a fairly formal method due to Buckingham (1914), which is commonly used in elementary texts.

The first step is to choose r repeating variables from the x's, where r is the rank of the matrix. The repeating variables must be linearly independent, so the submatrix of their dimensional exponents must have a nonzero determinant. In our example ρ, d, and U meet this condition and will be chosen as the repeating variables. The repeating variables will occur in all the pi variables. Therefore, if we want the dependent variable to occur in only one pi variable, it should not be chosen as a repeating variable.

Consider the r repeating variables plus one of the remaining x variables. Since there are only r independent dimensions, a nondimensional variable may be formed from these $r + 1$ variables. Taking the pressure p and the repeating variables ρ, d, and U, we form the first pi variable (using absolute sizes), that is, we construct $\hat{\Pi}_1$ as

$$\hat{\Pi}_1 = \hat{p}(\hat{\rho})^{\alpha}(\hat{d})^{\beta}(\hat{U})^{\gamma} \qquad 8.3.5$$

We seek values of the exponents α, β, and γ that make 8.3.5 nondimensional. Substituting 8.1.2 for each variable in 8.3.5 gives

$$\Pi_1 M^0 L^0 T^0 = pML^{-1}T^{-2} \cdot \rho^{\alpha}(ML^{-3})^{\alpha} \cdot d^{\beta}(L)^{\beta} \cdot U^{\gamma}(LT^{-1})^{\gamma}$$

$$= p\rho^{\alpha} d^{\beta} U^{\gamma} \cdot M^{1+\alpha} \cdot L^{-1-3\alpha+\beta+\gamma} \cdot T^{-2-\gamma} \qquad 8.3.6$$

This equation must hold for all choices of the measuring scales M, L, T;

therefore, the exponents of M, L, and T must be zero. Hence 8.3.6 becomes

$$\Pi_1 = p\rho^{\alpha} d^{\beta} U^{\gamma} \qquad 8.3.7$$

Equating the exponents of M, L, and T in 8.3.6 to zero yields a set of equations that determines the values of α, β, and γ:

$$\begin{aligned} &M \text{ exponent:} && 0 = 1 + \alpha && \Rightarrow \quad \alpha = -1 \\ &T \text{ exponent:} && 0 = -2 - \gamma && \Rightarrow \quad \gamma = -2 \\ &L \text{ exponent:} && 0 = -1 - 3\alpha + \beta + \gamma && \\ & && 0 = -1 + 3 + \beta - 2 && \Rightarrow \quad \beta = 0 \end{aligned} \qquad 8.3.8$$

(Note that the solution of the nonhomogeneous system above is guaranteed if the determinant of the coefficients of α, β, γ is nonzero. This is the same determinant used to show that the rank of the dimensional matrix was 3.) Thus, we find the first nondimensional variable is

$$\Pi_1 = \frac{p}{\rho U^2} \qquad 8.3.9$$

Taking each remaining x variable in turn, together with the r repeating variables, will produce the required $n - r$ nondimensional variables. They turn out to be

$$\Pi_2 = \frac{p_0}{\rho U^2}, \qquad \Pi_3 = \frac{\mathbf{x}}{d}, \qquad \Pi_4 = \frac{\mu}{\rho d U} \qquad 8.3.10$$

These are the variables that were listed in 8.2.4 as the nondimensional form of our result.

There is another method, which is somewhat simpler, called the method of scales. In this method we form units for the primary dimensions by using the repeating variables. This may be done by inspection in most instances. These units are "natural" measuring units for the specific problem. In the sphere problem the natural length unit is the diameter of the sphere. So we use the freedom to define a length unit to set

$$L = d \qquad 8.3.11$$

A natural mass scale is formed from the sphere diameter and the fluid density. Let

$$M = \rho d^3 \qquad 8.3.12$$

The time scale is formed using the fluid velocity and sphere diameter:

$$T = d/U \qquad 8.3.13$$

In essence we are organizing the repeating variables into groups so that the pi groups can be formed by inspection. Now the first nonrepeating variable p has the dimensions

$$p \qquad \frac{M}{LT^2}$$

In order to cancel these dimensions we divide by the mass scale and multiply by the length scale and the time scale squared. Hence

$$\Pi_1 = p \cdot \frac{d(d/U)^2}{(\rho d^3)} = \frac{p}{\rho U^2} \qquad 8.3.14$$

The process is continued using each nonrepeating variable in turn. The scale method is recommended only when the rank of the matrix is equal to the number of primary dimensions.

8.4 PUMP OR BLOWER ANALYSIS

Let us consider the analysis of a pump designed for use with incompressible liquids. The analysis is also valid, however we shall not prove it, for fans and blowers that transport gases. A schematic of the pump and a typical test setup is shown in Fig. 8-3. Before we list the variables we must agree on the choice of primary dimensions. We choose three primary dimensions: M, L, and T.

The second task is to list the variables that enter the problem. An accurate list is essential to obtain the correct answer. Leaving out a variable leads to an erroneously simple result, while including an extraneous variable usually leads

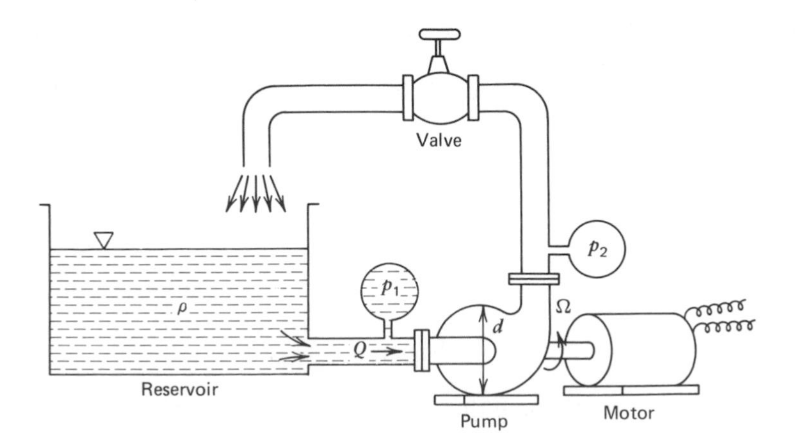

Figure 8-3 Test setup for pump performance.

to an extra pi group (but sometimes to a problen where it is impossible to nondimensionalize the extra variable). As an aid in listing the variables, let us imagine how a pump test would be performed. We want to be particularly careful about considering all the things that might be changed to cause a different flow situation in the pump. The size of the machine is characterized by the single dimension d. This may be taken as the impeller diameter or any other convenient dimensions. We assume that if the size of the machine changes, all of its dimensions change in proportion to d. A motor is coupled to the pump and turns the impeller at a constant speed. As test conditions change, it may be necessary to adjust the motor to maintain the constant speed. We assume the motor is of such a type that this may be done. With these stipulations, the geometry of the pump and the motion of all internal parts are fixed by the variables d and Ω.

The pump draws liquid in from a reservoir at a pressure p_1 and discharges it at a pressure p_2. A valve on the outlet is used to simulate the piping resistance and change the back pressure. As the back pressure changes, the volume flow through the machine also changes. In essence a series of tests with different valve positions gives the pressure rise vs. flow rate for the pump under fixed geometry and speed conditions. Changing the valve position will be considered as equivalent to controlling the flow rate Q as an independent variable and determining the pressure rise as the dependent variable. (Actually we could consider the position of the valve as an independent variable and the pressure rise as the dependent variable. That would be one problem. Then considering the valve position as determining the flow rate would be a second problem. Eliminating the valve position between these two problems leaves variables that refer only to the pump. Either the pressure or the flow rate could be considered as the independent variable replacing the valve position.)

Since the fluid is incompressible, no thermodynamic processes occur, and the fluid is characterized by its density ρ. There are two extra pieces of information that can be part of the problem. First, incompressible-flow theory tells us (or will tell us) that the level of pressure in a flow field is not important. That is, if the inlet pressure is raised a certain amount by increasing the level of fluid in the reservoir, the outlet pressure increases an equal amount. This fact is taken into account by considering $\Delta p = p_2 - p_1$ as a single variable and not p_2 and p_1 separately. The second extra assumption concerns friction. Does viscosity play an important role in determining the pressure in the pump? Again a knowledge of some of the general characteristics of fluid flows is useful. In most engineering situations viscous forces are much smaller than pressure forces, and unless the viscous forces act over a large area, they can usually be neglected in comparison with the pressure forces. We make the assumption that viscosity may be neglected and proceed to state the problem as

$$\begin{aligned} p_2 &= f(Q, \rho, d, \Omega) + p_1 \\ p_2 - p_1 \equiv \Delta p &= f(Q, \rho, d, \Omega) \end{aligned} \tag{8.4.1}$$

Now the pi theorem is applied to 8.4.1. The dimensional matrix is

	Δp	Q	ρ	d	Ω
M	1	0	1	0	0
L	-1	3	-3	1	0
T	-2	-1	0	0	-1

We choose ρ, d, and Ω as repeating variables. The last three columns have a nonzero determinant, so the rank is 3. This means we expect $5 - 3 = 2$ nondimensional variables. Let d be the length scale, ρd^3 the mass scale, and Ω^{-1} the time scale; then the two nondimensional variables are found to be

$$\Pi_1 = \frac{\Delta p}{\rho d^2 \Omega^2}, \qquad \Pi_2 = \frac{Q}{d^3 \Omega} \tag{8.4.2}$$

The nondimensional form of 8.4.1 becomes

$$\frac{\Delta p}{\rho d^2 \Omega^2} = f_1\left(\frac{Q}{d^3 \Omega}\right) \tag{8.4.3}$$

Experimental measurements on a pump of a given size are shown in Fig. 8-4a in dimensional terms. (Pressure is measured in terms of the "head", that is, $\Delta p/\rho g$ for water.) Figure 8-4b shows the same data in nondimensional form. Note that we could have drawn the characteristic curve for the pump by testing it at only one speed. One of the great powers of dimensional analysis is illustrated by this problem. By varying only two quantities in a test, Δp and Q, we can actually find the dependence for three additional variables, d, ρ, and Ω. Even though we did not test other fluids or different-size pumps, the data can be used to predict what would happen if we used a smaller pump or changed the fluid from water to oil.

Another interesting facet of the test is that we can consider the tests at different speeds as an evaluation of the effect of viscosity. In order to see this, let us consider what would happen if viscosity were important to the flow. Then, our functional assumption would be

$$\Delta p = f(Q, \rho, d, \Omega, \mu) \tag{8.4.4}$$

Dimensional analysis would yield the same answer as before except for the addition of a new variable, the Reynolds number based on the impeller tip velocity. The answer is

$$\frac{\Delta p}{\rho d^2 \Omega^2} = f\left(\frac{Q}{d^3 \Omega}, \text{Re}\right) \tag{8.4.5}$$

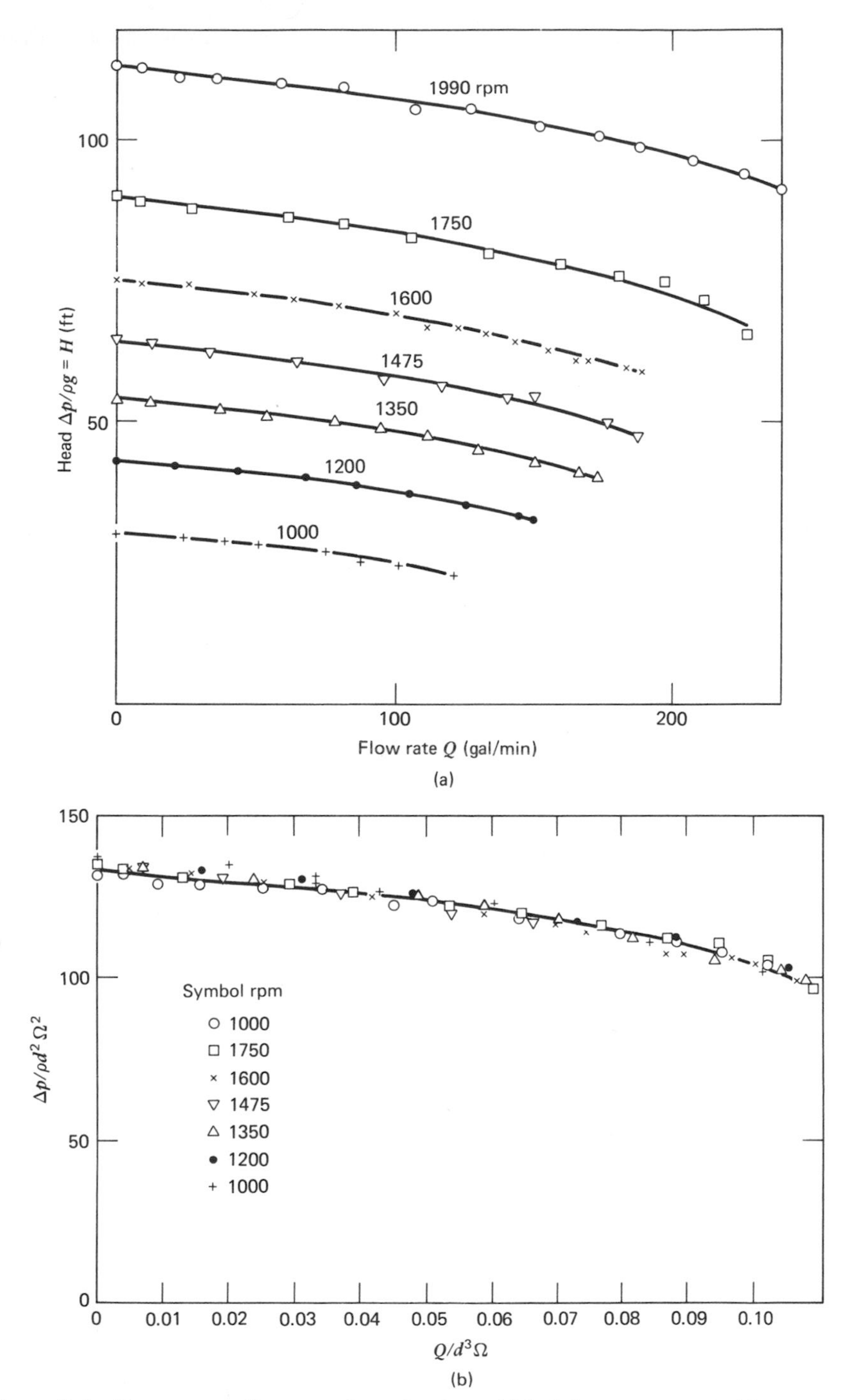

Figure 8-4 Pressure vs. flow rate for a backward-bladed pump (test fluid water): (a) dimensional variables, (b) nondimensional variables.

where

$$\mathrm{Re} \equiv \frac{\Omega d^2 \rho}{\mu} \qquad 8.4.6$$

Now from 8.4.6 we see that changes in the speed change the Reynolds number, which is equivalent to changing the viscosity of the fluid. The fact that all the different speed curves fall roughly together when plotted in Fig. 8-4 verifies that viscosity is not important in this case. We have actually tested the influence of viscosity while using only one fluid in the test.

Lest we give the false impression that pumps aways behave according to 8.4.3, we should point out that cavitation may occur. At very low pressures, in the neighborhood of $\frac{1}{30}$ atm, water will boil and produce pockets of vapor. The lowest pressure in the system occurs on the moving blades, and when cavitation occurs there, the pump characteristics change. Dimensional analysis including cavitation would include the vapor pressure of the liquid and lead to another dimensional parameter. Pumps, hydrofoils, ship propellers, and nozzle flows are all cases where cavitation is possible. For a review article on cavitation, see Arndt (1981).

8.5 NUMBER OF PRIMARY DIMENSIONS

There have been several attitudes towards the question of how many primary dimensions are required. The question is not as simple as it might first seem. Let us be more precise and ask the question: If we have a complicated continuum problem involving mechanics, thermodynamics, and electrodynamics, what is the minimum number of dimensions required to express the dimensional variables? That is, how many P's are needed in Bridgman's equation

$$\hat{x} = x P_1^a P_2^b P_3^c \cdots \qquad 8.5.1$$

for any variable of interest? It is a fact of experience that all of our physical concepts may be expressed in terms of three primary dimensions (see Sedov (1959) for more discussion of this point).

During the first part of this century some physicists thought that five dimensions were required. They added temperature and an electrical unit to M, L, and T. It is not the usual practice to express thermodynamic or electrical concepts in terms of mass, length, and time; but it can be done (energy and force are ideas common to all subjects). Since the number of primary dimensions plays such an important role in the pi theorem, it is natural to inquire about how the pi theorem changes if we use more or less than three primary dimensions. We begin with the customary discussion of mechanics and Newton's second law.

Newton's second law equates the forces to the mass times the acceleration:

$$F = m\frac{dv}{dt} \tag{8.5.2}$$

If we take the viewpoint that mass, length, and time are primary dimensions, then the dimensions of force are ML/T^2. A newton of force is just a shorter name for one kilogram meter per second squared. There is an alternate viewpoint where one assumes that there are four primary dimensions F, M, L, and T. If this viewpoint is taken, then Newton's second law must be written with a dimensional constant g_c:

$$F = \frac{m}{g_c}\frac{dv}{dt} \tag{8.5.3}$$

Because we overspecified the number of primary dimensions, we must introduce a compensating *dimensional unifier* into the mathematics. In the British engineering system where we use the pound-force, pound-mass, foot, and second, the value of g_c is

$$g_c = 32.17\frac{\text{lb}_m - \text{ft}}{\text{lb}_f - \text{sec}^2} \tag{8.5.4}$$

So the conclusion is that if we take the viewpoint that there are four primary dimensions, then a dimensional unifier appears in the governing laws and in all the answers.

The solution to a problem governed by Newton's law, for example the position of the body as a function of time, will then contain the dimensional unifier g_c:

$$x = f(t, m, \ldots, g_c)$$

In dimensional analysis any quantity that changes magnitude when the size of primary units are changed must be considered a variable. If, for example, we defined the "doublefoot" as a new length unit then g_c would be 16.1 lb_m-dbft/lb_f-sec. Thus, in a dimensional analysis with F, M, L, and T as primary dimensions, the dimensional matrix would look like

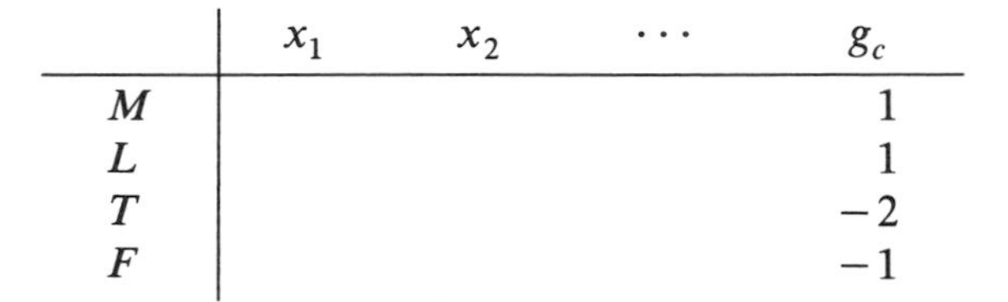

	x_1	x_2	$\cdots$	g_c
M				1
L				1
T				-2
F				-1

The effect on the pi theorem is nil. The rank of the matrix has been increased by one and the number of dimensional variables has also been increased by one.

Consider another, less obvious example. What if the problem involves an angle and we choose to measure the angle in degrees. The primitive definition of an angle is the length of an arc divided by the radius, a dimensionless quantity called the radian. However, if we insist on using the degree unit as a primary dimension, then all formulas will contain a unifying dimensional constant, $2\pi/360$ radians/degree. The dimensional matrix should be constructed as

	x_1	x_2	$\cdots$	$2\pi/360$
M				0
L				0
T				0
degree				-1

Whenever an additional primary dimension is added to the basic three, there must (in general) be a compensating dimensional unifier.

Although there is something special about the number of primary dimensions, there is nothing special about which dimensions are chosen for the primary role. For instance, we could use area instead of length. Then the dimensional exponent of a length would be $\frac{1}{2}$ and dimensional variables would be expressed as

$$\hat{x} = xM^aA^bT^c \qquad 8.5.5$$

As another example, consider using speed as a primary dimension. It would be related to the length and time dimensions by

$$S = LT^{-1} \qquad 8.5.6$$

Substituting into 8.1.2, which governs all variables, gives an equation of identical form but with different exponents:

$$\hat{x} = xM^aL^bT^c = xM^aL^{b+c}S^{-c}$$

$$= xM^aL^{b_1}S^{c_1} \qquad 8.5.7$$

The primary dimensions are now mass, length, and speed. Any transformation of the primary dimension is allowed as long as the form of Bridgman's equation is unchanged. That is, any product of the primary dimensions to any powers could be used as new primary dimensions. More complicated functions such as $\sin(L/T)$ or $\exp(M)$ are not acceptable.

It is customary for different fields of science to use different primary dimensions. Electricity and thermodynamics have different primary dimensions than those used in mechanics. This is, of course, the result of the separate historical developments of the subjects. In fact, the primary dimensions used in thermodynamics and electrodynamics are related to M, L, and T by products of powers.

Three primary dimensions are sufficient for any problem but are not always necessary. A problem in kinematics could be formulated using only L and T. If M were also included, then all entries on the M row of the dimensional matrix would be zero and the rank would automatically reduce from three to two. This is a somewhat trivial example but should be kept in mind.

There is another more subtle way in which the number of dimensions can be reduced. It is based on using a "universal" constant as the unit for a primary dimension. In order to see how this works, consider a problem with five variables that include the universal dimensional constant c, the speed of light in a vacuum. We may assume a set of primary dimensions of M, L, and S (speed), since it was demonstrated above how this set might be produced. The governing function is assumed to be

$$x_1 = f(x_2, x_3, x_4, c) \tag{8.5.8}$$

Let the dimensional matrix be

	x_1	x_2	x_3	x_4	c
M	1	0	1	0	0
L	1	1	−1	0	0
S	1	0	−1	−1	1

The number of variables and their exponents have been chosen to make the processes in the example concrete; the principles are the same irrespective of the exact values. Consider some simple combinations of the variables where c is used to eliminate the speed dimension from each variable. This is done by defining

$$\begin{aligned} X_1 &= \frac{x_1}{c} \qquad & X_2 &= x_2 \\ X_3 &= x_3 c \qquad & X_4 &= x_4 c \end{aligned} \tag{8.5.9}$$

substituting into 8.5.8 will give a new function

$$cX_1 = f\left(X_2, \frac{X_3}{c}, X_4 c, c \right)$$

or

$$X_1 = f_1(X_2, X_3, X_4, c) \tag{8.5.10}$$

Next we use the freedom to choose the unit for speed to set $S = \hat{c}$. Since $\hat{c} = cS$, this implies that $c = 1$ (the measurement unit for velocity is the speed of light). With $c = 1$ the function 8.5.10 is reduced to four variables. It now reads

$$X_1 = f_1(X_2, X_3, X_4, 1) \qquad 8.5.11$$

The dimensional matrix for the four new variables now has only two dimensions as follows:

	X_1	X_2	X_3	X_4
M	1	0	1	0
L	1	1	−1	0

We have essentially used the speed of light as a universal unit for all variables. The freedom to choose the speed unit has been given up in order to express variables in terms of two primary dimensions. The number of pi-variables predicted by the pi theorem is the same: four minus two instead of five minus three.

From another point of view what we have done is to partially nondimensionalize the problem. The mathematical operations are similar to those that will be used to prove the pi theorem in a later section. If we had continued the process to eliminate M and L, we would have had only two variables left. They would have been nondimensional.

The process of reducing the number of primary dimensions by using a physical constant as a unit is not often used. It was originally employed with the hope that some sort of simplification of a universal nature would occur (that is, three fundamental universal constants would govern all of nature). However, there are no universal dimensional constants that are relevant to all physical processes—this is true for microscopic as well as the macroscopic phenomena. If there are universal scales on the astronomical level, they have yet to be confirmed. We have not found any universal scales for the dynamics of the universe (though they have been proposed).

On the other hand, for the study of a limited class of problems where a certain dimensional constant always occurs, as in astronomy, it would be entirely proper to reduce the number of primary dimensions. It would however be incorrect and improper to list a dimensional constant, such as the speed of light, in a problem where it plays no role whatsoever.

At the outset of this section we spoke of different viewpoints on the question of the number of primary dimensions required for a general problem. Many writers take the viewpoint that the number of primary dimensions depends upon the problem being solved and whether we write the governing laws with a dimensional unifier. The question is further complicated by the fact that some special problems do not actually require a dimensional unifier (heat transfer in incompressible flow, for example). The structure of the equations that govern the subject may allow the dimensional unifier to be eliminated from the problem. When a dimensional unifier is not needed, it is because supplementary information about the physics has been introduced into the analysis. Thus, when one is allowed to introduce supplementary information during the process of listing the variables, one cannot make a general rule on the required number of primary dimensions. If we make a rule that supplementary information must be explicitly introduced into an analysis, three primary dimensions are sufficient, and the rank of the dimensional matrix will tell us if they are necessary.

*8.6 PROOF OF BRIDGMAN'S EQUATION

The equation $\hat{x} = xM^aL^bT^c$ is the cornerstone of dimensional analysis. It is not usually given a name. Bridgman (1922) seems to be the first to state it explicitly and offer proof, so we shall refer to it as Bridgman's equation.

In order to prove Bridgman's equation we adopt a different notation in this section. Subscripts will refer to different values of the same variable: y_1 is a specific value of y, and y_2 is another value of the same variable. Consider m as the mass variable. In accord with 8.1.1, we write

$$\hat{m} = mM \tag{8.6.1}$$

M is the size of a mass unit, m the number of mass units, and $\hat{m}$ the absolute size of the mass in question.

Next, we consider a variable y that depends in some way upon mass. Force, pressure, power, and density are examples. If y also depends upon length and time, we temporarily hold those variables constant. Assume

$$y = f(m) \tag{8.6.2}$$

For instance, y might be the density of a sphere with a certain volume. Our first step is to consider two spheres made of different substances, so that m_1 has one value and m_2 another. The numbers m_1 and m_2 are found using a certain mass scale M. Bridgman reasoned that there must be something intrinsic about a physical concept, in this case the density, that is independent of the measuring units. He proposed that the ratio of the densities of the two spheres should be a constant independent of the scale unit M. In mathematical terms

$$\frac{y_1}{y_2} = \frac{f(m_1)}{f(m_2)} = C \tag{8.6.3}$$

This is the major assumption: *The ratio of two definite values of any physical variable does not depend upon the size of the measuring units of the primary dimensions*. As far as we know, all our physical concepts satisfy this assumption.

Before we continue, let us take another example to illustrate 8.6.3. Suppose that $y = f(l)$ is the area of a rectangle. Next we consider two examples, say this piece of paper and a football field. We choose the foot as a length scale, measure the areas of the two rectangles, and form the ratio of these numbers. Our major assumption means that the ratio of areas would be the same number if we repeated the process using a meter as the unit instead of the foot.

To proceed with the proof, we put 8.6.3 in the form

$$f(m_1) = Cf(m_2) \tag{8.6.4}$$

Consider how a change in the mass unit M will affect the problem. To do this, hold $\hat{m}$ constant while m and M change. Differentiating 8.6.4 with respect to M yields

$$f'(m_1)\frac{dm_1}{dM} = Cf'(m_2)\frac{dm_2}{dM} \tag{8.6.5}$$

while from 8.6.1 we have the constraining equations

$$0 = m_1\,dM + M\,dm_1$$

$$0 = m_2\,dM + M\,dm_2$$

Substituting these equations into 8.6.5 and using 8.6.3 to replace C yields

$$\frac{m_1 f'(m_1)}{f(m_1)} = \frac{m_2 f'(m_2)}{f(m_2)} = \text{const} = k \tag{8.6.6}$$

This equation holds for all choices of m_1 and m_2 independently, and therefore each side is a constant.

Equation 8.6.6 is solved as follows (we can drop the subscript):

$$\frac{m}{f}\frac{df}{dm} = k$$

$$\ln\frac{f}{f_0} = k\ln\frac{m}{m_0} \tag{8.6.7}$$

$$\frac{y}{y_0} = \frac{f}{f_0} = \left(\frac{m}{m_0}\right)^k$$

The reference values are fixed by the notation that y_0 (the dimensional magnitude) is equal to $\hat{y}$ (the absolute size) when the scale unit M is equal to 1; that is,

$$\hat{m} = m_0 \qquad \text{when} \quad \hat{y} = y_0$$

Substituting this into 8.6.7 and noting that $\hat{m} = mM$, we have

$$\frac{y}{\hat{y}} = \left(\frac{m}{\hat{m}}\right)^k = M^{-k}$$

$$\hat{y} = yM^k \qquad 8.6.8$$

The exponent of the mass dimension is unrestricted.

The proof is completed by noting that the process may be repeated for the second and third variables of a more general function $y = f(m, l, t)$. The final result is Bridgman's equation,

$$\hat{y} = yM^{k_1}L^{k_2}T^{k_3} \qquad 8.6.9$$

This is the same equation as 8.1.2, where x replaces y and the constants are changed to a, b, and c.

*8.7 PROOF OF THE PI THEOREM

Before reading this section the student may want to review the pi theorem given in Section 8.3. It will not be restated here.

Consider a function that describes a physical process and contains n dimensional variables. Let x_1 be the dependent variable and write the function as

$$x_1 = f(x_2, x_3, \ldots, x_n) \qquad 8.7.1$$

Assume that all variables in 8.7.1 may be expressed in terms of three primary dimensions according to Bridgman's equation,

$$\hat{x}_i = x_i M^{a_i} L^{b_i} T^{c_i} \qquad 8.7.2$$

In addition to this equation, another major assumption is needed. We assume that the function 8.7.1 is valid for any measuring units we might choose. A function relating physical variables is valid for any choice of M, L, and T units. In particular, if we choose to use the "absolute" values, 8.7.1 becomes

$$\hat{x}_1 = f(\hat{x}_2, \hat{x}_3, \ldots, \hat{x}_n) \qquad 8.7.3$$

Another way of looking at this is that 8.7.1 should be valid when $M = 1$, $L = 1$, and $T = 1$.

The mass–length–time symbols will be used in the proof; however, they have no special properties, and any primary dimensions would suffice. We have already discussed how extra primary dimensions have no essential effect on the pi theorem because extra unifying dimensional constants must be added to the list of variables. The fact that three primary dimensions are sufficient to express all physical variables is the result of the intrinsic structure of physics and is not subject to proof.

The dimensional matrix for the problem contains the dependent variable as the first entry. If necessary we renumber the variables so that the rank of the matrix formed by x_2, x_3 and x_4 is 3 (the determinant is nonzero). We are going to prove the theorem for the case that $r = 3$. The matrix is

	x_1	x_2	x_3	x_4	...	x_n
M	a_1	a_2	a_3	a_4	...	a_n
L	b_1	b_2	b_3	b_4	...	b_n
T	c_1	c_2	c_3	c_4	...	c_n

The three variables x_2, x_3, x_4 will become new primary dimensions or scales. We define the new scales by the relations

$$S_2 \equiv \frac{\hat{x}_2}{x_2}, \qquad S_3 \equiv \frac{\hat{x}_3}{x_3}, \qquad S_4 \equiv \frac{\hat{x}_4}{x_4} \tag{8.7.4}$$

In terms of the M, L, T units the new scales are (by 8.7.2)

$$S_2 = M^{a_2}L^{b_2}T^{c_2}, \qquad S_3 = M^{a_3}L^{b_3}T^{c_3}, \qquad S_4 = M^{a_4}L^{b_4}T^{c_4} \tag{8.7.5}$$

In order to express a variable in the new S scales we need relations for M, L, and T in terms of the S's. To do this consider products of S_2, S_3, S_4 raised to some as yet undetermined exponents. First we find the powers that will produce a mass scale; that is, we solve the following equation for A_2, A_3, and A_4:

$$S_2^{A_2}S_3^{A_3}S_4^{A_4} = M^1L^0T^0$$

Substituting from equation 8.7.5, we have

$$M^1L^0T^0 = M^{a_2A_2}L^{b_2A_2}T^{c_2A_2}M^{a_3A_3}L^{b_3A_3}T^{c_3A_3}M^{a_4A_4}L^{b_4A_4}T^{c_4A_4} \tag{8.7.6}$$

Since M, L, and T are arbitrary, this equation is true only if the exponents sum to zero. This produces the system of linear equations for the unknowns A_2, A_3, A_4:

$$\begin{aligned} &\text{exponent of } M: && a_2A_2 + a_3A_3 + a_4A_4 = 1 \\ &\text{exponent of } L: && b_2A_2 + b_3A_3 + b_4A_4 = 0 \\ &\text{exponent of } T: && c_2A_2 + c_3A_3 + c_4A_4 = 0 \end{aligned} \tag{8.7.7}$$

A unique solution of this nonhomogeneous system is guaranteed by Cramer's rule if the determinant of the coefficients is nonzero. This condition is satisfied, since the coefficients of the system 8.7.7 are the same coefficients as appear in the dimensional matrix. We had arranged the variables so that this determinant was nonzero at the start.

By the same process coefficients for the L and T dimensions are found. When this is completed we have the following relations between the new scales and the old M, L, T scales:

$$M = S_2^{A_2} S_3^{A_3} S_4^{A_4}$$

$$L = S_2^{B_2} S_3^{B_3} S_4^{B_4} \qquad 8.7.8$$

$$T = S_2^{C_2} S_3^{C_3} S_4^{C_4}$$

The linear system of equations for the B's is the same as that for the A's except that the right-hand column of 8.7.7 is 0, 1, 0. Likewise, the system for the C's has the right-hand column 0, 0, 1.

All the dimensional variables in the problem may be expressed in terms of the new S dimensions by substituting 8.7.8 into 8.7.2:

$$\hat{x}_i = x_i \left(S_2^{A_2} S_3^{A_3} S_4^{A_4} \right)^{a_i} \left(S_2^{B_2} S_3^{B_3} S_4^{B_4} \right)^{b_i} \left(S_2^{C_2} S_3^{C_3} S_4^{C_4} \right)^{c_i}$$

This is Bridgman's equation with primary scales S_2, S_3, S_4, that is,

$$\hat{x}_i = x_i S_2^{\alpha_i} S_3^{\beta_i} S_4^{\gamma_i} \qquad 8.7.9$$

where the exponents are defined according to the relations

$$\alpha_i = a_i A_2 + b_i B_2 + c_i C_2$$

$$\beta_i = a_i A_3 + b_i B_3 + c_i C_3$$

$$\gamma_i = a_i A_4 + b_i B_4 + c_i C_4$$

Up to this point we have essentially done two things: we have chosen new scale units $S_2 = \hat{x}_2 / x_2$, S_3, and S_4, and we have shown that Bridgman's equation can be written in terms of the new scales. The dimensional matrix would now look like this:

	x_1	x_2	x_3	x_4	x_5	$\cdots$	x_n
S_2	α_1	1	0	0	α_5	$\cdots$	α_n
S_3	β_1	0	1	0	β_5	$\cdots$	β_n
S_4	γ_1	0	0	1	γ_5	$\cdots$	γ_n

We have essentially diagonalized the entries for the repeating variables.

Now we are ready to show that physical functions are homogeneous, and then use this fact to nondimensionalize the variables. Substituting 8.7.9 into 8.7.1 yields

$$\hat{x}_1 S_2^{-\alpha_1} S_3^{-\beta_1} S_4^{-\gamma_1}$$
$$= f\left(\hat{x}_2 S_2^{-1}, \hat{x}_3 S_3^{-1}, \hat{x}_4 S_4^{-1}, \hat{x}_5 S_2^{-\alpha_5} S_3^{-\beta_5} S_4^{-\gamma_5}, \ldots, \hat{x}_n S_2^{-\alpha_n} S_3^{-\beta_n} S_4^{-\gamma_n}\right) \tag{8.7.10}$$

Substituting 8.7.3 for $\hat{x}_1$ in the equation above produces

$$S_2^{-\alpha_1} S_3^{-\beta_1} S_4^{-\gamma_1} f(\hat{x}_2, \hat{x}_3, \ldots, \hat{x}_n) = f\left(\hat{x}_2 S_2^{-1}, \ldots, \hat{x}_n S_2^{-\alpha_n} S_3^{-\beta_n} S_4^{-\gamma_n}\right) \tag{8.7.11}$$

In mathematical terms a function is called *homogeneous of order k* if the substitution $\lambda x \to x$ produces λ^k times the original function,

$$\lambda^k f(x) = f(\lambda x) \tag{8.7.12}$$

The physical function 8.7.11 is homogeneous of degree α_1 in the x_2 variable, of degree β_1 in the x_3 variable, and of degree γ_1 in the x_4 variable. If we used different repeating variables and retrace the previous steps, we would find that the physical equation is homogeneous to some degree in all variables. Thus, all physical equations come from a special class of functions called homogeneous functions.

We could apply the term "dimensionally homogeneous" to the fact that physical equations are homogeneous functions of the variables chosen as repeating variables. However, various workers have their own special definitions of the phrase. Langhaar (1951) takes the assumption that physical equations are valid for all choices of M, L, and T as the proper definition. Most frequently the definition is that each additive term in an equation must have the same dimensions. This is really just a facet of the meaning we have given above.

There are just a few steps left to prove the pi theorem. Notice that the size of a unit in the new primary dimensions, S_2, S_3, and S_4, is arbitrary. Let us select

$$S_2 = \hat{x}_2, \qquad S_3 = \hat{x}_3, \qquad S_4 = \hat{x}_4 \tag{8.7.13}$$

This selection also means $x_2 = x_3 = x_4 = 1$; that is, we are using the values of x_2, x_3, and x_4 as scale units. Equation 8.7.10 now reads

$$\frac{\hat{x}_1}{\hat{x}_2^{\alpha_1} \hat{x}_3^{\beta_1} \hat{x}_4^{\gamma_1}} = f\left(1, 1, 1, \frac{\hat{x}_5}{\hat{x}_2^{\alpha_5} \hat{x}_3^{\beta_5} \hat{x}_4^{\gamma_5}}, \ldots, \frac{\hat{x}_n}{\hat{x}_2^{\alpha_n} \hat{x}_3^{\beta_n} \hat{x}_4^{\gamma_n}}\right) \tag{8.7.14}$$

The variables in this equation are all nondimensional. This is shown by considering 8.7.9 and substituting the scale definitions 8.7.4 to arrive at

$$\hat{x}_i = x_i\left(\frac{\hat{x}_2}{x_2}\right)^{\alpha_i}\left(\frac{\hat{x}_3}{x_3}\right)^{\beta_i}\left(\frac{\hat{x}_4}{x_4}\right)^{\gamma_i}$$

Rearranging yields

$$\frac{\hat{x}_i}{\hat{x}_2^{\alpha_i}\hat{x}_3^{\beta_i}\hat{x}_4^{\gamma_i}} = \frac{x_i}{x_2^{\alpha_i}x_3^{\beta_i}x_4^{\gamma_i}} \qquad 8.7.15$$

The values of the pi variables are independent of the choice of the M, L, and T units, that is $\hat{\Pi} = \Pi$. Therefore the variables in 8.7.15 are nondimensional, and we change the symbols in 8.7.14 accordingly. The new function has $n - r$ nondimensional variables:

$$\Pi_1 = f(1, 1, 1, \Pi_5, \Pi_6, \ldots, \Pi_n) \qquad 8.7.16$$

It is no longer a homogeneous function in these variables and has a completely arbitrary mathematical form. This completes the proof of the pi theorem for the case $r = 3$. The reader wishing to know the modifications necessary for r less than three may consult Langhaar (1951), Brand (1957), or Sedov (1959).

*8.8 VECTOR LENGTH SCALES (DISTINGUISHED DIRECTIONS)

During the formative stages of dimensional analysis, Williams (1892) proposed that different length scales could be used for different directions. The primary dimensions for each direction would be distinct, say L_x and L_y. They would be regarded as "apples and oranges" and could not be cancelled out any more than we would cancel M and T. By effectively increasing the number of primary dimensions, we would increase the rank of the matrix and thus would get fewer nondimensional variables, that is, a sharper result. Huntley (1953) gives many examples of this method. These methods are not founded on physical or mathematical theory, but are tricks for indirectly introducing supplemental information into a problem by increasing the number of primary dimensions.

Before we discuss the foundation of this method, an example will be worked to illustrate its use. Consider the problem of finding the range r of a projectile fired horizontally in a vacuum from a height h with velocity v. Conventional dimensional analysis proceeds as follows. We know from experience with the gravitational law that the acceleration of a falling object is constant; $a = g = f/m$. Therefore we assume

$$r = r(h, v, g) \qquad 8.8.1$$

The rank of the dimensional matrix is two, so two pi groups are required. Solving the problem gives the following result:

$$\frac{r}{h} = f\left(\frac{v^2}{hg}\right) \tag{8.8.2}$$

The actual answer to the problem is even simpler than 8.8.2 would indicate.

Now we rework this problem using the concept of distinguished directions. We consider L_x and L_y as separate noncancellable dimensions. The matrix becomes

	r	h	v	g
L_x	1	0	1	0
L_y	0	1	0	1
T	0	0	−1	−2

The rank is now three, indicating only one nondimensional parameter is needed. This gives the answer

$$\frac{r}{v}\sqrt{\frac{g}{h}} = \text{const} \tag{8.8.3}$$

—a sharper result than the conventional method produced.

At first glance the method seems to be founded on the physical principle that vector equations separate into components, and we could actually measure one direction in meters and the other in yards. This is not so. Recall that the theoretical basis of the pi theorem is Bridgman's dimensional equation

$$\hat{x} = xM^aL^bT^c \tag{8.8.4}$$

In order for the vector-length concept to work rigorously in the pi theorem, we would need a principle that said

$$\hat{x} = xM^aL_x^bL_y^cL_z^dT^e \tag{8.8.5}$$

This is not in general true.

The use of several primary length dimensions works when the physical process for one direction occurs independently of those in the other direction. To make it work we must be able to unambiguously specify that a length dimension is either L_x or L_y. It would be preferable to use the extra information in a more direct manner.

Let us rework the ballistics problem from a new viewpoint. Instead of seeking the range directly, we seek the time of flight. Next assume that the horizontal velocity does not influence the time of flight, since that is determined by the free fall. The functional form for the time of flight is

$$t = f(h, g) \tag{8.8.6}$$

By dimensional analysis, this problem has one pi group:

$$\Pi_1 = t\sqrt{\frac{g}{h}} = \text{const} \tag{8.8.7}$$

Now consider a second problem where the range is assumed to be a function of v and t. Obviously, it is $r = vt$; if we insist on using dimensional analysis on $r = f(v, t)$ we get

$$\Pi_2 = \frac{tv}{r} = \text{const} \tag{8.8.8}$$

Combining the results of these problems to eliminate t yields

$$\frac{r}{v}\sqrt{\frac{g}{h}} = \text{const} \tag{8.8.9}$$

This is the same result that was produced earlier using two length dimensions. By analyzing the problem without using two lengths, the supplementary assumptions have been made explicit.

*8.9 GENERALIZED MILLIKAN ARGUMENT

C. B. Millikan (1938) derived the logarithmic velocity profile for a turbulent boundary layer. This equation is valid over a large middle section of the boundary layer. It is not valid extremely close to the wall, nor is it valid in the outer portion of the boundary layer. The same equation also applies to turbulent flow in a pipe. The equation was known previously from experiments, and several semiempirical derivations (such as those by Prandtl and von Kármán) existed. Millikan's argument is important since it shows that the logarithmic law follows from dimensional analysis alone. The essence of the argument is that two different physical processes dominate in different regions of the flow. Viscosity dominates the profile near the wall, while inviscid turbulent motions dominate far away from the wall. Millikan showed how to obtain a considerable amount of information about the transition region, called the *overlap region*, where the two physical processes are about equally important (or equally unimportant).

In order to fix these ideas, we discuss the turbulent boundary layer problem in more detail. The region near the wall is dominated by viscosity, and the proper nondimensional velocity profile has a form called the *law of the wall*:

$$\frac{u}{u_*} = f\left(\frac{yu_*}{\nu}\right) \tag{8.9.1}$$

with

$$u_* \equiv \sqrt{\frac{\tau_\omega}{\rho}}$$

In this expression u is the velocity as a function of the distance y from the wall, and the shear stress at the wall is τ_ω. Figure 8-5 shows experimental data plotted in this form. The data give one curve near the wall, which becomes a straight line for $yu_*/\nu > 30$. The straight line is the logarithmic law of the overlap or transition region. At large distances from the wall the data deviate from the logarithmic law. The beginning of this deviation marks the end of the overlap region. Its location depends upon the Reynolds number $u_*\delta/\tau_\omega$ (δ is the boundary-layer thickness).

Far away from the wall, the velocity must be referenced to the free-stream velocity U, and the distance scaled by the boundary-layer thickness δ. In this region, the proper nondimensional form is known as the *defect law*,

$$\frac{u - U}{u_*} = f\left(\frac{y}{\delta}\right) \qquad 8.9.2$$

Figure 8-6 shows experimental data plotted in this form. The data correlate

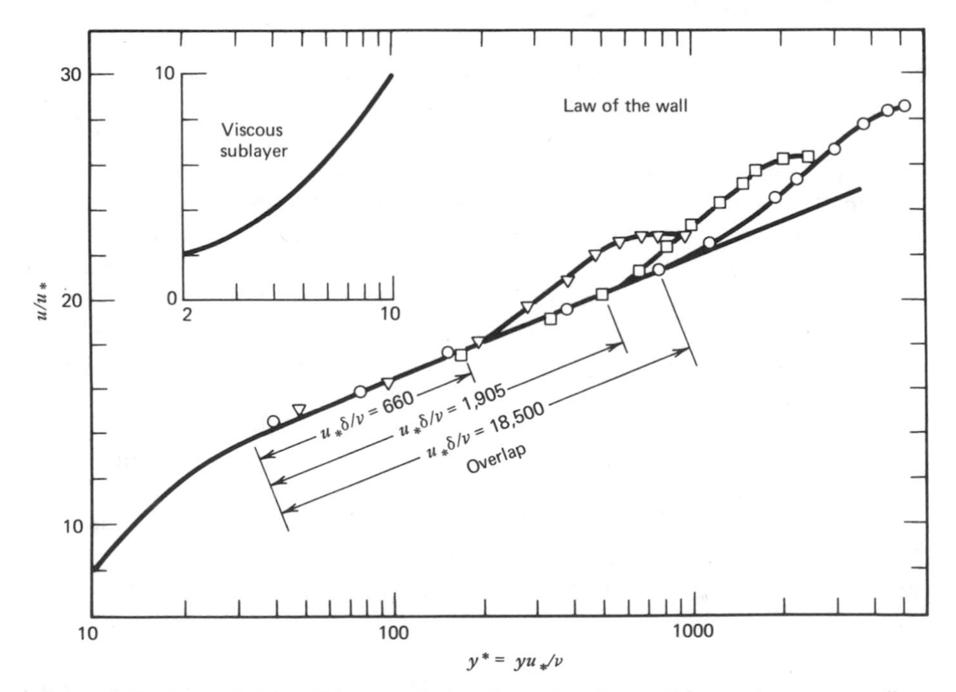

Figure 8-5 Law of the wall for a turbulent boundary layer with zero pressure gradient.

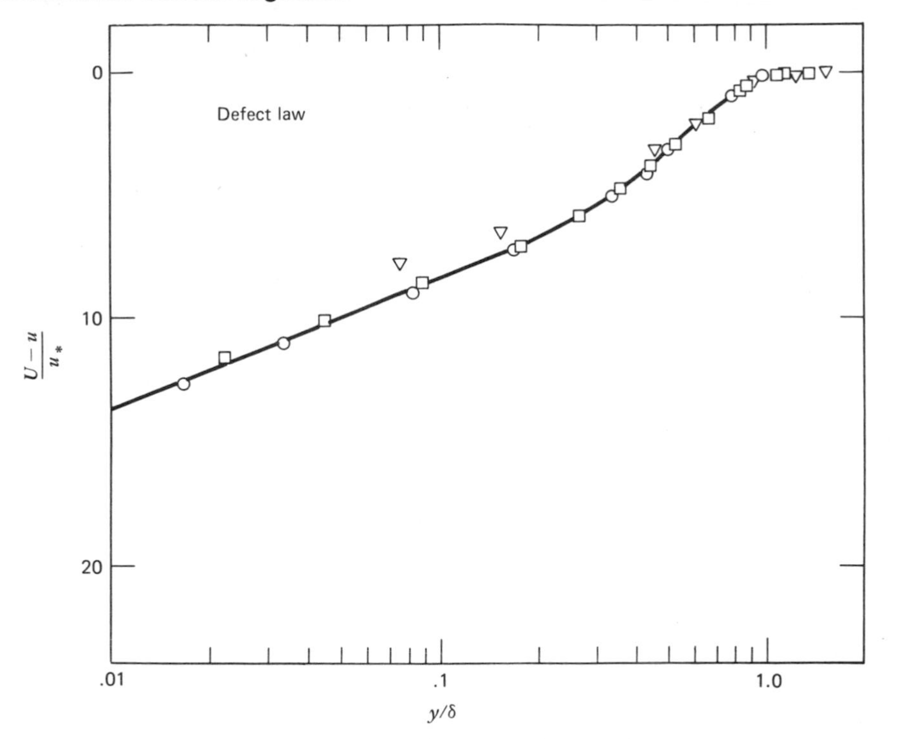

Figure 8-6 Defect law for a turbulent boundary layer with zero pressure gradient.

well in the outer region near $y/\delta = 1$ and becomes a logarithmic profile toward the wall. Very near the wall the data deviate. Again, where the deviation begins depends upon the Reynolds number $u_*\delta/\nu$.

The existence of an overlap region where both the law of the wall and the defect law are valid occurs at high Reynolds number, that is, in the limit $u_*\delta/\nu \to \infty$. From the mathematical standpoint the problem is one of singular perturbation with the Reynolds number as the perturbation parameter. The overlap region is the matching region of two matched asymptotic expansions. Millikan's argument can be interpreted as a dimensional analysis of the matching region. We now do the analysis in a very general way without any reference to a specific problem. Several cases will occur. One of these cases will apply to the turbulent boundary layer; others will apply to other problems.

Consider a physical problem where the dependent variable is y and the independent variable is x. Reference values for y may occur in the problem and are Y_1 and Y_2. A natural scale for y is y_0 and that for x is x_0. All of the foregoing variables are dimensional. A dimensionless perturbation parameter will be denoted by ϵ. We may assume the range on x is 0 to ∞, and that the perturbation parameter $\epsilon \to 0$. The restrictions on x and ϵ are not important, since it is always possible to redefine the origin and make transformations that will satisfy the specified requirements. The perturbation parameter ϵ is usually related to the scales x_0 and y_0; however, for the present purposes ϵ may be considered independent. The dimensional functional form is then

$$y = f(x; x_0, Y_1, Y_2, y_0, \epsilon, \ldots) \qquad 8.9.3$$

The dots at the end are to show that other variables may also occur but do not enter the analysis.

The limit $\epsilon \to 0$ is a singular perturbation, and we use dimensional analysis to find the form of the answer in the overlap region. If the overlap region is large (as it is in turbulent boundary layers), this will be a useful equation.

The nondimensional form for y in the outer region away from $x = 0$ is

$$F \equiv \frac{y - Y_1}{y_0} = F(\hat{x}, \epsilon, \ldots) \qquad 8.9.4$$

where the nondimensional independent variable is $\hat{x} = x/x_0$. The form of F in the limit $\epsilon \to 0$ is

$$F_0 = F(\hat{x}; \epsilon \to 0, \ldots) \qquad 8.9.5$$

Since by assumption this is a singular perturbation, F_0 is not a valid answer in the region near $x = 0$. In this inner region a rescaling is required. Near $x = 0$ we assume the proper nondimensional form is

$$f \equiv \frac{y - Y_2}{y_0 \epsilon^m} = f(x^*; \epsilon, \ldots) \qquad 8.9.6$$

The reference value for y has changed from Y_1 to Y_2, and the proper scaling is assumed to be $y_0 \epsilon^m$. There is no restriction on m. The new scaling for x is

$$x^* = \epsilon^{-n} \hat{x} \qquad 8.9.7$$

A scaling change on the independent variable is required in a singular perturbation problem, so $n \neq 0$.

Now if the problem has been organized properly, we can assume there is an overlap region where both forms 8.9.4 and 8.9.6 are valid. Equating these equations gives a relation valid in the overlap region,

$$F(\hat{x}, \epsilon) + \frac{Y_1(\epsilon)}{y_0} = \epsilon^m f(x^*, \epsilon) + \frac{Y_2(\epsilon)}{y_0} \qquad 8.9.8$$

Notice that the reference values may also be functions of ϵ. Equation 8.9.8 is differentiated with respect to $\hat{x}$ to produce

$$F' = \epsilon^m f' \frac{dx^*}{d\hat{x}} = \epsilon^{m-n} f'$$

But since, from 8.9.7, $\epsilon^n = (\hat{x}/x^*)$, we may rearrange this equation as follows:

$$\hat{x}^{(n-m)/n} F' = x^{*\,(n-m)/n} f' \qquad 8.9.9$$

The left-hand side is a function of $\hat{x}$ and ϵ, and the right-hand side a function of x^* and ϵ. This relation holds for the limit $\epsilon \to 0$, so

$$\hat{x}^{(n-m)/n} F_0' = x^{*\,(n-m)/n} f_0' = C_1 \qquad 8.9.10$$

C_1 must be a constant, since $\hat{x}$ and x^* are considered as independent. ($\hat{x}$ can be varied at constant x^* by letting ϵ change.) Thus, from 8.9.10 we have two differential equations:

$$dF_0 = C_1 \hat{x}^{-(n-m)/n}\, d\hat{x}$$
$$df_0 = C_1 x^{*\,-(n-m)/n}\, dx^* \qquad 8.9.11$$

The solution of these equations fall into two cases depending upon m.

The first case is when $m = 0$ and there is no scale change for the dependent variable. Many physical problems have this character. Then the solutions are

$$f_0 = C_1 \ln x^* + C_2$$
$$F_0 = C_1 \ln \hat{x} + C_3 \qquad 8.9.12$$

These equations are valid only in the overlap region. This is not all the information that may be obtained. Substituting back into 8.9.8 relates the constants and the reference values:

$$\frac{Y_2 - Y_1}{y_0} = C_3 - C_2 + C_1 n \ln \epsilon \qquad 8.9.13$$

This is the most general result.

The results above apply to the turbulent boundary layer. The variables are as follows:

$$m = 0, \quad n = -1, \quad Y_1 = U, \quad Y_2 = 0, \quad \epsilon = \text{Re} = \frac{u_* \delta}{\nu}$$

$$y = u, \quad x = y, \quad y_0 = u_*, \quad x_0 = \delta$$

$$\hat{x} = \frac{y}{\delta}, \qquad x^* = \frac{y u_*}{\nu}$$

$C_1 = k^{-1}$ (von Kármán constant)

$C_2 = B$ (roughness constant) 8.9.14

$C_3 = 2\Pi/k$ (pressure-gradient constant)

The equations 8.9.12 are the overlap-region expressions for the law of the wall

$$\frac{u}{u_*} = \frac{1}{k} \ln \frac{yu_*}{\nu} + B \qquad 8.9.15$$

and the *law of the wake*

$$\frac{u - U}{u_*} = \frac{1}{k} \ln \frac{y}{\delta} + \frac{2}{k}\Pi \qquad 8.9.16$$

The third equation from the analysis (8.9.13) is the friction law for a turbulent boundary layer,

$$\frac{U}{u_*} = B - \frac{2}{k}\Pi + \frac{1}{k} \ln \mathrm{Re} \qquad 8.9.17$$

The value of B changes with wall roughness (represented by the dots in 8.9.6), while Π is a nondimensional pressure-gradient parameter (represented by the dots in 8.9.4). The wall roughness influences only the inner-region equation and the pressure gradient influences only the outer-region equation. This is known from experiments and is not a result of the analysis.

Now return to 8.9.13 and consider another possibility. Assume that the references are equal ($Y_2 = Y_1$). Then C_1 must be zero (since ϵ changes independently) so that $C_2 = C_3$ and in the overlap region

$$F_0 = f_0 = C_2 \qquad 8.9.18$$

The majority of perturbation problems fall under this case, and therefore dimensional analysis is not very informative.

A second case for solutions to 8.9.11 is when the dependent variable must undergo a scale change in the inner region, that is, $m \neq 0$. The solutions are now

$$\begin{aligned} f_0 &= C_1 x^{*m/n} + C_2 \\ F_0 &= C_1 \hat{x}^{m/n} + C_3 \end{aligned} \qquad 8.9.19$$

and the auxiliary relation is

$$\frac{Y_1 - Y_2}{y_0} = C_3 - C_2 \epsilon^m$$

If we further require $Y_1 = Y_2$, then $C_2 = C_3 = 0$ and the final results are

$$\begin{aligned} f_0 &= C_1 x^{*m/n} \\ F_0 &= C_1 \hat{x}^{m/n} \end{aligned} \qquad 8.9.20$$

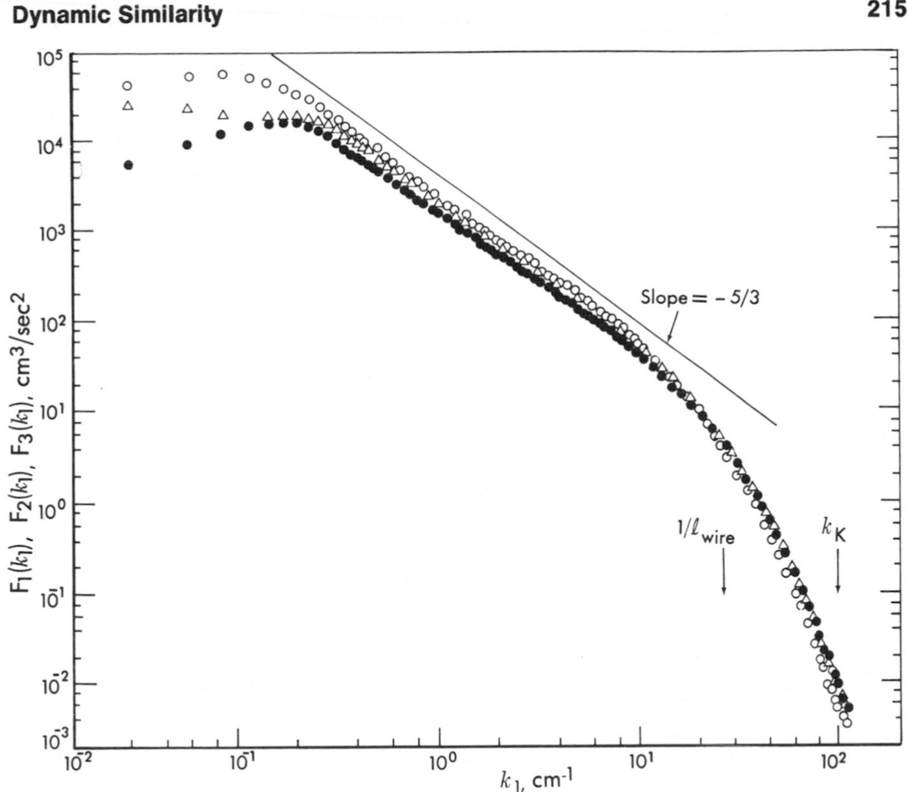

Figure 8-7 Spectrum of velocity fluctuations in a turbulent jet: ●, F_1 (longitudinal); ○, F_2 (lateral); △, F_3 (transverse). Wavenumber = k_1. From F. H. Champagne (1978).

This form of overlap is found in spectra where f_0 and F_0 are power density functions and x is the frequency or wavenumber. Kolmogorov (1941a, 1941b) found a relation of this type for the energy spectrum of turbulence. Figure 8-7 shows examples of spectra of turbulence taken by Champagne (1978). These data show the power spectra for the longitudinal $[F_1(k_1)]$ and lateral $[F_2(k_1)]$ velocity fluctuations as a function of the wavenumber k_1. The theoretical coefficient in 8.9.20 for this case is $m/n = -\frac{5}{3}$. The figure shows a large overlap region where this equation is valid.

8.10 DYNAMIC SIMILARITY

Two different physical problems are *dynamically similar* if the variables in one problem can be put in correspondence with the variables in the other. This is a very general statement, which includes even electrical–mechanical analogies. The simplest type of similarity occurs when two situations have different dimensional variables but the same nondimensional variables.

Consider a physical phenomenon governed by the equation

$$\Pi_1 = f(\Pi_2, \ldots, \Pi_k) \qquad 8.10.1$$

If two flows have the same values of the independent variables, then the dependent variables should be the same. Calling one flow the model and the other the prototype, then

$$\Pi_{im} = \Pi_{ip}, \qquad i = 1, 2, \ldots, k \qquad 8.10.2$$

There is an implicit assumption that both situations are governed by the same unique function. However, there are many instances in fluid mechanics where a unique answer is not obtained. For example, with the same pressure ratio across a converging–diverging nozzle there can be two different flow patterns. One is completely subsonic, while the other contains supersonic flow and shock waves. Which flow pattern occurs depends upon the past history of the imposed pressures. In general, failure of the uniqueness assumption is the exception rather than the rule.

The theory of the geometrically similar flow situations is easily handled by the pi theorem where the nondimensional variables of the two situations are equal. There are other types of similarity where the geometry differs between the two flows. In these situations we must look at the equations and boundary conditions governing the flow field. As an example of an analysis of this type we discuss the theory of flow over thin two-dimensional airfoils as shown on Fig. 8-8.

The origin of the equations and boundary conditions that govern the flow will not be given. The reader unfamiliar with them will not lose much in the way of understanding, as the procedure is essentially mathematical. In subsonic compressible flow a perturbation velocity potential ϕ is the major dependent variable. It is a function of x and y coordinates and the Mach number M. All of these variables are nondimensionalized in such a way that the airfoil is one unit long and the free-stream velocity is one. The convenient abbreviation $\beta \equiv (1 - M^2)^{1/2}$ is employed. The governing equation and boundary conditions are

$$\beta^2 \frac{\partial^2 \phi}{\partial x^2} + \frac{\partial^2 \phi}{\partial y^2} = 0 \qquad 8.10.3$$

At infinity the perturbation vanishes:

$$\left.\frac{\partial \phi}{\partial x}\right|_\infty = \left.\frac{\partial \phi}{\partial y}\right|_\infty = 0$$

and on the surface of the airfoil,

$$\frac{dy_s}{dx} = \left.\frac{\partial \phi}{\partial y}\right|_s$$

These equations are sufficient to solve for $\phi(x, y, \beta)$.

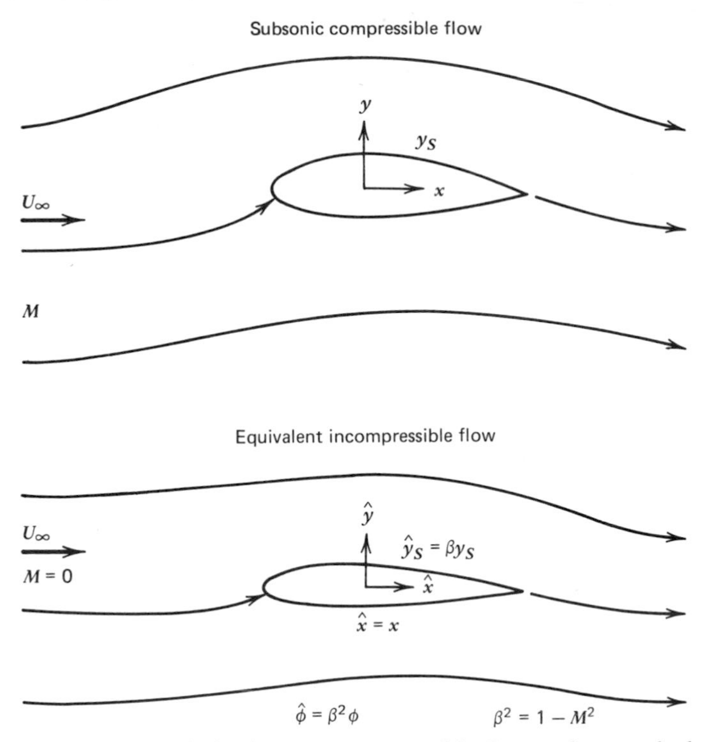

Figure 8-8 Dynamic similarity between compressible flow and an equivalent incompressible flow over a distorted model.

We consider the similarity between an incompressible flow where $\beta = 1$ and a compressible flow at a specified value of β. The shape of the airfoil will be different. When the Mach number is zero, the potential is ϕ_0 and is governed by the Laplace equation

$$\frac{\partial^2 \phi_0}{\partial x^2} + \frac{\partial^2 \phi_0}{\partial y^2} = 0 \qquad 8.10.4$$

The same boundary conditions apply to ϕ_0 as to ϕ.

Now, suppose that the compressible flow variables are transformed as follows:

$$\tilde{x} = x, \qquad \tilde{y} = \beta y, \qquad \tilde{\phi} = \beta^2 \phi \qquad 8.10.5$$

We shall show that the mathematical problem for the $\tilde{\phi}(\tilde{x}, \tilde{y})$ is identical to the incompressible problem. Substituting into the differential equation 8.10.3

yields

$$\beta^2 \frac{\partial^2(\tilde{\phi}/\beta^2)}{\partial \tilde{x}^2} + \frac{\partial^2(\tilde{\phi}/\beta^2)}{\partial \tilde{y}^2}\left(\frac{d\tilde{y}}{dy}\right)^2 = 0$$

$$\frac{\partial^2 \tilde{\phi}}{\partial \tilde{x}^2} + \frac{\partial^2 \tilde{\phi}}{\partial \tilde{y}^2} = 0 \qquad 8.10.6$$

The boundary conditions far from the body are

$$\frac{\partial(\tilde{\phi}/\beta^2)}{\partial \tilde{x}} = 0, \qquad \frac{\partial \tilde{\phi}}{\partial \tilde{x}} = 0$$

$$\frac{\partial(\tilde{\phi}/\beta^2)}{\partial \tilde{y}}\frac{d\tilde{y}}{dy} = 0, \qquad \frac{\partial \tilde{\phi}}{\partial \tilde{y}} = 0$$

On the surface of the body the boundary condition is

$$\frac{d}{d\tilde{x}}\left(\frac{\tilde{y}}{\beta}\right) = \frac{\partial}{\partial \tilde{y}}\left(\frac{\tilde{\phi}}{\beta^2}\right) \cdot \frac{d\tilde{y}}{dy}$$

$$\frac{d\tilde{y}}{d\tilde{x}} = \frac{d\tilde{\phi}}{d\tilde{y}}$$

The solution to this problem, $\tilde{\phi}(\tilde{x}, \tilde{y})$, is identical to $\phi_0 = \phi(x, y; \beta = 1)$. Thus, the compressible flow over an airfoil at Mach number M is related to the incompressible flow over an airfoil with a stretched shape. The length of the airfoil is the same (since $\tilde{x} = x$), but the surface coordinates are related by $\tilde{y} = \beta y$. This is an example of similarity between distorted models. The original nondimensional problem $\phi(x, y; \beta)$ has a mathematical structure that allows a new, smaller set of variables $\tilde{\phi}(\tilde{x}, \tilde{y})$. This new function is the same between the actual flow and the equivalent incompressible flow.

There is another physical result that this problem illustrates: that subsonic compressible flow over thin objects is qualitatively the same as incompressible flow. The streamlines and forces undergo only slight modifications as the Mach number increases. The exact magnitude and nature of the modification as well as the completely different phenomena which occur in transonic and supersonic flow is the subject of compressible-flow theory.

Distorted models are frequently used in hydraulics to model dams, river systems, and other large-scale problems. As with the compressible-flow example above, the local differential equations must be used to establish similarity of distorted models.

One of the largest hydraulic models ever constructed is a model of the Mississippi River basin built by the U.S. Army Corps of Engineers. Figure 8-9

Figure 8-9 Model of the Mississippi River. Photograph courtesy of U.S. Army Engineer Waterways Experiment Station, Vicksburg, Miss.

shows only a portion of the model. Both horizontal directions on the model have the same scale ratio 1 : 2000 (model : prototype). On the model this places Sioux City, South Dakota, about one-half mile from the mouth of the Mississippi River, an actual distance of about 1000 miles on the real river. Vertical distances on the model are scaled 1 : 100, giving a distortion factor of 20 : 1 (vertical : horizontal). The actual elevation of Sioux City is 1100 ft, so the corresponding point on the model would be just 11 ft higher than the mouth of the river. Through a detailed analysis of the governing equations one can determine that the time scales are such that 5.4 minutes on the model corresponds to one day on the river. With the time scale compressed in this way it is possible to trace the history of a hypothetical flood in a reasonably short experiment.

The model has had many uses. Primarily it has been used to evaluate the effect that dams, levees, and reservoirs have on the river flow, especially on the management of flood conditions. Upon several occasions the model has been used to forecast the progress of flood-fighting efforts. With the highly developed and managed system that the Mississippi has today, disastrous floods are much less frequent than in years past. Nevertheless, the model was recently called into action during a flood to predict the effect of a breached levee.

Another current use for the model is for public relations. Groups of civic leaders are brought to the model to see how proposed flood-control projects will benefit their regions.

The model has also been used to aid in developing a computer model of the river system. Fluid-dynamic events on the hydraulic model, the computer model, and the real river should ideally all agree. This means that some empirical coefficients in the computer model may actually be checked by running the hydraulic model at the desired conditions. In this way extreme flood conditions, which may never actually occur, can be verified.

A hydraulic model is very good in its ability to simulate geometric effects. Where the models are very poor is in their ability to faithfully reproduce friction or viscous effects (the Reynolds numbers of model and prototype are much different). On the model, additional friction is introduced by several methods. In the flood plains the high resistance of vegetation is simulated by wire-screen strips folded in accordion fashion. Where the main channel flow is fairly slow, an extra-rough concrete finish supplies enough friction; however, regions of fast flow require more artificial friction. In these places rectangular metal fingers protrude from the bottom and generate extra turbulence.

The last topic of this section is the internal type of similarity, wherein the variables at one point in a flow field are related to the variable at other positions in the same flow. Such flows are called *self-similar* and occur frequently in fluid mechanics. A classic example of a self-similar flow is the Rayleigh problem for the impulsive motion of an infinite flat plate. The velocity field at one time is simply related to the velocity field at any other time. Recall that the nondimensional form of the solution is

$$\frac{u}{u_0} = f\left(\frac{y}{\sqrt{\nu t}}\right)$$

All points y, t such that $y/\sqrt{\nu t} = \text{const}$ have the same velocity. This is essentially a simplification of the non-self-similar form

$$\frac{u}{u_0} = f\left(\frac{y}{\nu/u_0}, \frac{t}{\nu/u_0^2}\right)$$

Sometimes internal similarity can be found by dimensional analysis, and sometimes it cannot. Many examples of internal similarity will be forthcoming in later chapters.

8.11 NONDIMENSIONAL FORMULATION OF PHYSICAL PROBLEMS

In many instances we know the equations that govern a problem and can write out the relevant laws and conditions. The fact that solutions of physical problems must be dimensionally homogeneous is only contained implicity in the governing equations. It is often ignored as one finds the solution. This is a bad practice. If we recast the problem into nondimensional variables, we explicitly use the information that physical functions are dimensionally homogeneous. Boundary conditions and physical constants are used to nondimensionalize the dependent and independent variables. The nondimensional form of the problem will contain all the necessary variables. Inspection of these equations will reveal the nondimensional functions without using the pi theorem. Moreover, frequently there is information contained in the governing equations that reduces the number of nondimensional variables even more than the pi theorem would predict. The advantages of nondimensionalizing a problem are great; the problem has the fewest variables and the simplest mathematical structure when expressed in nondimensional variables.

Nondimensional variables may be thought of as variables whose scales or units of measurement come from the problem itself. In this sense they are natural scales. The standard units of measurement such as the meter, the kilogram, and the second have no special importance to any physical processes. The important scales (the S's of 8.7.4) come from the list of variables for the problem.

Let us consider for a moment the anatomy of a nondimensional variable. A nondimensional variable consists of three parts: the dimensional variable y, a reference value y_0, and the scale or unit y_s:

$$y^* = \frac{y - y_0}{y_s} \qquad 8.11.1$$

The first question to answer in composing nondimensional variables is about the reference y_0. Is the absolute value of the variable important to the problem, or only its value compared to a reference? For example, in heat conduction only differences in temperature are important; therefore we should look for a reference value at some point in the field. Once the reference has been decided upon, attention can be turned to the scale y_s. The scale is some combination of boundary conditions and/or dimensional constants that has the same dimensions as y. Its most important characteristic is that it measures the range that y takes on in the problem. If y takes on maximum and minimum values, we ask ourselves how large the difference in these values is. We do not need its exact value, but only a quantity that estimates it.

Pressure is a good example. Consider a compressible flow where the pressure changes by expansion and compression. The absolute magnitude of the pressure is important in this case. The work of the process depends upon the absolute level and not just the difference between initial and final values. The reference should be zero, and the proper nondimensional pressure is

$$p^* = \frac{p}{p_s} \qquad 8.11.2$$

The scale p_s is some specified pressure in the problem (say the initial pressure). Pressure in this case plays a thermodynamic role as well as providing a force. On the other hand, for incompressible flow the level of the pressure is not important. In this case some specified pressure in the flow is used as a reference p_0. Also, here the changes in pressure are the result of dynamic processes. Thermodynamic changes in pressure are absent. The unit of pressure is a characteristic kinetic energy (per unit mass) of the flow,

$$\tilde{p} = \frac{p - p_0}{\frac{1}{2}\rho V_0^2} \qquad 8.11.3$$

This form of pressure nondimensionalization leads to a mechanical pressure for fluid flow.

Perhaps the best way to demonstrate how to nondimensionalize a problem is through examples. We first consider the plane-Couette-flow problem worked in the previous chapter (note that the mathematical statement is complete only when all equations and boundary conditions are specified). The problem is

$$\frac{d^2u}{dy^2} = 0$$

$$y = 0, \qquad u = 0 \qquad 8.11.4$$

$$y = h, \qquad u = V_0$$

The first place to look for scales is in the boundary conditions. We have already eliminated any reference value for y by choosing the coordinate system on the lower wall. The range of y is from 0 to h, so h is the obvious scale. Let the nondimensional y variable be

$$y^* = \frac{y - 0}{h - 0} = \frac{y}{h} \qquad 8.11.5$$

By similar reasoning the velocity variable has a reference value 0 on the lower wall, and we expect the upper-wall velocity U to be the maximum:

$$u^* = \frac{u - 0}{V_0 - 0} = \frac{u}{V_0} \qquad 8.11.6$$

In terms of the new variables, the boundary condition are now pure numbers independent of measuring units:

$$y^* = 0, \qquad u^* = 0$$

$$y^* = 1, \qquad u^* = 1 \qquad 8.11.7$$

The differential equation transforms to

$$\frac{d^2u}{dy^2} = \frac{d^2(u^*V_0)}{dy^2} = V_0\frac{d^2u^*}{dy^*}\left(\frac{dy^*}{dy}\right)^2$$

$$\frac{d^2u^*}{dy^{*2}} = 0 \qquad 8.11.8$$

A lot of information can be found without solving the problem. Since there are no parameters to vary in 8.11.7 and 8.11.8, we conclude that $u^* = f(y^*)$. In this case the same result would be obtained by applying the pi theorem to $u = f(y; h, U)$. [If we had incorrectly included viscosity in the list of variables and set $u = f(y; h, U, \mu)$, we would have found that there was no way to nondimensionalize μ.] The solution to 8.11.8 is $u^* = y^*$.

As a second example, consider the flow in a slot that is driven by a constant pressure gradient. The problem is

$$\mu\frac{d^2u}{dy^2} = \frac{dp}{dx}$$

$$y = 0, \quad u = 0 \qquad 8.11.9$$

$$y = h \quad u = 0$$

The y variable is nondimensionalized as before: $y^* = y/h$. The boundary conditions show that $u = 0$ at the walls, and we find no information about how large u will become. The maximum value of u is determined by a balance between the pressure force and the viscous force. This information is contained within the differential equation itself. A convenient procedure is to assume an unknown velocity scale u_s:

$$\tilde{u} = \frac{u - 0}{u_s} \qquad 8.11.10$$

Substituting gives

$$\frac{\mu u_s}{h^2}\frac{d^2\tilde{u}}{dy^{*2}} = \frac{dp}{dx}$$

Now, the scale u_s is fixed so that the right-hand side is a pure number. Let u_s be chosen as

$$u_s = -\frac{h^2}{\mu}\frac{dp}{dx} \qquad 8.11.11$$

The problem in nondimensional variables becomes

$$\frac{d^2\tilde{u}}{dy^{*2}} = -1$$

$$y^* = 0, \qquad \tilde{u} = 0 \qquad 8.11.12$$

$$y^* = 1, \qquad \tilde{u} = 0$$

It is not important that the right-hand side was chosen to be one. What is important is that the size of the velocity profile is measured by

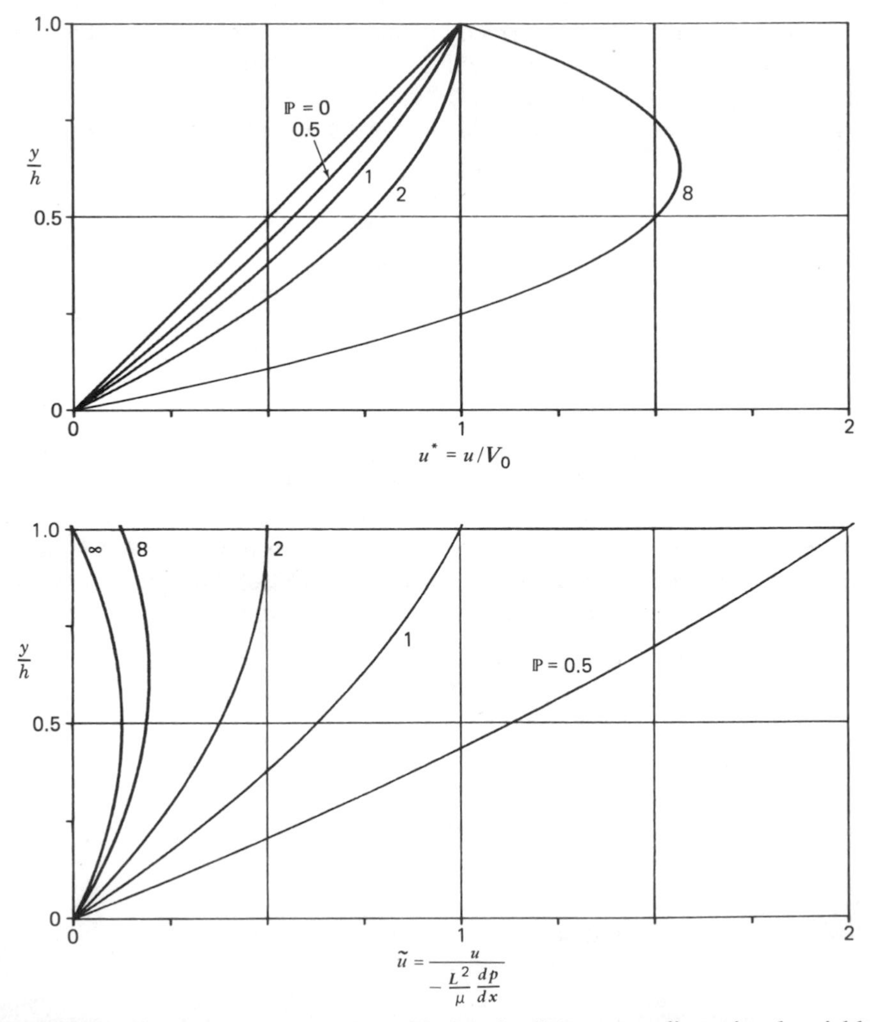

Figure 8-10 Combined Couette–Poiseuille flow in different nondimensional variables. The pressure-gradient parameter is $\mathbb{P}$.

$(h^2/\mu)(-dp/dx)$. The minus sign compensates for the fact that dp/dx is negative for a positive velocity. Again we have a problem without any parameters, so we know that the answer has the functional form $\tilde{u} = f(y^*)$. For the record, the solution is $\tilde{u} = \frac{1}{2}(\tilde{y}^* - y^{*2})$.

Next, we consider a combination of the two previous problems (Fig. 8–10). Couette flow with an imposed pressure gradient offers an example where two natural velocity scales appear in the same problem. In dimensional variables the problem is

$$\frac{d^2u}{dy^2} = \frac{1}{\mu}\frac{dp}{dx}$$

$$y = 0, \qquad u = 0 \qquad\qquad 8.11.13$$

$$y = h, \qquad u = U$$

If we choose to use the wall velocity V_0 as a scale, the problem transforms into

$$\frac{d^2u^*}{dy^{*2}} = \frac{h^2}{\mu V_0}\frac{dp}{dx} \equiv -\mathbb{P}$$

$$y^* = 0, \qquad u^* = 0 \qquad\qquad 8.11.14$$

$$y^* = 1, \qquad u^* = 1$$

The new parameter $\mathbb{P}$ defV_0ned above compares the pressure velocity scale with the wall velocity. The structure of the equations and boundary conditions indicates that the solution is of the form $u^* = f(y^*, \mathbb{P})$. From our previous work in Chapter 7 we know that the actual answer is

$$\frac{u}{V_0} = u^*(y^*; \mathbb{P}) = y^* + \tfrac{1}{2}\mathbb{P}(y^* - y^{*2}) \qquad\qquad 8.11.15$$

Now let us reconsider the problem, using u_s as the velocity scale. The problem is

$$\frac{d^2\tilde{u}}{dy^{*2}} = -1$$

$$y^* = 0, \qquad \tilde{u} = 0 \qquad\qquad 8.11.16$$

$$y^* = 1, \qquad \tilde{u} = \mathbb{P}^{-1}$$

The parameter $\mathbb{P}$ now appears in the boundary conditions, and the differential equation is free of parameters. The solution is still of the form $u^* = f(y^*, \mathbb{P})$

as shown below:

$$\tilde{u} = \frac{u}{-\dfrac{L^2}{\mu}\dfrac{dp}{dx}} = \frac{u^*}{\mathbb{P}} = \frac{1}{\mathbb{P}}y^* + \tfrac{1}{2}(y^* - y^{*2}) \qquad 8.11.17$$

This result is the same equation as 8.11.15 except the velocity variable has a different scale.

What if we have a solution and want to retrieve a special case by letting a parameter approach a certain value? For instance, $\mathbb{P} \to 0$ means that the pressure–velocity scale is small compared to the moving-wall scale. Allowing $\mathbb{P} \to 0$ in the form $\tilde{u}(y^*; \mathbb{P})$ gives an infinite answer. The velocity is not properly nondimensionalized for this limit. Allowing $\mathbb{P} \to 0$ in $u^*(y^*; \mathbb{P})$ produces the proper result:

$$u^* = y^* \qquad 8.11.18$$

which is the velocity profile for Couette flow without a pressure gradient.

If the two velocity scales are about the same size, the parameter $\mathbb{P}$ is a reasonable magnitude. In this case it makes no difference which scale is used for the velocity. If one scale is much larger than the other, then the dominant scale should be used in forming the nondimensional variables. If a limiting process is to be applied to an answer (or, more likely, to the problem itself before the answer is obtained), then the variables must be properly nondimensionalized for that limit process.

8.12 CONCLUSION

Dimensional analysis allow us to express mathematical or experimental results in their simplest form. The most complicated situations benefit the most from reducing the number of variables.

There are two fundamental ideas behind dimensional analysis. The first is that a measurement scale is not an intrinsic part of a physical quantity. The only intrinsic aspect is that the ratio of two different values is independent of the measurement system in which they are expressed. This fact led to Bridgman's equation. The second idea is that a law governing physical variables must be valid for all different measurement systems. Together these ideas underlie the pi theorem.

Even sharper results than given by the pi theorem are obtained if we can bring into play extra information provided by laws that govern the physics or by assumptions about the functional form of the result. The latter assumptions are based on either physical or purely intuitive arguments.

Nondimensional variables may be thought of as variables that are measured by a scale unit arising from the physical event itself. A proper selection of natural scales produces variables of a modest numerical size. The first, and sometimes the most difficult, step in organizing an analysis or in organizing experimental results is to find the proper scales for forming nondimensional variables.

PROBLEMS

8.1 (A) Rework the pump analysis using F, M, L, T as primary variables.

8.2 (A) Rework the pump analysis using M, S (speed), and T as primary variables.

8.3 (A) A list of variables for a problem has only one variable with the dimension of mass. In what two possible ways could the list be in error?

8.4 (A) Write out three unifying dimensional constants in common use. What is g_c in a system where kg-force, kg-mass, m, and sec are primary?

8.5 (B) Could the relation 1 pascal $\equiv$ 1 N/m^2 be a unifying dimensional constant?

8.6 (A) Write out four universal physical constants that are not usually employed as unifying dimensional constants.

8.7 (B) After making a list of variables for a problem, a worker states, "If I have left anything off of the list that just means I need another pi variable." When is this statement true and when is it untrue?

8.8 (A) A windmill is designed to operate at 20 rpm in a 15-mph wind and produce 300 kW of power. The blades are 175 ft in diameter. A model 1.75 ft in diameter is to be tested at 90-mph wind velocity. What rotor speed should be used and what power should be expected?

8.9 (B) A capillary tube of diameter d is vertical in a liquid with surface tension σ (M/T^2). How high will the column of liquid rise as a function of the other important parameters of the problem? Would an experiment done on earth predict the result of this experiment when done on the moon?

8.10 (B) Extend the derivation of Bridgman's equation to three variables; that is, consider $y = f(m, l, t)$.

8.11 (C) What modifications are required in the proof of the pi theorem if the rank of the matrix is two rather than three?

8.12 (C) Instant nondimensionalization of the Navier–Stokes equations is sometimes done in the following way. Suppose that the problem has characteristic U_0, L, ρ_0 specified. Imagine that these values are used as the measuring scales, and then set them equal to unity. Then any distance symbol x_i is measured in terms of L and is really x_i/L, v_i is really v_i/U_0, and ρ is really ρ/ρ_0. What is ν in terms of the measuring scales U_0, L, ρ_0? What is p in terms of U_0, L, ρ? If the flow is incompressible, what does the momentum equation reduce to?

8.13 (B) Consider a turbulent boundary layer in a zero pressure gradient ($\Pi = 0.6$). The Reynolds number is $u_* \delta/\nu = 1000$, and the wall is smooth ($B = 5.0$). Find the Reynolds number $U\delta/\nu$. If δ is 5 cm, find the y distances where the logarithmic region begins and where it ends (approximately). Repeat the calculation for $u_* \delta/\nu = 3000$.

8.14 (B) Consider an incompressible flow with constant properties where the viscous and the pressure forces are dominant (inertia terms are negligible). What is the proper nondimensional pressure if U_0 and L are scales from the boundary conditions?

8.15 (B) A closed cylinder with a flat end contains a disk that comes within a small distance h of the end. This space is of radius r_0 and is filled with oil. A shaft attached to the disk turns it with speed Ω. Formulate the differential equations and boundary conditions that govern the velocity profile in the cavity. Nondimensionalize the problem.

9 Compressible Couette Flow

In order to reach our goal of understanding when a flow may be considered as incompressible, we need to know some of the characteristics of compressible flows. The main purpose of the present chapter is to study the flow in a Couette apparatus for a fluid that has completely arbitrary thermodynamic equations. We want to pay particular attention to how the flow causes viscous heating and how this energy is conducted away.

From a practical standpoint, this problem is a model of how heat is generated in a bearing. The same effect occurs on high-speed airplanes and rockets because there is a layer of shear flow next to the surface. It is sometimes noted that this problem is an exact solution to the Navier–Stokes equation—that is, a solution to the full equations without assuming any transport properties or thermodynamic functions are constant. Illingworth (1950) was the first to give this analysis in his paper on solutions of the (compressible) Navier–Stokes equations.

9.1 COMPRESSIBLE COUETTE FLOW; ADIABATIC WALL

We consider only the plane-flow case, because the algebra is somewhat simpler and it contains the essential physics; there is no mathematical difficulty in solving the problem for the cylindrical case. Figure 9-1 gives a picture of the flow and defines the boundary conditions. Our first assumption is that none of the flow properties depend upon x, z, or t. The major dependent variables are

$$v_x = v_x(y), \qquad v_y = v_y(y)$$
$$T = T(y), \qquad \rho = \rho(y) \tag{9.1.1}$$

Enforcing the no-slip boundary conditions leads to the following:

$$v_x(0) = 0, \qquad v_y(0) = 0$$
$$v_x(h) = V_0, \qquad v_y(h) = 0 \tag{9.1.2}$$

Since the energy equation is now involved in the problem, we also need boundary conditions on the heat transfer. Assume that the upper wall is

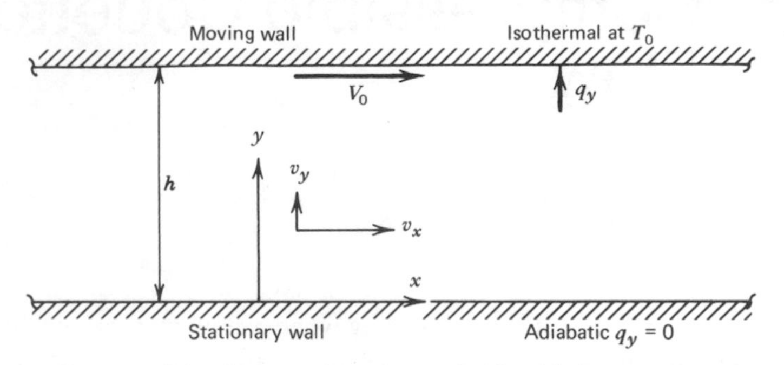

Figure 9-1 Compressible Couette flow for a fluid with known equation of state, viscosity law $\mu(T, p)$, and conductivity law $k(T, p)$.

isothermal and that the lower wall is adiabatic. This means

$$T(h) = T_0, \qquad q_y(0) = 0 \tag{9.1.3}$$

We might, as an alternative, have assumed that the lower wall was also isothermal at a different temperature T_1. There is a definite reason for choosing an adiabatic lower wall instead: this way we shall be able to pin down to what extent viscous dissipation influences the temperature profile. If we had allowed a constant temperature on the lower wall, then there would be a characteristic temperature scale $(T_0 - T_1)$ for conduction across the gap; the temperature profile might have been dominated by conduction, so that we would have lost the effect of viscous dissipation.

The governing equations will be considered one by one as they are simplified for this problem. First, the continuity equation under the assumption 9.1.1 reduces to

$$\frac{d}{dy}(\rho v_y) = 0 \tag{9.1.4}$$

This means that ρv_y must be a constant. Applying the boundary condition $v_y(0) = 0$ shows the constant to be zero. Therefore, we find that there is no y-direction motion.

In anticipation of the momentum equations, we calculate the viscous stresses employing the fact that $v_x(y)$ is the only velocity component ($v_y = 0$). They are

$$\tau_{xx} = \tau_{yy} = 0 \tag{9.1.5}$$

$$\tau_{yx} = \tau_{yx} = \mu \frac{dv_x}{dy} \tag{9.1.6}$$

Next we consider the y-direction momentum equation. It simplifies to

$$0 = \frac{dp}{dy} \tag{9.1.7}$$

implying that the pressure is a constant. This is an important result with respect to the thermodynamics of the fluid. A simple compressible substance has only two independent intensive properties, and we have just shown that one of them is constant. Since the pressure has only one value for the whole flow, we may consider all the transport properties as a function of the temperature only.

The x-direction momentum equation reduces to a statement that the net shear on a particle is zero:

$$0 = \frac{d\tau_{yx}}{dy} \qquad 9.1.8$$

This integrates to

$$\tau_{yx} = \text{const} = \tau_0 = \mu \frac{dv_x}{dy} \qquad 9.1.9$$

where τ_0 denotes the constant shear stress. We shall come back to this equation later on after a few other facts have been established from the energy equation.

The heat-flux vector under the assumption $T = T(y)$ becomes

$$q_x = 0, \qquad q_y = -k\frac{dT}{dy} \qquad 9.1.10$$

Employing this information along with our previous assumptions, we find that the energy equation 5.10.2 balances the net conduction away from a point with the energy generated at the point by viscous dissipation. In mathematical terms,

$$0 = -\frac{dq_y}{dy} + \tau_0 \frac{dv_x}{dy} \qquad 9.1.11$$

We can integrate this equation to give the relation

$$-q_y + \tau_0 v_x = C_1 \qquad 9.1.12$$

C_1 is determined to be zero by noting that both v_x and q_y are zero on the lower wall. Upon substituting 9.1.9 and 9.1.10 into 9.1.12 we find

$$k\frac{dT}{dy} + \mu v_x \frac{dv_x}{dy} = 0$$

or, written another way,

$$\frac{d}{dy}\left(\tfrac{1}{2}v_x^2\right) = -\frac{k}{\mu}\frac{dT}{dy} \qquad 9.1.13$$

Now we are in a position to use the fact that the transport properties may be considered as a function of temperature alone (the pressure is constant). Without even stating the relations for $k(T)$ and $\mu(T)$ explicitly, we can write the integral of 9.1.13 from $y = h$ to an arbitrary point y (T' denotes a dummy integration variable):

$$\tfrac{1}{2}v_x^2 - \tfrac{1}{2}V_0^2 = -\int_{T_0}^{T} \frac{k(T')}{\mu(T')}\,dT' \qquad 9.1.14$$

This is a somewhat unusual form. We have found v_x as a function of the temperature T. Since μ and k are always positive, the integral is an increasing function of T, and in principle we may find the inverse function $T(v_x)$.

To continue, let us once again consider 9.1.9:

$$\mu(T)\frac{dv_x}{dy} = \tau_0$$

Since T is a function of v_x, this may be written as

$$\mu(T(v_x))\,dv_x = \tau_0\,dy$$

Upon integration from $y = 0$, where $v_x = 0$, to an arbitrary y, v_x, we find the relation (v_x' denotes a dummy integration variable)

$$y = \frac{1}{\tau_0}\int_0^{v_x} \mu(T(v_x'))\,dv_x' \qquad 9.1.15$$

This is the velocity profile in an inverse form $y = y(v_x)$. The temperature profile can be found in principle by inverting 9.1.15 and substituting into 9.1.14. We find the constant τ_0 in 9.1.15 by integrating 9.1.15 all the way across the wall, that is,

$$\tau_0 = \frac{1}{h}\int_0^{V_0} \mu(T(v_x'))\,dv_x' \qquad 9.1.16$$

This concludes all the necessary relations. In principle, we have an exact solution. Equations 9.1.7, 9.1.14, 9.1.15, and 9.1.16 give us the velocity and two thermodynamic properties, p and T, at every point in the flow.

Note that the thermodynamic equations of state have not been specified, and indeed they do not appear in the solution. The energy equation of state (involving the heat capacity) is absent because there is no storage or convection of energy in the problem. The second equation of state is only necessary if we want to find the density from $\rho = \rho(p, T)$.

9.2 FLOW WITH POWER-LAW TRANSPORT PROPERTIES

As a specific example we compute the results when the viscosity and thermal conductivity are governed by power laws. Assume

$$\mu = \mu_0 \left(\frac{T}{T_0} \right)^n, \qquad k = k_0 \left(\frac{T}{T_0} \right)^n \tag{9.2.1}$$

The exponent n is arbitrary at this stage. It is chosen to make 9.2.1 fit the experimental data for a given substance. The reference constants μ_0 and k_0 are the values at the upper wall, where the temperature is T_0. Substitution of 9.2.1 into 9.1.14 gives an equation where n drops out,

$$v_x^2 - V_0^2 = -2 \frac{k_0}{\mu_0} \int_{T_0}^{T} dT'$$

Performing the integration and rearranging yields the temperature–velocity relation,

$$T = T_0 + \frac{1}{2} \frac{\mu_0}{k_0} \left(V_0^2 - v_x^2 \right) \tag{9.2.2}$$

The velocity profile will be found next. We take 9.1.15 and substitute 9.2.1 to get the integral

$$y = \frac{1}{\tau_0} \int_0^{v_x} \mu_0 \left(\frac{T(v_x')}{T_0} \right)^n dv_x'$$

$$= \frac{\mu_0}{\tau_0} \int_0^{v_x} \left[1 + \frac{1}{2} \frac{\mu_0}{k_0 T_0} \left(V_0^2 - v_x'^2 \right) \right]^n dv_x' \tag{9.2.3}$$

In order to proceed further, we must specify n. We consider two cases: $n = 0$, which implies constant properties, and $n = 1$, which is more like the behavior of a perfect gas.

For the case $n = 0$, integration of 9.2.3 produces

$$y = \frac{\mu_0}{\tau_0} v_x$$

Upon evaluating τ_0 at $y = h$, we obtain the answer previously given in Chapter 7, namely

$$\frac{v_x}{V_0} = \frac{y}{h} \tag{9.2.4}$$

The corresponding temperature profile is obtained by substituting 9.2.4 into 9.2.2. It is

$$T = T_0 + \frac{1}{2}\frac{\mu_0 V_0^2}{k_0}\left[1 - \left(\frac{y}{h}\right)^2\right] \qquad 9.2.5$$

The case $n = 1$ is only slightly more complicated. By performing algebra analogous to that above, we find the velocity equation is

$$\frac{y}{h} = \frac{v_x}{V_0}\,\frac{1 + \dfrac{\mu_0 V_0^2}{2k_0 T_0}\left[1 - \dfrac{1}{3}\left(\dfrac{v_x}{V_0}\right)^2\right]}{1 + \dfrac{\mu_0 V_0^2}{3k_0 T_0}} \qquad 9.2.6$$

In this equation we cannot solve explicitly for v_x. Because of this, the corresponding equation for the temperature cannot be stated as an explicit function of y, but must remain in the form 9.2.2.

9.3 NONUNIQUENESS OF NAVIER–STOKES SOLUTIONS

When we check the analytical results against experiments we find some very interesting things happen. The analysis is confirmed for low cylinder velocities, but a transition to a new flow regime occurs if the velocity is so high that the Taylor number is above a critical value of 1700. The *Taylor number*, for the case of inner cylinder rotating and outer cylinder stationary, is defined as

$$\mathrm{Ta} = \left(\frac{V_0 h}{\nu}\right)^2 \frac{h}{R_i}$$

G. I. Taylor (1923) (English physicist, 1886–1970) provided both the experimental measurements and the theoretical analysis for this transition phenomenon. Above the critical Taylor number the flow is a three-dimensional laminar flow with a series of spiral vortices. The vortices are somewhat like doughnuts stacked on top of each other as shown in Fig. 9-2. Taylor vortices are nearly square, and each rotates in a direction opposite to its neighbor. This is obviously a case where the Navier–Stokes equations admit more than one solution for the same geometry and boundary conditions. Our previous solution gave no indication that it would not apply for all Taylor numbers. It breaks down because the assumed mathematical form of the velocity—that is, $v_\theta(r)$, $v_r = 0$, $v_z = 0$—is invalid. Compressibility is actually not an issue here;

Figure 9-2 Taylor vortices show nonuniqueness of flow patterns. Note the different number of rings in each picture. Photograph from Burkhalter and Koschmieder (1973).

Taylor vortices occur in gases or liquids. Nor are they the end of the story: still higher modes of flow occur. In these modes the Taylor cells have a wavy structure. For very high rotation rates the flow within the vortices becomes turbulent. This flow is still the subject of active research, because it is an example of bifurcation theory. Further discussion is given in Section 22.9. Setting these more advanced problems aside, the main point to be taken from this section is that solutions of the Navier–Stokes equations are not necessarily unique.

9.4 CONCLUSION

Compressible Couette flow shows us the way in which a shear flow influences the temperature profile through viscous dissipation. The temperature profile has a maximum increase in temperature equal to $(\mu_0/2k_0)V_0^2$. Note that this is true for the most general fluids irrespective of n. It is instructive to cast 9.2.2 into a nondimensional form according to the philosophy of Section 8.11. We want a nondimensional variable of the form

$$T^* = \frac{T - T_{\text{ref}}}{T_{\text{scale}}}$$

Rearranging 9.2.2, we find the equation

$$\frac{T - T_0}{\mu_0 V_0^2/k_0} = \frac{1}{2}\left[1 - \left(\frac{v_x}{V_0}\right)^2\right]$$
$$T^* = \tfrac{1}{2}[1 - v^{*2}] \qquad 9.4.1$$

We see clearly that T_0 is only a reference value. The natural temperature scale is $\mu_0 V_0^2/k_0$. This is an estimate of the temperature rise that we can expect because of viscous dissipation from a velocity gradient of order V_0/h.

PROBLEMS

9.1 (A) Solve for the compressible flow in a slot with the stationary wall held at a constant temperature T_1 that is different from the upper wall temperature T_0. What specific form does the answer take for $n = 0$ in 9.2.1?

9.2 (A) A shaft 3 cm in diameter rotates at 30,000 rpm. A stationary collar around the shaft is 2 cm long and has a gap of 0.1 mm. Estimate the temperature of oil filling the gap.

9.3 (B) Solve the Couette-flow problem of Section 9.2 where the walls form an annulus with radii r_0 and r_1.

9.4 (B) Explain why the temperature of the adiabatic wall does not depend upon the thickness of the shear layer.

10 Incompressible Flow

Incompressible flow is a main subdivision of fluid mechanics. It includes within its boundaries a great many problems and phenomena that are found in engineering and nature. Flows of gases, as well as those of liquids, are frequently incompressible. The layman is usually surprised to learn that the pattern of the flow of air can be similar to that of water. From a thermodynamic standpoint, gases and liquids have quite different characteristics. As we know, liquids are often modeled as incompressible fluids. However, "incompressible fluid" is a thermodynamical term, whereas "incompressible flow" is a fluid-mechanical term. We can have an incompressible flow of a compressible fluid.

The main criterion for incompressible flow is that the Mach number be low ($M \to 0$). This is a necessary condition. In addition, other conditions concerning heat transfer must be satisfied. There are several different situations of heat transfer under which incompressible flow can occur. In this chapter a detailed study of these situations will be made. As part of this study we shall derive the equations which govern incompressible flow. Then we shall be in a position to observe some of the general features and characteristics of incompressible flow.

10.1 CHARACTERIZATION

The term "incompressible flow" is applied to any situation where changes in the density of a particle are negligible. A mathematical definition is

$$\frac{1}{\rho}\frac{D\rho}{Dt} = 0$$

From the continuity equation we have

$$\frac{1}{\rho}\frac{D\rho}{Dt} = -\partial_i v_i = \lim_{V_{\mathrm{MR}} \to 0} -\frac{1}{V_{\mathrm{MR}}}\frac{DV_{\mathrm{MR}}}{Dt}$$

This shows that equivalent definitions are that $\nabla \cdot \mathbf{v} = 0$ (the rate of expansion is zero) or $DV_{\mathrm{MR}}/Dt = 0$ (the rate of change of the volume of a particle is zero). Notice that all particles do not have to have the same density. The only requirement is that the density of each particle remain unchanged. In some

cases—for example, the ocean, where salt content and temperature are functions of depth—the density of adjacent particles changes but any one particle has a constant density. These *stratified* flows exhibit such interesting and unusual phenomena that they constitute a separate branch of fluid mechanics. We shall not study any stratified flows in this book. It will always be assumed that the density of all particles is the same.

The density of the fluid is also governed by a thermodynamic equation of state. For a general fluid we may write this as

$$\frac{1}{\rho}\frac{D\rho}{Dt} = \alpha\frac{Dp}{Dt} - \beta\frac{DT}{Dt} \tag{10.1.1}$$

where the isothermal compressibility is

$$\alpha(p,T) = \frac{1}{\rho}\frac{\partial\rho}{\partial p}\bigg|_T$$

and the bulk expansion coefficient is

$$\beta(p,T) \equiv -\frac{1}{\rho}\frac{\partial\rho}{\partial T}\bigg|_p$$

The functions α and β are thermodynamic variables that characterize the fluid. We want to study the most general type of fluid by leaving α and β unrestricted. When that is done, the right-hand side of 10.1.1 will be small only if the pressure and temperature changes are small enough. In turn, the magnitudes of these variables are governed by dynamic processes occurring in the flow field. The energy and momentum equations will play a major role in fixing the pressure and temperature. The advantage of writing the equation of state in the form 10.1.1 is that the flow-field effects are isolated in Dp/Dt and DT/Dt while the thermodynamic character of the fluid is isolated in α and β.

With dimensional analysis fresh in our minds, we should feel a little uneasy about the statement that temperature and pressure changes are to be small. Certainly we should not use dimensional variables which compare temperature and pressure with common units of measurement. We must nondimensionalize the pressure and temperature with scales that are determined by the dynamics of the flow. Since scales for nondimensionalizing variables are found in the boundary conditions and equations that govern the flow, we must be more specific about the problem statement.

10.2 INCOMPRESSIBLE FLOW AS A LOW-MACH-NUMBER FLOW WITH ADIABATIC WALLS

Consider the two flow situations depicted in Fig. 10-1. The external flow in Fig. 10-1a has specified values of velocity v_0, density ρ_0, and temperature T_0 far away from the body. The type of fluid is given so the thermodynamic functions $\alpha(p,T)$, $\beta(p,T)$, and $c_p(p,T)$ are also known in principle. Knowledge of the

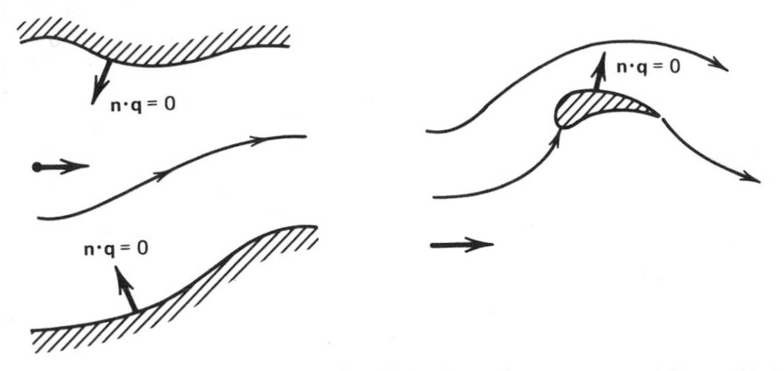

Figure 10-1 Incompressible flows with adiabatic walls: (a) external flow, (b) internal flow.

fluid also implies that equations for the transport coefficients $\mu(p, T)$ and $k(p, T)$ are available. We shall not need to specify the thermodynamic data in detail, but can perform the analysis for a general Newtonian fluid. In the case of an internal-flow problem (Fig. 10-1b), a similar specification is given at a certain reference location.

The body has a characteristic length L. It may also have other characteristic lengths, but they will not concern us, as they form geometric ratios with L when the problem is nondimensionalized. The no-slip condition on the solid surfaces requires that the velocity be zero. We also assume that the walls are adiabatic, so that no heat enters the flow through the walls. Later on in this chapter we shall do the problem again assuming a fixed wall temperature.

A list of all equations which govern the problem is as follows:

Thermodynamics:

$$\frac{1}{\rho}\frac{D\rho}{Dt} = \alpha\frac{Dp}{Dt} - \beta\frac{DT}{Dt} \tag{10.2.1}$$

with

$$\begin{aligned} \alpha &= \alpha(p, T), \qquad \beta = \beta(p, T), \qquad c_p = c_p(p, T) \\ \mu &= \mu(p, T), \qquad k = k(p, T) \end{aligned} \tag{10.2.2}$$

Continuity:

$$\frac{1}{\rho}\frac{D\rho}{Dt} = -\partial_i v_i \tag{10.2.3}$$

Momentum:

$$\rho\frac{Dv_i}{Dt} = -\partial_i p + \partial_j \tau_{ji} + \rho g_i \tag{10.2.4}$$

Net viscous stress:

$$\partial_j \tau_{ji} = -\tfrac{2}{3}\partial_i(\mu \partial_j v_j) + 2\partial_j(\mu \partial_{(j} v_{i)}) \tag{10.2.5}$$

Energy:

$$\rho c_p \frac{DT}{Dt} = \partial_i(k\partial_i T) + \Phi + \beta T\frac{Dp}{Dt} \qquad 10.2.6$$

where the viscous dissipation function Φ is defined by

$$\Phi \equiv \tau_{ij}\partial_{(j}v_{i)} = -\tfrac{2}{3}\mu(\partial_i v_i)^2 + 2\mu(\partial_{(i}v_{j)}\partial_{(j}v_{i)})$$

The boundary conditions at the reference position are

$$v_i = (v_0, 0, 0), \qquad T = T_0, \qquad \rho = \rho_0 \qquad 10.2.7$$

and on the walls

$$v_i = 0, \qquad n_i q_i = 0$$

A general fluid-flow problem for a given geometric arrangement would require the solution of 10.2.1 through 10.2.7. The dependent variables are ρ, p, T, and v_i. Note that all the equations are coupled together. For instance, the momentum equation contains terms in the density, pressure, and viscosity. These quantities depend upon the local temperature. The temperature in turn is governed by the energy equation, which contains the velocity in the convection terms. Even the most modern computers can deal effectively with these equations only for simple cases.

10.3 NONDIMENSIONAL PROBLEM STATEMENT

Not all the terms in 10.2.1 through 10.2.7 have the same importance in determining the flow solution. In order to determine which terms are large and which are small, we must cast the equations into nondimensional variables. Many of the nondimensional variables are formed in a straightforward manner using the boundary values. In this way we define

$$\begin{aligned}
x_i^* &= \frac{x_i}{L}, & t^* &= \frac{tv_0}{L} \\
v_i^* &= \frac{v_i}{v_0}, & \rho^* &= \frac{\rho}{\rho_0} \\
\alpha^* &= \frac{\alpha}{\alpha_0}, & \beta^* &= \frac{\beta}{\beta_0} \qquad 10.3.1 \\
c_p^* &= \frac{c_p}{c_{p0}}, & \mu^* &= \frac{\mu}{\mu_0} \\
k^* &= \frac{k}{k_0}, & F_i^* &= \frac{g_i}{v_0^2/L} = \hat{g}_i \mathrm{Fr}^{-2}
\end{aligned}$$

The temperature and pressure variables need some special consideration.

In incompressible flow, pressure will play the role of a force in the momentum equation. Since pressure occurs as a gradient in this equation, a reference level may be subtracted without any effect. That is,

$$\partial_i(p - p_0) = \partial_i p$$

The first step in finding the pressure scale is to substitute the definitions of 10.3.1 into the momentum equation. Then, we argue that both the pressure and inertia terms would be needed in a general incompressible-flow problem. Although there may be cases where one of the other terms is zero, there are certainly many incompressible flows where both terms are present. We temporarily use p_s to symbolize the proper scale, that is, let

$$p^* = \frac{p - p_0}{p_s}$$

Substituting into the momentum equation, we find that

$$\rho^* \frac{Dv_i^*}{Dt^*} = \frac{\rho v_0^2}{p_s} \frac{\partial p^*}{\partial x_i^*} + \text{viscous terms} + \text{body force}$$

This equation shows that the pressure term will be of the same order as the inertia terms if the nondimensional pressure is defined to be

$$p^* = \frac{p - p_0}{\rho_0 v_0^2} \qquad 10.3.2$$

When pressure changes in the flow are dominated by momentum effects, 10.3.2 is the proper nondimensional pressure variable.

Our experience with the Couette-flow (9.4.1) can help in formulating the nondimensional temperature. In the problems we have posed, one boundary value is a fixed temperature, which can serve as a reference, while the other boundary is adiabatic. If we consider the physical processes, we expect that heat (thermal energy) will be generated in the flow field by viscous dissipation. This heat will then be redistributed by conduction and carried to new places by convection. If we try the same nondimensional temperature that was used in Couette flow in the energy equation, we find that all three terms—convection, conduction, and dissipation—are of the same order. Thus we define

$$T^* = \frac{T - T_0}{\mu_0 v_0^2/k_0} = \frac{T - T_0}{\Pr v_0^2/c_{p0}} \qquad 10.3.3$$

In the second form, the Prandtl number ($\Pr = \mu_0 c_{p0}/k_0$) has been introduced. This is a standard dimensionless ratio used in heat transfer. Some workers prefer to use v_0^2/c_{p0} by itself as the temperature scale. The results will be the same in either case, because the Prandtl number is of moderate size (if we were dealing with flows where the Prandtl number took on extreme values, 0 or ∞, we would need to be more careful).

When the nondimensional variables defined above are substituted into the problem 10.2.1–10.2.7, several nondimensional groups occur. They are listed below:

$$\mathrm{Re} = \frac{\rho_0 L v_0}{\mu_0} \qquad \mathrm{Pr} = \frac{\mu_0 c_{p0}}{k_0} \qquad \gamma_0 = \frac{c_{p0}}{c_{v0}}$$

$$\mathrm{Fr}^2 = \frac{v_0^2}{gL} \quad M = \frac{v_0}{a_0} \quad A = \alpha_0 \rho_0 c_{p0} T_0 \quad B = \beta_0 T_0 \qquad 10.3.4$$

In formulating these nondimensional parameters we have used the relation for the speed of sound derived in Chapter 2 (2.17.8),

$$a_0^2 = \frac{\gamma_0}{\rho_0 \alpha_0} \qquad 10.3.5$$

The details of the substitution are simplified by noting that

$$M^2 = \frac{v_0^2}{a_0^2} = \frac{v_0^2 \rho_0 \alpha_0}{\gamma_0} \qquad 10.3.6$$

and that

$$\frac{v_0^2}{c_{p0} T_0} = \frac{M^2 \gamma_0}{A} \qquad 10.3.7$$

The final form of the mathematical problem statement in nondimensional variables is as follows:

Thermodynamics:

$$\frac{1}{\rho^*} \frac{D\rho^*}{Dt^*} = \gamma_0 M^2 \left\{ \alpha^* \frac{Dp^*}{Dt^*} - \frac{\mathrm{Pr}\, B \beta^*}{A} \frac{DT^*}{Dt^*} \right\} \qquad 10.3.8$$

Continuity:

$$\frac{1}{\rho^*} \frac{D\rho^*}{Dt^*} = -\partial_i^* v_i^* \qquad 10.3.9$$

Momentum:

$$\rho^* \frac{Dv_i^*}{Dt^*} = -\partial_i^* p^* + \partial_j^* \tau_{ji} + \hat{g}_i \mathrm{Fr}^{-2} \qquad 10.3.10$$

Viscous stress:

$$\partial_j^* \tau_{ji}^* = \frac{1}{\mathrm{Re}} \left\{ -\tfrac{2}{3} \partial_i^* \left(\mu^* \partial_j^* v_j^* \right) + 2 \partial_j^* \left(\mu^* \partial_{(j}^* v_{i)}^* \right) \right\} \qquad 10.3.11$$

Energy:

$$\rho^* c_p^* \frac{DT^*}{Dt^*} = \frac{1}{\text{Re}}\left\{\frac{1}{\text{Pr}}\partial_i^*(k^*\partial_i T^*) + \Phi^*\right\} + \beta^* B\left\{\frac{1}{\text{Pr}} + \frac{\gamma_0 M^2}{A} T^*\right\}\frac{Dp^*}{Dt^*} \qquad 10.3.12$$

where

$$\Phi^* = \frac{\Phi}{\mu_0 v_0^2/L^2} \qquad 10.3.13$$

Boundary conditions at the reference position:

$$v_i^* = (1,0,0), \qquad T^* = 0, \qquad \rho^* = 1 \qquad 10.3.14$$

Boundary conditions at the walls:

$$v_i^* = 0, \qquad n_i q_i^* = 0 \qquad 10.3.15$$

The thermodynamic functions α^*, β^*, μ^*, k^*, and c_p^* complete the problem. Assuming our guesses about the proper natural scales for the nondimensional variables are correct, then each variable is of order one. Furthermore, the nondimensionalizing process has introduced several parameters (M, γ_0, Pr, A, B, Fr, Re) into the equations. For any given flow problem these parameters have specific fixed values. If they are large or small, they magnify or diminish the effect of the terms in which they appear as coefficients.

At this stage we can see that incompressible flow will result when the right side of 10.3.8 becomes small, that is, in the limit $M^2 \to 0$. When $D\rho^*/Dt^* = 0$, the density of a particle is constant. This causes a domino effect in the remaining equations. In 10.3.9 the left side is zero, so that $\nabla^* \cdot \mathbf{v}^* = 0$. Terms containing $\nabla^* \cdot \mathbf{v}^*$ in the viscous stress and dissipation relations become small, as does the term preceded by M^2 in the energy equation. It also turns out that all the thermodynamic functions μ^*, k^*, c_p^*, α^*, and β^* are constant.

The thermodynamic functions depend upon the absolute magnitudes of the temperature and pressure. Let us consider the viscosity as an example:

$$\mu = \mu(p, T)$$

We know that at the reference state T_0, p_0 the viscosity is μ_0. Consider the nondimensional viscosity function

$$\mu^* = \frac{\mu}{\mu_0} = \mu^*\left(\frac{T}{T_0}, \frac{p}{p_0}\right) \qquad 10.3.16$$

This function may be expanded in a double Taylor series about $T/T_0 = 1$ and $p/p_0 = 1$:

$$\mu^* = 1 + \left.\frac{\partial \mu^*}{\partial (T/T_0)}\right|_{1,1} \frac{T - T_0}{T_0} + \left.\frac{\partial \mu^*}{\partial (p/p_0)}\right|_{1,1} \frac{p - p_0}{p_0} + \cdots \qquad 10.3.17$$

However, from the definition of T^* and 10.3.7 we have

$$\frac{T - T_0}{T_0} = \frac{\Pr \gamma_0}{A} M^2 T^* \qquad 10.3.18$$

and similarly, from the definition of p^* and 10.3.6,

$$\frac{p - p_0}{p_0} = \frac{\gamma_0 M^2 p^*}{p_0 \alpha_0} \qquad 10.3.19$$

In this equation $p_0 \alpha_0$ is a new dimensionless constant. Substituting 10.3.18 and 10.3.19 into 10.3.17 and allowing the Mach number to become small shows that

$$\mu^* = 1, \qquad \text{or} \quad \mu = \mu_0 = \text{const} \qquad 10.3.20$$

The same argument can be applied to all the other thermodynamic functions. The dynamic processes do not change the temperature or pressure enough to cause any appreciable change in the thermodynamic state. All the thermodynamic coefficients may be considered constants.

10.4 CHARACTERISTICS OF INCOMPRESSIBLE FLOW

The main criterion for incompressible flow is that the Mach number is low ($M \to 0$), that is, all velocities are small compared to the speed of sound. Recall that the speed of sound is given by 2.17.1:

$$a_0^2 = \left.\frac{\partial p}{\partial \rho}\right|_s \qquad 10.4.1$$

When a_0 appears in the incompressible-flow derivation, its role is not to tell how fast waves travel, but to indicate how much density change accompanies a certain pressure change. Pressure changes in the flow are of the order of $\rho_0 v_0^2$. A flow decelerated from $v = v_0$ at one location in the flow to $v = 0$ at another will undergo a pressure change $\Delta p = \frac{1}{2}\rho_0 v_0^2$ (neglecting viscosity). With these considerations the Mach number is interpreted as follows:

$$M^2 = \frac{v_0^2}{a_0^2} = v_0^2 \left.\frac{\partial \rho}{\partial p}\right|_s = \frac{\rho_0 v_0^2}{\rho_0} \left.\frac{\partial \rho}{\partial p}\right|_s$$

$$\approx \Delta p \frac{1}{\rho_0} \frac{\Delta \rho}{\Delta p} = \frac{\Delta \rho}{\rho_0} \qquad 10.4.2$$

M^2 is a measure of the size of density changes compared to the fluid density. As $M^2 \to 0$, density changes become only a small fraction of the fluid density.

Although $M^2 \to 0$ is required for incompressible flow, it is not the only requirement. Some flows where $M^2 \to 0$ are low-speed compressible flows. In these cases density changes are caused by temperature changes. In the analysis above this did not happen because the wall was adiabatic. There was not enough heat generated by viscous dissipation to cause large temperature changes.

We now list the governing equations for incompressible flow derived in the previous section:

Thermodynamics:

$$\frac{D\rho^*}{Dt^*} = 0$$

$$\rho^* = c_p^* = \alpha^* = \beta^* = \mu^* = k^* = 1 \qquad 10.4.3$$

Continuity:

$$\partial_i^* v_i^* = 0 \qquad 10.4.4$$

Momentum (recall from problem 3.16 that $2\partial_j \partial_{(j} v_{i)} = \partial_j \partial_j v_i$ if $\partial_i v_i = 0$):

$$\frac{Dv_i^*}{Dt^*} = -\partial_i^* p^* + \frac{1}{\text{Re}} \partial_j^* \partial_j^* v_i^* + \hat{g}_i \,\text{Fr}^{-2} \qquad 10.4.5$$

Energy:

$$\frac{DT^*}{Dt^*} = \frac{1}{\text{Re}} \left\{ \frac{1}{\text{Pr}} \partial_i^* \partial_i^* T^* + \partial_{(i}^* v_{j)}^* \partial_{(j}^* v_{i)}^* \right\} + \frac{B}{\text{Pr}} \frac{Dp^*}{Dt^*} \qquad 10.4.6$$

Boundary conditions at the reference location:

$$v_i^* = (1,0,0), \quad T^* = 0, \quad \rho^* = 1, \quad p^* = 0 \qquad 10.4.7$$

Boundary conditions at the walls:

$$v_i^* = 0, \qquad n_i q_i^* = 0 \qquad 10.4.8$$

Inspection of these equations reveals quite a lot about incompressible flow. First of all, the density and all thermodynamic coefficients are constants. A separate independent assumption that ρ, c_p, μ, and k are constant is not needed. It would be inconsistent to solve an incompressible-flow problem and allow viscosity to be a function of temperature. We would not obtain any greater accuracy (the viscosity would only change slightly), and the mathematics would be considerably more complicated.

Because the density and transport properties are constant, the continuity and momentum equations are decoupled from the energy equation. This is an extremely important result, as it means that we may solve for the three velocities and the pressure without regard for the energy equation or the temperature. The velocity field in incompressible flow is unaffected by heat transfer and thermal effects.

Since pressure is determined by the momentum equation, it plays the role of a mechanical force and not a thermodynamic variable. Moreover, pressure only occurs under a derivative and therefore, as we remarked previously, the level of the pressure is not important in incompressible flow. An incompressible-flow solution will determine $p^* = (p - p_0)/\rho v_0^2$ without any need to specify p_0. If in a given flow the reference p_0 is increased, the level of all pressures in the flow increases, so that p^* has the same values. The velocities and streamlines do not change when the pressure is increased.

Another important fact about incompressible flow is that only two parameters, the Reynolds number and the Froude number, Fr, occur in 10.4.5. The appearance of the Froude number is important in flows where there is a free surface. Open-channel flows, water waves, and the flow of liquid jets or sheets are examples of free-surface flows. In confined flows (i.e., those flows where the fluid occupies the entire region between walls or the entire region on the outside of a body), the gravity force produces an equivalent hydrostatic effect, which may be separated out of the flow problem. This will be discussed in detail in the next section. Meanwhile, we note that in incompressible confined flows, the Froude number does not appear explicitly in the problem.

For a given geometry, the character of a confined velocity field depends upon the single dimensionless number Re. Of course other parameters such as geometric ratios, velocity ratios, and so on may enter the problem through boundary conditions, but the Reynolds number is the only constant in the governing equations. As it appears in 10.4.5, the Reynolds number indicates the size of the viscous-force term relative to the other terms. Flow patterns change their character as the Reynolds number takes on different values. For this reason subsequent chapters dealing with flow patterns are organized according to the Reynolds number. It is only a slight exaggeration to say that the study of incompressible flow is a study of the Reynolds number.

Mathematically speaking, the momentum equation is nonlinear in the velocity v^*. The substantial derivative

$$\frac{Dv_i^*}{Dt^*} = \partial_0 v_i^* + v_j^* \partial_j^* v_i^* \qquad 10.4.9$$

contains v_i^* twice in the second term. This nonlinear term prevents the use of many of the standard mathematical techniques. It is also the cause of many interesting and unusual phenomena that occur in fluid mechanics. The equations are elliptic, and in general we specify the velocity around the surface of a region in order to determine the velocity field on the inside of the region.

Once the velocity and pressure are found, they may be substituted into the energy equation, leaving temperature as the sole dependent variable. The substantial derivative is now linear in T:

$$\frac{DT^*}{Dt^*} = \partial_0 T^* + v_j^* \partial_j^* T^* \qquad 10.4.10$$

Also notice that the temperature always occurs under a derivative everywhere in 10.4.6. This means that in incompressible flow, only changes in temperature with respect to some reference are important. As with pressure, the level of the reference temperature does not affect the solution. If ρ, μ, c_p, and k had not become constants, the absolute temperature would enter the problem through these variables. The actual temperatures in the flow do not differ from T_0 very much. This can be seen by recalling 10.3.18. When we reorganize the temperature variable T^* so that T is compared to the reference value T_0, we get

$$\frac{T - T_0}{T_0} = \frac{1}{A} \operatorname{Pr} \gamma_0 M^2 T^*$$

or

$$\frac{T}{T_0} = 1 + \frac{1}{A} \operatorname{Pr} \gamma_0 M^2 T^* \qquad 10.4.11$$

Since T^* is of order one, the actual temperatures are nearly the same as T_0. The variable T^* may be thought of as a correction of order M^2 to a uniform temperature. The problem of solving the energy equation really belongs to the subject of convective heat transfer, and we shall not pursue it after this chapter.

10.5 SPLITTING THE PRESSURE INTO HYDRODYNAMIC AND HYDROSTATIC PARTS

The body force due to gravity can be effectively eliminated from the momentum equation by defining a new pressure variable. In order to do this, we first imagine that the fluid is at rest (Fig. 10-2) and the pressure is p_0 at some reference height (this method is useful for flows without a free surface). The hydrostatic variations in pressure are governed by the momentum equation (in dimensional form) with zero velocity. It is

$$0 = -\frac{1}{\rho} \partial_i p_h + g_i \qquad 10.5.1$$

The solution of this equation gives p_h, the pressure that would exist at any point in a stationary fluid. Since the body force g_i is the same whether the fluid

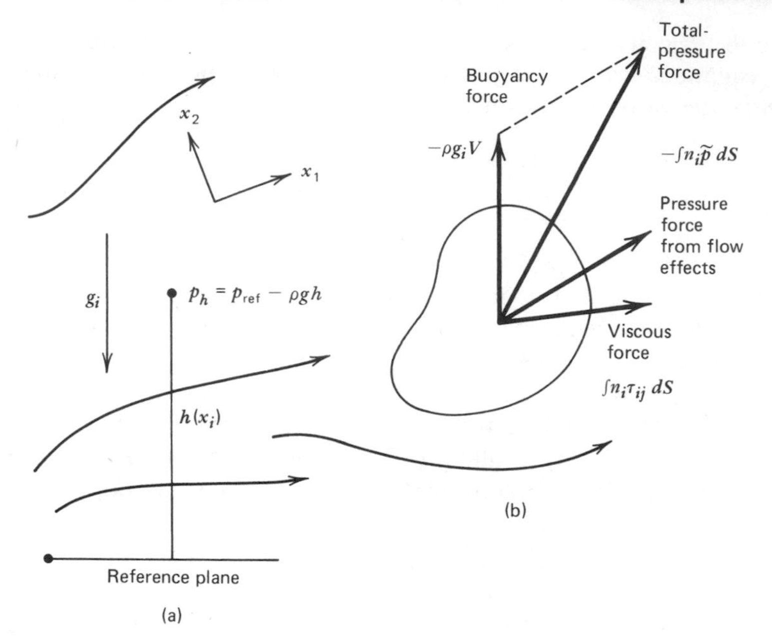

Figure 10-2 Separation of dynamic and hydrostatic effects: (a) hydrostatic component of the pressure; (b) net pressure force on a body consists of the buoyancy and the dynamic parts.

is moving or not, we can substitute for g_i in the momentum equations for the flowing fluid. This gives

$$\frac{Dv_i}{Dt} = -\frac{1}{\rho}\partial_i p + g_i + \nu \partial_j \partial_j v_i$$

$$= -\frac{1}{\rho}\partial_i (p - p_h) + \nu \partial_j \partial_j v_i$$

$$= -\frac{1}{\rho}\partial_i \tilde{p} + \nu \partial_j \partial_j v_i \qquad 10.5.2$$

where we define $\tilde{p} \equiv p - p_h$. Hence the pressure has in effect been split into two parts,

$$p = \tilde{p} + p_h \qquad 10.5.3$$

The pressure $\tilde{p}$ is the difference between the actual pressure in the flow and the hydrostatic pressure that would exist at that point if there were no flow. We can regard $\tilde{p}$ as the pressure resulting from the fact that the fluid is flowing. (We avoid the term "dynamic pressure" for $\tilde{p}$, as that term is used in engineering for the quantity $\frac{1}{2}\rho v_0^2$. Let $\tilde{p}$ be called the hydrodynamic pressure).

The local hydrostatic pressures are found from 10.5.1. In order to do this, recall that the gravity force has a magnitude g and a downward direction as shown in Fig. 10-2. Since gravity is a conservative force, it has a potential $-h(x_i)$, the distance above a horizontal reference plane:

$$g_i = g\hat{g}_i = -\partial_i[gh(x_i)] \qquad 10.5.4$$

Substituting into 10.5.1 and integrating produces the well-known equation for hydrostatic pressure,

$$p_h = p_0 - \rho g h \qquad 10.5.5$$

The combination ρg is denoted by γ in hydraulics and is called the specific weight. It is the force per unit volume. Since p_h is known, the only remaining question is to find $\tilde{p}$.

We may solve an incompressible flow and ignore the gravity force, that is, we use 10.5.2 as the momentum equation. This solution produces $\tilde{p}$, the hydrodynamic part of the pressure (and furthermore the results depend only on the Reynolds number). The true pressures in the flow include the hydrostatic pressure. Expressing 10.5.3 in the nondimensional form compatible with Section 10.3 gives

$$p^* = \frac{p - p_0}{\rho v_0^2} = \frac{\tilde{p} + p_h - p_0}{\rho v_0^2} = \tilde{p}^* - \mathrm{Fr}^{-2}\frac{h}{L}$$

Thus, in a confined flow the Froude number is merely a scale factor to indicate the level of the hydrostatic pressure component. Since the hydrostatic component has a simple universal solution, we really only need to find the dynamic portion of the pressure for any problem of interest. It is common practice to strike the gravity force from the momentum equation and let the symbol p stand for the hydrodynamic part of the pressure without any special notation.

The total pressure force on a body is frequently of interest. The net pressure force (Fig. 10-2b) is given by the integral

$$F_i^{(p)} = -\int n_i p \, dS \qquad 10.5.6$$

We introduce the pressure splitting into the integration by substituting 10.5.3 into 10.5.6. This gives

$$F_i^{(p)} = -\int n_i \tilde{p} \, dS - \int n_i p_h \, dS \qquad 10.5.7$$

The second integral is the net force on the body due to hydrostatic pressure—the buoyancy force that the body would experience if the flow were absent. The buoyancy force may be computed by substituting 10.5.5 and using

Gauss's theorem:

$$F_i^{(p)} = -\int n_i \tilde{p}\, dS - \int n_i p_0\, dS + \int n_i \rho g h\, dS$$

$$= -\int n_i \tilde{p}\, dS - p_0 \int n_i\, dS + \rho g \int \partial_i h\, dV$$

Noting that $\partial_i h = -\hat{g}_i$ from 10.5.4, we obtain

$$\underset{\substack{\text{total}\\ \text{pressure}\\ \text{force}}}{F_i^{(p)}} = \underset{\substack{\text{pressure}\\ \text{force due}\\ \text{to flow}}}{-\int n_i \tilde{p}\, dS} - \underset{\substack{\text{buoyancy}\\ \text{force}}}{\rho g_i V} \qquad 10.5.8$$

This is the familiar result that the buoyancy force is equal to the weight of fluid the body displaces and is in the direction opposite to the gravity vector. Equation 10.5.8 says that the buoyancy force and the pressure force due to flow may be separated in an unambiguous way in incompressible flow.

*10.6 MATHEMATICAL ASPECTS OF THE LIMIT PROCESS $M^2 \to 0$

From a mathematical viewpoint the complete compressible Navier–Stokes problem stated in Sections 10.2 and 10.3 has six dependent variables v_i, ρ, p, and T. Consider v_i as a typical variable. It is a function of position and a number of nondimensional constants,

$$v_i^* = f\left(x_i^*; M, \text{Re}, \text{Fr}, \text{Pr}, \gamma_0, A, B, p_0\alpha_0\right) \qquad 10.6.1$$

If we knew this complete answer we could set $M = 0$ and obtain the incompressible result

$$v_{i\text{IC}}^* = f\left(x_i^*; 0, \text{Re}, \text{Fr}\right) \qquad 10.6.2$$

The other parameters are not listed, because we found that they dropped out of the problem when $M^2 \to 0$.

Of course, the Mach number is never exactly zero for any flow. However, if 10.6.1 is not a strong function of M, the answer with $M = 0$ is a good approximation for flows with a small but nonzero M. As a general rule of experience, if $M < \frac{1}{3}$, then 10.6.2 will give a fairly close result. We usually do not know how fast 10.6.1 actually changes with M, as that depends on the exact shape of the body or the flow passage.

It is actually a much too complicated problem to find the complete answer to 10.6.1 and then set $M = 0$. The procedure we followed was to take the limit

$M \to 0$ in the governing equations and hope that the resulting equations would produce 10.6.2 when they were solved. This amounts to an interchange of differentiation and a limiting process. On a typical term the process would be

$$\lim_{M \to 0} \partial_i^* v_j^* \stackrel{?}{=} \partial_i^* \left(\lim_{M \to 0} v_j^* \right) = \partial_i^* v_{j\mathrm{IC}}^* \qquad 10.6.3$$

Moving the limiting process inside the derivative cannot always be justified mathematically. In this case, and in most other cases in fluid mechanics, it works out correctly.

Another danger in letting $M \to 0$ in the governing equations is that the variables may not have been nondimensionalized properly. As $M \to 0$ all variables must be nondimensionalized so that they are of order one, that is, they have finite values at $M = 0$. If a variable went to zero (or infinity) as $M \to 0$ and we did not know it, we would obtain the wrong equation (Fig. 10-3). In nondimensionalizing the variables we are actually making guesses as to how they vary with M in the neighborhood of $M = 0$.

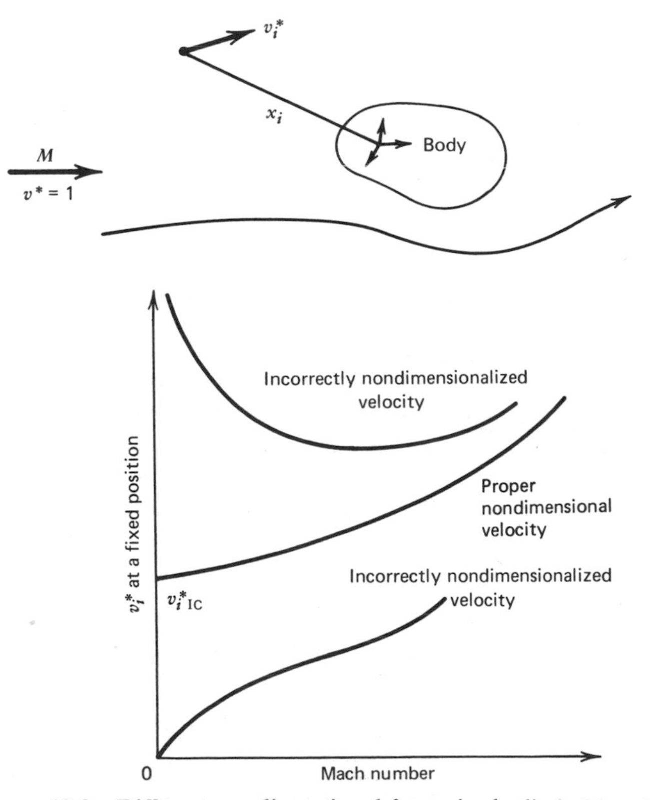

Figure 10-3 Different nondimensional forms in the limit $M \to 0$.

An example of improper nondimensionalization was given in Section 8.11. The problem for plane Couette flow with a pressure gradient gave the result

$$\tilde{u} = \frac{u}{-\dfrac{L^2}{\mu}\dfrac{dp}{dx}} = \frac{1}{\mathbb{P}} y^* + \tfrac{1}{2}\left(y^* - y^{*2}\right) \qquad 8.14.7$$

where

$$\mathbb{P} \equiv -\frac{L^2}{U\mu}\frac{dP}{dx}$$

The velocity $\tilde{u}$ is not properly nondimensionalized for the limit $\mathbb{P} \to 0$. In order to have a finite expression we must rearrange the equation and use a different nondimensional velocity,

$$u^* = \frac{u}{U} = \mathbb{P}\tilde{u} = y^* + \tfrac{1}{2}\mathbb{P}\left(y^* - y^{*2}\right)$$

The variable $\tilde{u}$ is infinite as $\mathbb{P} \to 0$. On the other hand, the limit would be valid if u^* were used.

*10.7 INVARIANCE OF INCOMPRESSIBLE-FLOW EQUATIONS UNDER UNSTEADY MOTION

The incompressible-flow equations satisfy a special invariance that allows some unsteady flows to be analyzed from a moving coordinate system. As an example, suppose a body is oscillating back and forth in a fluid. We could analyze this problem by using a coordinate system fixed in the body and by applying the usual incompressible-flow equations. There would be no special terms needed to account for the acceleration of the coordinate system.

We begin the proof by letting x_i, t be an inertial reference frame. As shown in Fig. 10-4, the origin of the noninertial frame is moving with a velocity $V_i(t)$, which is completely arbitrary in both magnitude and direction. However, rotation of the $\hat{x}_i$ system is not allowed. We will prove that the same equations govern the flow in the moving system as in the inertial system. The coordinates and velocities are related by the transformations below:

$$\hat{x}_i = x_i - \int_0^t V_i(t')\, dt'$$

$$\hat{t} = t \qquad 10.7.1$$

$$\hat{v}_i = v_i - V_i$$

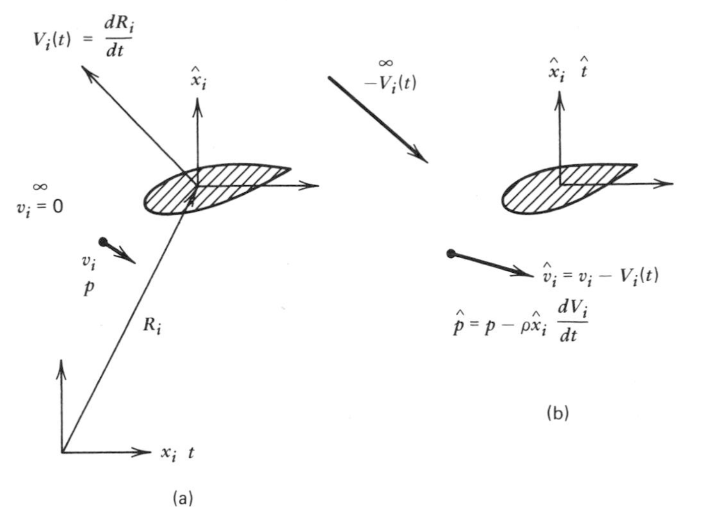

Figure 10-4 Invariance of incompressible equations under an unsteady translation: (a) body moving through an infinite fluid with translation $V_i(t)$, (b) equivalent flow in body-fixed coordinates.

The Galilean transformation is a special case when V_i is constant. From these equations the partial derivatives are computed as

$$\frac{\partial}{\partial x_i} = \frac{\partial}{\partial \hat{x}_i} \quad \text{and} \quad \frac{\partial}{\partial t} = \frac{\partial}{\partial \hat{t}} - V_i \frac{\partial}{\partial \hat{x}_i} \tag{10.7.2}$$

The continuity equation is unchanged by the transformation because

$$\frac{\partial v_i}{\partial x_i} = \frac{\partial}{\partial \hat{x}_i}(\hat{v}_i + V_i) = \frac{\partial \hat{v}_i}{\partial x_i} = 0 \tag{10.7.3}$$

The momentum equation in inertial coordinates is

$$\frac{\partial v_j}{\partial t} + v_i \frac{\partial v_j}{\partial x_i} = -\frac{1}{\rho}\frac{\partial p}{\partial x_j} + \nu \frac{\partial^2 v_j}{\partial x_i \partial x_i} \tag{10.7.4}$$

This is transformed into

$$\frac{\partial \hat{v}_j}{\partial \hat{t}} - V_i \frac{\partial \hat{v}_j}{\partial \hat{x}_i} + \frac{\partial V_j}{\partial \hat{t}} + \hat{v}_i \frac{\partial \hat{v}_j}{\partial \hat{x}_i} + V_i \frac{\partial \hat{v}_j}{\partial \hat{x}_i} = -\frac{1}{\rho}\frac{\partial p}{\partial \hat{x}_j} + \nu \frac{\partial^2 \hat{v}_j}{\partial \hat{x}_i \partial \hat{x}_i}$$

The two terms in V_i cancel. Now the unsteady-coordinate acceleration term can be rewritten as follows:

$$\frac{\partial V_j}{\partial \hat{t}} = \frac{dV_j}{dt} = \delta_{ij}\frac{dV_i}{dt} = \frac{\partial \hat{x}_i}{\partial \hat{x}_j}\frac{dV_i}{dt} = \frac{\partial}{\partial \hat{x}_j}\left(\hat{x}_i \frac{dV_i}{dt}\right) \tag{10.7.5}$$

In this form we may include this term with the pressure and define a new pseudopressure variable $\hat{p}$ according to

$$\hat{p} \equiv p + \rho \hat{x}_i \frac{dV_i}{dt} \qquad 10.7.6$$

The momentum equation now takes on the same form as in an inertial system, namely,

$$\hat{\partial}_0 \hat{v}_j + \hat{v}_i \hat{\partial}_i \hat{v}_j = -\frac{1}{\rho} \hat{\partial}_j \hat{p} + \nu \hat{\partial}_i \hat{\partial}_i \hat{v}_j \qquad 10.7.7$$

These equations are illustrated in Fig. 10-4, where a body is depicted moving through an infinite fluid without rotating. The fluid at infinity is at rest and a uniform pressure. The equivalent problem shown in (b) has a fixed body with the flow at infinity $V_i(t)$. The velocities in the two problems are related by 10.7.1 and the pressures by 10.7.6. The pressures in (b) consist of the pressure p due to the flow pattern of (a) and a pressure $\rho \hat{x}_i \, dV_i/dt$ required to accelerate the fluid. Since the fluid extends to infinity, the pressure at infinity will become infinite. It takes an infinite force to accelerate an infinite amount of fluid.

An alternate viewpoint is to retain the pressure equivalence between the two flows, that is, $p = \hat{p}$. Now the term $-dV_j/dt$ is added to the problem in $\hat{x}_i$ coordinates as an effective body-force term. This may be viewed as an imaginary body force required to accelerate the stream at infinity.

*10.8 LOW-MACH-NUMBER FLOWS WITH CONSTANT-TEMPERATURE WALLS

We have seen how $M \to 0$ with adiabatic walls is an example of incompressible flow. In may instances there is significant heat transfer through the walls. In this case we can isolate the flow situation by imagining that the walls are held at some fixed temperature T_w that is different from T_0. If the wall actually has a temperature distribution along its surface, then T_w is an estimate (such as the maximum) of the wall temperature. We may not be able to find the exact distribution of the wall temperature without solving a problem including the interior of the wall, but this is not important. As long as we can estimate T_w, we can assume some external agent exists which will give the proper wall temperatures on the flow boundaries. This new problem is dominated by different physical processes than the adiabatic wall problem. We must redo the nondimensional scales for the temperature and reanalyze the results. When this is done, we shall find that the flow is a compressible flow as long as $(T_0 - T_w)/T_0$ is finite. The further special case of small temperature differences will turn out to be an incompressible flow.

The problem now has a characteristic temperature scale $T_0 - T_w$, which is a driving force for the conduction of heat from the walls into the fluid. Since we expect that all temperatures will lie between these two values, the proper nondimensional temperature is

$$\hat{T} = \frac{T - T_0}{T_w - T_0} \qquad 10.8.1$$

The temperature variable T^* for the adiabatic wall problem is related to $\hat{T}$ by the equation

$$T^* = \frac{T_w - T_0}{T_0} \frac{A}{\gamma_0 \Pr M^2} \hat{T} \qquad 10.8.2$$

Since we expect no change in the nondimensional form of any other variables, we can substitute 10.8.2 into the adiabatic wall analysis (10.3.8 through 10.3.17) and again let the Mach number approach zero.

Temperature does not appear in the continuity and momentum equations, so the previous forms 10.3.9 and 10.3.10 remain unchanged. The equation of state is revised by substituting 10.8.2 into 10.3.8. This yields

$$\frac{1}{\rho^*} \frac{D\rho^*}{Dt^*} = \gamma_0 M^2 \alpha^* \frac{Dp^*}{Dt^*} - \beta^* B \left(\frac{T_w - T_0}{T_0} \right) \frac{D\hat{T}}{Dt^*} \qquad 10.8.3$$

When 10.8.2 is substituted into 10.3.12, the energy equation changes to

$$\rho^* c_p^* \frac{D\hat{T}}{Dt^*} = \frac{1}{\Pr \mathrm{Re}} \partial_i^* (k^* \partial_i^* \hat{T}) + \frac{\gamma_0 M^2}{A \, \mathrm{Re}} \left(\frac{T_0}{T_w - T_0} \right) \Phi^*$$

$$+ \frac{\beta^* B \gamma_0 M^2}{A} \left[\frac{T_0}{T_w - T_0} + \hat{T} \right] \frac{Dp^*}{Dt^*} \qquad 10.8.4$$

A typical transport property is recast in terms of $\hat{T}$ by inserting 10.8.1 and 10.3.19 into 10.3.17:

$$\mu^* = 1 + \frac{d(\mu/\mu_0)}{d(T/T_0)} \left(\frac{T_w - T_0}{T_0} \right) \hat{T} + \frac{d(\mu/\mu_0)}{d(p/p_0)} \frac{\gamma_0}{p_0 \alpha_0} M^2 p^* + \cdots \qquad 10.8.5$$

The boundary conditions at the reference position are

$$v_i^* = (1, 0, 0), \qquad \hat{T} = 0, \qquad \rho^* = 1 \qquad 10.8.6$$

and on the walls

$$v_i^* = 0, \qquad \hat{T} = 1 \text{ (or a known function)}$$

The limit $M^2 \to 0$ gives a low-speed compressible flow where the thermodynamic state equation shows that the density changes only because of the large changes in temperature:

$$\frac{1}{\rho^*}\frac{D\rho^*}{Dt^*} = -\beta^* B\left(\frac{T_w - T_0}{T_0}\right)\frac{D\hat{T}}{Dt^*} \qquad 10.8.7$$

The energy equation for the flow shows that convection and conduction determine the temperature field:

$$\rho^* c_p^* \frac{D\hat{T}}{Dt^*} = \frac{1}{\Pr \mathrm{Re}}\partial_i^*\left(k^* \partial_i \hat{T}\right) \qquad 10.8.8$$

Transport properties and thermodynamic functions such as c_p^* are not constants, but depend upon temperature:

$$\mu^* = 1 + \frac{d(\mu/\mu_0)}{d(I/T_0)}\left(\frac{T_w - T_0}{T_0}\right)\hat{T} + \cdots \qquad 10.8.9$$

The continuity and momentum equations no longer simplify, but must be considered with their temperature dependence. The complete set of equations is coupled together through the transport properties and the density. They apply to flows where the walls supply significant heating to the fluid. Problems of natural or free convection are of this type. In this case "significant" means large enough to cause the density to change.

The low-speed compressible-flow equations contain a new parameter $(T_w - T_0)/T_0$. This parameter compares the temperature changes in the flow, as measured by $T_w - T_0$, with the absolute temperature. In many flows of practical engineering interest this is a small number, which suggests that we consider the special cases where $(T_w - T_0)/T_0 \to 0$. When this limiting process is applied to 10.8.7 and 10.8.9, we again retrieve an incompressible flow with constant properties: low-Mach-number flow about a body with a small temperature difference. This flow is governed by the same equations as the adiabatic case, except that the energy equation is a little different. Equation 10.8.8 has only convection and conduction terms, while for the adiabatic case 10.4.6 also includes a viscous dissipation and a pressure term. Many heat-transfer problems studied in textbooks fall into the category where the simplified energy equation 10.8.8 may be used. These flows have the same general characteristics of incompressible flow that were discussed in connection with the adiabatic-wall case. A typical isothermal-wall problem with a small temperature difference would have $T_w - T_0 \approx 50°$ R (30° K) and an absolute temperature of 500° R (300° K); thus $(T_w - T_0)/T_0 \approx 0.1$.

The low-Mach-number flow over a body with a small temperature difference is a double limiting process. We found that incompressible flow results when

$$\lim_{\Delta T/T_0 \to 0}\left\{\lim_{M\to 0}[\text{Navier–Stokes}]\right\}$$

Notice that 10.8.4 contains expressions like $M^2T_0/\Delta T$, which go to zero in this limit. It is important to realize that if the order of limiting is interchanged, 10.8.8 does not result from 10.8.4. A true mathematical limit does not exist at the point $\Delta T/T_0 = 0$, $M^2 = 0$. Incompressible flow exists for any limit

$$\lim_{\substack{M\to 0 \\ \text{and} \\ \Delta T/T_0 = f(M)}} [\text{Navier–Stokes}]$$

where

$$\frac{M^2}{\Delta T/T_0} = \frac{M^2}{f(M)} \to 0$$

Most engineering cases meet these conditions. We now discuss those cases that do not.

Let us consider what happens if the temperature difference is extremely small, say five or ten degrees. For this we should do some careful thinking about the temperature scale for forming the nondimensional temperature. The temperature scale is our best guess as to how large temperature variations in the flow are going to be; it is an estimate of the maximum minus the minimum. If the profile is dominated by conduction from the walls, then the proper temperature scale is

$$\Delta T_{\text{isothermal wall}} = T_w - T_0$$

However, there is always viscous dissipation generating heat within the fluid. This effect might also increase the temperature a few degrees. So for very small $T_w - T_0$ our guess might be invalid. Recall that the adiabatic-wall temperature scale 10.3.3 is

$$\Delta T_{\text{adiabatic wall}} = \frac{\Pr v_0^2}{c_{p0}} = \frac{1}{A} \Pr \gamma_0 T_0 M^2$$

If we use air as an example $[\gamma_0/A = \gamma_0 - 1 = 0.4]$ and we let $T = 500°$ R and $M = 0.3$, then the resulting rise in temperature is $\Delta T_{\text{adiabatic}} = 13°$ R (7° C).

In order to decide which temperature scale to use, we can form the ratio

$$\frac{\Delta T_{\text{adiabatic}}}{\Delta T_{\text{isothermal}}} \approx \frac{\Pr v_0^2}{c_{p0}(T_w - T_0)}$$

$$= \frac{M^2 \gamma_0 \Pr}{A} \left(\frac{T_0}{T_w - T_0} \right) \qquad 10.8.10$$

If this variable is small, the isothermal analysis will be valid; if it is large, the adiabatic analysis will apply. It is of course possible that the two temperature scales are of the same order. In this case the temperature field is not dominated by either conduction or viscous dissipation. In principle we could use either temperature scale for this situation. In such cases the adiabatic scale and energy equation 10.4.6 are appropriate as they stand. This equation contains conduction, pressure, work, and dissipation, as the problem demands. The parameter we defined in 10.8.10 is related to the Eckert number found in the heat transfer literature. Since Pr is of order one for all common fluids (nonmetals), the Eckert number essentially indicates the relative influence of conduction and viscous dissipation in the heat-transfer process.

*10.9 THE ENERGY-EQUATION PARADOX

The incompressible flow energy equation we derived in the previous section balances the convection terms and the conduction terms.

$$\rho c_p \frac{DT}{Dt} = k\nabla^2 T \qquad 10.9.1$$

This equation is valid for a low-Mach-number flow (with a small imposed temperature difference). It was derived as a simplification of the complete thermal energy equation,

$$\rho c_p \frac{DT}{Dt} = k\nabla^2 T + \Phi + \beta T \frac{Dp}{Dt} \qquad 10.9.2$$

Now, as an alternative, we might have started the analysis with a different form of the energy equation in which c_v appears rather than c_p. The complete energy equation in terms of c_v is

$$\rho c_v \frac{DT}{Dt} = k\nabla^2 T + \Phi - T \left.\frac{\partial p}{\partial T}\right|_\rho \nabla \cdot \mathbf{v} \qquad 10.9.3$$

If we take an off-hand look at this equation and try to guess which terms could be neglected in incompressible flow, we would strike out the last term because $\partial_i v_i \approx 0$ and the viscous dissipation term because we have seen in our previous work that it is negligible for small Mach numbers. That would leave 10.9.3 as a balance between convection terms and conduction terms, but with an important difference. Equation 10.9.1 has c_p as a coefficient, while the simplified version of 10.9.3 has c_v as a coefficient. This is the paradox. The correct equation 10.9.1 implies that convection of enthalpy is balanced by heat conduction, whereas the appearance of c_v in 10.9.3 implies that internal energy is convected.

We cannot resolve this paradox by explaining that for *incompressible fluids* the differences between c_v and c_p vanish. While this is a true statement, it misses the point. The case under discussion is the incompressible flow of a *compressible fluid*. Gases are very compressible fluids, and c_p for them is distinctly different from c_v no matter what the flow situation is.

In order to resolve the paradox and convince ourselves that the analysis of the previous section that produces 10.9.1 is correct, we shall redo the analysis starting from 10.9.3. This will show that our off-hand guesses were wrong. When nondimensional variables are introduced into 10.9.3, we arrive at

$$\frac{\rho^* c_v^*}{\gamma_0} \frac{D\hat{T}}{Dt^*} = \frac{1}{\mathrm{Re}\,\mathrm{Pr}} \nabla^* \cdot (k^* \nabla^* \hat{T}) + \frac{T_0 M^2}{\Delta T A} \Phi^*$$

$$- \left(\hat{T} + \frac{T_0}{\Delta T} \right) \frac{B\beta^*}{A\alpha^*} \nabla^* \cdot \mathbf{v}^* \qquad 10.9.4$$

In computing the last term, the following thermodynamic identity has been used:

$$\left. \frac{\partial p}{\partial T} \right|_\rho = \frac{\beta}{\alpha} \qquad 10.9.5$$

The limit $M^2 \to 0$ applied to 10.9.4 removes the dissipation term, just as in Section 10.8. We are left with the equation,

$$\frac{\rho^* c_v^*}{\gamma_0} \frac{D\hat{T}}{Dt^*} = \frac{1}{\mathrm{Re}\,\mathrm{Pr}} \nabla^* \cdot (k^* \nabla^* \hat{T}) - \left(\hat{T} + \frac{T_0}{\Delta T} \right) \frac{B\beta^*}{A\alpha^*} \nabla^* \cdot \mathbf{v}^* \qquad 10.9.6$$

The second limit process, that for small temperature differences $\Delta T / T_0 \to 0$, cannot be applied to 10.9.6 without some rearranging. The difficulty is the term containing

$$\frac{T_0}{\Delta T} \nabla^* \cdot \mathbf{v}^*$$

This is recognized as an indeterminate form $\infty \cdot 0$, since we know that $\partial_i^* v_i^*$ will become zero in the limit $\Delta T / T_0 \to 0$. The continuity equation 10.3.9 and the state equation 10.8.7 show that

$$- \nabla^* \cdot \mathbf{v}^* = \frac{1}{\rho^*} \frac{D\rho^*}{Dt^*} = - \frac{\Delta T}{T_0} B\beta^* \frac{D\hat{T}}{Dt^*} \qquad 10.9.7$$

When 10.9.7 is substituted into 10.9.6 we find that the term with $\partial_i^* v_i^*$ may be taken to the right hand side. We now have

$$\left[\frac{\rho^* c_v^*}{\gamma_0} + \frac{B^2 \beta^{*2}}{A\alpha^*} \left(\frac{\Delta T}{T_0} + 1 \right) \right] \frac{D\hat{T}}{Dt^*} = \frac{1}{\mathrm{Pr}\,\mathrm{Re}} \nabla^* \cdot (k^* \nabla^* \hat{T}) \qquad 10.9.8$$

It takes just a little more work to show that the coefficient in square brackets is actually the proper c_p^* term. From thermodynamic theory we have the relation

$$c_p = c_v + \frac{T}{\rho}\frac{\beta^2}{\alpha} \qquad 10.9.9$$

The nondimensional form of this equation is

$$\rho^* c_p^* = \frac{\rho^* c_v^*}{\gamma_0} + \frac{B^2 \beta^{*2}}{A\alpha^*}\left(1 + \hat{T}\frac{\Delta T}{T}\right) \qquad 10.9.10$$

Comparing 10.9.10 with the bracket in 10.9.8, we get the final form. Thus, the final form of 10.9.8 does indeed have $\rho^* c_p^*$ preceding the substantial derivative of the temperature. Our more detailed analysis shows that 10.9.3 does reduce to 10.9.1 for incompressible flow. The paradox is explained by noting that when the limit $\Delta T/T_0 \rightarrow 0$ is applied to 10.9.3, the convection and conduction terms are just as small as the term containing $\partial_i v_i$. Thus, all three terms are important. A slight rearrangement of the $\partial_i v_i$ term using the continuity and state equations allows that term to be combined with the c_v term to produce the correct term where c_p is the coefficient of the substantial derivative.

We have illustrated the energy-equation paradox for incompressible flow with constant-temperature boundaries. The same paradox arises in the case of incompressible flow where the boundaries are adiabatic. To resolve the paradox in this case, one must nondimensionalize 10.9.3 using the temperature variable T^* appropriate to the adiabatic wall problem. Some algebraic steps similar to those above lead to the proper form of the energy equation 10.3.12 where c_p^* is the coefficient.

10.10 CONCLUSION

Incompressible flows require low Mach numbers for adiabatic flow boundaries and low Mach numbers plus small temperature differences for boundaries with prescribed temperatures. In either case the flows have effectively constant properties and the fluid-flow events are independent of the heat-transfer events. This is, in itself, a considerable simplification.

Only two nondimensional parameters, the Froude and Reynolds numbers, occur in the incompressible flow equation. In free-surface flows the Froude number is important; however, in confined flows its effect can be isolated into a hydrostatic pressure of no dynamic significance. The Reynolds number, on the other hand, is always a significant quantity in incompressible flow. The flow behaves differently and is dominated by different mechanisms as the Reynolds number changes.

PROBLEMS

10.1 (A) What are the values of α and β for air and water at standard conditions?

10.2 (A) For air, sketch a graph of $\mu(T)$ from 0 to 100°C. Replot μ/μ_0 as a function of T/T_0 for $T_0 = 30°C$. Consider air at 30°C flowing at $M = 0.3$. At a certain point in the flow $T^* = 1$. What is the value of μ/μ_0 at this point?

10.3 (A) Air at standard temperature and pressure is flowing at $M = 0.4$. Estimate the stagnation pressure. Estimate the density change between the free stream and the stagnation point if the flow is isentropic. What is the fractional change in density?

10.4 (B) A thin plastic garden hose has a nozzle at the end to control the flow from zero to the maximum, which is determined by the hydrant pressure. Is the flow in the hose independent of the level of pressure p_0 at the entrance to the nozzle?

10.5 (B) An incompressible flow has a certain velocity U_0 in a region of size L. If water and air are both used in this situation, which fluid is the more viscous?

10.6 (B) The pressure at the shoulder of a sphere, 1 ft in diameter, is measured as 2104.1 psf when it is moving at 100 ft/sec in standard air ($\rho = 0.00238$ slugs/ft^3, $p = 2116$ psf). Another sphere, 5 ft in diameter, is moving horizontally at a depth of 175 ft below the surface of a lake. Find the pressure at the upper and lower shoulder positions of the sphere if it is moving at 0.606 ft/sec.

10.7 (B) The center of a cylinder of radius r_0 oscillates according to $X_0 = A \sin \Omega t$ in a fluid that is at rest at infinity. If this problem is to be solved using a coordinate system fixed to the cylinder, what are the proper boundary conditions on the pressure and velocity? Assume the solution results in a surface pressure distribution $\hat{p}_s(\hat{\theta}) = F(\hat{\theta}) \cos \omega \hat{t}$. What are the true pressures on the surface?

11 Flows that are Exact Solutions of the Navier–Stokes Equations

The problems we investigate in this chapter are simplified situations that allow an exact mathematical answer. They will be of great use to us in learning how pressure and viscous forces exert their influence to produce different flow patterns. Consider the momentum equation for incompressible flow:

$$\partial_0 v_i + v_j \partial_j v_i = -\frac{1}{\rho}\partial_i p + \nu \partial_j \partial_j v_i$$

There are four different terms in this equation: local acceleration, convective acceleration, pressure forces, and viscous forces. Since the density is constant, we can incorporate it into the nondimensional pressure and it will be eliminated from the problem. The viscosity, on the other hand, will normally occur as a parameter in the solution, usually in the form of a Reynolds number. With two exceptions, all of the problems in this chapter are so simple that only the viscous term and one other term are nonzero. With only two nonzero terms one can always incorporate the viscosity into the definition of a nondimensional distance variable. Hence, ν no longer appears explicitly in the solution, and the velocity profile is independent of the Reynolds number. This is only a mathematical result. Experimentally, the solution may exist only in a certain range of Reynolds numbers, usually at low values, and a more complicated pattern or a turbulent flow is found for other Reynolds numbers.

On the mathematical side, this chapter offers examples of a variety of methods for solving partial differential equations. Separation variables, splitting and transforming dependent variables, similarity solutions, and finally a numerical technique for ordinary differential equations are illustrated. Whenever it is feasible, we give sufficient mathematical details so that the reader may supply the intermediate steps without undue effort. However, in a few instances the student will be asked to accept a result or be prepared for a lengthy mathematical exercise.

Another mathematical problem concerns notation. When we get down to the details of solving the Navier–Stokes equations, it is inconvenient to use the index notation (x_1, x_2, x_3, v_1, v_2, v_3, etc.). For a rectangular Cartesian coordinate system we shall change to the standard symbols x, y, z for the coordinates and u, v, w for the corresponding velocities. In cylindrical and spherical coordinates we shall use r, θ, z and r, θ, φ respectively. The velocity components in these systems will be v_r, v_θ, v_z and v_r, v_θ, v_φ.

11.1 PRESSURE-DRIVEN FLOW IN TUBES OF VARIOUS CROSS SECTIONS; AN ELLIPTICAL TUBE

Consider the flow in a tube of arbitrary cross section, and assume that the velocity has only one component, which is along the tube axis. This assumption may not be correct, even if we can find such a solution to the flow equations. There may also be other solutions to the flow equations, which contain other velocity components and secondary flows. The flow pattern we get experimentally may depend upon the transient events by which the flow is established. Although there are not a lot of detailed measurements, it is generally thought that the axial solutions are valid at low Reynolds numbers—for example, in the flow through the passageways of porous materials.

Let us proceed with the solution by noting that if u is the only velocity component, the momentum equations in directions perpendicular to the flow tell us that there is no pressure gradient in those directions. We may therefore conclude that the pressure is a function of x only. The x-direction momentum equation becomes

$$0 = -\frac{dp}{dx} + \mu\frac{\partial^2 u}{\partial y^2} + \mu\frac{\partial^2 u}{\partial z^2} \qquad 11.1.1$$

As in a circular pipe, the density drops out of the problem and we surmise that the velocity is determined solely by the pressure gradient and the absolute viscosity. Since dp/dx is a function of x alone and u is a function of y and z, the only way 11.1.1 can be satisfied is if dp/dx is constant. Therefore, we take dp/dx as a prescribed number. At either end of the tube we must have some external agent that provides the pressure difference and drives the flow. The mathematical statement of the problem is completed by the no-slip condition,

$$u(\text{wall}) = 0 \qquad 11.1.2$$

This problem has the same mathematical form as problems for the torsion of a rod or the deflection of a thin membrane loaded with a uniform pressure. Equation 11.1.1 is a Poisson equation and is of elliptic type. This means that if we change the shape of the boundary in one small region of the wall, it will affect the solution everywhere across the cross section.

Next, we undertake to nondimensionalize the problem. Let L be a characteristic dimension of the cross section. We use L to nondimensionalize the space variables:

$$y^* = \frac{y}{L}, \qquad z^* = \frac{z}{L} \qquad 11.1.3$$

There is no characteristic velocity in the problem, so we must form one from the constants in the differential equation. By trial and error we find that the differential equation will have a simple form if we define the nondimensional velocity as

$$u^* = \frac{u}{-\dfrac{L^2}{\mu}\dfrac{dp}{dx}} \qquad 11.1.4$$

(From another point of view one may find the characteristic velocity by asking: how can a quantity with dimensions L/T be formed from μ, L, and dp/dx?) In this type of problem the pressure force balances the net viscous force everywhere in the flow field (see 11.1.1 above), so it is appropriate that the ratio of dp/dx to the viscosity is a measure of the maximum velocity. The problem in nondimensional variables is

$$\frac{\partial^2 u^*}{\partial y^{*2}} + \frac{\partial^2 u^*}{\partial z^{*2}} = -1 \qquad 11.1.5$$

Specifying the shape of the cross section completes the problem.

We continue the solution for a tube that has an elliptical cross section as shown in Fig. 11-1. Solutions for a wide variety of shapes are given by Berker (1963). The equation for the wall location is

$$y_w^{*2} + Kz_w^{*2} = 1, \qquad \text{where} \quad K = \left(\frac{a}{b}\right)^2 \qquad 11.1.6$$

In this equation the semiaxes of the ellipse are a in the x direction and b in the y direction. We have chosen a as the characteristic length previously denoted by L.

One approach to solving a Poisson equation is to redefine the dependent variable so that it becomes a Laplace equation. Consider a new dependent variable

$$U = u^* + C_1 y^{*2} + C_2 z^{*2} \qquad 11.1.7$$

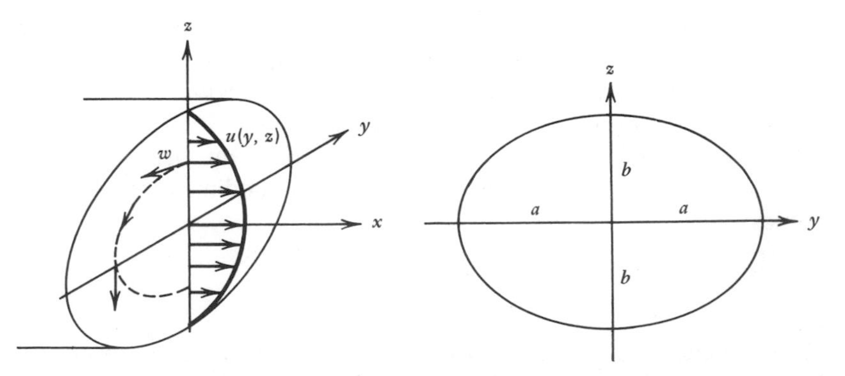

Figure 11-1 Parallel flow in an elliptical tube.

Computing the Laplacian gives

$$\nabla^2 U = \nabla^2 u^* + 2C_1 + 2C_2 \qquad 11.1.8$$

Evidently 11.1.5 will transform to $\nabla^2 U = 0$ if we require that

$$2C_1 + 2C_2 = 1 \qquad 11.1.9$$

The exact values of C_1 and C_2 are fixed by considering the boundary condition. On the wall this new variable U takes on the value

$$\begin{aligned} U(\text{wall}) &= C_1 y_w^{*2} + C_2 z_w^{*2} \\ &= C_1\left[y_w^{*2} + \frac{C_2}{C_1} z_w^{*2} \right] \end{aligned} \qquad 11.1.10$$

Comparing this with 11.1.6, we see that we can arrange for U(wall) to be the constant C_1 if we choose

$$K = \frac{C_2}{C_1} \qquad 11.1.11$$

Solving 11.1.10 and 11.1.9 gives the values of the constants as

$$C_1 = \frac{1}{2(1+K)}, \qquad C_2 = \frac{K}{2(1+K)} \qquad 11.1.12$$

The mathematical problem is now to solve $\nabla^2 U = 0$ for the boundary condition $U = C_1$ on the closed boundary consisting of the wall. One of the characteristics of the Laplace equation is that the maximum and the minimum values of the solution must occur on the boundary of the domain. Thus, we need look no further; the only solution is $U = C_1$ throughout the region enclosed by the ellipse. Setting 11.1.7 equal to C_1 and using 11.1.12, we find the velocity is given by

$$u^* = \frac{1}{2(1+K)}\left(1 - y^{*2} - Kz^{*2}\right)$$

The velocity has the same value at any point on an ellipse that has the same eccentricity as the tube wall:

$$y^{*2} + Kz^{*2} = C$$

The vorticity components are

$$\omega_z = \frac{1}{K+1} y^*, \qquad \omega_y = -\frac{K}{K+1} z^*$$

One can also show that the vortex lines (lines connecting the vorticity vectors in a tangential manner) are also ellipses and that the vorticity has a constant magnitude on any ellipse:

$$|\omega| = \frac{1}{K+1}\left(y^{*2} + Kz^{*2}\right)^{1/2}$$

Notice that the vorticity does not depend upon the Reynolds number; the only parameter that appears in the equation is a geometric parameter K. Integration of the velocity profile yields the volume flow rate

$$\frac{Q}{-\dfrac{a^4}{\mu}\dfrac{dp}{dx}} = \frac{\pi}{4}\frac{1}{K^{1/2}(K+1)}$$

11.2 STOKES OSCILLATING PLATE

This is one of the very first problems in which the Navier–Stokes equations were solved. It is one of several solved by Stokes (1845) during the course of his study of pendulum friction. This particular problem is a simplified one that illustrates the flow engendered by an oscillating boundary. A semiinfinite fluid is at rest initially and bounded below by a solid plane at $y = 0$. The problem is to find the motion of the fluid after the plate begins to oscillate in its own plane with a velocity given by

$$u_w = u_0 \sin \Omega t \tag{11.2.1}$$

If we assume the fluid has a single velocity component $u(y, t)$, the x-direction momentum simplifies to

$$\frac{\partial u}{\partial t} - \nu \frac{\partial^2 u}{\partial y^2} = 0 \tag{11.2.2}$$

The boundary conditions are that the fluid is initially quiescent, the velocity is bounded at infinity, and the no-slip condition applies at the plate surface:

$$u(y, t = 0) = 0 \tag{11.2.3}$$

$$u(y \to \infty, t) < \infty \tag{11.2.4}$$

$$u(y = 0, t) = u_0 \sin \Omega t \tag{11.2.5}$$

The solution to this problem can be expressed as the sum of a transient and a steady-state solution. In this section we give the steady-state solution, and in the following section the transient solution. The steady-state solution is a repetitive oscillation found by ignoring the initial condition.

Examining the boundary conditions we find that u_0 is a velocity scale and Ω^{-1} is a time scale. Therefore, we define nondimensional variables as

$$U = \frac{u}{u_0}, \qquad T = \Omega t \tag{11.2.6}$$

These variables are substituted into the differential equation together with the assumption $Y = y/\alpha$. We find that the problem will not contain any parameters if we choose α so that the nondimensional distance is

$$Y = \frac{y}{(\nu/\Omega)^{1/2}} \tag{11.2.7}$$

The problem statement in nondimensional variables is

$$\frac{\partial U}{\partial T} - \frac{\partial^2 U}{\partial Y^2} = 0 \tag{11.2.8}$$

$$U(Y, T = 0) = 0 \tag{11.2.9}$$

$$U(Y \to \infty, T) = 0 \tag{11.2.10}$$

$$U(Y = 0, T) = \sin T = \operatorname{Im} e^{iT} \tag{11.2.11}$$

The last boundary condition has been written as the imaginary part of a complex function. It is advantageous for us to consider U as a complex variable and then, in accord with 11.2.11, take the imaginary part as our answer.

The steady-state solution is sought by ignoring the initial boundary condition 11.2.9 and stipulating that the time dependence is an oscillation. We assume a solution of the form

$$U = f(Y)e^{iT} \tag{11.2.12}$$

When this form is substituted into the differential equation, we obtain the equation

$$(f'' - if)e^{iT} = 0 \tag{11.2.13}$$

Since e^{iT} is nonzero, we must require that the factor in parentheses be zero. This furnishes differential equation for f. A solution for f is assumed in the form

$$f = Ae^{aY}$$

where the constants may be complex. These assumptions will produce zero values of the factor in parentheses in 11.2.13 if

$$a = \pm\sqrt{i} = \pm\frac{1+i}{\sqrt{2}} \qquad 11.2.14$$

The second expression in 11.2.14 splits a into real and imaginary parts. We now have a solution to the differential equation:

$$U = Ae^{\pm(1+i)Y/\sqrt{2}}e^{iT} \qquad 11.2.15$$

The requirement that the answer be bounded at infinity dictates that the minus sign be chosen. The constant A is found by considering the boundary conditions. Setting $Y = 0$ in 11.2.15 and comparing with the wall condition 11.2.11, we find that the imaginary part of the equation represents the proper answer if $A = 1$. With these stipulations 11.2.15 becomes

$$U = \text{Im}\{e^{-(1+i)Y/\sqrt{2}}e^{iT}\}$$

or

$$U = e^{-Y/\sqrt{2}}\sin\left(T - \frac{Y}{\sqrt{2}}\right) \qquad 11.2.16$$

Figure 11-2 shows velocity profiles at various times.

The velocity profile is damped in the y direction by the first exponential term in 11.2.16. Let us denote the position $y = \delta$ as the place where the amplitude has decreased to 1% of the wall value. This occurs at about $Y/\sqrt{2} = 4.5$. Converting to dimensional variables, we find that the thickness of this region is

$$\delta = 4.5\left(\frac{2\nu}{\Omega}\right)^{1/2} \qquad 11.2.17$$

The depth to which viscosity makes itself felt is proportional to $\sqrt{\nu}$, just as it was in the Rayleigh problem of Chapter 7.

From the second term in 11.2.16 we have a wavelike behavior. A certain point on the wave is given when $T - Y/\sqrt{2}$ takes a specific value. In dimensional terms this point travels through space according to

$$y = C + \sqrt{2\nu\Omega}\,t \qquad 11.2.18$$

The wave velocity is $(2\nu\Omega)^{1/2}$. The mathematical solution may be thought of as damped viscous waves traveling away from the wall. This interpretation is really tied up with the oscillating boundary condition. The physical process that is occurring is viscous diffusion. The Rayleigh problem does not have a physical wave velocity, although one can trace the depth of penetration of the diffusion effect as a function of time. Note in Fig. 11-2 how the effect of the wall motion is delayed. When the wall reverses its motion and generates a net shear in the opposite direction at $T = \pi/2$, a net accelerating force from viscosity still exists deeper in the fluid, say $Y = 2$. Only after some time delay does the net shear force within the fluid change sign and begin to decelerate the fluid.

Another interpretation of this problem can also be made. We employ the fact, discussed in Section 10.7, that the equations governing incompressible flow are invariant under an "unsteady" Galilean transformation. This enables us to turn the problem around and say that the wall is stationary and the fluid far away is oscillating. An oscillating pressure gradient from external sources is needed to cause the free stream to oscillate.

The oscillating problem is analyzed as follows. Let the fluid velocity be $\overline{U}$. At $Y \to \infty$ the fluid is oscillating with a velocity

$$\overline{U}(Y \to \infty, T) = \sin T \qquad 11.2.19$$

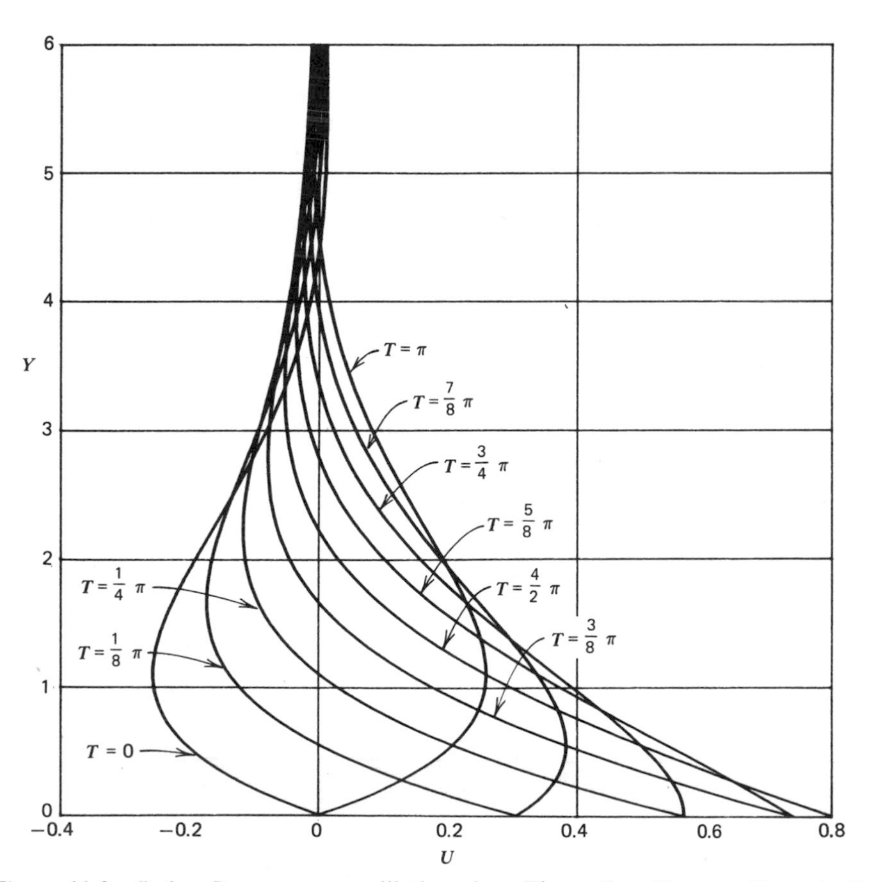

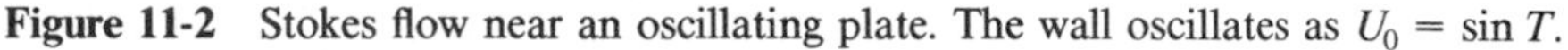
Figure 11-2 Stokes flow near an oscillating plate. The wall oscillates as $U_0 = \sin T$.

A pressure force $-dp/dx \propto \cos T$ would produce this motion. Because the wall is stationary,

$$\overline{U}(Y = 0, T) = 0 \tag{11.2.20}$$

According to Section 10.7 we can choose an $\hat{x}_i$ coordinate system that is fixed on a particle at infinity. The velocity U_∞ of the $\hat{x}_i$ system is the same as the fluid velocity 11.2.19, that is,

$$U_\infty = \sin T \tag{11.2.21}$$

Any fluid velocity measured in the new system $\hat{U}$ is related to the fluid velocity and coordinate-system velocity by

$$\hat{U} = \overline{U} - U_\infty \tag{11.2.22}$$

Hence, the boundary conditions 11.2.19 and 11.2.20 become

$$\begin{aligned} \hat{U}(\hat{Y} \to \infty, T) &= \sin T - \sin T = 0 \\ \hat{U}(\hat{Y} = 0, T) &= -\sin T \end{aligned} \tag{11.2.23}$$

Since we are assured by Section 10.7 that the governing equations in the new coordinates are unchanged in form, we may conclude that the new problem is simply Stokes's problem for a $-\sin T$ wall motion. The solution to the new problem must be the negative of 11.2.16, and the solution to the original problem is found by combining this result with 11.2.21 and 11.2.22. The final result is

$$\overline{U} = -\sin\left(T - \frac{Y}{\sqrt{2}}\right) e^{-Y/\sqrt{2}} + \sin T \tag{11.2.24}$$

This is the velocity profile for a uniform stream oscillating above a fixed wall. The first term is a viscous effect, and the second is the inviscid oscillation of the main stream.

At any given distance from the wall, 11.2.24 is the sum of two sine waves with the same frequency. It is always possible to represent this as a single wave with a different amplitude and a phase lag. Thus an equivalent form for 11.2.24 is

$$\overline{U} = A \sin(T + \Theta) \tag{11.2.25}$$

where

$$A = \left\{1 - 2e^{-Y/\sqrt{2}} \cos\left(\frac{Y}{\sqrt{2}}\right) + e^{-2Y/\sqrt{2}}\right\}^{1/2} \tag{11.2.26}$$

and

$$\Theta = \sin^{-1}\left\{\frac{e^{-Y/\sqrt{2}} \sin(Y/\sqrt{2})}{1 - e^{-Y/\sqrt{2}} \cos(Y/\sqrt{2})}\right\} \tag{11.2.27}$$

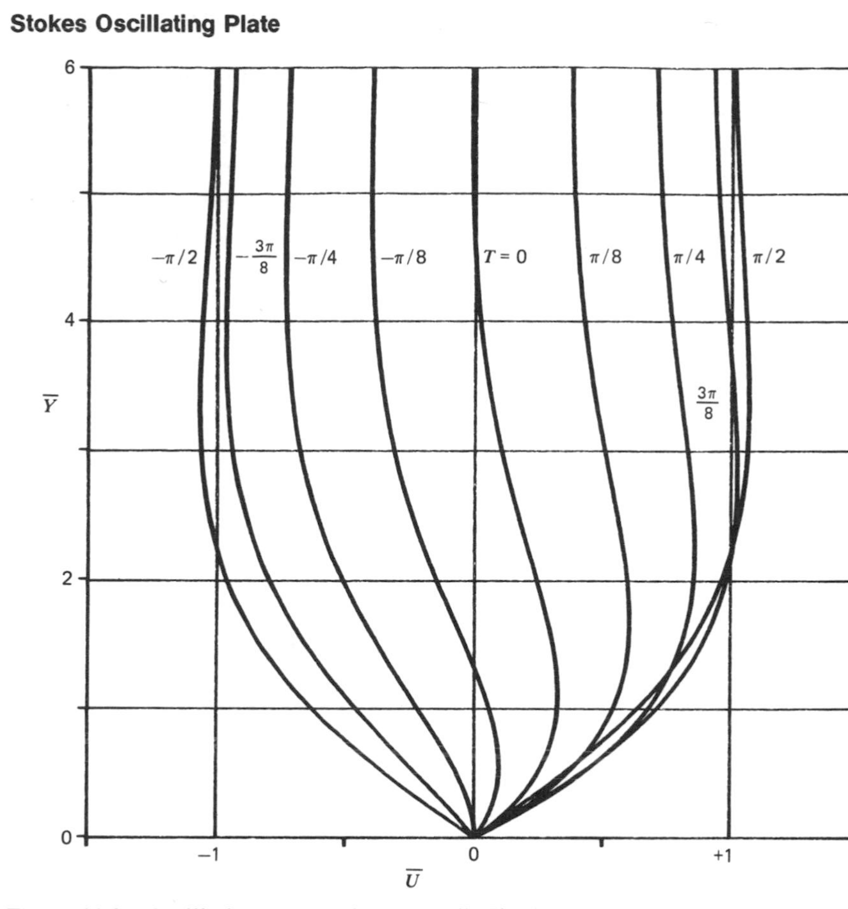

Figure 11-3 Oscillating stream above a wall. The free stream oscillates as $U_\infty = \sin T$.

One unexpected result is that the maximum amplitude of the oscillation is not at $Y \to \infty$ but at an intermediate point near the wall.

Figure 11-3 gives velocity profiles 11.2.24 for several times. The overshoot occurs at $Y \approx 3.2$; approximately one-half a viscous length away from the wall. To gain an insight into how the overshoot occurs, let us differentiate 11.2.24 to obtain

$$\frac{\partial \overline{U}}{\partial T} = \cos T - \cos\left(T - \frac{Y}{\sqrt{2}}\right) e^{-Y/\sqrt{2}} \qquad 11.2.28$$

$$\frac{\partial \overline{U}}{\partial T} = -\frac{d\overline{P}}{dX} + \frac{\partial^2 \overline{U}}{\partial \overline{Y}^2} \qquad 11.2.29$$

Written below 11.2.28 is the differential equation 11.2.29 that governs the flow. Since the pressure gradient is $-\cos T$, the terms in 11.2.28 and 11.2.29 correspond in the same order that they have been written ($\overline{P} \equiv p/(\rho V_0 \Omega L)$ where L is used in $X = x/L$). A nondimensional pressure gradient of $\cos T$

acts uniformly through the layer to drive the flow. At the wall $Y = 0$, the viscous force exactly counteracts the pressure force so that no motion occurs. As we move away from the wall, the oscillating viscous stresses die out according to the exponential factor. It is the intermediate region where the behavior is most complicated. The term $\cos(T - Y/\sqrt{2})$ contains a phase lag of $Y/\sqrt{2}$ compared to the term in $\cos T$. At the wall the phase lag is zero and the terms cancel. At small distances Y the term in $\cos(T - Y/\sqrt{2})$ peaks at later times, so the mismatch allows the fluid to accelerate. However, at certain distances from the wall the lag is so great that the viscous and pressure terms actually add together. The combined effect is to accelerate the fluid to higher velocities than those produced by the pressure force acting alone. We rationalize this by noting that the net viscous stress created at the wall diffuses into the flow and is attenuated at the same time. About half a cycle after the net viscous force was generated, viscous diffusion has carried it slightly from the wall, but it is still strong enough to aid the pressure force, which has now changed its direction. The combination of these forces accelerates the fluid to produce the overshoot. Two general mechanisms occurring in incompressible flow are illustrated by this problem: pressure forces are transmitted instantaneously through the fluid, while viscous forces are transmitted by viscous diffusion.

The overshoot phenomenon was originally observed by Richardson and Tyler (1929) in turbulent pipe flow. Sexl (1930) gave a laminar analysis showing the same effect for a round tube. It was originally known as the annular effect or as Richardson's annular effect. From our analysis above it is obvious that the effect is not related to the geometry and the term annular should be dropped.

As an application of the results of Stokes's analysis, we review a problem from acoustics. Consider a set of traveling acoustic waves in air bounded by a wall. The waves travel parallel to the wall and induce a velocity oscillation in the fluid away from the wall. Very near the wall, viscosity retards the motion, and the no-slip condition applies on the wall itself. Even though the velocity of the fluid far from the wall is determined by the compressible equations of acoustics, it turns out that the flow near the wall may be treated as if it were incompressible. Thus, we may apply 11.2.25 for the velocity profile near the wall. The frequency in 11.2.25 is, of course, the frequency of the sound, and the amplitude is the acoustic velocity amplitude, which depends upon the intensity of the sound. One quantity that does not depend upon the intensity is the thickness of the viscous region. For a sound frequency of 1000 Hz in air ($\nu = 0.15$ cm^2/sec) the thickness is found from 11.2.17 as $\delta = 6.5(\nu/\Omega)^{1/2} =$ 0.3 mm. In acoustics this thin region is known as the Stokes layer.

*11.3 TRANSIENT FOR STOKES OSCILLATING PLATE

At the initiation of the plate motion and during the first cycle of the oscillation, the velocity profile in the fluid differs from that given by the steady-state solution 11.2.16. During this period the solution consists of the sum of a transient and a steady-state solution,

$$U = U_s + U_t \qquad 11.3.1$$

The solution U_s is the one given in 11.2.16. In this section we shall find U_t, the transient solution.

Mathematically the problem for U_t is given by the diffusion equation

$$\frac{\partial U_t}{\partial T} - \frac{\partial^2 U_t}{\partial Y^2} = 0 \qquad 11.3.2$$

The fluid is initially quiescent; hence the transient answer must exactly cancel the steady-state solution at $T = 0$. From 11.2.16, we have

$$U_t(Y, T = 0) = -U_s(Y, T = 0)$$

$$= e^{-Y/\sqrt{2}} \sin \frac{Y}{\sqrt{2}} \qquad 11.3.3$$

It turns out that the only way to solve this problem in closed form is to represent the initial condition in complex-variable form (Panton 1968). Therefore we write 11.3.3 as

$$U_t(Y, T = 0) = \operatorname{Im}\left\{\exp\left[-\frac{(1 - i)Y}{\sqrt{2}}\right]\right\} \qquad 11.3.4$$

The other boundary conditions are

$$U_t(Y \to \infty, T) = 0 \qquad 11.3.5$$

and

$$U_t(Y = 0, T) = 0 \qquad 11.3.6$$

The last equation reflects the fact that the oscillating part of the boundary condition at the wall has already been satisfied by the steady-state solution.

There is a problem in heat conduction that is mathematically equivalent to our problem. Consider the unsteady heat conduction in a semiinfinite slab where an initial temperature profile 11.3.4 decays while the surface temperature is held at a constant value of zero (11.3.6). The solution to this problem is discussed in many texts and can be expressed as an integral. Carslaw and Jaeger (1947), in their classic text on heat-conduction solutions, give the solution as

$$U_t = (4\pi T)^{-1/2} \int_0^\infty f(\xi)\left\{e^{-(Y-\xi)^2/4T} - e^{-(Y+\xi)^2/4T}\right\} d\xi \qquad 11.3.7$$

Here the function $f(\xi)$ is the initial temperature profile, and for our problem it is the initial velocity profile. Upon substituting 11.3.4 for $f(\xi)$, the integral 11.3.7 is found to be tractable and is given for instance by Abramowitz and Stegun (1964). The result is

$$U_t(Y,T) = \mathrm{Im}\left\{ \tfrac{1}{2} e^{-CY/\sqrt{2} - iT} \operatorname{erfc}\left[\left(\tfrac{1}{2}T\right)^{1/2} \left(C - \frac{Y}{T\sqrt{2}} \right) \right] \right.$$

$$\left. - \tfrac{1}{2} e^{CY/\sqrt{2} - iT} \operatorname{erfc}\left[\left(\tfrac{1}{2}T\right)^{1/2} \left(C + \frac{Y}{T\sqrt{2}} \right) \right] \right\} \qquad 11.3.8$$

where $C = 1 - i$. The combined solution $U_s + U_t$ is given in Fig. 11-4. From the graphs we see that the transient is only significant during the first cycle of the oscillation.

There are several other types of plate motion that lead to exact results. For example, if the velocity increases as t^n for some integer power n, a solution may be found. All of these problems may be turned around and considered as a prescribed motion of an external stream over a fixed plate. They may also be added together to give a composite motion, because the problem is linear. Thus an arbitrary motion might be expressed as a polynomial in time, and each t^n term would contribute an exact answer to the combined motion. Another approach to the problem of an arbitrary plate motion is to represent the motion as a Fourier sine series. Then the solutions of this section and the

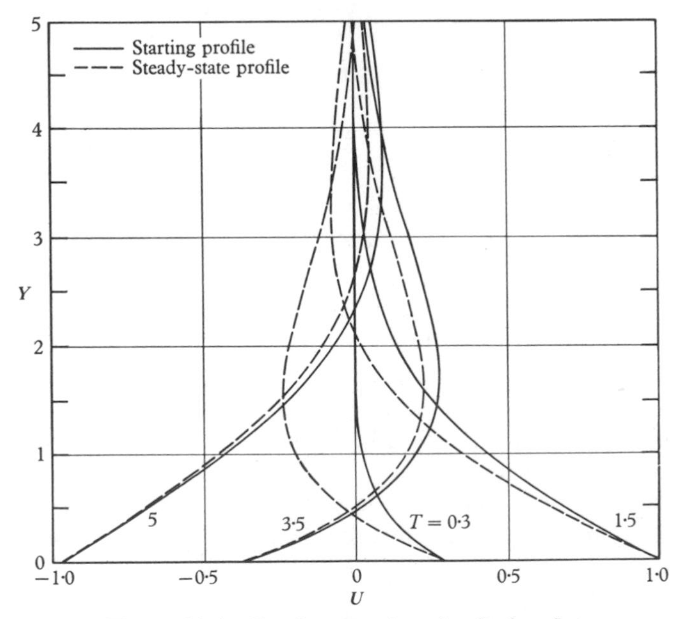

Figure 11-4 Starting transient for Stokes flow.

previous section can be found for each term in the series

$$u(y=0,t) = A\sin\Omega t + B\sin 2\Omega t + \cdots$$

This method can represent most motions with a small number of terms.

Finally we mention that an integral method (Duhammel's integral) is also available for direct numerical evaluations of this integral.

*11.4 STARTING TRANSIENT FOR PLANE COUETTE FLOW

We consider how the Couette flow in a slot is developed from the situation where the fluid and both walls are initially stationary. To start the flow, the lower wall, $y = 0$, is impulsively brought to the constant velocity u_0. The final Couette velocity profile

$$\frac{u}{u_0} = 1 - \frac{y}{h} \tag{11.4.1}$$

will be approached asymptotically as $t \to \infty$. Simplification of the x-direction momentum equation and the no-slip condition gives the differential equation and boundary conditions that govern the transient development. They are

$$\frac{\partial u}{\partial t} = \nu\frac{\partial^2 u}{\partial y^2}$$

$$\begin{aligned} u(y,t=0) &= 0 \\ u(y=0,t) &= u_0 \\ u(y=h,t) &= 0 \end{aligned} \tag{11.4.2}$$

We do not expect a similarity solution in this case, since there is a natural measuring scale for y in the problem. If we choose nondimensional variables

$$Y = \frac{y}{h}, \qquad U = \frac{u}{u_0} \tag{11.4.3}$$

and substitute into the differential equation, it will turn out that the proper time variable is

$$T = \frac{t}{h^2/\nu} \tag{11.4.4}$$

The physical interpretation of the time scale h^2/ν is the time it takes for viscous effects to diffuse a distance h. This could have been found by setting the viscous diffusion length equal to the gap height $h = \delta = \sqrt{\nu t}$, then solving for the time.

The problem now reads

$$\frac{\partial U}{\partial T} = \frac{\partial^2 U}{\partial Y^2}$$

$$U(Y,0) = 0$$

$$U(0,T) = 1$$

$$U(1,T) = 0 \tag{11.4.5}$$

Although this problem looks simple as it stands, there is one further change that will help. We split the problem into a transient component U_t and a steady-state component U_s:

$$U = U_s + U_t \tag{11.4.6}$$

so that we can use the separation-of-variables technique on U_t. The steady-state answer we already know to be

$$U_s = 1 - Y$$

The problem for the transient reads

$$\frac{\partial U_t}{\partial T} = \frac{\partial^2 U_t}{\partial Y^2}$$

$$U_t(Y,0) = -U_s(Y,0) = -1 + Y \tag{11.4.7}$$

$$U_t(0,T) = 0$$

$$U_t(1,T) = 0$$

This problem is now in a good form for a separation-of-variables solution: an initial profile at $T = 0$ and homogeneous boundary conditions at two points in space. The solution takes the form

$$U_t = \sum_{n=1}^{\infty} A_n \sin(n\pi Y)\exp(-n^2\pi^2 T) \tag{11.4.8}$$

Terms containing cosines are ruled out because they will not satisfy the boundary conditions. The initial condition is used to find the coefficients A_n. First set $T = 0$ to get

$$U_t(Y,0) = -1 + Y = \sum_{n=1}^{\infty} A_n \sin(n\pi Y) \tag{11.4.9}$$

The usual method for fitting a sine series to a function, in this case the function $Y-1$, is followed: the series is multiplied by $\sin m\pi Y$ and integrated from $Y=-1$ to $+1$. The orthogonality conditions selects out A_n as

$$A_n = 2\int_0^1 (Y-1)\sin n\pi Y\,dY \qquad 11.4.10$$

This integrates to

$$A_n = -\frac{2}{n\pi}$$

Thus, the answer is the infinite series

$$U = 1 - Y - 2\sum_{n=1}^{\infty}\frac{1}{n\pi}\sin(n\pi Y)\exp(-n^2\pi^2 T) \qquad 11.4.11$$

The series converges more rapidly for longer times. Also we see that the last component to die out is the one for which $n=1$.

Recall that T is scaled by the viscous diffusion time h^2/ν. In this expression we have left out a constant as it makes no real difference in the analysis. To be more precise, the diffusion distance in the Rayleigh plate problem was $\delta = 4\sqrt{\nu t}$, and so the time for diffusion through a distance h is really $t_{\text{Ray}} = h^2/16\nu$. This means that our nondimensional variable is

$$T = \frac{t}{h^2/\nu} = \frac{t}{16 t_{\text{Ray}}} \qquad 11.4.12$$

Hence when $T=\frac{1}{16}$, a time equal to one Rayleigh diffusion time has lapsed. The last component to die out, $n=1$, has a value $\exp(-\pi^2/16)=0.54$ of its original size after a time $T=\frac{1}{16}$. Thus, we find that the transient will die out in a time comparable to the time it takes for viscous diffusion to traverse the gap distance.

The series solution 11.4.11 does not converge rapidly for small times. An alternative solution leads to a more appropriate form (see Schlichting, 1950). The first few terms of the series are

$$U = \operatorname{erfc}\left(\frac{y}{2\sqrt{\nu t}}\right) - \operatorname{erfc}\left(\frac{2h-y}{2\sqrt{\nu t}}\right) + \operatorname{erfc}\left(\frac{4h+y}{2\sqrt{\nu t}}\right) + \cdots \qquad 11.4.13$$

Since $\operatorname{erfc}(\infty)=0$, the first term in 11.4.13 dominates for small values of time. This is the Rayleigh flat-plate solution. At small times the moving wall does not feel the presence of the far wall, and the flow near the moving wall is exactly the same as if the fluid extended to infinity.

We can also interpret this problem as an example of vorticity diffusion. The vorticity for this problem is

$$\omega^* = \frac{\omega}{U_0/h} = \frac{\partial U}{\partial Y}$$

Differentiation of the differential equation governing the velocity (11.4.2) yields an equation for the vorticity,

$$\frac{\partial \omega^*}{\partial T} = \frac{\partial^2 \omega^*}{\partial Y^2}$$

Initially there is no vorticity across the slot. At the instant the plate moves, the vorticity is infinite on the lower wall. The value of vorticity at the moving wall decreases as vorticity diffuses into the fluid. The final steady state is a uniform vorticity across the slot. We can differentiate our previous expression for U (11.4.11) to find ω^*:

$$\omega^* = -1 + 2\sum_{n=1}^{\infty} \cos(n\pi Y)\exp(-n^2\pi^2 T) \qquad 11.4.14$$

As with any Fourier series, differentiation produces a series that converges even slower than the original series. In fact, the expression for ω^* above diverges at $T = 0$.

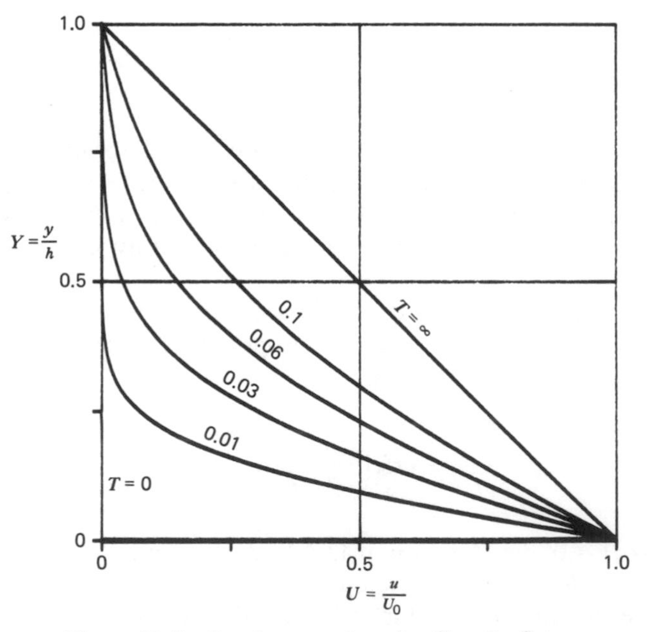

Figure 11-5 Starting transient for Couette flow.

For small times we can use the Rayleigh flat-plate solution as an approximation; for example,

$$U = \operatorname{erfc}\left(\frac{y}{2\sqrt{\nu t}}\right) = \operatorname{erfc}\left(\frac{Y}{2\sqrt{T}}\right)$$

$$\omega^* = -\frac{Y}{\pi T}\exp\left(-\frac{Y^2}{4T}\right)$$

A schematic plot of the vorticity at different times is shown in Fig. 11-5.

11.5 FLOW IN A SLOT WITH AN OSCILLATING PRESSURE GRADIENT

In Chapter 7 we found that flow in a slot with a constant pressure gradient resulted in a parabolic velocity profile. Here we consider that an extra oscillatory component is added to the steady gradient. The pressure is then

$$-\frac{1}{\rho}\frac{\partial p}{\partial x} = -\frac{1}{\rho}\frac{dp}{dx}\bigg|_0 + K\cos\Omega t \tag{11.5.1}$$

The coordinate system is taken with x in the flow direction and with its origin in the center of a channel of width $2h$.

The first step in an analytical solution is to assume that u is a function of y and t but not a function of x. Then the continuity equation simplifies to $\partial v/\partial y = 0$, and hence v is a function of x alone. Since v is zero on the wall, it must be zero everywhere.

The x-direction momentum equation simplifies to

$$\frac{\partial u}{\partial t} = -\frac{1}{\rho}\frac{dp}{dx}\bigg|_0 + K\cos\Omega t + \nu\frac{\partial^2 u}{\partial y^2} \tag{11.5.2}$$

Upon this equation we impose boundary conditions of no slip at the wall and assume the flow is symmetric about $y = 0$:

$$u(y = h, t) = 0, \qquad \frac{\partial u}{\partial y}(y = 0, t) = 0 \tag{11.5.3}$$

The initial condition is omitted, as we will seek only the steady-state result. The equation and boundary conditions are linear in u; so the answer may be separated into a part with the constant pressure gradient u_1 (the solution given in Chapter 7) and an oscillatory part u_2 that satisfies

$$\frac{\partial u_2}{\partial t} = K\cos\Omega t + \nu\frac{\partial^2 u_2}{\partial y^2} \tag{11.5.4}$$

u_2 obeys the same boundary conditions, 11.5.3. In future work we shall replace $\cos \Omega t$ by $e^{i\Omega t}$ and take the real part of the answer to satisfy 11.5.4.

Turning our attention to finding nondimensional variables, we choose

$$T = \Omega t, \qquad Y = \frac{y}{h}$$

The proper scale for the velocity is not so obvious. Let us denote it by α and substitute $U = u_2/\alpha$ into 11.5.4:

$$\frac{\Omega\alpha}{K}\frac{\partial U}{\partial T} = e^{iT} + \frac{\nu\alpha}{h^2 K}\frac{\partial^2 U}{\partial Y^2} \qquad 11.5.5$$

There are two ways to define α so that a coefficient will disappear from 11.5.5: $\alpha = K/\Omega$ or $\alpha = h^2K/\nu$. We choose the first form, which implies that the magnitude of the velocity depends on the amplitude and frequency of the oscillation. The problem now reads

$$\frac{\partial U}{\partial T} = e^{iT} + \frac{1}{\lambda^2}\frac{\partial^2 U}{\partial Y^2}$$

$$U(Y = 1, T) = 0, \qquad \frac{\partial U}{\partial Y}(Y = 0, T) \qquad 11.5.6$$

where

$$\lambda = \frac{h}{\sqrt{\nu/\Omega}}$$

is a nondimensional parameter that compares the slot height with the viscous diffusion length. We cannot eliminate λ from the problem; if we incorporated λ into a new distance variable ($\hat{Y} = \lambda Y$), we would have a differential equation free of parameters, but λ would appear explicitly in the boundary condition, for example, $U(\hat{Y} = \lambda,\ T) = 0$. It is usually better strategy to deal with a parameter in the equation than in the boundary condition.

We seek a steady-state oscillatory solution to 11.5.6 by trying to find an answer in the form

$$U(Y, T) = e^{iT}F(Y) \qquad 11.5.7$$

Substituting this into 11.5.6 produces the problem

$$iF = 1 + \lambda^{-2}F''$$

$$F(Y = 1) = 0, \qquad F'(Y = 0) \qquad 11.5.8$$

Next, the constant 1 is incorporated into a new dependent variable by the definition

$$\hat{F} = F + i \tag{11.5.9}$$

The problem now reads

$$\hat{F}'' = i\lambda^2 \hat{F}$$
$$\hat{F}(Y = 1) = i, \qquad \hat{F}'(Y = 0) = 0 \tag{11.5.10}$$

Solutions of 11.5.10 are $\exp(\pm\lambda\sqrt{i}\,Y)$ or equivalently a linear combination of exponentials such as sinh or cosh. We find that the sinh answer will not fit the boundary conditions at $Y = 0$, so we try

$$\hat{F} = A\cosh(\pm\lambda\sqrt{i}\,Y) \tag{11.5.11}$$

where A is determined by the boundary condition at $Y = 1$:

$$A = \frac{i}{\cosh(\pm\lambda\sqrt{i}\,)} \tag{11.5.12}$$

The answer is obtained by putting 11.5.12, 11.5.11, and 11.5.7 together. When this is done we find that the velocity is

$$U = \left[-i + i\frac{\cosh(\sqrt{i}\,\lambda Y)}{\cosh(\sqrt{i}\,\lambda)}\right]e^{iT} \tag{11.5.13}$$

The real part of this equation is the answer (note again that $\sqrt{i} = \pm(1 + i)/\sqrt{2}$). We also noticed that Y appears only as the combination $\hat{Y} = \lambda Y$, the very combination we chose to avoid at the beginning of the analysis.

The real part of 11.5.13 can be expressed in elementary functions; however, the result is lengthy. In order to write the answer completely we introduce the notation below:

$$\Lambda \equiv \frac{\lambda}{\sqrt{2}} = \frac{h}{\sqrt{2\nu/\Omega}} \tag{11.5.14}$$

and

$$C(x) \equiv \cosh x \cos x$$

$$S(x) \equiv \sinh x \sin x$$

$$M(Y; \Lambda) \equiv C(\Lambda Y)C(\Lambda) + S(\Lambda Y)S(\Lambda) \tag{11.5.15}$$

$$N(Y; \Lambda) \equiv C(\Lambda Y)S(\Lambda) - S(\Lambda Y)C(\Lambda)$$

$$J(\Lambda) \equiv C^2(\Lambda) + S^2(\Lambda)$$

The answer, the real part of 11.5.13, can now be written as

$$U = \left[1 - \frac{M(Y;\Lambda)}{J(\Lambda)}\right]\sin T - \frac{N(Y;\Lambda)\cos T}{J(\Lambda)} \qquad 11.5.16$$

These profiles must be superimposed upon the parabolic profile from the steady flow component.

The parameter $\Lambda = h/\sqrt{2\nu/\Omega}$ compares the slot width with the viscous diffusion length. As Λ approaches zero, the viscous diffusion length becomes much larger than h. Other things being equal, we can imagine that this occurs when $\Omega \to 0$ (a low-frequency limit). Perhaps a better interpretation is to consider Λ^2 as the ratio of the time for viscous effects to diffuse across the slot $(h^2/2\nu)$ to the period of a pressure oscillation $(1/\Omega)$. Then $\Lambda^2 \to 0$ means the viscous time is small compared to the oscillation period. In any event, the form of the velocity profile valid as $\Lambda \to 0$ is found from 11.5.16 by noting that $C(x) \sim 1$ and $S(x) \sim x^2$ as $x \to 0$. The result is

$$U = -\Lambda^2(1 - Y^2)\cos T \qquad 11.5.17$$

The same equation in dimensional variables is

$$\frac{u_2}{-Kh^2/\nu} = \tfrac{1}{2}\cos\Omega t\left[1 - \left(\frac{y}{h}\right)^2\right] \qquad 11.5.18$$

This is a quasi-steady-state result. The velocity profile is a parabola with the amplitude modified to correspond to the pressure gradient at that instant. Viscous diffusion is rapid enough to keep the profile in a quasi steady state.

The opposite extreme, $\Lambda \to \infty$, means that the viscous diffusion depth is small compared to the slot width. This is the high-frequency limit. For large values of x the proper approximations to 11.5.15 are

$$C(x) \sim \tfrac{1}{2}e^x\cos x$$
$$S(x) \sim \tfrac{1}{2}e^x\sin x \qquad 11.5.19$$

With these relations and employing some trigonometric identities, one can show that 11.5.16 becomes

$$U = \sin T - \sin(T - \eta)\,e^{-\eta} \qquad 11.5.20$$

Here a new distance variable η has been defined as

$$\eta \equiv \Lambda(1 - Y) = \frac{h - y}{\sqrt{2\nu/\Omega}} \qquad 11.5.21$$

The variable η compares the distance from the wall $(h - y)$ with the viscous diffusion distance.

The first term in 11.5.20 is the inviscid response to an oscillating pressure gradient. It satisfies 11.5.4 without the viscous term; that is, $\partial U/\partial T = \cos T$. As we move a few viscous lengths away from the wall, the second term drops out and the solution predicts that the flow will perform a simple oscillation with no y dependence. The second term in 11.5.20 is the same answer we found for Stokes's problem of an oscillating plate bounding an infinite fluid. When the viscous length is small, the flow behaves as if the opposite wall were absent. The flow is an inviscid oscillation of the bulk of the fluid with a Stokes layer at each wall.

11.6 DECAY OF AN IDEAL LINE VORTEX (OSEEN VORTEX)

An ideal vortex is a flow with circular streamlines where the particle motion is irrotational. Incompressible and irrotational flows are called *ideal flows*. The velocity profile of an ideal vortex obeys the equation

$$v_\theta = \frac{\Gamma}{2\pi r} \qquad 11.6.1$$

Here the constant Γ is the circulation defined by the integral 3.9.7; it indicates the strength of the vortex. As we shall see later chapters, an irrotational flow has no net viscous forces. Thus, 11.6.1 represents an inviscid flow (see Fig. 11-6a). At the origin, 11.6.1 indicates that the velocity becomes infinite. Such behavior is prohibited because continuity requires that the velocity be zero at the origin. In order to meet this requirement we must have a core region where the flow is rotational and viscous forces are important.

The mathematical problem that we shall solve in this section might be considered as an ideal vortex that, at time zero, is forced to obey the zero-velocity condition at the origin. The problem is much like the Rayleigh impulsive-plate problem in that the effects of viscosity diffuse through the fluid. In the present case the streamlines are curved, rather than straight as in the Rayleigh problem. Ultimately, viscous forces will completely destroy the vortex.

In order to begin the mathematical solution we assume that the velocity $v_\theta(r, t)$ is a function of r and t only. The continuity equation is satisfied by this assumption, and the θ-direction momentum equation simplifies to the following form:

$$\frac{\partial v_\theta}{\partial t} = -\nu \frac{1}{r^2}\frac{\partial (r v_\theta)}{\partial r} + \nu \frac{1}{r}\frac{\partial^2 (r v_\theta)}{\partial r^2} \qquad 11.6.2$$

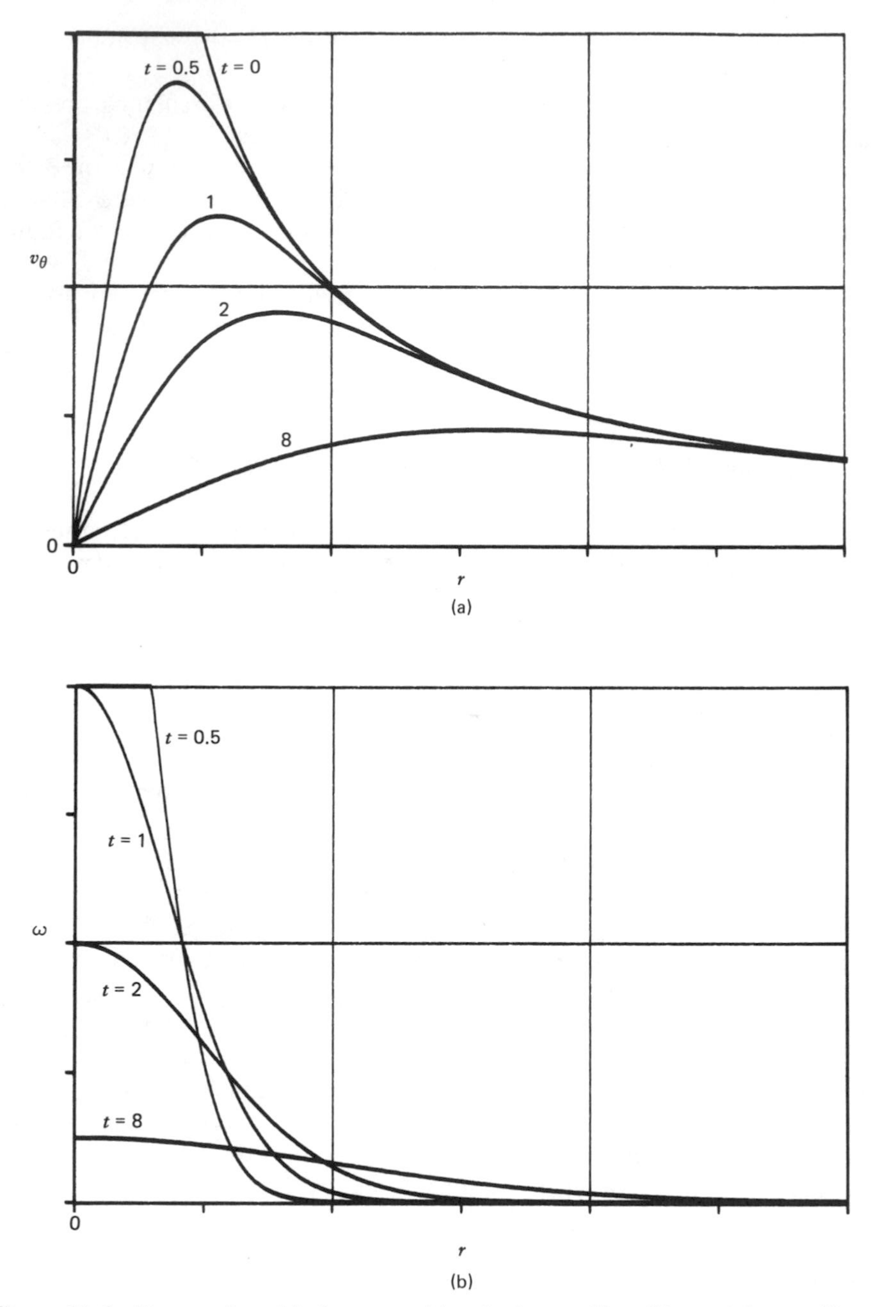

Figure 11-6 Decay of an ideal vortex: (a) velocity profiles, (b) vorticity profiles at corresponding times. Scales are arbitrary.

Boundaiy conditions are

$$v_\theta(r = 0, t) = 0$$

$$v_\theta(r \to \infty, t) \sim \frac{\Gamma}{2\pi r} \qquad 11.6.3$$

$$v_\theta(r, t = 0) = \frac{\Gamma}{2\pi r}$$

We note that neither r nor t has a natural measuring scale in the boundary conditions. This leads us to suspect that a similarity solution might be appropriate.

Before we construct the similarity variable, let us nondimensionalize the dependent variable. The boundary conditions and differential equation will have an especially simple form if we use $1/r$ in the velocity scale:

$$\gamma^* \equiv \frac{v_\theta}{\Gamma/2\pi r} = \frac{rv_\theta}{\Gamma/2\pi} \qquad 11.6.4$$

This is a slightly new twist. We are finding a similarity solution for a dependent variable that is itself scaled by one of the independent variables. The combination $\gamma = rv_\theta$ is called the *reduced circulation*. Mathematically the problem now consists of

$$\frac{\partial \gamma^*}{\partial t} = -\frac{\nu}{r}\frac{\partial \gamma^*}{\partial r} + \nu \frac{\partial^2 \gamma^*}{\partial r^2}$$

$$\gamma^*(r = 0, t) = 0 \qquad 11.6.5$$

$$\gamma^*(r \to \infty, t) \sim 1$$

$$\gamma^*(r, t = 0) = 1$$

At this point we have found a nondimensional dependent variable using information contained in the boundary conditions. We can find the similarity variable by using dimensional analysis (not all similarity variables may be found in this way). Assume a solution of the form $\gamma^* = \gamma^*(r, t, \nu)$. Dimensional analysis shows that two variables are required. Since one pi variable is γ^*, the other must be formed from r, t, and ν. Hence, with a little trial and error we find the similarity variable

$$\eta = \frac{r}{\sqrt{\nu t}} \qquad 11.6.6$$

The next task is to transform the problem into the variable η. Employing the notation

$$\gamma^* = f(\eta) \qquad 11.6.7$$

and applying the same mathematical procedures as in the Rayleigh problem, we find the transformed equation

$$f'' + \left(\frac{\eta}{2} - \frac{1}{\eta}\right)f' = 0 \qquad 11.6.8$$

This differential equation does not explicitly contain either r or t, so the similarity variable is valid as far as the differential equation is concerned. To be completely successful, the boundary conditions must also transform, and at the same time they must reduce from three to two in number. The boundary conditions in 11.6.5 transform into

$$\begin{aligned} \gamma^*(r = 0, t) &= f(\eta = 0) = 0 \\ \gamma^*(r \to \infty, t) &= f(\eta \to \infty) \sim 1 \\ \gamma^*(r, t = 0) &= f(\eta \to \infty) = 1 \end{aligned} \qquad 11.6.9$$

The equations constitute only two distinct boundary conditions, so 11.6.8 and 11.6.9 form a compatible problem.

Straightforward integration of 11.6.8 gives the reduced circulation as

$$\gamma^* = f(\eta) = 1 - e^{-\eta^2/4} \qquad 11.6.10$$

returning to dimensional variables, we find the velocity profile corresponding to this equation is

$$v_\theta = \frac{\Gamma}{2\pi r}\left(1 - e^{-r^2/4\nu t}\right) \qquad 11.6.11$$

or, in a slightly different form,

$$v_\theta = \frac{\Gamma}{2\pi\sqrt{\nu t}}\frac{1}{\eta}\left[1 - e^{-\eta^2/4}\right] \qquad 11.6.12$$

This solution is called an Oseen vortex (it is also called a Lamb vortex). Typical velocity profiles are plotted in Fig. 11-6 and (in similarity variables) in Fig. 11-7. Notice that v_θ is not a function of the similarity variable η, but either v_θ/r^{-1} or $v_\theta/t^{-1/2}$ is a function of η only.

The Oseen vortex profile is one member of a family of vortex profiles that satisfy the Navier–Stokes equations. For example, another profile discovered by G. I. Taylor is given by

$$v_\theta = \frac{\mathrm{H}}{8\pi}\frac{r}{\nu t^2}\exp\left(-\frac{r^2}{4\nu t}\right)$$

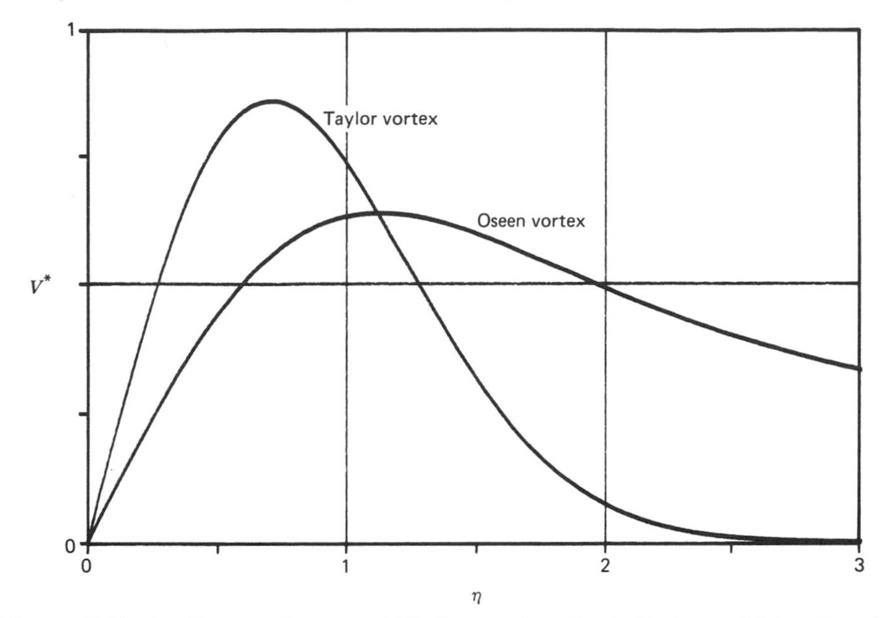

Figure 11-7 Profiles for Oseen and Taylor vortices in similarity variables. For the Oseen vortex $V^* \propto v_\theta/t^{-1/2}$, while for the Taylor vortex $V^* \propto v_\theta/t^{-3/2}$. In each case $\eta = r/\sqrt{\nu t}$.

The constant H in this relation is physically the amount of angular momentum in the vortex (the Oseen vortex contains an infinite amount of angular momentum). The similarity form of Taylors vortex is also shown in Fig. 11-7.

The vorticity in the Oseen vortex is found from the relation

$$\omega_z = \frac{1}{r}\frac{\partial}{\partial r}(rv_\theta) = \frac{\Gamma}{4\pi\nu t}e^{-\eta^2/4} \qquad 11.6.13$$

Sketches of the vorticity profiles at various times are shown in Fig. 11-6b. At any instant the distribution is a Gaussian bell curve. Where the vorticity is nonzero the flow is viscous, and where the vorticity is zero the flow remains like the original ideal vortex. The height of the curve falls off as t^{-1}, and the width increases by viscous diffusion as $(\nu t)^{1/2}$. The total vorticity in the flow may be computed either directly from 11.6.13 or by applying Stokes's theorem. The result shows that the total vorticity is a constant and equal to the circulation:

$$\int_0^\infty \omega_z 2\pi r\, dr = \Gamma \qquad 11.6.14$$

The equation above is an example of the interpretation of vorticity as the local circulation per unit area.

In terms of vorticity, this solution is analogous to a heat-conduction problem where a finite amount of energy Γ is concentrated in a line source at time zero. Subsequently, the heat is conducted radially away from the line. This causes the temperature ω_z to increase. As time goes on, the temperature ω_z decreases as the energy is dispersed further and further from the source. Even though the temperature eventually approaches zero, the total energy Γ is always the same constant value.

Next we turn to a practical application of this solution. Any lifting surface in an unbounded flow—a fan blade, a ship's propeller blade, or an airplane wing—has a trailing vortex, which forms at the tip. This vortex can be quite strong and concentrated. Frequently it is turbulent, and in the case of a large aircraft it may last for several minutes. Squire (1965) was able to model the decay of these vortices by using 11.6.12. He proposed, first, that the kinematic viscosity be replaced by an effective turbulent viscosity, which would be a constant for a given vortex, but would change as a function of the vortex strength Γ/ν (a vortex Reynolds number). Second, he proposed to relate the decay time to the distance behind the aircraft divided by the aircraft speed:

$$t \to \frac{z - z_0}{V_\infty} \tag{11.6.15}$$

We must include an arbitrary constant z_0 as an effective origin for the ideal vortex. The detailed process by which the vortex is formed at the wing tip produces a vortex that is already in some stage of decay. Although there are more sophisticated models of a turbulent vortex, it turns out that Squire's model is reasonably good. The predicted decay of the maximum velocity ($\sim z^{-1/2}$) and growth of the core ($\sim z^{1/2}$) are fairly accurate when the proper effective origin z_0 is used in the model.

11.7 STAGNATION-POINT FLOW (HIEMENZ FLOW)

The next flow we discuss is a local flow solution—one that is good in only a small part of the entire flow field. Consider a two-dimensional body in an infinite stream as shown in Fig. 11-8. In subsequent chapters we shall show that the flow in the neighborhood of the stagnation point has the same character irrespective of the shape of the body (as long as the flow is two dimensional). A high Reynolds number is necessary for this to be true. The neighborhood where this solution is valid may not be very large, but it is nevertheless a finite size. You might imagine that you are at the stagnation point and begin shrinking in size. Soon the surface, as far as you are concerned, becomes flat. You can't see the details of the flow as it approaches or what happens after it turns and goes away, however, in a small neighborhood near your vantage point the flow is much like Fig. 11-8.

In order to analyze the problem, we assume a flat wall with a two-dimensional flow $u(x, y)$, $v(x, y)$ which obeys the Navier–Stokes equations and the

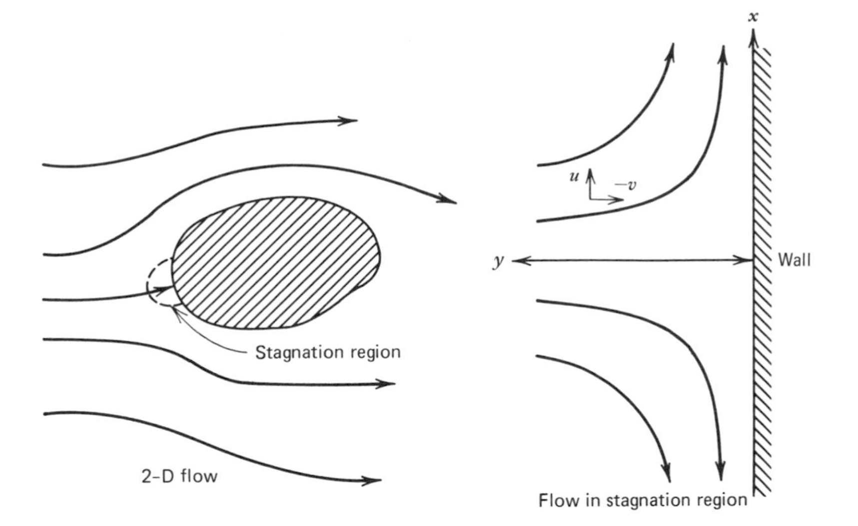

Figure 11-8 Stagnation-point flow; a local solution.

no-slip boundary conditions,

$$u(x, y = 0) = 0$$

$$v(x, y = 0) = 0$$

Far away from the wall, the flow approaches as if it is slowing down linearly, that is,

$$v(x, y \to \infty) = -ay + b = -a\left(y - \frac{b}{a}\right) \qquad 11.7.2$$

Also far away from the wall and as we go along the wall in x, the flow is accelerating as

$$u(x, y \to \infty) = ax \qquad 11.7.3$$

The reason these boundary conditions are appropriate is not obvious. Equations 11.7.2 and 11.7.3 are actually the solution for an inviscid flow near a stagnation point (the solution used in the example in Chapter 4). The only effect that viscosity has in the inviscid far field is to shift the apparent location of the wall by the amount b/a as shown by 11.7.2. The constant a in these equations is proportional to the free-stream velocity far away from the body divided by a characteristic length of the body:

$$a = \frac{\alpha U_\infty}{L} \qquad 11.7.4$$

The constant of proportionality α depends on the exact shape of the body.

It will turn out that the separation-of-variables assumption

$$u = xf'(y) \qquad 11.7.5$$

gives the proper form for the answer. The reason that the derivative of f is used in 11.7.5 instead of the completely equivalent form $u = xF(y)$ is that the continuity equation

$$\frac{\partial u}{\partial x} + \frac{\partial v}{\partial y} = 0$$
$$f'(y) + \frac{\partial v}{\partial y} = 0 \qquad 11.7.6$$

integrates to

$$v = -f(y) + C(x)$$

Since $v(x,0) = 0$, the function of integration $C(x)$ turns out to be zero, and

$$v = -f(y) \qquad 11.7.7$$

Hence, 11.7.5 and 11.7.7 will satisfy the continuity equation, and the two unknowns u and v are replaced by one unknown $f(y)$. The no-slip condition is satisfied if

$$f'(y=0) = 0 \quad \text{and} \quad f(y=0) = 0 \qquad 11.7.8$$

The boundary condition far away, 11.7.2, and 11.7.3 require that

$$f'(y \to \infty) \to a$$
$$f(y \to \infty) \to ay - b \qquad 11.7.9$$

These are essentially the same condition, as the first may be integrated to give the second.

When the assumptions above are inserted into the x-momentum equation, we obtain

$$u\frac{\partial u}{\partial x} + v\frac{\partial u}{\partial y} = -\frac{1}{\rho}\frac{\partial p}{\partial x} + \nu\frac{\partial^2 u}{\partial y^2}$$

$$-\frac{1}{\rho}\frac{\partial p}{\partial x} = x\{(f')^2 - ff'' - \nu f'''\}$$

If we denote the terms in braces as

$$-H(y) \equiv (f')^2 - ff'' - \nu f''' \qquad 11.7.10$$

the momentum equation is

$$-\frac{1}{\rho}\frac{\partial p}{\partial x} = -xH(y)$$

Partial integration yields

$$\frac{1}{\rho}p = \tfrac{1}{2}x^2H(y) + K(y) \qquad 11.7.11$$

The function $K(y)$ is the pressure along the stagnation streamline at $x = 0$. In anticipation of the y-momentum equation, we compute the derivative

$$-\frac{1}{\rho}\frac{\partial p}{\partial y} = -\tfrac{1}{2}x^2H' - K'$$

Putting this, together with our previous assumptions, into the y-momentum equation yields

$$\begin{gathered} v\frac{\partial v}{\partial y} = -\frac{1}{\rho}\frac{\partial p}{\partial y} + \nu\frac{\partial^2 v}{\partial y^2} \\ ff' = -\tfrac{1}{2}x^2H' - K' - \nu f'' \end{gathered} \qquad 11.7.12$$

This equation cannot be true for arbitrary x and y unless $H' = 0$. Hence, H is constant, and 11.7.10 is the differential equation that governs the problem. When 11.7.10 is evaluated as $y \to \infty$ with the assumption that $f''(\infty)$ and $f'''(\infty)$ are zero (a requirement that the flow smoothly approach the free-stream conditions), we find

$$-H = a^2$$

The complete problem for $f(y)$ now reads

$$\begin{gathered} (f')^2 - ff'' - \nu f''' = a^2 \\ f(0) = 0 \\ f'(0) = 0 \\ f'(\infty) = a \end{gathered} \qquad 11.7.13$$

We have waited till the last possible moment before nondimensionalizing. Since a has dimensions $1/T$, and ν has dimensions L^2/T, a length scale is $\sqrt{\nu/a}$ and a velocity scale is $\sqrt{\nu a}$ (for the v velocity; the u-velocity scale is ax in view of 11.7.3). Therefore, it is appropriate to define new variables as follows:

$$\eta = \frac{y}{\sqrt{\nu/a}}$$
$$F = \frac{f}{\sqrt{\nu a}} \qquad 11.7.14$$

When this choice of variables is substituted into the problem 11.7.13, we obtain

$$(F')^2 - FF'' - F''' = 1$$
$$F(0) = 0$$
$$F'(0) = 0 \qquad 11.7.15$$
$$F'(\infty) = 1$$

The problem is now free of all parameters and therefore can be solved once and the solution used for all stagnation points. That is, the same solution may be applied for all different bodies and flow velocities as characterized by $a = \alpha U_\infty / L$ and for all different fluid viscosities as characterized by ν.

The mathematical problem given as 11.7.15 is a third-order nonlinear ordinary differential equation. It does not have a closed-form solution. The boundary conditions are applied at two different points, $\eta = 0$ and $\eta \to \infty$. For this reason it is called a two-point boundary-value problem.

One of the purposes of this section is to learn how problems of this type are solved using a standard computer program, which solves a system of coupled first-order differential equations. Appendix E lists a short program using the Runge–Kutta method as organized by Fehlberg (1969). The first step in arranging the problem for the computer is to convert the third-order differential equation into three first-order differential equations. Consider the new dependent variables defined by

$$Y_1 = F, \qquad Y_2 = F', \qquad Y_3 \equiv F'' \qquad 11.7.16$$

The original differential equation 11.7.15 now becomes

$$Y_3' = Y_2^2 - Y_3 Y_1 - 1 \qquad 11.7.17$$

and, from the definitions above, two other equations are

$$Y_2' = Y_3$$
$$Y_1' = Y_2$$

Thus, the third-order equation is replaced by the three ordinary equations given above. In general, this method can be used to replace an nth-order ordinary differential equation by n first-order differential equations of the form

$$Y_i' = f_i(Y_1, Y_2, \ldots, Y_n) \qquad i = 1, 2, \ldots, n$$

The functions f_i in this equation are unrestricted.

Boundary data for the computer equation must be given at an "initial" location (let $\eta = x$ be the computer independent variable). From 11.7.15 we find our boundary conditions are

$$Y_1(0) = 0, \qquad Y_2(0) = 0, \qquad Y_2(\infty) = 1 \tag{11.7.18}$$

We have no known condition for $Y_3(0)$, but instead there is a known value for $Y_2(\infty)$. The most popular way around this difficulty is to assume a value for $Y_3(0)$, solve the problem, then compare the result for $Y_2(\infty)$ with that required by the original boundary condition. This is called a shooting method, and is sometimes automated so that the computer makes a new guess for $Y_3(0)$ based upon the last error in $Y_2(\infty)$.

A second question arises because the integration should extend over the infinite domain $X[0, \infty]$. We must pick a finite value of X and call it $X = \infty$. In general this value depends upon the answer, and we must either do an asymptotic analysis as $X \to \infty$ or watch the solution and make sure the answer has stabilized at the chosen "$X = \infty$" point.

Graphs of the u and v velocity profiles are given in Fig. 11-9. The v velocity is a function of y only, and the slope dv/dy is zero at the wall. This latter fact is a general result; the velocity component normal to a wall and its derivatives are zero at a solid wall for any flow. Far away from the wall, the v velocity increases linearly as required by 11.7.2. From the solution one can evaluate the constant b/a in 11.7.2. In nondimensional form 11.7.2 is

$$\lim_{\eta \to \infty} (\eta - F) = \frac{b/a}{\sqrt{\nu/a}} = 0.648 \tag{11.7.19}$$

This is the effective displacement of the wall, b/a, in nondimensional form.

The horizontal velocity profile has the same shape at each location in x, while the magnitude of the profile increases in proportion to x. As a result, the distance from the wall at which u is 99% of the free-stream value (11.9.3) is a constant; $u/ax \geq 0.99$ when $\eta \geq 2.4$. We denote this thickness by δ:

$$\begin{aligned} \frac{\delta}{\sqrt{\nu/a}} &= \eta_{99} = 2.40 \\ \delta &= 2.40\sqrt{\nu/a} \end{aligned} \tag{11.7.20}$$

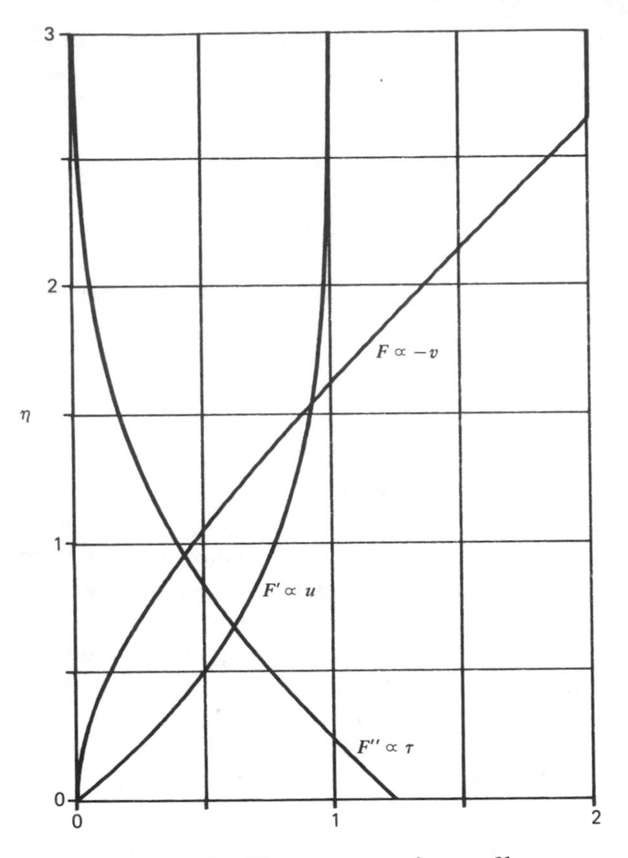

Figure 11-9 Hiemenz stagnation profile.

The shear stress and vorticity are confined within this region, which has a thickness proportional to $\sqrt{\nu}$.

As a concrete example let us consider the stagnation region on a circular cylinder. For a body of this shape, it is known from inviscid flow theory that the constant $a = 2U/r_0$ where U is the free-stream speed and r_0 the cylinder radius. Hence, the thickness of the stagnation region is

$$\frac{\delta}{r_0} = 2.4\sqrt{\frac{\nu}{2U_\infty r_0}} = 2.4\,\mathrm{Re}^{-1/2} \qquad 11.7.21$$

The physical thickness of the viscous layer compared to the radius decreases as the Reynolds number increases. This confirms the statement made at the beginning of this section—the analysis will be good for high Reynolds numbers. For Re = 1000 we have $\delta/r_0 = 0.076$ ($r_0 = 5$ cm, $U = 15$ cm/sec in air, $\nu = 0.15$ cm/sec, $\delta = 0.38$ cm), and for Re = 10^5 we have $\delta/r_0 = 0.0076$ (5 cm, 15 m/sec, 0.038 cm).

The pressure in the flow can be evaluated by returning to 11.7.11:

$$\frac{1}{\rho}p = \tfrac{1}{2}x^2H + K(y)$$

Previously H was determined to be $-a^2$, and $K(y)$ can be determined by integrating 11.7.12. The result for $K(y)$ is

$$K = -\tfrac{1}{2}f^2 - \nu f' + C$$

The constant C is found by setting the pressure at $x = 0$, $y = 0$ equal to the stagnation pressure p_0. Inserting these facts into the equation above gives

$$\frac{p_0 - p}{\rho} = \tfrac{1}{2}a^2x^2 + \tfrac{1}{2}\nu aF^2 + \nu aF' \qquad 11.7.22$$

A better physical understanding of this equation is obtained if we recall that 11.7.4 states that a is proportional to the free-stream speed divided by a body dimension. When $a = \alpha U_\infty/L$ is inserted into 11.7.22, we find the nondimensional pressure as

$$\frac{p_0 - p}{\tfrac{1}{2}\rho U_\infty^2} = \alpha^2\left(\frac{x}{L}\right)^2 + \frac{\alpha}{\text{Re}}[F^2 + 2F'] \qquad 11.7.23$$

where

$$\text{Re} = \frac{U_\infty L}{\nu}$$

As long as we are close to the surface, F and F' are of order one; hence in this region the second term becomes negligible as the Reynolds number becomes large (the condition for the analysis to apply). This means that the pressure near the wall is nearly constant across the viscous-dominated region. Equation 11.7.23 is essentially the Bernoulli equation in the free stream, since 11.7.3 and 11.7.4 show that $\alpha x/L = u(x, y \to \infty)/U_\infty$. The fact that the pressure is constant *across* the viscous layer is a general result, which we shall find is valid for all boundary layers.

This problem was first analyzed by Hiemenz (1911), and improved calculations have been done by many people over the course of years. A similar problem for the stagnation point on an axisymmetric blunt body can also be solved exactly (Homann 1936). Howarth (1951) solved the general two-dimensional stagnation-point problem where flow comes toward the point in the y direction and leaves in the x and z directions according to $u(x, y \to \infty, z) = a_1x$ and $w(x, y \to \infty, z) = a_2z$. These problems will be taken up in Chapter 20.

11.8 VON KÁRMÁN'S VISCOUS "PUMP"

This problem concerns a very large flat disk rotated at speed Ω in a semiinfinite fluid as shown in Fig. 11-10a. Attention is focused on the flow on one side of the disk in the local region near the axis of rotation. At the surface of the disk, the no-slip condition requires that the fluid rotate with the same velocity as the disk. Viscous effects diffuse away from the disk and induce a rotation in nearby fluid in the same manner as Rayleigh's impulsive plate. However, there is no pressure gradient in the radial direction to balance the centrifugal force. Once particles have been accelerated by the plate, they are also flung out in a radial flow. Continuity demands that we replace the outward moving fluid. This is accomplished by an axial flow toward the disk from the quiescent fluid far from the disk. Fluid is pumped from the far stream toward the disk, where viscous forces induce a swirl; then the resulting centrifugal effect produces a radial flow.

The solution to this problem is fairly complex. First, let us note that $v_\theta = r\Omega$ for the fluid at the disk while $v_\theta = 0$ far away from the disk. This suggests the appropriate nondimensional form for v_θ would be

$$G(z) \equiv \frac{v_\theta}{r\Omega} \qquad 11.8.1$$

von Kármán proposed that G was a function of z alone. It also turns out that the radial velocities induced by the centrifugal force should scale the same way. We let F be the nondimensional radial velocity defined by

$$F(z) \equiv \frac{v_r}{r\Omega} \qquad 11.8.2$$

Since viscous diffusion operates in the z direction, we use the viscous length $\sqrt{\nu/\Omega}$ to nondimensionalize that direction:

$$z^* \equiv \frac{z}{\sqrt{\nu/\Omega}} \qquad 11.8.3$$

The proper scaling for the axial velocity v_z can be found by substituting into the continuity equation. Consider the continuity equation

$$\frac{1}{r}\frac{\partial}{\partial r}(rv_r) + \frac{\partial v_z}{\partial z} = 0$$

and define

$$H(z) \equiv \frac{v_z}{\sqrt{\nu\Omega}} \qquad 11.8.4$$

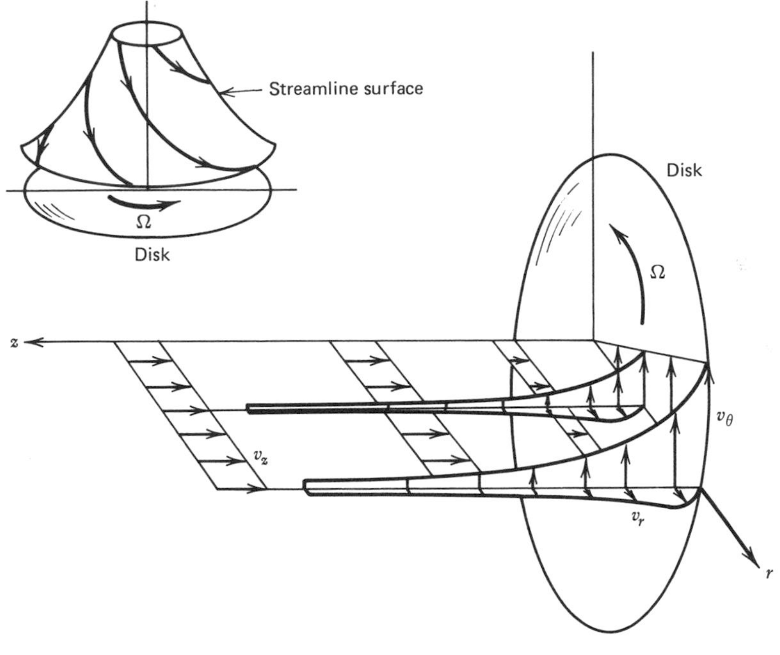

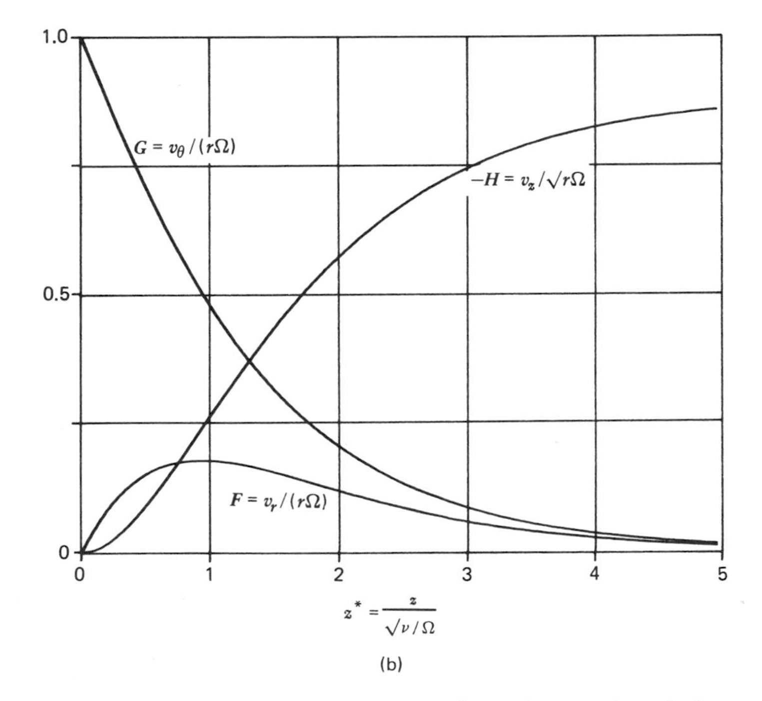

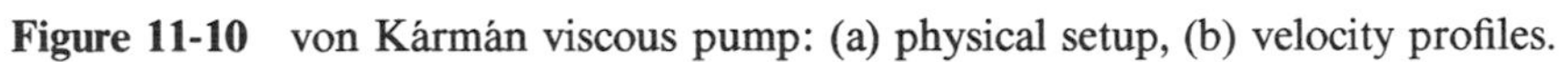

Figure 11-10 von Kármán viscous pump: (a) physical setup, (b) velocity profiles.

The continuity equation in nondimensional terms becomes

$$2F + H' = 0 \tag{11.8.5}$$

—a relation between the radial flow and the axial flow.

A second differential equation is found from the r-momentum equation. Here we write only terms that are not zero:

$$v_r \frac{\partial v_r}{\partial r} + v_z \frac{\partial v_r}{\partial z} - \frac{v_\theta^2}{r} = \nu \frac{\partial^2 v_r}{\partial z^2}$$

With the assumptions above, this becomes

$$F^2 + HF' - G^2 = F'' \tag{11.8.6}$$

Not only does the centrifugal force G^2 cause a radial flow, but two minor effects, z-direction convection and viscous stress, also play a role.

The θ-momentum equation is similarly complicated. It is

$$v_r \frac{\partial v_\theta}{\partial r} + v_z \frac{\partial v_\theta}{\partial z} + \frac{v_r v_\theta}{r} = \nu \frac{\partial^2 v_\theta}{\partial z^2}$$

Transforming to nondimensional variables gives

$$FG + HG' + FG = G'' \tag{11.8.7}$$

The major effect in this equation is the convection HG' competing with the viscous diffusion G''.

After defining the nondimensional pressure

$$P(z) \equiv \frac{p}{\rho \Omega \nu}$$

the last equation is obtained from the z-momentum equation. It is

$$v_z \frac{\partial v_z}{\partial z} = -\frac{1}{\rho}\frac{\partial p}{\partial z} + \nu \frac{\partial^2 v_z}{\partial z^2}$$

$$-HH' = -P' + H''$$

A better form is obtained if we eliminate H'' by using the continuity equation:

$$2HF - 2F' = P' \tag{11.8.8}$$

We now have three coupled ordinary differential equations, 11.8.5, 11.8.6, and 11.8.7, for the unknowns F, G, and H, while, in principle, P need not be determined from 11.8.8 until the first three equations are solved.

Boundary conditions are imposed as follows:

at $z^* = 0$

$$\begin{aligned} v_\theta &= r\omega, & G(0) &= 1 \\ v_r &= 0, & F(0) &= 0 \\ v_z &= 0, & H(0) &= 0 \\ p &= 0, & P(0) &= 0 \end{aligned} \qquad 11.8.9a$$

at $z^* \to \infty$,

$$\begin{aligned} v_r &= 0, & F(\infty) &= 0 \\ v_\theta &= 0, & G(\infty) &= 0 \end{aligned} \qquad 11.8.9b$$

This is another two-point boundary-value problem and may be solved by computer computations. When 11.8.5, 11.8.6, 11.8.7, and 11.8.8 are written as six first-order differential equations, we gain F' and G' as additional variables. Initial guesses $F'(0)$ and $G'(0)$ replace the boundary conditions at $z \to \infty$. Rogers and Lance (1960) found that the initial condition $F'(0) = 0.510233$ and $G'(0) = -0.514922$ give solutions that closely satisfy 11.8.9.

Graphs of the solutions are given in Fig. 11-10b. The swirling flow near the plate is confined to the region out to about $z^* = 5.5$. Once again we denote the thickness of the viscous region as δ. Using 11.8.3, we find

$$\delta = 5.5\sqrt{\frac{\nu}{\Omega}}$$

which might be interpreted as the characteristic length for the viscous diffusion that occurs in one revolution of the disk. The main counteracting effect is the flow of fluid toward the plate. From the solution we find $H(\infty) = 0.886$; thus the disk produces a volume flow per unit area of

$$v_z(\infty) = 0.886\sqrt{\nu\Omega}$$

The more viscous fluids display a better pumping effect. The average distance the flow moves in the z direction in one revolution is proportional to

$$L = \frac{v_z}{\Omega} \sim \sqrt{\nu\Omega} \sim \sqrt{\frac{\nu}{\Omega}}$$

Because the flow balances viscous diffusion and convection, this length and δ have the same parametric dependence.

In order to get a feeling for the size of this effect, let us assume that a disk is spinning in air ($\nu = 0.15$ cm/sec) at 1200 rpm (125 rad/sec). The axial flow will be at a velocity

$$v_z(\infty) = 0.896\sqrt{0.15 \times 125} = 3.85 \text{ cm/sec}$$

and the viscous region will be

$$\delta = 5.5\sqrt{\frac{0.15}{125}} = 0.19 \text{ cm}$$

At higher speed, the thickness δ becomes even smaller while the axial flow velocity increases.

The radial flow occurs because there is no pressure gradient in the r direction to force the particles into circular motion. An interesting variation of the problem is where the flow at infinity rotates as a solid body and the disk is stationary. This is like a vortex core interacting with a solid wall. In this case there must be a radial pressure gradient at infinity to maintain the solid-body-like rotation. Viscous forces near the wall slow the fluid down and destroy the balance between the pressure gradient and the centrifugal inertia force. In this region the pressure gradient accelerates the flow inward toward the center. Continuity then produces an outward flow along the z axis. This flow is known as the Bödewadt (1940) problem. The solution has some of the characteristics of a tornado intersecting with the ground; however, tornado-like solutions in general can be very much more complicated.

Both of these problems are special cases of the general situation where the fluid at infinity and the disk may be assigned different rotation speeds. Another set of problems involves disks set finite distances apart.

11.9 CONCLUSION

Viscous diffusion has been the major theme of this chapter. Shear stresses generated at a wall take some time to diffuse into the interior. The important physical property of the controlling viscous diffusion is the kinematic viscosity ν (L^2/T). The depth of penetration of viscous diffusion in a time t is given by $\sqrt{\nu t}$.

Pressure forces, on the other hand, are transmitted instantaneously in an incompressible flow because the effective speed of sound is infinite. For example, consider the problem where the fluid oscillates back and forth above a fixed wall. The pressure gradient along the wall is established instantaneously by some external agent. Viscous effects diffusing from the wall have a much slower time scale. We found an unexpected overshoot in the velocity profile because the pressure and viscous forces could combine at a certain distance from the wall. The disparity in time scales for transmitting pressure forces (instantaneous) and for transmitting viscous forces (diffusive) is responsible for many striking phenomena in fluid mechanics. We shall see these mechanisms again in later chapters.

PROBLEMS

11.1 (B) The cross section of a tube is an equilateral triangle with sides of length l and a horizontal base. Flow in the tube is produced by an imposed pressure gradient dp/dx. Verify that the velocity profile is given by

$$u(y, z) = \frac{1}{2\sqrt{3}\,\mu l}\left(-\frac{dp}{dx}\right)\left(z - \frac{\sqrt{3}}{2}l\right)(3y^2 - z^2)$$

where the coordinate origin is at the apex of the triangle with z bisecting the angle and positive downward, and y is horizontal. Check that the flow rate is

$$Q = \frac{\sqrt{3}}{320}\frac{l^4}{\mu}\left(-\frac{dp}{dx}\right)$$

11.2 (A) Waves in shallow water induce an oscillatory motion that extends to the bottom. The motion is parallel to the bottom and sinusoidal. Estimate the thickness of the viscous effect caused by the no-slip condition at the bottom when the wave period is one second.

11.3 (B) Solve for the velocity profile above a plate that oscillates in its own plane according to $u(0, t) = u_0 \cos \Omega t$. Choose the nondimensional y variable as $Y = y/\sqrt{2\nu/\Omega}$.

11.4 (B) Consider an infinite stream oscillating according to $u(t) = u_0 \sin \Omega t$. What pressure gradient would cause this oscillation? A solid wall is inserted into the flow so that it is parallel to the motion. What is the shear stress on the wall? What is the phase of the shear stress with respect to the velocity $u(y \to \infty, t)$? What is the y location, as a function of time, where the particle acceleration is a maximum (of either sign)? How much of the acceleration is due to pressure and how much to viscosity?

11.5 (A) Flow in a slot of width $2h$ is driven by a pressure gradient $[dp/dx]_0 + K \cos \Omega t$. How does the average flow rate for this situation compare with the flow rate produced by the steady gradient $[dp/dx]_0$?

11.6 (A) Flow in a pipe has a steady component and an oscillatory component. Develop a criterion to determine when quasi-steady-state assumptions can be used to find the wall shear stress.

11.7 (C) Consider a steady swirling flow that is unbounded. In cylindrical coordinates the flow moves away from the origin (in both directions) according to $v_z = az$, where a is a constant. The swirl component v_θ is a function only of the radial position r. With this information and the continuity equation, find the equation for the radial velocity v_r. At $r \to \infty$, v_r may become unbounded in order to supply the flow that is leaving along the z axis. Next, simplify the θ-momentum equation. Introduce the nondimensional variables $\gamma^* = 2\pi r v_\theta/\Gamma$, where Γ is a constant and $r^* = r/\sqrt{2\nu/a}$. Compare the resulting differential equation with 11.6.8, and find the solution. You have now found v_z, v_r, and v_θ. What role would the r- and z-momentum equations play in completing the solution?

11.8 (B) The function $\gamma^*(r, t, \nu)$ determined by 11.6.5 leads to the nondimensional form $\gamma^*(\eta)$, where $\eta = r/\sqrt{\nu t}$. What does dimensional analysis predict about the function $v_\theta(r, t, \Gamma, \nu)$? Why does the first form lead to a sharper result?

11.9 (B) Fluid is contained in a slot of width h. Find the velocity profile if the lower wall oscillates sinusoidally in its own plane while the upper wall is fixed.

11.10 (B) The flat bottom surface under a liquid of depth h is moved in its own plane according to $u = u_0 \sin \Omega t$. Find the velocity and vorticity profiles in the liquid.

11.11 (B) Find the pressure field for an Oseen vortex. (Hint: Look for integrals that cancel.) What is the pressure at the origin?

11.12 (A) Find the exact relations for the maximum velocity and its position (the core radius) as functions of time for the Oseen vortex.

11.13 (A) Solve for the velocity profile in the stagnation-point flow. Tabulate $F(\eta)$, $F'(\eta)$ and $F''(\eta)$ for $\eta = 0.0, 0.2, 0.4, \ldots, 3.0$. Use $\eta_\infty = 8$.

11.14 (B) Air flows around a cylinder of 5-cm radius at 15 cm/sec (the Reynolds number is 1000), with the free-stream velocity perpendicular to the axis. Find the dimensional u and v components of the velocity at a point 0.15 cm away from the surface and 0.5 cm away from the symmetry plane on the upwind side of the cylinder. Find the shear stress on the wall at a point 0.5 cm away from the symmetry plane.

11.15 (C) A solid stationary wall exists at $z = 0$. Far away from the wall the fluid is rotating as if it were a solid body: $v_\theta = r\Omega$ as $z \to \infty$ (Bödewadt's problem). Solve for the velocity profiles. [This computation tends to blow up at infinity. The published values for the starting conditions are $F'(0) = -0.941971$ and $G'(0) = 0.772886$ (Rogers and Lance 1960).]

12 Stream Functions and the Velocity Potential

Two very useful ideas will be introduced in this chapter: the stream function and the velocity potential. These quantities have physical interpretations, and, perhaps more importantly, they are frequently used as dependent variables in the solution of flow problems. In this role they replace the velocity components. The stream function and the velocity potential exist only for specific types of flows that meet certain kinematic restrictions.

A stream function exists when a flow has symmetry with respect to a coordinate system and also has zero rate of expansion everywhere:

$$\frac{1}{V_{\mathrm{MR}}}\frac{dV_{\mathrm{MR}}}{dt} = \nabla \cdot \mathbf{v} = 0$$

Since the remainder of this book deals with incompressible flows, this second criterion is always met.

The velocity potential, on the other hand, does not require any symmetry of the flow field, but it imposes a much stronger requirement on the particle motion, namely, that the vorticity is zero throughout the flow:

$$\boldsymbol{\omega} = \nabla \times \mathbf{v} = 0$$

In the next chapter we shall find that viscous forces may be neglected in irrotational flows. Because of this, the velocity potential is useful in inviscid flows.

12.1 STREAM FUNCTION FOR PLANE FLOWS

Streamlines in a fluid are lines that are everywhere tangent to the velocity vectors. An equation that would describe such lines in a plane, two-dimensional flow may be written in the form

$$\psi = \psi(x, y) \qquad 12.1.1$$

where ψ is called the *stream function*. When ψ is constant, 12.1.1 is a relation between x and y that describes a streamline. If ψ is given a new value, the relation $\psi(x, y) = \psi$ describes a different streamline.

The stream function is more useful than you might at first suspect. All the properties of the flow—the velocities, the pressure, and so on—may be related to the it. The stream function is a single scalar unknown that can yield a complete description of the flow.

Let us start the mathematical development by noting that if ψ exists, it must obey the differential relation

$$d\psi = \frac{\partial \psi}{\partial x} dx + \frac{\partial \psi}{\partial y} dy \qquad 12.1.2$$

Next, we propose the important definition that relates ψ and the velocities. At this stage the definitions are tentative; we must demonstrate later that they are appropriate and mathematically proper. Let

$$u = \frac{\partial \psi}{\partial y}, \qquad v = -\frac{\partial \psi}{\partial x} \qquad 12.1.3$$

[Note that an arbitrary constant (or function of time) may be added to ψ without affecting the velocities.] Substituting into 12.1.2 gives

$$d\psi = -v\, dx + u\, dy \qquad 12.1.4$$

This is a Pfaffian form ($dF = M\, dx + N\, dy$), and we know from mathematics that it is "exact"—that is, ψ is really a function of x and y—if and only if the following derivatives are equal ($\partial M/\partial y = \partial N/\partial x$):

$$-\frac{\partial v}{\partial y} = \frac{\partial u}{\partial x} \qquad \left(\frac{\partial^2 \psi}{\partial y\, \partial x} = \frac{\partial^2 \psi}{\partial x\, \partial y} \right) \qquad 12.1.5$$

The equation above is the continuity equation. Thus, the definitions 12.1.3 are acceptable and the stream function exists as a result of the fact that $\nabla \cdot \mathbf{v} = 0$. This also means that a stream function will automatically satisfy the continuity equation.

From 12.1.4 it is easy to show that when ψ is constant, the resulting equation describes a streamline. If ψ is a constant, $d\psi$ is zero and 12.1.4 becomes

$$\frac{v}{u} = \left. \frac{dy}{dx} \right|_{\psi = \text{const}} \qquad 12.1.6$$

This equation states that the velocity vector is tangent to the curve $\psi = $ const.

The second major characteristic of the stream function is that the numerical difference in ψ between two streamlines is equal to the volume flow rate between those streamlines. In order to prove this, consider two streamlines with values ψ_A and ψ_B as shown in Fig. 12-1. Two points A and B are chosen and connected by any smooth path. The volume flow between the streamlines

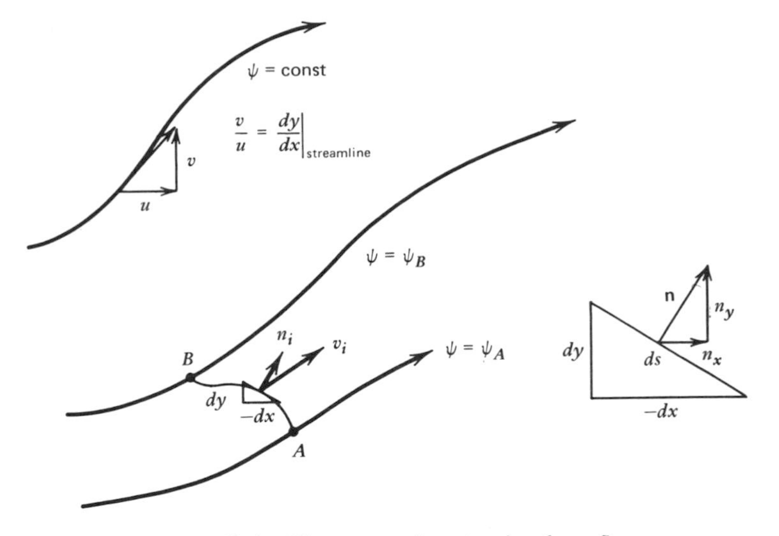

Figure 12-1 The stream function in plane flow.

is

$$Q_{AB} = \int_{A-B} n_i v_i \, ds = \int_{A-B} (n_x u + n_y v) \, ds \qquad 12.1.7$$

where n is normal to the integration element ds. By geometry we have the relations $n_x \, ds = dy$ and $n_y \, ds = -dx$. With these relations, 12.1.7 becomes

$$Q_{AB} = \int_{A-B} (u \, dy - v \, dx) = \int_{A-B} d\psi$$

$$Q_{AB} = \psi_B - \psi_A \qquad 12.1.8$$

The flow rate between streamlines is the difference in their stream-function values. This equation is also unaffected by the addition of an arbitrary constant to ψ.

There are two other important relations involving the stream function. The first is a relation between it and the vorticity. In plane (two-dimensional) flow the vorticity has only one nonzero component:

$$\omega = \omega_z = \varepsilon_{zjk} \partial_j v_k = -\frac{\partial u}{\partial y} + \frac{\partial v}{\partial x} \qquad 12.1.9$$

When 12.1.3 are introduced into the equation above, we find

$$-\omega_z = \frac{\partial^2 \psi}{\partial x^2} + \frac{\partial^2 \psi}{\partial y^2} = \nabla^2 \psi \qquad 12.1.10$$

In later chapters we shall use this equation in a computation scheme where ψ and ω are the major dependent variables. This equation takes on added importance in irrotational flows. Then ω is zero, and ψ satisfies the Laplace equation.

The major equation involving ψ is derived from the momentum equations. In plane, two-dimensional flow these are

$$\frac{\partial u}{\partial t} + u\frac{\partial u}{\partial x} + v\frac{\partial u}{\partial y} = -\frac{1}{\rho}\frac{\partial p}{\partial x} + \nu\frac{\partial^2 u}{\partial x^2} + \nu\frac{\partial^2 u}{\partial y^2} \qquad 12.1.11a$$

$$\frac{\partial v}{\partial t} + u\frac{\partial v}{\partial x} + v\frac{\partial v}{\partial y} = -\frac{1}{\rho}\frac{\partial p}{\partial y} + \nu\frac{\partial^2 v}{\partial x^2} + \nu\frac{\partial^2 v}{\partial y^2} \qquad 12.1.11b$$

The pressure is eliminated from these equations by the following process: differentiate 12.1.11a with respect to y so that $\partial^2 p/\partial y\,\partial x$ occurs; similarly differentiate 12.1.11b with respect to x, so that $\partial^2 p/\partial x\,\partial y$ occurs, subtract the two equations to cause these terms to cancel. Using 12.1.9 to identify the vorticity, we then have the following equation:

$$\frac{\partial \omega}{\partial t} + u\frac{\partial \omega}{\partial x} + v\frac{\partial \omega}{\partial y} = \nu\nabla^2\omega \qquad 12.1.12$$

Finally, 12.1.3 and 12.1.10 are used to eliminate the velocities and the vorticity. The equation then becomes

$$\frac{\partial}{\partial t}\nabla^2\psi + \frac{\partial \psi}{\partial y}\frac{\partial}{\partial x}\nabla^2\psi - \frac{\partial \psi}{\partial x}\frac{\partial}{\partial y}\nabla^2\psi = \nu\nabla^4\psi \qquad 12.1.13$$

where (subscripts denote differentiation)

$$\nabla^4\psi \equiv \psi_{xxxx} + 2\psi_{xxyy} + \psi_{yyyy}$$

This equation has only ψ as an unknown and offers a starting point for solution of any incompressible, two-dimensional flow. Its advantage is that it is a single equation with one unknown; its disadvantage is that it is fourth-order. This contrasts with our previous use of two momentum equations and the continuity equation. These equations are of second order but have u, v, and p as unknowns.

As an example of the use of the stream function, we again consider the steady flow in a slot driven by pressure gradient. The coordinate system is placed on the lower wall. Assuming that $v = -\partial\psi/\partial x = 0$ means that ψ is a function of y alone which reduces 12.1.13 to

$$0 = \frac{d^4\psi}{dy^4} \qquad 12.1.14$$

The solution may be found immediately as

$$\psi = C_0 + C_1 y + C_2 y^2 + C_3 y^3 \tag{12.1.15}$$

Next, we formulate the boundary conditions. Any solid wall is a streamline and therefore has a constant value of ψ. Also, since ψ is only determined to within an arbitrary constant, we can fix the value on one wall to any value we choose. Thus we let

$$\psi(y = 0) = 0 \tag{12.1.16}$$

The value on the upper wall must be the flow rate in light of 12.1.8; hence

$$\begin{aligned} Q &= \psi(y = h) - \psi(y = 0) \\ Q &= \psi(y = h) \end{aligned} \tag{12.1.17}$$

When we specify $\psi = \text{const}$ on a wall, it is equivalent to saying that the derivative of ψ along the wall, which is by definition the velocity component normal to the wall, is zero. The remaining boundary conditions are found from the no-slip conditions:

$$u(y = 0) = 0 = \left.\frac{d\psi}{dy}\right|_0 \tag{12.1.18}$$

$$u(y = h) = 0 = \left.\frac{d\psi}{dy}\right|_{y=h} \tag{12.1.19}$$

With these four conditions we can determine the constants in 19.1.15. The result is

$$\frac{\psi}{Q} = 3\left(\frac{y}{h}\right)^2 - 2\left(\frac{y}{h}\right)^3 \tag{12.1.20}$$

It is common practice to plot streamlines with equal increments in ψ. Since equal increments imply equal flow rates between the lines, one can get a rough idea of the velocity. The average velocity between streamlines is inversely proportional to the distance between them.

Velocities are obtained from ψ using the definitions 12.1.3:

$$\begin{aligned} \frac{u}{Q/h} &= \frac{\partial(\psi/Q)}{\partial(y/h)} \\ \frac{u}{Q/h} &= 6\frac{y}{h} - 6\left(\frac{y}{h}\right)^2 \end{aligned} \tag{12.1.21}$$

When ψ is used in a problem, it naturally brings in the flow rate Q as a boundary condition, and information about the pressures is not needed. We must integrate the momentum equation to find the pressures. In this case the x-momentum equation simplifies to

$$0 = -\frac{dp}{dx} + \mu\frac{d^2u}{dy^2}$$

$$\frac{dp}{dx} = \mu\frac{d^2(uh/Q)}{d(y/h)^2}\frac{Q}{h^3} = -12\frac{\mu Q}{h^3} \qquad 12.1.22$$

The pressure-drop–flow-rate equation gives the equivalence between dp/dx and Q as parameters in the profiles.

*12.2 STREAMLINES AND STREAMSURFACES FOR A THREE-DIMENSIONAL FLOW

In this section we take a more general approach to the problem of describing streamlines. Consider a three-dimensional flow where the space coordinates are x, y, and z and the corresponding velocity components are u, v, and w. Since, by definition, a streamline is everywhere tangent to the velocity, the projected slopes of a streamline are given by

$$\left.\frac{dy}{dx}\right|_{\text{str}} = \frac{v}{u}, \qquad \left.\frac{dy}{dz}\right|_{\text{str}} = \frac{v}{w}, \qquad \left.\frac{dx}{dz}\right|_{\text{str}} = \frac{u}{w} \qquad 12.2.1$$

A more compact form for these equations is

$$\frac{dx}{u} = \frac{dy}{v} = \frac{dy}{w} \quad \text{or} \quad \mathbf{v} \times \mathbf{t}\, ds = 0 \qquad 12.2.2$$

For a given velocity field, the solutions of this set of equations are the streamline trajectories.

Several aspects of streamlines should be noted. First, when the velocity is zero, there is no unique direction for the streamline. If the streamline should happen to split or branch, it must do so at places where the velocity is zero. Such points are called stagnation points. As an example, consider the streaming flow around a blunt axisymmetric body. The streamline on the axis approaches the nose, and the velocity becomes zero. This streamline then splits into an infinite number of streamlines, which follow the surface of the body. The viscous no-slip condition actually requires that the velocity be zero everywhere on the body. One might legitimately ask, "Are there streamlines on a solid surface?" In this instance we must broaden the definition of a streamline slightly. If one takes a line normal to the surface and considers the velocity vector at various positions on this line, there is a well-defined direction in the limit of approaching the wall. This direction determines the wall streamline (sometimes known as the limiting streamline). At the nose of the

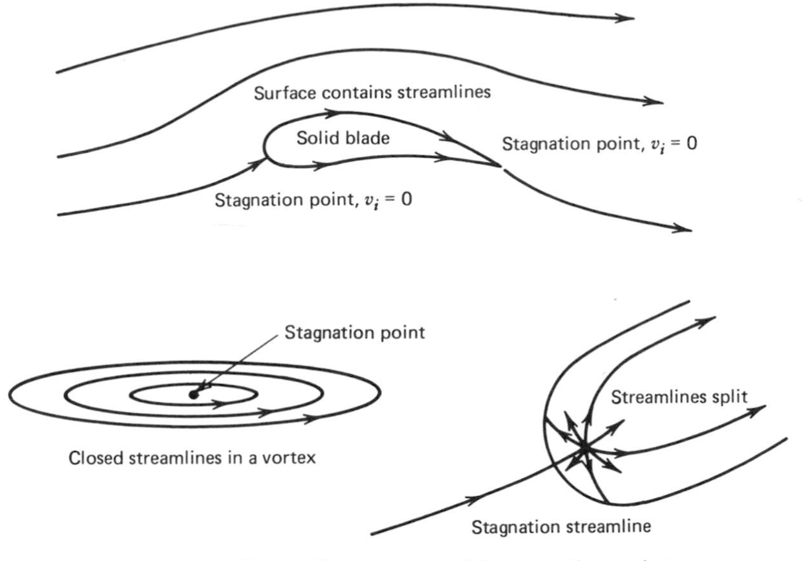

Figure 12-2 Streamline patterns with stagnation points.

body a special direction does not exist for the stagnation streamline, (Fig. 12-2). As another example, consider the trailing edge of an airfoil with a finite angle. The wall streamlines from the top and bottom surfaces come together at an angle, then leave the surface in a streamline that divides the upper flow from the lower flow. At this stagnation point (or, more properly, stagnation line) the velocity must be zero. Stagnation points are not limited to surfaces and may occur in the interior of the fluid as well. If the velocity is zero anywhere, then it is possible for the streamline to split.

It is often said that in incompressible flow, streamlines cannot end within the fluid. They either come from and return to infinity, or they form closed loops. It is also possible, however, for them to emanate from surfaces, though only if they are stagnation streamlines. These results can be understood by considering a small streamtube (which does not contain a stagnation streamline in its interior). The streamtube consists of two ends, A_1 and A_2, and a side surface. The velocity vector and the surface normal are always perpendicular on the side surface, while on the end surfaces they are nearly parallel. Assuming there are no sources within the streamtube, we apply Gauss's theorem to the continuity equation:

$$0 = \int \partial_i v_i \, dV = \int n_i v_i \, dS$$

$$0 = \int_{A_1} n_i v_i \, dS + \int_{A_2} n_i v_i \, dS$$

From this equation we argue that streamtubes can never end within the fluid. If a finite integral exists for the surface A_1, then A_2 could only vanish if the velocity became infinite. An infinite velocity represents an unrealistic situation that violates the continuum assumption; therefore, A_2 is always finite. It is, of course, possible that streamtubes could form closed loops. The argument also applies to streamlines as they can be considered as the limit of a streamtube of small area.

There are several ways to describe a line that is imbedded in space. The method most useful for our purposes is to regard the line as the intersection of two independent surfaces f and g (Fig. 12-3):

$$f = f(x, y, z) \qquad g = g(x, y, z) \tag{12.2.3}$$

As f and g take on different values, we are describing different streamlines. These surfaces will be called streamsurfaces and are regarded as solutions to 12.2.1. For any velocity field, mathematicians tell us that it is possible to find two such sets of independent surfaces f and g as long as $\mathbf{v}$ is not zero. This is a local result. We can find the surfaces for any point in the flow, but we cannot necessarily find one set that will work for the whole flow. Most flows are simple enough that this is not a problem. Another aspect of 12.2.3 is that the surfaces are not unique. Any surface h described by a function of f and g,

$$h = h(f, g) \tag{12.2.4}$$

is also a streamsurface. We could replace either f or g in 12.2.3 with the new function h. Since the walls or solid surfaces containing a flow must also contain a set of streamlines, this result means that it is always possible to make one set of surfaces in 12.2.3 have a member that coincides with the walls bounding the flow.

At this point we digress for a moment to discuss some properties of surfaces. It is necessary to have this background material to understand the

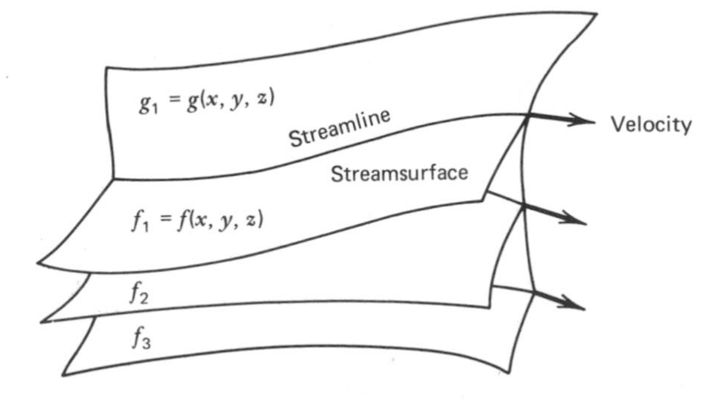

Figure 12-3 Streamsurfaces f_1 and g_1 intersect to define a streamline.

development of the stream function. The equation

$$f = f(x, y, z) \qquad 12.2.5$$

describes a surface in space for each value of f. The equation is a numbering system for each surface as well as an equation to determine the shape of the surface. For instance, any other function constructed from f produces the same set of surfaces, but assigns a different number to each; that is $F = F(f)$ gives the same surfaces as f.

As an illustration, consider the planes described by $f = x - y$. Now $f = 2$ is a certain plane from this set. However, the equation $F = (x - y)/3$ gives the same plane when $F = \frac{2}{3}$, or the equation $F = e^{(x-y)^2}$ gives the same plane when $F = e^4$. Thus, the shape of the surfaces, which is the only property important for the intersection of two surfaces, and the numbering system for the surfaces are somewhat independent.

Another property of a surface is its normal vector. At each point of the surface there is a normal vector given by ∇f. Although the direction of this gradient is always normal, its magnitude depends upon the numbering system chosen for the surfaces. When we state that two sets of surfaces are independent, as in 12.2.3, we are requiring that their normals not be parallel. Thus, f and g are independent if $\nabla f \times \nabla g \neq 0$.

Now we can return to the central question: how are the functions f and g related to the velocity? Since ∇f and ∇g are perpendicular to the streamsurfaces, they are also perpendicular to the velocity. Hence, the product $\nabla f \times \nabla g$ will be in the direction of the velocity as shown in Fig. 12-4a. The velocity itself can be obtained if we adjust the magnitude by a function h:

$$\mathbf{v} = h(x, y, z)\nabla f \times \nabla g \qquad 12.2.6$$

Up to this point the discussion has been very general and applies to any vector field. Actually 12.2.6 does not represent much progress, as the three components of the vector $\mathbf{v}$ are replaced by three functions f, g, and h on the right-hand side.

The continuity equation in incompressible flow states that the rate of expansion is always zero: $\nabla \cdot \mathbf{v} = 0$ ($\mathbf{v}$ is a solenoidal vector field). From this it is possible to prove that one may choose f and g in 12.2.6 in such a way as to make $h = 1$. This gives

$$\mathbf{v} = \nabla f \times \nabla g \quad \text{if} \quad \nabla \cdot \mathbf{v} = 0 \qquad 12.2.7$$

The equation above is the important relation between the streamsurfaces and the velocity. It corresponds to the relations 12.1.3 for plane flows. The surfaces f and g are not uniquely determined, and several choices are possible. With regard to the numbering system, we can choose that for only one set. The numbering for the other set must be determined so as to give the proper

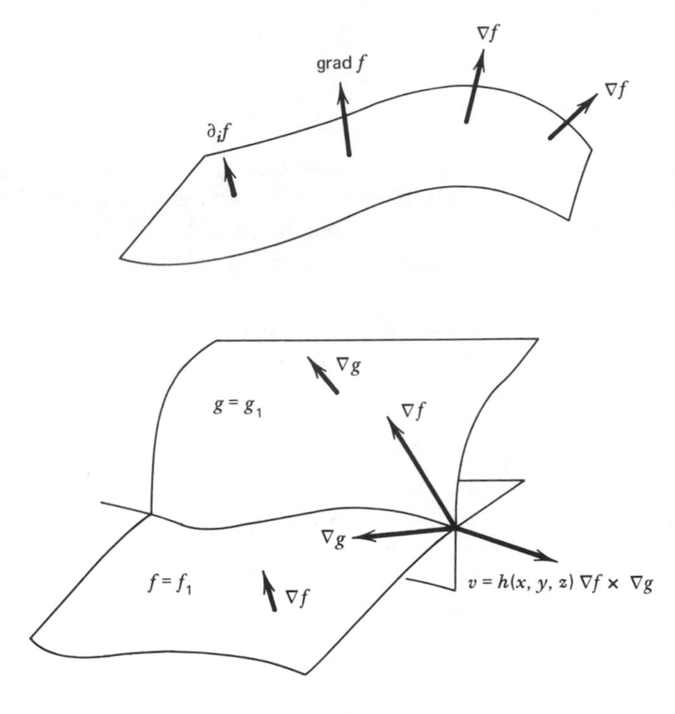

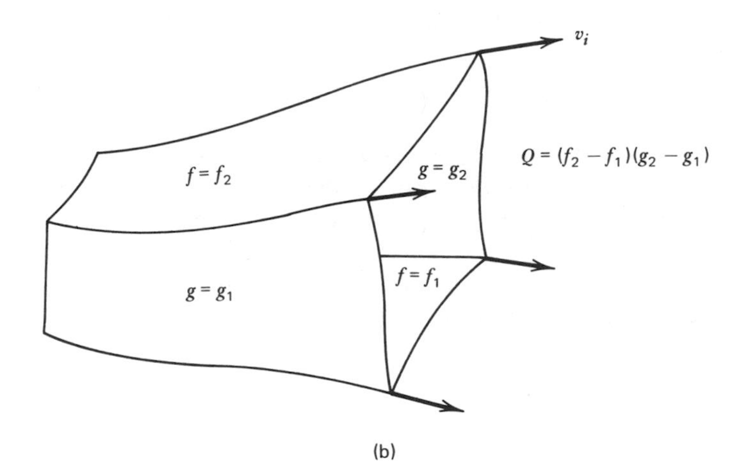

Figure 12-4 Relationship between streamsurfaces and flow properties: (a) $\mathbf{v} = h\ \nabla f \times \nabla g$, (b) $Q = (f_2 - f_1)(g_2 - g_1)$.

magnitude for the velocity through 12.2.7. In other words, if a change in the numbering system is made for g, then a compensating change must occur in both the shape and numbering system of f so that 12.2.7 is still true.

The streamsurfaces f and g are also related to the flow rate. Consider the streamtube formed by the surfaces f_1, f_2, g_1, and g_2 as shown in Fig. 12-4b. We simply state the flow-rate equation without proof (see Yih 1969):

$$Q = (f_2 - f_1)(g_2 - g_1) \qquad 12.2.8$$

Note that this relation and 12.2.7 remain unchanged if one adds an arbitrary constant to either f or g.

In order to illustrate how these equations work, we return to the plane-flow case in rectangular coordinates. There the proper choices are $g = z$ and $f = \psi$. Then 12.2.7 becomes (subscripts take on values x, y, z)

$$v_i = \varepsilon_{ijk} \partial_j \psi \partial_k (z) = \varepsilon_{ijk} \partial_j \psi \delta_{kz}$$

$$v_i = \varepsilon_{ijz} \partial_j \psi$$

$$v_x = \varepsilon_{xjz} \partial_j \psi = \frac{\partial \psi}{\partial y}$$

$$v_y = \varepsilon_{yjz} \partial_j \psi = -\frac{\partial \psi}{\partial x}$$

The flow rate is computed for a unit depth in the z direction:

$$Q = (\psi_2 - \psi_1)(z_2 - z_1) = \psi_2 - \psi_1 \qquad 12.2.9$$

These formulas correspond to the relations given in the previous section.

The incompressible continuity equation $\nabla \cdot \mathbf{v} = 0$ allows a general flow with three independent velocity components to be represented in terms of two streamsurfaces. When, in addition, a flow possesses a symmetry such that only two velocity components are nonzero, it is possible to choose g as a set of coordinate planes and thus reduce the problem to one of finding the remaining streamsurfaces $f = \psi$. Flows of primary interest are plane flows in either rectangular or cylindrical coordinates and axisymmetric flows in either cylindrical or spherical coordinates. Equations for these flows are given in Appendix D.

There is another method of formulating the stream function, which is complementary in that it gives us more information and allows us to connect the stream function with the vorticity. Once this is done, we may find the general equation governing the stream function by simply substituting into the dynamic equation governing vorticity (Chapter 13).

Consider a vector function $\mathbf{B}$ defined by the relation

$$\mathbf{B} \equiv f \nabla g = \psi \nabla g \qquad 12.2.10$$

which gives ψ in terms of $\mathbf{B}$. We compute the curl of $\mathbf{B}$:

$$\nabla \times \mathbf{B} = \nabla \times (f \nabla g)$$

$$= \nabla f \times \nabla g + f \nabla \times \nabla g$$

Since the last term is always zero for any scalar function g, we find that an equivalent expression to 12.2.7 is

$$\mathbf{v} = \nabla \times \mathbf{B} \qquad \text{if} \quad \nabla \cdot \mathbf{v} = 0 \qquad 12.2.11$$

The vector $\mathbf{B} \equiv \psi \nabla g$ is called the *vector potential* (not to be confused with the velocity potential). Equation 12.2.10 does not uniquely define $\mathbf{B}$. We may also specify that $\mathbf{B}$ be chosen so that $\nabla \cdot \mathbf{B} = 0$ ($\mathbf{B}$ is solenoidal as well as $\mathbf{v}$). This is also true in our application of f and g to flows that are symmetric with respect to an orthogonal coordinate system. ($\nabla \cdot \mathbf{B} = \psi \nabla^2 g + \nabla \psi \cdot \nabla g$: for all coordinate systems in Appendix E, g is taken as a coordinate plane and the numbering system is chosen so that $\nabla^2 g = 0$. Moreover, the symmetry of the flow and the orthogonality of the coordinates imply that $\nabla \psi$ is perpendicular to ∇g.)

The first important relation is found by computing the vorticity and using a vector identity,

$$-\boldsymbol{\omega} = -\nabla \times \mathbf{v} = -\nabla \times (\nabla \times \mathbf{B})$$

$$= \nabla^2 \mathbf{B} - \nabla(\nabla \cdot \mathbf{B})$$

or, since $\nabla \cdot \mathbf{B} = 0$,

$$-\boldsymbol{\omega} = \nabla^2 \mathbf{B} \qquad 12.2.12$$

This is the general formula that corresponds to 12.1.9 in the previous section. You will find the proper simplified form of this equation in each of the stream-function tables. [Recall that $\nabla^2 \mathbf{B}$ stands for the vector $\nabla \cdot \nabla \mathbf{B}$. This is important because in some coordinate systems and for certain components, it is not true that $(\nabla^2 \mathbf{B})_{\text{component}} = \nabla^2 (\mathbf{B}_{\text{component}})$].

The stream function satisfies the continuity equation by virtue of the way in which it is constructed. For an equation to govern ψ, we must look to the momentum equations. Actually, the equation we need is the vorticity transport equation. It is derived by taking $\nabla \times$ the momentum equation. This is done in the next chapter. The equation is

$$\frac{\partial \boldsymbol{\omega}}{\partial t} + \mathbf{v} \cdot \nabla \boldsymbol{\omega} = \boldsymbol{\omega} \cdot \nabla \mathbf{v} + \nu \nabla^2 \boldsymbol{\omega} \qquad 12.2.13$$

Substitution of $-\boldsymbol{\omega} = \nabla^2 \mathbf{B}$, $\mathbf{v} = \nabla \times \mathbf{B}$, and $\mathbf{B} = \psi \nabla g$ produces a single fourth-order equation for ψ. The only restriction is that the flow must be incompressible and that it must have symmetry with respect to an orthogonal coordinate system. In all symmetric flows the vector **B** has only one nonzero component, the vorticity has only one nonzero component, and the vorticity is perpendicular to the velocity. This makes plane and axisymmetric flows somewhat special since 12.2.12 and 12.2.13 have only one nontrivial component. These equations form the basis for many numerical solution methods where the vorticity and stream function are the major unknowns.

12.3 VELOCITY POTENTIAL AND UNSTEADY BERNOULLI EQUATION

Most students are familiar with the fact that a conservative force field may be represented by a potential. The gravity force is the best-known example, with the weight force per unit mass given by

$$F_i = g\,\partial_i(h)$$

where g is the acceleration of gravity and h is the vertical distance above a reference plane. Under special circumstances the velocity field itself may be represented by a *velocity potential*. The potential and the velocity are related by

$$v_i = \partial_i\phi \qquad 12.3.1$$

where ϕ may have an arbitrary constant added without a change in v_i. This equation replaces three unknown velocity components by a single unknown scalar function ϕ. In the light of this simplification one can guess that the conditions under which 12.3.1 exists are very restrictive. It turns out that the necessary and sufficient condition for a velocity potential to exist is that the flow is irrotational,

$$\boldsymbol{\omega} = \nabla \times \mathbf{v} = 0 \qquad 12.3.2$$

A flow field that lacks vorticity is a very special situation where the particles have never experienced a net viscous force; it is an inviscid flow. (The terms potential flow and inviscid flow are almost synonymous and are frequently used interchangeably.) The great advantage of the velocity potential is that it may be used in three-dimensional flows; no special symmetry is required. The great disadvantage is that it only works for inviscid flows.

The velocity potential is frequently used in compressible flow and acoustics. However, our interest is its application to incompressible flows. The major equation that governs ϕ for incompressible flows is found by substituting 12.3.1 into the continuity equation. The result is

$$0 = \partial_i v_i = \partial_i\partial_i\phi = \nabla^2\phi \qquad 12.3.3$$

The velocity potential satisfies the Laplace equation. Solutions to the Laplace equation are termed *harmonic functions*. There is a vast amount of mathematical information about this equation, which goes under the name potential theory. More will be said about the characteristics of potential flows in a later chapter. At this point it is sufficient to note that two kinematic conditions, zero vorticity and zero expansion, have led to an equation for a single unknown that will determine the velocity field.

If the velocity is determined from purely kinematic considerations, what role does the momentum equation play? The momentum equation 5.7.19 can be written using problems 3.16 and 3.17 together with the identity $\nabla^2 \mathbf{v} = -\nabla \times \boldsymbol{\omega}$ as

$$\frac{\partial \mathbf{v}}{\partial t} + \nabla\left(\tfrac{1}{2}v^2 + \frac{p}{\rho}\right) = \mathbf{v} \times \boldsymbol{\omega} - \nu \nabla \times \boldsymbol{\omega} \qquad 12.3.4$$

The terms on the right must be zero because the vorticity is zero. The unsteady term is changed by inserting 12.3.1:

$$\frac{\partial \mathbf{v}}{\partial t} = \frac{\partial}{\partial t}(\nabla \phi) = \nabla\left(\frac{\partial \phi}{\partial t}\right)$$

Hence 12.3.4 may be written

$$\nabla\left(\frac{\partial \phi}{\partial t} + \tfrac{1}{2}v^2 + \frac{p}{\rho}\right) = 0$$

This integrates to the unsteady Bernoulli equation for irrotational flow,

$$\frac{\partial \phi}{\partial t} + \tfrac{1}{2}v^2 + \frac{p}{\rho} = C(t) \qquad 12.3.5$$

The "constant" C is a function of time, which must be determined from boundary information. Assuming that the velocities have been determined as discussed above, the Bernoulli equation tells us what pressure forces are required to produce those motions.

12.4 GROWTH OF A GAS BUBBLE IN A LIQUID

We shall discuss two examples of the application of the velocity potential to flows. The first concerns a sphere imbedded in an infinite fluid. The sphere surface undergoes a prescribed expansion or contraction $R(t)$ as depicted in Fig. 12-5. This problem, and variations of it, are relevant to boiling, to cavitation [one of the first applications to cavitation was by Rayleigh (1917)], and to underwater explosions. We assume that the radial expansion or contraction of the sphere produces a purely radial flow through the action of pressure

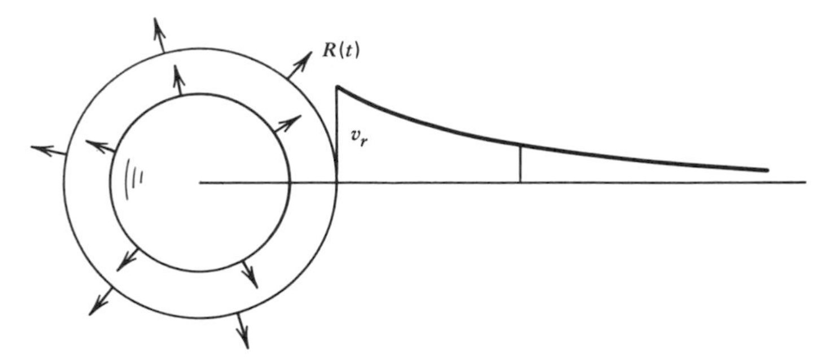

Figure 12-5 Expansion of a sphere in an infinite fluid.

forces. Pressure forces cannot impart any rotational motion to the particles. Rotation can only be started by unbalanced shear stresses (these facts will be proved in later chapters). In such a motion the velocity will be related to the potential by

$$v_r = \frac{\partial \phi}{\partial r} \qquad 12.4.1$$

In spherical coordinates, the governing equation 12.3.3 for $\phi(r, t)$ is

$$\nabla^2 \phi = \frac{1}{r}\frac{\partial}{\partial r}\left(r^2 \frac{\partial \phi}{\partial r}\right) = 0 \qquad 12.4.2$$

Boundary conditions are that the fluid is stationary at infinity and that it follows the sphere motion at $r = R(t)$:

$$v_r = \frac{\partial \phi}{\partial r}(r \to \infty, t) = 0 \qquad 12.4.3$$

$$v_r = \frac{\partial \phi}{\partial r}(r = R, t) = \dot{R}(t) \qquad 12.4.4$$

From 12.4.2 and 12.4.4 we find that

$$r^2 \frac{\partial \phi}{\partial r} = C(t) = R^2 \dot{R} \qquad 12.4.5$$

In light of 12.4.1, this is really an equation for the fluid velocity; that is,

$$v_r = \left(\frac{R}{r}\right)^2 \dot{R} \qquad 12.4.6$$

The velocity decays as r^{-2}. A second integration of 12.4.5 yields

$$\phi = -\frac{R^2 \dot{R}}{r} + \phi_\infty \qquad 12.4.7$$

where ϕ_∞ may be taken as zero to fix the arbitrary constant.

Next, we imagine that the pressure far away from the bubble has a constant value (that is, the pressure minus the hydrostatic contribution is a constant). With the results above, Bernoulli's equation 12.3.5 becomes

$$-\frac{1}{r}\left[R^2\ddot{R} + 2R(\dot{R})^2\right] + \frac{1}{2}\left(\frac{R^2\dot{R}}{r^2}\right) + \frac{p}{\rho} = \frac{p_\infty}{\rho} \qquad 12.4.8$$

The pressure approaches the pressure at infinity as r^{-1}. A quantity of importance is the pressure at the bubble surface:

$$\frac{p(R) - p(\infty)}{\rho} = R\ddot{R} + \tfrac{3}{2}\dot{R}^2 \qquad 12.4.9$$

Rayleigh used this result to find the time it takes for a cavitation bubble to collapse (see Knapp, Daily, and Hammitt 1970). For this problem it is assumed that the bubble is somehow formed at time zero without any internal pressure. With this restriction, 12.4.9 can be integrated to find the time of collapse.

12.5 FLOW IN AN IDEAL VORTEX—CIRCULATION

As a second example we shall find the velocity profile for the ideal vortex. We ask what types of swirling, steady, irrotational flows are possible with cylindrical symmetry. The flows we seek have only one velocity component,

$$v_\theta(r) = \frac{1}{r}\frac{\partial\phi}{\partial\theta} \quad \text{and} \quad v_r = \frac{\partial\phi}{\partial r} = 0 \qquad 12.5.1$$

The Laplace equation in cylindrical coordinates is

$$\nabla^2\phi = \frac{1}{r}\frac{\partial}{\partial r}\left(r\frac{\partial\phi}{\partial r}\right) + \frac{1}{r^2}\frac{\partial^2\phi}{\partial\theta^2} = 0$$

The first term is zero in light of the assumption that $v_r = \partial\phi/\partial r = 0$. The second term can be integrated to obtain

$$\phi = f(r)\theta + g(r) \qquad 12.5.2$$

Again, using $\partial\phi/\partial r = 0$ shows that $f(r)$ and $g(r)$ must be constants. Furthermore, we may arbitrarily set $\phi = 0$ at $\theta = 0$, so that $g = 0$. The final result is

$$\phi = C\theta \qquad 12.5.3$$

The corresponding velocity profile is

$$v_\theta = \frac{C}{r} \qquad 12.5.4$$

The constant C remains arbitrary and describes the strength of the vortex. This profile is singular at the origin, as v_θ becomes infinite. One can verify that the vorticity is zero.

The ideal vortex will help us demonstrate an important aspect of potential flows. Let us formally define the *circulation* Γ as the closed line integral

$$\Gamma \equiv \oint_C v_i t_i \, ds \qquad 12.5.5$$

The line element, with length ds and unit tangent vector t_i, is considered positive if the circuit is traversed in a counterclockwise direction. The integrand $v_i t_i$ is the velocity component along the line element, as shown in Fig. 12-6. By the theorem of Stokes we may transform the line integral 12.5.5 into a area integral over any surface A that has the circuit C as the boundary:

$$\Gamma = \int_A n_i \varepsilon_{ijk} \partial_j v_k \, dS \qquad 12.5.6$$

or in terms of the vorticity,

$$\Gamma = \int_A n_i \omega_i \, dS \qquad 12.5.7$$

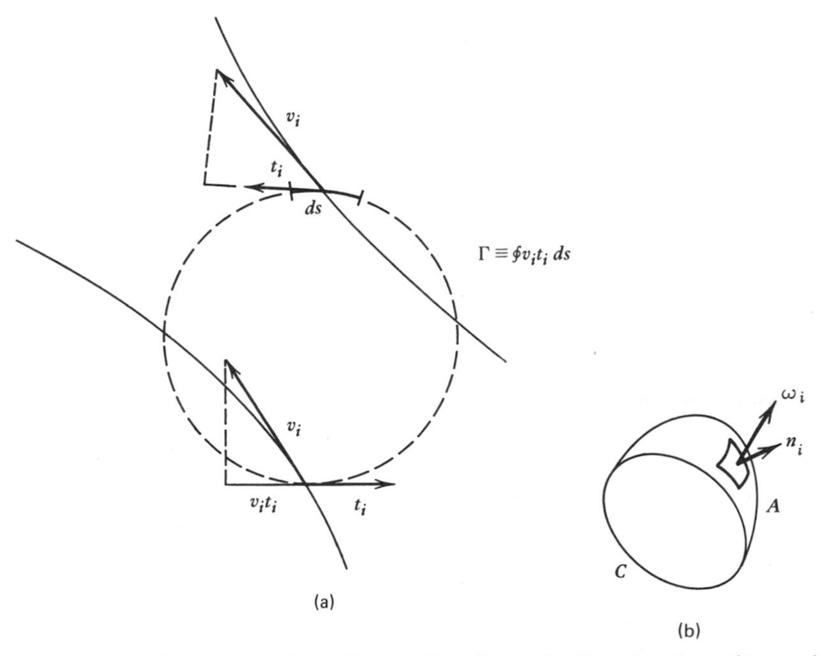

Figure 12-6 Circulation: (a) defined as a line integral, (b) related to the vorticity "flow" across a surface.

The equation above shows that Γ is equal to the total vorticity flux across A. Consider a circular circuit around the ideal vortex; then 12.5.6 is the same for a circuit of any radius:

$$\Gamma = \oint_0^{2\pi} v_\theta r \, d\theta = C \int_0^{2\pi} d\theta = 2\pi C \qquad 12.5.8$$

The ideal vortex has a constant circulation for any path that includes the origin. The constant $C = \Gamma/2\pi$ in 12.5.3 indicates the strength of the vortex in terms of its circulation.

Since potential flows must have zero vorticity, an offhand application of 12.5.7 would indicate that $\Gamma = 0$, instead of the finite value found by a direct calculation. The discrepancy lies in the fact that the ideal vortex is singular; the solution is invalid at the origin, where v_θ becomes infinite. The finite value of Γ in 12.5.7 is explained by saying that ω becomes infinite at the origin in just such a way as to make the integral finite.

Another aspect of the potential vortex is the fact that $\phi = C\theta$ is not a continuous function. Along the line $\theta = 0$, ϕ jumps by $2\pi C$. Notice that the difference in ϕ at two points in the flow can be found as follows:

$$\phi_2 - \phi_1 = \int_1^2 d\phi = \int_1^2 \partial_i \phi \, dx_i$$

$$= \int_1^2 v_i \, dx_i = \int_1^2 v_i t_i \, ds \qquad 12.5.9$$

When a closed circuit is made, the integral 12.5.9 becomes the circulation. If the circuit does not loop around the origin, then $\phi_1 = \phi_2$ and the circulation is zero. If the circuit does include the origin in its interior, the circulation integral indicates the amount that ϕ jumps as one crosses the line where ϕ is discontinuous. The fact that ϕ is discontinuous is perfectly acceptable so long as the velocities, which are related to the derivative of ϕ, are continuous.

12.6 CONCLUSION

The stream function and velocity potential offer us alternative variables to use in describing fluid flows. Through their mathematical relationships with the velocity, $\mathbf{v} = \nabla\psi \times \nabla g = \nabla \times \mathbf{B}$ and $\mathbf{v} = \nabla\phi$, they can give a complete description of the flow. Only in the case of flows with symmetry, however, is the stream function useful. We need the symmetry so that g surfaces may be taken as coordinate surfaces and are therefore known. The velocity potential, on the other hand, is a natural choice for flows without vorticity. In fact, the terms potential flow and irrotational flow are used interchangeably because $\omega = 0$ is a necessary and sufficient condition for ϕ to exist.

PROBLEMS

12.1 (A) Find the stream function and the velocity potential for the ideal flow toward a plane stagnation point. The velocity components are $u = ax$ and $v = -ay$. Plot several streamlines using equal increments in ψ.

12.2 (B) Find the stream function for a stream oscillating above a fixed plate. The velocity is given by 11.2.24.

12.3 (A) Find the stream function for the asymptotic suction profile $u = u_0(1 - e^{-yV_0/\nu})$ which occurs when a streaming motion u_0 goes over a porous plate with a sucking velocity V_0. Sketch several streamlines with equal increments in ψ.

12.4 (B) Consider a uniform stream from left to right with a speed U. Find the stream function for this flow in all four coordinate systems of Appendix D.

12.5 (B) An infinitely small point source of fluid exists at the origin. The flow away from the source is purely radial and is irrotational. Find the stream function in spherical coordinates for this flow.

12.6 (B) Find the velocity potential for the flow of Problem 5.

12.7 (B) Convert the answer to Problem 5 into cylindrical coordinates, using the relations between coordinates. Check the velocity components in the axisymmetric system with Appendix D.

12.8 (B) The Hiemenz stagnation-point flow has velocity components $u = xf'(y)$ and $v = -f(y)$. How is f related to the two-dimensional stream function ψ? What equation governs ψ for this flow? Express this equation in terms of f, and compare it with 11.7.13.

12.9 (A) Find the velocity potential for a uniform stream. Solve the problem for each of the coordinate systems of Appendix D.

12.10 (A) Find the velocity potential for Problem 5.

12.11 (B) The stream function for flow over a circular cylinder is $\psi = Ur\sin\theta\,(1 - r_0^2/r^2)$. Find the pressure distribution on the surface.

12.12 (B) Find the pressure distribution on the surface of Hill's spherical vortex.

12.13 (C) Prove the flow-rate equation $Q = (f_2 - f_1)(g_2 - g_1)$.

12.14 (B) Consider a bubble of radius r_0 without any internal pressure. Find the time of collapse by numerical integration. Also compute pressure profiles in the liquid for several stages of the bubble collapse.

12.15 (C) Consider the cylinder-piston problem of Fig. 5–9. Assume ideal flow and set the velocity potential ϕ to zero at the piston face. Sketch $U(x)$ and $\phi(x)$ for one dimensional flow through the cylinder, orifice, and into the jet at several distinct times. What does $\phi(x)$ look like if the transition region where U goes from U_p to U_0 is infinitely thin? Find $\partial\phi/\partial t$ at the jet and apply Bernoulii's equation from the piston to the jet. With the aid of 5.14.4 find the force on the end wall.

13 Vorticity Dynamics

The momentum equations for incompressible flow focus our attention on the velocity and pressure as the major items of interest. Interactions that occur in a flow field are explained in terms of inertia, pressure forces, gravity forces, and viscous forces. These basic concepts are the elements at our disposal in interpreting fluid-dynamic events. In this chapter we shall broaden our outlook. In many instances it is advantageous to interpret the events in a flow in terms of the vorticity and the dynamic events that are interacting to give a certain vorticity distribution.

The existence of vorticity generally indicates that viscous effects are important. This is because fluid particles can only be set into rotation by an unbalanced shear stress. Vorticity dynamics, roughly speaking, offers a method to separate a flow into viscous and inviscid effects. It is especially valuable in cases where there is only a weak interaction between viscous and inviscid effects.

13.1 VORTICITY

In Chapter 4 we defined vorticity as

$$\boldsymbol{\omega} = \nabla \times \mathbf{v} \qquad 13.1.1$$

Vorticity has several physical interpretations. The most common is that vorticity measures the solid-body-like rotation of a material point P' that neighbors the primary material point P. Since the motion is like a solid-body rotation, the rotation velocity increment of P' with respect to P is

$$d\mathbf{v}^{(r)} = \tfrac{1}{2}\boldsymbol{\omega} \times d\mathbf{r} \qquad 13.1.2$$

where $\boldsymbol{\omega}$ is the vorticity at P and $d\mathbf{r}$ is the distance increment from P to P'.

Several other slightly different interpretations may be attached to vorticity. For instance, we know that at each point one may find a set of orthogonal principal axes. Particles on these axes have no shearing deformation, and their instantaneous motion is translation, expansion, and rotation. Therefore, a second interpretation is that vorticity is a measure of the instantaneous rotation rate of the fluid on the principal axes.

Our next interpretation is a more vivid physical picture. We imagine that a small spherical piece of fluid about P is instantaneously frozen. The frozen ball would then translate and rotate as a result of the previous motion of its particles. It we give the frozen sphere the same angular momentum about P that the unfrozen particles had, the rotation will occur at a speed $\omega/2$ and the angular momentum will be given by the product of the vorticity and the moment of inertia of the sphere,

$$L = \tfrac{1}{2} I \omega \qquad 13.1.3$$

The calculation that leads to this result is valid only for a sphere; the same statement cannot be made for a frozen ellipsoidial particle. Another slight difficulty with the frozen-ball concept occurs if we try to apply it at a solid wall. The no-slip condition at the wall means that the particles are not translating; however, they are undergoing a rotation. To compute the rotation of the particle P on the wall we must look at the particle P' a small distance away within the fluid. Note that the velocity derivatives are discontinuous at the wall, so we must compute the vorticity of the fluid particles by using derivatives only on the fluid side. If we wanted to use the frozen-ball idea at the wall, we would need to imagine that the fluid is extended into the wall in such a way that the velocity derivatives are continuous. We might imagine that the wall consists of an array of marbles, which are rotating but remain at the same location on the wall.

The final interpretation is in a somewhat different vein. It connects vorticity and circulation. Circulation, you will recall, was defined as the line integral around a closed circuit,

$$\Gamma = \oint_C t_i v_i \, ds \qquad 13.1.4$$

From Stokes's theorem we have the equivalent expression

$$\Gamma = \int_A n_i \omega_i \, dS \qquad 13.1.5$$

where A is any surface having C as its boundary. In this form we interpret the integrand as the circulation per unit area:

$$n_i \omega_i = \frac{d\Gamma}{dS} \qquad 13.1.6$$

We may say that the vorticity is the circulation per unit area for a surface perpendicular to the vorticity vector.

13.2 KINEMATIC RESULTS CONCERNING VORTICITY

Many of the ideas we have associated with the velocity field may be adapted to apply also to the vorticity field. For example, a vortex line is defined as a line that is everywhere tangent to the vorticity vectors. In general, vortex lines are distributed throughout the flow. Sometimes regions of the flow are idealized to have zero vorticity and hence have no vortex lines. For example, in the high-Reynolds-number flow over a wing, the vortex lines are concentrated near the surface and in the wake behind the wing. The flow away from these regions is idealized as irrotational. On the other hand, in the flow in a pipe the vortex lines are rings, which exist throughout the fluid.

In this section we discuss some geometric restrictions that vorticity and vortex lines must obey. Because of its very definition as the curl of the velocity, we know that the "rate of expansion" for vorticity must be zero. That is, the vector identity $\nabla \cdot \nabla \times \mathbf{v} = 0$ takes the following form when the vorticity is identified:

$$\nabla \cdot \boldsymbol{\omega} = 0 \qquad 13.2.1$$

Before we discuss the consequences of 13.2.1, it is useful to investigate the geometry of vortex lines on a solid wall.

On a solid, stationary wall the no-slip condition requires that the velocity be zero. Nevertheless, we are able to define a wall streamline by the following limiting process. Consider a smooth wall, and erect a local coordinate system at a point P on the wall. The wall will lie in the x, z plane with y as the normal direction as shown in Fig. 13-1. A flat wall is assumed for simplicity; the argument is also valid for curved walls. The continuity equation at P is

$$\frac{\partial u}{\partial x} + \frac{\partial v}{\partial y} + \frac{\partial w}{\partial z} = 0 \qquad 13.2.2$$

Since u and w are zero all along the wall, $\partial u/\partial x = 0$ and $\partial w/\partial z = 0$. From 13.2.2 this implies that $\partial v/\partial y = 0$ on a solid wall. This fact is often useful in itself; however, we shall use it in a Taylor's expansion. Taylor's series for the

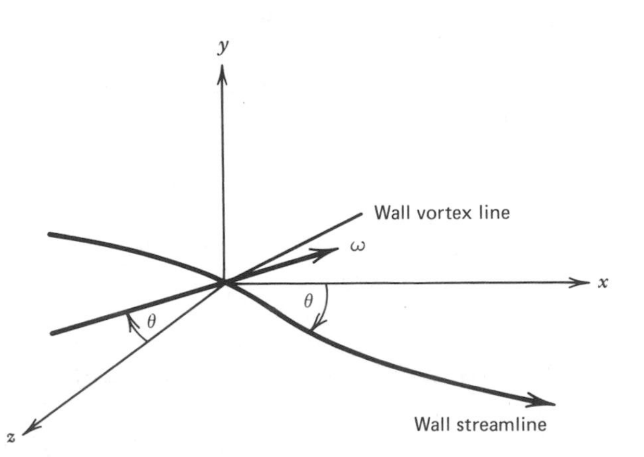

Figure 13-1 Relationship between streamlines, vorticity, and shear stress on a solid wall.

velocity components as we leave the wall in the y direction are

$$u = 0 + \left.\frac{\partial u}{\partial y}\right|_0 y + \cdots$$

$$v = 0 + 0 + \left.\frac{\partial^2 v}{\partial y^2}\right|_0 \frac{y^2}{2} + \cdots \qquad 13.2.3$$

$$w = 0 + \left.\frac{\partial w}{\partial y}\right|_0 y + \cdots$$

Now we are in a position to find the direction of the wall streamline:

$$\left.\frac{dz}{dx}\right|_{\text{streamline}} = \tan\theta = \lim_{y\to 0}\frac{w}{u} = \frac{\left.\frac{\partial w}{\partial y}\right|_0}{\left.\frac{\partial u}{\partial y}\right|_0} \qquad 13.2.4$$

where θ is the angle the wall streamline makes with the x axis in the plane of the wall. We can also conclude that the streamline lies in the wall because the streamline angles in the y, x and y, z planes—that is, the limits of v/u and v/w —are both zero.

With this information about the surface streamline, we now turn to the vortex lines. The vorticity components at the point P on the wall are calculated as follows:

$$\omega_x = \frac{\partial w}{\partial y} - \frac{\partial v}{\partial z} = \left.\frac{\partial w}{\partial y}\right|_0$$

$$\omega_y = \frac{\partial u}{\partial z} + \frac{\partial w}{\partial x} = 0 \qquad 13.2.5$$

$$\omega_z = \frac{\partial v}{\partial x} - \frac{\partial u}{\partial y} = -\left.\frac{\partial u}{\partial y}\right|_0$$

The vorticity component perpendicular to the wall is zero, so we know that this vector also lies in the wall (this result is true for curved surfaces as well). An additional fact of importance is that wall vortex lines are always perpendicular to the wall streamlines. This is found be direct calculation of the slope together with the previous result 13.2.4:

$$\left.\frac{dz}{dx}\right|_{\text{vortex line}} = \frac{\omega_z}{\omega_x} = \frac{-\left.\frac{\partial u}{\partial y}\right|_0}{\left.\frac{\partial w}{\partial y}\right|_0} = \frac{-1}{\left.\frac{dz}{dx}\right|_{\text{streamline}}} \qquad 13.2.6$$

Away from the wall, vortex lines and streamlines are not necessarily perpendicular. The primary exceptions, which include 95% of all flows that are analyzed, are two-dimensional and axisymmetric flows where $\mathbf{v}$ and $\boldsymbol{\omega}$ are perpendicular. In a general three-dimensional flow, the vorticity and velocity vectors are not perpendicular except when the wall is approached.

Everywhere on the surface of a body, the streamlines and vortex lines are orthogonal. When the vorticity is nonzero, a unique direction for the vortex line is assured. If a vortex line leaves the surface, it can only do so at a point (or line) where the vorticity is zero. Just as in the case of streamlines, it is necessary to have $\boldsymbol{\omega} = 0$ at any place where a vortex line splits and goes in several directions.

The fact that vorticity is a solenoidal vector ($\nabla \cdot \boldsymbol{\omega} = 0$) means that vortex lines and vortex tubes must obey the same rules that streamlines and streamtubes obey as a consequence of the fact that $\nabla \cdot \mathbf{v} = 0$. Namely: vortex tubes cannot end within the fluid; they must either form closed loops, extend to infinity, or intersect a wall at a place where the vorticity is zero. Another statement that results from the solenoidal condition is derived by considering the integral of $\nabla \cdot \boldsymbol{\omega} = 0$ over any volume and applying Gauss's theorem:

$$0 = \int \partial_i \omega_i \, dV = \int n_i \omega_i \, dS \qquad 13.2.7$$

This equation may be applied to a vortex tube with end caps A_1 and A_2. On the surface of the vortex tube $n_i \omega_i = 0$, so the only contributions come from the end caps. Equation 13.2.7 implies that the integral over any cross section of a vortex tube is constant:

$$-\int_{A_1} n_i \omega_i \, dS = \int_{A_2} n_i \omega_i \, dS = \Gamma \qquad 13.2.8$$

The integral of $n_i \omega_i$ across a vortex tube is called the *strength* of the vortex tube. Equation 13.2.8 says that the strength of a vortex tube must be constant, and from 13.1.5 it is equal to the circulation of any circuit around the vortex tube.

It is well to note that all the results of this section apply to the flow at an instant. In steady flow it is perhaps natural to consider vortex lines and vortex tubes fixed in space. In other words, an identity is given to a vortex tube that goes through the same points in space. We often talk the same way about streamtubes in a steady flow. A streamtube through the same points in space is thought to retain its identity as time goes on. The streamtube is identified by two values of the stream function ψ (or definite positions of the streamsurfaces f and g). However, this is not the only viewpoint. The streamfunction is only defined up to an arbitrary "constant", which may be a function of time. Thus, in principle we can add an arbitrary function of time to ψ. Now, even though the flow is steady, the numbering system for the streamlines keeps changing with time. A specific number for ψ no longer identifies a streamline that goes through the same points in space. Thus, even in steady flow there is a certain arbitrariness to the identity of a streamtube. The same may be said about vortex tubes. In fact, we shall find that sometimes (in inviscid flow) it is advantageous as well as permissible to imagine that vortex lines in a steady flow are not stationary, but move along with the fluid velocity. More discussion concerning this will be given later. The major point of the present discussion is that 13.2.8 applies at any instant to any vortex tube. At the next instant in time the definition of the vortex tube can change in any manner we choose.

13.3 THE VORTICITY EQUATION

The dynamic equation that governs vorticity is derived from the momentum equation. We start with the momentum equation for incompressible flow,

$$\partial_0 v_i + v_j \partial_j v_i = -\frac{1}{\rho} \partial_i p + \nu \partial_j \partial_j v_i \qquad 13.3.1$$

Into this equation we substitute the vector identity

$$v_j \partial_j v_i = \partial_i\left(\tfrac{1}{2} v_j v_j\right) + \varepsilon_{ijk} \omega_j v_k \qquad 13.3.2$$

The resulting equation is differentiated with ∂_q and multiplied by ε_{pqi} to yield

$$\partial_0\left(\varepsilon_{pqi} \partial_q v_i\right) + \varepsilon_{pqi} \partial_q \partial_i\left(\tfrac{1}{2} v_j v_j\right) + \varepsilon_{pqi} \partial_q\left(\varepsilon_{ijk} \omega_j v_k\right)$$

$$= -\frac{1}{\rho} \varepsilon_{pqi} \partial_q \partial_i p + \nu \varepsilon_{pqi} \partial_j \partial_j \partial_q v_i \qquad 13.3.3$$

Consider this equation term by term. The first term can be identified as the time derivative of the vorticity. The second term is zero because ε_{pqi} is antisymmetric and $\partial_q \partial_i$ is symmetric; the same can be said for the pressure term on the right-hand side. Moreover, note that the last term contains the vorticity. The term we skipped is expanded to yield (the last line below is obtained by noting that $\partial_k v_k$ and $\partial_j \omega_j$ are always zero)

$$\varepsilon_{pqi} \varepsilon_{ijk} \partial_q\left(\omega_j v_k\right) = \partial_k\left(\omega_p v_k\right) - \partial_j\left(\omega_j v_p\right)$$

$$= v_k \partial_k \omega_p - \omega_j \partial_j v_p \qquad 13.3.4$$

Collecting these results yields the final vorticity transport equation:

$$\partial_0 \omega_i + v_j \partial_j \omega_i = \omega_j \partial_j v_i + \nu \partial_j \partial_j \omega_i$$

or, in symbolic notation,

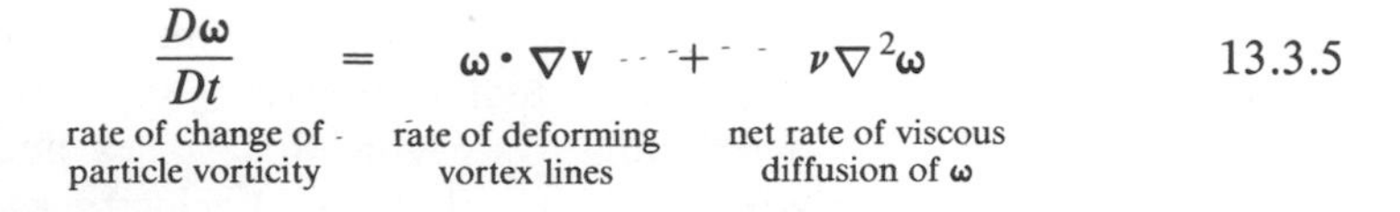

$$\frac{D\boldsymbol{\omega}}{Dt} = \boldsymbol{\omega} \cdot \nabla \mathbf{v} + \nu \nabla^2 \boldsymbol{\omega} \qquad 13.3.5$$

This equation is almost as important to fluid mechanics as the momentum equation itself.

One of the most important things about the vorticity equation 13.3.5 is not what appears, but what does not appear—namely, the pressure. The usefulness of vorticity in interpreting fluid flow problems is that vorticity tracks only the effect of viscous forces; pressure and gravity forces will not directly change it. The physical reason behind this has to do with the fact that vorticity is an indicator of solid-body rotation. Pressure forces and gravity forces act through the center of mass of a particle and cannot produce a rotation. On the other hand, shear stresses act tangentially at the surface of a particle and, if they are unbalanced, will generate vorticity.

The intimate connection between unbalanced shear stresses, or viscous action, and vorticity is made even clearer by noting that the viscous term in the momentum equation can be written as

$$\nabla \cdot \boldsymbol{\tau} = \mu \nabla^2 \mathbf{v} = -\mu \nabla \times \boldsymbol{\omega}$$

An unbalanced shear stress can only exist when the vorticity is nonzero. As a general rule the existance of vorticity means that a particle is, or at least in its past history was, subjected to net viscous forces.

As an aside one should note that 13.3.5 is not the proper vorticity equation for a stratified flow of an incompressible fluid. The density gradient in these flows implies that the center of mass of a particle does not coincide with its geometric center. Since pressure forces act through the geometric center, they can now generate rotational motion. This is one of the distinctive characteristics of stratified flows. Internal vorticity generation processes occur in the oceans and the atmosphere when density differences are significant.

13.4 VORTICITY DIFFUSION

In this section we discuss the physical meaning and interpretation of $\nu \nabla^2 \boldsymbol{\omega}$, the last term in the vorticity equation of 13.3.5. It shows that vorticity can diffuse through a flow by viscous action in the same way that momentum diffuses. There is also an analogy with heat transport. Recall that the thermal-energy equation, after simplification for incompressible flow with constant-temperature boundaries, is

$$\rho c_p \frac{DT}{Dt} = k \nabla^2 T \qquad 13.4.1$$

Comparing this equation with 13.3.5, we see that there is an analogy between vorticity and temperature in plane two-dimensional flows. These flows have only one vorticity component ω_z and since $\boldsymbol{\omega} \cdot \nabla \mathbf{v}$ is always zero, the equation governing ω_z is of the same form as 13.4.1. The fact that vorticity takes on negative values is immaterial, as the temperature could include an arbitrary reference level without changing the governing equation.

All of the problems in Chapters 7 and 11 obey this analogy except for von Kármán's problem and pressure-driven flow in a tube of arbitrary cross section, neither of which is a plane flow. In most of the steady-state problems —pressure-driven flow in a slot, Couette flow, and the falling film—the vorticity obeys the simplified equation

$$0 = \frac{\partial^2 \omega}{\partial y^2}$$

The vorticity is distributed so that the vorticity "flux", in analogy with the heat flux q, is constant:

$$\frac{\partial \omega}{\partial y} = \text{const}$$

The unsteady problems—Rayleigh's problem, Stokes's problem, the oscillating pressure gradient in a slot, and the vortex decay problems—are classical diffusion problems. They all obey a vorticity equation of the form

$$\frac{\partial \omega}{\partial t} = \nu \nabla^2 \omega$$

The rate of change of vorticity at a point is equal to the net diffusion flux into the point. Note especially that the viscous diffusion of momentum and of vorticity have the same diffusivity constant ν.

We learned from these problems that the depth of penetration of "free diffusion" obeyed a relation of the form

$$\delta \propto \sqrt{\nu t}$$

The viscous diffusion length is independent of the size of the vorticity pulse that occurs at the boundary. For example, in Rayleigh's problem the magnitude of the impulsive velocity given to the plate does not affect the diffusion length.

The plane stagnation-point problem of Chapter 11 shows a balance between viscous diffusion and convection. In this flow convection occurs by both u and v flow velocities. The proper equation is

$$u \frac{\partial \omega}{\partial x} + v \frac{\partial \omega}{\partial y} = \nu \frac{\partial^2 \omega}{\partial y^2}$$

Diffusion of vorticity in the y direction is counteracted by the downward convection from the negative v velocity and the outward convection by the u velocity. A constant thickness of the vortical region $\delta \approx \sqrt{\nu/a}$ results. Again the viscous thickness is proportional to $\sqrt{\nu}$.

13.5 VORTICITY INTENSIFICATION BY STRAINING VORTEX LINES

Having reviewed viscous diffusion, let us now turn our attention to the first term on the right-hand side of 13.3.5, namely $\boldsymbol{\omega} \cdot \nabla \mathbf{v}$. This term represents the generation or destruction of vorticity by stretching or turning the vortex lines. In order to back up this interpretation let us consider a material line with two points dr_j apart. The motion of dr_j with time was computed in Chapter 4 (4.6.3) as

$$\frac{\partial(dr_j)}{\partial t} = dr_i\, \partial_i v_j \qquad 13.5.1$$

This equation has exactly the same form as 13.3.5 when the viscous term in the latter equation is ignored. The vorticity vector ω_j plays a role analogous to the material line vector dr_j. Stretching a vortex line produces vorticity in the same way that stretching a material line produces length. Such a process is, of course, reversible: the contraction of a vortex line decreases the vorticity. There is also another effect in the term $\boldsymbol{\omega} \cdot \nabla \mathbf{v}$: that of turning by angular strain. The vector nature of ω_i (and dr_i) means that turning a vortex line creates vorticity in one direction at the expense of vorticity in another direction.

We may see this more clearly by noting that in the term $\boldsymbol{\omega} \cdot \nabla \mathbf{v}$, the velocity gradient may be replaced by the rate-of-strain tensor, that is,

$$\omega_i \partial_i v_j = \omega_i \partial_{(i} v_{j)} \qquad 13.5.2$$

(Proof of this equation follows by substituting 3.5.13 for the velocity gradient:

$$\omega_i \partial_i v_j = \omega_i \left[\partial_{(i} v_{j)} + \tfrac{1}{2} \varepsilon_{kij} \omega_k \right]$$

The last term is zero because $\omega_i \omega_k$ is symmetric and ε_{kij} is antisymmetric.) To illustrate, we write out the x component of the vorticity equation 13.3.5, and insert 13.5.2

$$\frac{D\omega_1}{Dt} = \omega_1 \partial_{(1} v_{1)} + \omega_2 \partial_{(2} v_{1)} + \omega_3 \partial_{(3} v_{1)}$$

The first term on the right-hand side is ω_1 vorticity generated by stretching the vortex line with the extension rate $\partial_{(1} v_{1)}$, while the second and third terms are ω_1 vorticity generated by turning the vortex line with the angular deformations $\partial_{(2} v_{1)}$ and $\partial_{(3} v_{1)}$.

In making the interpretation that the $\boldsymbol{\omega} \cdot \nabla \mathbf{v}$ term represents vortex-line turning and stretching, we have tacitly assumed that the vortex line at any instant is a material line moving with the fluid velocity. This is, in fact, true in inviscid flows, where the viscous diffusion is zero. For this case the analogy between the motion of a material line governed by 13.5.1 and that of a vortex line governed by 13.3.5 is exact. In viscous flows, where the diffusion term is nonzero, we must qualify our interpretation because vortex lines cannot be given a identity and treated as moving with the fluid. Under these circumstances we can say that the term generates vorticity *as if* the vortex line were moving as a material line.

As noted earlier, the stretching and turning mechanism is absent in all plane flow. In such flows the vorticity vector is perpendicular to the velocity vector, so the vortex lines are perpendicular to the plane of the flow. Since everything is uniform in the vortex-line direction, the lines have a constant length and do not turn. Other flows where $\boldsymbol{\omega} \cdot \nabla \mathbf{v}$ is always zero include the unidirectional flows of the first example of Chapter 11. In these flows the vortex lines form closed loops perpendicular to the straight streamlines. They do not turn or expand.

One method to get information about the *magnitude* of the vorticity, a method that disregards the vector nature of ω_i, is to consider the dot product $\omega_i \omega_i$. This is analogous to considering the kinetic energy $\frac{1}{2} v_i v_i$ rather than the momentum v_i. The equation that governs $\frac{1}{2}\omega_i \omega_i$ is derived by forming the dot product of ω_i with 13.3.5. The result is

$$\partial_0\left(\tfrac{1}{2}\omega_i\omega_i\right) + v_j\partial_j\left(\tfrac{1}{2}\omega_i\omega_i\right) = \omega_i\omega_j\partial_{(j}v_{i)} + \nu\omega_i\partial_j\partial_j\omega_i \qquad 13.5.3$$

$$= \omega_i\omega_j\partial_{(j}v_{i)} + \nu\partial_j\partial_j\left(\tfrac{1}{2}\omega_i\omega_i\right) - \nu\partial_j\omega_i\partial_j\omega_i \qquad 13.5.4$$

The left side is the substantial derivative of $\frac{1}{2}\omega_i\omega_i$ as usual. The first term on the right-hand side is the generation of vorticity by stretching the vortex lines. This is brought out more clearly if we write the vorticity as a unit vector α_i times a scalar magnitude ω. The second term is now

$$\omega_i\omega_j\partial_{(j}v_{i)} = \omega^2\alpha_i\alpha_j\partial_{(j}v_{i)}$$

Comparing with 4.7.2, we see that this term is ω^2 times the rate of extension in the direction of the vortex line. From this we surmise that any turning or angular deformation (shear strain) does not increase the magnitude of the vorticity. The only motion that increases the vorticity is the extension strain, which is along the vortex line.

The viscous effect in 13.5.3 indicates that the net diffusion of vorticity, $\nu\omega_i\partial_j\partial_j\omega_i$, will change the magnitude of the vorticity. This same effect is represented in 13.5.4 as equal to the difference of two terms. The first is the diffusion of the quantity $\frac{1}{2}\omega_i\omega_i$ itself, and the second is the viscous destruction of $\frac{1}{2}\omega_i\omega_i$. The second term is analogous to viscous dissipation.

13.6 HILL'S SPHERICAL VORTEX

There is an interesting example of a flow in which the stretching of vortex lines plays a dominant role. Hill's (1894) spherical vortex is a model of the internal flow in a gas bubble moving in a liquid, or a droplet of an immiscible liquid moving through another liquid. Because of the motion outside the bubble, an internal circulation is set up. Figure 13-2 gives a diagram of the geometry and sets a cylindrical coordinate system moving with the bubble so that the internal flow is steady. The vortex lines for this flow are circular loops around the z axis. As the flow carries vortex lines to positions of larger radius, the loops increase in length in direct proportion to the radius. Due to the vortex-line stretching effect, the vorticity is therefore proportional to the radius:

$$\omega_\theta = Cr = \frac{5U}{R}\frac{r}{R} \qquad 13.6.1$$

The constant C has been given the value $5U/R$ for reasons that will surface later (it turns out that Hill's vortex solution satisfies the complete Navier–Stokes equations including the nonzero viscous terms).

Let us examine the vorticity equation 13.3.5 as it applies to Hill's vortex. Only the θ component has nonzero vorticity. The terms in 13.3.5 are as follows:

Convection $$\frac{D\omega_\theta}{Dt} = v_r\frac{\partial \omega_\theta}{\partial r} = Cv_r$$

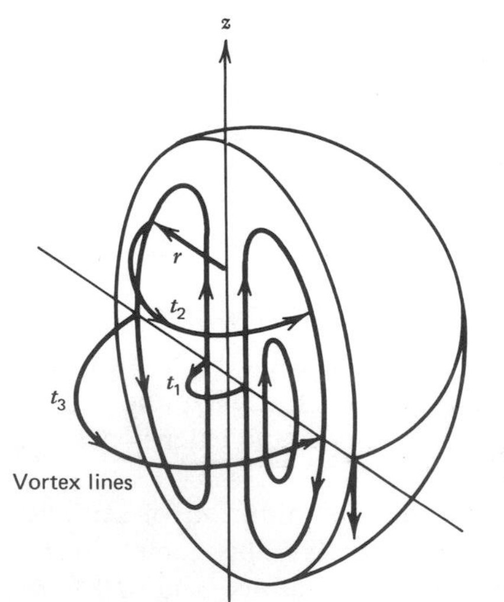

Figure 13-2 Hill's spherical vortex. Vortex lines are complete circles; those shown are for the same fluid particles at three different times. Kidney shaped lines are streamlines. Left half of sphere is cut away for clearity.

Stretching $$[\boldsymbol{\omega} \cdot \nabla \mathbf{v}]_\theta = \omega_\theta \frac{v_r}{r} = C v_r$$

Diffusion $$[\nu \nabla^2 \boldsymbol{\omega}]_\theta = \nu \frac{\partial}{\partial r}\left[\frac{1}{r}\frac{\partial}{\partial r}(r\omega_\theta)\right] = 0$$

Thus, the vorticity balance is between convection and stretching without any net viscous diffusion.

Although this is a completely viscous problem, there is no net diffusion of vorticity and thus we can imagine that the vortex lines are material lines that move with the fluid. The increase in vorticity is wholly the result of generation by vortex-line stretching. The fact that v_r does not have to be specified emphasizes that any movement of the circular vortex lines is allowed as long as the vorticity is proportional to the circumference of the loop.

Let us digress for a moment to give the complete details of Hill's solution. The stream function for this flow is given by

$$\psi = \frac{UR^2}{2}\left(\frac{r}{R}\right)^2\left[1 - \left(\frac{z}{R}\right)^2 - \left(\frac{r}{R}\right)^2\right] \qquad 13.6.2$$

Thus ψ is zero on the axis of the bubble ($r = 0$) and over the spherical surface given by

$$\left(\frac{r}{R}\right)^2 + \left(\frac{z}{R}\right)^2 = 1$$

The flow goes along the axis toward a stagnation point at $z = +1$ and then around the outer portion of the bubble to converge at another stagnation point at $z = -1$. From the formulas of the previous chapter, the velocities are computed to be

$$v_z = U\left[1 - \left(\frac{z}{R}\right)^2 - 2\left(\frac{r}{R}\right)^2\right]$$

and

$$v_r = U \cdot \left(\frac{r}{R}\right)\left(\frac{z}{R}\right)$$

At the bubble surface, the magnitude of the velocity is

$$\left(v_r^2 + v_z^2\right)^{1/2} = U \cdot \left(\frac{r}{R}\right)$$

The stagnation points at the poles have zero velocity, of course, while at the equator, the maximum velocity U is attained. In general, U is a constant that is determined by the flow on the outside of the bubble. The flow field on the outside of the bubble can be found for the two special cases of low and high Reynolds numbers (Sections 19.8 and 21.5).

13.7 WALL PRODUCTION OF VORTICITY

A fixed solid wall is the source of the vorticity that enters the flow. The first fact of importance is the direct connection between the viscous shear stress on the wall and the vorticity. We restrict the discussion to Newtonian fluids. At a solid wall a Newtonian fluid has no normal viscous force; the viscous stress vector lies in the wall and is the same direction as the wall streamline.

In order to relate the vorticity and the wall shear stress, we consider a flat wall with a coordinate system originating at a point P on the wall. We use the same notation as in Section 13.2, but here we orient the coordinate system so that the streamline is exactly in line with the x axis at the point P (Fig. 13-3).

The results of Section 13.2 showed that on the wall many velocity derivatives are zero:

$$\frac{\partial u}{\partial x} = 0, \qquad \frac{\partial w}{\partial z} = 0, \qquad \frac{\partial v}{\partial y} = 0 \tag{13.7.1}$$

By virtue of choosing the streamline angle θ to be zero, we find from 13.2.4 that

$$\frac{\partial w}{\partial y} = 0 \tag{13.7.2}$$

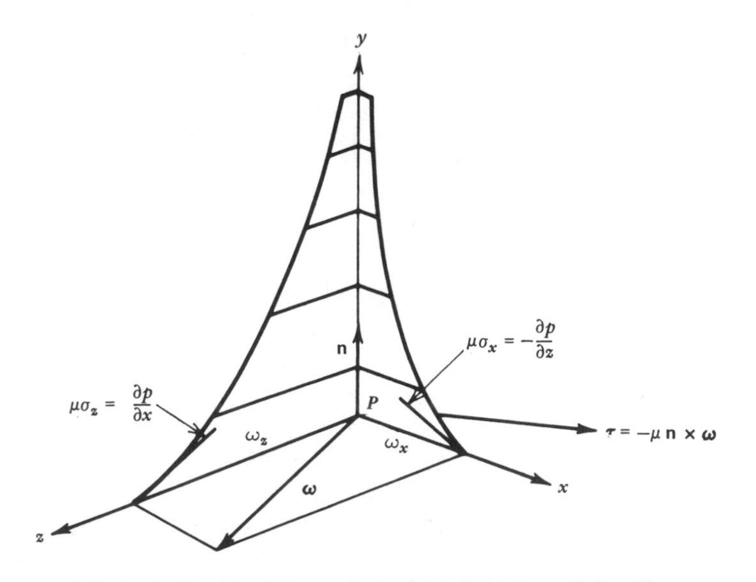

Figure 13-3 Vorticity and vorticity flux at a solid wall.

The relations of 13.7.1 and 13.7.2 are inserted into a computation of the viscous stress on the wall. The result is

$$F_{j\,\text{viscous}} = n_i \tau_{ij}$$
$$F_{x\,\text{viscous}} = n_y \tau_{yx} = \mu \frac{\partial u}{\partial y} \qquad 13.7.3$$

Now from 13.2.5 we note that $\omega_z = -\partial u/\partial y$ $(\omega_x = \omega_y = 0)$. Hence, 13.7.3 becomes

$$F_{x\,\text{viscous}} = -\mu \omega_z \qquad 13.7.4$$

The wall vorticity is directly proportional to the wall shear stress. A direct calculation in an arbitrary coordinate system will show that 13.7.4 is a special case of the general expression

$$\mathbf{F}_{\text{viscous}} = \mathbf{n} \cdot \boldsymbol{\tau} = -\mu \mathbf{n} \times \boldsymbol{\omega} \qquad 13.7.5$$

The numerical values of the wall shear and the vorticity are directly related, with the viscosity as the proportionality constant. In terms of the temperature analogy we can think of the wall shear stress as the "vorticity temperature". When the wall shear is high, the vorticity is large.

Just as the temperature of a wall does not indicate how much energy is leaving the wall, the wall vorticity does not indicate how much vorticity is entering the flow. Recall that the heat-flux equation says that the heat flux across a plane with orientation n_i is given by $n_i q_i$, where q_i is the heat-flux vector. By analogy, we define the vorticity flux σ_i as the inner product with the "vorticity flux tensor", that is,

$$\sigma_i \equiv -n_j \partial_j \omega_i \qquad 13.7.6$$

(Note the analogy with the stress and the stress tensor.) The vector σ_i is the flux of i vorticity across a plane with normal n_j. [This viewpoint is due to Lighthill (1963).]

We can find out some information about σ_i at the wall from the momentum equation. Consider the momentum equation in the form

$$\partial_0 v_i + \partial_i \left(\tfrac{1}{2} v^2 + \frac{p}{\rho} \right) = -\varepsilon_{ijk} \omega_j v_k - \nu \varepsilon_{ijk} \partial_j \omega_k \qquad 13.7.7$$

Evaluating this equation at the wall where $v_i = 0$ yields

$$\partial_i p = -\mu \varepsilon_{ijk} \partial_j \omega_k \qquad 13.7.8$$

The components of this equation in a local wall coordinate system relate the pressure gradient to the vorticity flux across the wall into the fluid. The components are

$$\frac{\partial p}{\partial x} = -\mu \frac{\partial \omega_z}{\partial y} = \mu \sigma_z$$
$$\frac{\partial p}{\partial z} = \mu \frac{\partial \omega_x}{\partial y} = -\mu \sigma_x \qquad 13.7.9$$

A pressure gradient along the surface is necessary to sustain a flux of vorticity into the fluid. The equations 13.7.9 are the key relations, that quantify the flux of vorticity from the wall. To reiterate, a vorticity flux from a wall results from pressure gradients along the wall.

The third flux of vorticity across the wall is σ_y. This may be found by evaluating the equation $\nabla \cdot \boldsymbol{\omega} = 0$ at the wall. The result is

$$\sigma_y = -\frac{\partial \omega_y}{\partial y} = \frac{\partial \omega_x}{\partial x} + \frac{\partial \omega_z}{\partial z} \qquad 13.7.10$$

Although ω_y is zero at the wall, there may be a flux of ω_y vorticity out of the wall. This flux depends upon the distribution of ω_x and ω_z on the wall itself.

Up to this point a major theme has been that pressure does not influence vorticity. More precisely, it does not do so directly. The pressure-gradient–vorticity-flux relation in 13.7.9 gives us a coupling whereby pressure forces associated with inviscid motions can introduce vorticity into the fluid. Previously we had indicated that the major mechanism for generating vorticity was the torque produced by an unbalanced shear stress. A little physical thinking will convince us that this is also true at the wall. For particles at the wall the momentum equation reduces to

$$0 = -\frac{1}{\rho}\partial_i p + \partial_j \tau_{ji} \qquad 13.7.11$$

Since particles at the wall are restrained from gaining linear momentum, any pressure gradient must be exactly canceled by the unbalanced shear. Hence in this special situation, the vorticity-producing stresses can be replaced by the pressure gradient. Fluid particles at a wall cannot have a linear translational velocity, but they do indeed have a "rotational" velocity.

As a final topic in this section a few remarks will be made about the effect of wall motion on vorticity generation. If the wall itself has vorticity, as in the case a spinning ball or a spinning disk (von Kármán problem), then additional vorticity exists at the surface, and the vortex lines no longer lie in the plane of the wall. Such problems must be considered individually.

Another interesting case is when the wall has translational motion $V(t)$ only, without any rotation. In Section 10.7 we found that the very same incompressible flow equations apply in a moving coordinate system as in an inertial coordinate system. The only change was that an effective pressure $\hat{p}$ must be introduced. When a coordinate system fixed to the wall is used to find the vorticity, and in particular to compute the wall vorticity flux, the effective pressure must be used. It can be calculated from 10.7.6 as follows:

$$\frac{\partial \hat{p}}{\partial \hat{x}_i} = \frac{\partial p}{\partial x_i} - \rho \frac{dV_i}{dt} \qquad 13.7.12$$

where p is the true pressure gradient on the translating body and $\rho \, dV_i/dt$ is an effective contribution from the translation motion. From 13.7.12 we see that the body-motion term contributes to the pressure gradient in the same direction as the acceleration; that is, the component of dV_i/dt perpendicular to the wall contributes to the normal wall pressure gradient, while the component along the wall contributes to the pressure gradient along the wall. The importance of these facts is brought out when we note that the vorticity-flux equations 13.7.9 are not concerned with the pressure gradient normal to the wall. Therefore, acceleration of the wall in the normal direction does not directly contribute to the vorticity flux (however, this motion may set up a true pressure gradient along the wall, which does contribute). An expanding sphere generates an irrotational potential flow (Chapter 12) because the wall motion is always in the normal direction. A plane wall, such as a piston or loudspeaker, moving only in the normal direction, would likewise generate a potential motion. On the other hand a sliding wall, such as the Rayleigh flat plate, generates vorticity because of the wall motion in its own plane. The actual dp/dx in the fluid is zero, but the effective gradient in the moving coordinates is

$$\frac{\partial \hat{p}}{\partial \hat{x}_i} = -\rho \frac{dV_i}{dt} \qquad 13.7.13$$

Vorticity flux generation occurs only during the acceleration phase. Once a steady velocity is attained, the vorticity flux becomes zero. An impulsive motion can be regarded as a finite amount of vorticity that is dumped into the flow at the initial instant.

13.8 HOW VORTICITY DISTRIBUTIONS ARE ESTABLISHED

Two examples will be discussed in this section. First, suppose that a certain internal flow is to be started from rest. For definiteness envision a long tube that accepts fluid from a large reservoir. The other end of the tube is connected to a pump or to a blower as the case may be. Initially the stationary fluid particles have no vorticity. Starting the pump or blower causes a pressure difference between the reservoir and the tube. This pressure difference is responsible for accelerating the fluid. Since both $\nabla^2\boldsymbol{\omega} = 0$ and $\boldsymbol{\omega} \cdot \nabla \mathbf{v} = 0$, the vorticity equation for the major portion of the flow is

$$\frac{D\boldsymbol{\omega}}{Dt} = 0 \qquad 13.8.1$$

With zero vorticity, the stretching and diffusion mechanisms cannot operate, and the particles initially move irrotationally. Equation 13.8.1 is the basis for Helmholtz's (1867) statement (for the flow of an effectively inviscid fluid): *No element of the fluid that was not originally in rotation is made to rotate.* The viscous mechanism is the only way to impart vorticity to an initially irrotational motion. A wall is a source of vorticity in the same sense that a hot wall is a source of heat.

The fully developed steady flow is established in the following manner. The time scale to establish a flow by pressure forces is determined by the speed of sound of the fluid. In incompressible flow the speed is effectively infinite. Hence, we may imagine that the irrotational flow developed by the pressure forces is established impulsively. In reality, the time it takes to accelerate the moving parts of the blower or pump determines how rapidly the pressure gradient is applied. At any given pump rotation rate during startup, the inviscid portion of the flow is quasisteady.

Since pressure forces can only generate an irrotational flow, a fluid particle must receive its initial rotation by viscous diffusion. We may consider that the situation within the tube is like Rayleigh's problem in which the flow stream is started impulsively while the wall is at rest. A thin layer of high vorticity is generated at the wall (at time zero the layer has infinite vorticity and zero thickness) and subsequently diffuses into the main fluid.

Viscous diffusion, as we have seen in previous examples, is a relatively slow process. Using Rayleigh's analogy (Section 7.5) we estimate that it would take a time $t_{\text{vis}} = h^2/(3.6\nu)$ for viscous diffusion to cross a tube of width $2h$. If h is 10 cm and the fluid is air, then the diffusion time is about 1 min. For vegetable oil ($\nu = 1.1 \text{ cm}^2/\text{sec}$) it is 7 sec. Another characteristic time in this problem is the time it takes the fluid to flow through the tube. The flow time is $t_{\text{flow}} = L/U$, where L is the length of the tube and U the fluid velocity. The relationship between these time scales determines the final steady flow. When the diffusion time is short compared to the flow time, the vorticity generated at the walls will diffuse completely across the tube as the particles slowly move down the tube. The final flow pattern occurs when a balanced vorticity distribution is established according to the complete vorticity equation 13.3.5. In the steady flow a pressure drop down the tube is required to balance the viscous forces. This means that a flux of vorticity continues to enter the flow from the walls according to 13.7.9. For a plane slot with a fully developed parabolic profile, the vorticity distribution is linear (7.2.5). The flux of vorticity is thus constant. Vorticity flux from one wall passes through the fluid and exits through the opposite wall. This is entirely analogous to a linear temperature distribution with a constant heat flux out of one wall and into the opposite wall. Each term in the vorticity equation—convection, line stretching, and diffusion—is identically zero.

Flow in a round tube is slightly different. A parabolic velocity profile in cylindrical coordinates leads to a conical increase in vorticity from the centerline, $\omega_\theta = 2rv_\theta/R^2$. This in turn implies a constant flux of vorticity through the fluid. Again, all terms in the vorticity transport equation are identically zero. In particular, the net vorticity diffusion term is zero. At first it might seem paradoxical that the net vorticity diffusion is zero, while the tube has a constant vorticity flux from the wall. It is, of course, impossible for a tube to have a constant heat flux from the wall and a steady temperature profile. However, this problem does not have plane geometry, and it does not obey the analogy between temperature and vorticity. The net diffusion term for vorticity in tube flow is

$$\nabla^2 \boldsymbol{\omega} = [\nabla^2 \boldsymbol{\omega}]_\theta = \frac{\partial}{\partial r}\left(\frac{1}{r}\frac{\partial}{\partial r}(r\omega_\theta)\right) = 0$$

The corresponding term in the heat equation is

$$\nabla^2 T = \frac{1}{r}\frac{\partial}{\partial r}\left(r\frac{\partial T}{\partial r}\right) = 0$$

The vorticity ω_θ cannot be made analogous to the temperature, because $[\nabla^2 \boldsymbol{\omega}]_\theta \neq \nabla^2 \omega_\theta$.

The opposite extreme, a short flow time compared to the diffusion time, means that vorticity is convected down the tube and never gets very far from the walls before the pump or blower is encountered. Such a flow would be typical of a wind tunnel with an irrotational flow in the core and thin boundary layers on the walls. The boundary layers are always growing as convection sweeps the vorticity introduced at the wall downstream. The time to establish this flow is measured by the particle flow time.

The second qualitative example is the external flow over an airfoil. Assume the flow is strictly two-dimensional, resulting in a vorticity vector that is always perpendicular to the velocity as shown in Fig. 13-4. Furthermore, the Reynolds number is assumed large. This means that the vorticity diffusion is primarily normal to the wall. A local coordinate system with $y = 0$ on the surface of the airfoil and x in the flow direction is assumed. The origin is placed at the stagnation point so that the positive x-axis is on the upper surface. The curvature of this coordinate system is not important since our arguments are only qualitative in nature.

The stagnation point is a point of zero shear, and hence by 13.7.5 zero vorticity. As the flow accelerates away from the stagnation point on the upper surface, the shear stress becomes positive, and the vorticity, again through 13.7.5, becomes negative. In this region the pressure drops, and we have a flux

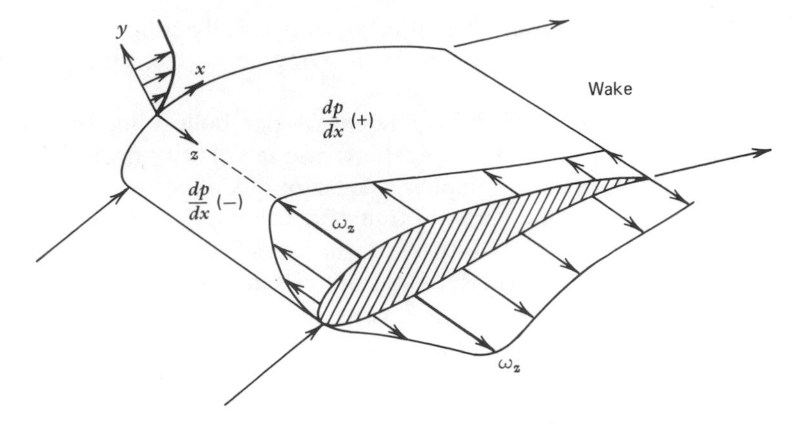

Figure 13-4 Vorticity distribution over an airfoil. Section cut through chord allows vorticity on lower surface to be shown.

of negative vorticity from the wall,

$$\mu\sigma_z = -\mu\frac{\partial\omega_z}{\partial y} = \frac{\partial p}{\partial x} < 0$$

The surface acts as a source to generate negative vorticity in the flow. Somewhere near the front of the airfoil the pressure reaches a minimum, followed by a gentle increase as the flow proceeds toward the trailing edge. In this region $\partial p/\partial x$ is positive, meaning that the wall acts as a sink to absorb some of the negative vorticity from the flow. The wall flux is positive (negative vorticity diffusing toward the wall). Notice that the maximum vorticity now occurs within the flow, as the sign of $\partial\omega_z/\partial y$ is negative at the wall. This process continues until the trailing edge is reached.

On the bottom side of the airfoil similar processes occur, except the x coordinate is now decreasing in the flow direction and the signs of the events switch. The pressure gradient accelerating the flow generates positive vorticity, while the subsequent decelerating pressure gradient creates a sink for positive vorticity. When the trailing edge is reached, the upper and lower streams must merge. At this point there is a discontinuity in the vorticity. This discontinuity is quickly washed out as it goes downstream. The negative vorticity from the upper surface and the positive vorticity from the lower surface merge into the wake. These regions diffuse together to destroy the wake. It is usually permissible in practical cases to assume that the vorticity has not diffused very far from the surface by the time the flow reaches the trailing edge. A calculation of the net vorticity across the wake shows it to be zero:

$$\text{average } \omega_z \text{ at trailing edge} = \int_{\delta_L}^{\delta_U}\omega_z\,dy = \int_{\delta_L}^{\delta_U}\frac{\partial u}{\partial y}\,dy = [u]_{\delta_L}^{\delta_U} = 0 \qquad 13.8.2$$

(In this calculation we neglect a small contribution to the vorticity from $\partial v/\partial x$. This contribution is very small for high Reynolds numbers, and in any event it dies out as the wake profile decays.) Another interesting fact is that the net flux of vorticity from the surface of the airfoil is zero. This is proved by integrating the vorticity flux over the surface from the lower side trailing edge x_L to the upper side trailing edge x_U:

$$\text{total flux of } \omega_z \text{ from airfoil} = \int_{\text{TE lower}}^{\text{TE upper}} \sigma_z \, dx = -\int_{\text{TE lower}}^{\text{TE upper}} \frac{\partial \omega_z}{\partial y} dx$$

$$= \frac{1}{\mu} \int_{x_L}^{x_U} \frac{dp}{dx} dx = \frac{1}{\mu} [\, p \,]_{x_L}^{x_U} = 0 \qquad 13.8.3$$

The last step above makes use of the fact that the pressure of the merging steams is the same. The sources and sinks of vorticity over the surface of the airfoil must cancel each other out.

Although the airfoil no longer puts out a net vorticity flux, there is a net vorticity within the flow. If we integrate the vorticity in the region outside the airfoil out to a radius R and then let $R \to \infty$, we find

$$\int \omega_z \, dA = \oint_R t_i v_i \, ds = \Gamma \qquad 13.8.4$$

—a finite number equal to the circulation. The net nonzero vorticity is inserted into the flow during the transient process by which the flow is established. In the transient process the flow does not leave the trailing edge smoothly, and the so-called "starting vortex" is formed. Figure 13-5 depicts a starting vortex formed by impulsively moving the airfoil. The starting vortex contains the same net amount of vorticity as the airfoil but with the opposite sign. A circulation loop going around the airfoil and including the starting vortex has $\Gamma = 0$.

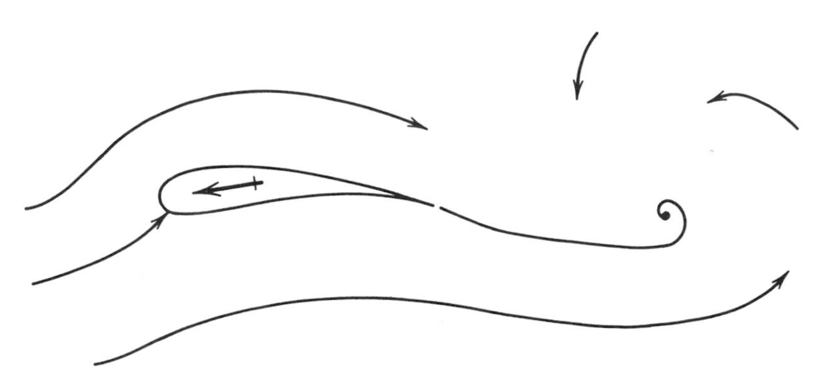

Figure 13-5 Vortex generated during the starting transient when an airfoil just begins to move.

13.9 INVISCID MOTION OF LINE VORTICES

Vortex lines are essentially instantaneous concepts. At any given time we can connect the vorticity vectors together to form the vortex lines. Giving an identity to a line in the sense that we can follow the time history of a specific vortex line is a separate issue. In this section results that apply to the inviscid motion of vortex lines will be given. For this special case an identity may be assigned to vortex lines, and in fact, they can be considered as material lines that move with the fluid.

The idea of inviscid motion of vortex lines might at first seem paradoxical, since vorticity must be generated by unbalanced viscous forces. However, there are many instances when the net viscous force produces vorticity in a transient process and the subsequent flow occurs as an inviscid flow carrying vorticity. A smoke ring is such an example. After the ring is generated, the vorticity is confined to a thin ring, but the fluid motion occurs in a much larger region. Even a region which is permeated with vorticity (so to speak) may behave in an inviscid manner in certain instances. Since viscous diffusion is such a slow process, events on a time scale much shorter than the diffusion time may be considered as inviscid. For such events vortex lines are material lines and move with the flow.

Let us consider a specific example. Assume that two potential line vortices of opposite rotation are a distance $2h$ apart. We view the line vortex as a collection of vortex lines, or a vortex tube, which has been shrunk to zero area. There must actually be a core of finite size, but it is not important for the present problem. The flow everywhere outside the core is irrotational and therefore inviscid. By Helmholtz's theorem, which we shall prove later, the vortex lines may be regarded as material lines, and hence they move with the local fluid velocity. A single vortex by itself has no tendency to move, but two counterrotating vortices propel each other through the fluid. (The problem for the velocity potential is linear in the potential and also in the velocity components. Superposition can therefore be employed.) The velocity field due to vortex A has a magnitude v_A given by

$$v_A(\mathbf{r}) = \frac{-\Gamma}{2\pi|\mathbf{r} - \mathbf{r}_A|}, \qquad \mathbf{r} \neq \mathbf{r}_A \tag{13.9.1}$$

This equation is valid everywhere except at the core position $\mathbf{r} = \mathbf{r}_A$, where the viscous forces are important and $v_A(r_A) = 0$. Vortex A has a negative circulation, while vortex B, with a positive circulation, is a distance $2h$ away. We regard vortex B as a material line and require that it move with the local fluid velocity. The velocity of vortex B is the motion induced at B by the velocity field of vortex A:

$$v_A(\mathbf{r} = \mathbf{r}_B) = V_B = -\frac{\Gamma}{4\pi h} \tag{13.9.2}$$

Likewise, the core of vortex A is propelled by the flow set up by vortex B.

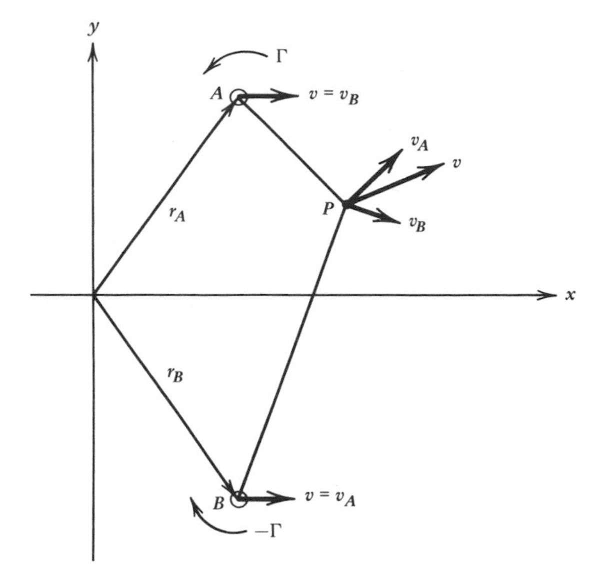

Figure 13-6 Motion of a pair of line vortices by self-induction.

Figure 13-6 shows the fluid motion set up by the vortices. Thus we observe that the vortices set up by the stroke of a canoe paddle are not stationary but propel themselves through the water in a direction opposite the motion of the boat. The same mechanism is responsible for the motion of a smoke ring.

When a pair of self-propelled vortices approach a solid wall, a very interesting series of events happen. The inviscid events are easily described by replacing the wall with a mirror-image set of vortices. The symmetry of the arrangement makes the x axis a streamline, which as far as inviscid flow is concerned can be taken as a wall. As the vortex pair approaches the wall, the image vortices become important, and the velocity induced by them pushes the vortices along the wall. The closer to the wall the vortex comes, the faster the image vortex will propel it along. One can show by calculation that the trajectories of the vortices form a curve known as the "cross curve".

This argument can be extended to apply to a ring vortex encountering a wall. Now, as the vortex nears the wall and begins to move outward, it must stretch. Stretching results in a proportional increase in the vorticity. For example, one can observe the increase in rotation of a smoke ring as it nears a wall.

13.10 KELVIN'S THEOREM AND HELMHOLTZ'S THEOREM

Two theorems will make the ideas we discussed above precise. Kelvin's theorem concerns the change in circulation of a set of material particles. From it we can proceed to deduce Helmholtz's theorem that vortex lines in an inviscid flow are material lines (this is the second of three major results about vortex lines by Helmholtz). Consider the circulation around a certain loop, and follow the material particles as they move. The symbol Γ_{ml} will emphasize that as time proceeds the same material loop is to be used in the calculation. It will be an aid in the mathematics if we consider distances along the material curve to be a function of a parameter m. As m goes from m_1 to m_2 we proceed around the curve and return to the starting point. From the definition of circulation we have

$$\Gamma_{\text{ml}} = \oint_C t_i v_i \, ds = \oint_C v_i \, dr_i = \int_{m_1}^{m_2} v_i \frac{\partial r_i}{\partial m} \, dm$$

The quantities v_i and r_i are formulated in Lagrangian variables: $v_i(r_0, \hat{t})$ and $r_i(r_0, \hat{t})$ [or, with m as a parameter specifying the curve, $r_i(r_0(m), \hat{t})$]. The time rate of change of the circulation is computed as follows:

$$\frac{d\Gamma_{\text{ml}}}{d\hat{t}} = \frac{d}{d\hat{t}} \int_{m_1}^{m_2} v_i \frac{\partial r_i}{\partial m} \, dm$$

$$= \int_{m_1}^{m_2} \frac{\partial v_i}{\partial \hat{t}} \frac{\partial r_i}{\partial m} \, dm + \int_{m_1}^{m_2} v_i \frac{\partial}{\partial \hat{t}} \left(\frac{\partial r_i}{\partial m} \right) dm$$

The second integral turns out to be zero (see 4.1.8), since

$$\int v_i \frac{\partial (dr_i)}{\partial \hat{t}} = \int v_i \, dv_i = \int d\left(\tfrac{1}{2} v_i v_i\right) = 0$$

The beginning and end points of the circuit are identical and have the same velocity. We are left with

$$\frac{d\Gamma_{\text{ml}}}{d\hat{t}} = \oint \frac{\partial v_i}{\partial \hat{t}} \, dr_i$$

Converting the expression to Eulerian variables gives

$$\frac{D\Gamma_{\text{ml}}}{Dt} = \oint \frac{Dv_i}{Dt} \, dx_i$$

Further simplification can be found if we substitute the momentum equation for incompressible flow. This produces

$$\frac{D\Gamma_{\text{ml}}}{Dt} = \oint \left[-\partial_i \left(\frac{p}{\rho} \right) + \nu \partial_j \partial_j v_i \right] dx_i$$

Noting that the pressure term integrates to zero yields Kelvin's theorem,

$$\frac{D\Gamma_{\text{ml}}}{Dt} = \oint \nu \partial_j \partial_j v_i \, dx_i \qquad 13.10.1$$

The circulation around a material loop of particles changes only if the net viscous force on those particles gives a nonzero integral. The special case of inviscid flow gives the result that the circulation of a material loop never changes:

$$\frac{D\Gamma_{\text{ml}}}{Dt} = 0 \qquad 13.10.2$$

From this key result it is possible to derive Helmholtz's theorem.

Consider a simple closed loop A in the fluid, which defines a vortex tube at a certain instant. On the surface of the vortex tube we choose another simple material loop B in such a way that it does not enclose the vortex tube. The surface of the tube enclosed by loop B contains the vortex lines and by 13.1.5 it has zero circulation. After some time has lapsed, the material particles that form B have moved to a new position. According to Kelvin's theorem the circulation around B must still be zero. Since the choice of the curve defining B is arbitrary, we allow it to shrink to zero size on the original vortex tube. For each choice of loop B, $\Gamma = 0$, and Γ continues to be zero at the second instant in time. The only way this can happen is if the surface described by this process remains a vortex surface. If at the second instant in time there is any portion of the surface that is punctured by a vortex line, we choose a circuit around that portion and find that it must have a nonzero circulation by 13.1.5. This contradicts Kelvin's theorem, since the circuit must have zero circulation at the original time.

Any loop A that goes around the vortex tube has a certain circulation Γ. Kelvin's theorem tells us that at any later instant, the loop will continue to have the same circulation. These considerations lead us to the conclusion that as long as the flow behaves in an inviscid manner, we may regard a vortex tube as a set of material particles. Limiting the vortex tube to zero area gives Helmholtz's second theorem: *The elements which at any time belong to one vortex line, however they may be translated, remain on one vortex line*. This means it is permissible to give an identity to each vortex line and imagine that vortex lines move with the fluid. Note that this is just a permissible description; in a given steady flow it might also be advantageous to envision vortex lines as fixed in space.

*13.11 INTEGRAL VORTICITY EQUATIONS

By following the procedures of Section 5.16 we may derive the global equation governing vorticity in a control volume, the surface of which moves with an arbitrary velocity w_i. The result is

$$\underbrace{\frac{d}{dt}\int \omega_i \, dV}_{\substack{\text{rate of change}\\ \text{of } \omega_i \text{ within}\\ \text{the control}\\ \text{volume}}} = \int \Bigg[\underbrace{n_j(w_j - v_j)\omega_i}_{\substack{\text{rate } \omega_i \text{ is}\\ \text{carried in}\\ \text{by fluid}\\ \text{flow}}} + \underbrace{n_j \omega_j v_i}_{\substack{\text{rate vortex}\\ \text{lines are}\\ \text{stretched}\\ \text{by } v_i}} + \underbrace{\nu n_j \partial_j \omega_i}_{\substack{\text{viscous}\\ \text{diffusion}\\ \text{of vorticity}\\ \text{across the surface}}} \Bigg] dS \qquad 13.11.1$$

This equation in several respects has a closer analogy with the momentum equation than with the thermal-energy equation. First of all, ω_i is a vector, so that in the integral over the volume, vorticity in one direction may cancel vorticity in the other direction. The total amount of vorticity in a smoke ring is zero because the direction of ω_i changes in such a way that the integral vanishes.

Another aspect of 13.11.1 is that only surface integrals appear on the right-hand side; there are no "body force" mechanisms to generate vector vorticity. The first surface integral is convection of vorticity into the region by the relative flow velocity $v_j - w_j$. We have seen this effect so many times that it needs no further comment. The second term is the result of the local vortex turning and stretching mechanisms. Any turning and stretching within the control volume generates "length" in both directions, and the vector nature of the equation causes these increments to cancel out. The only way a "net" length is added to a vortex line is if the ends of the line (the points where the vortex line crosses the control volume) move away from each other because of different velocities v_i. The last term is the viscous-diffusion flux of ω_i into the region. Momentum analogies for the last two terms would be surface pressure forces and viscous forces.

Sometimes it is also useful to interpret a flow using the global equation for $\frac{1}{2}\omega_i\omega_i$. This suppresses the vector nature of vorticity and gives us a measure, albeit a nonlinear measure, of the amount of vorticity within a region irrespective of its direction. The global form of 13.5.4 may be found as

$$\frac{d}{dt}\int \tfrac{1}{2}\omega^2 \, dV = \int \left[n_i(w_i - v_i)\tfrac{1}{2}\omega^2 + \nu n_j \partial_j\left(\tfrac{1}{2}\omega^2\right) \right] dS$$

$$+ \int \left[\omega_i \omega_j \partial_{(j} v_{i)} - \nu \partial_j \omega_i \partial_j \omega_i \right] dV \qquad 13.11.2$$

The surface integrals show that ω^2 may increase within a region by convection and by a viscous flux of ω^2. The volume integrals display the fact that the competition between vorticity production by stretching and destruction by viscosity must be viewed as local processes. These volume integrals cannot be converted to integrals over the surface of the control volume.

Let us consider, from the standpoint of vorticity, another of the basic problems of fluid mechanics, the semiinfinite flat plate in a uniform stream. We consider the high-Reynolds-number case where diffusion only occurs normal to the plate. The flat plate does not have a pressure gradient along its surface, and therefore there is no flux of vorticity from the wall. Very far away from the plate the free stream is irrotational. Consider a control volume containing the plate and with two stations a short distance apart on faces ① and ②. The global vorticity equation 13.11.1 for this region is

$$0 = -\int_{①} u\omega_z \, dy + \int_{②} u\omega_z \, dy$$

The convection of vorticity across each section is the same. Where does this vorticity originate? Its source is in the neighborhood of the leading edge. We might think of the leading edge as having the same effect as an airfoil whose nose radius is allowed to approach zero. Even when the flat plate is sharp, there are pressure gradients near the leading edge, which are the source of the vorticity. In the limit, the upper surface has a negative line source of vorticity at the leading edge. The lower surface has a corresponding positive line source. If we revert to the temperature analogy, the flat plate constitutes a line sink of energy on the upper leading edge and a line source on the underside. In other words, the leading edge is a doublet with its axis in the vertical direction. The plate itself is an adiabatic wall. Vorticity conducted from the doublet at the front is carried downstream by the fluid flow.

13.12 CONCLUSION

Vorticity vectors and their associated vortex lines give us additional concepts with which to interpret fluid flow patterns. This viewpoint emphasizes viscous effects, since pressure does not play a direct role in the equation governing vorticity dynamics. Pressure nevertheless does have some important influences in at least two respects: (a) pressure forces help determine the velocity field, which in turn convects the vortex lines and stretches them; (b) pressure gradients along a solid wall are directly related to the flux of vorticity from the wall into the fluid. Moreover, the wall is the only place where vorticity can originate.

Helmholtz's theorem, which allows us to think of vortex lines as stringing together a set of material particles, is applicable whenever viscous diffusion (i.e., the net viscous force) is negligible. Such flows are ideal flows carrying vorticity. In these flows, stretching vortex lines increases the vorticity in direct proportion to the increase in the length of the vortex line.

PROBLEMS

13.1 (A) A disk of radius R is spinning about its axis at a speed Ω. What is the vorticity of the particles at $r = 0$, $r = \frac{1}{2}R$, and $r = R$?

13.2 (A) Burgers' vortex in cylindrical coordinates has the velocity components $v_r = -ar$, $v_z = az$, and $v_\theta = (\Gamma/2\pi r)[1 - \exp(-ar^2/2\nu)]$ What is the vorticity field for this flow?

13.3 (B) Compute the vorticity for the von Kármán pump problem. Leave your answer in terms of the functions F, G, and their derivatives. What relations of F and G determined the fluid vorticity at the wall? Contrast with Problem 1.

13.4 (A) Compute each term in the vorticity equation 13.3.5 for Problem 2.

13.5 (B) Prove relation 12.7.5 between the wall shear stress and the vorticity for a curved wall.

13.6 (B) Find the equation of the "cross-curve" which marks the path of two counter-rotating vortices as they approach a wall.

13.7 (B) Consider an airfoil that is stationary in an infinite fluid. At time zero, the airfoil starts to move with a constant speed U_0. How does the vorticity and circulation in the starting vortex compare to that over the airfoil?

13.8 (A) Compare the vorticity distributions for Stokes' oscillating plate and the oscillating freestream above a fixed plate. The velocity profiles are 11.2.16 and 11.2.24.

14 Kinematic Decomposition of Flow Fields

Previously we introduced the decomposition of fluid motion into categories of translation, solidlike rotation, and deformation. This is a local picture of the flow that is valid for the motion of particles in a vanishingly small neighborhood. In this chapter we seek to decompose the entire velocity field into parts that have kinematic significance. This can be accomplished in several different ways. We shall review two different methods. Helmholtz's decomposition will be discussed first, as it is the most popular and best known. The second method uses Monge's potentials, which also go under the name "Clebsch variables".

*14.1 GENERAL APPROACH

We seek to divide the velocity field into two parts as follows:

$$\mathbf{v} = \mathbf{v}^{(\omega)} + \mathbf{v}^{(\phi)} \tag{14.1.1}$$

The first part, $\mathbf{v}^{(\omega)}$, is the rotational component and accounts for all of the vorticity in the flow. As a consequence, the second part, $\mathbf{v}^{(\phi)}$, is irrotational. Mathematically these statements imply that

$$\boldsymbol{\omega} = \nabla \times \mathbf{v} = \nabla \times \mathbf{v}^{(\omega)} \tag{14.1.2}$$

$$0 = \nabla \times \mathbf{v}^{(\phi)} \tag{14.1.3}$$

Recall that the necessary and sufficient condition for the existence of a velocity potential is simply that the flow is irrotational. Hence, we call the second part the potential component. It is related to a potential ϕ by

$$\mathbf{v}^{(\phi)} = \nabla\phi \tag{14.1.4}$$

The decomposition 14.1.1 is not unique. For a given velocity field $\mathbf{v}$, we could choose any potential flow whatsoever and subtract it from the real flow to arrive at $v^{(\omega)}$. To make the decomposition unique we need to apply more conditions.

In the local description of fluid motion, discussed in Chapter 4, we considered $d\mathbf{v}$, the velocity of a material particle P' with respect to the primary particle P. This velocity increment was further divided into rotational and straining components; $d\mathbf{v} = d\mathbf{v}^{(r)} + d\mathbf{v}^{(s)}$. Notice that the rotational component is denoted by $d\mathbf{v}^{(r)}$, while in 14.1.1 the rotational component is denoted by $\mathbf{v}^{(\omega)}$. These velocities are not related. As a matter of fact, $d\mathbf{v}^{(r)}$ is an inexact differential and hence cannot be integrated to produce a function. Equation 14.1.1 is simply a splitting where one component produces the vorticity when it is differentiated. Many splittings have this property.

*14.2 HELMHOLTZ'S DECOMPOSITION

This decomposition bears Helmholtz's name because he employed it in his famous paper on vortex-line behavior. As with most ideas, it developed in stages with several contributors. In fact, Stokes published a key result in a slightly different form prior to the paper of Helmholtz.

Let us impose the requirement that $\mathbf{v}^{(\omega)}$ have zero divergence (that is, $\mathbf{v}^{(\omega)}$ is solenoidal:

$$\nabla \cdot \mathbf{v}^{(\omega)} = 0 \qquad 14.2.1$$

This cuts down the choices for $\mathbf{v}^{(\omega)}$, but it still does not produce a unique decomposition. To show this, we take any harmonic function Φ ($\nabla^2\Phi = 0$) and let $\nabla\Phi$ be another velocity potential. Now, consider a certain decomposition denoted by subscripts 1, and add and subtract $\nabla\Phi$:

$$\begin{aligned} \mathbf{v} &= \mathbf{v}_1^{(\omega)} + \mathbf{v}_1^{(\phi)} \\ &= \mathbf{v}_1^{(\omega)} + \nabla\Phi + \mathbf{v}_1^{(\phi)} - \nabla\Phi \\ &= \mathbf{v}_2^{(\omega)} + \mathbf{v}_2^{(\phi)} \end{aligned}$$

The combination $\mathbf{v}_2^{(\omega)} = \mathbf{v}_1^{(\omega)} + \nabla\Phi$ still satisfies all the requirements for $\mathbf{v}^{(\omega)}$, including 14.2.1, and the combination $\mathbf{v}_2^{(\phi)} = \mathbf{v}_1^{(\phi)} - \nabla\Phi$ is still a potential flow.

Equation 14.2.1 has the effect of placing the expansion motions in the potential component. We denote the rate of expansion by Δ. For a given velocity field we have

$$\Delta \equiv \nabla \cdot \mathbf{v} = \nabla \cdot \mathbf{v}^{(\phi)} = \nabla^2\phi \qquad 14.2.2$$

To keep the discussion general we allow the flow to be compressible, so the effective source distribution Δ is not zero. In incompressible flow $\Delta = 0$ and the potential ϕ will become a harmonic function.

For the sake of finding a solution to 14.2.2, let us assume that Δ is a known distribution of sources. Textbooks in mathematics show that the solution to

Poisson's equation 14.2.2 is given by

$$\phi(\mathbf{x}) = \Phi - \frac{1}{4\pi}\int \frac{\Delta' \, dV'}{|\mathbf{r}|} \qquad 14.2.3$$

The corresponding velocity is

$$\mathbf{v}^{(\phi)}(\mathbf{x}) = \nabla\Phi + \frac{1}{4\pi}\int \frac{\mathbf{r}\Delta' \, dV'}{|\mathbf{r}|^3} \qquad 14.2.4$$

In these equations $\mathbf{x}$ is the position of interest, while $\mathbf{x}'$ is the position of the integration element dV' and the point where Δ' is evaluated. The vector $\mathbf{r}$ is defined as $\mathbf{x} - \mathbf{x}'$. The given distribution of sources determines the integral in 14.2.4. The gradient of any harmonic function Φ may be added to the source effect. Φ is usually chosen to satisfy any boundary conditions.

Turning now to the rotational component $\mathbf{v}^{(\omega)}$, we note the result from tensor analysis (Phillips 1933, Brand 1957, Aris 1962, Batchelor 1967), that any vector field that satisfies $\nabla \cdot \mathbf{v}^{(\omega)} = 0$ (14.2.1) may be represented by a vector potential $\mathbf{B}$. The representation is such that

$$\mathbf{v}^{(\omega)} = \nabla \times \mathbf{B} \qquad 14.2.5$$

A unique choice for $\mathbf{B}$ is made on mathematical grounds. The vector identity

$$\nabla^2 \mathbf{B} = -\nabla \times (\nabla \times \mathbf{B}) + \nabla(\nabla \cdot \mathbf{B})$$

will simplify to

$$\nabla^2 \mathbf{B} = -\boldsymbol{\omega} \qquad 14.2.6$$

if we make the assumption $\nabla \cdot \mathbf{B} = 0$. The solution to 14.2.6 is

$$\mathbf{B} = \frac{1}{4\pi}\int \frac{\boldsymbol{\omega}'}{|\mathbf{r}|} dV' \qquad 14.2.7$$

This leads to the well-known Biot–Savart law:

$$\mathbf{v}^{(\omega)}(x) = -\frac{1}{4\pi}\int \frac{\mathbf{r} \times \boldsymbol{\omega}'}{|\mathbf{r}|^3} dV' \qquad 14.2.8$$

A known vorticity distribution produces a specific rotational component $v^{(\omega)}$ from 14.2.8. This formula is arrived at by requiring that $\nabla \cdot \mathbf{B} = 0$.

The expressions 14.2.8 and 14.2.4 give a definite decomposition of the velocity field. The ambiguity noted at the beginning amounts to adding the potential of a harmonic function to 14.2.7 and subtracting the same from 14.2.4.

With the mathematical assumptions that **v** is continuous and has bounded derivatives, the integrals in 14.2.4 and 14.2.8 are well defined and yield well-behaved solutions. The decomposition is valid in the global sense. We have no difficulties in principle in applying the equations to the whole flow field.

*14.3 LINE VORTEX AND VORTEX SHEET

Consider a small cylindrical region of radius r_0 which contains vorticity (Fig. 14-1). The vorticity is directed along the z axis and extends from z_1 to z_2. Outside of this region the vorticity is zero. Recall that the strength of a vortex tube must be constant; hence we let

$$\Gamma = \int_0^{r_0} \omega_z \, dA \qquad 14.3.1$$

It is not physically possible for the vortex lines to exist only between z_1 and z_2; however, we shall consider only this piece to find the influence of a section of finite length.

Without loss of generality we compute the velocity at the point P a radial distance R from the origin. The vortex induces a velocity v_θ at this point. When R is large compared to the vortex-tube radius r_0, the integral 14.2.8 becomes

$$\begin{aligned} v_\theta &= \frac{\Gamma}{4\pi} \int_{z_1}^{z_2} \frac{R\, dz}{(R^2 + z^2)^{3/2}} = \frac{\Gamma}{4\pi R} \left[\frac{z_2}{\left(R^2 + z_2^2\right)^{1/2}} - \frac{z_1}{\left(R^2 + z_1^2\right)^{1/2}} \right] \\ &= \frac{\Gamma}{4\pi R}(\cos\alpha_2 - \cos\alpha_1) \end{aligned} \qquad 14.3.2$$

From this equation we see that the velocity in the end plane of a semi-infinite vortex is $v_\theta = \Gamma/(4\pi R)$. This result has application to the downwash velocity

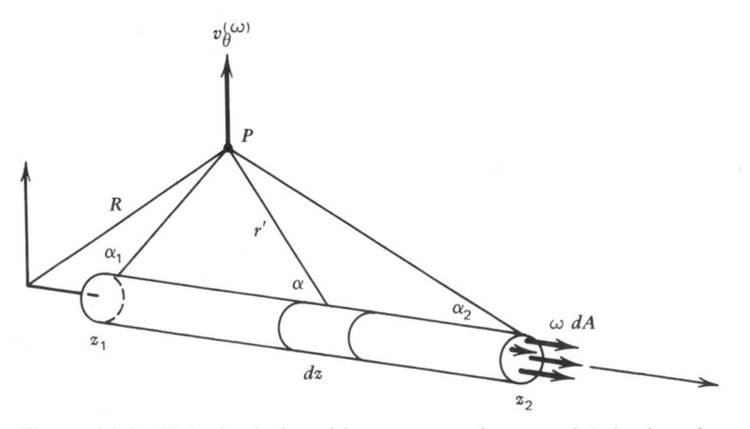

Figure 14-1 Velocity induced by a vortex element of finite length.

at the wing caused by a vortex filament in the wake. A line vortex that is infinite in both directions results in the customary formula for an ideal vortex; $v_\theta = \Gamma/(2\pi R)$. The location of the origin becomes immaterial in this instance.

As a second example, consider a plane vortex sheet as shown in Fig. 14-2. A uniform vorticity $\boldsymbol{\omega}$ exists in the region $-h < x_3 < h$, where h is very small. Integration across this layer produces a constant,

$$K = \int_{-h}^{h} \omega_2 \, dx_3 \qquad 14.3.3$$

For a vortex tube dx_1 by $2h$ in cross section, the incremental circulation is

$$d\Gamma = K \, dx_1 \qquad 14.3.4$$

Thus, K is the circulation per unit length of the vortex sheet. Everywhere outside the sheet the vorticity is zero.

Since the sheet is infinite, we need only consider points on the x_3 axis to find how the velocity changes with distance from the sheet. The Biot–Savart law 14.2.8 for these points yields

$$\begin{aligned} v_2(0,0,x_3) &= -\frac{K}{4\pi} \iint_{-\infty}^{\infty} \frac{x_1' \, dx_1' \, dx_2'}{\left(x_1'^2 + x_2'^2 + x_3'^2\right)^{3/2}} = 0 \\ v_1(0,0,x_3) &= \frac{K}{4\pi} \iint_{-\infty}^{\infty} \frac{x_3' \, dx_1' \, dx_2'}{\left[x_1'^2 + x_2'^2 + x_3'^2\right]^{3/2}} = \pm \frac{K}{2} \end{aligned} \qquad 14.3.5$$

The plus sign is for $x_3 > 0$ and the minus sign for $x_3 < 0$. The velocity field associated with a vortex sheet is a uniform flow parallel to the sheet and

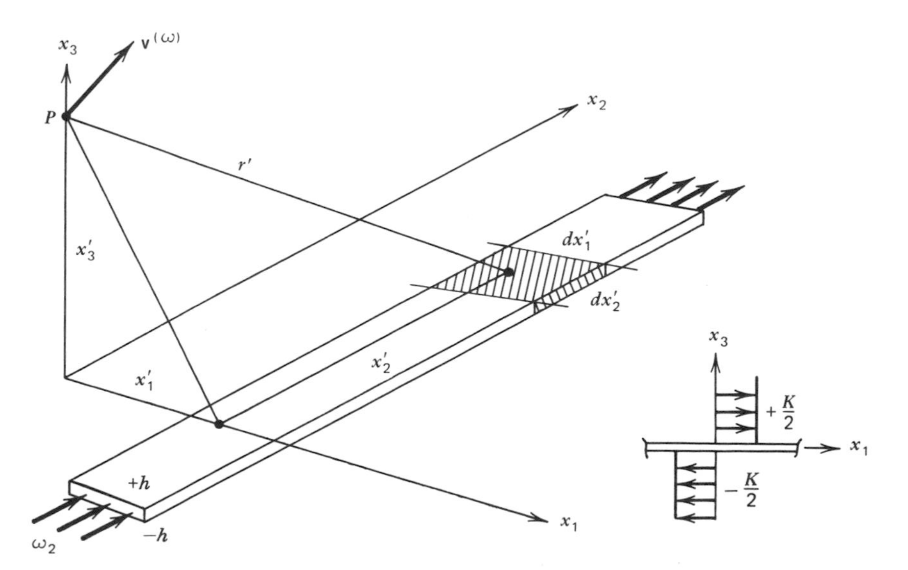

Figure 14-2 Velocity induced by a vortex sheet.

perpendicular to the vorticity vector. The sheet separates two uniform streams, the upper with velocity $K/2$ and the lower with velocity $-K/2$. The jump in tangential velocity as the sheet is crossed is the sheet strength K.

The interpretation of K as the circulation density $d\Gamma/ds$ is especially relevant to thin-airfoil theory. Consider a small section of a thin airfoil where dx_1 is along the chord direction. The boundary layer on the upper surface has a positive vorticity as the velocity goes from zero at the wall to U_{upper} outside the boundary layer. On the lower surface the vorticity in the boundary layer is negative with the velocity attaining a value U_{lower} outside the boundary layer. Integration across both layers gives a net strength $K = U_{\text{upper}} - U_{\text{lower}}$ at any chord position. Since $K = d\Gamma/dx$, the integration of $K\,dx_1$ from leading edge to trailing edge yields the total circulation around the airfoil, Γ. In this way the ideal flow over an airfoil may be represented as a uniform stream with a vortex sheet of varying strength inserted to represent the foil (thickness effects may be represented by sources and sinks in 14.2.4, but are usually negligible).

In the examples above we have flows where the vorticity is concentrated in thin regions. The Biot–Savart law is not a cause–effect relationship in the sense that a force produces an acceleration. The relationship is purely kinematic. For this reason people frequently say that a vorticity distribution "induces" a certain remote velocity. We say a vortex ring is propelled by self-induction. A similar kinematic relation occurs when a small pipe injects water in the center of a large, deep pool. The source of mass requires a flow velocity be induced at remote positions through 14.2.4. This is a kinematic requirement. In a similar manner changes in a vorticity distribution must be accompanied by a remote effect through 14.2.8.

When a flow is restricted by walls, as most flows are, the decomposition into rotational and potential parts is accomplished by imagining the fluid extends beyond the walls and has fictitious sources and vortices that will produce the proper result within the real flow. This introduces another ambiguity in that many distributions of sources and vortices within the walls can produce the same flow confined between the walls (suggestions are contained in Richardson and Cornish, 1977).

*14.4 COMPLEX-LAMELLAR DECOMPOSITION

A second method of decomposing the velocity field into a potential part $\mathbf{v}^{(\phi)}$ and a rotational part $\mathbf{v}^{(\omega)}$ is to choose $\mathbf{v}^{(\omega)}$ to be a complex-lamellar field. This somewhat awkward term was applied by Kelvin and finds favor with workers in rational mechanics. (Potential flows were called lamellar by Kelvin.) A complex-lamellar vector field is one that becomes a potential flow if it is divided by an integrating function σ. That is, a potential function χ exists such that

$$\frac{\mathbf{v}^{(\omega)}}{\sigma} = \nabla\chi \qquad 14.4.1$$

Substituting into 14.1.1 and using 14.1.4, we see that the three "potentials" σ, χ, and ϕ determine the velocity:

$$\mathbf{v} = \sigma \nabla \chi + \nabla \phi \tag{14.4.2}$$

These potentials are sometimes called Clebsch's variables, as he used them in an early application to fluid mechanics. The decomposition 14.4.2 is not unique; several combinations of σ, χ, and ϕ can be found that give the velocity field.

Another completely equivalent definition of a complex-lamellar vector is that it always is perpendicular to its own curl. Hence,

$$\begin{gathered} \mathbf{v}^{(\omega)} \cdot (\nabla \times \mathbf{v}^{(\omega)}) = 0 \\ \mathbf{v}^{(\omega)} \cdot \boldsymbol{\omega} = 0 \end{gathered} \tag{14.4.3}$$

In any flow where the total velocity is perpendicular to the vorticity (plane flows and axisymmetric flows), this condition is satisfied by the velocity itself. For these flows $\mathbf{v} = \mathbf{v}^{(\omega)} = \sigma \nabla \chi$ and $\mathbf{v}^{(\phi)} = 0$ constitute an acceptable representation.

The most interesting aspect of the decomposition of 14.4.2 is revealed by computing the vorticity,

$$\boldsymbol{\omega} = \nabla \times \mathbf{v}^{(\omega)} = \nabla \sigma \times \nabla \chi \tag{14.4.4}$$

Note the similarity between this equation and 12.2.7. In 12.2.7 the incompressibility condition $\nabla \cdot \mathbf{v} = 0$ allowed the velocity to be given by streamsurfaces f and g such that $\mathbf{v} = \nabla f \times \nabla g$. Vorticity always meets the condition $\nabla \cdot \boldsymbol{\omega} = 0$, so a similar representation is possible. Equation 14.4.4 means that surfaces of $\sigma = \text{const}$ and $\chi = \text{const}$ are vortex surfaces containing the vortex lines. This is easily shown by proving that the normal at the surface σ, that is $\nabla \sigma$, is perpendicular to the vorticity vector: $\nabla \sigma \cdot \boldsymbol{\omega} = \nabla \sigma \cdot \nabla \sigma \times \nabla \chi = 0$. A triple vector product containing the same vector twice is always zero. Similarly $\nabla \chi \cdot \boldsymbol{\omega} = 0$ establishes that χ surfaces also contain the vortex lines.

Since σ and χ surfaces contain the vortex lines, their intersection describes a vortex line. Thus, this decomposition offers a method of identifying vortex lines in a viscous flow. There are, however, two drawbacks, which need to be immediately pointed out. The decomposition 14.4.2 may not be globally valid. It may happen that a certain set of σ, χ, and ϕ surfaces cannot be extended over the whole flow. For example, this happens if the vortex lines have knots (in the same way, the stream-function surfaces are not globally valid if the streamlines have knots). Second, the identity of the vortex lines involves an arbitrary choice of reference surfaces.

To bring out some of the details of this decomposition, consider a flow over a wall as shown in Fig. 14-3. Assume the velocity, and hence the vorticity, are

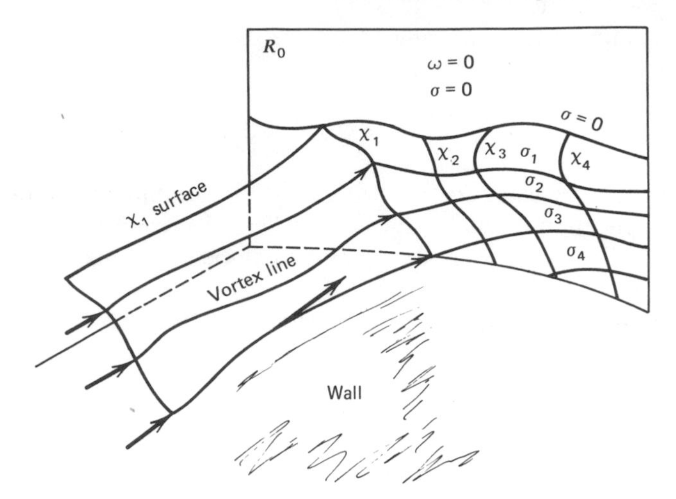

Figure 14-3 Clebsch variables for a concentrated region of vorticity.

known. Also assume the vorticity is nonzero near the wall, becomes zero on a certain surface within the flow, and is zero at all positions beyond this surface. Let the flow be regular in such a way that we may choose a certain reference plane R_0 that is pierced by all the vortex lines. All the σ and χ surfaces will also cut this plane. We may take any arbitrary set of curves on R_0 as lines where χ is constant. These lines on R_0 can be extended along the vortex lines throughout the flow to form the χ surfaces. Choosing a family of χ curves on R_0 determines the χ surfaces through the entire flow when all vortex lines pierce R_0.

In the region where $\omega = 0$ we must have $\nabla\sigma = 0$, since $\boldsymbol{\omega} = \nabla\sigma \times \nabla\chi$ and $\nabla\chi \neq 0$. Hence $\sigma = \sigma_0$ is constant in these regions (or possibly a function of time). If we take $\sigma_0 = 0$, then the component $v^{(\omega)}$ will also be zero in this region. Unlike the Helmholtz decomposition, the present decomposition can be arranged to give $\mathbf{v}^{(\omega)} = 0$ when $\omega = 0$.

We continue by explaining how it is possible in principle to compute σ and ϕ. Let dx by a differential line element which lies in a χ = const surface. Now

$$\boldsymbol{\omega} \times d\mathbf{x} = \nabla\chi \, d\sigma \qquad \text{on} \quad \chi = \text{const}$$

This vector is perpendicular to the χ surface. Take a helping unit vector that does not lie in the χ surface. For example, the surface normal $\nabla\chi$ would be suitable. We are assured the dot product with $\boldsymbol{\omega} \times d\mathbf{x}$ is nonzero:

$$\nabla\chi \cdot \boldsymbol{\omega} \times d\mathbf{x} = \nabla\chi \cdot \nabla\chi \, d\sigma$$

Integrating from the region where $\boldsymbol{\omega} = 0$ to any point x yields the function σ as

$$\sigma = \int^{\mathbf{x}} \frac{\nabla\chi \times \boldsymbol{\omega} \cdot d\mathbf{x}}{\nabla\chi \cdot \nabla\chi} \qquad \text{on} \quad \chi = \text{const} \tag{14.4.5}$$

The irrotational potential is found by integrating

$$d\phi = \nabla\phi \cdot d\mathbf{x} = (\mathbf{v} - \sigma\nabla\chi)\cdot d\mathbf{x}$$

Since the right-hand side is known, the decomposition is complete.

A different choice for the family of curves on R_0 that define the χ surfaces leads to a different potential–complex-lamellar decomposition. With the new choice of curves, $\mathbf{v}^{(\omega)}$ and $\mathbf{v}^{(\phi)}$ will be different.

We return now to the question of identifying vortex lines as the intersection of σ and χ surfaces with specific numbers. Obviously the first requirement is that we have a unique decomposition defined by a certain choice of χ surfaces. Recall that in Section 12.2 we found that the shape of a family of surfaces $\chi = \text{const}$ and the numbering system for the surfaces can be chosen separately. Any function $\hat{\chi} = g(\chi)$ simply renumbers the same family of surfaces. Suppose we renumber the surfaces (even in a steady flow we could do this as a function of time). What effect would this have on the decomposition? If $\hat{\chi} = g(\chi)$, then $\nabla\hat{\chi} = g'\nabla\chi$. Substituting into 14.4.5 reveals that $\hat{\sigma} = (1/g')\sigma$. Thus, surfaces $\sigma = \text{const}$ are not the same as surfaces $\hat{\sigma} = \text{const}$. The intersection of the σ and χ surfaces is not the same vortex line as the intersection of the $\hat{\sigma}$ and $\hat{\chi}$ surfaces with the same numbers. The history of a vortex line is ambiguous to this extent. However, even though the σ family is changed to a new set of surfaces by a simple renumbering of the χ surfaces, the vector decomposition remains the same. This is shown by noting that

$$\mathbf{v}^{(\omega)} = \sigma\nabla\chi = \hat{\sigma}\hat{\nabla}\chi$$

A numbering change of the χ surfaces gives the same decomposition but with a new family of σ surfaces and the same set of χ surfaces.

The arbitrariness in following vortex lines noted above is not at odds with Helmholtz's theorem that vortex lines follow the material particles in an inviscid flow. This theorem means only that one is *allowed* to choose the σ and χ surfaces so that an intersection follows the fluid particle [see Sudarshaw and Mukundew (1974) for a proof]. Other choices are also possible. In a steady inviscid flow we might also envision that the vortex lines are fixed in space.

*14.5 CONCLUSION

Many different decompositions of the velocity field are possible. The two most widely applied methods have been outlined.

Helmholtz's decomposition has global validity and many nice characteristics. The different components may be expressed as integrals of the kinematic properties of the given velocity field. The potential component $\mathbf{v}^{(\phi)}$ results from integrals of the source distribution $\nabla \cdot \mathbf{v}$, while the vortical component $\mathbf{v}^{(\omega)}$ results from integrals of $\nabla \times \mathbf{v}$. The ambiguity in this decomposition amounts to a harmonic function.

The second decomposition uses Monge's potential function σ, χ, and ϕ. The potential component $\mathbf{v}^{(\phi)} = \nabla\phi$ is irrotational, but $\nabla \cdot \mathbf{v}^{(\phi)} \neq 0$. Not only does the vortical component $\mathbf{v}^{(\omega)} = \sigma\nabla\chi$ yields the vorticity through $\boldsymbol{\omega} = \nabla\sigma \times \nabla\chi$, but this equation also means that the σ and χ surfaces intersect to determine the vortex lines. This physical interpretation is perhaps the most interesting aspect of this decomposition.

PROBLEMS

14.1 (C) Consider a closed region where **B** is given by 14.2.7. By computing $\nabla \cdot \mathbf{B}$ from 14.2.7 show that the condition $\nabla \cdot \mathbf{B} = 0$ implies that the following surface integral over the region is zero,

$$0 = \frac{1}{4\pi} \int \frac{\mathbf{n} \cdot \boldsymbol{\omega}'}{|\mathbf{r}|} dS'$$

14.2 (B) Consider a square vortex of side L and strength Γ. What is the velocity induced on one side by the other three sides? Are there any singular positions? Would the square vortex remain square?

14.3 (A) A line vortex of strength Γ extends from $x = 0$ to $x = +\infty$. What is the velocity profile as a function of the distance R away from the vortex at the positions $x = +\infty$, $x = 0$, and $x = -\infty$?

14.4 (A) Consider a point in space where the velocity is in the x direction and the vorticity is at 45° in the x-z plane. For the complex lamellar decomposition, sketch a picture of $\mathbf{v}$, $\boldsymbol{\omega}$, $\mathbf{v}^{(\omega)}$, $\mathbf{v}^{(\phi)}$, and $\mathbf{v} \cdot \boldsymbol{\omega}$ giving arbitrary magnitudes to the vectors. Sketch surfaces σ, χ, and ϕ = constant as they go through this point in space.

14.5 (A) Prove that 14.4.4 follows from 14.4.1.

14.6 (B) Consider the Monge potentials σ, χ, ϕ in a large region where $\boldsymbol{\omega} = 0$. How does this restrict the functions σ, χ, and ϕ and what does this imply about $\mathbf{v}^{(\phi)}$ and $\mathbf{v}^{(\omega)}$?

15 Flows at Moderate Reynolds Numbers

Flows where inertia, pressure forces, and viscous forces are all significant occur at moderate Reynolds numbers. Moderate Reynolds numbers cannot be defined as a precise range of values. For one thing, the numerical values depend upon the particular flow situation and the way the Reynolds number itself is defined. For flow over a circular cylinder, moderate Reynolds numbers are from $VD/\nu = 0.5$ to about 200. For flow into the entrance of a slot, moderate Reynolds numbers might be 0.5 to 50.

The other difficulty in specifying an exact range for moderate Reynolds numbers is that there is no abrupt change in the flow at either end of the spectrum. Low-Reynolds-number flows, where inertia effects are unimportant, make a rather smooth transition into moderate-Reynolds-number flows. Likewise, the confinement of viscous effects to the wall regions occurs progressively at the high end of the moderate-Reynolds-number regime. To assign a particular number to these transitions is a matter of personal judgement.

Regardless of the difficulties in defining them precisely, we can characterize moderate-Reynolds-number flows as those where both pressure and viscous forces contribute importantly to fluid accelerations *over the major extent of the flow field*. We need this last proviso in order to exclude boundary layers.

The momentum and vorticity equations for moderate Reynolds number flows are

$$\frac{D\mathbf{v}}{Dt} = -\nabla p + \frac{1}{\mathrm{Re}} \nabla^2 \mathbf{v}$$

and

$$\frac{D\boldsymbol{\omega}}{Dt} = \boldsymbol{\omega} \cdot \nabla \mathbf{v} + \frac{1}{\mathrm{Re}} \nabla^2 \boldsymbol{\omega}$$

The variables in these equations have been nondimensionalized using a characteristic velocity, a characteristic length, and the density as in Chapter 11. The highest-order terms, $\nabla^2 \mathbf{v}$ and $\nabla^2 \boldsymbol{\omega}$, give these equations an elliptic nature. We therefore expect that conditions in one part of the flow will influence all other parts and, in particular, there will be an upstream influence. An important distinguishing feature of these flows is that Re occurs as a parameter. As the Reynolds number changes we not only get a different balance of terms, but in many cases the flow patterns take on different and unusual forms.

15.1 SOME UNUSUAL FLOW PATTERNS

Figure 15-1 shows a circular cylinder as it joins a flat wall, much like a bridge support intersecting the river bottom or an airplane wing joining the fuselage. Intuition would lead us to expect that the flow would neatly divide and pass around the obstacle with a minimum of complication. In fact, this does not happen quite so simply. One of the major features of this quite complex flow is a horseshoe-shaped vortex that is looped around the cylinder in the front and

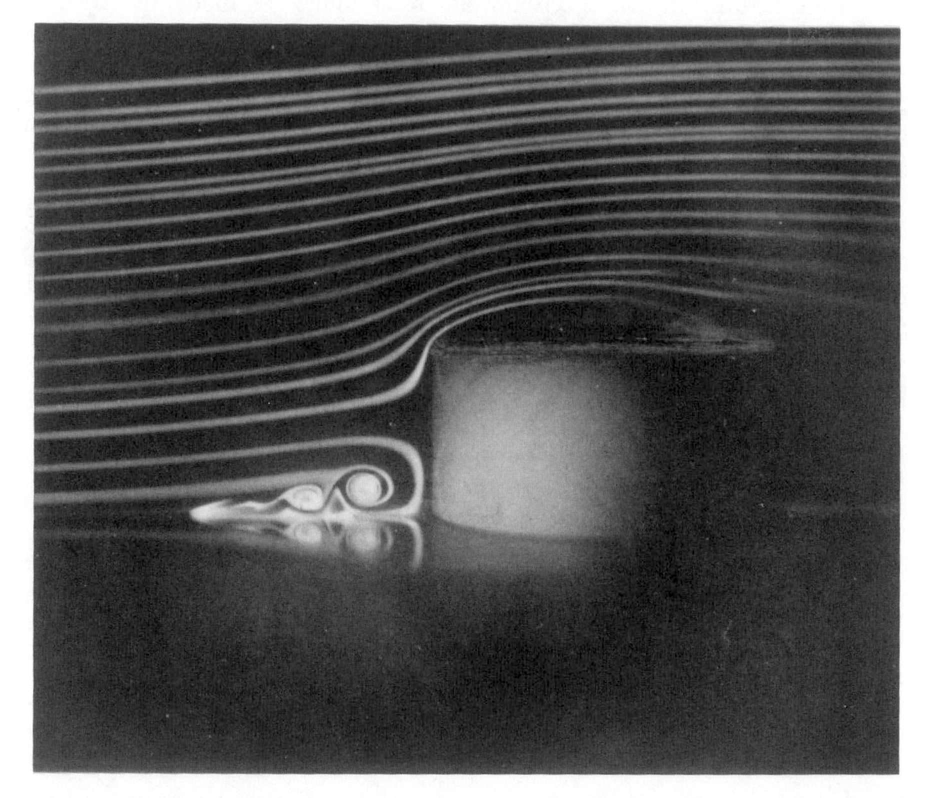

Figure 15-1 Several horseshoe vortices formed at the junction of a cylinder and a wall are made visible by smoke and illuminated on the centerline (from Thwaites, 1960). The boundary layer in the approach flow contains vorticity that is ultimately organized into the horseshoe vortex.

trails off behind. The vortex is next to the wall and continuously entrains more fluid as it proceeds around the cylinder and downstream. The intense velocities of these vortices are responsible for scouring the river bottom around the sides of bridge supports and for making the distinctive patterns in the snow around telephone poles. Figure 15-1 is a photograph of the vortices illuminated on the centerplane of the cylinder. At this Reynolds-number there are not one, but several horseshoe vortices adjacent to each other. The same pattern also exists at higher Reynolds numbers, except that the number of vortices decreases and their size becomes much smaller as shown in Fig. 15-2.

It is difficult to imagine the complicated interaction of inertia, pressure forces, and viscous forces that produces this flow. Although a detailed explanation is not possible, we can observe some major effects. One important aspect is the shear region on the wall far ahead of the cylinder. The thickness of this shear layer and the fact that it contains a lot of transverse vorticity are essential to the development of the organized vortices. Far away from the shear

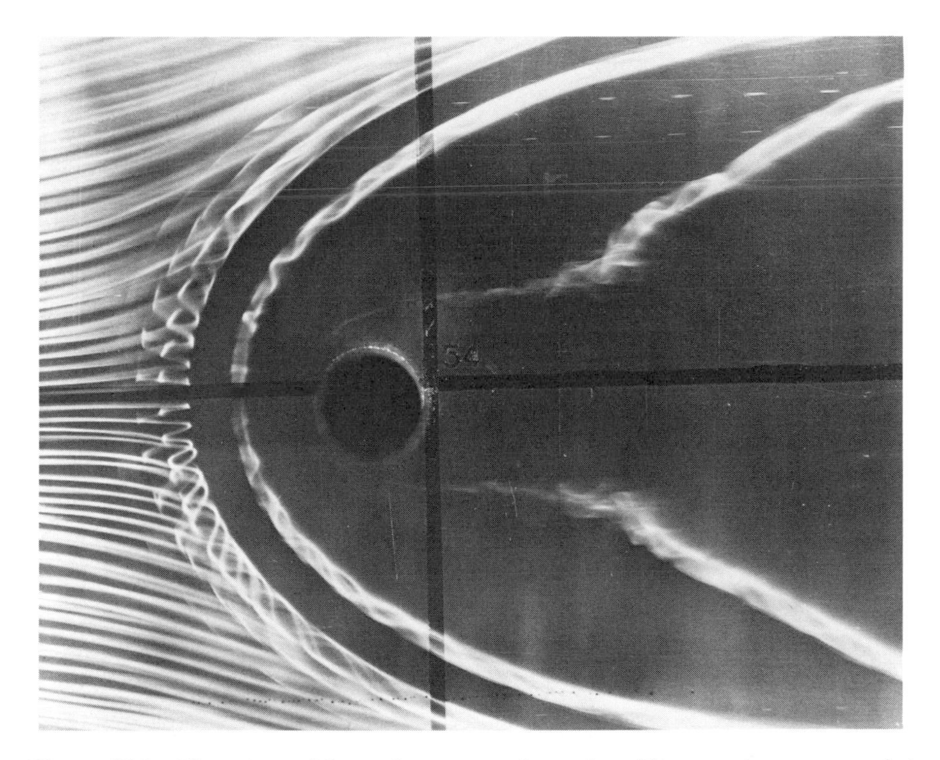

Figure 15-2 Plan view of horseshoe vortex formation. Photograph courtesy of A. Thomas, Lockheed-Georgia Co. This flow is at a higher Reynolds number than Fig. 15-1, giving an unsteady pattern where the vortices pinch off and reconnect downstream of the cylinder.

layer the flow is simply pushed away from the cylinder and made to go around either side by pressure forces. This motion is accompanied by straining deformation of the particles. Next to the wall, within the shear layer, there is a transverse vorticity component. When this vorticity undergoes a straining motion, the term $\boldsymbol{\omega} \cdot \nabla \mathbf{v}$ in the vorticity equation 13.3.5 turns the vorticity vector to give $\boldsymbol{\omega}$ a large streamwise component and also intensifies it by stretching. If the flow is at a modest Reynolds number, the viscous forces of one vortex can induce another vortex of the opposite sign next to it. The final flow pattern is a delicate balance of inertia, pressure forces, and viscous forces.

As a second example consider the flow under a sluice gate as shown in Fig. 15-3. Again our intuition would lead us to the wrong flow pattern. We might imagine that the flow would smoothly drain under the gate from all parts of the upstream area. In particular, the flow could smoothly accelerate down the gate, with the pressure dropping until at the exit the pressure would be atmospheric and the velocity correspondingly high. Figure 15-3 is a photograph of the streamlines that actually occur. A separated region exists near the top of the gate and extends downward for a considerable distance. The circular

Figure 15-3 Vortex at the top of a sluice gate. The flow is visualized with fluorescent dye and a slit of light on the center plane. Photograph courtesy of D. G. Bogard and L. N. Goenka, University of Texas.

Figure 15-4 Corner vortices forming from the wall boundary layer as the flow dives under a sluice gate. Air bubbles are carried downstream in the vortex core. Photograph by the author and L. N. Goenka.

flow within this region is sometimes accompanied by an outward flow toward the side walls, but this is not necessary.

Another unexpected pattern occurs at the corners where the gate and the walls meet. Near the wall a vortex is formed, which has its axis nearly vertical at the surface. The vortex core is marked by a stream of bubbles in the photograph (Fig. 15-4). The core trails under the gate and turns in the streamwise direction as it enters the tailwater flow. A little reflection will reveal the analogy between this vortex and the horseshoe vortices of the first example.

Figure 15-5 reveals the flow pattern in a fluid contained in a closed cylindrical tank where the top is rotated. Through the viscous pump effect, the rotating top causes an outward flow. The flow then proceeds down the sides and inward along the bottom. As the flow merges to rise in the center, a vortex is formed with its axis on the centerline of the tank. Dye has been introduced into the core of the vortex through a hole in the bottom. A short way up from the bottom the vortex "bursts". A free-standing separation bubble containing a torodial flow is clearly seen in the figure. The rising flow goes around this bubble as it proceeds toward the top.

Figure 15-5 "Vortex burst" in a closed cylinder whose top is rotated. Dye seeps in at the bottom and flows toward the top along the centerline. Photograph courtesy of M. P. Escudier, BBC, Baden, Switzerland.

15.2 ENTRANCE FLOWS

The events that occur in the entrance region of a tube or slot connected to a large reservoir were described previously in Section 7.1. At high Reynolds numbers, the flow was divided into several parts. Near the entrance the fluid is accelerated from the reservoir into the tube by pressure forces. This produces a velocity profile at the entrance to the straight section that is flat except for thin viscous regions next to the walls. Downstream from the entrance there is a hydrodynamic entrance region where the flat profile is transformed into the parabolic profile. The last region is characterized by a fully developed profile at any downstream station.

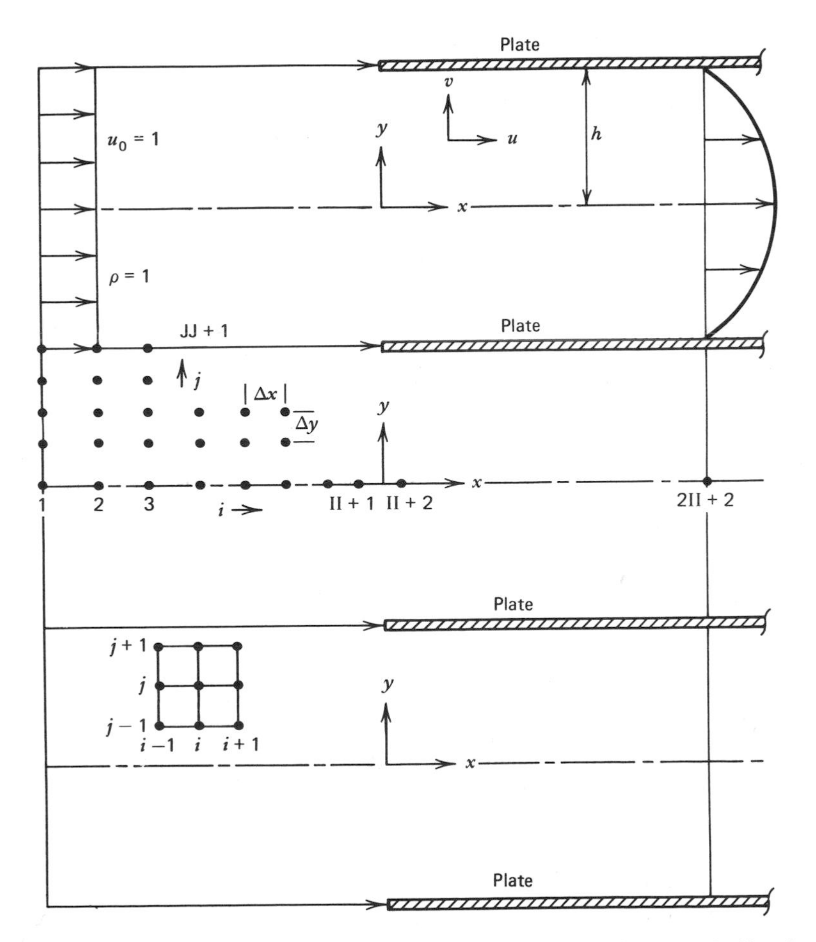

Figure 15-6 Flow into a cascade of thin plates. The flow in each cell is similar; in this diagram top cell shows the coordinate system, the middle cell shows the i, j finite-difference grid for one half cell, and the bottom cell contains a typical finite-difference molecule.

This same flow at a moderate Reynolds number cannot be so easily divided into separate parts. There is upstream influence as the diffusion of vorticity extends into the entrance and even a slight distance into the reservoir itself. A uniform profile at the beginning of the straight section is no longer an acceptable assumption. Because of the elliptic nature of the equations, we need to formulate the problem with the reservoir, the entrance, and part of the tube connected together.

Such a flow is sensitive to geometry, and we need to specify the exact shape of the entrance contour. As an academic problem that illustrates moderate-Reynolds-number effects and yet avoids assigning a definite shape to the entrance, we consider the flow into a cascade of channels. Figure 15-6 depicts the arrangement. The essential idea is to contrive a physical arrangement where 100% of the flow coming from a uniform stream at infinity is accepted into the tube. Physically, we must have some external method to suck the flow into the cascade. Once it is there, vorticity is generated on the walls. Because the Reynolds number is low, the vorticity may diffuse out of the entrance and modify the flow in front of the entrance. Geometric symmetry allows us to assume that the streamlines that stagnate at the edges of the plates are straight. Far downstream within the cascade, the velocity profile develops into the parabolic profile of fully developed flow.

15.3 ENTRANCE FLOW INTO A CASCADE OF PLATES: COMPUTER SOLUTION BY THE STREAM-FUNCTION–VORTICITY METHOD

Numerical methods are well suited to moderate-Reynolds-number flows. A computer can take on the added complication of having all terms in the equation present without undue effort. Although it is quite feasible to solve problems using the pressure and velocities as unknown, this particular problem is most easily formulated in terms of the stream function and vorticity. Recall that the vorticity–stream-function method applies only when the flow is plane or axisymmetric. Emmons (1949) was one of the first to use this method in a finite-difference solution. This particular problem was first done numerically by Wang and Longwell (1964).

This example also affords an opportunity to introduce some rudiments of finite-difference procedures. Its major purpose is nevertheless the physics of entrance flow. In the end we shall observe how the flow patterns change as the Reynolds number takes on different values.

Figure 15-6 shows a cascade of plates with a spacing $2h$. A coordinate system is placed along the centerline of one channel with the origin aligned with the edge of the plates. Far upstream, the flow is uniform with velocity u_0 and a density ρ. We assume that the flow pattern has the same symmetry as the geometry and select one half of one channel as the solution region. A simple way to nondimensionalize is to let the half width h be the length unit; that is, let $h = 1$, choose $u_0 = 1$ as a velocity unit, and let $\rho = 1$ be the density unit (from the standpoint of dimensional analysis we are choosing units for length, velocity $v = L/T$, and density $\rho = M/L^3$, instead of the traditional M, L, T). All lengths in the problem may be regarded as nondimensionalized by h, all velocities by u_0, the vorticity by u_0/h, and the stream function by $u_0 h$. The Reynolds number is $\text{Re} = u_0 2h/\nu = 2/\nu$.

The equations that govern the flow are the vorticity equation for the component $\omega_z = \omega$,

$$\frac{\partial \omega}{\partial t} = -\frac{\partial}{\partial x}(u\omega) - \frac{\partial}{\partial y}(v\omega) + \frac{2}{\text{Re}}\nabla^2\omega \tag{15.3.1}$$

the stream-function–vorticity equation,

$$\nabla^2\psi = -\omega \tag{15.3.2}$$

and the stream-function–velocity relations

$$u = \frac{\partial \psi}{\partial y}, \qquad v = -\frac{\partial \psi}{\partial x} \tag{15.3.3}$$

The variables ψ and ω are governed by coupled second-order equations. We could eliminate the velocity relations 15.3.3 from the problem; however, because of their physical importance it is useful to keep them as an intermediate calculation.

We shall solve the problem as if the flow were time-dependent. Then, once the solution becomes steady, we shall take the steady-state solutions as the answer. The general scheme will be to assume a flow field and calculate values of ω for the next instant in time using 15.3.1 with the right side known. The new ω values are then carried to 15.3.2, where the corresponding ψ values are found. In this way the calculation procedures decouple the equations. Velocities are determined from 15.3.3 to complete the step at this time level. Taking the new flow properties back to 15.3.1 starts the process over again. When the values of ω at the new time are nearly the same as those at the last time, we stop the process and consider the steady-state flow solved. It is of interest to note that the initial conditions are not necessarily a realistic flow pattern.

Since ψ and ω are the primary variables, we need to specify boundary and initial conditions on these variables. The far stream is uniform [$u(x \to -\infty, y) = 1$] and without vorticity; hence

$$\begin{aligned} \omega(x \to -\infty, y) &= 0 \\ \psi(x \to -\infty, y) &= \int \frac{\partial \psi}{\partial y} dy + F(x) = \int_0^y u\, dy + F(x) \end{aligned} \tag{15.3.4}$$

But, since $u(x \to -\infty, y)$ is unity, we have

$$\psi(x \to -\infty, y) = y \tag{15.3.5}$$

where the arbitrary function $F(x)$ has been chosen so that the centerline $y = 0$ is the streamline:

$$\psi(x, y = 0) = 0 \tag{15.3.6}$$

The centerline is also a line of symmetry with $\partial u/\partial y = 0$ and $v = 0$. These conditions require that the vorticity be zero, that is,

$$\omega(x, y = 0) = -\frac{\partial u}{\partial y} + \frac{\partial v}{\partial x} = 0 \tag{15.3.7}$$

The position $y = 1$ consists of a stagnation streamline for $x < 0$ and a solid wall for $x > 0$. Along both sections the stream function is constant and equal to the nondimensional flow rate. Hence, we have

$$\psi(x, y = 1) = \frac{Q}{u_0 h} = 1 \tag{15.3.8}$$

The stagnation streamline is also a line of symmetry; so in a manner similar to 15.3.6 we find that for $x < 0$

$$\omega(x < 0, y = 1) = 0 \tag{15.3.9}$$

Vorticity is generated on the solid wall $x > 0$, but we do not know exactly how much. The most we know is that because $v = 0$ on the wall, 15.3.2 reduces to

$$\omega(x > 0, y = 1) = -\left.\frac{\partial^2 \psi}{\partial y^2}\right|_{y=1} \tag{15.3.10}$$

This is another place where ψ and ω are coupled in the problem. Far downstream the flow becomes a fully developed parabolic profile. Hence, as $x \to \infty$ the velocity is

$$u = \tfrac{3}{2}(1 - y^2), \qquad v = 0$$

The corresponding stream-function equation is

$$\psi(x \to \infty, y) = \tfrac{3}{2}y - \tfrac{1}{2}y^3 \tag{15.3.11}$$

and the vorticity is

$$\omega(x \to \infty, y) = 3y \tag{15.3.12}$$

Equations 15.3.4 through 15.3.10 are the boundary conditions for the steady-flow problem.

The mathematical problem we have laid out above needs to be rewritten in a finite-difference form. Two issues arise in this process. First, does the finite-difference algorithm converge? In many cases, what looks like a reasonable scheme does not converge. The second issue is accuracy: do the answers from the computer give a good approximation of the answer to the continuous problem? The reader may find more information on these points in Roache (1972).

The first step in converting the problem to a form suitable for finite-difference calculation is to define a grid for the domain. An arbitrary point on the grid will be labeled i, j. In the y direction, grid points run uniformly from $j = 1$ to $j = \mathrm{JJ} + 1$ (Fig. 15-6). Thus, the increment in y is

$$\Delta y = \frac{1}{\mathrm{JJ}} \tag{15.3.13}$$

and the y position of the point j is

$$y = (j - 1)\,\Delta y \tag{15.3.14}$$

The x direction presents a slight problem in that the range is $-\infty$ to $+\infty$. It is necessary in finite-difference solutions to approximate an infinite domain with a finite one. We choose L as the length of the plate and define the computation region as $-L \le x \le L$. Boundary conditions will be applied at $\pm L$. We hope that L has been chosen large enough so that it has negligible effect on the answer. The index i will run from $i = 1$ at $x = -L$ to $i = 2\,\mathrm{II} + 2$ at $x = L$. By making sure that i ends in an even number, we force the leading edge to be between the grid points $i = \mathrm{II} + 1$ and $i = \mathrm{II} + 2$. By this trick we avoid the $x = 0$ point where the vorticity is infinite. The increment in x is the length of the field divided by the number of intervals,

$$\Delta x = \frac{2L}{2\,\mathrm{II} + 1} \tag{15.3.15}$$

An arbitrary point is then

$$x = \left(i - \mathrm{II} - \tfrac{3}{2}\right)\Delta x \tag{15.3.16}$$

The mesh aspect ratio is

$$\beta \equiv \frac{\Delta x}{\Delta y} = \frac{\mathrm{JJ}}{\mathrm{II} + \frac{1}{2}} L \tag{15.3.17}$$

By way of summary, observe that the grid is defined by three numbers II, JJ, and L. With these numbers Δx, Δy, and β may be determined.

Next, we take on the task of translating the differential equations into difference equations. Consider an arbitrary point i, j and all points in the immediate neighborhood as shown in Figure 15-6. We denote the value of some arbitrary function f at a given point by subscripts. For any y level, a

Taylor expansion gives f_{i+1} as

$$f_{i+1} = f_i + \left.\frac{\partial f}{\partial x}\right|_i \Delta x + \left.\frac{\partial^2 f}{\partial x^2}\right|_i \frac{\Delta x^2}{2} + \cdots \tag{15.3.18}$$

The value f_{i-1} is given in a similar manner as

$$f_{i-1} = f_i + \left.\frac{\partial f}{\partial x}\right|_i (-\Delta x) + \left.\frac{\partial^2 f}{\partial x^2}\right|_i \frac{(-\Delta x)^2}{2} + \cdots \tag{15.3.19}$$

A *centered finite-difference formula* for the second derivative is found by adding 15.3.18 and 15.3.19 and solving:

$$\left.\frac{\partial^2 f}{\partial x^2}\right|_{ij} = \frac{1}{\Delta x^2}\left[f_{i+1,j} - 2f_{ij} + f_{i-1,j}\right] \tag{15.3.20}$$

Similarly, for the y direction at any x level i we have

$$\left.\frac{\partial^2 f}{\partial y^2}\right|_{ij} = \frac{\beta^2}{\Delta x^2}\left[f_{i,j+1} - 2f_{ij} + f_{i,j-1}\right] \tag{15.3.21}$$

The sum of 15.3.20 and 15.3.21 is the Laplacian in finite-difference form.

One of the equations we want to solve is 15.3.2,

$$0 = \nabla^2\psi + \omega$$

The finite-difference form of this equation at the point i, j is found by substituting 15.3.20 and 15.3.21 and multiplying by Δx^2. The result is

$$\begin{aligned} 0 &= \psi_{i+1,j} + \psi_{i-1,j} + \beta^2\psi_{i,j+1} + \beta^2\psi_{i,j-1} - 2(\beta^2 + 1)\psi_{ij} \\ &\quad + \omega_{ij}\Delta x^2 \\ 0 &= D(\psi; \omega) \end{aligned} \tag{15.3.22}$$

where $D(\psi; \omega)$ is defined as the right-hand side of the first line. In the decoupled problem we assume that we know the values of ω_{ij} and that all the ψ values are to be found.

One of many methods of solving 15.3.22 for the values of ψ is by iteration. Let ψ^n be the last known value and ψ^{n+1} the next estimate. An iteration formula (where F is some number) is constructed as

$$\psi^{n+1} = \psi^n + \frac{F}{2(\beta + 1)} D(\psi^n; \omega) \tag{15.3.23}$$

When $D(\psi; \omega)$ is zero, 15.3.22 is satisfied and no change in ψ should occur. Equation 15.3.23 says that ψ should be changed in accordance with how far away we are from satisfying $D(\psi; \omega) = 0$. This method is known as *successive overrelaxation* (SOR). F is called the relaxation parameter. If $1 \leq F \leq 2$, then the method is convergent. Theory also shows that convergence can be optimized on a rectangular domain if one chooses

$$F = \frac{2}{\xi}\left(1 - \sqrt{1 - \xi}\right)$$

where

$$\xi = \frac{1}{(\beta^2 + 1)^2}\left[\cos\frac{\pi}{L_x/\Delta x} + \beta^2 \cos\frac{\pi}{L_y/\Delta y}\right]$$

Of course an iteration method never satisfies 15.3.22 exactly. In application we must establish a convergence criterion. For a chosen error E_ψ the iteration is stopped when

$$D(\psi, \omega) < E_\psi \qquad \text{for all } i, j \qquad 15.3.24$$

Equation 15.3.24 must be satisfied at all points in the domain.

There is one last trick in applying the iteration formula 15.3.23. Consider for a moment how the calculation for ψ^{n+1} would proceed. Begin with the line $i = 1$ in Fig. 15-6. This is the far stream, where boundary data $\psi = y$ are specified; so we move immediately to $i = 2$. In order, we compute 15.3.23 for $j = 2$ to $j = \mathrm{JJ}$. Next we go to $i = 3$ and again sweep across the slot in j. For the sake of argument, say we are computing for the point $i = 3$, $j = 4$. We already know ψ for all $i < 3$ and for $i = 3$ if $j < 4$. Some of these points are included in $D(\psi; \omega)$ for the computation of $\psi_{3,4}^{n+1}$. We can use the updated values of ψ whenever they are available by redefining the operator $D(\psi; \omega)$ as

$$\begin{aligned} D^*(\psi, \omega) \equiv \{&\psi_{i+1,j}^n + \psi_{i-1,j}^{n+1} + \beta^2\psi_{i,j+1}^n \\ &+ \beta^2\psi_{i,j-1}^{n+1} - 2(\beta^2 + 1)\psi_{ij}^n + \omega_{ij}^n \Delta x^2\} \end{aligned} \qquad 15.3.25$$

Using D^* in the iteration formula 15.3.23 speeds the convergence somewhat and also allows us to use only one storage array for ψ. We do not need to have storage for ψ^n and for ψ^{n+1}, but only for current values of ψ, that is, ψ^{n+1} or ψ^n as the case may be.

The vorticity equation 15.3.1 is expressed in finite-difference form using similar arguments. First we multiply by Δt to get

$$\frac{\partial \omega}{\partial t}\Delta t = \Delta t\left\{-\frac{\partial}{\partial x}(u\omega) - \frac{\partial}{\partial y}(v\omega) + \frac{2}{\mathrm{Re}}\nabla^2\omega\right\} \qquad 15.3.26$$

All terms on the right-hand side are considered known at the last time step n. We can explicitly compute a new set of ω_{ij}^{n+1} for the new time $n+1$ using the formula

$$\frac{\partial \omega}{\partial t}\Delta t = \omega_{ij}^{n+1} - \omega_{ij}^{n} \qquad 15.3.27$$

For the Laplacian term on the right-hand side of 15.3.26 we use 15.3.20 and 15.3.21:

$$\frac{2\,\Delta t}{\mathrm{Re}}\nabla^2\omega = \frac{2\,\Delta t}{\mathrm{Re}\,\Delta x^2}\Big\{\omega_{i+1,j}^{n} + \omega_{i-1,j}^{n} + \beta^2\omega_{i,j+1}^{n} + \beta^2\omega_{i,j-1}^{n} - 2(\beta^2+1)\omega_{ij}^{n}\Big\} \qquad 15.3.28$$

All of the ω's in this expression are old values at time level n.

Special care is needed for the convective terms in 15.3.26. We have at least three choices of ways to estimate the first-derivative forms needed in the convective terms. If we truncate 15.3.18 and solve for $\partial f/\partial x|_i$, we get a *forward-difference formula*,

$$\left.\frac{\partial f}{\partial x}\right|_i = \frac{1}{\Delta x}(f_{i+1} - f_i) \qquad 15.3.29$$

If we truncate 15.3.19 and solve, we get a *backward-difference formula*,

$$\left.\frac{\partial f}{\partial x}\right|_i = \frac{1}{\Delta x}(f_i - f_{i-1}) \qquad 15.3.30$$

In forming both of the equations above we have neglected a term of order Δx^2. The *centered-difference formula* is found by subtracting 15.3.19 from 15.3.18 and solving. This leads to an expression that is accurate through order Δx^2:

$$\left.\frac{\partial f}{\partial x}\right|_i = \frac{1}{2\,\Delta x}(f_{i+1} - f_{i-1})$$

Experience has shown that there is not one best formula, but that we should change the difference formula in accordance with the direction of the flow. We want to carry information into the point ij from points upstream of ij (streamlines are subcharacteristics of the equations). The *upwind-differencing formulas* are

$$\Delta t\,\frac{\partial}{\partial x}(u\omega) = \frac{\Delta t}{\Delta x}\times\begin{Bmatrix}(u\omega)_{ij}^{n} - (u\omega)_{i-1,j}^{n} & \text{if} \quad u>0\\ (u\omega)_{i+1,j}^{n} - (u\omega)_{ij}^{n} & \text{if} \quad u<0\end{Bmatrix} \qquad 15.3.31$$

$$\Delta t\,\frac{\partial}{\partial y}(v\omega) = \frac{\beta\,\Delta t}{\Delta x}\times\begin{Bmatrix}(v\omega)_{ij}^{n} - (v\omega)_{i,j-1}^{n} & \text{if} \quad v>0\\ (v\omega)_{i,j+1}^{n} - (v\omega)_{ij}^{n} & \text{if} \quad v<0\end{Bmatrix} \qquad 15.3.32$$

Using upwind differencing gives an unconditionally stable computation scheme for the vorticity equation.

Another expression that the student will run across in the literature on finite differences is *artificial viscosity*. This is a numerical effect that occurs because we have truncated the second-order terms in formulating 15.3.31 and 15.3.32. This error causes an effect that is equivalent to modifying the viscosity coefficient of the second-order diffusion terms; hence the name.

Another noteworthy point is that there are two ways to write the convective terms:

$$\frac{\partial}{\partial x}(u\omega) + \frac{\partial}{\partial y}(v\omega) \quad \text{or} \quad u\frac{\partial \omega}{\partial x} + v\frac{\partial \omega}{\partial y}$$

When written as finite-difference equations these forms are not equivalent, because the approximating equations throw away slightly different parts. Experience has shown that in general the version on the left-hand side, the so called *conservative form*, is to be preferred.

The final vorticity equation is a relation for ω_{ij}^{n+1} obtained by substituting 15.3.27, 15.3.28, 15.3.31, and 15.3.32 into 15.3.26. Only known values of ω^n, u^n, and v^n occur on the right-hand side. After each new vorticity value is computed, it is compared with the old value to see if a steady state has been reached. The calculation is stopped once all values change less than a specified amount E_ω:

$$\max_{ij} \left| \omega_{ij}^{n+1} - \omega_{ij}^{n} \right| < E_\omega \qquad 15.3.33$$

We now have, in principle, the methods for finding ψ and ω for the problem.

The velocities must be calculated as an intermediate step after the ψ_{ij}'s are determined. At all interior points the centered-difference formula for 15.3.3 gives

$$u_{ij} = \frac{\psi_{i,j+1} - \psi_{i,j-1}}{2\,\Delta x/\beta}, \qquad v_{ij} = -\frac{\psi_{i+1,j} - \psi_{i-1,j}}{2\,\Delta x} \qquad 15.3.34$$

These formulas cannot be used along the boundaries, because one ψ point would be outside the computation region. Along the centerline, the forward-difference version is used:

$$u_{i1} = \frac{\psi_{i,2} - 0}{\Delta x/\beta}, \qquad v_{i1} = 0 \qquad 15.3.35$$

And along the stagnation streamline, the backward-difference formula is applied; that is, for $i = 2, \text{II} + 1$,

$$u_{i,\text{JJ}+1} = \frac{1 - \psi_{i,\text{JJ}}}{\Delta x/\beta}, \qquad v_{i,\text{JJ}+1} = 0 \qquad 15.3.36$$

The values of u on the inflow boundary, the outflow boundary, and the wall are specified conditions and do not change during the calculation.

You should not expect the calculation to produce complete consistency between ψ, ω, u, and v. In different parts of the problem different types and levels of approximation are used. For example, if a backward-difference formula similar to 15.3.36 is used on the wall, a value $u \neq 0$ will result. On the other hand, the centered-difference formula could be applied at the wall if we were to place a phantom point inside the wall. At the phantom point the value of ψ would be defined as $\psi_{i,\mathrm{JJ}+2} = \psi_{i,\mathrm{JJ}}$, so 15.3.34 would produce $u = 0$, the proper value.

Values of ψ are fixed on all boundaries, and values of ω are fixed on all boundaries except the wall. At the wall ψ and ω are related by 15.3.10, an equation for which we need a finite-difference equivalent. Consider the following Taylor expansion of ψ from the wall ($j = \mathrm{JJ} + 1$) to the first interior point $j = \mathrm{JJ}$:

$$\psi_{i,\mathrm{JJ}} = \psi_{i,\mathrm{JJ}+1} + \left.\frac{\partial \psi}{\partial y}\right|_{i,\mathrm{JJ}+1}(-\Delta y) + \left.\frac{\partial^2 \psi}{\partial y^2}\right|_{i,\mathrm{JJ}+1}\frac{(-\Delta y)^2}{2} + \cdots$$

Substituting $\psi_{\mathrm{wall}} = 1$, $u_{\mathrm{wall}} = \partial\psi/\partial y = 0$, $\omega_{\mathrm{wall}} = -\partial^2\psi/\partial y^2$ and solving gives

$$\omega_{i,\mathrm{JJ}+1} = (1 - \psi_{i,\mathrm{JJ}})\frac{2\beta^2}{\Delta x^2} \qquad \text{for} \quad j = \mathrm{II}+2,\ldots,2\mathrm{II}+1 \qquad 15.3.37$$

This relation determines ω at the wall for known values of ψ.

A flow chart of the computer program is given in Fig. 15-7. After the customary statements to dimension the variables and set the format, the input parameters are specified. Next, the known boundary conditions on ψ, ω, u and v are fixed and the initial conditions for the interior points are specified. The program in Appendix F employed initial conditions $u = 1$, $v = 0$, $\omega = 0$, and $\psi = y$ on all interior points, while the initial wall vorticity was taken as $\omega = 3$. The main calculation consists of two loops. The outside loop of the program solves the vorticity equation 15.3.26 at all interior points. These vorticity values are used as known quantities in the inner loop for the stream-function equation 15.3.23. This equation needs only to be supplied with ω on interior points. The inner iteration loop on ψ is complete once 15.3.24 has been satisfied. Next, velocities at all interior points, the centerline, and the stagnation streamline are found by using 15.3.34, 15.3.35, and 15.3.36. The final calculation uses 15.3.37 to evaluate the wall vorticity. This completes one pass through the program. The resulting flow pattern is taken back to the top of the loop, where we begin again by calculating the interior vorticity at the next time step. Once successive vorticity values are within the tolerance set by 15.3.33, the calculation is halted and the last values sent to the output.

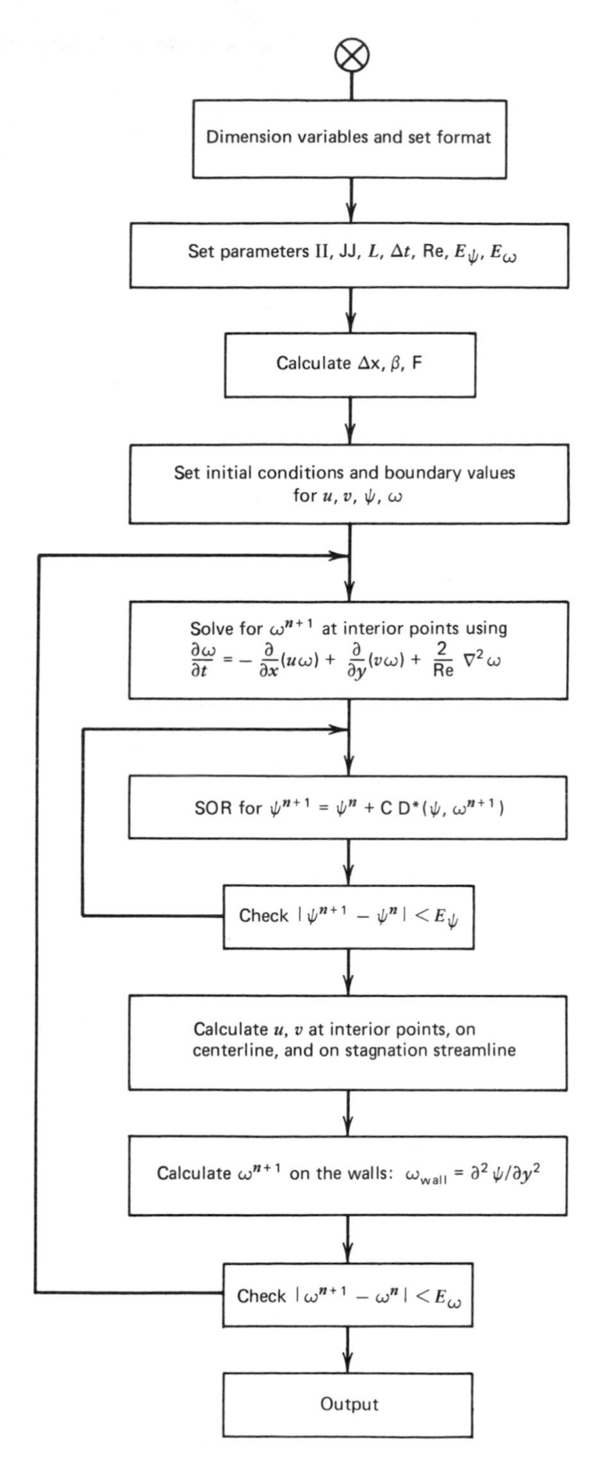

Figure 15-7 Flow chart for ψ–ω computer solution. A sample program is given in Appendix F.

The stability and accuracy of the solution are determined by the choices we make about the mesh size and the time step. The diffusion of vorticity is not accurately described if the mesh size becomes too large. We need several mesh points to accurately resolve any steep velocity gradients. A useful rule of thumb is that the Reynolds number of grid cells should be somewhat less than 10. This translates into a mesh spacing

$$\Delta y < \frac{10}{\text{Re}}, \qquad \text{Re} = \frac{u_0(2h)}{\nu} \tag{15.3.38}$$

In this equation Δy is nondimensionalized by $2h$, the same length as used in Re. For small Reynolds numbers, meshes much finer that 15.3.38 are used for better resolution.

Stability considerations will fix the largest value of the time step that you can choose. It has already been noted that the equation for ψ is convergent for any mesh size as long as $1 \leq F \leq 2$. The vorticity equation offers a more stringent stability criterion. Roache (1972) gives the maximum time step for stability as

$$\Delta t < \frac{1}{\dfrac{|u|}{\Delta x} + \dfrac{|v|}{\Delta y} + \dfrac{4}{\text{Re}}\left(\dfrac{1}{\Delta x^2} + \dfrac{1}{\Delta y^2}\right)} \tag{15.3.39}$$

Note that neglecting the terms containing u and v gives a larger estimate of Δt. As Ames (1977) points out, this time step is only a guide: some problems may require larger or smaller steps than 15.3.39 would indicate. This seems to be especially true at the higher Re, where smaller steps are needed.

There is a side issue that needs a little further discussion. Recall that in the far stream, the specified boundary conditions were $\psi = y$ and $\omega = 0$. The first of these conditions is derived from the fact that $u = 1$. The point of caution is that the second condition $\omega = 0$ does not also imply that $v = 0$ in the far stream. The vorticity definition

$$\omega = -\frac{\partial u}{\partial y} + \frac{\partial v}{\partial x}$$

shows that at the upstream position we have

$$0 = \frac{\partial v}{\partial x}$$

This does not also require that $v = 0$. The solution we obtain using $u = 1$, $\omega = 0$ will in general have a nonzero value of v. In formulating the problem we imagine that the far stream is approached smoothly with $u = 1$, $v = 0$, and all space derivatives equal to zero, so that $\omega = 0$. This could be the case, but mathematically we can only specify two things. The solution determines everything else.

15.4 ENTRANCE FLOW INTO A CASCADE OF PLATES: PRESSURE SOLUTION

The ψ–ω method of solution does not give any information about the pressure field. A separate calculation for the pressure is needed. There are two major ways this can be approached. The first method is to do a numerical quadrature of the relation

$$dp = \frac{\partial p}{\partial x} dx + \frac{\partial p}{\partial y} dy \qquad 15.4.1$$

where the derivatives are given by the momentum equations,

$$\frac{1}{\rho}\frac{\partial p}{\partial x} = -u\frac{\partial u}{\partial x} - v\frac{\partial u}{\partial y} + \nu\nabla^2 u \qquad 15.4.2$$

$$\frac{1}{\rho}\frac{\partial p}{\partial y} = -u\frac{\partial v}{\partial x} - v\frac{\partial v}{\partial y} + \nu\nabla^2 v \qquad 15.4.3$$

Since u and v are known at every point, the derivatives on the right-hand sides of 15.4.2 and 15.4.3 can be expressed in finite-difference formulas. In an alternate procedure, the right-hand sides can be formulated entirely in terms of the stream function, which is also known.

In many instances we are only seeking a drag force and therefore only need the pressure on the surface of the body. The quadrature of 15.4.1 is an effective method in these problems.

In the next section, the pressure on the upper and lower flow boundaries will be given. These pressures were found by quadrature of 15.4.1 in a simplified form. On these streamlines $dy = 0$, so 15.4.1 becomes

$$dp = \frac{\partial p}{\partial x} dx = \rho\left[-\frac{\partial}{\partial x}\left(\tfrac{1}{2}u^2\right) + \nu\nabla^2 u\right] dx$$

By noting that $\nabla^2 u = -\partial\omega/\partial y$, this equation reduces to

$$\frac{1}{\rho}(p - p_\infty) = -\tfrac{1}{2}\left(u^2 - u_\infty^2\right) - \nu\int_\infty^x \frac{\partial \omega}{\partial y} dx \qquad 15.4.4$$

For these special streamlines the viscous effect is all that needs to be integrated numerically (special care must be taken at the leading edge, where $\partial\omega/\partial y$ is infinite).

The second method of finding the pressure is to find a numerical solution to the differential equation for the pressure field. The equation is

$$-\frac{1}{\rho}\nabla^2 p = \left(\frac{\partial u}{\partial x}\right)^2 + 2\left(\frac{\partial u}{\partial y}\frac{\partial v}{\partial x}\right) + \left(\frac{\partial v}{\partial y}\right)^2 \qquad 15.4.5$$

With a known velocity field, 15.4.5 is a Poisson equation and can be solved by exactly the same method used to solve the stream-function equation 15.3.2. The boundary conditions on the pressure are more complicated, however.

In most instances we do not know the pressure on the complete boundary (unless we solve 15.4.1), so the problem is one of mixed boundary conditions. Along any solid wall the pressure gradient is given by 13.7.8. When this relation is evaluated in the normal direction, we have

$$\frac{\partial p}{\partial n} = \mu \frac{\partial \omega}{\partial s}$$

where n is normal and s is along the wall. On all flow boundaries we know the value of either the pressure or the pressure gradient.

15.5 ENTRANCE FLOW INTO A CASCADE OF PLATES: RESULTS

Figure 15-8 shows the streamline patterns in the entrance for several values of the Reynolds number. These results were obtained from the computer program in Appendix F. At any Reynolds number, streamlines that are equally spaced

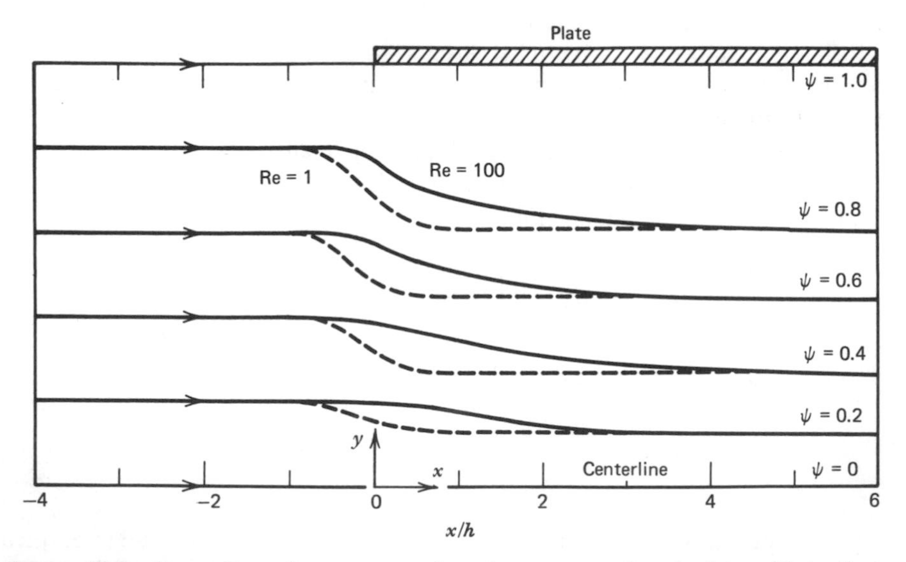

Figure 15-8 Streamlines for entrance flow into a cascade of plates. Note that x distances are a factor of 1 : 5 shorter than y distances.

far upstream become pinched toward the center as the fully developed profile is reached far downstream. The final location of the streamlines is always the same. The pinching process itself is the only thing that changes as the Reynolds number varies. At low Re, the pinching begins outside the plates and is completed a short distance into the channel. As Re increases, this pattern shifts downstream and extends. As a matter of fact, as Re becomes large, the length it takes to establish the final profile grows in direct proportion to Re.

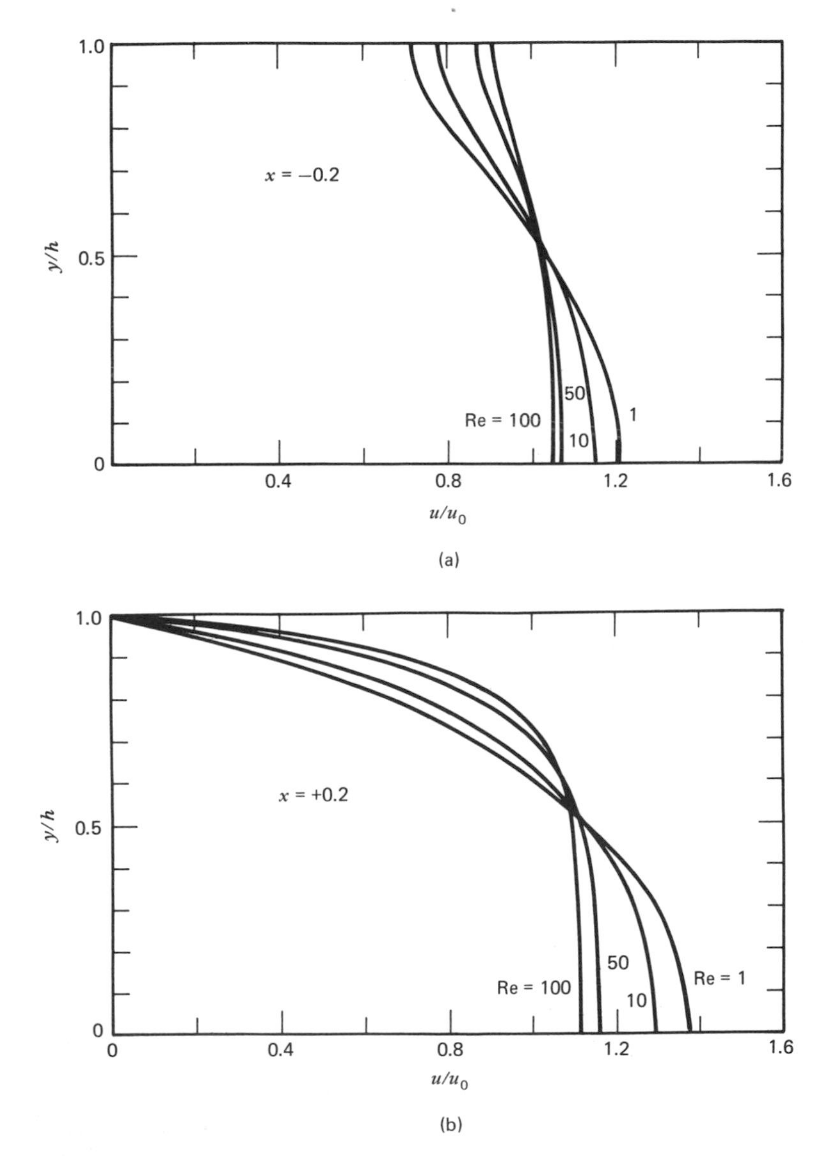

Figure 15-9 Velocity profiles for entrance flow: (a) $x = -0.2$, (b) $x = +0.2$.

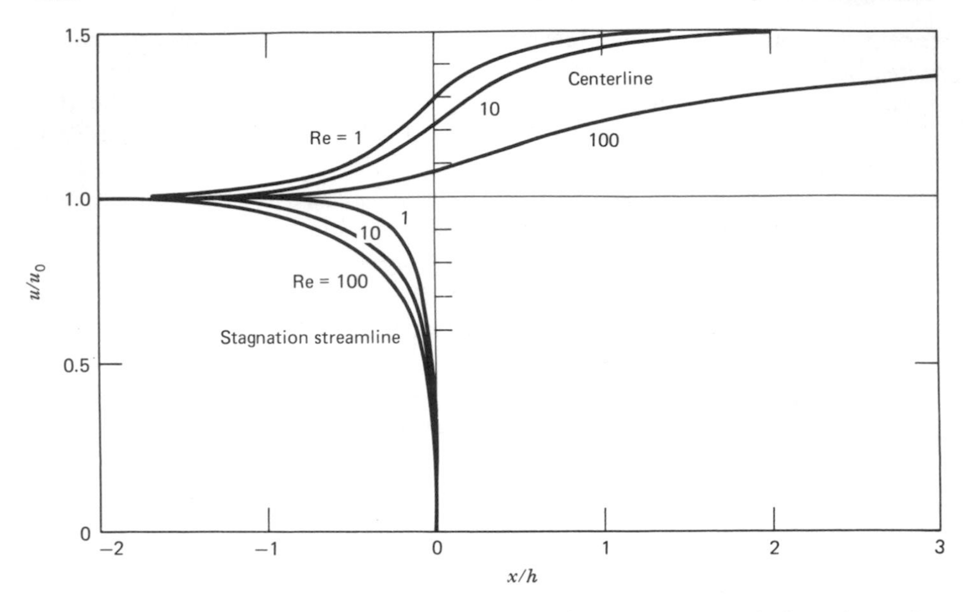

Figure 15-10 Variation of the velocity along the centerline and that along the stagnation streamline for the entrance flow problem.

Velocity profiles at several x stations are given in Fig. 15-9. From the profiles in front of the plate it is evident that the upstream influence is much stronger at the lower Reynolds numbers. In fact, when Re = 100 there is practically no upstream influence; the u velocity is still very close to one even at $x = -0.2$. Profiles at the downstream stations confirm that the fully developed parabolic profile is established very rapidly for low Re, while a much longer entrance length is required at high Re. A different perspective of these same trends is given in Fig. 15–10, where the velocity on the centerline is plotted as a function of x. This figure also gives the velocity on the stagnation streamline. The extent to which viscous diffusion can progress against the oncoming stream is seen to be greater as the Reynolds number becomes lower.

Figure 15-11 displays the pressure coefficient along the boundaries of the flow. On the centerline, the pressure always decreases. At first the decrease is an inviscid effect; the pressure gradient is needed to accelerate the core flow. The pressure continues to drop beyond the entrance length in order to balance the viscous forces retarding the flow. The final steady-state pressure gradient is $-12/\text{Re}$.

The pressure on the stagnation streamlines increases as we approach the leading edge because the flow is slowing down. In a purely inviscid process the pressure force would slow the flow to zero velocity and the pressure coefficient would be unity. Then, the stagnation pressure, as given by 15.4.4 with $\nu = 0$

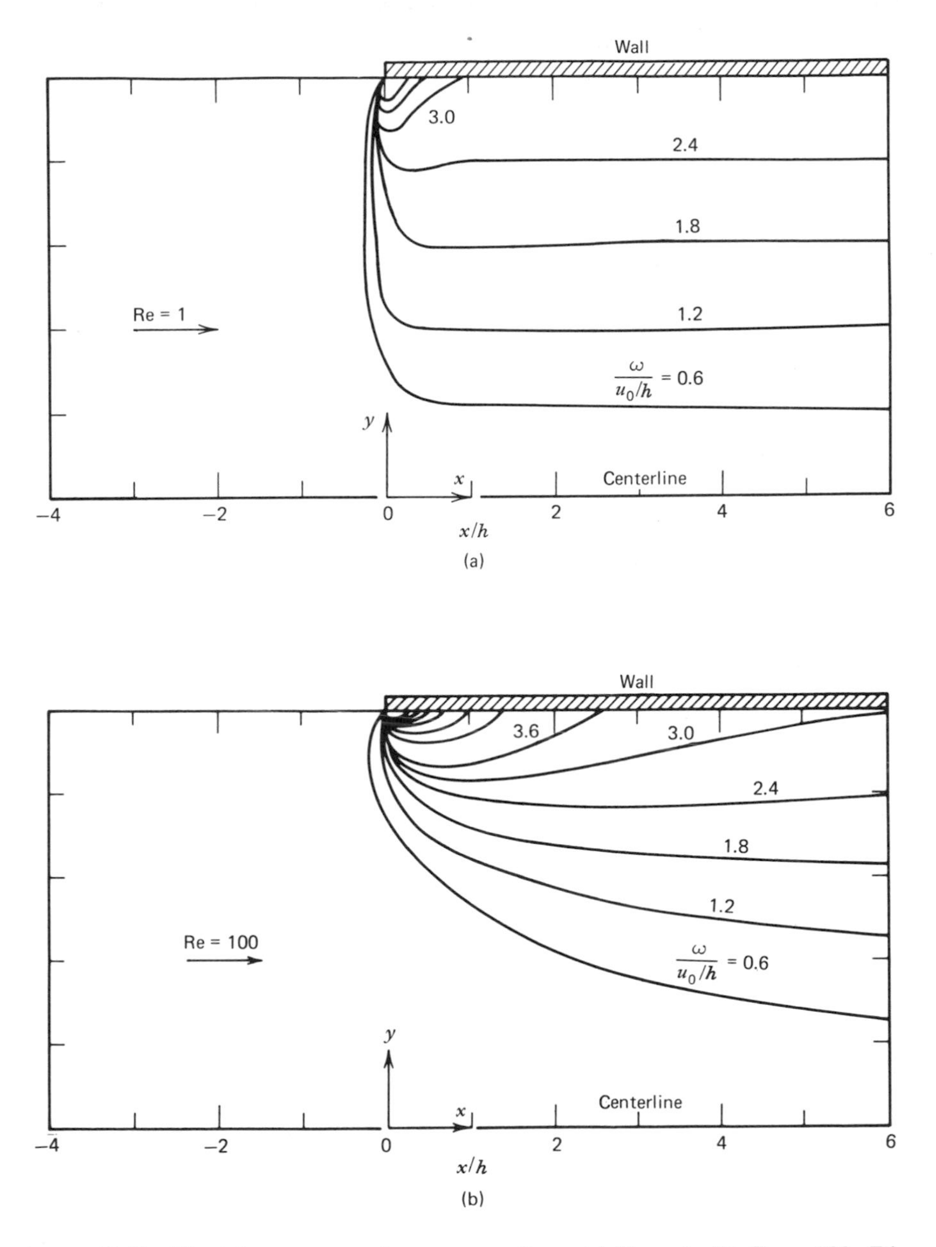

Figure 15-11 Vorticity contours for entrance flow: (a) Re = 1, (b) Re = 100. Distortion is $x : y$; 1 : 5.

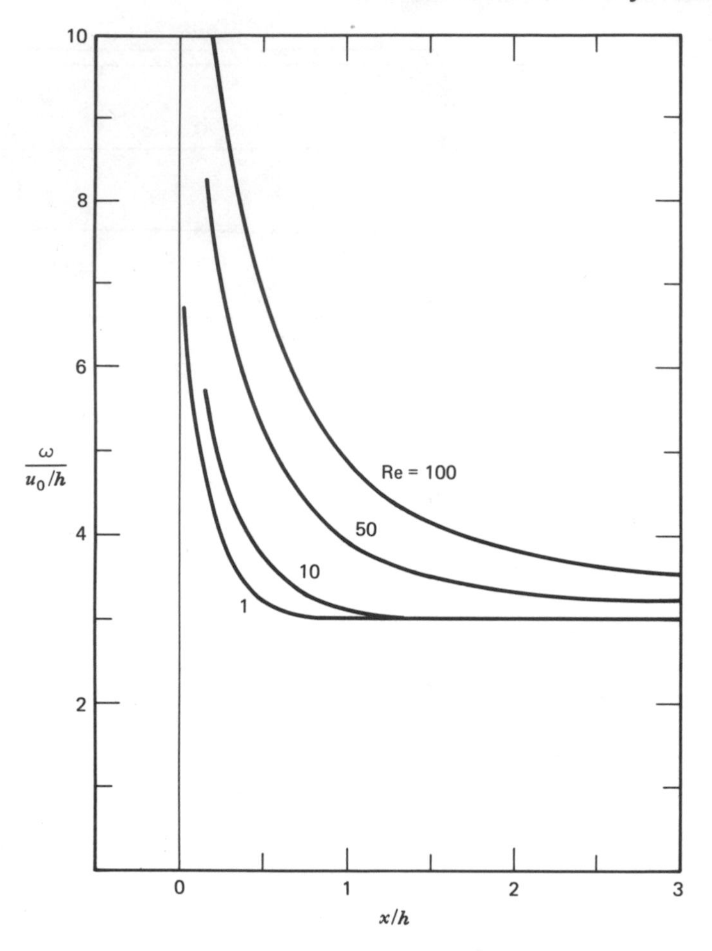

Figure 15-12 Vorticity at the wall for entrance flow.

and $u = 0$, would simply be the static pressure plus the dynamic pressure $\frac{1}{2}\rho u_0^2$. This is essentially what occurs at the high Reynolds numbers. At the other extreme, the lower-Re flows show an effect that is contrary to intuition. The viscous forces accelerate the fluid particles on the stagnation streamline. The pressure gradient in this case must retard the particles against the accelerating effect of viscosity. The net result is that the final pressure at the stagnation point is much higher than that for inviscid flow.

Contour plots of the vorticity are shown in Fig. 15-12. The initial vorticity in the far stream is zero. By symmetry, the vorticity on the centerline and the stagnation streamline are also zero. The plate, especially near the leading edge, acts as a source of vorticity. Recall that the vorticity is singular at a sharp leading edge. Approaching the leading edge along the stagnation streamline shows the vorticity is zero, while approaching the leading edge along the wall leads to an infinite vorticity (Fig. 15-13). Mathematically the leading edge is a vorticity doublet putting out positive vorticity on one side and negative vorticity on the other.

Recall that generation of vorticity on a solid wall is related to the pressure gradient along the wall. The flux of vorticity out of the wall (in dimensional

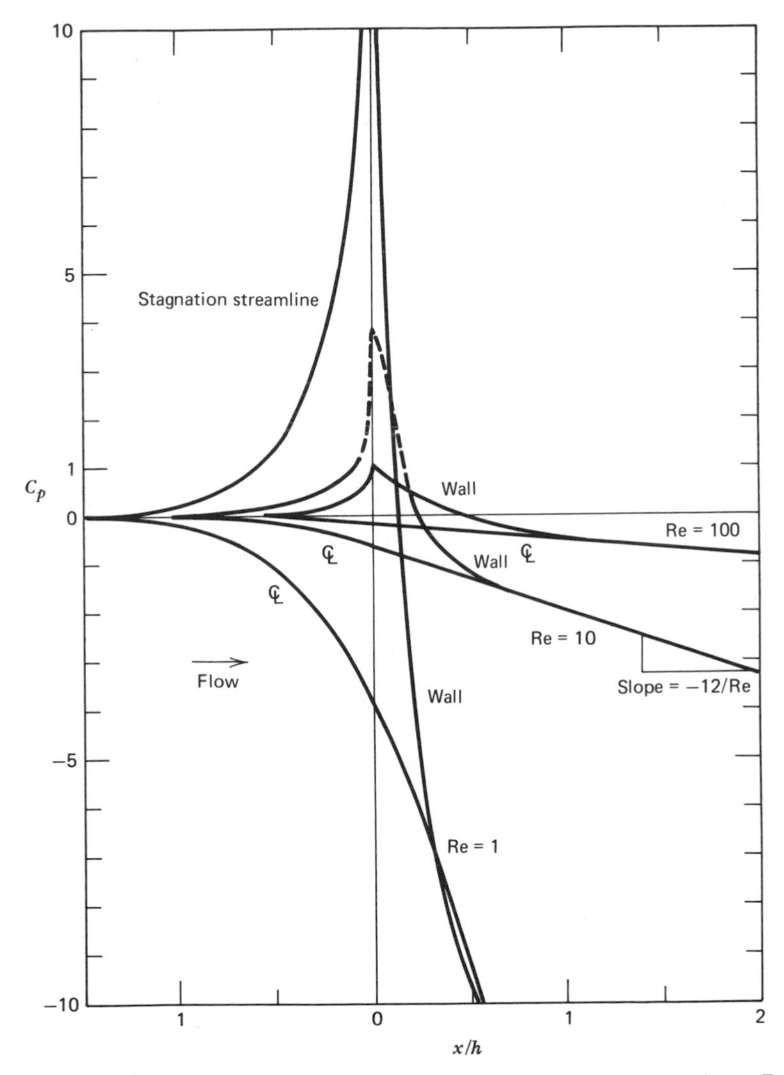

Figure 15-13 Pressure on the centerline and the wall for entrance flow. Pressure coefficient is $C_p \equiv (p - p_\infty)/\frac{1}{2}\rho u_0^2$.

variables) is

$$\left.\frac{\partial \omega}{\partial y}\right|_{\text{wall}} = \left.\frac{1}{\mu}\frac{\partial p}{\partial x}\right|_{\text{wall}}$$

Far downstream, the pressure gradient becomes constant and the flux of vorticity from the wall also becomes constant. The vorticity flux out of the upper wall is absorbed at the lower wall (see Section 13.6).

Within the fluid the value of the vorticity is determined by a balance of diffusion and convection. This being a plane flow, there is no vortex-line stretching. When the fully developed flow is established, the vorticity equation shows that both effects, the net convection term and the net diffusion term, are identically zero. Within the entrance region these two effects compete to distribute the vorticity that is generated at the walls. As the Reynolds number increases, the vorticity generated at the walls becomes higher and tends to stay closer to the wall for a longer distance. In the entrance region the downstream convection is faster than the cross-stream diffusion.

15.6 FLOW AROUND A CIRCULAR CYLINDER

A circular cylinder mounted perpendicular to a steady stream is a flow situation of fundamental interest. It gives us a chance to observe the flow patterns that develop around a smooth body with a finite thickness.

In an actual test we might mount a cylinder across the test section of a wind tunnel. The walls of the test section constrain the flow and make the streamlines conform to the wall at a finite distance away from the cylinder. The first question is: Does the presence of the walls modify the flow in a substantial way? Can we make the walls so far away that the flow around the cylinder can be isolated and the walls thought of as infinitely far? The answer depends upon the Reynolds number. If the Reynolds number is zero, the answer is no; the walls will always influence the flow when they are a finite distance away. However, for any Reynolds number $\text{Re} > 0$, it is thought that the flow may be isolated and the presence of the walls ignored. The flow disturbance caused by the cylinder does not extend to infinity either upstream or on the sides, but is confined to a wake, which trails off downstream and grows slowly. The wake is a region where the velocity is still lower than the free-stream value u_0. Viscous stresses cause the wake to spread out, and they also accelerate the fluid to bring the velocity back up toward u_0.

The net force of the fluid upon the cylinder is a drag force aligned with the flow direction. There is an important relation between the drag on the cylinder and a property of the wake called the momentum thickness. We shall find this relationship by an integral analysis. Consider the fixed control region shown in Fig. 15-14. It is rectangular, one unit in depth, and far enough away from the cylinder so that the pressure at both ends is atmospheric. A hole in the central part of the control region surrounds the cylinder. The upstream flow into the region is uniform at a value u_0, and at the wake on the downstream boundary the flow has a profile $u = u_w = u_w(y)$. Because the volume flow through the downstream boundary is less than that entering through the upstream boundary, there must also be an outflow across the sides of the control region. The velocity on the sides is $u = u_0$, v unknown. We denote the mass flow across both sides as $\dot{m}$, and begin the analysis with the integral continuity equation for a steady flow, 5.13.2:

$$\int_{\text{FR}} \rho v_i n_i \, dS = 0 \qquad 15.6.1$$

When the assumptions above are introduced into 15.6.1, we have

$$-\rho u_0 l + \dot{m} + \rho \int_{-l/2}^{+l/2} u_w \, dy = 0 \qquad 15.6.2$$

We shall need this relation for $\dot{m}$ in the next step.

The x-momentum equation for a steady flow through a fixed region is 5.14.1,

$$\int_{\text{FR}} [\rho n_i v_i u + n_x p - n_i \tau_{ix}] \, dS = 0 \qquad 15.6.3$$

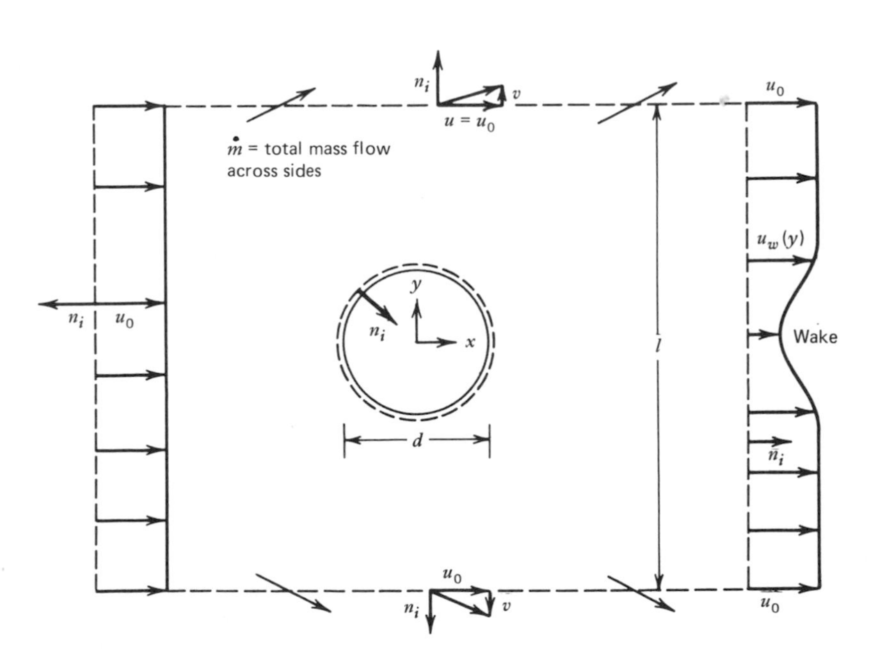

Figure 15-14 Drag analysis for flow around a two-dimensional object.

In applying 15.6.3 the viscous forces on the flow boundaries (τ_{xx} on the upstream and downstream boundaries and τ_{yx} on the sides) are assumed to be zero. The pressure and viscous forces on the surface of the cylinder are by definition the drag force; that is,

$$F_D \equiv \int_{\text{cyl}} [n_x p - n_i \tau_{ix}] \, dS \qquad 15.6.4$$

These facts, together with previous assumptions, allow 15.6.3 to reduce to

$$-\rho u_0^2 l + \dot{m} u_0 + \rho \int_{-l/2}^{+l/2} u_w^2 \, dy + F_D = 0$$

Substituting from 15.6.2 for $\dot{m}$ and rearranging produces

$$F_D = \rho u_0^2 d \int_{-l/2d}^{+l/2d} \left[\frac{u_w}{u_0} - \left(\frac{u_w}{u_0} \right)^2 \right] d\left(\frac{y}{d} \right) \qquad 15.6.5$$

Here we have introduced the diameter of the cylinder as a characteristic length. Note that the integrand in 15.6.5 goes to zero as y becomes large. Hence, we let $l \to \infty$ and define the *momentum thickness* of the wake as

$$\theta = d \int_{-\infty}^{\infty} \left[\frac{u_w}{u_0} - \left(\frac{u_w}{u_0} \right)^2 \right] d\left(\frac{y}{d} \right) \qquad 15.6.6$$

The momentum thickness is an integral property of the wake profile. The drag is found from 15.6.5 as

$$F_D = \rho u_0^2 \theta \qquad 15.6.7$$

The customary way to nondimensionalize the drag is by dividing by the kinetic energy of the flow, $\frac{1}{2}\rho u_0^2$, and the cross-sectional area of the body, $1 \times d$. Equation 15.6.7 expressed as a drag coefficient is

$$C_D = \frac{F_D}{\frac{1}{2}\rho u_0^2 d} = 2\frac{\theta}{d} \qquad 15.6.8$$

The drag coefficient as a function of Reynolds number is given as Fig. 15-15.

We summarize the picture at this stage as follows. The effect of a cylinder on the flow far away is concentrated in the wake. The velocity in the wake, u_w, gradually approaches the free-stream velocity; however, it must do so in such a way that there is always a deficit in momentum. The momentum deficit is measured by the momentum thickness, which is constant as the wake decays. The drag force on the cylinder is directly proportional to θ.

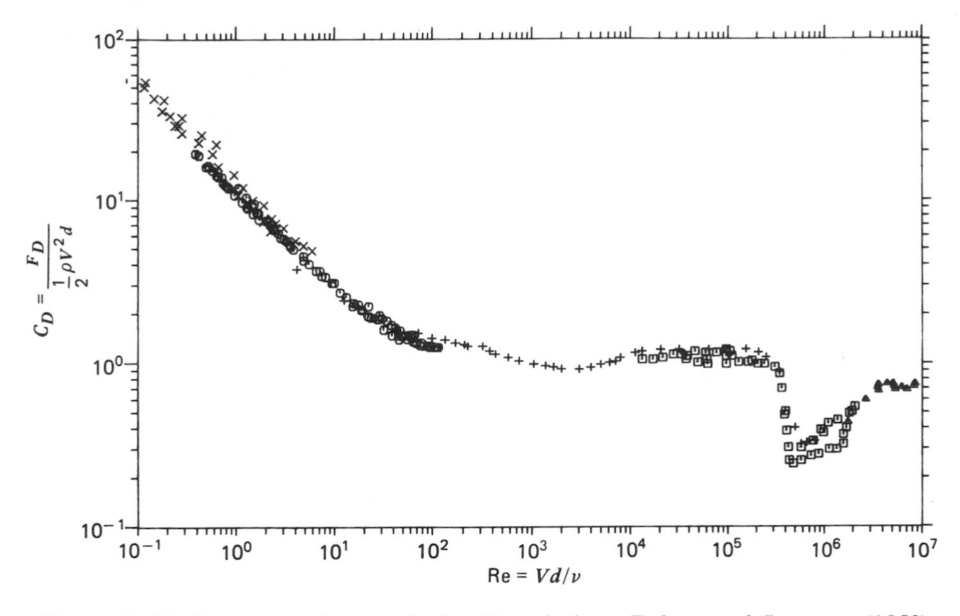

Figure 15-15 Drag curve for a cylinder. Data is from Delany and Sorenson (1953), Finn (1953), Roshko (1961), Tritton (1959), and Wieselsberger (1921).

In the case where the wake becomes turbulent—the most common case in practice—the arguments above are still valid if the velocity u_w is taken as a time-averaged value. The only other effect is that the decay is dominated by turbulent stresses instead of viscous stresses.

Next, we move closer to the cylinder and examine the rich variety of flow patterns that have been observed. Several of these patterns are shown schematically in Fig. 15-16 and in flow visualization experiments of subsequent Figures.

At low Reynolds numbers (Fig. 15-16a and Fig. 15-17a), the flow tends to divide and reunite smoothly approaching a symmetric pattern fore and aft as the Reynolds number decreases. The drag is quite high, as shown in Fig. 15-15. A change occurs in the flow patterns at about Re = 4. The flow separates on the downstream side, and two steady standing eddies are formed (Fig. 15-16c and Fig. 15-17b). These eddies are stable and remain attached to the body.

When the Reynolds number is about 40 the next flow pattern develops. The wake behind the cylinder becomes unstable. Oscillations in the wake grow in amplitude and finally roll up into discrete vortices with a very regular spacing. This trail of vortices in the wake is known as the *Karman vortex street*. The vortices travel downstream at a speed slightly less than u_0. They are not turbulent, and the flow near the cylinder remains steady with two attached eddies. If we placed a velocity-measuring instrument in the wake, it would show a regular oscillation with one cycle corresponding to the distance between

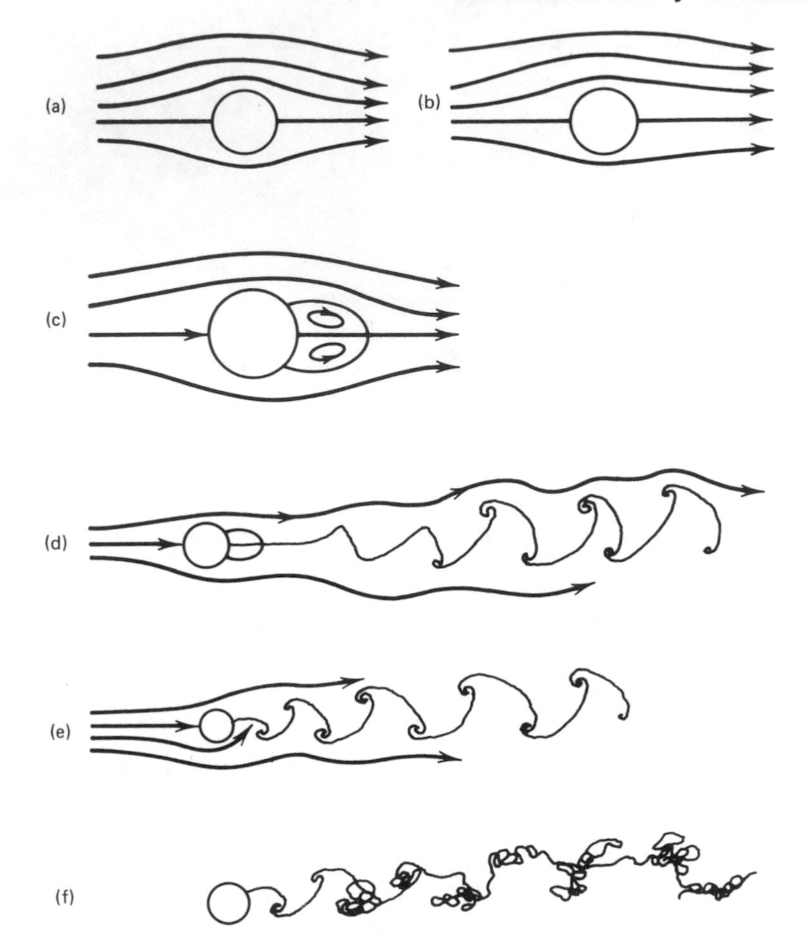

Figure 15-16 Flow regimes for a cylinder: (a) Re = 0, symmetrical; (b) 0 < Re < 4; (c) 4 < Re < 40, attached vortices; (d) 40 < Re < 60–100, Kármán vortex street; (e) 60–100 < Re < 200, alternate shedding; (f) 200 < Re < 400, vortices unstable to spanwise bending;

two vortices of the same sign. The frequency of this oscillation, f, when nondimensionalized by the diameter and the free-stream speed, is called the *Strouhal number*. It is defined as

$$\mathrm{St} \equiv \frac{fd}{u_0}$$

The Strouhal number varies slightly with Re but is roughly 0.2 over a wide range in Re.

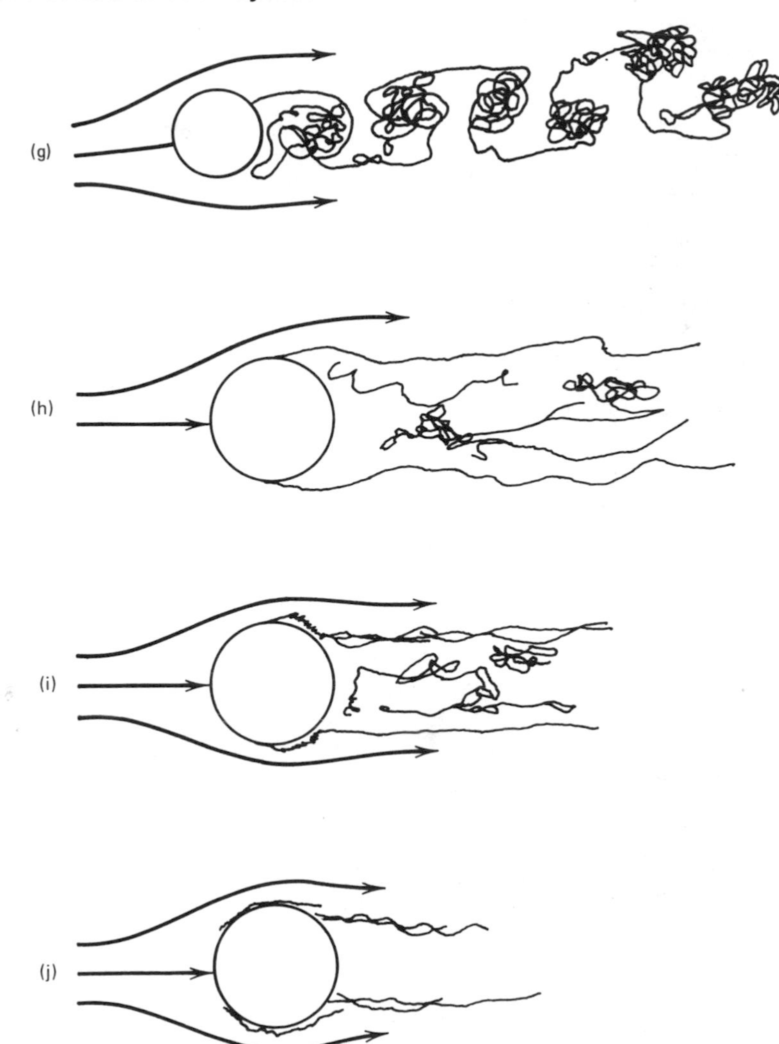

Figure 15-16 Flow regimes for a cylinder: (continued) (g) $400 < \text{Re}$, vortices turbulent at birth; (h) $\text{Re} < 3 \times 10^5$, laminar boundary layer separates at 80°; (i) $3 \times 10^5 < \text{Re} < 3 \times 10^6$, separated region becomes turbulent, reattaches, and separates again at 120°; (j) $3 \times 10^6 < \text{Re}$, turbulent boundary layer begins on front and separates on back.

As the Reynolds number increases, the vortex street forms closer to the cylinder, until finally the attached eddies themselves begin to oscillate. Ultimately the attached eddies give way to eddies that alternately form and then shed. Depending upon the details of the experiment, this first occurs at a Reynolds number somewhere between 60 and 100. Figure 15-18a shows the vortex street development. As one goes further downstream the circular motion

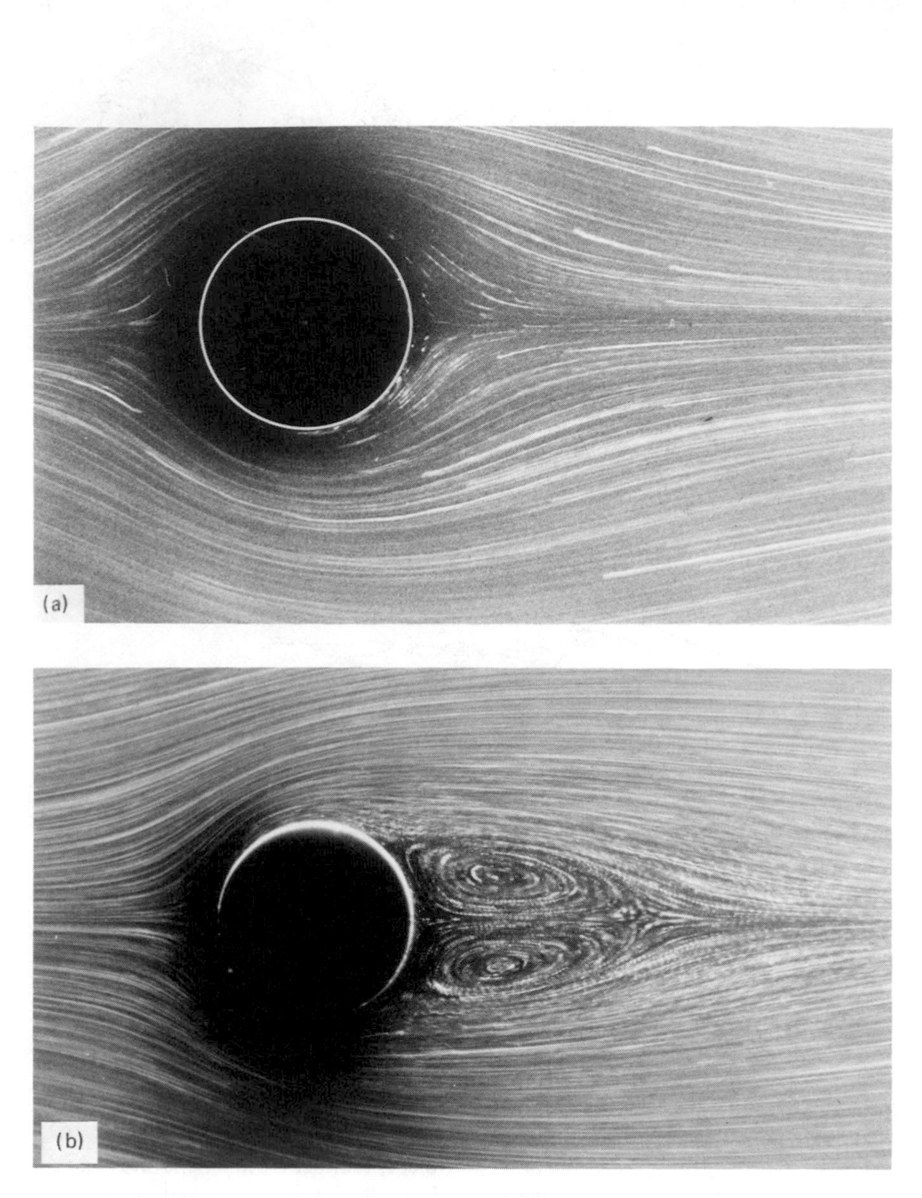

Figure 15-17 Flow over the circular cylinder is from left to right: (a) slight asymmetry at Re = 1.54, but flow is still attached; (b) standing vortices at Re = 26. From S. Taneda (1979).

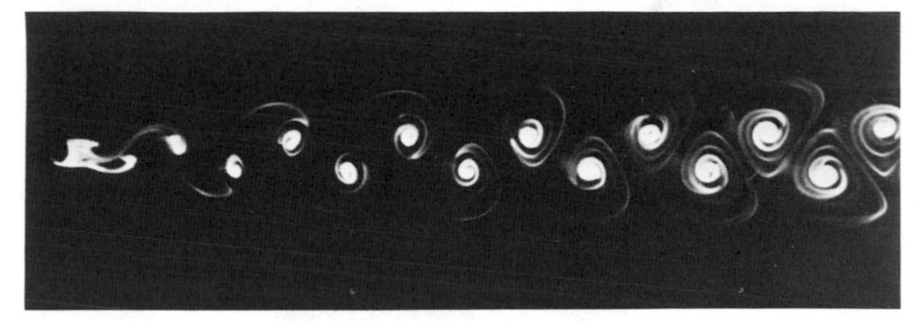

Figure 15-18a Development of the Kármán vortex street at Re = 105. Photograph courtesy of S. Taneda, Kyushu University, Japan.

of the vorticies is stopped by viscous forces. In an experiment such as Fig. 15-18a it is difficult to see when this happens as the flow visualization marker retains its distinctive pattern even after the vorticies have stopped. The effect of introducing a smoke marker at different positions in the wake is illustrated in Fig. 15-18b. When the smoke wire, which produces several streaklines, is located at $x/d = 150$ or 200, there is no trace of the vortex street. However, at the same location in the experiments where the smoke is introduced at earlier stations, a pattern is still seen. These patterns are fossils of events that occur where the smoke was introduced. Cimbala, Nagib, and Roshko have not only shown the vortex street decay, but they have also vividly demonstrated how our eyes can be deceived by flow visualization patterns.

We now have an unsteady flow near the cylinder, and the drag force oscillates with the formation of each eddy. In addition, the top-to-bottom asymmetry of the flow gives rise to an oscillating lift force. As the flow forms a clockwise eddy, it rushes past the top of the cylinder somewhat faster than the flow across the bottom. This causes the pressure on the top to be less, resulting in a lift force toward the top of the page. When the clockwise eddy breaks away, the opposite pattern develops on the bottom and the lift force reverses its direction. The shedding process is very regular and coherent in the spanwise direction.

Oscillations in the lift and drag forces on bluff bodies sometimes take on great importance. Figure 15-19 shows an oil platform with spiral strakes attached to the legs (shown in the jacked-up position). If the bending frequency of the legs is nearly the same as the shedding frequency of the vortices, the oscillating force may, over the course of several cycles, build up to destructive magnitude. The purpose of the spiral strakes is to break up the spanwise coherence of the vortices by forcing them to tear away at different times along the length of the leg. Then different parts of the leg are in different phases of the force oscillation, and a destructive motion is avoided. Similar phenomena

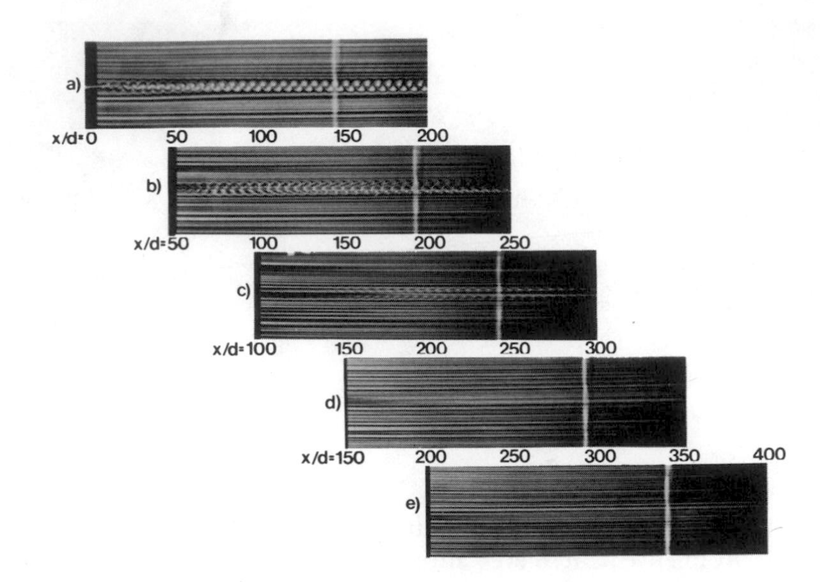

Figure 15-18b Wake of a circular cylinder at Re = 93. A smoke wire shows different patterns when inserted at different locations. Research described in Cimbala, Nagib, and Roshko (1981).

can occur on transmission lines, heat-exchanger tubes, and even suspension bridges. You may have seen the widely distributed movie of the collapse of the Tacoma Narrows bridge (the last of about 50 such major bridge accidents). Here, a torsional mode of oscillation in the bridge structure synchronized with the vortex shedding to destroy the bridge.

At a Reynolds number of 200 the vortex street becomes unstable to bends in the axial (spanwise) direction. As one goes further downstream these bends grow and the wake ultimately becomes turbulent. In the range $200 < \mathrm{Re} < 400$ the Strouhal number loses its regular, well-defined character. Somewhere in the neighborhood of Re = 400 the vortices themselves become turbulent. The turbulence within the vortices gives them a different velocity profile and restores the spanwise coherence. This restabilizes the Strouhal number, and it returns to its value 0.2.

Over the higher Reynolds-number range (except for $3 \times 10^5 < \mathrm{Re} < 3 \times 10^6$) the vortex shedding becomes somewhat irregular and in flow-visualization experiments it is difficult to see vortex shedding. Nevertheless, a time history of the velocity at any point has a large spectral component at the Strouhal frequency of 0.2.

The time-averaged drag coefficient for the cylinder (Fig. 15-15) drops to a value of about one at Re ~ 100–200 and then remains relatively constant with

Figure 15-19 A "jack-up rig" for offshire drilling. Note spiral strakes on the legs at the top of the picture. Photograph taken by Bethlehem Steel, and supplied by O. Griffin Naval Research Lab.

increasing Re. This indicates that pressure forces dominate the drag; viscous forces are negligible. From here on we are in the high Re range. Viscous forces and vorticity are confined near the surface of the cylinder in a boundary-layer region. The abrupt drop in the drag coefficient at $\text{Re} = 3 \times 10^5$ is due to another change in the flow pattern. Although this is a high-Re phenomena, we discuss it in a qualitative way for the sake of completeness.

Below $\text{Re} = 3 \times 10^5$, the boundary layer on the cylinder is laminar and separates on the front half of the cylinder (80°) with a shallow angle as depicted in Fig. 15-16. The pressures in the separated region on the downstream side are nearly constant, but much lower than the free-stream pressure.

This causes the high drag. The *critical Reynolds number* 3×10^5 marks the point where the laminar boundary itself becomes unstable just after it separates. In a very short distance the shear layer becomes turbulent and then reattaches to the cylinder. The actual thickness of the boundary layer, the separation bubble, and the reattachment zone is greatly exaggerated in Fig. 15-16 so that we may see it. The turbulent boundary layer itself separates from the cylinder at about the 120° position. The net result is that the area of the large separation region has decreased and the pressure in this region has almost come back to the free-stream value. Accordingly, a dramatic drop in the drag (over 70%) is realized.

With a slight further increase in Re to about 3×10^6, the drag increases again. As far as experiments have gone, this is the final flow pattern. The boundary layer now becomes turbulent on the front half of the cylinder while it is still attached. Separation of the turbulent layer occurs a little earlier than before, and the pressure is somewhat lower. As a result, the drag is moderately increased.

Boundary-layer separation and transition to turbulence are sensitive to many things. If the surface is rough or the free stream contains a little turbulence, the critical Reynolds number will change slightly. Even the presence or absence of sound can change the critical Reynolds number.

Figure 15-20 Subcritical flow over a sphere is shown at Re = 15,000. Laminar separation occurs forward of the equator. Photograph courtesy of H. Werlé, ONERA, Catillon, France.

The pattern of flows described above is common for smooth, bluff bodies; only the values of the transition Reynolds number changes. Elliptical cylinders, spheres, ellipsoids, and so on all show similar behavior. Figures 15-20 and 15-21 show flow-visualization experiments on a sphere at subcritical and supercritical Reynolds numbers.

Many moderate-Reynolds-number calculations of the flow patterns have been done. Almost all of them use a ψ–ω numerical method, more or less like the one we studied in the entrance-flow problem. Thom (1933) did the first published calculation at Re = 10 and 20. Many, but not all, of the flow patterns discussed above have been reproduced by the numerical solutions. Figures 15-22 and 15-25 show streamline patterns from calculation by Fornberg (1980). At Re = 2, some fore–aft asymmetry of the flow can already be detected. Higher-Re solutions show the attached eddies and delineate their growth together with the forward movement of the separation point as Re increases. This particular calculation method has forced symmetry. Instabilities in the wake and vortex shedding are prevented by computing only half of the flow. Nevertheless, the results are of theoretical interest as a solution, albeit an

Figure 15-21 Supercritical flow at Re = 30,000. Normally this flow is subcritical, but a small trip wire has induced transition to a turbulent boundary layer. Separation is now downstream of the equator, and the wake is smaller. Photograph from ONERA by H. Werlé (1980).

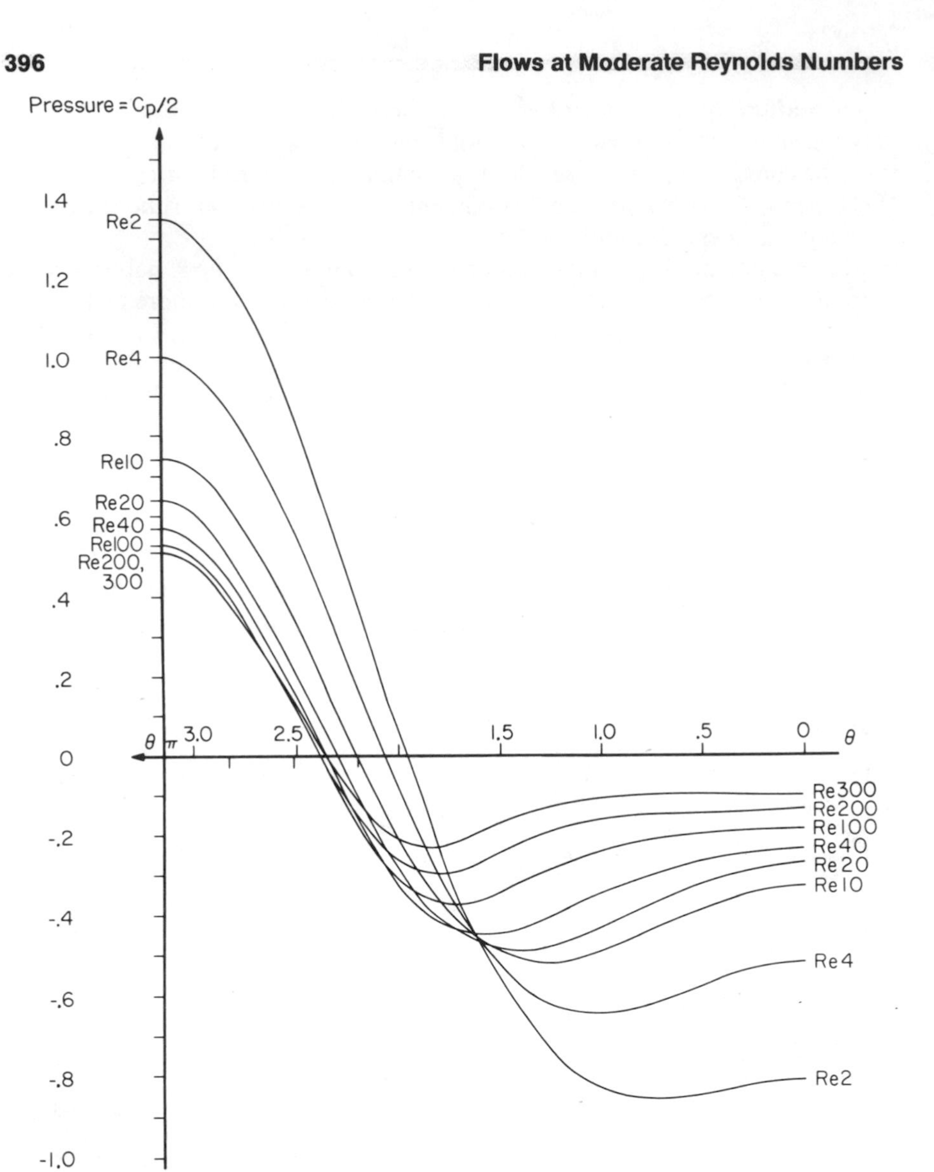

Figure 15-22 Pressure distribution over a cylinder. Pressure axis is $(p - p_\infty)/(\rho u_\infty^2)$. From Fornberg (1980).

unstable solution, of the Navier–Stokes equations. The surface pressures from these calculations are given in Fig. 15-23, and the vorticity results are given in Figs. 15-24 and 15-25. When Re = 2, vorticity diffuses some distance away from the cylinder with only a slight asymmetry caused by convection. As the Reynolds number increases the intensity of the vorticity increases—an indication of sharper velocity gradients near the cylinder. A most pronounced effect

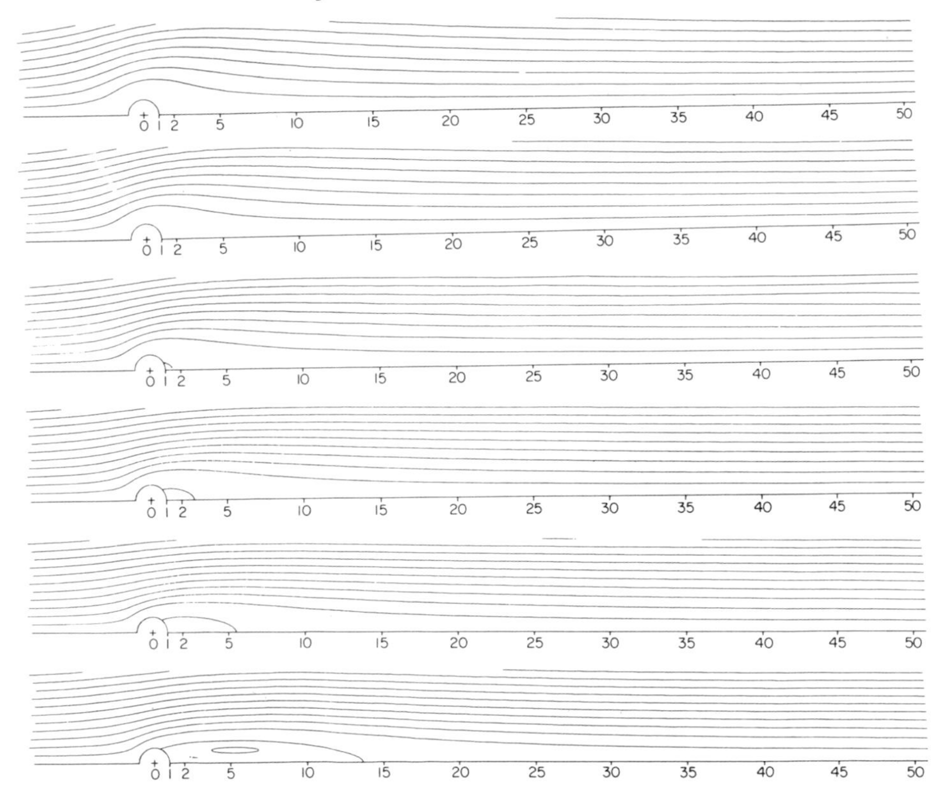

Figure 15-23 Streamlines for flow over a cylinder. Reynolds numbers from top to bottom are 2, 4, 10, 20, 40, and 100. From Fornberg (1980).

is that the vorticity is concentrated near the front and sides of the cylinder and swept downstream into the wake. These patterns show that convection is becoming more important than diffusion as the Reynolds number increases.

Numerical calculations for this problem have been carried out at some very high Reynolds numbers. One interesting calculation is by Thoman and Szewczyk (1968). They made time-dependent calculations without forcing symmetry into the solution. Shortly after the flow was started, the cylinder was given a short rotation to introduce an asymmetry into the vorticity. Subsequently the flow developed into the alternating vortex-shedding mode, which continued as long as the calculation was run. Results at Re = 40,000 are shown in Fig. 15-26. At this Re the real flow would have turbulent vortices and small flow structures that the computations, because of the coarse mesh size, could not hope to resolve. In spite of this, and in spite of questions regarding the finite computation region and artificial-viscosity errors (which the authors themselves cautioned about), the calculations have the proper overall qualitative behavior. Not only is the vortex shedding exhibited, but the Strouhal

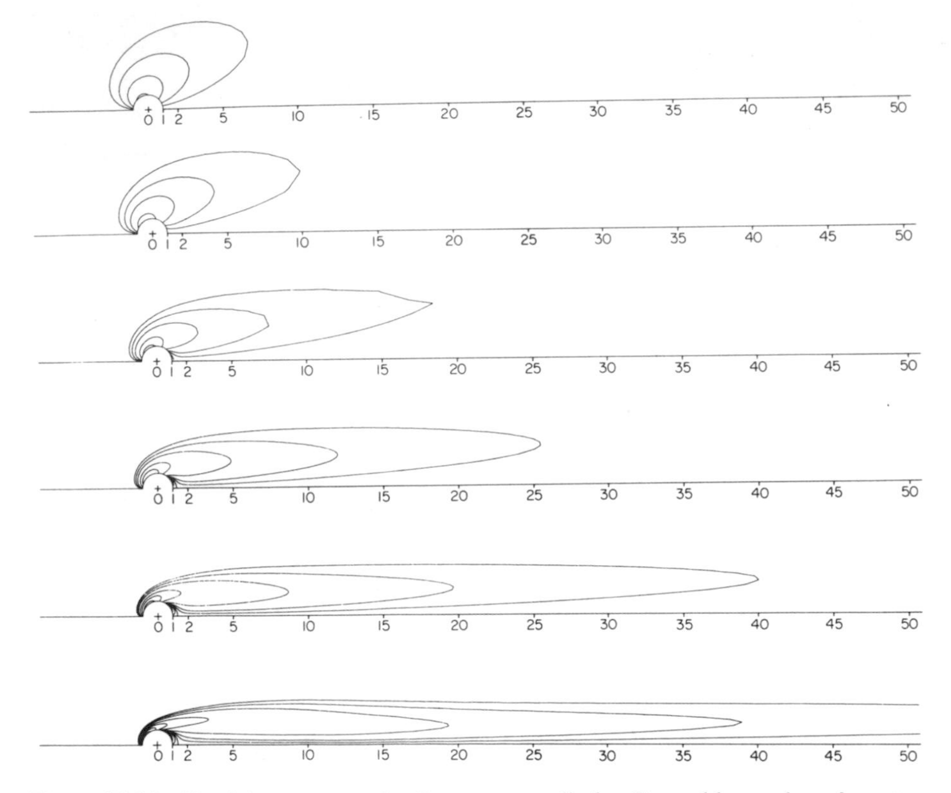

Figure 15-24 Vorticity contours for flow over a cylinder. Reynolds numbers from top to bottom are 2, 4, 10, 40, and 100. From Fornberg (1980).

number has nearly the experimental value. These calculations demonstrate that many characteristics of vortex shedding are not sensitive to viscosity. It is a phenomena dominated by inviscid mechanisms. This conclusion is in agreement with experiments, since they show that the Strouhal number is not a function of the Reynolds number.

Performing truly accurate numerical calculations becomes more difficult as the Reynolds number is increased. The difficulties can be grouped into three classes: adequate resolution of small, sharp changes in the solution, adequate size of the computation region for unbounded problems, and adequate convergence rate for the numerical iterations. If the flow has turbulent regions, there is the added problem of a proper turbulence model in the computations.

The first difficulty demands that the computation mesh be refined in those areas where the gradients are severe; this is especially difficult if the locations are unknown at the outset. Putting more mesh points into a calculation can soon tax the computer storage, especially in a three-dimensional problem. The computer storage capacity also bears on the second difficulty, adequate size of the computation region for unbounded domains. Transforming the unbounded

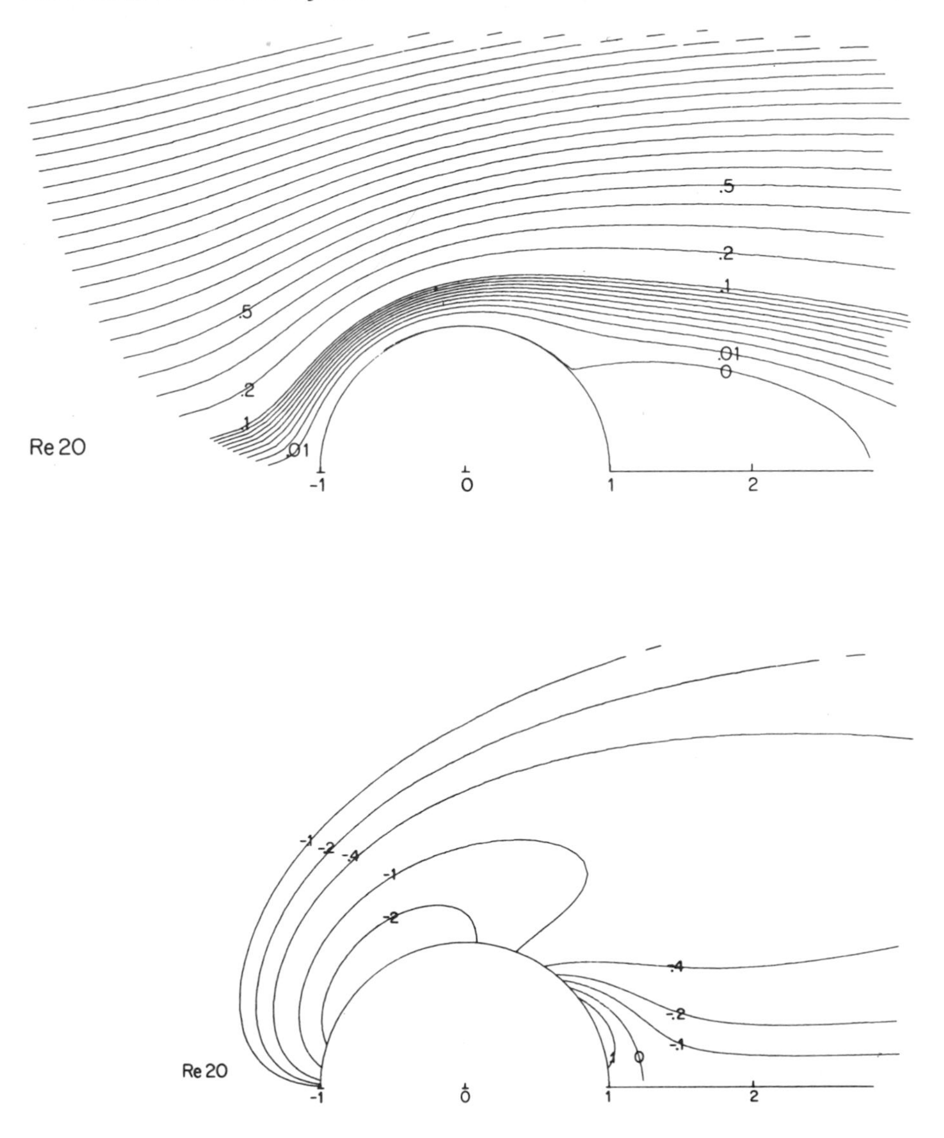

Figure 15-25 Details of streamlines and vorticity contours at Re = 20. From Fornberg (1980).

domain into a finite domain does not really solve this problem, but it helps. Although gross qualitative behavior may be found with small domains, very large domains are needed to get even three-place accuracy. Fornberg (1980) used a domain with a distance of 300 cylinder diameters to the outer boundary. The last difficulty is that the rate of convergence of the numerical schemes tends to deteriorate as the Reynolds number becomes large. The exact manner

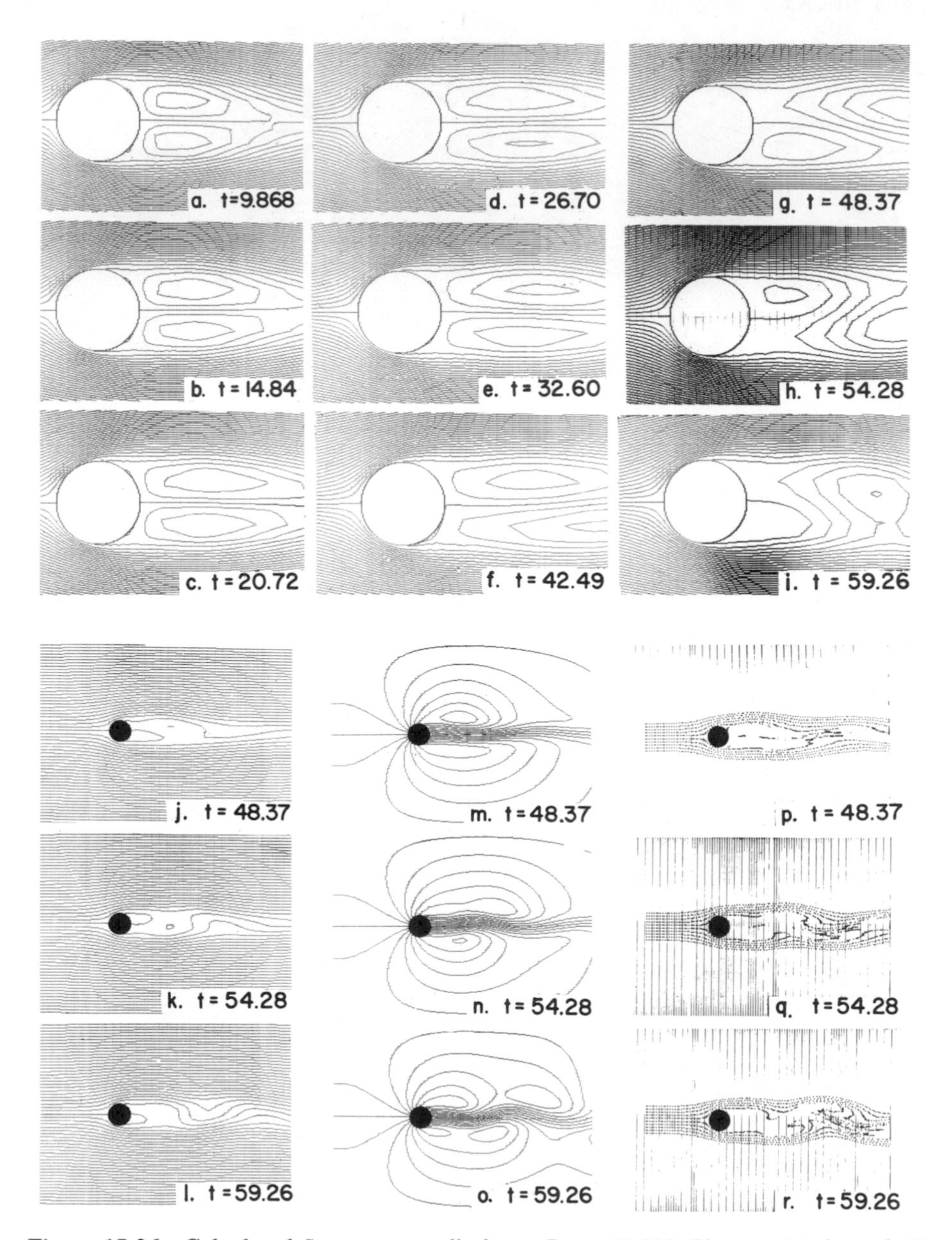

Figure 15-26 Calculated flow over a cylinder at Re = 40,000. Pictures (a) through (i) are close-up views of streamlines as vortex shedding begins. Pictures (j), (k), and (l) show streamlines at long distances; (m), (n), and (o), streamlines viewed by an observer moving with the freestream fluid; (p), (q), (r), the particle paths. From Thoman and Szewczyk (1968).

in which the convective terms are linearized and the equations decoupled when they are transcribed into a numerical form determines the convergence rate for the iteration process. For example, at Reynolds numbers approaching 300, Fornberg's calculation shows that the attached vortices are becoming slightly shorter. This unexpected result is a numerical problem. In subsequent calculations with 64-bit accuracy, Fornberg found the vortex length increased linearly with Re. The moral is that very accurate calculations at high Re need extreme care. Even large scale features such as vortex length can be incorrectly given by apparently valid calculations.

15.7 CONCLUSION

Moderate-Reynolds-number flows are characterized by the fact that inertia, pressure (and/or gravity), and viscous forces are all important. In some cases this competition produces complicated and intricate flow patterns.

Computers are very good at predicting and confirming flow patterns at moderate Re. When all the terms in the governing equations are about the same size, the error made in the numerical processes will be smaller than all the terms. As the Reynolds number takes on either extreme value, zero or infinity, some terms in the governing equations become very small. Under these conditions the errors in the large terms and the true size of small terms are comparable. As a result, the small-term effect is lost. You might raise the question: does this really make any difference? Won't the small effect average itself out and be unimportant anyway? Sometimes this is true, but in fluid mechanics it is frequently not true. A small term may accumulate to give a large effect. For example, a very small force applied to a satellite over the course of a year makes a drastic change in the orbit. The problem of doing valid numerical computations at extreme Reynolds numbers, especially the practical case of high Re, is a current research challenge in fluid mechanics.

PROBLEMS

15.1 (B) The velocities for ideal flow around a cylinder are given by 18.6.3. Compute the strain rate tensor and find the principal axes. Consider a line vortex of negligible strength that is perpendicular to the freestream and to the cylinder axis. The vortex is carried into the cylinder and wraps around it. Where would it suffer the most intense stretching?

15.2 (A) Run the entrance flow program of Appendix F for Re = 0.01 and 35.

15.3 (A) The "driven cavity" problem consists of a rectangular box with the lid sliding at a prescribed velocity U_0. Modify the program of Appendix F to solve this problem for a box twice as deep as it is wide. Run the program for several moderate Reynolds numbers.

15.4 (B) Do the driven cavity problem for the case where both the bottom and top lids move in the same direction at U_0. Do not assume a symmetric flow pattern to start the problem.

15.5 (B) A parabolic flow exists at the entrance of a slot of width h and length $10h$. Half way through the slot a flat plate divides the flow so that the upper channel is $2h/3$ and the lower is $h/3$. At the exit of these channels the upper section has 3/4th of the inlet flow. Solve for the pattern at Re = 1 and 10.

15.6 (C) Change the conditions of Problem 15.5 so that the exit pressures of the two channels are the same. What is the division of flow between the passages?

15.7 (B) A rectangular cavity $h \times 2h$ has an opening $0.2h$ on the short side next to the corner and an opening of $0.3h$ on the opposite side at the opposite corner. Flow comes in the first opening at a constant speed U_0 and is withdrawn from the $0.3h$ opening with a constant velocity. Solve for the flow pattern in the cavity when Re = 0.5.

15.8 (C) A slot $2h$ high by $6h$ long is open on the left end and closed on the right end. A splitter plate divides the channel as it extends along the centerline from the left to a distance $2h$ from the closed end. The flow enters the top channel with a uniform parabolic velocity profile, turns at the closed end, and exits going back along the lower channel. Assume an exit profile and solve for the flow when Re = 2.

15.9 (C) Construct a scheme for a finite difference approximation of the equation $\omega = \nabla^2\psi$ at a 90° corner that is concave. Repeat for a convex corner.

15.10 (C) Solve for the flow in a channel of width h that has a 90° right angle. Run the program at Re = 0.1, 1, and 10.

15.11 (A) Consider a slot where the walls are $y = \pm h + \pm A \sin(2\pi x/L)$. Find transformation $X(x, y)$, $Y(x, y)$ that will make the wavy slot into a rectangle in the X, Y domain.

15.12 (B) Transform the equation $\nabla^2\psi = \omega$ into the X, Y variables of 15.11.

15.13 (B) Transform the vorticity equation into the X, Y variables of 15.11.

15.14 (B) A two-dimensional body with characteristic length L is in a stream of velocity U_0. Far downstream the wake velocity is

$$u = u_0 - \frac{21\nu}{\sqrt{xd}} \exp\left(- \frac{y^2 u_0}{\nu x} \right)$$

Find the drag force per unit length and the momentum thickness.

16 Asymptotic Analysis Methods

We are frequently interested in the behavior of a flow as a certain parameter or variable approaches a limiting value. For example, consider a jet issuing into a large room. What is the flow like far away from the mouth of the jet? As another example, we might ask what would be the effect on Poiseuille flow if the pipe were slightly flattened instead of round. We have already seen how incompressible flow may be considered as a flow where the Mach number, a parameter, approaches zero. This is an example of an asymptotic theory: a class of flow problems that have some common characteristics as $M \to 0$.

In this chapter we will deal with several specific fluid-flow problems. These problems will be solved using asymptotic methods, also called perturbation methods. In between the sample problems are sections that give some of the formal mathematics relevant to asymptotic methods.

Before we take up the first problem, we note that the variable or parameter that approaches a limit always is (or can be made into) a nondimensional variable. The expression "far away from a jet" must refer to the distance divided by the radius of the jet exit. A "slightly flattened" pipe must mean the wall location is slightly out of round compared to the pipe radius; and, of course, a low Mach number means that velocities are small compared to the speed of sound.

16.1 OSCILLATION OF A GAS BUBBLE IN A LIQUID

One of the early investigations of gas bubbles was in order to explain the musical sounds that occur when bubbles are formed as the result of water flowing over the rocks in small streams, or a jet falling into a pool. The process of forming the bubble also sets it into oscillation. One important question is: what is the natural frequency of a bubble in a liquid? In an engineering context we might be interested in this problem with regard to the flow of a bubbly mixture or with regard to the attenuation of sound by the bubbles in the wake of a ship's propeller. A transient wave in a bubbly mixture will be strongly influenced by bubble oscillations.

In Section 12.4 we derived an equation for the change in the radius of a cavity $R(t)$ while the density of the surrounding fluid was ρ and the steady pressure at infinity was p_0. The equation is

$$R\ddot{R} + \tfrac{2}{3}\dot{R}^2 - \frac{1}{\rho}(p - p_0) = 0 \qquad 16.1.1$$

We take the pressure p at the liquid surface to be the same as that within the cavity. This assumption ignores surface tension, which becomes important in this problem when R is less than 10^{-2} cm. Let the bubble contain a perfect gas in such an amount that if the pressure in the bubble were p_0, the volume would be V_0 and the radius R_0. For isentropic bubble oscillations we have the relationship

$$\frac{p}{p_0} = \left(\frac{V_0}{V}\right)^{\gamma} = \left(\frac{R_0}{R}\right)^{3\gamma} \tag{16.1.2}$$

Substitution of 16.1.2 into 16.1.1 gives an equation that governs $R(t)$:

$$R\ddot{R} + \tfrac{2}{3}\dot{R}^2 - \frac{p_0}{\rho}\left[\left(\frac{R}{R_0}\right)^{-3\gamma} - 1\right] = 0 \tag{16.1.3}$$

Every term of this oscillator equation is nonlinear. In order to set a definite problem, we assume that the bubble radius is initially displaced from the equilibrium position but has zero initial velocity. That is,

$$\begin{aligned} R(0) &= R_0(1 + \epsilon) \\ \dot{R}(0) &= 0 \end{aligned} \tag{16.1.4}$$

Here ϵ is a nondimensional parameter for the initial displacement. The complete problem above is so complicated that a closed form-solution is very difficult (see Plesset and Prosperetti (1979)).

Let us lower our sights a little and seek an answer to the problem for oscillations of small amplitude. Mathematically we want the asymptotic behavior of $R(t)$ as the parameter $\epsilon \to 0$. To start the analysis assume the answer has the form

$$R(t) = R_0 + \epsilon R_1(t) + \epsilon^2 R_2(t) + \cdots \tag{16.1.5}$$

Without any initial displacement, the solution is a constant $R = R_0$. The functions $R_1(t)$ and $R_2(t)$ are to be found. In 16.1.5 we have explicitly assumed how the answer depends upon ϵ. An assumption of this type is the hallmark of a perturbation method.

Next, we substitute 16.1.5 into the differential equation. Taking each term in the equation separately yields

$$R\ddot{R} = \epsilon R_0\ddot{R}_1 + \epsilon^2[R_0\ddot{R}_2 + R_1\ddot{R}_1] + \cdots \tag{16.1.6}$$

$$\tfrac{2}{3}\dot{R}^2 = \epsilon^2 \tfrac{2}{3}\dot{R}_1^2 + \cdots \tag{16.1.7}$$

The pressure term requires some extra work. We must simplify the expression

$$\left(\frac{R}{R_0}\right)^{-3\gamma} = \left(1 + \epsilon\frac{R_1}{R_0} + \epsilon^2\frac{R_2}{R_0} + \cdots\right)^{-3\gamma}$$

This expression is expanded using the binomial expansion

$$(1 + x)^n = 1 + nx + \frac{n(n-1)}{2}x^2 + \frac{n(n-1)(n-2)}{3!}x^3 + \cdots \qquad 16.1.8$$

This is valid for any n as long as x is small. Applying this result with $n = -3\gamma$ and x taken as the two terms containing ϵ and ϵ^2 gives

$$-\frac{p_0}{\rho}\left[\left(\frac{R}{R_0}\right)^{-3\gamma} - 1\right] = \frac{p_0}{\rho}\left\{\epsilon 3\gamma\frac{R_1}{R_0} + \epsilon^2 3\gamma\left[\frac{R_2}{R_0} - \frac{3\gamma + 1}{2}\left(\frac{R_1}{R_0}\right)^2\right]\cdots\right\} \qquad 16.1.9$$

Given enough paper, one could write out the terms of the differential equation, 16.1.6, 16.1.7, and 16.1.9, and regroup them according to the powers of ϵ. The equation would look like this:

$$\epsilon[\cdots] + \epsilon^2[\cdots] + \cdots = 0$$

Since ϵ is an independent parameter, we argue that each bracket must be zero by itself. Thus, each bracket produces a differential equation. The equations below are generated by collecting together terms of like powers of ϵ in 16.1.6, 16.1.7, and 16.1.9:

Coefficient of ϵ:

$$R_0\ddot{R}_1 + \frac{3\gamma p_0}{\rho R_0}R_1 = 0 \qquad 16.1.10$$

Coefficient of ϵ^2:

$$R_0\ddot{R}_2 + \frac{3\gamma p_0}{\rho R_0}R_2 = -R_1\ddot{R}_1 - \tfrac{2}{3}\dot{R}_1^2 + \frac{3\gamma(3\gamma + 1)p_0}{2\rho}\left(\frac{R_1}{R_0}\right)^2 \qquad 16.1.11$$

We now have a sequence of problems. The first contains $R_1(t)$ as the unknown. Once this solution is found, it may be substituted into 16.1.11, and then $R_2(t)$ is the only remaining unknown.

Initial conditions for these equations are found by substituting the assumed form of the answer 16.1.5 into the conditions 16.1.4:

$$R(0) = R_0(1+\epsilon) = R_0 + \epsilon R_1(0) + \epsilon^2 R_2(0) + \cdots$$

$$\dot{R}(0) = 0 = \epsilon \dot{R}_1(0) + \epsilon^2 \dot{R}_2(0) + \cdots$$

Again, since ϵ is independent, we conclude that the initial conditions on R_1 and R_2 are

$$\begin{aligned} R_1(0) &= R_0, \qquad \dot{R}_1(0) = 0 \\ R_2(0) &= 0, \qquad \dot{R}_2(0) = 0 \end{aligned} \qquad 16.1.12$$

In principle the perturbation expansion 16.1.5 results in a series of problems for R_1, R_2, and so on. Usually only the first term or the first two terms are wanted. If we have organized the problem properly, the dominant physics will be in these terms.

The solution to 16.1.10 with initial condition 16.1.12 is found to be

$$\frac{R_1}{R_0} = \cos \omega_0 t \qquad 16.1.13$$

with

$$\omega_0 = \frac{1}{R_0}\left(\frac{3\gamma p_0}{\rho}\right)^{1/2} \qquad 16.1.14$$

Hence a bubble, in the first approximation, is a linear oscillator without any damping. The resonant-frequency equation of the bubble, as given by 16.1.14, was first proposed by Minnaert (1933) using a different method. It turns out that a bubble one inch in diameter oscillates at middle C on the musical scale. Smaller bubbles have correspondingly higher tones.

16.2 ORDER SYMBOLS, GAUGE FUNCTIONS, AND ASYMPTOTIC EXPANSIONS

Suppose that for some physical or mathematical purpose we want to know what a certain function $f(\epsilon)$ is like. Futhermore $f(\epsilon)$ is complicated in that it is not a familiar elementary function like ϵ^2, $\epsilon^{2/3}$, e^ϵ, or $\sin \epsilon$. It may be defined only in terms of an integral or as the solution of a differential equation. Obviously, if we are to characterize a function by approximating it with elementary functions, we can do so only over a limited range in ϵ. First off, we restrict our question to finding the characteristics of $f(\epsilon)$ in the neighborhood of the point ϵ_0. We take ϵ_0 to be zero without loss of generality: a transformation $\hat{\epsilon} = \epsilon - \epsilon_0$ will shift the neighborhood of interest to the origin (in case $\epsilon_0 = \infty$, the inversion formula $\hat{\epsilon} = 1/\epsilon$ can be used).

What is the shape of $f(\epsilon)$ near the origin? One answer, if $f(\epsilon)$ is analytic, is the Taylor series

$$f(\epsilon) = f(0) + f'(0)\epsilon + f''(0)\frac{\epsilon^2}{2} + \cdots$$

$$= A + B\epsilon + C\epsilon^2 + \cdots \qquad 16.2.1$$

Now, this may not be the best answer. For instance, if $f = \epsilon^{4/3}$ the Taylor series is an unnecessarily crude description of the function. Next, consider a slight generalization of the Taylor expansion where ϵ^n is replaced by a sequence of elementary functions $\delta_i(\epsilon)$. The function $f(\epsilon)$ is now represented by

$$f(\epsilon) = a + b\,\delta_1(\epsilon) + c\,\delta_2(\epsilon) + d\,\delta_3(\epsilon) + \cdots \qquad \text{as} \quad \epsilon \to 0 \qquad 16.2.2$$

In this equation the $\delta_i(\epsilon)$ are called *gauge functions*. Each gauge function must be smaller than the preceding one in the following sense:

$$\lim_{\epsilon \to 0} \frac{\delta_1(\epsilon)}{1} = 0$$

$$\lim_{\epsilon \to 0} \frac{\delta_2(\epsilon)}{\delta_1(\epsilon)} = 0 \qquad 16.2.3$$

and in general

$$\lim_{\epsilon \to 0} \frac{\delta_{i+1}(\epsilon)}{\delta_i(\epsilon)} = 0 \qquad \text{for all } i$$

A sequence δ_i satisfying these relations is called an *asymptotic sequence*.

Order symbols are a shorthand notation to express how two functions compare. The function $f(\epsilon)$ is order of $g(\epsilon)$ as $\epsilon \to 0$ if a nonzero number A exists ($0 < |A| < \infty$) such that

$$\lim_{\epsilon \to 0} \frac{f(\epsilon)}{g(\epsilon)} = A \qquad 16.2.4$$

This is written using the order symbol O as

$$f(\epsilon) = O[g(\epsilon)] \qquad 16.2.5$$

The size of A in 16.2.3 is immaterial. Hence, there is no connection between "order" and "order of magnitude". If $f = 10^4 g$, then f and g differ by four orders of magnitude, but they are still of the same order in the above sense. In any physical problem that has been properly nondimensionalized, the number A will be of reasonable size, so the order of magnitude of A does not really concern us. In asymptotic analysis we are concerned about how the shape of f compares to the shape of g. The statement

$$f(\epsilon) \sim A g(\epsilon), \qquad \epsilon \to 0$$

is certainly more informative than knowing the values of $f(0)$ and $g(0)$. Some examples of the use of order symbols are given below. For $\epsilon \to 0$,

$$\sin\epsilon = O[\epsilon], \qquad \tan\epsilon = O[\epsilon]$$

$$\sin^2\epsilon = O[\epsilon^2], \qquad J_0(\epsilon) = O[1]$$

$$\sin 2\epsilon = O[\epsilon], \qquad (1+\epsilon)^n - 1 - n\epsilon = O[\epsilon^2]$$

$$1000 = O[1], \qquad \sinh\epsilon = O[1]$$

If the value of A in 16.2.4 is zero, then $f(\epsilon)$ is said to be of smaller order than $g(\epsilon)$. This is written with a small o:

$$f(\epsilon) = o[g(\epsilon)]$$

The same thing in longhand is

$$\lim_{\epsilon \to 0} \frac{f(\epsilon)}{g(\epsilon)} = 0 \qquad 16.2.6$$

An important example is

$$\exp\left[-\frac{A}{\epsilon}\right] = o[\epsilon^n] \quad \text{for all } n \qquad \text{as} \quad \epsilon \to 0$$

Using the order symbols we can now write 16.2.1 in the form

$$f(\epsilon) = a + b\,\delta_1(\epsilon) + c\,\delta_2(\epsilon) + O[\delta_3(\epsilon)] \qquad \text{as} \quad \epsilon \to 0$$

or

$$f(\epsilon) = a + b\,\delta_1(\epsilon) + c\,\delta_2(\epsilon) + o[\delta_2(\epsilon)] \qquad \text{as} \quad \epsilon \to 0$$

An *asymptotic expansion* of $f(\epsilon)$ is a finite number of terms in a series using a chosen set of gauge functions. An asymptotic expansion is written as

$$f(\epsilon) \sim a + b\,\delta_1(\epsilon) + c\,\delta_2(\epsilon) \qquad \text{as} \quad \epsilon \to 0 \qquad 16.2.7$$

The sign $\sim$ means *asymptotically equal to*. We should not use an equal sign, because the three terms on the right only approximate f for any finite ϵ. The coefficients in the asymptotic expansion are defined formally by

$$\lim_{\epsilon \to 0} f(\epsilon) = a$$

$$\lim_{\epsilon \to 0} \frac{f(\epsilon) - a}{\delta_1(\epsilon)} = b \qquad 16.2.8$$

$$\lim_{\epsilon \to 0} \frac{f(\epsilon) - b\,\delta_1(\epsilon) - a}{\delta_2(\epsilon)} = c$$

and so on. In practice the coefficients are usually determined by using some equation that governs $f(\epsilon)$. The term *asymptotic series* is used when the gauge functions are chosen as $1, \epsilon, \epsilon^2, \ldots, \epsilon^n$ (or $1, \epsilon^{-1}, \epsilon^{-2}, \epsilon^{-3}, \ldots, \epsilon^{-n}$ if $\epsilon \to \infty$). A Taylor series is an asymptotic series where the coefficients are also known to be derivatives of the function.

An asymptotic expansion does not necessarily converge to $f(\epsilon)$ as a large number of terms are taken (originally, indeed, asymptotic series were simply known as divergent series). This is not a disadvantage. Convergence of a series is a property of the tail end of the series. It tells nothing about how close a finite number of terms may be to the function in question. An asymptotic expansion, on the other hand, may do what we want—closely approximate the value of a function—with only a few terms. Never mind that after a few terms in the expansion we find that the partial sum begins to grow more rapidly and ultimately diverges. [There is, of course, the question of how many terms are needed to give the best answer. No universal answer to this question can be given. The interested reader should consult Bender and Orszag (1978) for more information on this aspect.]

To emphasize the difference between convergence and usefulness as an approximation, consider the example of the Bessel function $J_0(x)$. The absolutely convergent series for $J_0(x)$ is

$$J_0(x) = 1 - \frac{x^2}{2^2} + \frac{x^4}{2^2 4^2} - \frac{x^6}{2^2 4^2 6^2} + \cdots + (-1)^n \frac{x^{2n}}{2^2 4^2 \cdots (2n)^2} \qquad 16.2.9$$

This series converges for all values of x. Now, an asymptotic expansion of $J_0(x)$ as $x \to \infty$ is

$$J_0(x) \sim \sqrt{\frac{2}{\pi x}} \cos\left(x - \tfrac{1}{4}\pi\right), \qquad x \to \infty \qquad 16.2.10$$

For small x, 16.2.9 is useful, but for $x = 4$ the one-term expansion 16.2.10 gives three-place accuracy, while eight terms of 16.2.9 are needed to obtain the same accuracy. Perhaps equally as important, the asymptotic expansion immediately gives us an idea about the shape of $J_0(x)$, while the pushing and pulling of the alternating signs in 16.2.9 give no such clue.

An asymptotic expansion is not unique. For a certain function $f(\epsilon)$ and a given set of gauge function $\delta_i(\epsilon)$, we obtain one expansion. However, the choice of gauge functions can be changed and then a different asymptotic expansion for $f(\epsilon)$ results. Herein lies one of the powers of asymptotic methods. The proper choice of gauge functions can lead to a very good approximation for $f(\epsilon)$ with only a few terms. In some problems the choice is not critical; the series of gauge functions $1, \epsilon, \epsilon^2, \ldots, \epsilon^n$ is fine. In other more difficult problems only a special set of gauge functions (say $1, \epsilon^{1/3}, \epsilon^{2/3}, \epsilon^1, \ldots$ or $1, \epsilon, \epsilon \ln \epsilon, \epsilon^2, \ldots$) will do the job. The best choice, or proper choice as the case may be, of gauge functions is one of the art aspects of asymptotic methods. It requires us to guess what the shape of f is as $\epsilon \to 0$.

16.3 INVISCID FLOW OVER A WAVY WALL

In this example we study the potential flow over a wavy wall. The solution is approximate in two respects. First, the flow does not really slip over the wall; a vorticity layer must be created next to the wall. We assume that this layer is very thin, so that it has no effect on the flow. Secondly, the solution will be expanded using the wall height as the perturbation parameter. Hence, we are investigating walls with small waviness. Mathematically this problem will illustrate a perturbation analysis where it is necessary to "transfer the boundary condition". Thin-airfoil theory in aerodynamics is based on this same mathematical technique.

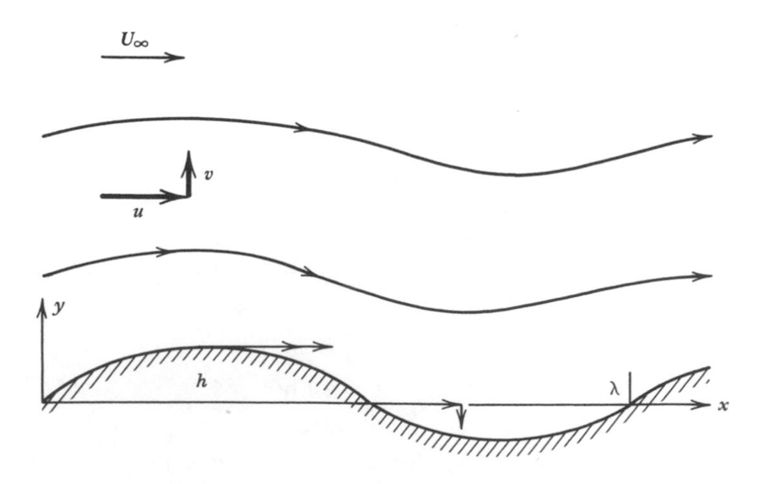

Figure 16-1 Inviscid flow over a wavy wall.

Figure 16-1 depicts the problem and nomenclature. The height of the wave is h, the length λ, and the free-stream speed U_∞. The velocity components are $\hat{u}$ in the $\hat{x}$ direction and $\hat{v}$ in the $\hat{y}$ direction, and the velocity potential is $\hat{\phi}$. Nondimensional variables are defined as follows:

$$\phi = \frac{\hat{\phi}}{U_\infty \lambda}, \qquad x = \frac{\hat{x}}{\lambda}, \qquad y = \frac{\hat{y}}{\lambda} \tag{16.3.1}$$

The perturbation parameter is the nondimensional wall amplitude,

$$\epsilon = \frac{h}{\lambda} \tag{16.3.2}$$

Recall that the velocity potential is related to the velocity components by $v_i = \partial_i \phi$; that is,

$$u = \frac{\partial \phi}{\partial x}, \qquad v = \frac{\partial \phi}{\partial y} \tag{16.3.3}$$

where ϕ is governed by the Laplace equation (derived from continuity: $\partial_i v_i = 0$ and $v_i = \partial_i \phi$)

$$\phi_{xx} + \phi_{yy} = 0 \tag{16.3.4}$$

The boundary condition is that there is no flow through the wall: $n_i v_i = 0$. An equivalent statement is that the direction of the velocity is tangent to the wall:

$$\frac{dy_w}{dx} = \left.\frac{v}{u}\right|_{y=y_w} \tag{16.3.5}$$

where

$$y_w = \epsilon \sin 2\pi x \tag{16.3.6}$$

The second boundary condition is the free-stream condition

$$y \to \infty, \qquad u = 1 \tag{16.3.7}$$

By using 16.3.2 we can cast the boundary conditions in terms of the velocity potential

$$\begin{aligned} &y \to \infty, \qquad \phi_x = 1 \\ &y = \epsilon \sin 2\pi x, \qquad \phi_x \cdot 2\pi\epsilon \cos 2\pi x = \phi_y \end{aligned} \tag{16.3.8}$$

The perturbation analysis is begun by assuming that the answer is an asymptotic series

$$\phi \sim \phi^0 + \epsilon \phi^1 \tag{16.3.9}$$

When this is substituted into 16.3.4 we find

$$\phi^0_{xx} + \phi^0_{yy} + \epsilon\left[\phi^1_{xx} + \phi^2_{yy}\right] = 0 \qquad 16.3.10$$

But since ϵ is independent, both ϕ^0 and ϕ^1 separately satisfy the Laplace equation. Because our original equation is linear, we see that all approximations satisfy the same equation. If our original equation had been nonlinear, the first approximation, ϕ^0, might be governed by a linear or nonlinear equation; however, all higher approximations would obey linear (usually nonhomogeneous) equations.

Substituting 16.3.9 into the free-stream boundary condition produces

$$1 = \phi_x(x, y \to \infty) = \phi^0_x(x, y \to \infty) + \epsilon\phi^1_x(x, y \to \infty)$$

or, since ϵ is independent,

$$\begin{aligned} \phi^0_x(x, y \to \infty) &= 1 \\ \phi^1_x(x, y \to \infty) &= 0 \end{aligned} \qquad 16.3.11$$

The wall condition requires more careful work. The condition is

$$\begin{aligned} &\left[\phi^0_x(x, y = y_w) + \phi^1_x(x, y = y_w)\epsilon\right] 2\pi\epsilon\cos 2\pi x \\ &\qquad -\left[\phi^0_y(x, y = y_w) + \phi^1_y(x, y = y_w)\right] = 0 \end{aligned} \qquad 16.3.12$$

Since y_w is also a function of ϵ, we have not yet displayed the dependence on ϵ explicitly in the equation above. For example, $\phi^0_x(x, y = y_w)$ depends on ϵ through the variable y. We solve this difficulty by expanding each term above in a Taylor series about $y = 0$. The term for ϕ^0_x is

$$\begin{aligned} \phi^0_x(x, y = y_w) &= \phi^0_x(x,0) + \phi^0_{xy}(x,0)\, y_w + \cdots \\ &= \phi^0_x(x,0) + \phi^0_{xy}(x,0)\epsilon\sin 2\pi x + \cdots \end{aligned}$$

When each term in 16.3.12 is treated in this manner we find

$$\begin{aligned} &\left[\phi^0_x(x,0) + \phi^0_{xy}(x,0)\epsilon\sin 2\pi x + \phi^1_x(x,0)\epsilon + \phi^1_{xy}(x,0)\epsilon^2\sin 2\pi x\right] 2\pi\epsilon\cos 2\pi x \\ &\quad -\left[\phi^0_y(x,0) + \epsilon\phi^0_{yy}(x,0)\epsilon\sin 2\pi x + \phi^1_y(x,0)\epsilon + \phi^1_{yy}(x,0)\epsilon^2\sin 2\pi x\right] = 0 \end{aligned}$$

Now, we may group together the coefficients of each power of ϵ and equate to zero. The wall boundary condition then becomes transferred back to the basic surface $y = 0$:

$$\phi_y^0(x,0) = 0 \tag{16.3.13}$$

$$\phi_x^0(x,0)2\pi\cos 2\pi x - \phi_y^1(x,0) = 0 \tag{16.3.14}$$

After we solve the problem for ϕ^0 using 16.3.13 as a boundary condition, we can substitute into 16.3.14 to get a boundary condition on ϕ^1. Equations 16.3.10, 16.3.11, 16.3.13, and 16.3.14 constitute decoupled problems for ϕ^0 and ϕ^1.

The solution for ϕ^0 is easily found to be

$$\phi^0 = x \tag{16.3.15}$$

This is simply a uniform stream flowing past the wall. The ϕ^1 problem is

$$\phi_{xx}^1 + \phi_{yy}^1 = 0$$

$$y = 0, \qquad \phi_y^1 = 2\pi\cos 2\pi x$$

$$y \to \infty, \qquad \phi_x^1 = 0$$

Separation of variables shows that the solution is

$$\phi^1 = -\cos 2\pi x\, e^{-2\pi y} \tag{16.3.16}$$

Hence, the complete answer is

$$\phi \sim x - \epsilon\cos 2\pi x\, e^{-2\pi y} \tag{16.3.17}$$

and from this the velocities are

$$u = \phi_x \sim 1 + 2\pi\epsilon\sin 2\pi x\, e^{-2\pi y}$$

$$v = \phi_y \sim 2\pi\epsilon\cos 2\pi x\, e^{-2\pi y}$$

The flow is a uniform stream perturbed by the presence of the waviness in the wall. Because the flow is inviscid and irrotational, the Bernoulli equation may be used to find the pressure:

$$\frac{p - p_\infty}{\frac{1}{2}\rho U_\infty^2} \sim 1 - \frac{\hat{u}^2 + \hat{v}^2}{U_\infty^2}$$

$$\sim 4\pi\epsilon\sin 2\pi x\, e^{-2\pi y} + O[\epsilon^2] \tag{16.3.18}$$

The pressure is in phase with the u velocity component.

Frequently we are interested in values of the flow variables at the wall. These may be obtained by substituting the position $y = 0$, the transferred position of the wall, into the answer. When this is done we obtain wall values to the same order of accuracy with which we have solved the problem. The surface pressure and velocities are

$$\begin{aligned} u_{\text{wall}} &= 1 + 2\pi\epsilon \sin 2\pi x \\ v_{\text{wall}} &= 2\pi\epsilon \cos 2\pi x \\ \frac{p_{\text{wall}} - p_\infty}{\frac{1}{2}\rho U_\infty^2} &= 4\pi\epsilon \sin 2\pi x \end{aligned} \tag{16.3.19}$$

Sketches of the results are shown in Fig. 16-1. The maximum velocities are at the crest of the wave. An adverse pressure gradient decelerates the flow as it goes toward the valley, where the minimum velocity is reached. The magnitude of the velocity perturbations depends directly on the height of the wall, ϵ. This was built into our solution by the assumption that ϵ was the proper gauge function. The fact that we were able to fit the problem together using this assumption gives us confidence that this is a reasonable choice. Note that perturbations die out away from the wall as $\exp(2\pi\hat{y}/\lambda)$. The dominant influence in this term is the wavelength λ.

*16.4 NONUNIFORM EXPANSIONS

When a function of two variables is expanded in one of the variables—for example, $f(x, \epsilon)$ is expressed as an asymptotic expansion in ϵ—the expansion may not be good for all values of the other variable. Then we say the expansion is not *uniformly valid* in x. Nonuniform expansions are very frequent in physical problems. Problems that lead to nonuniform expansions are called *singular perturbation* problems.

Consider the function

$$u = f(x, \epsilon) = \frac{x^2(1 - \epsilon)}{x^2 + \epsilon} \tag{16.4.1}$$

A graph of this function is displayed in Fig. 16-2 with solid lines. An asymptotic expansion of u might be

$$u \sim 1 - \epsilon\left(\frac{1 + x^2}{x^2}\right) \tag{16.4.2}$$

This expression is given by dashed lines on Fig. 16-2. For $\epsilon = 0$ the result is exact, but for any $\epsilon > 0$ the expansion 16.4.2 does not come close to the true

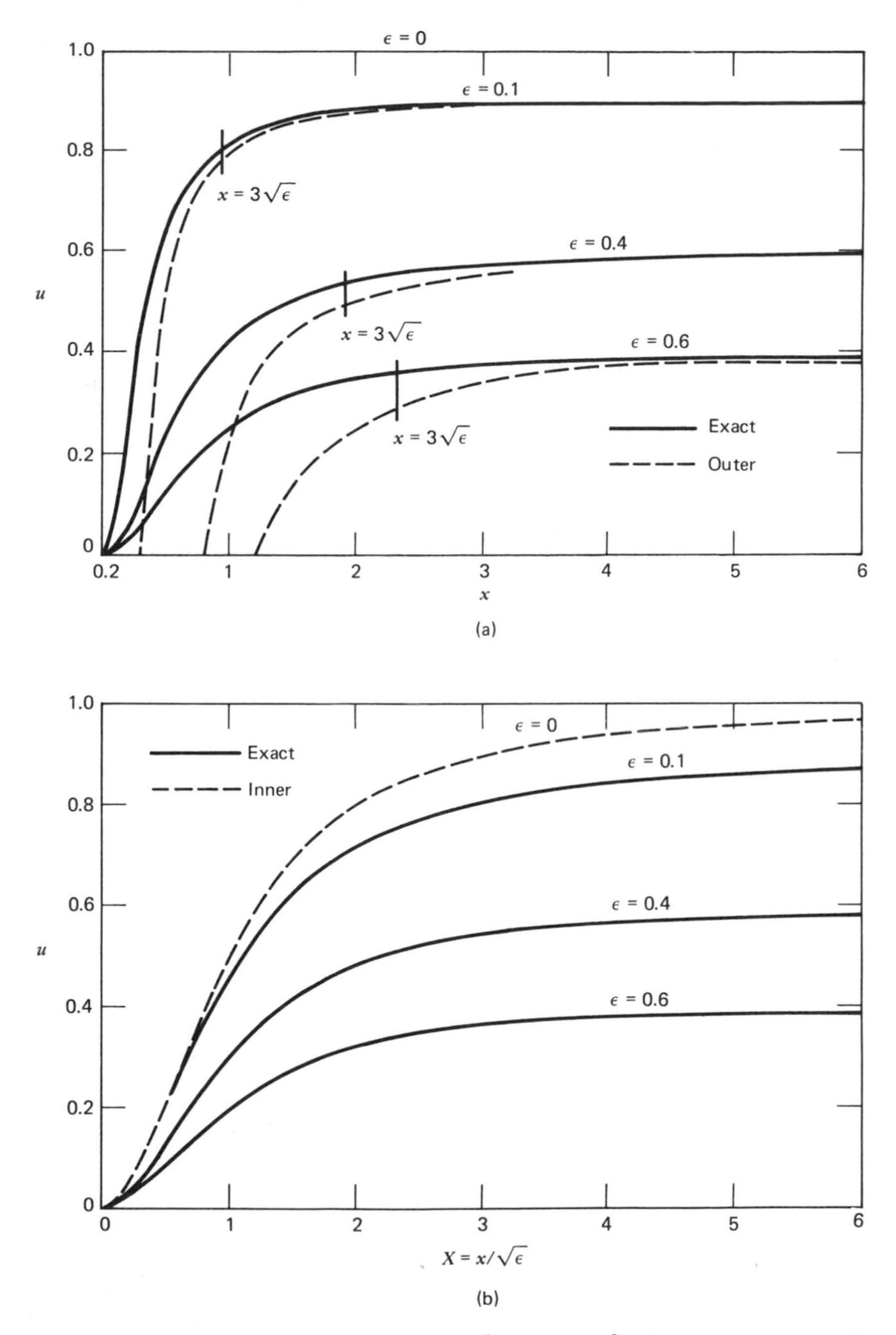

Figure 16-2 Asymptotic expansions of $u = x^2(1 - \epsilon)/(x^2 + \epsilon)$: (a) outer expansions for different values of ϵ, (b) inner expansions where the inner variable is $X \equiv x/\epsilon^{1/2}$.

result 16.4.1 when we are in the neighborhood of $x = 0$. Equation 16.4.2 is a nonuniform expansion because it fails at $x = 0$. Adding more terms to 16.4.2. does not correct this difficulty.

We can construct an expansion for $u = f(x, \epsilon)$ that will be valid at $x = 0$. To do this we first change variables so that the neighborhood of $x = 0$ is enlarged as $\epsilon \to 0$. Let

$$X = \frac{x}{g(\epsilon)}$$

The function $g(\epsilon)$ goes to 0 as $\epsilon \to 0$. It must be a measure of the size of the nonuniform region at $x = 0$. The usual way to determine $g(\epsilon)$ is to look at the expansion 16.4.2 and find the dominant behavior as $x \to 0$. We ask ourselves the question: when will the second term in 16.4.2 become comparable with the final term? At $x \to 0$ the second term is dominated by ϵ/x^2. Hence, we guess that the size (in x) of the nonuniform region is $O[\epsilon^{1/2}]$. On Fig. 16-2a the position $x_0 = 3\sqrt{\epsilon}$ has been marked for each curve (the number 3 has no special significance except that the curves were reasonably close at this value). In the light of these facts we choose $g(\epsilon) = \epsilon^{1/2}$, so that

$$X = \frac{x}{\sqrt{\epsilon}} \qquad 16.4.3$$

The points $x = 3\sqrt{\epsilon}$ on Fig. 16-2 are also $X = 3$. In terms of the new variables 16.4.1 becomes

$$u = F(X, \epsilon) = (1 - \epsilon)\frac{X^2}{X^2 + 1} \qquad 16.4.4$$

Transforming $u = f(x, \epsilon)$ into $u = F(X, \epsilon)$ does not cause any loss of information. Now we make an asymptotic expansion of u for $\epsilon \to 0$ with X fixed,

$$u \sim \frac{X^2}{X^2 + 1}, \qquad \epsilon \to 0 \qquad 16.4.5$$

The one-term expansion 16.4.4 corresponds to the dashed line of Fig. 16-2b. Our new expansion is valid at $x = 0$.

It so happens that the two-term expansion of $F(X, \epsilon)$ is identical to the function itself. Thus the two-term inner expansion is uniformly valid and 100% accurate. This is an unusual occurrence. In fact, most of the time the expansion of $F(X, \epsilon)$ will itself be invalid at $X \to \infty$. In this event we would need two expansions to describe u: one good for $x > 0$, and one good for the neighborhood of $x = 0$.

*16.5 MATCHED ASYMPTOTIC EXPANSIONS: FRIEDRICH'S PROBLEM

This problem is a singular perturbation that displays a boundary layer. The governing differential equation (a slightly modified version of Friedrich's original problem) is

$$\epsilon \frac{d^2u}{dy^2} + \frac{du}{dy} = -\tfrac{3}{2}(1 - 3\epsilon)e^{-3y} \qquad 16.5.1$$

This equation is similar to the boundary-layer momentum equation where u is an x-direction velocity, ϵ is analogous to $1/\text{Re}$, and the right-hand side represents the missing terms with x derivatives. For the boundary conditions we choose conditions that model a flow problem:

$$\begin{aligned} u(y = 0) &= 0 \\ u(y \to \infty) &= 1 \end{aligned} \qquad 16.5.2$$

The $y = 0$ condition is analogous to the no-slip condition at a solid wall, while the $u = 1$ condition represents a specified free stream far away from the wall.

We seek a solution to Friedrich's problem for $\epsilon \to 0$ and y fixed, that is,

$$u = f(y, \epsilon) \sim f^0(y) + O[\epsilon]$$

Letting $\epsilon \to 0$ in 16.5.1 and 16.5.2 produces the equation and boundary conditions governing f^0:

$$\begin{aligned} \frac{df^0}{dy} &= -\tfrac{3}{2}e^{-3y} \\ f^0(0) &= 0 \\ f^0(\infty) &= 1 \end{aligned} \qquad 16.5.3$$

Notice that the highest-order term in 16.5.2 is dropped when the limit $\epsilon \to 0$ is applied. As a consequence we can no longer satisfy both boundary conditions. If we choose to satisfy one boundary condition, the other will not be met, and the answers become singular at that location. Losing the highest derivative always leads to a nonuniform expansion, but the converse is not true: nonuniform expansions can still happen when the highest derivative is retained. There is no general way to know which boundary condition should be satisfied; we can try one and then the other to see which answer makes the most sense. In this case the proper boundary condition to satisfy is the one at $y = \infty$. Solving 16.5.3 with $f^0(y \to \infty) = 1$ gives

$$u \sim f^0 = 1 + \tfrac{1}{2}e^{-3y} \qquad 16.5.4$$

This is called the outer expansion of $u(y, \epsilon)$.

Equation 16.5.4 gives $u \approx 1.5$ at $y = 0$. We must conclude that the expansion is singular at $y = 0$ since the boundary condition $u(y = 0) = 0$ is not satisfied. This deficiency is corrected by changing variables and constructing an inner expansion. Let a new space variable be defined as

$$Y = \frac{y}{g(\epsilon)} \qquad 16.5.5$$

In contrast with the example of Section 16.4, where we found $g(\epsilon)$ by observing how fast the answer became unbounded at $x = 0$, our outer expansion 16.6.4 is well behaved at $y = 0$. We must use a different and (since we do not know the true answer) indirect method to find $g(\epsilon)$. When the transformation 16.5.5 is substituted into 16.5.1, we obtain a differential equation to govern $u(Y, \epsilon)$. It is

$$\epsilon \frac{d^2u}{dY^2} + g\frac{du}{dY} = -\tfrac{3}{2}g^2(1 - 3\epsilon)e^{-3gY} \qquad 16.5.6$$

Now, we argue that the inner expansion $u \sim F(Y, \epsilon)$ should be governed by a different equation than the outer expansion. In particular we are interested in retaining the second-derivative term at least. Hence, we choose $g(\epsilon) = \epsilon$. (The choice $g = O[1]$ gives the same problem as before with no magnification of y in the neighborhood $y = 0$, whereas $g = O[\epsilon^2]$, say $g = \epsilon^2$, gives an overly simple equation $d^2u/dY^2 = 0$.) Letting $\epsilon \to 0$ in 16.6.6 gives the problem

$$\frac{d^2F^0}{dY^2} + \frac{dF^0}{dY} = 0 \qquad 16.5.7$$

We impose the wall boundary condition on this problem:

$$F^0(Y = 0) = 0 \qquad 16.5.8$$

If this were a physical problem, we would have a balance between certain physical terms in the outer region 16.5.3, and a balance between different effects in the inner region 16.5.7.

Our outer solution $f^0(y, \epsilon)$ has already satisfied the far boundary condition $u(y \to \infty) = 1$, so we do not impose this same condition on F. Integration of 16.5.6 and application of 16.5.7 results in the answer

$$u \sim F^0(Y) = A(1 - e^{-Y}) \qquad 16.5.9$$

The constant A is undetermined.

The solution to a singular perturbation problem may be represented by two *matched asymptotic expansions*. The word "matched" indicates the philosophy by which the constant A in 16.5.9 is found.

For one-term inner and outer expansions such as we have here, the simplest matching principle is that the outer answer as $y \to 0$ (the nonuniform region) is equal to the inner answer as $Y \to \infty$:

$$f^0(y \to 0) = F^0(Y \to \infty) \qquad 16.5.10$$

As applied to our problem, 16.5.10 becomes (see 16.5.4 and 16.5.9)

$$\tfrac{3}{2} = A$$

Matching essentially replaces a boundary condition. This is the matching rule that applies to the u velocity in a boundary layer: the inviscid velocity at the wall is equal to the boundary-layer velocity at infinity.

Figure 16-3 shows a graph of the various approximations and also the exact answer. In order to explain how a complicated problem might be solved, we have avoided stating the exact solution. It turns out to be

$$u = \tfrac{3}{2}\frac{1 - e^{-y/\epsilon}}{1 - e^{-1/\epsilon}} - \tfrac{1}{2}(1 - e^{-3y})$$

By performing the outer limit ($\epsilon \to 0$, y fixed) and the inner limit ($\epsilon \to 0$, Y

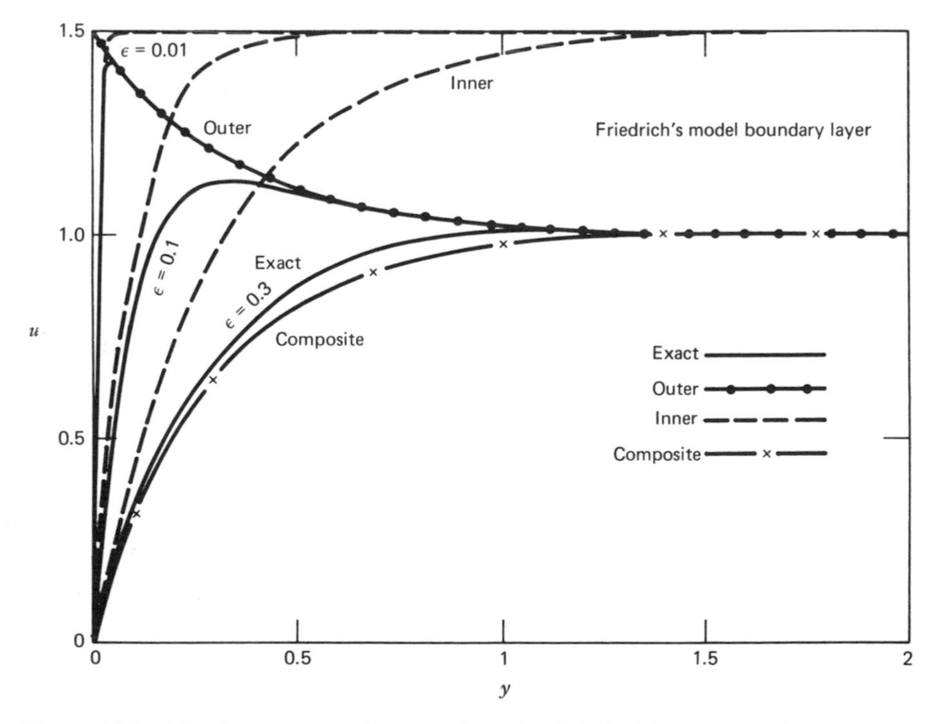

Figure 16-3 Matched asymptotic expansions for Friedrich's problem. For $\epsilon = 0.3$ a composite expansions is also given.

fixed) on this function you can verify that the expansions we obtained, 16.5.4 and 16.5.9, are correct.

*16.6 THE MATCHING PROCESS

Two functions are patched together by choosing a certain point, making the values of the functions agree at that point, and making an arbitrary number of the derivatives of the functions agree at the chosen point. We do not patch asymptotic expansions together. For example, in Friedrich's problem we had an outer expansion

$$u \sim f^0(y, \epsilon), \qquad \epsilon \to 0$$

which was valid away from $y = 0$, and an inner expansion

$$u \sim F^0(Y, \epsilon) \qquad \epsilon \to 0$$

which was valid near $y = 0$. From Fig. 16-3 we can see that for the case $\epsilon = 0.3$ neither of these approximations is very good in the region $0.2 < y < 0.6$. It is only asymptotically as $\epsilon \to 0$ that the accuracy of these expansions becomes good. It is true that two or three terms in the expansions would usually give better accuracy, but this is beside the point. We have no good place to choose as a patching point.

The idea behind matching two asymptotic expansions is that as $\epsilon \to 0$, there is a region in y where both expansions represent the true function. The *matching region*, or overlap region, is where the inner and outer expansions have the same functional form as $\epsilon \to 0$. (A given asymptotic expansion does not represent a unique function. Several functions may have the same asymptotic expansion. They form a class of functions that are asymptotically equal. Thus, it is not strictly correct to say that the inner and outer expansions are equal to the same function in the overlap region.)

As a way of illustrating both the need for matching and the technique of matching, we shall solve Friedrich's problem to higher orders and introduce Van Dyke's (1964) matching rule.

Consider again Friedrich's problem, and assume the outer expansion has three terms (superscripts on f indicate different functions),

$$u \sim f^0(y) + \epsilon f^1(y) + \epsilon^2 f^2(y) \qquad 16.6.1$$

Substituting this into the differential equation 16.5.1 and the boundary conditions 16.5.2, and equating the coefficients of each power of ϵ to zero results in the following set of problems:

Coefficient $O[1]$:

$$\frac{df^0}{dy} = -\tfrac{3}{2}e^{-3y}, \qquad f^0(\infty) = 1 \qquad 16.6.2$$

Coefficient $O[\epsilon]$:

$$\frac{df^1}{dy} = -\frac{d^2f^0}{dy^2} + \tfrac{9}{2}e^{-3y}, \qquad f^1(\infty) = 0 \tag{16.6.3}$$

Coefficient $O[\epsilon^2]$:

$$\frac{df^2}{dy} = -\frac{d^2f^1}{dy^2}, \qquad f^2(\infty) = 0 \tag{16.6.4}$$

The solution to 16.6.2 we previously found as

$$f^0 = 1 + \tfrac{1}{2}e^{-3y} \tag{16.6.5}$$

Substituting for f^0 in 16.6.3 shows that $df^1/dy = 0$, so the next solution is identically zero: $f^1 = 0$. Continuing, we see that 16.6.4 requires $df^2/dy = 0$ and hence

$$f^2 = 0$$

The three-term expansion has only f^0 as a nonzero contribution.

Now, we turn our attention to the inner expansion. In terms of the inner variable $Y = y/\epsilon$, the differential equation is

$$\frac{d^2u}{dY^2} + \frac{du}{dY} = -\tfrac{3}{2}\epsilon(1 - 3\epsilon)e^{-3\epsilon Y} \tag{16.6.6}$$

One boundary condition is the wall condition

$$u(Y = 0) = 0 \tag{16.6.7}$$

We do not have a second boundary condition for this problem. The lost boundary condition is replaced by the requirement that the inner and outer expansions be matched in the overlap region. In this way the outer solution supplies the missing boundary condition to the inner solution. This also implies a natural order to the solution: the first-term outer problem can be solved by itself, the first-term inner receives a boundary condition from the first term outer, the two-term outer may need information from the one-term inner, the two-term inner gets a boundary condition from the two-term outer, and so on.

Assume the inner expansion has the form

$$u \sim F^0(Y) + F^1(Y)\epsilon + F^2(Y)\epsilon^2, \qquad \epsilon \to 0, \quad Y \text{ fixed} \tag{16.6.8}$$

The now familiar procedure of substitution and equating the coefficient to zero gives the following sequence of problems:

Coefficient $O[1]$:

$$\frac{d^2F^0}{dY^2} + \frac{dF^0}{dY} = 0, \qquad F^0(0) = 0$$

The solution is

$$F^0 = A_0(1 - e^{-Y}) \tag{16.6.9}$$

Coefficient $O[\epsilon]$:

$$\frac{d^2F^1}{dY^2} + \frac{dF^1}{dY} = -\frac{9}{2}, \qquad F^1(0) = 0$$

The solution is

$$F^1 = A_1(1 - e^{-Y}) - \tfrac{3}{2}Y \tag{16.6.10}$$

Coefficient $O[\epsilon^2]$:

$$\frac{d^2F^2}{dY^2} + \frac{dF^2}{dY} = \frac{9}{2}(1 + Y), \qquad F^2(0) = 0$$

The solution is

$$F^2 = A_2(1 - e^{-Y}) + \tfrac{9}{4}Y^2 \tag{16.6.11}$$

Each solution contains an unknown constant because we have only one boundary condition to apply to each problem. The constants must be determined by matching.

Van Dyke's matching rule is a practical recipe for matching two expansions. While it has not yet been connected with the mathematical theory of matching, it offers a simple method that is easily applied (only in rare occasions does it fail). First we consider the question "How does the three-term outer expansion behave in the inner region?" To find out, we write out the outer expansion,

$$u \sim 1 + \tfrac{1}{2}e^{-3y} + 0 \cdot \epsilon + 0 \cdot \epsilon^2$$

and reexpress it in the inner variable using $y = \epsilon Y$:

$$u \sim 1 + \tfrac{1}{2}e^{-3\epsilon Y}$$

Expanding this for fixed Y as $\epsilon \to 0$ gives us the part of the outer expansion that is dominant in the inner region. Hence, we expand the exponential in the expression above and obtain

$$u \sim 1 + \frac{1}{2}\left[1 + 3\epsilon Y + \frac{9\epsilon^2 Y^2}{2} + \cdots\right]$$

Next, we truncate this at three terms in ϵ:

$$u \sim \tfrac{3}{2} - \epsilon\tfrac{3}{2}Y + \epsilon^2\tfrac{9}{4}Y^2 \qquad 16.6.12$$

The notation for the process above is

$$\left[f_0^{(n)}\right]_i^{(m)}$$

This is read: m-term inner expansions of the n-term outer expansion. By this process we have extracted the part of the outer expansion that is controlling in the inner region. To avoid confusion, the student should note that $f_0^{(n)}$ is the n-term outer expansion of f, while f^0, f^1, f^2 are the coefficients of that expansion with powers of ϵ equal to 0, 1, and 2 respectively.

The matching principle is to do the same for the inner expansion and equate the results. Mathematically the principle is

$$\left[f_0^{(n)}\right]_i^{(m)} = \left[F_i^{(m)}\right]_0^{(n)} \qquad 16.6.13$$

Either the integers m and n are taken the same, or m is taken as one larger than n. In our example above $m = n = 3$ was chosen.

Let's see how the matching principle works in Friedrich's problem. To do this we need to compute $[F_i^{(3)}]_0^{(3)}$. The three-term inner expansion is

$$u \sim A_0(1 - e^{-Y}) + \epsilon\left[A_1(1 - e^{-Y}) - \tfrac{3}{2}Y\right] + \epsilon^2\left[A_2(1 - e^{-Y}) + \tfrac{9}{4}Y^2\right] \qquad 16.6.14$$

Rewritten in the outer variable using $Y = y/\epsilon$,

$$u \sim A_0(1 - e^{-y/\epsilon}) + \epsilon\left[A_1(1 - e^{-y/\epsilon}) - \frac{3}{2}\frac{y}{\epsilon}\right]$$

$$+\epsilon^2\left[A_2(1 - e^{-y/\epsilon}) + \frac{9}{4}\frac{y^2}{\epsilon^2}\right]$$

Expanded for $\epsilon \to 0$, y fixed, yields

$$u \sim A_0 - \tfrac{3}{2}y + \tfrac{9}{4}y^2 + \epsilon A_1 + \epsilon^2 A_2 + O[e^{-1/\epsilon}]$$

Recall that the exponential term $e^{-1/\epsilon}$ goes to zero faster than any power of ϵ:

$$e^{-1/\epsilon} = o[\epsilon^n] \quad \text{for all } n, \qquad \epsilon \to 0$$

These "exponentially small" terms are the first to drop out, as they go at the very end of any useful set of gauge functions. Next, u is truncated at three terms in ϵ:

$$u \sim A_0 - \tfrac{3}{2}y + \tfrac{9}{4}y^2 + \epsilon A_1 + \epsilon^2 A_2 \qquad 16.6.15$$

According to the matching rule we equate 16.6.15 and 16.6.12. To do this we need to use the same variable, either y or Y, and so we convert 16.6.15 back to the inner variable:

$$u \sim A_0 + \epsilon[A_1 - \tfrac{3}{2}Y] + \epsilon^2[A_2 + \tfrac{9}{4}Y^2] \qquad 16.6.16$$

Comparing 16.7.16 and 16.7.12 determines the constants in the inner solution as

$$A_0 = \tfrac{3}{2}, \qquad A_1 = 0, \qquad A_2 = 0 \qquad 16.6.17$$

Matching the common parts of the two expansions replaces the boundary condition that was lost in the inner problems.

The theoretical basis of matching is a little more complicated than Van Dyke's matching rule would indicate. In case the interested student investigates this subject further, we outline what he is likely to find. Three types of variables and limits are defined. First, we have the outer limit $\epsilon \to 0$ with y fixed. This leads to the outer expansion

$$u = f(y, \epsilon) \sim \tilde{f}(y, \epsilon)$$

The tilde means an asymptotic expansion. An inner variable has a general form

$$Y = \frac{y}{g(\epsilon)} \qquad \text{where} \quad g(\epsilon) \to 0 \quad \text{as } \epsilon \to 0$$

that is,

$$g(\epsilon) = o[1]$$

The function, when expressed in inner variables and expanded for $\epsilon \to 0$, Y fixed, has the form

$$u = F(Y, \epsilon) \sim \tilde{F}(Y, \epsilon)$$

In addition to the inner and outer variables there are intermediate variables defined by

$$\mathscr{Y} = \frac{y}{h(\epsilon)} \qquad 16.6.18$$

where $h(\epsilon) \to 0$ as $\epsilon \to 0$ but at a slower rate than $g(\epsilon)$. If $g(\epsilon) = \epsilon$, then $h(\epsilon)$ might be $\epsilon^{1/2}$. The functions h and g obey the relation

$$\frac{h(\epsilon)}{g(\epsilon)} \to \infty \qquad \text{as} \quad \epsilon \to 0$$

The intermediate limit, $\epsilon \to 0$ with $\mathscr{Y}$ fixed, is where the inner and outer expansions have their common region of validity. The theory indicates that when the expansions are expressed in intermediate variables they are *asymptotically equal*. Asymptotically equal means that with respect to a chosen set of gauge functions $\Delta_n(\epsilon)$, the difference between the expansions is smaller than order $\Delta_n(\epsilon)$ for all n, that is,

$$\tilde{f}(y = \mathscr{Y}h(\epsilon), \epsilon) - \tilde{F}(Y = h(\epsilon)\mathscr{Y}/g(\epsilon), \epsilon) = o[\Delta_n(\epsilon)]$$

$$\epsilon \to 0, \quad \mathscr{Y} \text{ fixed}, \quad \text{for all } n \qquad 16.6.19$$

In many cases the form of the expression above is unchanged if the inner variable is used instead of an intermediate variable.

*16.7 COMPOSITE EXPANSIONS; ACCURACY

The inner and outer expansions have different regions of validity and also a common form in the overlap region. In some instances it is desirable to have an expansion that is uniformly valid over the entire region. Consider what would happen if we simply added the two expansions, that is,

$$f_0^{(m)}(y, \epsilon) + F_i^{(n)}\left(Y = \frac{y}{\epsilon}, \epsilon\right)$$

If y is small, then F is approximately equal to the true function while f takes on values typical of the overlap region. If y is large, then f is approximately equal to the true function while F takes on values typical of the overlap region. Those values common to both expansions, the values of the overlap region, are found from $[f_0^{(m)}]_i^{(n)}$, or $[F_i^{(n)}]_0^m$ in light of the matching rule 16.6.13. Hence a composite expansion for f is formed by subtracting $[f_0^{(m)}]_0^{(n)}$ (or $[F_i^{(n)}]_0^m$) from the sum of the inner and outer expansions:

$$u \sim f_{\text{comp}}^{(m,n)} = f^{(m)} + F^{(n)} - \left[f_0^{(m)}\right]_i^{(n)} \qquad 16.7.1$$

For the middle values of y, both f and F have values typical of the overlap region, so 16.7.1 is still a good approximation to the real function.

We construct a composite expansion for Friedrich's problem using $m = n = 1$. The outer expansion (16.6.5) is

$$f_0^{(1)} = f^0 = 1 + \tfrac{1}{2}e^{-3y}$$

The inner expansion (16.6.9) is

$$F_i^{(1)} = F^0 = \tfrac{3}{2}(1 - e^{-y/\epsilon})$$

Using the same steps as those used to arrive at 16.6.12, we find

$$\left[f_0^{(1)}\right]_i^{(1)} = \tfrac{3}{2}$$

The composite expansion is

$$u \sim f_c^{(1,1)} = 1 + \tfrac{1}{2}e^{-3y} - \tfrac{3}{2}e^{-y/\epsilon} \qquad 16.7.2$$

A comparison of this composite expansion with the exact answer is given in Fig. 16-3. Accuracy of 5% is obtained for $\epsilon = 0.3$ and all values of y. (A far better result than if one uses either the inner or the outer expansion.)

The accuracy of an asymptotic expansion is an important practical consideration. How large can the perturbation parameter become before a uniformly valid expansion becomes inaccurate? There is no general answer to this question. In each situation the answer depends on the nature of the true function and the particular expansion used to represent it (that is, the gauge functions chosen). If the true answer is the equation of a straight line, say $f = 6 + \epsilon y^2$, then two terms in a series give the exact answer for all values of ϵ. Even though the process of producing this relation uses the limit $\epsilon \to 0$, the result is valid for values of ϵ that we normally consider far from zero. One guiding rule is that the accuracy is at least as good as the size of the next neglected term in the expansion. To apply this rule, however, requires that we compute the next term, which we might not want to do in a complicated practical problem.

Our intuition about when a parameter is large or small should be used with caution. Consider the following: is the value $\epsilon = 0.333$ close enough to zero for an expansion about $\epsilon \to 0$ to apply? If we considered all possible values of ϵ from 0 to infinity, then we might suppose $\epsilon = 0.333$ reasonably close to zero. Now, consider an expansion about $\epsilon \to \infty$. Would $\epsilon = 3$ be a reasonable number for which such an expansion would be valid? Normally our linear minds regard 3 as a long way from infinity. Actually an expansion about $\epsilon \to \infty$ is completely equivalent to an expansion about $\hat{\epsilon} \to 0$ with $\hat{\epsilon} \equiv 1/\epsilon$. The point $\epsilon = 3$ with $\epsilon \to \infty$ is equivalent to the point $\hat{\epsilon} = 0.333$ with $\hat{\epsilon} \to 0$. Our intuition is deceiving. In the above sense, anyway, there are "just as many" points between 0 and 1 as there are between 1 and infinity. In Section 16.2 we compared a Taylor series for the Bessel function $J(x)$ about the point $x = 0$ with an asymptotic expansion about $x \to \infty$. The two representatives were compared at $x = 4$, where the asymptotic expansion was shown to be much superior. Another way of looking at this is that $x = 4$ for an expansion $x \to \infty$ is really the point $\hat{x} = \frac{1}{4}$ in an expansion of $\hat{x} \equiv 1/x \to 0$. This is not far from zero.

16.8 CONCLUSION

The method of matched asymptotic expansion is only one of a variety of perturbation methods. It is particularly suited to several important fluid-flow problems and as a viewpoint for the theoretical division of fluid flows. The reader interested in learning about other methods should consult the books by Van Dyke (1964), Nafey (1973), Cole and Kevorkian (1981), and Bender and Orszag (1978).

Expansions are essentially a method to simplify problems and break out the two or three most dominant aspects of the physics. Frequently very useful analytic expressions are formed by these methods. For example, asymptotic methods show that the drag on a sphere is given by

$$C_D = \frac{24}{\text{Re}}\left\{1 + \tfrac{3}{16}\text{Re} + \tfrac{9}{160}\text{Re}^2 \ln \text{Re} + O[\text{Re}^2]\right\}, \qquad \text{Re} \to 0$$

In a computer solution to this same problem, it would be necessary to make a series of runs at different Reynolds numbers and then curve-fit for a drag equation. The theoretical and physical simplifications provided by an asymptotic analysis would be hidden in the computer. On the other hand, many flow problems are very complex and cannot be simplified; a large number of competing events occur without any dominant physics. Here asymptotic methods, by their nature, are inappropriate. The computer then becomes the most powerful and useful approach. Perturbation methods offer a great amount of flexibility in extracting the major elements from a problem. In this respect perturbation methods require considerable guesswork, insight, and creativity. Workers using perturbation methods are essentially doing mathematical engineering.

PROBLEMS

16.1 (A) Which of these functions goes to infinity faster when $x \to \infty$; $f_1 \sim x$ or $f_2 \sim \ln x$?

16.2 (A) Prove that when $x \to 0$ that $\exp(-1/x) \to 0$ faster than x^n for any n no matter how large.

16.3 (A) Find the limit as $\epsilon \to 0$ for the function

$$f(y, \epsilon) = \frac{3}{2} \frac{[1 - \exp(-y/\epsilon)]}{[1 - \exp(-1/\epsilon)]} - \frac{1}{2}[1 - \exp(-3y)]$$

Change the function by letting $Y = y/\epsilon$ and find the limit of $F(Y, \epsilon)$ as $\epsilon \to 0$. In each case compare the iterated limits

$$\lim_{y \to 0} \lim_{\epsilon \to 0} f(y, \epsilon) \quad \text{and} \quad \lim_{\epsilon \to 0} \lim_{y \to 0} f(y, \epsilon)$$

16.4 (A) Solve the bubble oscillator problem to find $R_2(t)$. (Hint: In solving the nonhomogeneous differential equation, change squares of trig functions into trig functions of the double angle.)

16.5 (C) An oscillator perturbation will become invalid at large times if a term like $t\cos(\omega t)$ occurs. This is called a secular term. Show that secular terms will occur in $R_3(t)$ of the bubble oscillator problem.

16.6 (A) Consider a slightly wavy wall of a channel given by

$$y = \pm h \pm A\left[\sin\left(\frac{2\pi x}{\lambda}\right) + B\sin\left(\frac{4\pi x}{\lambda + c}\right)\right]$$

Solve for the viscous flow in this channel.

16.7 (B) Do the viscous wavy wall analysis using asymptotic expansions in the variables u, v, and p.

16.8 (B) The upper surface of a thin body ($\epsilon = h/L \to 0$) is given by the parabola $y_w/h = 1 - (x/L)^2$ for $-L \le x \le L$. The streaming flow u_0 is perturbed so that $u = u_0 + \epsilon u_1$ and $v = \epsilon v_1$. Transfer the wall boundary condition given below to $y = 0$.

$$\left.\frac{dy_w}{dx}\right|_{y_w} = \left.\frac{v}{u}\right|_{y_w}$$

16.9 (A) Consider the differential equation below with boundary conditions $f(0) = 0$, $f(1) = 1$;

$$\epsilon f'' + f' = a$$

Make an expansion for $f(x, \epsilon)$ for $\epsilon \to 0$. Locate the singular behavior and make a matched asymptotic expansion for the inner region.

16.10 (A) Construct a multiplicative composite expansion for 16.9.

17 Characteristics of High-Reynolds-Number Flow

Many, probably the vast majority, of engineering flows are at high Reynolds numbers. It is not unusual for the flow in a pipe to have Reynolds number of 10^5, or the flow over the wing of even a small airplane to have Reynolds number 10^6. In this chapter we investigate some of the main characteristics of high-Reynolds-number flows. We shall find that the flow can be divided into two parts: an inviscid flow in the major part of the flow region, and a boundary layer near the walls. Boundary layer principles also apply to thin regions of high shear (shear layers) within the main flow region.

Equations for both inviscid flow and boundary layers will be derived and discussed in this chapter. The purpose in doing this, rather than placing the discussions in the separate chapters on inviscid flow and boundary layers, is to emphasize that these subjects are not distinct, but that they hold complementary positions in the theory of fluid mechanics.

17.1 PHYSICAL MOTIVATION

Viscous diffusion of momentum, or of vorticity, is a slow process: to be specific, it is slow compared to convection. Let us again (cf. Section 13.8) consider a duct of length L and characteristic diameter D (Fig. 17-1). The flow comes from a very large reservoir connected to the entrance. On the other end of the duct a fan or pump is placed in order to produce the flow. Pressure forces are responsible for accelerating the flow into the duct, and by their very nature they cannot generate a net viscous force or vorticity. Likewise the center flow of the tube does not contain any net shear stress, because only pressure forces have acted on these particles. Once the flow is moving, the no-slip condition at the wall causes shear stresses at the wall. The imbalance in shear stress that occurs is transferred toward the center through viscous diffusion. Using Rayleigh's argument, we estimate that the thickness of the viscous effect is $\delta \sim \sqrt{\nu t}$. In order to estimate the viscous thickness at the end of the duct we insert the flow time for t ($t_{\text{flow}} = L/U$). The final thickness of the viscous region is compared with the duct diameter:

$$\frac{\delta}{D} \sim \left(\frac{\nu}{DU}\frac{L}{D}\right)^{1/2} = \left(\frac{1}{\text{Re}}\frac{L}{D}\right)^{1/2} \tag{17.1.1}$$

If this estimate is valid, we can expect viscous effects to be confined to a

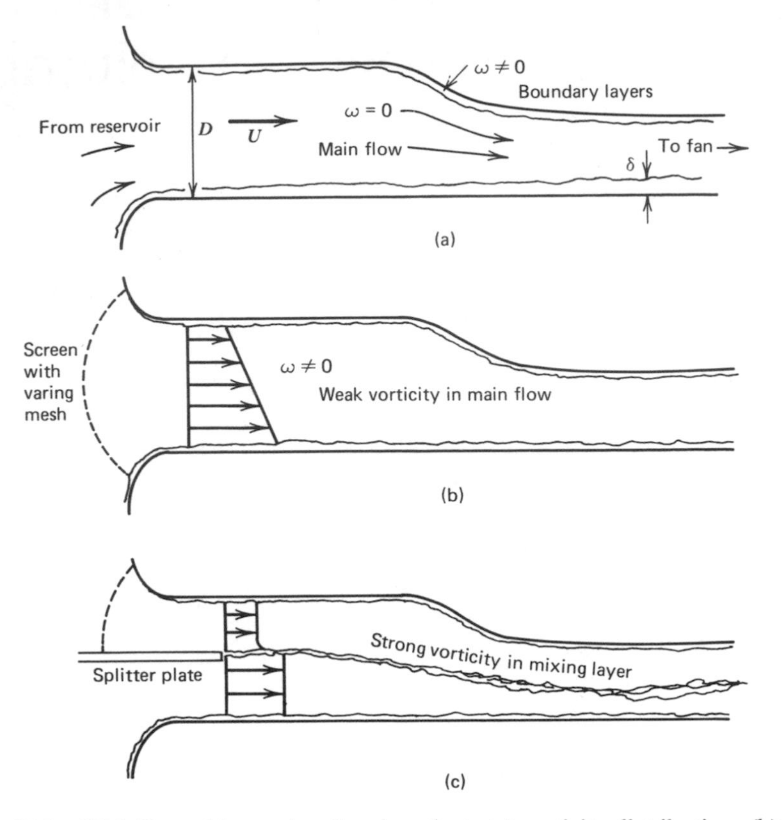

Figure 17-1 High-Reynolds-number flow in a duct: (a) vorticity distribution; (b) weak vorticity in inviscid main flow; (c) strong vorticity in shear layer, which must be treated as a boundary layer.

vanishingly thin region next to the wall as the Reynolds number becomes infinite. For any fixed value of L/D one could produce a high enough Reynolds number so that $\delta/D \to 0$.

We can also make the same argument for the external flow over a body. Consider a two-dimensional flow over an airfoil. Since the airfoil may be quite thin compared to its chord length L, we compare the viscous diffusion thickness δ with L. Using the same logic as above, we find

$$\frac{\delta}{L} \sim \left(\frac{\nu}{UL}\right)^{1/2} = \mathrm{Re}^{-1/2} \qquad 17.1.2$$

A high airfoil Reynolds number indicates that the diffusion of vorticity is confined to a thin layer next to the surface of the airfoil. Furthermore, the viscous wake of the airfoil (the region downstream where the viscous regions from the top and the bottom merge) will be thin for a distance downstream of the order of L_w. Say $L_w = 100L$; then

$$\frac{\delta_{\text{wake}}}{L} \sim \left(\frac{L_w}{L}\frac{\nu}{UL}\right)^{1/2} = 10\,\text{Re}^{-1/2}$$

The factor of 10 is really not important as Re → ∞. Thus, the viscous effects in the wake of the airfoil are also confined to a thin region. The nice thing about a wake is that the viscous-diffusion effects proceed in both directions and the viscous part of the wake destroys itself in a distance of the order of several L. Thus, in this case, the downstream flow is again free of viscous effects.

The actual situation is slightly different from what we have described above. At high Reynolds numbers the viscous layers become turbulent. In spite of this, they still remain thin, and the principle of separating the flow into viscous and inviscid regions remains valid. The turbulent wake of the airfoil will also destroy itself; both the mean velocity and the turbulence will eventually disappear.

There is one event that can invalidate the arguments above: separation. If a flow goes smoothly over the walls and always continues in the same direction, the boundary layer concept is valid and the viscous regions are thin. However, sometimes the flow separates and leaves the wall, carrying with it the vorticity and viscous effects of the boundary layer. A large region of backflow or recirculating flow exists downstream of the separation and frequently leads to an unsteady turbulent wake. When this happens it is no longer true that viscous effects and vorticity diffusion are confined to thin regions near the walls, or that the main body of the flow is inviscid. Although some specific numerical calculations of separated flows have been made, they still remain an unresolved area in fluid mechanics.

17.2 INVISCID MAIN FLOWS

As the Reynolds number becomes large, the viscous regions become vanishingly thin, so that almost the entire flow is governed by inviscid equations. The proper scales for these flows are the same as those used in Chapter 10 for the incompressible-flow equations: U, a characteristic velocity; L, a characteristic length of flow path or body; and ρ, the fluid density. The nondimensional continuity and momentum equations are 10.4.4 and 10.4.5:

$$\partial_i^* v_i^* = 0 \qquad 17.2.1$$

$$\frac{Dv_i^*}{Dt} = -\partial_i^* p^* + \frac{1}{\text{Re}}\partial_j^*\partial_j^* v_i^* \qquad 17.2.2$$

The simplified equations when Re → ∞, known as Euler's Equations, show that acceleration results solely from pressure forces:

$$\frac{Dv_i^*}{Dt^*} = -\partial_i^* p^* \qquad 17.2.3$$

The principle of dominant balance says that the solution of this equation should approximate the complete solution of 17.2.2 with Re → ∞.

The second important point about 17.2.3 is that we have lost the highest-order term, $\partial_j \partial_j v_i$. This means that we can no longer stipulate as many boundary conditions as for a viscous flow. The boundary condition that must be given up is the no-slip condition at a solid wall. In general there is no way to deduce mathematically which boundary condition must be given up. From past experience we know that the proper approach to inviscid flow is to hold on to the condition that the velocity normal to the wall is zero. Thus the proper boundary condition for inviscid flow is

$$[n_i v_i]_{\text{solid wall}} = 0 \qquad 17.2.4$$

This makes the wall into a surface containing streamlines. The velocity component along the wall is nonzero, except at stagnation points, and is determined by the inviscid solution. We realize that at the wall this solution is not correct, in the sense that it does not satisfy the no-slip condition.

Mathematically we may view the inviscid flow as the first term in an asymptotic expansion of the exact answer. The expansion parameter is Re → ∞. The problem is a singular perturbation with the nonuniform region (i.e., the place where the answer is incorrect) next to the walls. A boundary layer exists in these regions in order to complete the solution and satisfy the no-slip condition.

The vorticity in an inviscid flow is nondimensionalized as

$$\omega^* = \frac{\omega}{U/L} \qquad 17.2.5$$

With this definition the nondimensional vorticity equation becomes

$$\frac{D\omega_i^*}{Dt^*} = \omega_j^* \partial_j^* v_i^* + \frac{1}{\text{Re}} \partial_j^* \partial_j^* \omega_i^*$$

As Re → ∞, viscous diffusion becomes negligible, and we obtain the inviscid equation governing vorticity,

$$\frac{D\omega_i^*}{Dt^*} = \omega_j^* \partial_j^* v_i^* \qquad 17.2.6$$

The vorticity of a material particle changes by turning and stretching of the vortex lines. From Chapter 13 we recall that Helmholtz's theorem allows us to treat the vortex lines as material lines moving with the fluid particles. As a line stretches, the vorticity increases; as it shrinks, the vorticity decreases.

The existence of vorticity indicates that viscous forces are, or at least have been, active. In the case of inviscid flow we must choose the perfect tense in this statement. Viscous forces are not important in an inviscid flow carrying vorticity, but somewhere in the past history of the particle motion, viscous forces were active in order to generate the vorticity. As an example suppose that we place a fine mesh screen in front of the duct entrance in Fig. 17-1b. The screen is in a low-velocity region, and the screen wires have a moderate to low Reynolds number. Hence, as the flow goes through the screen, viscous forces are important. Next, imagine that the screen has a very fine mesh on one side gradually changing to a coarse mesh on the other side. The pressure drop across the screen will be the same everywhere, but the viscous forces will reduce the flow velocity where the screen has its finest mesh. The result is that the flow at the entrance to the duct has a velocity profile with nonzero vorticity. The flow within the duct is an inviscid flow carrying vorticity. Stretching and turning are now the only mechanisms to change the vorticity.

Note that the scale of the vorticity in 17.2.6 must be U/L, which is relatively small. Let us consider a flow that has a larger scale for the vorticity. In Fig. 17-1b a splitter plate has been added to the entrance along with a second screen, which has a different mesh. Now, the flow contains two regions of weak vorticity with a thin layer of stronger vorticity in the shear layer formed downstream of the plate. The main flow consists of two parts with a thin shear layer between them. The shear layer is really another type of boundary layer and can be treated in a manner similar to wall boundary layers.

A special class of inviscid flows occurs when the vorticity is zero. A flow that is both inviscid and irrotational is called an *ideal flow*, or equivalently a *potential flow* (note that irrotational flow implies inviscid flow but not the other way around). The discussion in the remainder of this section is limited to ideal flows.

The velocity field of an ideal flow is completely determined by two kinematic considerations: the rate of particle expansion and the rate of particle rotation are both zero. Mathematically these conditions are

$$\nabla \cdot \mathbf{v} = 0 \qquad 17.2.7$$

$$\nabla \times \mathbf{v} = \boldsymbol{\omega} = 0 \qquad 17.2.8$$

The solution of these equations is most easily found by using the velocity potential defined by

$$\mathbf{v} = \nabla \phi \qquad 17.2.9$$

The velocity potential ϕ exists if and only if the flow is irrotational ($\omega = 0$). The equation for ϕ is found by substituting 17.2.9 into 17.2.7. The result is the Laplace equation,

$$\nabla^2\phi = 0 \qquad 17.2.10$$

Boundary conditions appropriate for the solution of 17.2.10 are either to specify ϕ or to specify the normal derivative $\mathbf{n} \cdot \nabla\phi$ around a closed region. The second condition is used in fluid mechanics, as physically it corresponds to the velocity normal to the boundary:

$$[n_i v_i]_{\text{boundary}} = n_i \partial_i \phi \qquad 17.2.11$$

If the boundary is within the fluid, we must know the flow velocity across it. If the boundary is a stationary solid wall, then $n_i v_i = 0$.

We shall not go deeply or systematically into the mathematical properties of the Laplace equation. Many mathematics books cover the subject adequately. As we study several specific potential flows in the next chapter, we shall bring up the required mathematical results as they are needed. Nevertheless, some general characteristics should be noted.

Potential flows are dominated by the geometry. The dimensions of length and time, but not that of mass, occur in 17.2.7 through 17.2.11. The shape and locations of the walls of a duct or a closed body completely establish the velocities and streamlines. Except possibly for variables describing the boundary region (the aspect ratio of an ellipse, for example), the problem contains no parameters (the characteristic values U and L are absorbed in the nondimensional variables). The fact that geometry controls the flow pattern is emphasized by noting that the velocity may be found without ever using the momentum equation. With known velocities, the momentum equation, after integration to form the Bernoulli equation, is used to find the pressure field. The actual fluid density (and mass dimension) become important at this time.

Potential flows are elliptic in their mathematical classification. Any change in a boundary condition is felt instantaneously at all points in the fluid. The influence is of course greatest at points closest to the place where the change was made. Any change goes upstream as well as downstream with equal intensity. If we have a sphere about which there is a streaming flow, the flow anticipates the presence of the sphere and moves aside to go around it. The flow then closes and proceeds downstream. The presence of the sphere is felt equally upstream and downstream, yielding a symmetric flow pattern. Next, let us imagine that the sphere pulsates with a sinusoidal motion about the mean radius in addition to the streaming flow. Because the governing equation and the boundary conditions are linear, we may consider the velocity solution as the sum of the steady-flow solution and the solution for an oscillating sphere in a infinite medium (a problem we solved in Chapter 15). Furthermore, the pulsating effect is transmitted instantaneously throughout the flow. Their are no storage effects, as time derivatives are absent from 17.2.7 through 17.2.11. As a result, the solution depends only on the instantaneous position and velocity of the boundaries. The past history of the boundary motion has no influence on the flow pattern. History effects occur in boundary layers, and history effects occur at moderate Reynolds numbers, but ideal flows have no memory of previous states of motion.

17.3 PRESSURE CHANGES IN FLOWS

In this section we investigate the momentum equations in streamline coordinates with unit vectors **t**, **n**, and **b** for the tangential, normal, and binormal directions (see Section 4.9). The velocity has only a tangential component:

$$v_t = v, \qquad v_n = 0, \qquad v_b = 0 \tag{17.3.1}$$

and the vorticity has components (4.9.4)

$$\begin{aligned} \omega_t &= v(\mathbf{t} \cdot \nabla \times \mathbf{t}) \\ \omega_n &= \frac{dv}{db} \\ \omega_b &= \frac{v}{R} - \frac{dv}{dn} \end{aligned} \tag{17.3.2}$$

Now, consider the steady-flow momentum equation written in the form used in 12.3.4;

$$\partial_i\left(\frac{p}{\rho} + \tfrac{1}{2}v^2\right) = \varepsilon_{ijk}v_j\omega_k + \nu\partial_j\partial_j v_i \tag{17.3.3}$$

(The pseudo pressure variable includes the gravity potential; that is, $p/\rho \to p/\rho + gh$ in problems where the hydrostatic pressure is important.) It is also convenient to define the "total head" as the sum of pressure, kinetic energy, and potential energy,

$$H \equiv \frac{p}{\rho} + \tfrac{1}{2}v^2 + gh \tag{17.3.4}$$

These terms have the dimensions of energy per unit mass. In some engineering disciplines it is customary to use H/g as an alternate definition of the total head. This head has the dimension of length.

There are several different situations under which we can integrate 17.3.3. To derive our first Bernoulli equation let us take the component of 17.3.3 along a streamline by forming the dot product of **t** with 17.3.3. The term $\mathbf{v} \times \boldsymbol{\omega}$ is always perpendicular to the streamline (or to a vortex line); hence $\mathbf{t} \cdot (\mathbf{v} \times \boldsymbol{\omega})$ is zero. The result of integrating 17.3.3 along a streamline is

$$\int_1^2 t_i \partial_i H \, ds = \nu \int_1^2 t_i \partial_j \partial_j v_i \, ds$$

Since $t_i \partial_i H \, ds = dH$, we obtain

$$H_2 - H_1 = \nu \int_1^2 t_i \partial_j \partial_j v_i \, ds \qquad 17.3.5$$

This equation states that the net viscous force in the streamline direction is the only thing that changes the total head. If the viscous force accelerates the flow, then the head H increases. On the other hand, if the viscous force decelerates the flow, then H decreases.

Next, let us suppose that the flow is inviscid and steady, but that it has a nonzero vorticity. The vortex lines are material lines in this flow, and together with the streamlines they form a set of surfaces (sometimes called Lamb surfaces), as shown in Fig. 17-2. Since $\mathbf{v} \times \boldsymbol{\omega}$ is everywhere perpendicular to this surface and the viscous term is negligible by assumption, we can integrate 17.3.3 anywhere on the surface to find that

$$H \equiv \frac{p}{\rho} + gh + \tfrac{1}{2}v^2 = \text{const} \qquad \text{on a } \psi\text{–}\omega \text{ surface} \qquad 17.3.6$$

Each ψ–ω surface may have a different value of the constant.

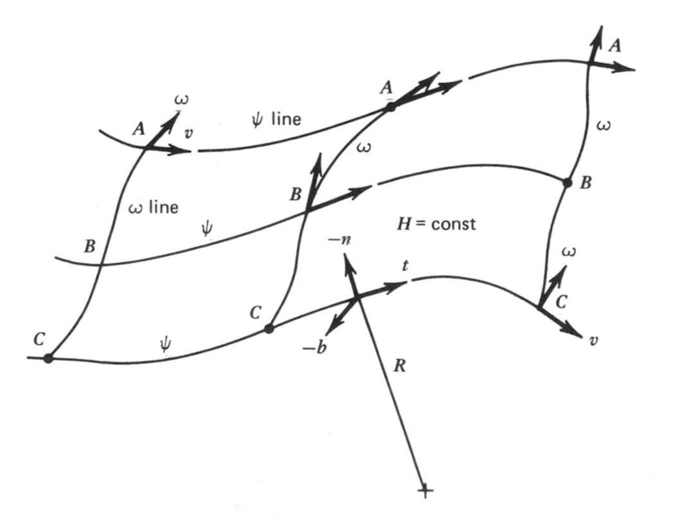

Figure 17-2 Bernoulli equation has the same constant on a Lamb surface (ψ–ω surface).

To interpret 17.3.6 consider the flow along a streamline. If the pressure and height of the streamline are such that $p/\rho + gh$ is constant, then the hydrostatic distribution prevails and $\frac{1}{2}v^2$ is a constant. The fluid particles are not being accelerated as they go along the streamline. On the other hand, a difference in $p/\rho + gh$ between two points on the streamline indicates an imbalance from the hydrostatic situation and a resulting fluid acceleration. We do not need to know the exact history of the acceleration, because the net pressure force ∇p has an effective potential in p and the net gravity force ρg_i has a potential ρgh. The integrated momentum equation is essentially a statement that the change in kinetic energy per unit mass, $\frac{1}{2}v^2$, between two points is the difference in the potentials $p/\rho + gh$.

The next special case is steady, inviscid, irrotational flow. The $\mathbf{v} \times \boldsymbol{\omega}$ term is now zero, and we can integrate 17.3.3 anywhere in the flow. One constant value of H is valid for the entire flow:

$$H = \frac{p}{\rho} + \tfrac{1}{2}v^2 + gh = \text{const} \qquad 17.3.7$$

This equation is the dynamic equation from which the pressure may be found after the velocities are determined as described in the previous section. It is also possible to find a slightly more general Bernoulli equation for inviscid, irrotational, but *unsteady* flow. That equation was given as 12.3.5.

We started the development above by considering a steady, viscous flow with vorticity and integrating the momentum equation along a streamline. Now we return to 17.3.3 and consider the component of the equation in the normal direction. When the $\mathbf{v} \times \boldsymbol{\omega}$ term is evaluated using 17.3.1 and 17.3.2, we find that 17.3.3 becomes

$$n_i \partial_i \left[\frac{p}{\rho} + \tfrac{1}{2}v^2 + gh \right] = -\frac{v^2}{R} + \frac{d}{dn}\left(\tfrac{1}{2}v^2\right) + \nu n_i \partial_j \partial_j v_i$$

or

$$\frac{d}{dn}\left(\frac{p}{\rho} + gh \right) = -\frac{v^2}{R} + \nu n_i \partial_j \partial_j v_i \qquad 17.3.8$$

Recall that R is the radius of curvature and n is positive toward the center of curvature.

One of the most useful arguments we can make from 17.3.8 proves that the hydrostatic pressure equation applies to flows whenever the streamlines are straight and parallel. In the region where the streamlines are straight, the radius of curvature is infinite. Furthermore the **n**, **b**, coordinates system may be taken as a rectangular system with an arbitrary choice of the normal and binormal directions. The viscous term in 17.3.8 is zero since v_n is zero:

$$n_i \partial_j \partial_j v_i = \partial_t \partial_t v_n + \partial_n \partial_n v_n + \partial_b \partial_b v_n = 0$$

Hence, 17.3.8 produces the hydrostatic-pressure–height relationship,

$$\frac{p}{\rho} + gh = \text{const} \qquad \text{across streamlines} \tag{17.3.9}$$

In a steady viscous flow, the assumption of straight parallel streamlines implies that the hydrostatic pressure equation applies in the plane *across* the streamlines. The constant of integration may change with position along the streamlines. For example, in the parallel flow in a straight tube, the constant decreases in the flow direction as the pressure drops.

A second special case of 17.3.8 occurs when the flow is inviscid. Then the equation becomes

$$\frac{d}{dn}\left(\frac{p}{\rho} + gh\right) = -\frac{v^2}{R} \tag{17.3.10}$$

When the streamlines are curved, the centrifugal force required to turn the flow is supplied by a pressure gradient. The pressure decreases toward the center of curvature. This is the reason the pressure in the center of a vortex is much lower than the pressure in the surrounding fluid.

The momentum equation in the binormal direction will be considered next. Working out the $\mathbf{v} \times \boldsymbol{\omega}$ term for this direction reduces 17.3.3 to

$$b_i \partial_i \left[\frac{p}{\rho} + \tfrac{1}{2}v^2 + gh\right] = \frac{d}{db}\left(\tfrac{1}{2}v^2\right) + \nu b_i \partial_j \partial_j v_i$$

or

$$\frac{d}{db}\left[\frac{p}{\rho} + gh\right] = \nu b_i \partial_j \partial_j v_i \tag{17.3.11}$$

Changes in $p/\rho + gh$ in the binormal direction are brought about only by viscous forces. The special case of steady inviscid flow allows 17.3.11 to be integrated to yield

$$\frac{p}{\rho} + gh = \text{const} \qquad \text{on a binormal line} \tag{17.3.12}$$

If, in addition, the flow is irrotational, we can subtract 17.3.12 from the Bernoulli equation 17.3.7 to prove that the velocity has a constant magnitude on a binormal line. We can use this fact to advantage in high-Reynolds-number ideal flows. Consider a solid body with an ideal inviscid flow over the surface. At the surface, the tangent and binormal directions must be in a plane tangent to the surface. Hence, the streamlines and binormal lines form an orthogonal net along the body surface. According to the results above, the pressure $(p/\rho + gh)$ and velocity are both constant on a binormal line. In principle, measurement of the surface pressure allows one to find the lines of constant pressure. Along these lines the inviscid velocity is constant (given by Bernoulli's equation) and the streamline direction is perpendicular to the surface isobars. In practice, the resolution of the pressure field is usually too coarse to give the flow direction accurately.

17.4 BOUNDARY LAYERS

At high Reynolds numbers, the viscous effects are confined to thin regions. Although the regions are thin, it is very important to know the details of the flow within them. Many processes of engineering interest—such as shear-stress, heat, and mass transfer—are controlled by the viscous regions. The term *boundary-layer theory* applies to regions next to walls, mixing layers between two portions of the flow moving at different speeds, thin wakes behind streamlined bodies, and even jets of fluid discharging into large reservoirs. The essential characteristics of these regions are that they are thin and that they have steep velocity gradients that make the viscous effects important. Frequently these regions become turbulent, but this does not invalidate the boundary-layer concept, which need only be modified to include the turbulent characteristics. (That, however, is no small problem.)

As we look at a flow from the outside, so to speak, we see that the vorticity or viscous effects are concentrated into thinner and thinner regions as the Reynolds number increases. The first approximation is that the flow is a completely inviscid flow enclosed by the geometry of the walls. The defect in this picture is that the inviscid flow cannot satisfy the no-slip boundary conditions. It only has enough flexibility to produce streamlines that follow the wall. The velocity on the wall streamline cannot be specified, but is determined as part of the inviscid solution. This error next to the wall always exists, no matter how high the Reynolds number becomes. Boundary-layer theory is a complement to inviscid-flow theory for the purpose of correcting the flow near the walls.

Next we derive the boundary-layer equations and discuss the proper boundary conditions. For definiteness consider the boundary layer on a solid wall in a two-dimensional flow as shown in Fig. 17-3. The wall will be taken as smooth and continuous, with a radius of curvature that is always large compared to the boundary-layer thickness. We erect a *boundary-layer coordinate system* where the surface $y = 0$ conforms to the body; the y axis is normal to the body, and the x axis is along the body. In order to emphasize the physical aspects, we make two simplifications in the derivation. First, we deal only with a two-dimensional flow, so that only two velocity components are nonzero. Second, we ignore terms in the equations that come from curvature in the coordinate systems. A more detailed derivation would show that these are indeed negligible (e.g. see Rosenhead 1963). As far as the boundary layer is concerned, the world is flat, but three-dimensional. Figure 17-4 shows the boundary layer as it is unwrapped from the body.

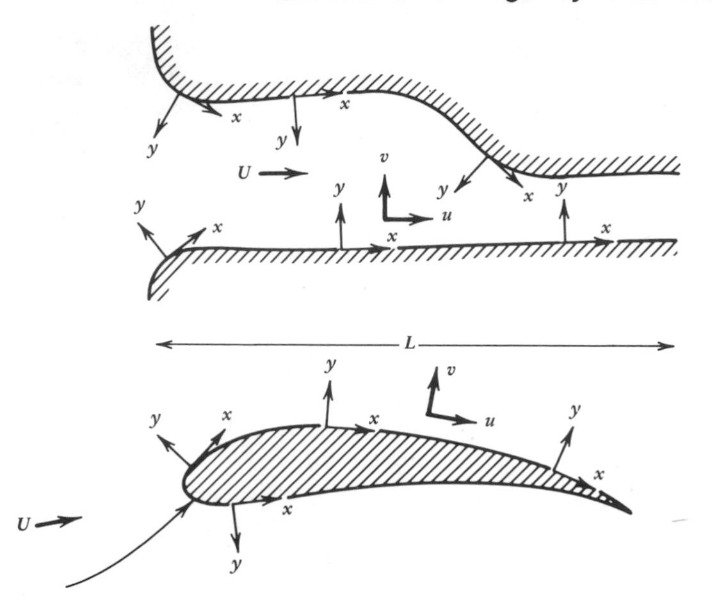

Figure 17-3 Boundary-layer coordinates conform to the wall.

Throughout the derivation keep in mind that we are making a correction to the inviscid flow so that the no-slip condition may be satisfied. We call the thickness of the boundary layer δ, and note the important fact that as $\text{Re} \to \infty$, δ is approaching zero. The reason inviscid-flow theory fails near the wall is that the proper scale for viscous effects is not L. The natural scale for the y direction is the distance δ. Let us see what happens to the u velocity component along the wall as we go across the boundary layer. At the wall the no-slip condition means that the velocity is always zero. On the inviscid side the slip velocity along the wall is determined by the inviscid solution. This velocity is zero at stagnation points and rises to values somewhat greater than U, the inviscid velocity scale, as a maximum. With this information we are in a

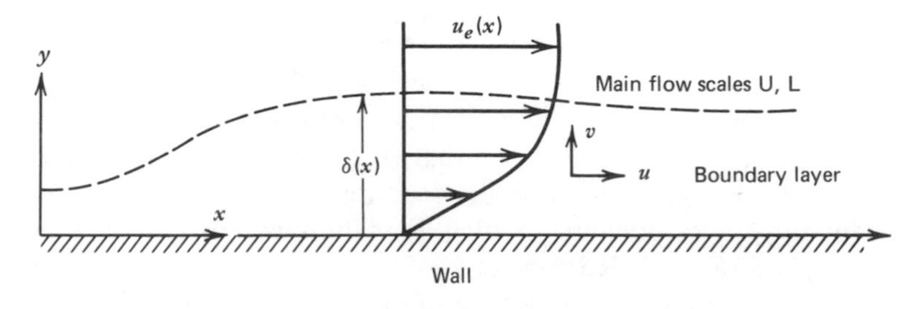

Figure 17-4 Boundary layer unwrapped from the wall.

position to estimate what the derivatives will be as we cross the boundary layer. A term like $\partial u/\partial y$ would be of the order

$$\frac{\partial u}{\partial y} \sim \frac{U-0}{\delta-0} = \frac{U}{\delta} \qquad 17.4.1$$

In inviscid theory a term like the one above would be estimated as

$$\frac{\partial u}{\partial y} \sim O\left[\frac{U}{L}\right]$$

This is the reason the inviscid theory failed at the wall; it contained an unreasonably low estimate of $\partial u/\partial y$.

So far we have argued that the boundary layer variables should have the following scales:

tangential velocity u:	scale U (same as inviscid)	
tangential distance x:	scale L (same as inviscid)	17.4.2
normal distance y:	scale δ (boundary layer thickness)	

The boundary-layer thickness is really unknown, except that we assume it approaches zero as Re goes to infinity.

We would expect v, the vertical or normal velocity in the boundary layer, to become zero as the boundary layer becomes thin. This is true, but it is not a precise enough estimate for our purposes. Let us introduce an unknown scale α for the vertical velocity. The nondimensional normal velocity will be

$$v^* = \frac{v}{\alpha} \qquad 17.4.3$$

Now we examine the continuity equation for the boundary layer. Below this equation we estimate the size of the terms:

$$\begin{array}{ccc} \dfrac{\partial u}{\partial x} & + \quad \dfrac{\partial v}{\partial y} & = 0 \\ O\left[\frac{U}{L}\right] & O\left[\frac{\alpha}{\delta}\right] & \end{array} \qquad 17.4.4$$

One general rule of incompressible fluid mechanics is that you should never drop a term from the continuity equation. In other words, don't let the flow gain or lose mass. (If we posed the theory in terms of the stream function, this rule would automatically be enforced.) Applying this principle means that α must be of order

$$\alpha \sim \frac{\delta}{L} U \qquad 17.4.5$$

When this is true, both terms in the continuity equation are the same size and no approximation to that equation occurs. The boundary layer is governed by the complete continuity equation 17.4.4.

The y-direction momentum equation will be considered next. We write the equation, and then below each term our guess as to its size. We do not know the proper pressure scale, so we introduce an unknown scale P. The equation is

$$\begin{array}{ccccccccc} u\dfrac{\partial v}{\partial x} & + & v\dfrac{\partial v}{\partial y} & = & -\dfrac{1}{\rho}\dfrac{\partial p}{\partial y} & + & \nu\dfrac{\partial^2 v}{\partial x^2} & + & \nu\dfrac{\partial^2 v}{\partial y^2} \\ O\left[U\dfrac{1}{L}\dfrac{\delta}{L}U\right] & & O\left[\dfrac{\delta}{L}U\dfrac{1}{\delta}\dfrac{\delta}{L}U\right] & & O\left[\dfrac{P}{\rho\delta}\right] & & O\left[\dfrac{\nu}{L^2}\dfrac{\delta}{L}U\right] & & O\left[\dfrac{\nu}{\delta^2}\dfrac{\delta}{L}U\right] \end{array} \quad 17.4.6$$

Reorganizing the orders into nondimensional form produces

$$O\left[\left(\frac{\delta}{L}\right)^2\right] + O\left[\left(\frac{\delta}{L}\right)^2\right] = O\left[\frac{P}{\rho U^2}\right] + O\left[\frac{1}{\mathrm{Re}}\left(\frac{\delta}{L}\right)^2\right] + O\left[\frac{1}{\mathrm{Re}}\right]$$

For a moment let us assume that the pressure scale is the inviscid scale ρU^2 and see what happens. With this assumption, terms in the y-momentum equation will have the following orders as $\mathrm{Re} \to \infty$:

$$O[0] + O[0] = O[1] + O[0] + O[0]$$

Thus, only one term is of order one, and the boundary-layer y-momentum equation reduces to a statement that

$$0 = \frac{\partial p}{\partial y} \quad \Rightarrow \quad p = p(x) \qquad 17.4.7$$

From this we conclude that the pressure is constant across a boundary layer. The pressure on the inviscid side of the layer is fixed by the inviscid flow. Whatever value occurs is impressed across the boundary layer without change. Since the proper inviscid scale for the pressure is ρU^2 and this same pressure is carried across the boundary layer, ρU^2 is also the proper boundary-layer pressure scale. Our assumption that $P = \rho U^2$ leads to approximations that are consistent and reasonable.

The fact that the pressure is constant across the boundary layer is an important result. It means that pressure forces on a body are solely the result of the inviscid flow (a geometry-dominated flow). They are not modified by the boundary layer. As an example of an application of this principle, let us consider the lift force on an airfoil. It is a direct result of pressure forces. At high Reynolds numbers the boundary layers become so thin that the pressure forces, and hence the lift force, are determined by the inviscid flow. Further increase in the Reynolds number will not change the lift force (when nondimensionalized by inviscid scales). This principle is the basis of wind-tunnel tests. It is impractical to always make the Reynolds number of the model equal to that of the actual prototype. But if one tests at a sufficiently high Reynolds number, the lift no longer changes and the test will predict the lift of the prototype at any higher Reynolds number. This argument does not apply to the drag force, because drag is a combination of both viscous and pressure forces. The argument also fails for bluff bodies, because they have regions of separated flow where boundary-layer theory itself is invalid.

Now we turn to the x-direction momentum equation for the boundary layer,

$$u\frac{\partial u}{\partial x} + v\frac{\partial u}{\partial y} = -\frac{1}{\rho}\frac{dp}{dx} + \nu\frac{\partial^2 u}{\partial x^2} + \nu\frac{\partial^2 u}{\partial y^2}. \qquad 17.4.8$$

The size of each term is estimated below:

$$O\left[\frac{U^2}{L}\right] + O\left[\frac{\delta}{L}U\frac{U}{\delta}\right] = O\left[\frac{1}{\rho}\frac{\rho U^2}{L}\right] + O\left[\frac{\nu}{L^2}U\right] + O\left[\frac{\nu}{\delta^2}U\right]$$

Clearing U/L so that the terms are nondimensional produces

$$O[1] + O[1] = O[1] + O\left[\frac{1}{\mathrm{Re}}\right] + O\left[\left(\frac{\delta}{L}\right)^{-2}\frac{1}{\mathrm{Re}}\right] \qquad 17.4.9$$

If we apply the limit $\mathrm{Re} \to \infty$ to 17.4.9, the next to last term definitely vanishes and the first three terms definitely stay. The question concerns the last viscous term. It has an order that is an indeterminate form,

$$\frac{1/\mathrm{Re}}{(\delta/L)^2} \sim \frac{0}{0} \to ? \qquad 17.4.10$$

We have been operating up to this point on the assumption that $\delta/L \to 0$ as $\mathrm{Re} \to \infty$, but we have not specified how fast that happens. There are three distinct possibilities: If $(\delta/L)^2$ goes to zero slower than $1/\mathrm{Re}$, then the ratio 17.4.10 becomes zero (this is in fact just a mathematical statement of what "slower" really means); if $(\delta/L)^2$ goes to zero faster than $1/\mathrm{Re}$, then the ratio is infinity; finally, if $(\delta/L)^2$ goes to zero at the same rate as $1/\mathrm{Re}$ then the ratio is a finite number. Let us consider the ramifications of each of these possibilities, one at a time.

If

$$\frac{1/\mathrm{Re}}{(\delta/L)^2} \to 0 \quad \text{as} \quad \mathrm{Re} \to \infty$$

then both viscous terms in the momentum equation are small. We are left with the same momentum equation as for the inviscid flow. Our boundary layer is governed by inviscid equations, and we cannot satisfy the no-slip conditions. This possibility must be thrown out, as it does not afford us enough flexibility to correct the inviscid flow.

The second possibility is if

$$\frac{1/\text{Re}}{(\delta/L)^2} \to \infty \qquad \text{as} \quad \text{Re} \to \infty \tag{17.4.11}$$

This case requires us to reorganize 17.4.9 by multiplying by $(\delta/L)^2$ Re. The terms now have the following orders:

$$O\left[\frac{(\delta/L)^2}{1/\text{Re}}\right] + O\left[\frac{(\delta/L)^2}{1/\text{Re}}\right] = O\left[\frac{(\delta/L)^2}{1/\text{Re}}\right] + O\left[\left(\frac{\delta}{L}\right)^2\right] + O[1]$$

Applying 17.4.11 to the relation above leads to the following result as $\text{Re} \to \infty$:

$$O[0] + O[0] = O[0] + O[0] + O[1]$$

For this case the momentum equation governing the boundary layer is the single term

$$0 = \frac{\partial^2 u}{\partial y^2} \tag{17.4.12}$$

The solution to this equation can be given immediately. It is

$$\begin{aligned} \frac{\partial u}{\partial y} &= C_1(x) \\ u &= C_1(x)y + C_2(x) \end{aligned} \tag{17.4.13}$$

One boundary condition is that $u = 0$ at $y = 0$; this mans that $C_2 = 0$. This solution says that the boundary layer has a constant shear stress and a linear velocity profile. Such a solution will not smoothly match the inviscid flow, as it has a discontinuity in the shear stress. For these reasons we reject the second possibility.

The last possibility, and the one correct choice, is that the ratio 17.4.10 is finite. This means

$$\frac{\delta}{L} \sim \sqrt{\frac{1}{\text{Re}}} \tag{17.4.14}$$

From 17.4.9 and 17.4.14 we find that the momentum equation for the boundary layer is

$$u\frac{\partial u}{\partial x} + v\frac{\partial u}{\partial y} = -\frac{1}{\rho}\frac{dp}{dx} + \nu\frac{\partial^2 u}{\partial y^2} \qquad 17.4.15$$

This equation differs from the inviscid momentum equation by retaining one viscous term. It offers us sufficient flexibility to meet the no-slip condition at the wall and to match smoothly to the inviscid flow.

An alternate approach to finding 17.4.14 is to use the Rayleigh argument—the diffusion of vorticity is $\delta \sim \sqrt{\nu t}$ and t is the flow time $t = L/U$. A physical argument such as this is often useful in pointing out the proper path to take in a complicated analysis.

The boundary-layer momentum equation 17.4.15 has a much different character than the inviscid flow equations. The y direction is dominated by the viscous diffusion term, the highest-order term in the equation. The term $u\,\partial u/\partial x$ on the left hand side of the equation together with the viscous term $\nu\,\partial^2 u/\partial y^2$, give the problem a parabolic mathematical character.

We have already encountered an example of a parabolic problem when we studied Rayleigh's problem for the impulsive motion of a fluid above a fixed flat plate:

$$\frac{\partial u}{\partial t} = \nu\frac{\partial^2 u}{\partial y^2}$$

$$\begin{aligned} u(y, t = 0) &= u_{\text{initial}} = 0 \\ u(y = 0, t) &= u_{\text{wall}} = 0 \\ u(y \to \infty, t) &= u_{\text{external}} = U \end{aligned} \qquad 17.4.16$$

In particular, notice that the boundary conditions are prescribed on an open domain in the y, t plane: an initial condition $u = 0$, and the values of u at two space points $y = 0$ and $y \to \infty$. The boundary-layer equation 17.4.15 has a similar character to 17.4.16 except that x in the boundary layer takes on the role of time in Rayleigh's problem. This analogy is mathematically correct, and we can use 17.4.16 as a guide to the proper boundary conditions for the boundary-layer equations.

The usual boundary-layer boundary conditions are an initial profile,

$$u(x = x_0, y) = u_{\text{initial}} = u_{\text{in}}(y) \qquad 17.4.17$$

the no-slip condition at the wall,

$$\begin{aligned} u(x, y = 0) &= 0 \\ v(x, y = 0) &= 0 \end{aligned} \qquad 17.4.18$$

and an external flow condition,

$$u(x, y \to \infty) = u_{\text{external}} = u_e(x) \qquad 17.4.19$$

We cannot impose a downstream boundary condition. That is prohibited for a parabolic equation. Also, a prescribed initial profile for v cannot be given, because, in principle, the continuity and momentum equations together with the initial velocity $u(y)$ could be solved to find $v(y)$. Similarly, to specify $v(x, y \to \infty)$ would overdetermine the problem. Whatever value the solution produces for $v(x, y \to \infty)$ must be accepted.

The external flow condition also deserves special comment. The exact value of $u_e(x)$ is to be determined by matching the boundary layer to the inviscid flow. As we look from the inviscid flow toward the boundary layer and allow the Reynolds number to increase, the thickness of the boundary layer decreases toward zero. Our first approximation for the inviscid flow was to neglect the thickness of the boundary layer and find a flow that slips over the surface of the body. Thus, we now argue that the external velocity of the boundary layer should be the inviscid flow evaluated at the wall,

$$u_e(x) = u_{\text{inviscid}}(\text{wall}) \qquad 17.4.20$$

This means that we must know the inviscid flow before we attempt to analyze the boundary layer.

Let us continue by considering things from the boundary-layer side. The proper nondimensional y variable for the boundary layer is scaled by the boundary-layer thickness:

$$y^* = \frac{y}{\delta} = \frac{y}{L/\sqrt{\text{Re}}} \qquad 17.4.21$$

The scale unit for the boundary layer becomes smaller and smaller as the Reynolds number becomes large. If you wanted to watch the events within the boundary layer, you would have to shrink yourself down in size as the Reynolds number increased. A constant unit in y^* occupies a smaller and smaller fraction of L as Re increases; $y/L = y^*/\sqrt{\text{Re}}$. The next question is: where is the inviscid flow in terms of boundary-layer variables? As far as the boundary layer is concerned, you must go out to $y^* \to \infty$ before you get to the inviscid flow. To make things clearer we can even write subscripts on u and y in 17.4.19 to show that these are boundary-layer variables. Equations 17.4.19 and 17.4.20 tell us precisely how to get the boundary condition on u by matching the boundary layer and the inviscid flow

$$u_{\text{bl}}(x, y_{\text{bl}} \to \infty) = u_e(x) = u_{\text{inviscid}}(\text{wall}) \qquad 17.4.22$$

The external velocity for the boundary layer—that is, the velocity as the boundary-layer distance approaches infinity—is equal to the inviscid-flow velocity evaluated at the inviscid coordinates corresponding to the body surface. This is the matching between the inviscid flow and the boundary layer.

The inviscid flow supplies a major boundary condition to the boundary layer through 17.4.22. It also determines the pressure within the boundary layer. Since the inviscid streamline on the wall must obey Bernoulli's equation, we have that the boundary layer pressure is given by

$$p(x) + \tfrac{1}{2}u_e^2(x) = \text{const} \qquad 17.4.23$$

The boundary-layer momentum equation requires that we know the pressure gradient. From 17.4.23 we find

$$-\frac{1}{\rho}\frac{dp}{dx} = u_e\frac{du_e}{dx} \qquad 17.4.24$$

This expression may also be derived by evaluating the boundary-layer momentum equation, 17.4.15, as $y \to \infty$. The assumption that the boundary layer smoothly matches the inviscid flow implies that $\partial u/\partial y$, $\partial^2 u/\partial y^2$, and all other y derivatives of u become zero as $y \to \infty$. Equation 17.4.24 results from applying these conditions to 17.4.15.

From an analytical point of view, the inviscid-flow problem is solved first. Then we can evaluate the velocity on the wall, $u_e(x)$, for use in the boundary-layer calculation. Experiments may be substituted for a knowledge of the inviscid flow. To do this one makes pressure measurements on the surface of the body. In principle this pressure is the inviscid pressure, and through Bernoulli's equation 17.4.23 the inviscid velocity $u_e(x)$ is determined.

Before we discuss the physical character of boundary layers it is good to collect together in one place a complete mathematical statement of the problem. The boundary-layer equations are

$$\begin{aligned}
&\frac{\partial u}{\partial x} + \frac{\partial v}{\partial y} = 0 \\
&u\frac{\partial u}{\partial x} + v\frac{\partial u}{\partial y} = u_e\frac{du_e}{dx} + \nu\frac{\partial^2 u}{\partial y^2} \\
&u(x = x_0, y) = u_{\text{in}}(y) \\
&u(x, y = 0) = 0 \\
&v(x, y = 0) = 0 \\
&u(x, y \to \infty) = u_e(x)
\end{aligned} \qquad 17.4.25$$

Two pieces of information are needed to complete the problem: the initial velocity profile $u_{in}(y)$ and the external velocity variation $u_e(x)$. Also note that unlike the inviscid flow, where the velocity field depended only upon geometry, the boundary-layer equations contain ν, the kinematic viscosity, as a parameter.

The parabolic nature of the boundary-layer problem means that a signal will travel across the layer at infinite speed. For example a small pulsation at the wall or an injection of fluid at the wall instantaneously changes the entire velocity profile across the boundary layer. In the x direction, the direction along the wall, events are convected with the flow velocity. Thus, whatever disruption the wall pulsation causes is not felt downstream until a later time when fluid that was at the pulsed point arrives downstream. This means that boundary layers contain a history dependence that comes from the initial profile. (Note that from the mathematical standpoint the initial profile can be at any place we choose. We can start the boundary-layer calculation at an arbitrary position. The velocity profile at that place becomes the initial profile.)

Let us consider a flat plate with a block initial profile and the external velocity $u_e = U$, which is constant over the plate (Fig. 17-5a). The boundary layer grows in a regular manner as the flow proceeds along the plate. Now, we wish to compare this problem with a second situation where a rounded leading edge is attached to the plate. Around the nose of the plate there are pressure gradients, and only after we pass the position x_0 does the pressure become uniform and u_e take on a constant value. At x_0 a certain initial velocity profile exists, which is not the block profile. We have constructed two situations where u_e is the same but the initial profiles are different. As we go downstream, these boundary layers have different velocity profiles as a result of their different initial conditions. The boundary layer remembers its initial flow situation. The further downstream we go, the less difference we can detect between the two boundary layers. The effect of the initial condition gradually dies out. At positions that are the same distance downstream, the two boundary layers will never be exactly the same. However, the careful observer would notice that if we shifted the origin of the rounded-nose boundary layer so that an effective position $\bar{x}_0$ was used as the origin, then the two boundary layers would approach the same profile as $x \to \infty$: the effect of the initial profile is equivalent, at downstream positions, to a virtual shift in the origin of the boundary layer.

Boundary layers transfer effects only in the downstream direction; there can be no upstream influence in a boundary layer. For instance imagine that a flat plate forms one wall of a flow channel of constant area. The external velocity u_e is constant in this case, just as in the previous example. Now, a second test is run where a large hump is attached to the opposite wall as shown in Fig. 17-5. In this case the inviscid flow must accelerate as it goes through the area constriction, causing an increase in the velocity u_e on the flat wall. Suppose for the sake of argument that the increase in u_e begins at a certain location x_1. This is not actually true, because inviscid flows are sensitive to all boundary conditions, but it is not an unreasonable approximation. Now, if we calculate

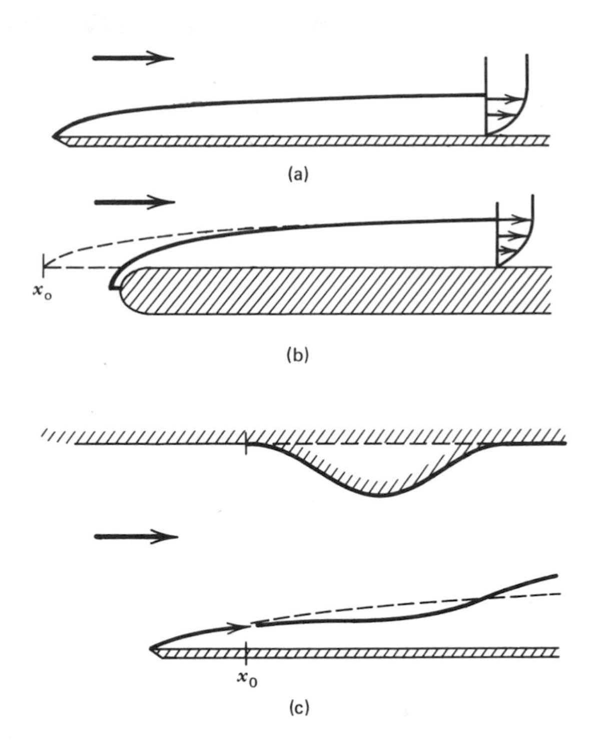

Figure 17-5 Boundary layers do not have any upstream influence: (a) The boundary layer develops on a flat plate. (b) For a boundary layer on a rounded nose, the development downstream is equivalent to (a) with a displaced effective origin. (c) The boundary layer in a flat channel develops on the dashed line. A bump on the opposite wall causes a different development after x_0.

the boundary condition for the new $u_e(x)$, we shall find exactly the same solution up to the position x_1. Nothing has changed in either $u_{in}(y)$ or $u_e(x)$ up to x_1, and since the mathematical nature of the boundary layer prohibits any upstream influence, the solution must be the same. Only when $u_e(x)$ starts changing does the boundary layer become different from the original problem. We shall find that when $u_e(x)$ increases, the boundary layer tends to thin out; when $u_e(x)$ decreases, the boundary layer thickens rapidly.

17.5 CONCLUSION

Unseparated flow at high Reynolds number may be divided into two complementary flows: an inviscid flow and a boundary layer. Geometry, that is, the shape of the bounding walls and their motion, determines the inviscid flow pattern. This flow then drives the boundary layer through two effects: the boundary layer-velocity at infinity must match the inviscid velocity over the wall, and the pressure gradient in the inviscid flow is directly imposed within the boundary layer. The special qualities of typical inviscid flows will be given in the next two chapters; Chapter 20 concerns boundary-layer theory.

PROBLEMS

17.1 (A) A jet of water traveling at a relative velocity of 50 ft/sec encounters the blade of a Pelton wheel turbine. The flow path over the blade is 4 in. long. Estimate the thickness of the boundary layer at the end of the blade.

17.2 (B) Air at room temperature flows in a tube 10 cm in diameter. The tube ends 1 cm from a flat wall that is perpendicular to the flow. An end flange 30 cm in diameter is flat so that the flow turns comes radially out along the wall exiting to the atmosphere. The flow in the tube is uniform at 15 m/sec. Estimate the thickness of the boundary layers on the wall and the flange.

17.3 (B) Compute the pressure variation along the flange and in the pipe. Estimate the pressure behavior in the region where the flow turns.

17.4 (B) Find the form of each of the following equations that is appropriate for a two dimensional boundary layer: $\boldsymbol{\omega} = \nabla \times \mathbf{v}$, $\nabla \cdot \boldsymbol{\omega} = 0$, $\nabla^2\psi = \omega_z$, $D\boldsymbol{\omega}/Dt = \boldsymbol{\omega} \cdot \nabla\mathbf{v} + \nu\nabla^2\boldsymbol{\omega}$.

17.5 (B) Consider a fan blade operating at a high Reynolds number. Low pressures exist on the upper surface. They are especially low in the root region but not quite so low as the tip is approached. Why will contour lines of constant pressure on top of the blade have cusp shapes?

17.6 (B) A cone with a 45° half angle and 20-cm height is in an inverted position and filled with water. The apex is open to the atmosphere. Find the pressure at all points on the cone surface if the cone is moved upward according to $U = 2 + 5t + 3t^3$ where U is in m/sec when t is in seconds. What is the pressure distribution at $t = 3$ sec?

18 Ideal Flows in a Plane

Inviscid flows occur at high Reynolds numbers UL/ν. The student may hear it said that inviscid flows are situations where the fluid is frictionless or has zero viscosity. These statements are not meant to be taken literally. The question is, does the flow pattern have any significant unbalanced viscous forces? If not, the equations that govern the flow are those obtained by setting $\mu = 0$ in the Navier–Stokes equations. This is completely equivalent to letting the Reynolds number become infinite with a fixed velocity scale U and a fixed length scale L.

Ideal flow is a special type of inviscid flow where the vorticity is zero. The particles in an ideal flow have never experienced an unbalanced shear stress and therefore are not rotating. The unbalanced pressure and gravity forces that produce the flow cannot induce any particle rotation.

Let us consider some typical situations where ideal flow is a reasonable assumption. As the first example, consider a wing mounted in a uniform airstream. The oncoming flow is without vorticity, and thus the major portion of the flow contains irrotational motion. The vorticity that is generated at the surface of the wing is confined to a thin viscous wake and to two vortices coming from the tips of the wing. A second example is an internal flow where a large reservoir supplies an irrotational flow to a duct or channel. Again the flow keeps the vorticity confined to the walls, and the main flow is an ideal flow. In this case, for any given Reynolds number VD/ν, the duct may become so long that the vorticity diffuses away from the wall to contaminate a significant region of the cross section. When this happens the flow is no longer ideal.

If flow separation occurs, either on an external flow or within an internal flow, a finite portion of the flow is occupied by the wake or recirculation region. Consequently the flow is no longer completely irrotational. A strict application of the inviscid-flow boundary-layer theory fails in these cases. Nevertheless, there is a large portion of the flow where the motion remains inviscid and irrotational. The flow in these regions still obeys the ideal-flow equations. Because these equations are elliptic, the wake region exerts an upstream influence. From one viewpoint, the inviscid flow is over an effective body shape, which is the real body plus the influence of the wake. The difficulty with this approach is that the size and shape of the wake are not known beforehand. Several attempts have been made to model such flows with computer solutions. A necessary ingredient is an assumption about the nature of the wake and the way it interacts with the main flow. A general theory of this type does not exist and may not even be possible.

In this chapter we restrict ourselves to the study of unseparated ideal flows. In fact, all of the flows will be two-dimensional, a mathematical convenience adopted for simplicity. The two-dimensional simplification is not critical, as the types of flow patterns we study are also elements of three-dimensional flow patterns.

18.1 PROBLEM FORMULATION FOR PLANE IDEAL FLOWS

Ideal flows are dominated by geometry. The position and shape of the body or the confining walls determines the flow pattern. A solution for the streamlines and the velocity field may be found from the two kinematic requirements that the particles do not rotate (zero vorticity) and that they do not expand (zero divergence). The condition $\boldsymbol{\omega} = 0$ is all that is required mathematically for the existence of a velocity potential. The potential ϕ is defined by

$$v_i = \partial_i \phi \qquad 18.1.1$$

Substituting 18.1.1 into the second kinematic requirement, $\nabla \cdot \mathbf{v} = 0$, shows that the velocity potential is governed by the Laplace equation,

$$\nabla^2 \phi = 0 \qquad 18.1.2$$

A function that satisfies the Laplace equation is called a harmonic function.

Since the flow pattern is completely determined by kinematics, what role does the momentum equation have? The momentum equation can be integrated to yield the pressures. The unsteady form of Bernoulli's equation (see Section 12.3 for a derivation) is

$$\frac{\partial \phi}{\partial t} + \frac{p}{\rho} + \tfrac{1}{2}\mathbf{v} \cdot \mathbf{v} + gz = C(t) \qquad 18.1.3$$

After the velocity potential is found, everything is known in this equation except the pressure. We might characterize ideal flow by the following statement: in an ideal flow the pressure adjusts itself according to Bernoulli's equation so that the fluid is accelerated to those values of velocity dictated by the geometry of the boundaries.

There is an alternative method of formulating the problem, which applies only to two-dimensional flows, either plane or axisymmetric. The stream function, which we studied in Section 12.1. applies to these flows, and it also obeys the Laplace equation

$$\nabla^2 \psi = 0 \qquad 18.1.4$$

Hence, ψ as well as ϕ is a harmonic function. Recall that for a plane flow, the velocity components are related to the stream function by

$$u = \frac{\partial \psi}{\partial y}, \qquad v = -\frac{\partial \psi}{\partial x} \tag{18.1.5}$$

Note that 18.1.4 is again a kinematic condition. It is a simplification of the mathematical identity $\nabla^2\psi = \omega_z$.

Most of the analysis we shall do in this chapter uses complex-variable theory, which is a very powerful mathematical method to find solutions of the Laplace equation. Unfortunately it is restricted to the plane two-dimensional case, where the Laplace equation takes the form

$$\frac{\partial^2 \phi}{\partial x^2} + \frac{\partial^2 \phi}{\partial y^2} = 0, \qquad \frac{\partial^2 \psi}{\partial x^2} + \frac{\partial^2 \psi}{\partial y^2} = 0$$

Complex variables do not work for axisymmetric flows because the Laplace equation has a different form in cylindrical or spherical coordinates. In the remainder of this section we shall study the special complex-variable nomenclature that is used in ideal flows.

We begin by letting the physical plane in which the flow occurs be represented by the complex variable z. The relation between z and the Cartesian variables x, y and with the polar coordinates, r, θ is

$$z = x + iy = re^{i\theta} \tag{18.1.6}$$

Flow properties may be expressed as complex functions of z. For example, the *complex potential* $F = F(z)$ is defined as

$$F = F(z) = \phi(x, y) + i\psi(x, y) \tag{18.1.7}$$

The real part of F is the velocity potential, and the imaginary part is the stream function. The motive behind this definition is the fact that any analytic function of a complex variable has real and imaginary parts that are each solutions to the Laplace equation.

As an example consider the analytic function

$$\begin{aligned} F &= iz^2 \\ &= i(x + iy)^2 = i(x^2 + 2ixy + i^2y^2) \\ &= -2xy + i(x^2 - y^2) \\ &= \phi + i\psi \end{aligned}$$

The real part of F, $\phi = -2xy$, and the imaginary part of F, $\psi = x^2 - y^2$, are both harmonic functions because they satisfy $\nabla^2\phi = 0$ and $\nabla^2\psi = 0$. Another example is

$$F = \sin z = \sin(x + iy)$$

$$= \sin x \cosh y + i \cos x \sinh y$$

Hence

$$\phi = \sin x \cosh y$$

and

$$\psi = \cos x \sinh y$$

are both harmonic functions satisfying $\nabla^2\phi = 0$ and $\nabla^2\psi = 0$.

Calculus operations on an analytic function of a complex variable can be performed using the same rules that one uses for real variables. The derivative of F is known as the *complex velocity*. It is

$$W(z) \equiv \frac{dF}{dz} \tag{18.1.8}$$

We interpret 18.1.8 by a special computation of dF/dz. Consider F as a function of x, y and allow $dz = dx + i\,dy$ to be taken only in the dx direction, that is,

$$W = \frac{\partial F}{\partial x} = \frac{\partial \phi}{\partial x} + i\frac{\partial \psi}{\partial x}$$

Equations 18.1.1 and 18.1.5 show that this relation may be interpreted in terms of the velocity components as

$$W = u - iv \tag{18.1.9}$$

or in polar form, where q is the magnitude of the velocity and α its angle,

$$W = qe^{-i\alpha} \tag{18.1.10}$$

The complex velocity is actually the complex conjugate of the velocity vector. The complex potential and the complex velocity are the essential ideas that relate the flow quantities to the theory of complex variables.

The student will find an introduction to the elements of complex variables in Churchill et al. (1974). In the remainder of this section we point out some of the algebra that is especially useful in the analysis of flows.

The dependent and independent variables in a complex function can be regarded as two-dimensional vectors in a plane. Geometrically the function

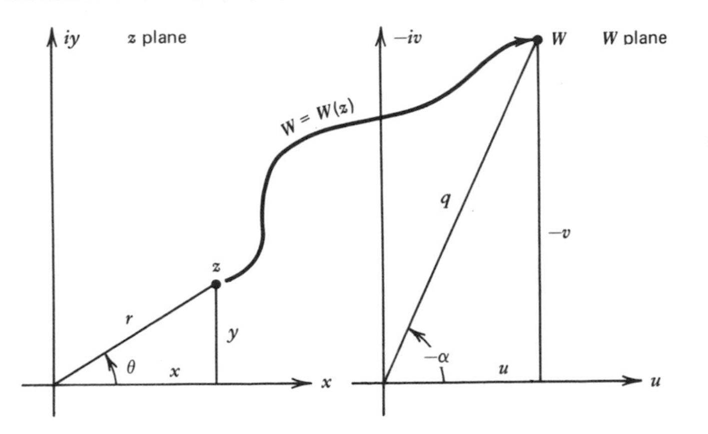

Figure 18-1 A complex function as a mapping from the z plane to the W plane. The example shows the complex velocity W as a function of the position in the real plane z.

$W = W(z)$ is thought of as a mapping or transformation from the z plane to the W plane (Fig. 18-1). Each point in the z plane represents the tip of a vector and has, through the function $W(z)$, an associated point or vector in the W plane. The physical flow occurs in the z plane, while the W plane consists of the velocity vectors.

The complex conjugate of $z = x + iy$ is found by replacing i with $-i$; the complex conjugate of z is $\bar{z} = x - iy$. Next, consider the complex function $W = W(z)$. If we want the complex conjugate $\overline{W}$, we replace i with $-i$ everywhere it occurs in the function $W(z)$. This means that $z \rightarrow \bar{z}$ and that any place where i appears explicitly, it is replaced by $-i$. For example, if $W(z) = z + i \sin z$, then $\overline{W} = \overline{W}(\bar{z}) = \bar{z} - i \sin \bar{z}$. Notice that for $W = W(z)$ we denote the complex conjugate as $\overline{W} = \overline{W}(\bar{z})$. A notation $W(\bar{z})$ would only imply that we set $z \rightarrow \bar{z}$ in the function $W(z)$. Likewise, the notation $\overline{W}(z)$ might be mistaken to mean that we replaced i with $-i$ in the function $W(z)$ but did not change z.

One of the uses of the complex conjugate is to find the magnitude of the vector. In the case of the velocity, we have that the flow speed q is given by

$$q^2 = W\overline{W} \tag{18.1.11}$$

Another useful mathematical expression is illustrated in Fig. 18-2. Frequently it is advantageous to work with polar coordinates r, θ. The corresponding velocity components are related to the rectangular components by the equations

$$\begin{aligned} u &= v_r \cos \theta - v_\theta \sin \theta \\ v &= v_r \sin \theta + v_\theta \cos \theta \end{aligned} \tag{18.1.12}$$

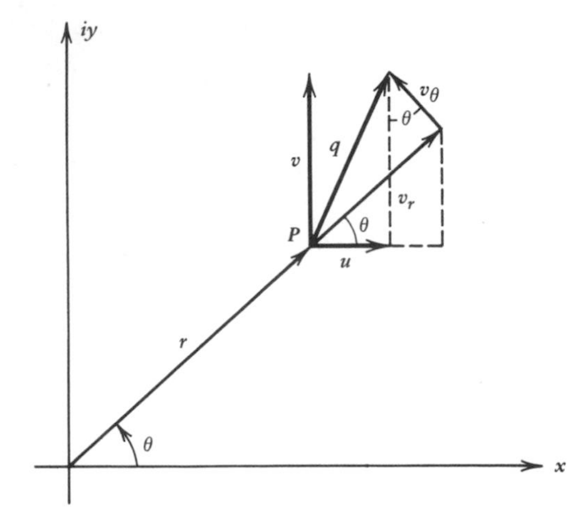

Figure 18-2 Velocity components displayed as a vector at the point P in real space in rectangular and cylindrical coordinates.

To obtain an expression for the complex velocity, we substitute these expressions into $W = u - iv$. This yields

$$W = [v_r(r,\theta) - iv_\theta(r,\theta)]e^{-i\theta} \tag{18.1.13}$$

Working in rectangular coordinates, one finds the velocity components u and v by substituting $z = x + iy$ into $W = W(z)$ and separating the result into its real and imaginary parts. Working in polar coordinates, one substitutes $z = re^{i\theta}$ into $W = W(z)$ and arranges the equation in the form 18.1.13. In this form the velocity components v_r and v_θ may be identified.

18.2 SIMPLE PLANE FLOWS

Since any analytic function represents some ideal flow, we can turn the problem around. Instead of choosing a definite flow geometry to analyze, we look at simple mathematical functions and see if they represent a practical flow situation.

We begin by noting that a constant may be added to F without changing the velocities. This is a reflection of the fact that ϕ and ψ may have an arbitrary constant added without changing the velocity. The next simplest function to a constant is $F \propto z$. In particular we let the proportionality constant be a complex number, so that

$$F = (Ue^{-i\alpha})z \tag{18.2.1}$$

For this potential, the complex velocity is

$$W = \frac{dF}{dz} = Ue^{-i\alpha} = U\cos\alpha - iU\sin\alpha$$

and therefore from 18.1.10 we find

$$u = U\cos\alpha, \qquad v = U\sin\alpha \tag{18.2.2}$$

This represents a uniform stream of magnitude U flowing at an angle of attack α with respect to the negative x axis. The stream moves from left to right.

Next, we consider the case where F is a power function of z. Let A and n be real constants, and take

$$F = Az^n \tag{18.2.3}$$

Realistic flow patterns occur if $n \geq \frac{1}{2}$ (Fig. 18-3). For these flows the complex velocity is

$$W = \frac{dF}{dz} = nAz^{n-1}$$

It is most convenient to use polar coordinates to interpret this expression; hence we substitute $z = re^{i\theta}$ and arrange the result as

$$\begin{aligned} W &= nAr^{n-1}e^{in\theta}e^{-i\theta} \\ &= nAr^{n-1}[\cos n\theta + i\sin n\theta]e^{-i\theta} \end{aligned}$$

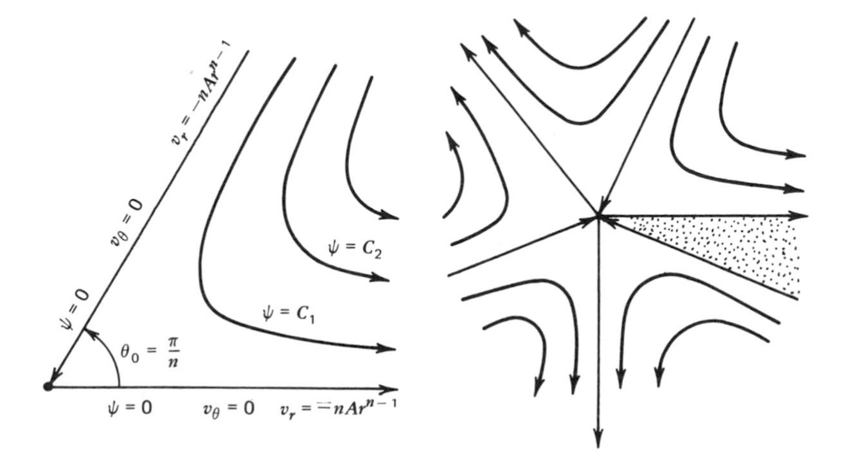

Figure 18-3 Flows with the potential $F = Az^n$.

Comparison with 18.1.13 shows that

$$v_r = nAr^{n-1}\cos n\theta$$
$$v_\theta = -nAr^{n-1}\sin n\theta \qquad 18.2.4$$

To aid in interpreting this flow pattern, it is useful to have the stream function. Expanding 18.2.3 into polar form and using the definition 18.1.7, we find that

$$\phi = Ar^n \cos n\theta \qquad 18.2.5$$

$$\psi = Ar^n \sin n\theta \qquad 18.2.6$$

Equation 18.2.6 shows that ψ will be zero for all values of r on lines from the origin at angles where $\sin n\theta = 0$. These rays are at angles $\theta = k\pi/n$ for integer values of k. Along these radial lines 18.2.4 shows that $v_\theta = 0$, and since $\cos 2\pi k = (-1)^k$, the radial velocity is

$$v_r = (-1)^k nAr^{n-1} \qquad 18.2.7$$

In particular, on the positive x axis ($k = 0$), the equation above shows that v_r is positive, indicating an outflow along the x axis. The next line on which $\psi = 0$ is for $k = 1$, giving $\theta = \pi/n$. Here the flow v_r is negative as it comes toward the origin. The flow within the wedge $\theta = 0$, π/n first comes toward the origin and then flows away along the x axis. This repetitive pattern may not come out even as we approach $\theta = 2\pi$. The values of W for $\theta = 2\pi$ and $\theta = 0$ are not necessarily the same. For $\theta \geq 2\pi$ we begin a second sheet of the function. By restricting the values of θ to $0 \leq \theta < 2\pi$ we have one unique function with a branch cut along $\theta = 0, 2\pi$.

An ideal flow allows the fluid to slip along the wall. In an ideal-flow solution, any streamline in the pattern may be taken to represent a solid wall. Figure 18-4 shows several different choices of walls for the same ideal-flow solution. Figure 18-5 shows some typical interpretations for different values of the parameter n. Two of the most important interpretations are the plane stagnation point where $n = 2$ and the flow over a sharp pointed wedge. In the

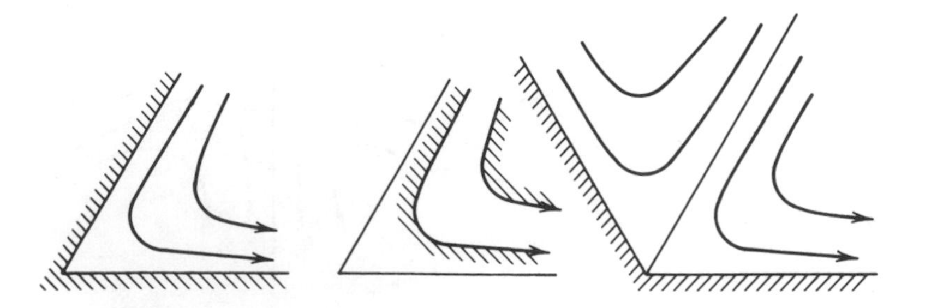

Figure 18-4 Any streamline may be taken as a wall in ideal flow.

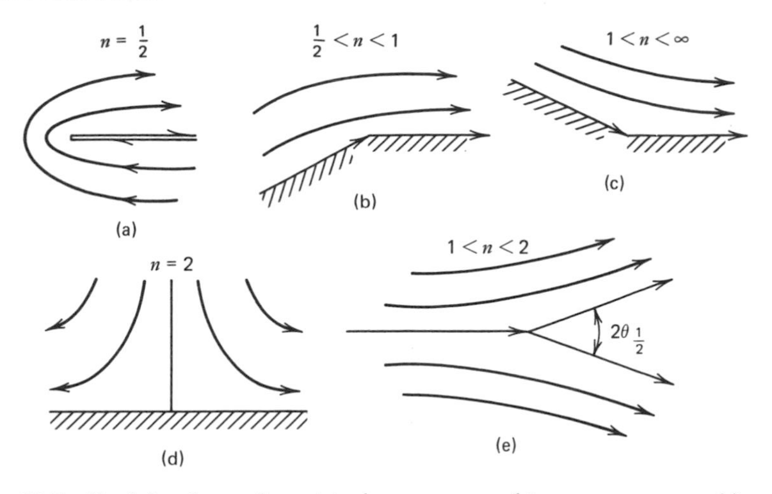

Figure 18-5 Useful values of n: (a) sharp corner, (b) convex corner, (c) concave corner, (d) $n = 2$ two flow segments giving a plane stagnation point, (e) two segments with $1 < n < 2$ giving the flow over a wedge.

latter case, the wedge angle is

$$\alpha = 2\pi\left(\frac{n-1}{n}\right) = 2\theta_{1/2} \tag{18.2.8}$$

We would expect these solutions to be only a local approximation for a region that is a part of a much larger flow. For instance, the sharp wedge might be the most forward portion of a body. The ideal-flow solution for an infinite wedge would be valid for some small neighborhood near the nose of the body.

The corner flows are also local solutions. If any ideal flow has a corner, there is some small region near the corner where only the angle of the corner is important. In this region the local flow corresponds to a portion of the solution with $F = Az^n$. A characteristic common to all convex corners is that the velocity becomes infinite and the pressure drops to minus infinity. These flows contain a physically unrealistic singularity at the corner. In an actual flow, viscous regions would exist so that the corner would effectively be rounded out. If a convex corner is very acute, a small separated region within the viscous flow might occur.

Concave corners always have a stagnation point where the velocity is zero and the pressure is a finite stagnation value. As the flow comes into the corner, the pressure rises. In the next chapter we shall find that this situation causes the viscous boundary layer to thicken, and in some cases there may even be a separated region in these corners. If the viscous region is still thin, we can still apply the ideal-flow solution to streamlines that are slightly away from the walls.

We conclude this section with a mathematical observation. It can be proved that the only ideal flow that has a finite velocity over the whole plane is the uniform stream. Note that the wedge-shaped flows, on the other hand, have at least one point with an infinite velocity. At a convex corner the infinite velocity occurs at the corner itself, while at a concave corner the infinite velocity occurs at infinity.

18.3 LINE SOURCE AND LINE VORTEX

The complex potential for a line source located at the position $z = z_0$ is

$$F = \frac{m}{2\pi} \ln(z - z_0) \qquad 18.3.1$$

Without loss in generality we shift the origin to $z = 0$ and express the potential in polar form:

$$F = \frac{m}{2\pi} \ln(re^{i\theta}) = \frac{m}{2\pi}[\ln r + i\theta] \qquad 18.3.2$$

Hence, by 18.1.7 we find

$$\phi = \frac{m}{2\pi} \ln r, \qquad \psi = \frac{m}{2\pi}\theta \qquad 18.3.3$$

The velocity components may be found by differentiation of 18.3.3, employing the relations of Appendix D, or by computation of the complex velocity and the use of 18.1.13. The latter method yields

$$W = \frac{dF}{dz} = \frac{m}{2\pi z} = \frac{m}{2\pi r}e^{-i\theta}$$

Comparison with 18.2.13 shows that

$$v_r = \frac{m}{2\pi r}, \qquad v_\theta = 0 \qquad 18.3.4$$

This is purely radial flow, either into a sink (when m is negative) or away from a source (when m is positive).

A sketch of the flow appears in Fig. 18-6. The velocity becomes infinite at the origin as $1/r$; hence this point is unrealistic. The strength of the source is given by the constant m. A physical interpretation of m is obtained if we compute the volume flow. Taking any surface around the origin with unit depth, the flow rate is

$$\text{volume flow per unit depth} = \int n_i v_i \, ds$$

$$\frac{Q}{L} = \int n_r v_r \, ds = \int_0^{2\pi} \frac{m}{2\pi r} r \, d\theta$$

$$= m$$

The constant m is the volume flow from the source per unit length.

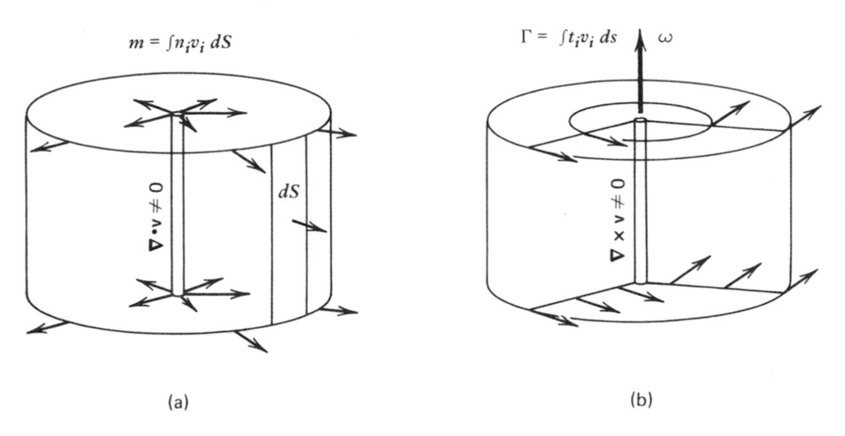

Figure 18-6 Singularities in ideal flow: (a) line source, (b) line vortex.

The calculation of the volume flow m can be made in another manner by using Gauss's theorem:

$$m = \int n_i v_i \, ds = \int \partial_i v_i \, dV$$

In incompressible flow $\nabla \cdot \mathbf{v} = 0$, so the fact that the volume integral is not zero means that this condition is not met throughout the whole flow. The singularity in $\mathbf{v}$ at the origin is a source of zero diameter. At that point the flow does not obey the condition $\nabla \cdot \mathbf{v} = 0$.

The ideal line vortex is very similar in mathematical form to the source. Its complex potential is also the logarithmic function, but with an imaginary constant:

$$F = -i\frac{\Gamma}{2\pi} \ln(z - z_0) \qquad 18.3.5$$

The velocity components for this potential turn out to be

$$v_r = 0, \qquad v_\theta = \frac{\Gamma}{2\pi r} \qquad 18.3.6$$

This flow swirls in a counterclockwise direction for positive values of Γ and decreases in magnitude as $1/r$—the same type of decrease observed for the source.

Again the origin has an infinite velocity, so we must exclude that point from any realistic flow. In a real vortex, the core velocity drops to zero through a viscous region where the vorticity is nonzero, a fact that was noted previously.

The strength of a vortex is given by the circulation Γ. The circulation was defined (Section 12.5) to be the counterclockwise line integral of the tangential velocity component around a closed path,

$$\Gamma \equiv \oint t_i v_i \, dl \qquad 18.3.7$$

A negative value of Γ means the vortex is swirling clockwise.

Recall that the circulation and the fluid vorticity are connected by an important integral relation. If Stokes's theorem is applied to 18.3.7, we find that

$$\Gamma \equiv \oint_C t_i v_i \, dl = \int_S n_i \omega_i \, ds \qquad 18.3.8$$

In this expression S is any simple surface bounded by the closed path C. Since Γ is not zero, there is some point in the flow (in this case, the origin) where the vorticity is not zero. The vorticity goes to infinity at the origin in such a way that the integral over the surface 18.3.8 has a finite value.

Ideal flows satisfy the two kinematic conditions of no expansion ($\nabla \cdot \mathbf{v} = 0$) and no particle rotation ($\boldsymbol{\omega} = \nabla \times \mathbf{v} = 0$). A source is a flow that violates the first condition at one point, while a vortex is a flow that violates the second condition. These flows are the two basic types of singularities.

18.4 FLOW OVER A NOSE OR A CLIFF

The equations and boundary conditions that govern ϕ, ψ, and v_i in ideal flows are linear. Thus, two potentials or velocities may be added together to produce a new flow pattern. In the present example we take a source of strength m and add a uniform stream. This will result in the flow pattern shown in Fig. 18-7. If we take the streamline that divides the source flow from the stream flow as a solid wall, then this pattern represents the flow over a two-dimensional body with a certain streamline shape. The body extends to infinity, where it has a half thickness h. The flow over this nose is typical of the flow over the front of

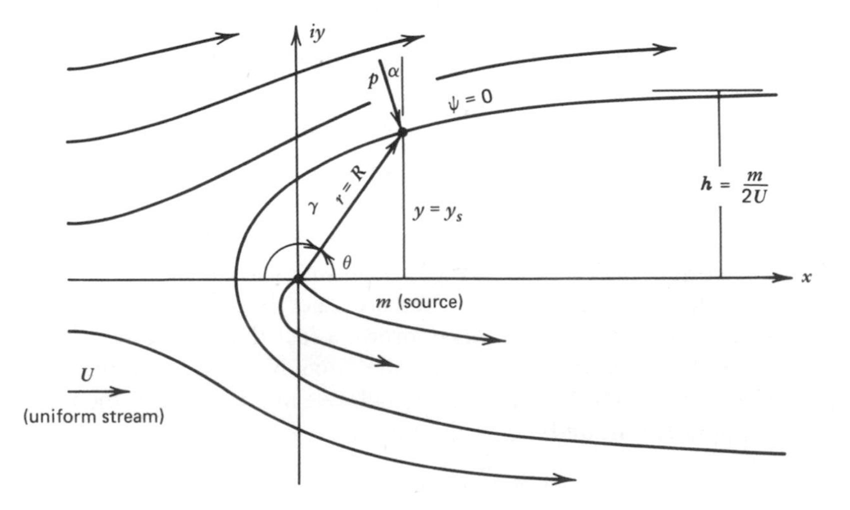

Figure 18-7 Flow over a nose or a cliff: a source in a uniform stream.

any smooth two-dimensional object, such as a fan blade, a wing, a contoured bridge piling, or a strut. If we take only the upper half of the flow, we can imagine that the pattern represents the flow coming from a plane (or the ocean) and passing over a cliff. Of course the cliff has a peculiar shape, but again the flow is typical of any such shape.

Another characteristic of ideal flows is that they are reversible. Simply changing the sign of the velocity boundary condition changes the sign of all velocities while leaving the magnitudes undisturbed. Now the flow travels in the opposite direction along the same streamlines as before. The pressure remains the same. In the present case we could interpret the flow as one coming off a cliff onto the plane.

The complex potential for the flow is the sum of the potentials for a source and uniform stream:

$$F = Uz + \frac{m}{2\pi} \ln z + iC \tag{18.4.1}$$

A constant iC has been included so that we may adjust the zero of the stream function to correspond to the streamline separating the source and stream flows. A real constant could be included to adjust the velocity potential, but we have no need to do this. By direct calculation from 18.4.1, or by adding the results of the previous two sections, we find that the stream function is

$$\psi = Uy + \frac{m}{2\pi}\theta + C \tag{18.4.2}$$

The complex velocity for this flow is the derivative of 18.4.1,

$$W = \frac{dF}{dz} = U + \frac{m}{2\pi z} \tag{18.4.3}$$

Any stagnation point will have $W = 0$, so from the equation above we have one such point located at

$$\begin{gathered} z = -\frac{m}{2\pi U} \\ x = -\frac{m}{2\pi U}, \qquad y = 0 \end{gathered} \tag{18.4.4}$$

This is the point on the negative x axis where the velocity of the stream, U, is exactly balanced by the velocity of the source, $-m/2\pi x$.

The streamline though the stagnation point is adjusted to zero by substituting $y = 0$, $\theta = \pi$, $\psi = 0$ into 18.4.2. This gives $C = -m/2$, and the stream-function equation becomes

$$\psi = Uy - \frac{m}{2\pi}(\pi - \theta) \tag{18.4.5}$$

When $\psi = 0$, we have an equation for the surface of the body,

$$y_s = \frac{m}{2\pi U}\gamma \qquad 18.4.6$$

where

$$\gamma \equiv \pi - \theta \qquad 18.4.7$$

The parametric form above, with the angle γ measured clockwise from the negative x-axis, has some advantages. A substitution $y_s = R\sin\theta = R\sin\gamma$ into 18.4.6 produces the polar form of the equation for the surface as

$$R = \frac{m}{2\pi U}\frac{\gamma}{\sin\gamma} \qquad 18.4.8$$

From 18.4.6 we determine the half width of the body by setting $\gamma = \pi$. The result is

$$h = \frac{m}{2U} \qquad 18.4.9$$

The point on the surface directly above the origin ($\gamma = \pi/2$) has $y_s = m/4U$; the surface has risen to half its final height.

One of the things we want to investigate is the pressure distribution over the surface and the drag it might produce—a sort of frontal drag typical of such bodies. We begin by finding the speed q from 18.4.3:

$$q^2 = W\overline{W} = \left[U + \frac{m}{2\pi z}\right]\left[U + \frac{m}{2\pi\bar{z}}\right]$$

Simplifying and inserting $z = -re^{i\gamma}$ produces

$$q^2 = U^2 - \frac{mU}{\pi r}\cos\gamma + \left(\frac{m}{2\pi}\right)^2\frac{1}{r^2} \qquad 18.4.10$$

To obtain the surface velocities we introduce 18.4.8 and arrive at

$$q_s^2 = U^2\left[1 - \frac{2}{\gamma}\sin\gamma\cos\gamma + \frac{1}{\gamma^2}\sin^2\gamma\right] \qquad 18.4.11$$

This form is suitable for computing the pressure.

The Bernoulli equation for steady flow is

$$p + \tfrac{1}{2}\rho q^2 = p_\infty + \tfrac{1}{2}\rho U^2 \qquad 18.4.12$$

In ideal flow it is more or less natural to refer the pressure to the value at infinity; recall that the level of all pressures in an incompressible flow increases directly with the reference pressure. The natural scale for the pressure is the dynamic pressure. Hence, the nondimensional pressure or *pressure coefficient* (also known as the Euler number) is defined as

$$C_p \equiv \frac{p - p_\infty}{\frac{1}{2}\rho U^2} \qquad 18.4.13$$

Inserting the Bernoulli equation into this definition results in the simple formula

$$C_p = 1 - \frac{q^2}{U^2} \qquad 18.4.14$$

Substituting 18.4.11 gives the equation for the surface pressures:

$$C_p = \frac{2}{\gamma}\sin\gamma\cos\gamma - \frac{1}{\gamma^2}\sin^2\gamma \qquad 18.4.15$$

A special subscript for the surface is not used, as most of the time we use only C_p as a surface quantity. In the equation above, C_p is expressed as a function of the angle γ. This is not really the best form.

In boundary-layer theory the distance along the body surface is a natural coordinate. Figure 18-8 shows the geometry for computing s, the distance

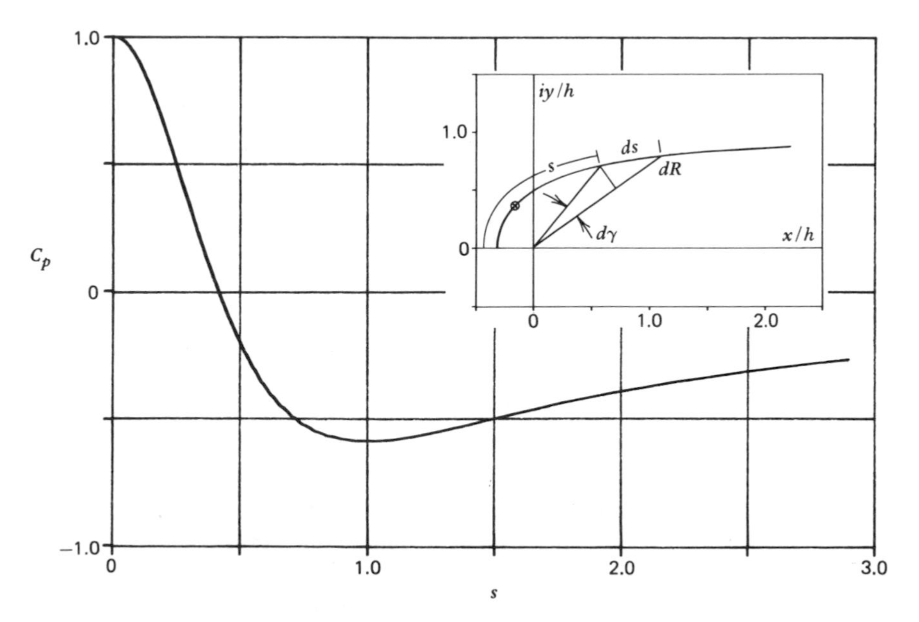

Figure 18-8 Pressure distribution on a smooth nose.

coordinate along the surface. The triangle pictured shows that

$$(ds)^2 = (R\,d\gamma)^2 + (dR)^2 \tag{18.4.16}$$

R is eliminated in this expression in favor of γ by employing 18.4.8. Some additional algebra leads to

$$ds = \frac{h}{\pi \sin^2 \gamma}\left[\gamma^2 - 2\gamma \sin\gamma \cos\gamma + \sin^2\gamma\right]^{1/2} d\gamma \tag{18.4.17}$$

Integration of this expression produces $s = s(\gamma)$, which, together with $C_p = C_p(\gamma)$ from 18.4.15, allows one to plot $C_p(s)$ as in Fig. 18-8. The graph shown was obtained by using a computer to integrate 18.4.17.

At the nose of the body, or at the base of the cliff, the stagnation pressure occurs and $C_p = 1$ as a result. Because of symmetry, the slope of the curve at the stagnation point is zero, but as we move around the body, the fluid accelerates rapidly and the pressure drops accordingly. The velocity becomes equal to the free-stream velocity at $s = 0.42h$ ($\gamma = 1.17$ radians). This point is noted in Fig. 18-8; it marks the end of the region where the pressure is higher than p_∞. The fluid continues to accelerate and C_p continues to drop. A minimum is reached at about $s = 0.98h$ ($\gamma = 2.04$), where $C_p = -0.585$. The corresponding maximum surface speed is $q_{\max} = 1.26U$. This low-pressure region continues over the remainder of the surface, and we approach p_∞ from below as $s \to \infty$.

It is easy to use streamline patterns to surmise the trends in the pressure. Pressure change results from two effects: streamline convergence and streamline curvature. Converging streamlines indicate an increasing velocity and thus a decreasing pressure. Diverging streamlines, as in the region of the stagnation point, indicate an increasing pressure. Streamline curvature indicates a low pressure toward the center of curvature and a higher pressure toward the outside. The concave curvature near the shoulder of the body indicates a low pressure in this region.

The pressure and velocity patterns on this body are typical of patterns on the nose of any well-rounded body. The velocity around the shoulders always increases above the free-stream value, and the pressure always decreases below the free-stream value. The maximum and minimum values depend somewhat upon the exact contour of the surface, but the trend is always the same.

It is a general mathematical result that solutions of the Laplace equation have maximum and minimum values on the boundaries. Starting with this fact, it may be proved that the maximum velocity in an ideal flow always occurs on the surface of the body. By Bernoulli's theorem this point also has the minimum value of pressure in the flow. On the other hand, the smallest magnitude the velocity can have is zero, a stagnation point. Stagnation points may occur within the fluid as well as on the surface of a body.

Since we are taking this example as typical of the flow on the forward portion of a two-dimensional body, the question arises of the drag that might be attributed to this portion of the body. If a body of this same shape is immersed in a still fluid, the pressure is everywhere p_∞ and there is no net drag force. In order to find the frontal drag due to the motion, we take the pressure $p - p_\infty$ on the surface element ds and resolve it into a component in the flow direction (see Fig. 18-7),

$$F_D = \int_0^\infty (p - p_\infty) \sin \alpha \, ds$$

The drag coefficient per unit depth is defined (note that $dy = \sin \alpha \, ds$) as

$$C_D \equiv \frac{F_D}{\frac{1}{2}\rho U^2 h} = \int_0^h \frac{p - p_\infty}{\frac{1}{2}\rho U^2} \frac{dy}{h} \qquad 18.4.18$$

Substituting 18.4.6, 18.4.8, and 18.4.15 into this integral gives

$$C_D = \frac{1}{2}\int_0^\pi \left[\frac{2}{\gamma} \cos \gamma \sin \gamma - \frac{1}{\gamma^2} \sin^2 \gamma \right] d\gamma$$

$$= \frac{1}{2}\int_0^\pi d\left[\gamma^{-1} \sin^2 \gamma\right] = \tfrac{1}{2}\left[\gamma^{-1} \sin^2 \gamma\right]_0^\pi = 0 \qquad 18.4.19$$

Applying L'Hôspital's rule to this indeterminate form shows that the drag is zero. The drag force caused by the high pressure at the front of the body is exactly canceled out by the thrust force on the shoulders, where the pressure is less than the free-stream value. The fact that the frontal drag on a smooth slender body in ideal flow is zero is a striking result. It is valid for any smooth-shaped body; we could simulate any shape by placing sources and sinks at several positions along the x axis, and the same zero-drag result would come out.

Our calculation has been made for a body that extends to infinity. Let us consider how this result might appear for a long slender body of finite length. We might guess that far away from the nose the pressure returns to nearly the free-stream value. At some distance downstream we may suppose that the body is terminated and has a blunt base. If the pressure in the base region is p_∞, then the result above will apply and the drag force will be zero. Unfortunately, our guess that the base pressure is nearly equal to the free-stream pressure is not true. If the body has a blunt base, the flow separates and a turbulent wake is formed. The base pressure is usually fairly uniform, but lower than the free-stream pressure. Hence, actual slender bodies have a finite drag, which is almost solely the result of the base drag, the frontal drag being negligible.

Since ideal flows are reversible, why don't we change the sign of the flow and consider that the pattern represents the flow at the end of a slender body? The body contour would end with the same shape as the previous nose shape. The reason this cannot be done is that the flow would need to penetrate a high-pressure region at the rear stagnation point. The ideal flow can in

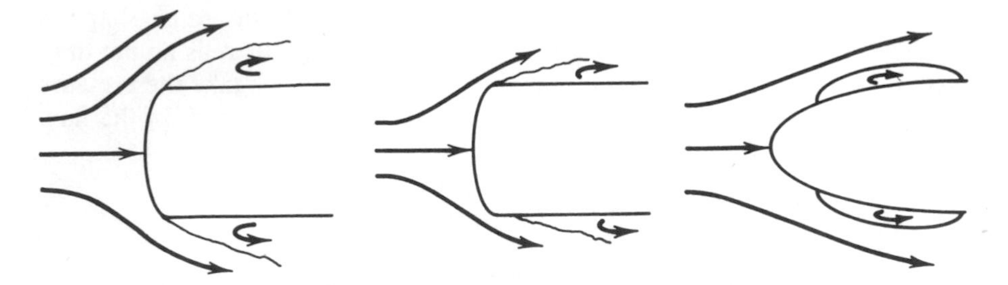

Figure 18-9 Nose drag occurs when flow separates from the sides. The last frame shows a separation bubble on a smooth nose. The size of the bubble is exaggerated.

principle do this, but the boundary layer cannot. The particles in the boundary layer have been slowed down by wall friction and do not have sufficient momentum to move against the high pressure gradient. The flow separates and once again we have a base region, which produces a drag force.

Separation at the rear of a body can be delayed if we decrease the height very slowly and form a sharp point at the end. Most airfoils have this shape. The slowly decreasing thickness allows the pressure to rise slowly, and more importantly, the streams from the upper and lower sides may merge smoothly without decreasing to zero velocity (the stagnation point is submerged within the viscous boundary layers). Shapes made in this way can have very low drag coefficients: measured values less than $C_D = 0.1$ (based on the cross section) are not uncommon for airfoil shapes.

The zero-frontal-drag principle does not apply to shapes with sharp corners such as those shown in Fig. 18-9. When the flow separates from the shoulders, we lose the low-pressure region that provides the suction force. As a result, the high-pressure region at the nose is not counterbalanced, and the body has a substantial frontal drag. Separation always occurs if the surface has a sharp corner and is common for bodies with sharp curvature near the shoulders.

Even very smooth bodies have a region of increasing pressure once the minimum has been reached (recall Fig. 18-8). The increasing pressure is a likely candidate for separation, depending upon the state of the boundary layer. In many cases the flow separates and then reattaches without forming a wake. Such regions are called *separation bubbles* and are very thin. We have exaggerated the thickness in the sketch so that the bubble may be seen. As far as the drag is concerned, this type of separation has a negligible effect.

18.5 DOUBLETS

The line source and line vortex are two types of singular points in the flow where the velocity is infinite and one of the kinematic conditions, either $\nabla \cdot \mathbf{v} = 0$ or $\nabla \times \mathbf{v} = 0$, is violated. There are other types of singular points that satisfy the Laplace equation everywhere except at one point. The singularity we treat in this section is called a *doublet*. Perhaps the most enlightening way to introduce the doublet is to consider it as the superposition of a source and a sink that are brought close together. Let a source of strength m be located on the negative x-axis at a point ϵ away from the origin, as shown in Fig. 18-10. The complex potential for the source is

$$F = \frac{m}{2\pi} \ln(z + \epsilon)$$

Next, a sink of equal strength $-m$ is placed at a position ϵ on the positive x axis. For this combination of a source and a sink, the potential is

$$F = \frac{m}{2\pi} \ln\left(\frac{z + \epsilon}{z - \epsilon}\right)$$

$$F = \frac{m}{2\pi} \ln\left(\frac{1 + \epsilon/z}{1 - \epsilon/z}\right) = \frac{m}{2\pi} \ln\left[\left(1 + \frac{\epsilon}{z}\right)\left(1 - \frac{\epsilon}{z}\right)^{-1}\right] \qquad 18.5.1$$

As the source and the sink come close together, ϵ becomes small. The binomial expansion (16.2.8) says that for $\epsilon \to 0$

$$\left(1 - \frac{\epsilon}{z}\right)^{-1} = 1 + \frac{\epsilon}{z} + O\left[\left(\frac{\epsilon}{z}\right)^2\right]$$

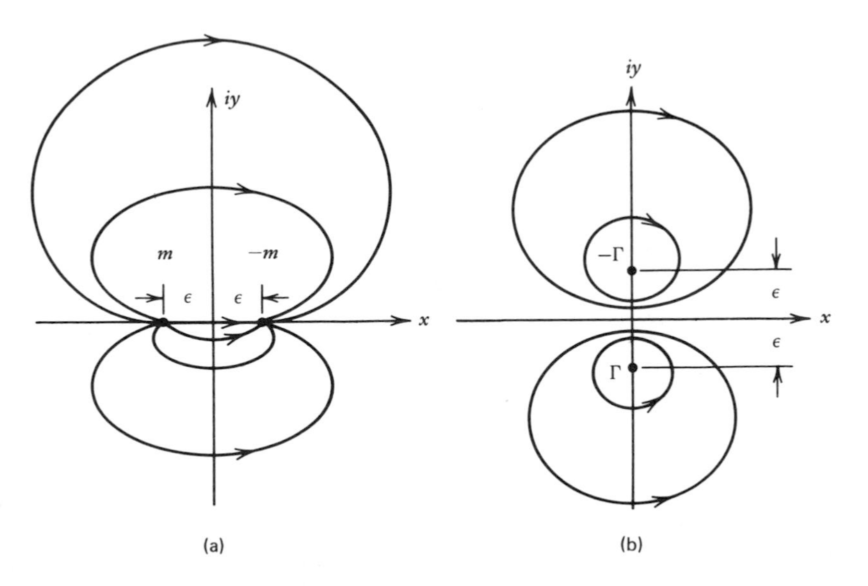

Figure 18-10 Doublet as the limit of: (a) source + sink, $\epsilon \to 0$, $m\epsilon = \mu$; (b) vortices on y axis, $\epsilon \to 0$, $\Gamma\epsilon = \mu$.

Inserting this expression in 18.5.1 above gives

$$F = \frac{m}{2\pi}\ln\left\{\left(1+\frac{\epsilon}{z}\right)\left(1+\frac{\epsilon}{z}+O\left[\left(\frac{\epsilon}{z}\right)^2\right]\right)\right\}$$

$$= \frac{m}{2\pi}\ln\left\{1+2\frac{\epsilon}{z}+O\left[\left(\frac{\epsilon}{z}\right)^2\right]\right\}$$

Next, we note that for small x the following expansion of $\ln x$ is valid:

$$\ln x = (x-1) - \tfrac{1}{2}(x-1)^2 + \cdots$$

Using this expansion results in the following relation for F:

$$F = \frac{m\epsilon}{\pi z} + O\left(\frac{\epsilon}{z}\right)^2$$

If we allow the source and sink to approach each other ($\epsilon \to 0$) with a constant strength m, the result is $F = 0$; they swallow each other. However, suppose instead that as the source and sink approach each other their strengths m are increased in such a way that the product $m\epsilon = \mu$ is a constant. The constant μ is called the strength of the doublet. The complex potential for a doublet is

$$F = \frac{\mu}{\pi z} \qquad 18.5.2$$

One can show that the streamlines for a doublet are circles through the origin given by the equation

$$x^2 + \left(y + \frac{\mu}{2\psi}\right)^2 = \left(\frac{\mu}{2\psi}\right)^2 \qquad 18.5.3$$

The centers of the circles are at $y = \pm\mu/2\psi$; when ψ is positive, the circles are in the lower half plane, and when ψ is negative, they are in the upper half plane. Figure 18-10a shows the pattern.

From 18.5.2 the complex velocity is found to be

$$W = \frac{dF}{dz} = -\frac{\mu}{\pi z^2}$$

$$= -\frac{\mu}{\pi r^2}e^{-i2\theta} \qquad 18.5.4$$

The influence of a doublet at long distances is smaller than that of a source (or a vortex), in the sense that the velocity induced by the doublet decreases as $1/r^2$, compared to $1/r$ for a source or a vortex.

The doublet has a direction associated with it, as well as a strength. We placed the sink to the right of the origin and the source on the left. Switching these positions would merely change the sign of μ. Consider what would happen if we placed the source on the negative y axis and the sink on the positive iy axis. In 18.5.1, the argument $(z+\epsilon)/(z-\epsilon)$ would change to $(z+i\epsilon)/(z-i\epsilon)$. A doublet aligned with the iy axis would have the potential

$$F = \frac{i\mu}{\pi z} \qquad 18.5.5$$

An arbitrary orientation for a doublet is constructed by replacing i by a unit vector in the desired direction ($i \to -e^{i\beta}$, where β is the doublet orientation).

The streamline pattern of a doublet was derived by considering a source and a sink brought together while the strength increased in such a way that $m\epsilon = \mu$ was a constant. There is another physical interpretation of a doublet, which is equally as valid and deserves some remarks. A doublet may be considered by merging two line vortices of opposite circulation in such a way that the strength times the separation distance, $\epsilon\Gamma$, is a constant. If we place the vortices on the positive and negative axes, the doublet is aligned with the iy axis, that is, 18.5.5 results. A vortex doublet is the same as a source–sink doublet turned through 90°.

From a mathematical standpoint the source and the vortex are the strongest singularities; they have potentials proportional to $\ln z$. The doublet is mathematically the derivative of the source or the vortex, having a potential $\propto 1/z$. One may continue this process and form higher-order singularities. The next singularity is constructed from a source–sink combination on the x axis paired with another on the iy axis. This is called a quadrupole singularity. In the same nomenclature, sources and vortices are known as monopole singularities, and the doublet is called a dipole singularity. We have very little use in fluid mechanics for the quadrupole and higher singularities.

18.6 CYLINDER IN A STREAM

The ideal flow about a circular cylinder with a uniform stream perpendicular to the cylinder axis is given by the superposition of a doublet and a stream. The potential is

$$F = Uz + U\frac{r_0^2}{z} \qquad 18.6.1$$

The doublet strength is taken as Ur_0^2, where r_0 is the radius of the circle. Computing the complex velocity gives

$$W = U - U\frac{r_0^2}{z^2} \qquad 18.6.2$$

Setting 18.6.2 equal to zero shows that stagnation points are located at

$z = \pm r_0$; that is, $x = \pm r_0$, $y = 0$. These are the most forward and aft points on the cylinder. With a little further algebra the velocity components may be found. Using cylindrical coordinates we have

$$W = U\left(1 - \frac{r_0^2}{r^2}e^{-i2\theta}\right) = U\left(e^{i\theta} - \frac{r_0^2}{r^2}e^{-i\theta}\right)e^{-i\theta}$$

$$= U\left[\left(1 - \frac{r_0^2}{r^2}\right)\cos\theta + i\left(1 + \frac{r_0^2}{r^2}\right)\sin\theta\right]e^{-i\theta}$$

Comparing this form with 18.1.13 shows that the velocity components are

$$v_r = U\left(1 - \frac{r_0^2}{r^2}\right)\cos\theta$$

$$v_\theta = -U\left(1 + \frac{r_0^2}{r^2}\right)\sin\theta \qquad 18.6.3$$

On the cylinder surface $r = r_0$ these equations show that $v_r = 0$ and hence

$$q = v_\theta = -2U\sin\theta \qquad 18.6.4$$

The velocity at the maximum-thickness point of the cylinder is twice the free-stream value.

Figure 18-11 shows the pattern of streamlines for this flow. We disregard the doublet flow on the inside of the circle $r = r_0$ and imagine that a solid cylinder replaces this portion of the flow. A remarkable feature is the symmetry of the flow upstream and downstream of the cylinder. The symmetry of the geometry results in a symmetry of the streamlines.

The pressure force on the surface of the cylinder is obtained by substituting 18.6.4 into 18.4.13:

$$C_p \equiv \frac{p - p_\infty}{\frac{1}{2}\rho U^2} = 1 - \left(\frac{q}{U}\right)^2$$

$$C_p = 1 - 4\sin^2\theta \qquad 18.6.5$$

A plot of this function is given in Fig. 18-11. At the forward and aft stagnation points $C_p = 1$. At the maximum-thickness point, the point where $q = 2U$, we find that the pressure has dropped to $C_p = -3$, which is 3 dynamic pressure units lower than atmospheric pressure. The pressure distribution has a pattern similar to that found for flow over a rounded nose shape in Section 18.5: high pressure at the stagnation region followed by low pressure at the shoulder of the body. For the nose the minimum value of C_p was -0.585, whereas for the cylinder the lowest value is -3. If we compute the drag, we find that, because of the symmetric pressure distribution, the force on the front half cancels that on the rear half to produce zero drag.

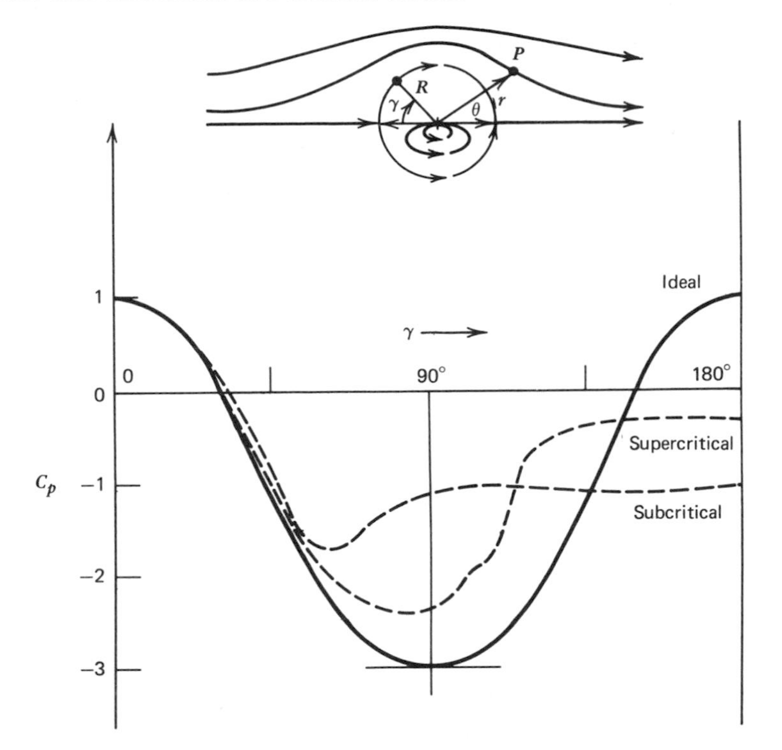

Figure 18-11 Pressure coefficient for streaming flow over a cylinder. Typical experimental trends for subcritical and supercritical Reynolds numbers are shown.

The ideal flow about a cylinder is not a realistic solution (though we shall find an important use for this solution in connection with the flow over airfoils). As discussed in Chapter 14, the flow at high Reynolds number always separates on the sides of the cylinder. Nevertheless, the solution does give reasonable results for the front portion of the cylinder in the neighborhood of the stagnation region. From Fig. 18-11 we can see that the actual surface pressures and the ideal values given by 18.6.5 agree for a distance up to $\theta \approx 60°$. Thereafter the separated wake flow causes a drastic change in the main flow, so the ideal solution is no longer valid.

18.7 CYLINDER WITH CIRCULATION IN A UNIFORM STREAM

The flow discussed in the previous section is not a unique solution. We can construct other ideal-flow solutions that also represent the flow on the outside of a cylinder. Consider the flow given by the potential

$$F = Uz + U\frac{r_0^2}{z} - \frac{i\Gamma}{2\pi}\ln\frac{z}{r_0} \tag{18.7.1}$$

This potential is the superposition of a uniform stream, a doublet, and a vortex of strength Γ located at the origin. An imaginary constant $(i\Gamma \ln r_0)/2\pi$ has been incorporated into the potential so that the streamline on the surface of the cylinder remains the $\psi = 0$ streamline.

The addition of the vortex changes the flow pattern everywhere except at the surface of the cylinder and at infinity. That is, the streamline that represents the cylinder is still a circle; the values of the surface velocity are, of course, changed. Figure 18-12 shows several flow patterns for different values of the circulation constant. Note that the circulation constant has been changed to $\Gamma_a = -\Gamma$. It will turn out that this flow is relevant to the flow about wings and airfoils. Aeronautical engineers, in order to make the lift on an airfoil positive, define Γ as the negative of the definition previously given. Let us compute the complex velocity from 18.7.1 and at the same time insert $\Gamma = -\Gamma_a$. The result is

$$W = \frac{dF}{dz} = U - U\frac{r_0^2}{z^2} + i\frac{\Gamma_a}{2\pi z} \qquad 18.7.2$$

Since we may assign any value we choose to Γ_a, there are an infinite number of ideal flows for the streaming motion over a cylinder.

The velocity components v_r, v_θ are found by inserting $z = re^{i\theta}$ into 18.7.2 and organizing the equation in the form 18.1.13—the same method we used in

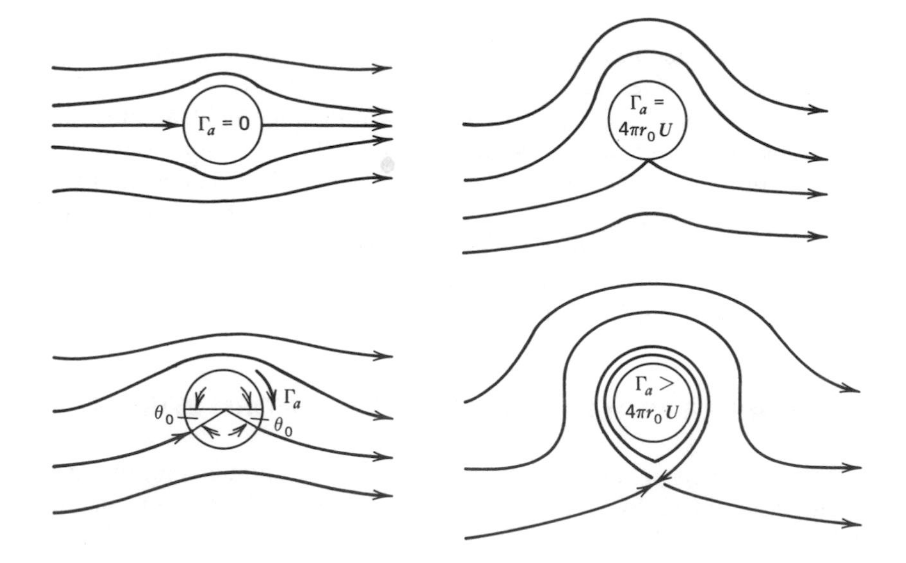

Figure 18-12 Nonuniqueness of flow over a cylinder. The circulation constant Γ_a must be specified to determine a unique flow.

the previous section. When this is done, the components are found to be

$$v_r = U\left(1 - \frac{r_0^2}{r^2}\right)\cos\theta$$

$$v_\theta = -U\left(1 + \frac{r_0^2}{r^2}\right)\sin\theta - \frac{\Gamma_a}{2\pi r} \qquad 18.7.3$$

The vortex only affects v_θ.

Stagnation points on the cylinder are located by setting $r = r_0$ and $v_\theta = 0$. The result is an equation for the θ_0 location of the stagnation points,

$$\sin\theta_0 = -\frac{\Gamma_a}{4\pi r_0 U} \qquad 18.7.4$$

Without any circulation, the stagnation points are symmetrical at $\theta_0 = 0, 2\pi$, as we noted in the previous section. When Γ_a is less than $4\pi r_0 U$ the right side of 18.7.4 will be a negative number smaller than one. This results in stagnation points on the lower portion of the cylinder as shown in Fig. 18-12. In the case where $\Gamma_a > 4\pi r_0 U$, the stagnation points cannot be found from 18.7.4. In this case a single stagnation point moves away from the surface of the cylinder and occurs within the flow. (Recall that it is only the maximum speed that must occur on the body. The minimum speed, $q = 0$, may occur anywhere in the flow.) If the stagnation point is within the fluid, there is a portion of fluid that is trapped next to the surface and continually rotates around the cylinder.

Ideal-flow theory allows the fluid to slip over the surface of a body. In principle, any of the solutions for various Γ would apply to a stationary solid cylinder. Recall that the circulation is related to the integral of the vorticity over an area enclosed by the circuit. Consider the circuit shown in Fig. 18-13,

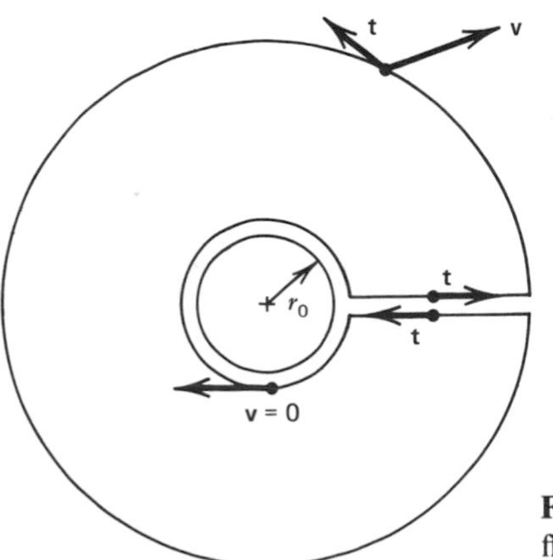

Figure 18-13 Circulation-integral circuit for flow over a cylinder.

which goes around the cylinder surface, has two coincident lines moving away from the cylinder, and is finally closed by a counterclockwise loop within the flow itself. Stokes's theorem says that

$$\Gamma = \oint_C \mathbf{v} \cdot \mathbf{t}\, dl = \int_A \mathbf{n} \cdot \boldsymbol{\omega}\, dA \qquad 18.7.5$$

For the real flow, the no-slip condition on the cylinder surface means that $\mathbf{v} \cdot \mathbf{t} = 0$ on the inside loop. On the lines connecting the two loops $\mathbf{v} \cdot \mathbf{t}$ changes sign as one integrates toward or away from the cylinder. Hence, the only contribution to Γ is the loop far out in the flow itself. This loop is in an inviscid-flow region and thus has the ideal-flow circulation. The right side of 18.7.5 is the area integral of the normal vorticity component. Since the vorticity is nonzero only in the boundary layers next to the surface, we see that the circulation constant of the ideal flow is in fact determined by the vorticity distribution in the boundary layers. Therefore the particular value of Γ and the associated ideal flow are determined by an integrated effect from the viscous flow near the body.

18.8 LIFT AND DRAG ON TWO-DIMENSIONAL SHAPES

The ideal lift and drag force on any cylindrical body, no matter what the cross-section shape, can be related to the complex potential. From a practical standpoint, we might as well limit our thoughts to cylinders with an airfoil shape. Any bluff shape would have a wake of finite thickness, and this would invalidate the theory. The entire flow must be an ideal flow for this theory to apply. Boundary layers and wakes must be vanishingly thin.

It will turn out that the drag force is always zero and that the lift force is directly proportional to the circulation constant Γ_a. The exact relation for the lift force is

$$F_L = \rho U \Gamma_a \qquad 18.8.1$$

This equation is called the Kutta–Joukowski law after the two people who independently discovered it.

To prove the statements above, we consider a body of arbitrary cross section as shown in Fig. 18-14. The flow around this body is an ideal flow without any separation. Hence, the viscous forces are zero and the pressure force on the body may be divided into a lift component and a drag component. For an increment of area ds these components are

$$\begin{aligned} dF_L &= p\, dx \\ dF_D &= -p\, dy \end{aligned} \qquad 18.8.2$$

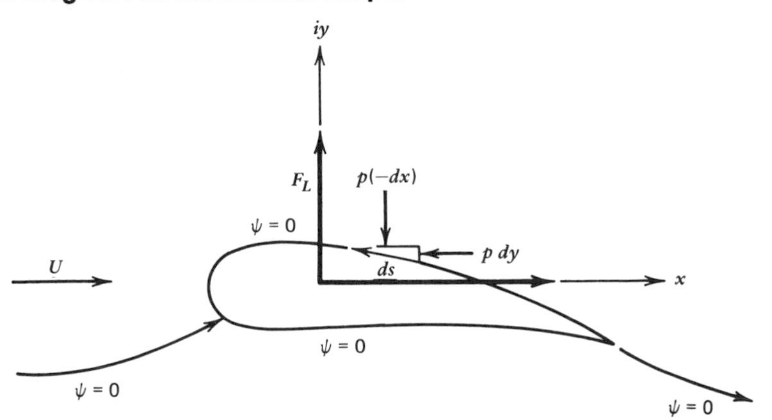

Figure 18-14 Lift and drag forces defined with respect to the flow direction at infinity. Ideal flow: $F_D = 0$, $F_L = \rho U \Gamma_a$.

We know from Chapter 17 that the pressure is constant across a boundary layer. This means that we shall get the proper lift and drag components that arise from the pressure forces if the ideal-flow values are used in 18.8.2.

Next, we form a complex vector for the conjugate of the force on the cylinder. It is

$$\begin{aligned} d(F_D - iF_L) &= -p\,dy - ip\,dx \\ &= -ip\,d\bar{z} \end{aligned} \qquad 18.8.3$$

Integration of this expression around the contour in a counterclockwise manner will yield an expression for the total lift and drag. In this integration the pressure on the surface may be evaluated using Bernoulli's equation

$$\begin{aligned} p &= p_0 - \tfrac{1}{2}\rho q^2 = p_0 - \tfrac{1}{2}\rho W\overline{W} \\ &= p_0 - \tfrac{1}{2}\rho \frac{dF}{dz}\frac{d\overline{F}}{d\bar{z}} \end{aligned}$$

Substituting this into 18.8.3 yields

$$d(F_D - iF_L) = -ip_0\,d\bar{z} + \tfrac{1}{2}i\rho \frac{dF}{dz}\,d\overline{F}$$

The surface contour C is a streamline $\psi = \text{const}$, and since $F = \phi + i\psi$, it follows that $dF = d\phi$ is real, and $dF = d\overline{F}$. Inserting this fact and noting that the integral of the constant p_0 around a closed contour is zero, we have the equation

$$F_D - iF_L = i\frac{\rho}{2}\int_C \left(\frac{dF}{dz}\right)^2 dz$$

$$= i\frac{\rho}{2}\int_C W^2\, dz \qquad 18.8.4$$

This formula is known as the *Theorem of Blasius*. The restriction of 18.8.4 to a contour on the surface of the body may be relaxed. Complex-variable theory shows that any contour of an analytic function gives the same answer as long as it loops around the same singular points of the function. If W has no singularities (sources, vortices, doublets, etc.) outside the cylinder, then the contour may be enlarged to $z \to \infty$ without difficulty. Since streaming motions around airfoils are of this type, we can use a contour at infinity to evaluate the lift and the drag.

Contour integrals of analytic functions are most easily evaluated by using the residue theorem, which states that

$$\int_C W^2\, dz = 2\pi i \sum_k R_k \qquad 18.8.5$$

where R_k are the residues of the function W^2. Residues are given by a Laurent expansion of the function. Any streaming motion must have a complex potential of the following form as $z \to \infty$:

$$F = Uz + \frac{1}{2\pi}(m + i\Gamma_a)\ln z + (a + ib)\frac{1}{z} + \cdots \qquad 18.8.6$$

The first term is a streaming motion aligned with the x axis. This orientation was chosen because the definitions of lift and drag are made with respect to the flow direction, and the formulation 18.8.4 implies that the stream is aligned in this way. The second term contains a source of strength m and a vortex of strength Γ. (A closed body would have a source strength of zero but a circulation constant of unknown value.) The next term represents a doublet with arbitrary direction. The remaining terms would be quadrupoles and higher-order singularities.

The complex potential corresponding to 18.8.6 is

$$W = U + \frac{i\Gamma_a}{2\pi z} + O\left[\frac{1}{z^2}\right]$$

and W^2 is given by

$$W^2 = U^2 + i\frac{\Gamma_a U}{2\pi z} + O\left[\frac{1}{z^2}\right]$$

By definition, the residue of W^2 is the coefficient of the $1/z$ term, $i\Gamma_a U/2\pi$. Combining this result with 18.8.5 and 18.8.4 yields

$$F_D = 0 \qquad 18.8.7$$

$$F_L = \rho U \Gamma_a \qquad 18.8.8$$

The first of these equations states that any airfoil has a zero drag force. The second relation shows that the lift force increases directly as the circulation. This formula is a little misleading in that the circulation around an airfoil is not just a function of the size and shape but it also increases directly with the free-stream velocity. Hence F_L is actually proportional to U^2. Equation 18.8.8 is very important in that it points out that lift can only result from a flow that has a circulation, or vortex-like behavior, in the far field away from the body.

The prediction of zero pressure drag for a two-dimensional airfoil shape is fairly accurate. The drag of actual streamline shapes is very small and is largely caused by the viscous friction on the surface. One thing that we should take warning of is the fact that 18.8.7 applies strictly to unseparated flow about airfoil shapes that are infinitely long. This equation does not apply to shapes that are finite in length such as an actual wing or fan blade. A finite-length wing, even in ideal theory, has a drag force caused by the pressure. This extra drag is called the induced drag. We shall consider it in Section 19.10.

The Kutta–Joukowski formula shows that the circulation Γ_a is the most important property of the flow in determining the lift. Recall the result of the previous section concerning circular cylinders: there was no unique answer until we specified a value for Γ_a. Choosing Γ_a determines a specific flow pattern and at the same time establishes the value of the lift through the Kutta–Joukowski law.

18.9 MAGNUS EFFECT

The flow over a nonrotating cylinder does not look much like the ideal-flow solution for $\Gamma = 0$ even though the lift force is zero. In this case the flow separates because fluid particles in the boundary layer do not have enough momentum to penetrate into the high-pressure region at the back of the cylinder. This situation changes somewhat if the cylinder is rotated. Since the no-slip condition demands that the fluid next to the wall move with the wall velocity, the boundary-layer profiles and the separation points are greatly modified. On the top side of the cylinder, the wall and flow velocities are in the same direction. These particles have extra momentum, and as a result the flow proceeds around this side further before it separates. The opposite effect occurs on the other side, as one can see in the photographs in Fig. 18-15.

The asymmetric flow around a rotating cylinder leads to a lift force. This is called the Magnus effect. The rotation parameter $r_0\Omega/U$ compares the surface

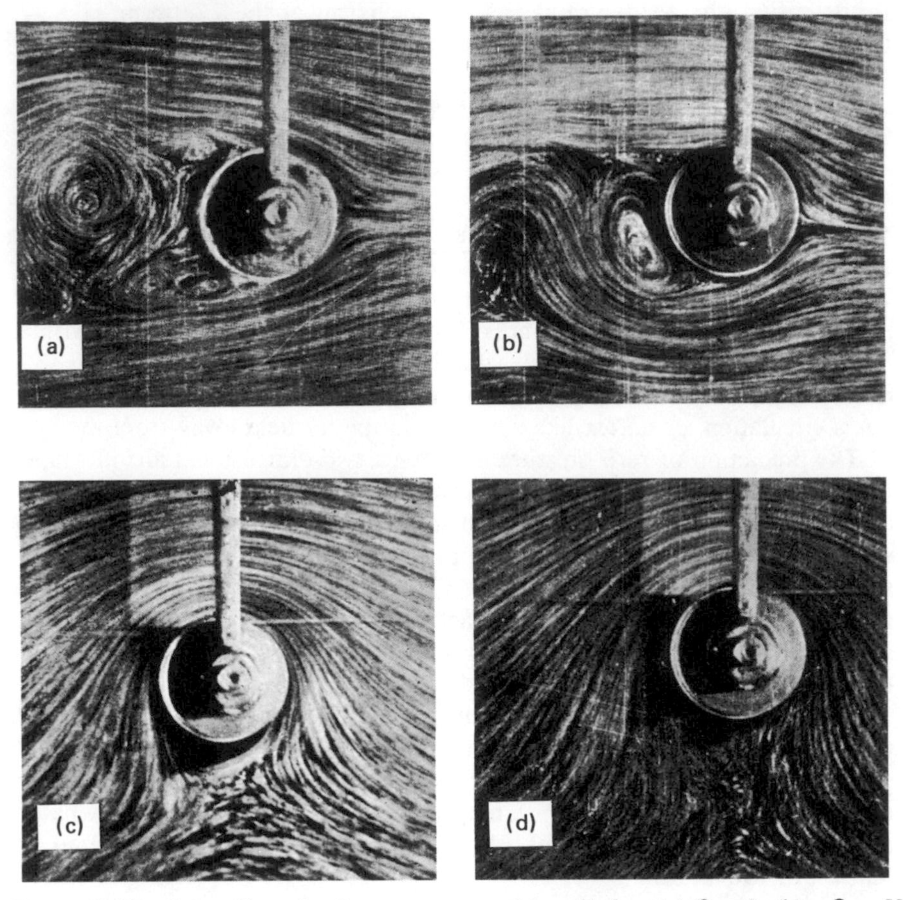

Figure 18-15 Streamlines for flow over a rotating cylinder: (a) $\Omega = 0$, (b) $r_0\Omega = U$, (c) $r_0\Omega = 4U$, (d) $r_0\Omega = 6U$. Photographs from Prandtl and Tietjens (1934).

velocity with the free-stream speed. As the rotation increases, the amount of lift also increases. First the increase is almost linear; then, around $r_0\Omega/U = 3$, a slower rate of increase begins.

At the higher rotation rates, separation can actually be suppressed, giving a flow pattern that is very much like the ideal flow pattern for a single stagnation point. In this regard one should compare Fig. 18-12 with $\Gamma_a = 4\pi r_0 U$ and Fig. 18-15 with $r_0\Omega/U = 4$. The question naturally arises whether the analysis could also predict the lift force, since the actual and theoretical flow patterns are so much alike. In order to predict the lift force using the theory of the previous two sections, we need to make a connection between the circulation Γ_a and the cylinder rotation Ω. Several arguments and analyses have been proposed to make this connection.

The Magnus problem was extensively investigated by Prandtl; he regarded it as a fundamental question relevant to the production of lift. Prandtl started by examining the particular ideal-flow solution that has only one stagnation point, that is, the case with $\Gamma_a = 4\pi r_0 U$. In this case the ideal velocity at the wall varies from zero to a maximum of $4U$. Prandtl reasoned that if the cylinder were rotated at this speed ($\Omega = 4U/r_0$), the boundary layer would never have fluid with a lower momentum than the ideal flow. Therefore, the boundary layer would not separate, and the ideal flow pattern, including the lift, would be realized. Prandtl thought that any higher rotation rate would give the same lift but with a slightly different boundary layer. This theory produces a lift value that is a little too high.

The Magnus lift force is not always in the same direction. At some Reynolds numbers (those near the critical value for transition from laminar to turbulent boundary layer), the Magnus effect can actually be negative. This happens only for very low rotation rates. This effect is thought to be the result of turbulent reattachment on the upwind-moving side and laminar separation on the downwind-moving side. Unless the flow is at the proper Reynolds number, the negative lift does not occur. Usually the rotation simply shifts the separation positions asymmetrically and creates a lift force in the expected direction.

18.10 CONFORMAL TRANSFORMATIONS

A very useful geometric interpretation of an analytic function is to consider that the function maps points from the plane of the independent variable $z = x + iy$ into points on a plane of the dependent variable $\zeta = \xi + i\eta$. For every point z, the function

$$\zeta = \zeta(z)$$

gives a point in the ζ plane. From calculus we know that the mapping is one-to-one as long as the derivative $d\zeta/dz$ is not zero. Places where $d\zeta/dz = 0$ are called *critical points* of the mapping.

Let us consider as an example the mapping given by the function

$$\zeta = z + \frac{r_0^2}{z} \qquad 18.10.1$$

where r_0 is a real constant. One of the best ways of visualizing a mapping is to draw some lines in the z plane and trace the corresponding lines mapped into the ζ plane. Special choices of lines in one plane usually give simple geometric patterns in the other plane. Figure 18-16 shows how the region in the z plane outside of the circle $z = r_0$ maps into the entire ζ plane. The points A, B, C, D are drawn in each plane. Substituting $z = r_0 e^{i\theta}$ into 18.10.1 will give ζ coordinates corresponding to the circle. They are

$$\zeta = r_0 e^{i\theta} + r_0 e^{-i\theta} = 2r_0 \cos\theta$$

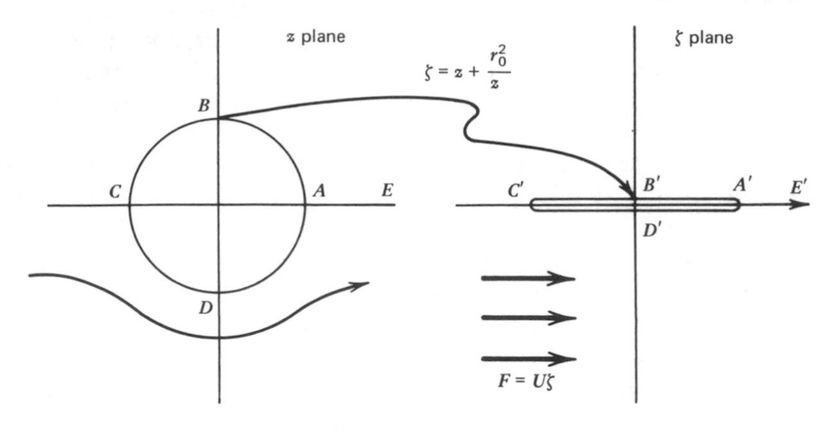

Figure 18-16 Conformal mapping of a cylinder to a flat plate.

As θ ranges from 0 to 2π, the corresponding points in the ζ plane are $\eta = 0$, $\xi = 2r_0 \cos\theta$. Note that the points B and D in the z plane map to the same point in the ζ-plane. This does not violate the one-to-one property of the mapping: the circle $z = r_0$ merely gives a boundary in the z plane for a region that covers the entire ζ plane. Points on the inside of the circle map to the ζ-plane in such a way that they also cover the entire plane. The point $z = 0$ maps to the "point at infinity" ($\zeta = \infty$) according to 18.10.1.

The name *conformal transformation* denotes the fact that the angle formed by the intersection of two lines in the z plane is unchanged when these lines are transformed into the ζ plane. In Fig. 18-17 the point P in the z plane has a certain line going through it. Along this line the differential increment dz may

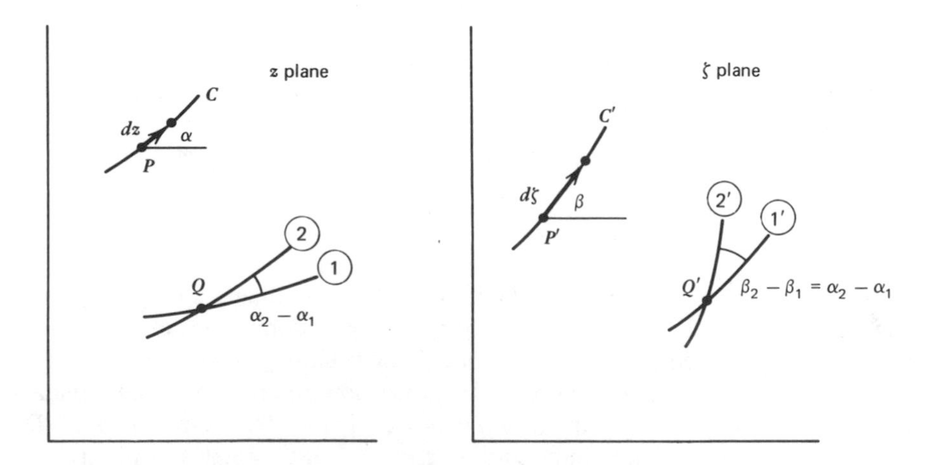

Figure 18-17 Curves mapped by a conformal transformation $\zeta(z)$ retain the same included angle.

be given in polar form as

$$dz = |dz|e^{i\alpha}$$

where α is the inclination of the line at P. A certain transformation $\zeta(z)$ maps P and the increment dz into the point P' in the ζ plane and the corresponding differential increment $d\zeta$:

$$d\zeta = |d\zeta|e^{i\beta}$$

Dividing these expressions gives

$$\frac{d\zeta}{dz} = \zeta'(z) = \frac{|d\zeta|}{|dz|}e^{i(\beta-\alpha)} \qquad 18.10.2$$

This equation presents the derivative of $\zeta(z)$ in polar form; the magnitude of $\zeta'(z)$ is the stretching factor $|d\zeta/dz|$ of a line though P, while the angle $\beta - \alpha$ is the amount of rotation the line is subjected to when it is transformed from the z to the ζ plane. Since $\zeta'(z)$ is independent of the direction of dz, all lines through P are stretched by the same amount and rotated through the same angle. Hence, any two lines through P are mapped into lines rotated by the same amount, and retain the same included angle in the ζ plane.

A critical point in the transformation occurs when $\zeta'(z) = 0$. Both the real and imaginary parts of $\zeta'(z)$ are then zero, implying that the magnitude is zero and the direction is undetermined. At critical points, lines through P may be rotated by different amounts as they are transformed onto the ζ plane. In the example furnished by 18.10.1 and illustrated in Fig. 18-16, the critical points are $z = \pm r_0$. The lines AB and AE are rotated by different amounts and the angle between these lines is not preserved during the transformation to the ζ plane. Note that a single line in one plane may split into several lines in the other plane at a critical point. Splitting may occur only at critical points, where the transformation is not one-to-one.

Although we shall not prove the fact, it should be stated for the record that any analytic function $F(z)$ is a conformal transformation, except at singular points of the function and at the critical points defined by $F'(z) = 0$. For example ln z is conformal except at the origin, where it is singular, while sin z is conformal except where the derivative cos z is zero.

When conformal mapping is employed for ideal-flow problems, we shall always let z be the physical plane where the body is drawn in its true shape. In this plane we have the streamlines given by $\psi(x, y)$ and the potential lines by $\phi(x, y)$. These lines map to new positions in the ζ plane, where there are corresponding lines $\psi(\xi, \eta)$ and $\phi(\xi, \eta)$. The complex potential in the physical plane is

$$F = F(z) = \phi(x, y) + i\psi(x, y) \qquad 18.10.3$$

When the inverse transformation $z = z(\zeta)$ is substituted, we call the new function $\hat{F}(\zeta)$:

$$F = F(z(\zeta)) = \hat{F}(\zeta) = \hat{\phi}(\xi, \eta) + i\hat{\psi}(\xi, \eta) \qquad 18.10.4$$

From this we see that it is perfectly reasonable to talk about an ideal flow in the z plane, $\phi(x, y)$, and an equivalent flow in the ζ plane, $\hat{\phi}(\xi, \eta)$.

In our example (Fig. 18-16) we considered how a cylinder in the z plane would transform into an infinitely thin plate between A and C in the ζ plane. Next, let us assume that the flow in the ζ plane is a uniform stream from right to left. This flow has the complex potential

$$F = \hat{F}(\zeta) = U\zeta \qquad 18.10.5$$

When the flat plate is transformed into a circle, we find the corresponding potential by substituting 18.10.1 into 18.10.5. The result is

$$F = U\left(z + \frac{r_0^2}{z}\right) \qquad 18.10.6$$

Thus, we have a new way of looking at the ideal flow over a cylinder. We imagine that a conformal transformation maps the cylinder into a flat plate; then the flow over the plate is a streaming motion given by 18.10.5. Equation 18.10.6 may be thought of as a composite of 18.10.5 and 18.10.1.

The usefulness of conformal transformation is that a complicated geometry may be mapped into a simple figure in one step, and then a simple flow pattern may be imagined in the transformed plane.

Streamlines and potential lines in one plane map into streamlines and potential lines in the transformed plane. The velocities, on the other hand, are modified. The original complex velocity is

$$W(z) = \frac{dF}{dz}$$

while in the ζ plane, the velocity of the equivalent flow is

$$\hat{W}(\zeta) = \frac{d\hat{F}}{d\zeta}$$

The chain rule applied to $F(z) = \hat{F}(\zeta(z))$ yields

$$\frac{dF}{dz} = \frac{d\hat{F}}{d\zeta}\frac{d\zeta}{dz}$$

This can be rewritten as

$$W(z) = \hat{W}(\zeta)\frac{d\zeta}{dz} \qquad 18.10.7$$

The derivative of the mapping $\zeta(z)$ indicates the velocity ratio between the z and ζ planes. One important fact we see from 18.10.7 is that critical points of the transformation are always stagnation points $W(z) = 0$ of the flow in the real plane. At these points the streamlines may branch. Futhermore stagnation points in the z plane are not necessarily stagnation points in the ζ plane as $\hat{W}(\zeta)$ is not necessarily zero.

So far we have dealt exclusively with the geometric aspects of conformal transformations. How can we be sure that an ideal flow in the z plane obeys the proper equations for a flow in the ζ plane? Recall that the basic fact we have been using is that all analytic functions $F(z)$ have real and imaginary parts that satisfy the Laplace equation. For $F = \phi + i\psi$ we have

$$\frac{\partial^2 \phi}{\partial x^2} + \frac{\partial^2 \phi}{\partial y^2} = 0, \qquad \frac{\partial^2 \psi}{\partial x^2} + \frac{\partial^2 \psi}{\partial y^2} = 0$$

Therefore any $F(z)$ represents a flow pattern of some sort where $\psi =$ const is a streamline and the velocity across a streamline is zero ($\partial\phi/\partial n = 0$). In order to establish that a flow in the z plane is also a flow in the ζ plane, we need another mathematical fact. It may be proved that an analytic function of an analytic function is another analytic function; that is, if $F(z)$ and $z(\zeta)$ are both analytic functions, then $F(z(\zeta))$ is an analytic function of the variable ζ. Hence, $F = F(z(\zeta))$ has real and imaginary parts that satisfy the Laplace equation in terms of ξ and η. That is,

$$F = F(z(\zeta)) = \hat{\phi}(\xi, \eta) + i\hat{\psi}(\xi, \eta)$$

$$\frac{\partial^2 \hat{\phi}}{\partial \xi^2} + \frac{\partial^2 \hat{\phi}}{\partial \eta^2} = 0, \qquad \frac{\partial^2 \hat{\psi}}{\partial \xi^2} + \frac{\partial^2 \hat{\psi}}{\partial \eta^2} = 0$$

The functions $\phi(x, y)$, $\psi(x, y)$, which satisfy the Laplace equation in x and y, also satisfy the Laplace equation in ξ and η when $x, y \to \xi, \eta$ under any conformal transformation.

Let us illustrate the facts discussed above with a concrete example. Recall that the analytic function

$$F = \sin z = \phi + i\psi$$

has real and imaginary parts $[\sin(x + iy) = \sin x \cosh y + i \cos x \sinh y]$

$$\phi = \sin x \cosh y$$

$$\psi = \cos x \sinh y$$

Previously (Section 18.1), we verified that these functions satisfy the Laplace equation. Next, we consider the conformal transformation given by the analytic function

$$\zeta = z^{1/2}$$

Under this transformation we find that F becomes

$$F = \hat{F}(\zeta) = \sin \zeta^2$$

$$= \sin(\xi + i\eta)^2$$

$$= \sin\left[(\xi^2 - \eta^2) + i2\xi\eta\right]$$

$$= \phi(\xi, \eta) + i\hat{\psi}(\xi, \eta)$$

A little further algebra separating the sine function into its real and imaginary parts shows that

$$\hat{\phi} = \sin(\xi^2 - \eta^2)\cosh 2\xi\eta$$

$$\hat{\psi} = \cos(\xi^2 - \eta^2)\sinh 2\xi\eta$$

It may be shown that these functions satisfy the Laplace equation in terms of ξ and η. An analytic function of an analytic function is an analytic function.

We have discussed how the functions $\hat{\phi}$ and $\hat{\psi}$ satisfy the proper differential equation that governs ideal flow, the Laplace equation. A few remarks should be made about the boundary conditions. In ideal flow we deal with boundary conditions of the type $\psi = \text{const}$ on a specified wall (or correspondingly $\partial\phi/\partial n = 0$) or $U = \partial\phi/\partial n = \text{const}$ on a certain curve. This type of boundary condition retains its form under a conformal transformation. Therefore we can be assured that the complete ideal-flow problem—the differential equation and the boundary conditions—remains the same in both planes. The only reason this fact is mentioned is that in some fields of science the boundary condition $\partial f/\partial n = cf$ is encountered. With this type of boundary condition, the constant c changes as f is transformed from the z plane to the ζ plane.

18.11 JOUKOWSKI TRANSFORMATION; AIRFOIL GEOMETRY

The study of ideal flow over two-dimensional cylindrical objects, (circular cylinders, elliptic shapes, flat plates, and a certain type of airfoil shape) is simplified by using the conformal-transformation technique. In the previous section we used the example of a circular cylinder in the real plane mapped into a flat plate under the transformation $\zeta = z + r_0^2/z$. If we simply rename the planes—that is, let the flat plate be in the physical plane and the cylinder be in the transformed plane—the transformation equation is $z = \zeta + r_0^2/\zeta$. We introduce the Joukowski transformation as a generalization of this form where c replaces r_0:

$$z = \zeta + \frac{c^2}{\zeta} \tag{18.11.1}$$

The constant c can be any real number. The inverse transformation is

$$\zeta = \tfrac{1}{2}z \pm \left[\left(\tfrac{1}{2}z\right)^2 - c^2\right]^{1/2} \tag{18.11.2}$$

It is important to keep in mind that the constant c is now arbitrary and is not necessarily a cylinder radius in the ζ plane.

Figure 18-18 shows the flat plate $ABCD$ in the z plane as it maps into a circle of radius $r_0 = c$ in the ζ plane. If we let the flow in the ζ plane [i.e., the complex potential $F(\zeta)$] be the flow over a circular cylinder, then the corresponding flow $F(z)$ in the z plane will be a flow over a flat plate. As a second example, consider the ellipse $EGFH$. It turns out that this ellipse in the z plane also maps into a circle in the ζ plane. (If the transformation constant is c and the radius of the cylinder is R_0, then the major and minor axes of the ellipse are $R_0 + c^2/R_0$ and $R_0 - c^2/R_0$.) Now, if we let the flow in the ζ plane be that for a circle of radius R_0, while the transformation constant is c, the flow in the real plane will be that over an elliptic cylinder. The student may find the details of these flow patterns in any of the supplementary texts recommended for this chapter.

The useful characteristic of the Joukowski transformation is that it sends a certain airfoil-like figure in the z plane into a circle in the ζ plane. It turns out that the center of the circle is off axis in the ζ plane (Fig. 18-19). Hence, the

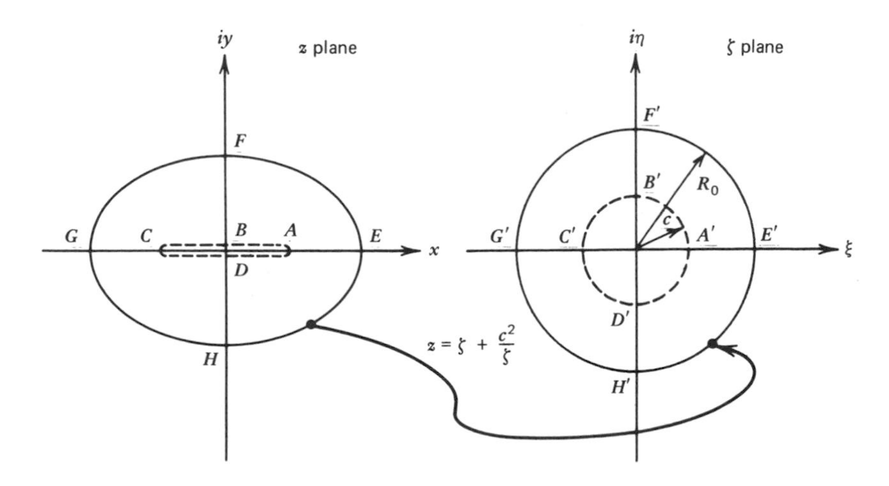

Figure 18-18 Joukowski transformation of an ellipse to a circle.

complex potential $F(\zeta)$ for flow over an off-axis circle in the ζ plane represents the flow over an airfoil shape in the real z plane. This is a very important problem as it offers a key to the flow pattern over airfoil shapes. From these solutions we can discover how various geometric parameters of the airfoil influence the lift. Of course, an airfoil does not necessarily have the same shape as a Joukowski airfoil. In fact, real airfoils are not Joukowski shapes; their contours are determined by considering viscous effects and how boundary-layer separation may be avoided. Nevertheless, it is useful to study Joukowski airfoils, as the general trends for ideal flow over these shapes and for flow over actual airfoil shapes are the same.

In the remainder of this section we shall investigate the geometry of Joukowski airfoils. The major question is how points on an off-center circle in the ζ plane transform into an airfoil shape in the z-plane. An exact closed-form equation for the airfoil contour in the z-plane does not exist. This is not a serious difficulty, as practical airfoils are thin, and we can find approximate equations for airfoil contours for this case. Consider the nomenclature defined for the circle on Fig. 18-19. Let the center of the circle of radius r_0 in the ζ plane be at the point ζ_0. This is given by the polar form

$$\zeta_0 = me^{i\delta} \tag{18.11.3}$$

The circle cuts the ξ axis at c, where c is the constant in the Joukowski transformation. An arbitrary point on the circle is a distance b from the origin at an angle γ,

$$\zeta_{\text{cir}} = be^{i\gamma}$$

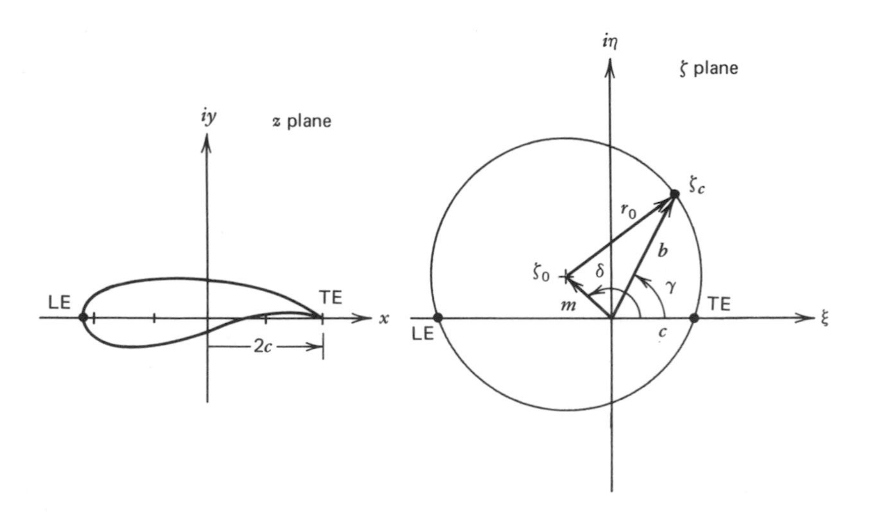

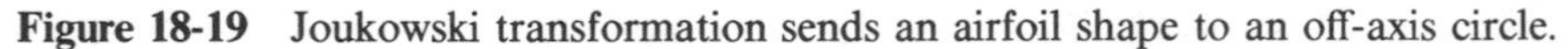

Figure 18-19 Joukowski transformation sends an airfoil shape to an off-axis circle.

Our first task is to find a relation for b as a function of the angle γ. This can be done if the displacement of the center of the circle is small. To be precise we define

$$\epsilon \equiv \frac{m}{c} \qquad 18.11.4$$

and seek asymptotic expressions valid as $\epsilon \to 0$.

Consider the law of cosines applied to the triangle formed by m, r, and b in Fig. 18-19:

$$r_0^2 = b^2 + m^2 - 2mb\cos(\delta - \gamma) \qquad 18.11.5$$

Dividing 18.11.5 by c, introducing 18.11.4, and rearranging yields

$$\frac{r_0}{c} = \frac{b}{c}\left[1 - 2\epsilon\frac{c}{b}\cos(\delta - \gamma) + \left(\frac{c}{b}\right)^2 \epsilon^2\right]^{1/2}$$

The binomial series is used to expand the bracket for small ϵ. The result is

$$\frac{r_0}{c} = \frac{b}{c}\left[1 - \epsilon\frac{c}{b}\cos(\delta - \gamma) + O[\epsilon^2]\right] \qquad 18.11.6$$

We set this equation aside for a moment. Next, consider the triangle formed by m, c, and another radial line with length r_0. The law of cosines for this triangle is

$$r_0^2 = m^2 + c^2 - 2mc\cos\delta$$

Dividing by c, inserting the definition $\epsilon = m/c$, and expanding by the binomial theorem gives

$$\frac{r_0}{c} = 1 - \epsilon\cos\delta + O[\epsilon^2] \qquad 18.11.7$$

Equating 18.11.6 and 18.11.7 yields

$$\frac{b}{c} = 1 + \epsilon[\cos(\delta - \gamma) - \cos\delta] + O[\epsilon^2]$$

$$= 1 + \epsilon B + O[e^2] \qquad 18.11.8$$

where B stands for the bracketed expression on the first line. By introducing the trigonometric formula for $\cos(\delta - \gamma)$, the bracket in 18.11.8 is expressed as

$$B = [\sin\gamma\sin\delta - \cos\delta\,(1 - \cos\gamma)] \qquad 18.11.9$$

Equations 18.11.8 and 18.11.9 are the desired expression for b as a function of the angle γ. These relations describe the circle in the ζ plane. The coordinate of any point on the circle is a function of the parameter γ through the equations

$$\frac{\zeta_{\text{cir}}}{c} = \frac{b}{c} e^{i\gamma}$$

$$\approx [1 + \epsilon B(\gamma)] e^{i\gamma} \qquad 18.11.10$$

The airfoil coordinates in the physical plane are denoted by $z_s = x_s + iy_s$. They are found from the Joukowski transformation evaluated on the circle:

$$z_s = \zeta_{\text{cir}} + \frac{c^2}{\zeta_{\text{cir}}}$$

Substituting 18.11.10 into the equation above and dropping terms of order ϵ^2 yields

$$\frac{x_s}{c} + i\frac{y_s}{c} = (1 + \epsilon B)(\cos\gamma + i\sin\gamma)$$

$$+ (1 + \epsilon B)^{-1}(\cos\gamma - i\sin\gamma)$$

Noting that $(1 + \epsilon B)^{-1} = 1 - \epsilon B + O[\epsilon^2]$, this expression simplifies to the equations

$$\frac{x_s}{4c} = \frac{\cos\gamma}{2}$$

$$\frac{y_s}{4c} = \frac{\epsilon B(\gamma)\sin\gamma}{2} \qquad 18.11.11$$

These equations are parametric relations for the airfoil coordinates in terms of γ. From 18.11.11 we find that the trailing edge of the airfoil is at $x_s = 2c$, while the leading edge is $x_s = -2c$. The length $4c$ is the chord of the airfoil, which is denoted by l. Thus, the Joukowski transformation constant c is one-fourth of the airfoil chord:

$$c = \frac{l}{4} \qquad 18.11.12$$

It is also useful to note that the leading and trailing edges of the airfoil correspond to the points in the ζ plane where the ζ axis cuts the circle ($\gamma = 0, \pi$).

The y equation for the airfoil surface, 18.11.11, can be expressed in terms of x_s. To do this we note from 18.11.11 that

$$\cos\gamma = \frac{x_s}{2c} = \frac{2x_s}{l}$$

Therefore by trigonometry

$$\sin \gamma = \pm\sqrt{1 - \left(\frac{2x_s}{l}\right)^2}$$

When the two expressions above are inserted into 18.11.11 together with 18.11.9 and 18.11.12, the following equation results (see Fig. 18-19):

$$\frac{y_s}{l} = \frac{\epsilon}{2}\left\{\sin \delta \left[1 - \left(\frac{2x_s}{l}\right)^2\right] \pm \cos \delta \left(1 - \frac{2x_s}{l}\right)\left[1 - \left(\frac{2x_s}{l}\right)^2\right]^{1/2}\right\} \qquad 18.11.13$$

In this form the equation of the airfoil, $y_s(x_s)$, consists of two parts. The first part is the *camber line*, given by the first term within the braces. The second part, the term following the $\pm$ sign, adds and subtracts a *thickness distribution* to the camber line. The camber line reaches its maximum at $x_s = 0$; we denote this maximum as h. Inserting it into 18.11.13 gives

$$\frac{h}{l} = \frac{\epsilon}{2} \sin \delta \qquad 18.11.14$$

The thickness distribution is zero at the leading and trailing edges and reaches a maximum, denoted by $t/2$, at $x_s/l = -\frac{1}{4}$. Introducing this fact into 11.11.13 shows that

$$\frac{t}{2l} = \frac{\epsilon}{2} \cos \delta \cdot \frac{3\sqrt{3}}{4} \qquad 18.11.15$$

By using the notation h for maximum camber and t for maximum thickness, 18.11.13 can be rewritten in the following form:

$$\frac{y_s}{l} = \frac{h}{l}\left[1 - \left(\frac{x_s}{l/2}\right)^2\right] \pm \frac{2}{3\sqrt{3}}\frac{t}{l}\left(1 - \frac{x_s}{l/2}\right)\left[1 - \left(\frac{x_s}{l/2}\right)^2\right]^{1/2} \qquad 18.11.16$$

This is the Joukowski profile equation linearized for small values of thickness and camber.

The nomenclature we introduced in splitting the Joukowski airfoil equation into a chord line, a camber line, and a thickness distribution is not the only accepted method for defining an airfoil shape. Figure 18-19 gives another method of constructing the definitions. First, one lays out the chord line of the proper length. The leading and trailing edges are the ends of the chord line. A camber line, sometimes also called the mean line, is marked off at specified distances from the chord line. There are no restrictions on the camber-line shape other than that it must begin and end at the leading and trailing edges. Joukowski airfoils described by 18.11.16 have a parabolic camber line with the maximum at the 50% chord position. Other airfoils have different shapes for the camber line. The airfoil shape is completed by adding a thickness distribution at equal distance above and below the camber line. In the alternate method, the thickness distribution must be added on a line perpendicular to the local slope of the camber line. If an airfoil begins with a lot of camber and also has a large leading-edge radius, then the airfoil is actually longer than the chord length and the leading edge is not the furthest forward position. In 18.11.16 we imply that the thickness distribution is added to the camber line in a direction perpendicular to the chord line instead of perpendicular to the mean line. This is mathematically the most convenient method, and for small camber, which incidentally most airfoils have, the two definitions are equivalent.

Joukowski airfoils have the general shape of all airfoils. They have a parabolic camber line, which is adjusted through the parameter h/l, and they have a certain thickness distribution, which is adjusted through the parameter t/l. If t/l is taken as zero, then the airfoil reduces to a camber line without any thickness: a model for a cloth sail or a sheet-metal fan blade. If h/l is taken as zero, the airfoil is symmetric. Symmetric airfoils have been used on stunt planes that fly upside down as much as they do right side up. If t/l and h/l are both zero, the airfoil becomes a flat plate. One characteristic of all Joukowski airfoils is that they have a cusp at the trailing edge. Actual airfoils have a sharp trailing edge with a finite wedge angle. A typical Joukowski profile is shown in Fig. 18-20.

We have in 18.11.16 a complete description of the airfoil shape in the z plane. The Joukowski transformation maps this shape into a circle in the ζ plane. The geometry of the circle is described by c (the position where it cuts the ξ axis) and ζ_0 (the position of the center with respect to the ζ origin). By way of summary we note that the position c is given by 18.11.12:

$$\frac{c}{l} = \frac{1}{4}$$

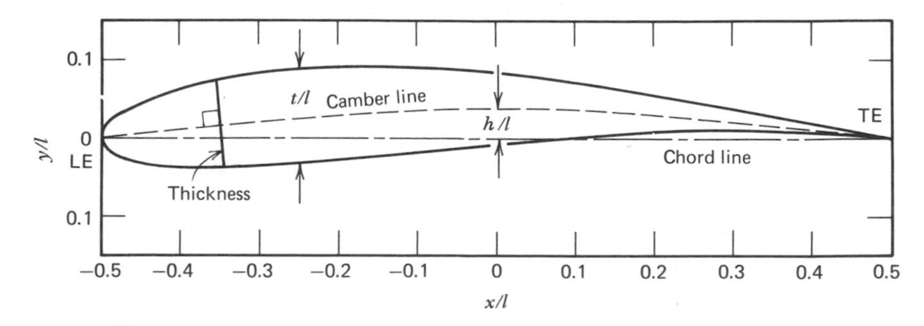

Figure 18-20 Joukowski airfoil with camber $h = 0.04l$ and thickness $t = 0.12l$. Thickness is defined as perpendicular to the camber line, but is equal to vertical distance for small camber.

The location of the center is given by 18.11.14 and 18.11.15, which we put in a more convenient form using 18.11.3:

$$\frac{m}{c}\sin\delta = 2\frac{h}{l} \qquad 18.11.17$$

$$-\frac{m}{c}\cos\delta = \frac{4}{3\sqrt{3}}\frac{t}{l} \qquad 18.11.18$$

The camber and thickness of the airfoil are related to the vertical and horizontal displacement of the center respectively.

18.12 KUTTA CONDITION

The ideal flow about a cylinder or other simple two-dimensional body is not unique because the region is doubly connected. A single arbitrary circulation constant must be specified in order to obtain a unique ideal flow pattern. In the case of a triply-connected region, say two cylinders or two airfoils side by side, two circulation constants must be specified. For bluff bodies such as elliptic or circular cylinders there is no method to determine the circulation constant. This is not very important, because such bodies have large wake regions and ideal flow cannot be applied anyway. Airfoils, on the other hand, offer a situation where ideal-flow theory is very useful, and a method to determine the circulation constant is required.

Figure 18-21 gives several ideal flow patterns for airfoil shapes at the same angle of attack. Each of these patterns has a different circulation constant and, according to the Kutta–Joukowski theorem, a different lift force. The *Kutta condition* (also known as Joukowski's hypothesis) is the assumption that the flow cannnot go around the sharp trailing edge, but must leave the airfoil so that the upper and lower streams join smoothly at the trailing edge. There is only one flow pattern and one circulation value that will do this. Formally stated, the Kutta condition says that the proper circulation constant for the flow over an airfoil is the value that causes the velocity to leave the trailing edge in a direction that bisects the angle formed by the upper and lower surfaces. An equivalent statement is that the velocity at the trailing edge cannot be infinite.

The Kutta condition is not subject to proof. It is a rule of thumb that works fairly well for most airfoils. In our theoretical analysis of flows we have, in a certain sense, oversimplified the problem by entirely neglecting viscosity and the no-slip condition. Stokes's theorem, 18.7.5, shows that circulation is equal to the integral of the vorticity in the boundary layers on the airfoil (a two-dimensional wake has no net vorticity). Thus in some complicated way the viscous effects in the boundary layers actually go hand in hand with the circulation. At high Reynolds numbers the details of the viscous effects are no

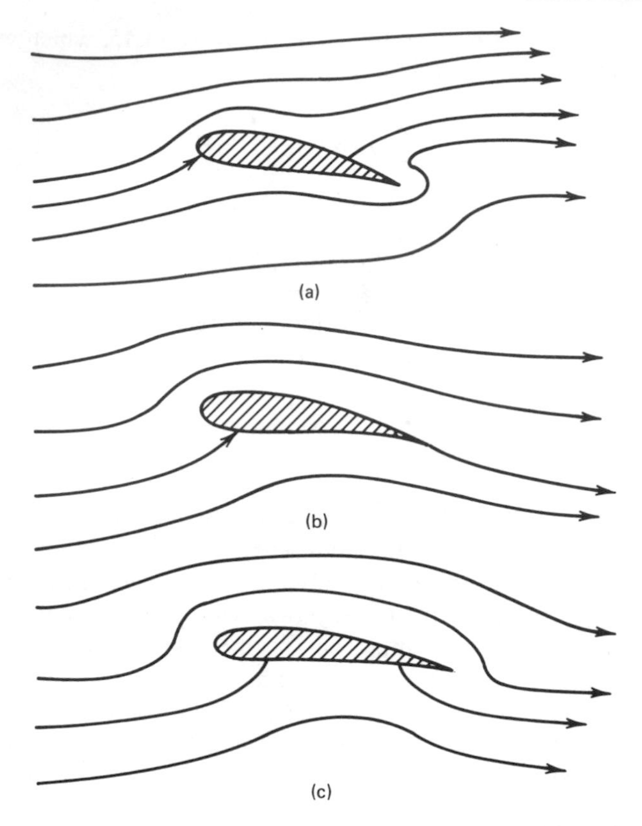

Figure 18-21 The Kutta condition requires that the streamline bisect the trailing edge as it leaves the airfoil.

longer important and their effect on the main inviscid flow can be distilled into the Kutta condition. For these reasons it is sometimes said that the Kutta condition is the result of viscosity.

When two inviscid flow streams merge, as they do at the trailing edge of an airfoil, the pressure must be the same on either side of the streamline; it cannot be discontinuous. Since in the case of an airfoil the two streams have the same Bernoulli constant, the velocity at the trailing edge has the same value for the upper and lower streams. Figure 18-21 shows two streams merging from a cusp trailing edge such as a Joukowski airfoil would have. The velocity there is generally slightly lower than the free-stream velocity. In the case of a finite angle at the trailing edge we might envision flow patterns where one stream turns through a larger angle than the other. Such patterns are in fact impossible in ideal flow. Recall that the wedge-flow solution of Section 18.2 has stagnation points when the flow is turned through such an angle and that the velocity increases as you move away from the stagnation point. The way in which the velocity increases depends upon the angle through which the flow is turned (see 18.2.7). If the top flow and the bottom flow were turned through different angles, then the velocity and pressure on the merging streamline would not match as we move away from the trailing edge. The only acceptable ideal-flow solution when the streamline leaves the trailing edge is one for which it bisects the wedge angle.

Also shown in Fig. 18-21 are solutions that go around the trailing edge and flow up the other side. The pressure at the corner of such a flow is minus infinity, and the velocity becomes infinite. This behavior is ruled out by the Kutta hypothesis. Although infinite velocities are ruled out at a trailing edge, they are allowed at the leading edge of a flat plate or cambered airfoil with zero thickness.

In reality, the trailing edge of an airfoil is hidden beneath viscous boundary layers. The top boundary layer is usually somewhat thicker than the bottom layer, a condition that progresses as more lift is obtained. Thus, the inviscid flow does not actually see a sharp trailing edge at all, but a geometry modified by the thickness of the boundary layers. In spite of these difficulties in detail, the Kutta condition is one of the major working assumptions in any airfoil theory. As a first approximation it gives a remarkably good estimate of the lift as long as separation does not occur.

18.13 FLOW OVER A JOUKOWSKI AIRFOIL

The geometry of a Joukowski airfoil is specified by the angle of attack α, the camber ratio h/l, and the thickness ratio t/l. We have no control over the distributions of camber and thickness, as all Joukowski airfoils have the same distributions. It turns out that this is not a critical simplification, as the events that produce lift are fairly insensitive to these distributions. This is especially true at modest angles of attack.

The flow around a Joukowski airfoil is found using the ideas of the previous three sections. In Section 18.10 the idea of conformally transforming one flow field into a much simpler flow field was introduced. Section 18.11 gave us a specific transformation, the Joukowski transformation, which transforms an airfoil shape in the z plane into a circular cylinder in the ζ plane. Since in 18.7 we already have in hand the solution for flow over a cylinder, the only remaining step is to reinterpret this flow after it is transformed back into the z plane for the airfoil. The obstacle to this procedure is the fact that there are an infinite number of ideal flows over a circular cylinder and we must pick one. This difficulty is overcome by invoking the Kutta condition to select that flow which leaves the trailing edge smoothly with a finite velocity.

The cylinder in the ζ plane is shown in Fig. 18-22. Recall that the trailing edge of the airfoil maps to the point $\zeta = c$ on the circle and that the thickness and camber determine the center position of the circle, denoted by ζ_0. To apply the Kutta condition, we must arrange the circulation constant so that the flow

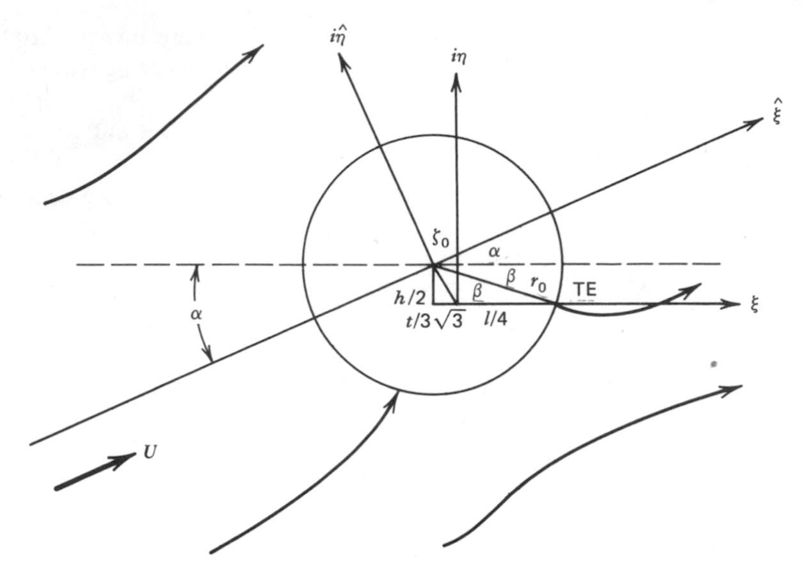

Figure 18-22 Circular cylinder off center and at angle of attack α.

leaves the circle at the point marked *TE*; this point will be the rear stagnation point for the flow in the ζ plane.

Before we can write down the complex potential, we need one more detail. The real airfoil is at an angle of attack α (the angle between the chord and freestream). What influence will this have in the ζ plane? Looking at the Joukowski transformation 18.11.1, we see that as $\zeta \to \infty$, $z \sim \zeta$. Since the functions of the complex potentials in the ζ and z planes are related by

$$F = \hat{F}(\zeta) = \hat{F}(\zeta(z)) = F(z)$$

at infinity we may set $z = \zeta$ to show

$$\hat{F}(z) \sim F(z), \qquad z \to \infty \tag{18.13.1}$$

At infinity the flows have exactly the same form in either plane. This result leads us to conclude that the angle of attack for the cylinder must be the same as for the airfoil.

Returning to Fig. 18-22, let us introduce a $\hat{\zeta} = \hat{\xi} + i\hat{\eta}$ coordinate system that is aligned with the flow at infinity and has its origin at the center of the circle. The relation with the ζ coordinates is given by a translation of ζ_0 and a rotation through the angle α, that is,

$$\hat{\zeta} = (\zeta - \zeta_0)e^{-i\alpha} \tag{18.13.2}$$

The complex potential for flow about a cylinder with circulation Γ_a is given by 18.7.1. In the current notation, the potential is

$$F = U\left(\xi + \frac{r_0^2}{\xi}\right) + i\frac{\Gamma_a}{2\pi}\ln\frac{\xi}{r_0} \qquad 18.13.3$$

Inserting 18.13.2 gives the potential in terms of ζ,

$$F = U\left[(\zeta - \zeta_0)e^{-i\alpha} + r_0^2 e^{i\alpha}(\zeta - \zeta_0)^{-1}\right] + i\frac{\Gamma_a}{2\pi}\ln\left(\frac{\zeta - \zeta_0}{r_0}\right) \qquad 18.13.4$$

A constant term that arises from $\ln e^{i\alpha}$ has been dropped. This merely changes the numbering system for the potentials. Equation 18.13.4 together with the inverse Joukowski transformation 18.11.2 defines the potential $F(z)$ in the real plane.

Let us take a careful look at 18.13.4 and see how all the symbols are related to quantities that refer to the airfoil geometry and to the flow in the z plane. The symbols U and α are obviously the free-stream speed and angle of attack. ζ is a parametric variable related to z through 18.11.2, where in turn $c = l/4$ (18.11.12) introduces the airfoil chord l. The center displacement of the circle is (18.11.17 and 18.11.18)

$$\begin{aligned} \zeta_0 &= m\cos\delta + im\sin\delta \\ &= -\frac{t}{3\sqrt{3}} - i\frac{h}{2} \end{aligned} \qquad 18.13.5$$

The only symbols in 18.13.4 that remain to be interpreted are the circle radius r_0 and the circulation Γ_a.

Figure 18-22 shows some of the details of the circle geometry. In this figure the right triangle that involves r_0 yields the relation

$$r_0^2 = \left(\frac{h}{2}\right)^2 + \left(\frac{l}{4} + \frac{t}{3\sqrt{3}}\right)^2$$

The airfoil has small h and t, so to the same degree of approximation with which we found the airfoil coordinates, we find

$$r_0 = \frac{l}{4} + \frac{t}{3\sqrt{3}} \qquad 18.13.6$$

Likewise, as $m \to 0$, the angle β in Fig. 18-22 is

$$\beta = \frac{h/2}{l/4} = \frac{2h}{l} \qquad 18.13.7$$

We need to know this angle in order to find the circulation Γ_a.

In our study of flow over a cylinder with circulation, we found that the stagnation points were moved away from the flow axis (the $\hat{\xi}$ axis) as the circulation increased. The angular position was given by 18.7.4. In order to satisfy the Kutta condition, we need to position the stagnation point at an angle $\alpha + \beta$ in Fig. 17-22. Substituting $\alpha + \beta$ for $-\theta_0$ in 18.7.4 yields

$$\sin(\alpha + \beta) = \sin\left(\alpha + \frac{2h}{l}\right) = \frac{\Gamma_a}{4\pi r_0 U} \tag{18.13.8}$$

This equation, together with 18.13.6, gives Γ_a in terms of the airfoil geometry and flow parameters. This completes the interpretation of the complex potential 18.13.4 in terms of variables related to the airfoil.

Velocities that occur around the airfoil are related to the velocities at corresponding points on the cylinder by 18.10.7:

$$\begin{aligned} W(z) &= \hat{W}(\zeta)\frac{d\zeta}{dz} \\ &= \frac{d\hat{F}(\zeta)}{d\zeta} \bigg/ \frac{dz}{d\zeta} \end{aligned} \tag{18.13.9}$$

Performing the operations indicated above, we arrive at

$$u - iv = \left\{U\left[e^{-i\alpha} - r_0^2 e^{i\alpha}(\zeta - \zeta_0)^{-2}\right] + i\frac{\Gamma_a}{2\pi r_0}\left(\frac{r_0}{\zeta - \zeta_0}\right)\right\}\left[1 - \left(\frac{c}{\zeta}\right)^2\right]^{-1} \tag{18.13.10}$$

This expression is left with ζ as a parameter. The velocities u and v are in the z plane at a position given by $z(\zeta)$ in 18.11.1. The factor in braces is the velocity in the ζ plane, while the bracket on the second line is the derivative of the Joukowski transformation. This is always nonzero except at the two critical points $\zeta = \pm c$.

One critical point lies within the circle, and the other lies at the rear stagnation point on the circle. At this point both W and $dz/d\zeta$ are zero, leading to an indeterminate form in 18.13.10. This expression can be evaluated using L'Hôspital's rule. For small values of thickness and camber, the velocity at the trailing edge is found to be

$$[u - iv]_{\text{TE}} = U\left[1 - \frac{4t}{3\sqrt{3}\,l}\right]\left[1 + i\frac{2h}{l}\right] \tag{18.13.11}$$

The velocity is slightly smaller in magnitude than the free-stream velocity (an effect due to the thickness) and is directed downward in alignment with the camber line. Figure 18-23 is a sketch of how the streamlines look in the physical plane, while Fig. 18-24 displays the corresponding surface pressures.

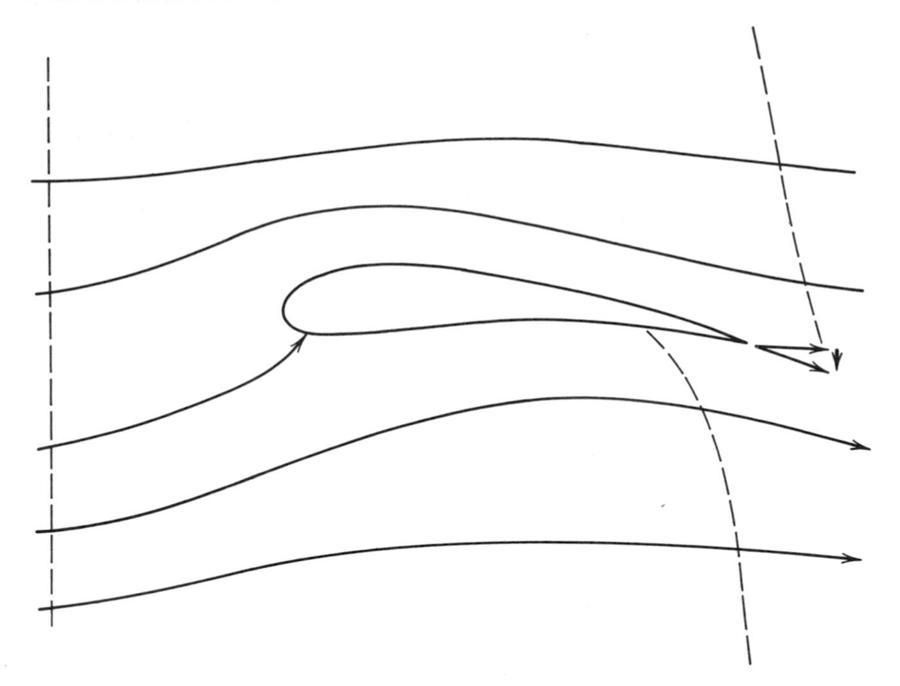

Figure 18-23 Flow over a Joukowski airfoil. Note that particles flow over the top faster than over the bottom, so that a time line is split in two as it flows over the foil.

The forward stagnation point on the cylinder maps into the forward stagnation point on the airfoil. From Fig. 18-23 we note that this position is below the leading edge. The position of the stagnation point is given approximately by

$$\left[\frac{x_s}{l/2}\right]_{\text{stag}} = -\tfrac{1}{2} + \alpha^2 \qquad 18.13.12$$

The corresponding y_s position can be found from 18.11.16. As the angle of attack increases, the stagnation point moves further away from the leading edge. The flow on the lower surface accelerates away from the stagnation point and generally has a velocity somewhat lower than the free-stream velocity. Correspondingly, through Bernoulli's equation, the pressure on the lower surface is slightly higher than the free-stream values, and hence an upward force results.

The flow that goes over the upper surface first accelerates from the stagnation point as it moves around the leading edge. In fact, if the airfoil has zero thickness (a cambered plate), the velocity at the leading edge is infinite. With nonzero thickness the velocity is not infinite, but it does reach high values and

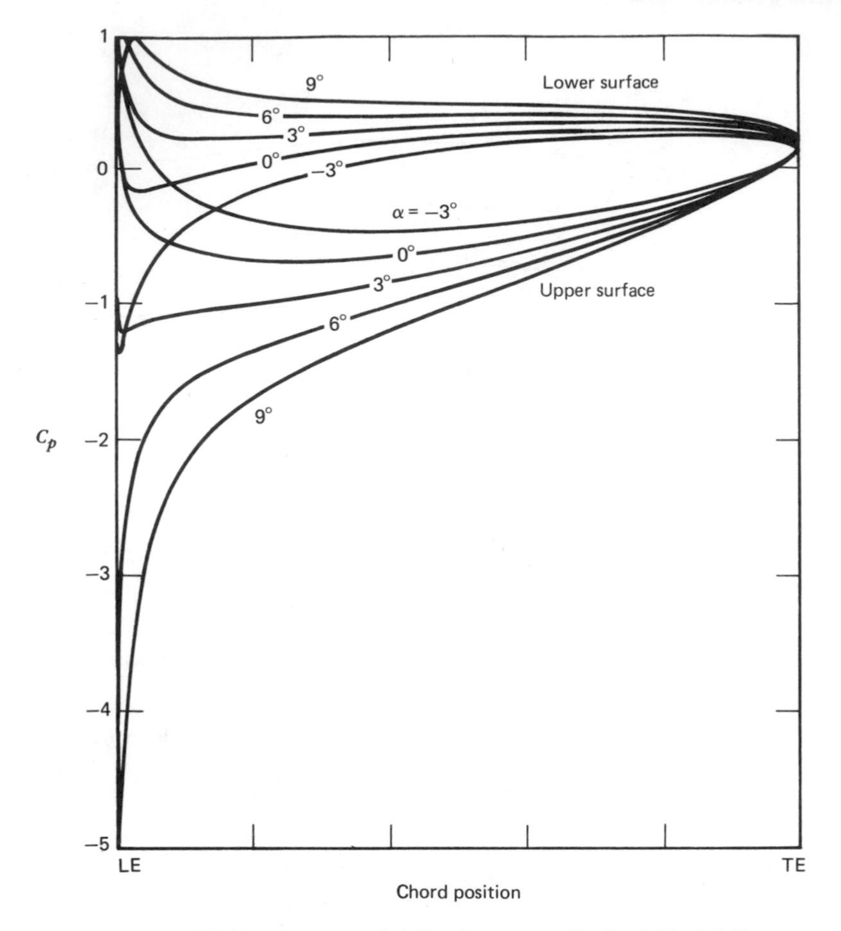

Figure 18-24 Pressure distribution over a Joukowski airfoil.

associated low pressures, as shown in Fig. 18-24. The extremely low pressures are not sustained very long, as the flow is quickly decelerated to more modest speeds. Recall that we found a similar velocity overshoot on a nose shape (Section 18.4). Having the nose at a nonzero angle of attack intensifies the overshoot on an airfoil. Large velocities in this region are undesirable, as the subsequent deceleration may lead to boundary-layer separation and airfoil stall (a type of stall known as leading-edge stall). The nose of a real airfoil is contoured to avoid separation by controlling the excessively low pressures near the nose. Figure 18-25 shows the streamlines over a typical airfoil in two tests at different angles of attack.

Low pressures on the upper surface persist over the major portion of the surface, and typically they make a much larger contribution to the lift than those on the lower surface. Of course, the pressure at the trailing edge on the

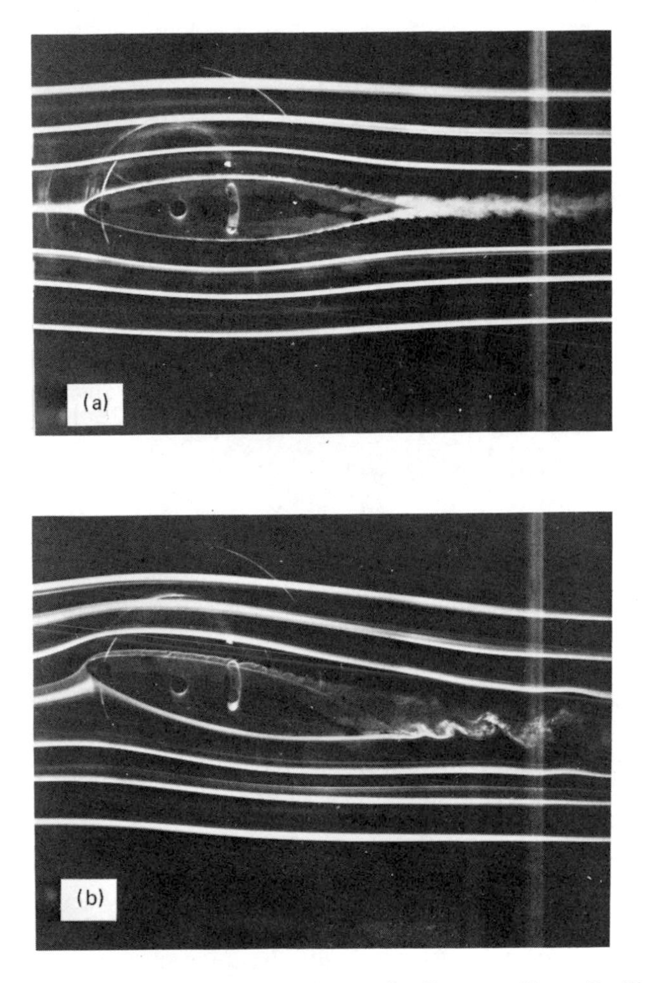

Figure 18-25 Flow over an airfoil shown by smoke filaments. From Batill and Mueller (1980).

upper and lower surfaces must match. Since this pressure is theoretically only slightly higher than free stream, the flow on the upper surface must gradually decelerate to reach this value. Again, we have an adverse pressure, which tends to cause the boundary layer to separate. Although the pressure gradient is not as great as in the nose region, the adverse gradient exists over a longer region. Increasing the angle of attack of the airfoil accentuates these effects, and ultimately the airfoil will stall—first at the trailing edge, and progressively further up the surface as the angle of attack is increased. This type of stall is known as trailing-edge stall.

18.14 AIRFOIL LIFT

When the circulation constant for the Joukowski airfoil, 18.13.8, is substituted into the Kutta–Joukowski lift law 18.8.1, and r_0 and β are replaced by airfoil parameters through the use of 18.13.6 and 18.13.7, we arrive at

$$F_L = \rho U^2 4\pi \left(\frac{l}{4} + \frac{t}{3\sqrt{3}} \right) \sin\left(\alpha + 2\frac{h}{l} \right) \qquad 18.14.1$$

In realistic situations $\alpha + 2h/l$ is small, so the sine term may be approximated by its argument. With this simplification, the lift coefficient per unit span is

$$C_L = \frac{F_L}{\frac{1}{2}\rho U^2 l} = 2\pi \left(1 + \frac{4}{3\sqrt{3}} \frac{t}{l} \right) \left(\alpha + 2\frac{h}{l} \right)$$

$$\approx 2\pi \left(\alpha + \frac{2h}{l} \right) \qquad 18.14.2$$

By far the most important geometric influences on lift are the amount of camber and the angle of attack.

A comparison with experimental results is shown in Fig. 18-26. Equation 18.14.2 predicts that the lift curve slope $dC_L/d\alpha$ should be 2π plus a small effect due to thickness. In practice the effect of thickness is very slight, and in some cases results show it even decreases the lift curve slope. For this reason it is usual to ignore the t/l term in 18.14.2. Indeed, one of the most popular and useful assumptions in airfoil computer codes is to ignore the thickness effects entirely. With this approximation all Joukowski profiles have a lift-curve slope of 2π. Figure 18-26 shows that the actual values are too low; however, the slope is nearly correct. It is customary to blame the slightly lower lift on the fact that the boundary layers, especially the thick one on the upper surface, allow the flow to leave the trailing edge at an angle smaller than the Kutta condition requires.

Lift is essentially an integrated effect of the local pressures over the airfoil, and as such, it is not very sensitive to the detailed shape of the airfoil. Indeed, thin-airfoil theory for an arbitrary camber distribution still predicts that the lift curve slope will be 2π. This is in general agreement with experiments. At low Re, typical lift-curve slopes are sometimes as low as 5.5, but above $\text{Re} = 3 \times 10^6$ practically all airfoils give a value of $dC_L/d\alpha$ between 6 and 2π. Even a flat plate or a cambered plate gives lift curves in good agreement with this theory, in spite of the fact that these surfaces are known to have separation bubbles at their leading edges.

The lift in 18.14.2 increases directly as the sum of the angle of attack and the maximum camber parameter h/l. To get more lift we need only increase the angle of attack. We might also propose to increase the lift by adding camber to the airfoil; however, another viewpoint is that camber merely shifts

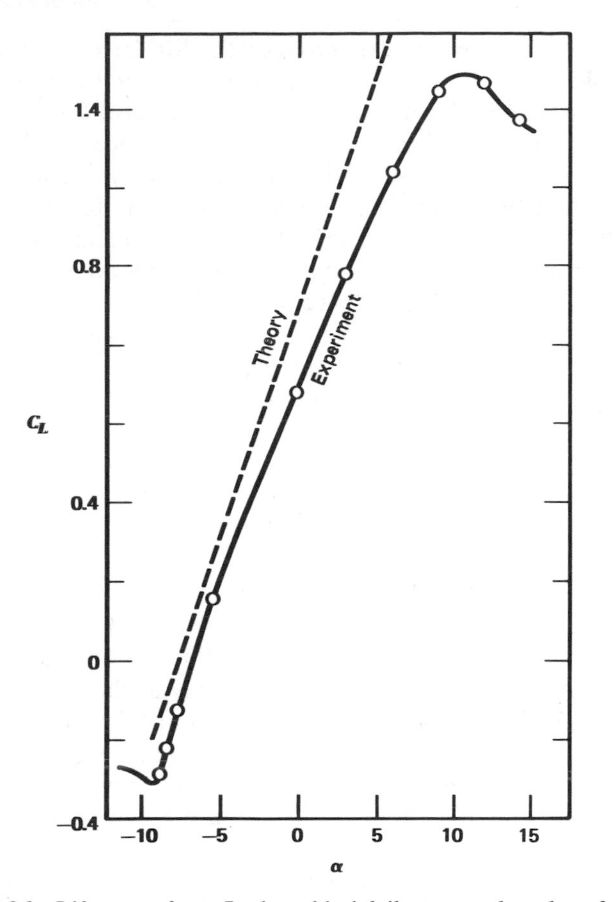

Figure 18-26 Lift curve for a Joukowski airfoil at several angles of attack.

the angle of attack at which zero lift occurs. A cambered airfoil produces some lift at $\alpha = 0$. If we are interested in large values of lift, then the maximum C_L is important. The angle of attack at which the maximum occurs may not be very important. The maximum C_L is determined by stall, which in turn is related to viscous effects. Surprisingly enough, boundary-layer separation at very high Reynolds numbers is independent of Re (another example of a viscous effect becoming independent of the viscosity value). It is generally true that additional camber gives a higher $C_{L\max}$. The student should also be aware the $C_{L\max}$ depends on the boundary-layer behavior; hence the details of the camber and thickness distributions, especially the nose radius, are very important in determining $C_{L\max}$. For a flat plane, $C_{L\max}$ is about 0.75, common airfoil shapes have $C_{L\max}$ of about 1.4, and modern high-lift designs can product $C_{L\max}$ values of almost 2. The reader who needs more information on airfoil theory can consult the introductory books by Bertin and Smith (1979) or Kuethe and Chow (1976); a more advanced treatment is given by Schlichting and Truckenbrodt (1979).

The fluid-mechanics events that produce the lift force deserve some additional discussion. The lift force in the theory above is the result of pressure forces on the surfaces of the airfoil. The reason that ideal-flow theory works so well is that viscous forces have practically no influence on the lift. Viscous forces are not only relatively small, but they are nearly aligned with the flow direction and therefore primarily contribute to the drag force.

The geometry of an airfoil, primarily its camber and angle of attack, sets up an ideal flow, which leaves the trailing edge smoothly according to the Kutta condition. This flow field has high velocities on the upper surface and low velocities on the lower surfaces. When the airfoil is at an angle of attack, the stagnation point moves to the underside of the airfoil and the fluid particles actually have a longer path over the top of the airfoil than on the underside. Even though particles traversing the upper surface have further to go, they still arrive at the trailing edge before companion particles on the underside. If we were to mark a line of particles in front of the wing, as shown in Fig. 18-23, and follow that line as the flow goes over the airfoil, we would find that the line was broken after the streams rejoined. Particles from the upper stream have moved ahead of particles from the lower stream. Of course this effect dies out as we move away from the airfoil. (The calculation of the transit time of a particle is complicated by the fact that the dividing streamline has a stagnation point where the velocity is zero. A particle exactly at the stagnation point has zero velocity and never leaves in either direction. To get around this we imagine two particles are a small distance ϵ on either side of the stagnation point. At time zero we release both particles and compute the time it takes them to reach the trailing edge. As $\epsilon \to 0$ there is a finite difference in the two transit times.)

Let us now leave the flow near the airfoil and look at the effects in the far field. Previously (18.13.1) it was argued that the flow at infinity for the Joukowski airfoil and for the transformed circle are the same, since $z \sim \zeta$ as $z \to \infty$. This is actually true for any airfoil. The flow far away from any airfoil is the same as the ideal flow over a circular cylinder with the same circulation constant. The velocities in the far field are found by setting $\zeta = z$ in 18.13.10 and retaining only the dominant parts as $z \to \infty$. The result is

$$u - iv = Ue^{-i\alpha} + i\frac{\Gamma_a}{2\pi z} + O[z^{-2}], \qquad z \to \infty \tag{18.14.3}$$

The $e^{-i\alpha}$ may be dropped if z is taken in a coordinate system aligned with the flow U as shown in Fig. 18-27. The flow 18.14.3 consists of a uniform stream and a vortex. Far upstream, the airfoil induces an upward flow as the fluid anticipates the presence of the airfoil. Far downstream, the flow is a downwash of equal magnitude. On the top and the bottom the vortex adds to and subtracts from the uniform stream. This in turn causes a lower pressure above the airfoil and a higher pressure below. The fact that these effects die out at the

Figure 18-27 Control-volume analysis of lift: (a) circular region, (b) rectangular region with top and bottom at infinity, (c) rectangular region with sides at infinity.

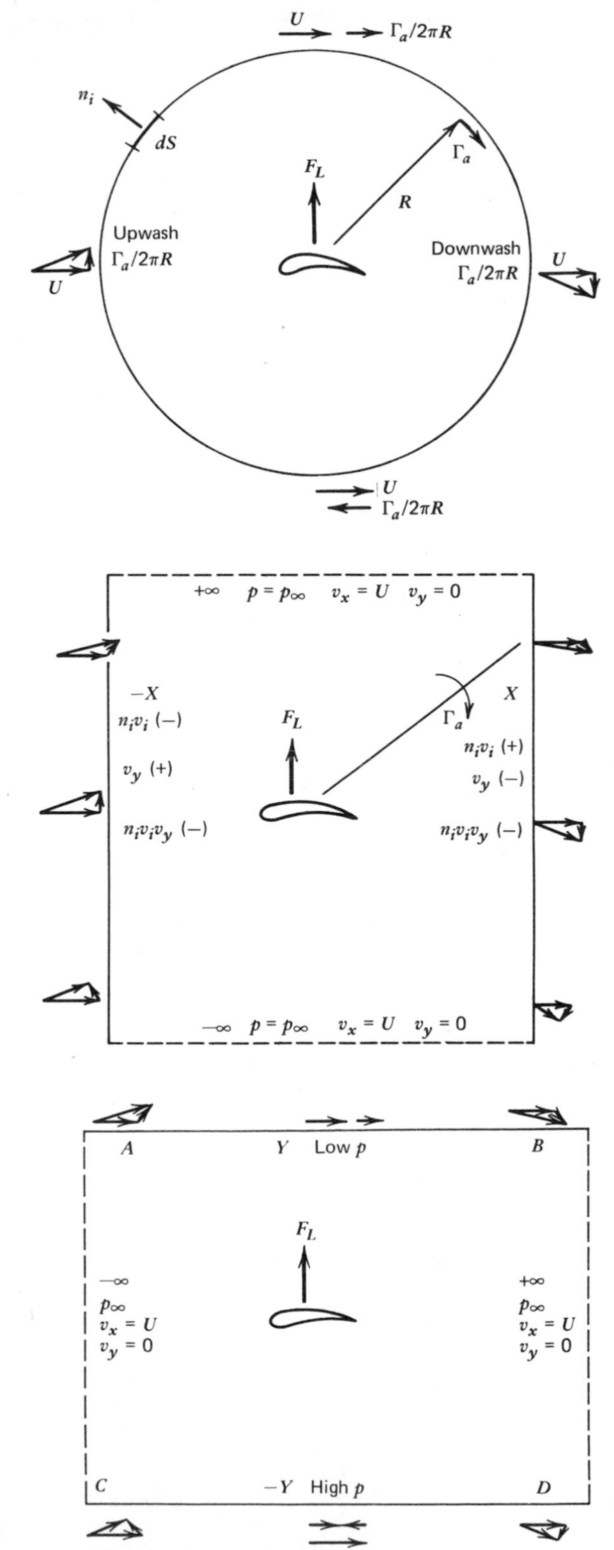

U
$\Gamma_a/2\pi R$
n_i
dS
Γ_a
F_L
R
Upwash
$\Gamma_a/2\pi R$
U
Downwash
$\Gamma_a/2\pi R$
U
(a)
U
$\Gamma_a/2\pi R$
$+\infty$ $p = p_\infty$ $v_x = U$ $v_y = 0$
$-X$
$n_i v_i$ (−)
v_y (+)
$n_i v_i v_y$ (−)
F_L
Γ_a
X
$n_i v_i$ (+)
v_y (−)
$n_i v_i v_y$ (−)
(b)
$-\infty$ $p = p_\infty$ $v_x = U$ $v_y = 0$
A
Y Low p
B
F_L
$-\infty$
p_∞
$v_x = U$
$v_y = 0$
$+\infty$
p_∞
$v_x = U$
$v_y = 0$
C
$-Y$ High p
D
(c)

same rate in all directions illustrates the elliptic character of ideal flows. The influence of the airfoil extends equally far upstream, downstream, up, and down the top, and to the bottom.

If an infinitely long airfoil could operate in an unbounded fluid, the airfoil would experience a lift force. It is of interest to ask ourselves the following questions: Does the production of this force leave a permanent mark on the flow? Is there a residual effect that can be identified as the source of the lift? (Recall that an analysis in Section 15.6 showed that the drag of a nonlifting body causes a momentum defect in the wake flow.)

In order to answer this question, recall the integral momentum equation 5.14.1. Applied to a steady inviscid flow without a body force, the equation is

$$0 = -\int_{\text{FR}} \rho n_i v_i v_j \, dS - \int_{\text{FR}} n_i p \, dS \qquad 18.14.4$$

Consider the fixed region, shown in Fig. 18-27a, that surrounds the airfoil as one boundary, and whose other boundary is a large cylinder of radius R. At the airfoil surface the fluid momentum effect is zero, since $n_i v_i$ is zero. The pressure integral over the airfoil surface is the lift force F_L. With these considerations, 18.14.4 shows that the lift is given by two integrals over the surface at R:

$$F_L = -\int_R \rho n_i v_i v_y \, dS - \int_R n_y p \, dS \qquad 18.14.5$$

In this equation, v_i is the vector velocity, which has components v_x, v_y, and the outward normal has components n_x, n_y. Since the velocities at the far boundary (18.14.3) differ slightly from the free-stream value, the pressure does also. By Bernoulli's equation we have

$$p = p_\infty + \tfrac{1}{2}\rho\left[U^2 - \left(v_x^2 + v_y^2\right)\right]$$

It is most convenient to evaluate the velocities 18.14.3 in polar coordinates $z = \text{R}e^{i\theta}$ and then substitute into Bernoulli's equation. The result is (to order $1/R$)

$$p \sim p_\infty + \tfrac{1}{2}\rho \frac{\Gamma_a U}{\pi R} \sin\theta \qquad 18.14.6$$

The details of the integration will be omitted. The important conclusion is that if 18.14.6 for the pressure and 18.14.3 for the velocity components are substituted into the momentum equation 18.14.5, each integral contributes exactly one-half to the lift. That is,

$$F_L = \underbrace{\tfrac{1}{2}\rho U \Gamma_a}_{\substack{\text{change in } y \\ \text{momentum} \\ \text{of fluid at} \\ \text{boundary } R}} + \underbrace{\tfrac{1}{2}\rho U \Gamma_a}_{\substack{\text{pressure} \\ \text{force on} \\ \text{boundary } R}} \qquad 18.14.7$$

As $R \to \infty$ the pressure and momentum effects contribute equally to producing the lift.

The result above is somewhat misleading, as it depends on the shape of the control volume. What we will really learn in these several paragraphs is that flow effects die out very slowly and that integrals of these effects over large areas give results that depend on the choice of control region. Figure 18-27 shows two alternative choices of the control region. The first (b) is a region with two fixed planes at X and $-X$. These planes are so far removed from the airfoil that the velocity approximation 18.14.3 is valid. The top and bottom are at $Y = \pm\infty$, where the flow is uniform and the pressure is p_∞. If we applied the momentum equation to this region, we would find no contribution from the top and bottom surfaces. On the other hand the side surfaces contribute only a momentum effect, since the pressure force is in the flow direction. Hence, we find that the lift force is

$$F_L = 2\rho \int_{-\infty}^{\infty} v_x v_y \, dy \sim 2\rho \int_{-\infty}^{\infty} U v_y \, dy$$

$$\sim 2\rho \int_{-\infty}^{\infty} \frac{x}{x^2 + y^2} \frac{\Gamma_a U}{2\pi} \, dy = \rho U \Gamma_a \qquad 18.14.8$$

In contrast with our previous conclusion, the lift is now attributed to the momentum effect of the upwash and downwash velocities before and after the airfoil.

To complete the picture we will take the other extreme. Consider a control region at far finite positions Y and $-Y$ for the horizontal positions and infinite positions for the upstream and downstream planes (Fig. 18-27c). The velocity at the X planes is uniform, so the flow here can make no contribution to the lift force. In Fig. 18-27c the top and bottom planes are shown with the velocities depicted. Momentum in the y direction that is carried into the bottom forward portion C of the control region is passed out on the top forward portion A without change. Likewise, momentum coming in the top aft surface B leaves in the same amount through the lower aft surface D. Hence by symmetry, the momentum-convection process can contribute nothing to the lift. The lift is entirely the result of pressure effects. The high pressure on the lower surface and an equally low pressure on the upper surface are the forces on the control region that produce the lift.

The situation of an isolated airfoil in an infinite stream is an artificial problem where we hope to isolate the fluid-mechanical events so that they may be studied. Our attempts to divide the cause of lift into momentum and pressure effects showed that the shape of the analysis region changed the result. In effect, this says to us that there is no single cause–effect relation for an unbounded flow. In any practical problem we should consider the actual situation in which the airfoil operates. Let us assume that the airfoil is placed in a large wind tunnel. In this case the walls of the tunnel confine the flow and force it to be straight and uniform at remote upstream and downstream locations and also along the upper and lower walls near the model. Hence the pressure on the walls above and beneath the airfoil deviates from the free-stream value, so that a net force is produced. The flow produces an upward force on the airfoil and an equal downward force on the tunnel walls.

*18.15 SCHWARZ–CHRISTOFFEL TRANSFORMATION

H. A. Schwarz and E. B. Christoffel, two German mathematicans, independently discovered a conformal transformation that will map the region inside a given polygon to the upper half plane. The polygon in question must be a simple closed polygon, but can have an arbitrary number of sides.

Figure 18-28 shows several examples of simple closed polygons. As is seen from the figure, a simple closed polygon can have one or more sides at infinity. Indeed, polygons with some sides at infinity are the most useful ones for fluid-mechanics applications. The strict definition of a simple closed polygon is that every point in the plane is either an interior point (any two interior points may be connected by a curve that never crosses a boundary), an exterior point, or a boundary point. This rules out figures where the boundary crosses itself and proceeds to form another polygon on the outside of the previous figure.

If we take the polygon in the z plane, we find that the transformation maps boundary points of the polygon to the real axis in the ζ plane. This is one of the major aspects of the Schwarz–Christoffel transformation. As shown in the figure, a polygon boundary at infinity in the z plane may or may not map to finite points on the ζ plane. Likewise the point at infinity in the ζ plane is

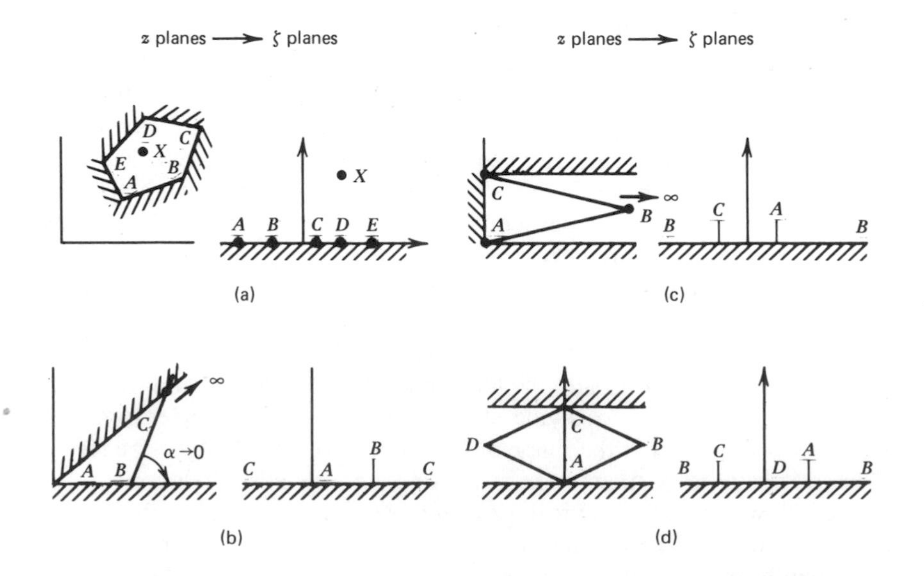

Figure 18-28 Schwarz–Christoffel transformations for several typical polygons.

frequently mapped to a finite point in the z plane, and always to a point on the polygon boundary. Let us define the vertices of the polygon as the points $A, B, C, \ldots$ in the z plane, where the interior angles are $\alpha, \beta, \gamma, \ldots$. Under a Schwarz–Christoffel transformation these points are mapped to points on the real axis in the ζ plane. Figure 18-28 displays this nomenclature. It is customary to take A, B, C in the counterclockwise sense, proceeding along the boundary with the interior on the left; then A', B', C' occur in the same sense on the real axis of the ζ plane.

The transformation is expressed in the form of a differential equation, which must be integrated for any given polygon. The equation that defines the transformation $z = f(\zeta)$ is

$$\frac{dz}{d\zeta} = K(\zeta - \xi_a)^{\alpha/\pi - 1}(\zeta - \xi_b)^{\beta/\pi - 1}(\zeta - \xi_c)^{\gamma/\pi - 1} \cdots \qquad 18.15.1$$

In this equation K is an arbitrary complex constant, and $\xi_a, \xi_b, \xi_c, \ldots$ are the transformed locations of the vertices in the ζ plane. One very important fact in using 18.15.1 is that the terms corresponding to a vertex at infinity in the ζ plane are omitted. If vertex B, located at finite z or at $z \to \infty$, is to be sent to $\xi_b \to \pm\infty$ in the ζ plane, then the term $(\zeta - \xi_b)$ is left out of 18.15.1. The reason for this is fully explained in the following example.

To illustrate the use of the transformation, consider an infinite slit of width π. Figure 18-28d shows the slit as the limiting form of a quadrangle as points B and D go to infinity. The interior angles take on the following limiting values:

Vertex	z	Angle	(Angle/π − 1)
A	0	π	0
B	∞e^{i0}	0	−1
C	$i\pi$	π	0
D	$\infty e^{i\pi}$	0	−1

Equation 18.15.1 for this case is

$$\frac{dz}{d\zeta} = K(\zeta - \zeta_a)^0(\zeta - \xi_b)^{-1}(\zeta - \xi_c)^0(\zeta - \xi_d)^{-1}$$

Let us choose to send the point B to $\xi_b \to \infty$. Then we can exclude the term $(\zeta - \xi_b)^{-1}$ from the transform expression. [This may be rationalized by noting that the term $(\zeta - \xi_b)^{-1}$ is dominated by ξ_b. If $dz/d\zeta$ is to be finite as $\xi_b \to \infty$, then K/ξ_b must be finite. When we omit $(\zeta - \xi_b)^{-1}$ from the equation, we are essentially redefining the constant K.]

The transform equation reduces to

$$\frac{dz}{d\zeta} = K(\zeta - \xi_d)^{-1}$$

which integrates to

$$z = K \ln(\zeta - \xi_d) + L \qquad 18.15.2$$

This equation maps the degenerate quadrangle $ABCD$ to the upper half plane with the boundary points on the ξ axis. The only stipulation so far is that $\xi_b \to \infty$. We may still choose the mapped positions for two more vertices. Choosing $\xi_d = 0$, the transformation becomes

$$z = K \ln \zeta + L \qquad 18.15.3$$

For the second choice, we note that $z_a = 0$ and set $\zeta_a = \xi_a = 1$. This yields

$$0 = K \ln 1 + L$$

$$0 = L$$

We still must find the constant K. Note that when ζ is on the positive real axis between A' and B' (i.e., $\zeta = \text{R}e^{i0}$ with $R > 1$), z must be real and positive. Thus 18.15.3 becomes real $= K \cdot$ real $+ 0$, and we conclude that K is a real number. Since point D' has been sent to the origin ($\xi_d = 0$), point C' must lie on the negative ξ axis and can be given by $\zeta_c = |\xi_c|e^{i\pi}$. Substituting this into 18.15.3 yields

$$0 + i\pi = K \ln|\xi_c| + iK\pi$$

Equating real and imaginary parts shows that $K = 1$ and that $\xi_c = -1$. The final transformation is given by

$$z = \ln \zeta \qquad 18.15.4$$

In arriving at this result we have been able to choose the ζ-plane locations of three vertices. The only restriction is that the vertices are on the real axis and retain the proper counterclockwise order. If one of the vertices is sent to infinity this is counted as one of the arbitrary choices. The ζ location of the fourth vertex cannot be specified. In the present case the fourth vertex turns out to be located at $\xi_c = -1$. The number of arbitrary choices for vertex locations in the ζ plane is the same for all polygons irrespective of the number of sides.

*18.16 FLOW AROUND A SQUARE BODY OR OVER A STEP

Consider a plane flow that encounters a square-faced body. The face is at $x = 0$ in the z plane and extends to infinity in the negative x direction. This problem is similar to the rounded-nose problem of Section 18-4 except this nose is square and sharp. Figure 18-29 gives a sketch of the flow. In light of the symmetry of the body, we elect to solve for only the upper half of the flow. This configuration is a step and offers another interpretation of the flow.

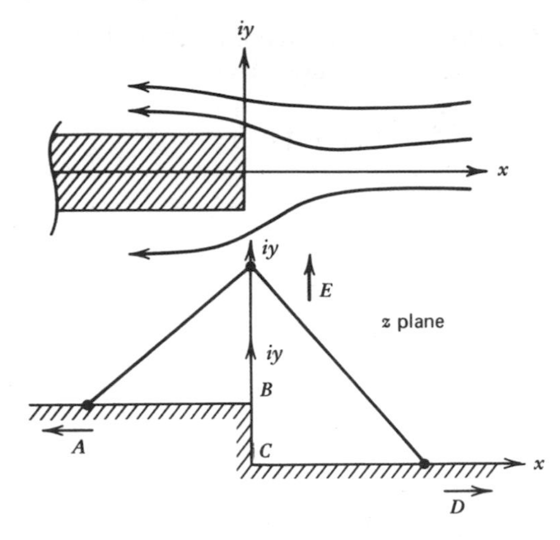

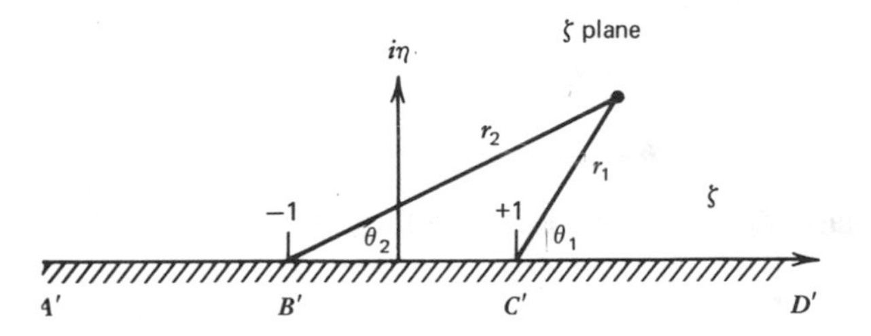

Figure 18-29 Flow over a square nose or over a step.

In order to consider this external-flow problem as the flow in a polygon, we denote the vertices of a figure $ABCDE$ and allow the points A, D and E to go to infinity. The following table gives the points and the important information for the transform equation:

Point	z	θ	$\theta/\pi - 1$	ζ
A	$-\infty$	$\pi/4$	Not needed	∞
B	ih	$3\pi/2$	$\frac{1}{2}$	-1
C	0	$\pi/2$	$-\frac{1}{2}$	$+1$
D	$+\infty$	$\pi/4$	Not needed	∞
E	$+i\infty$	$\pi/2$	Not needed	∞

Exponents for points sent to infinity in the ζ plane are omitted in the transform equation. We have chosen to send point B to $\zeta = -1$ and point C to $\zeta = +1$. The height of the step, h, is retained explicitly in the problem as the z location of point B.

The Schwarz–Christoffel transformation is found by simplifying 18.15.1 according to the tabulated information. The results is

$$\frac{dz}{d\zeta} = K(\zeta + 1)^{1/2}(\zeta - 1)^{-1/2}$$

$$= K\frac{\zeta + 1}{(\zeta^2 - 1)^{1/2}} \qquad 18.16.1$$

This may be integrated to yield

$$z = K\left\{\sqrt{\zeta^2 - 1} + \ln\left[\zeta + \sqrt{\zeta^2 - 1}\right]\right\} + L \qquad 18.16.2$$

Substituting for point C,

$$0 = K \ln 1 + L$$

$$0 = L$$

Substituting for point B,

$$ih = K\ln(-1) = K\ln e^{i\pi} = iK\pi$$

$$h = K\pi$$

Thus the mapping from the z plane to the ζ plane is given by

$$z = \frac{h}{\pi}\left[\sqrt{\zeta^2 - 1} + \ln\left(\zeta + \sqrt{\zeta^2 - 1}\right)\right] \qquad 18.16.3$$

The step in the z plane has been flattened out onto the real axis in the ζ plane. Note that when $\xi \to \infty$, z also approaches ∞.

In the ζ plane the flow pattern is a uniform stream from right to left. The complex potential for this flow is

$$F = \hat{F}(\zeta) = -U_1\zeta \qquad 18.16.4$$

where U_1 is a positive constant. (Inserting a negative U_1 would mean that the flow would go from left to right.) Velocities are changed as we transform from one plane to the other. The complex velocity in the z plane is

$$W(z) = \hat{W}(\zeta)\frac{d\zeta}{dz} = \frac{\hat{W}(\zeta)}{dz/d\zeta}$$

Using 18.16.4 and 18.16.1, we evaluate the expression above to find

$$W = -U_1\frac{\pi}{h}\frac{\sqrt{\zeta^2 - 1}}{\zeta + 1} \qquad 18.16.5$$

The velocity in the z plane at ∞ will be called $-U$. Since $\zeta \to z$ as $z \to \infty$, we find from 18.16.5 that

$$U = U_1 \frac{\pi}{h} \tag{18.16.6}$$

Hence, in terms of U, 18.16.5 becomes

$$W = -U \frac{\sqrt{\zeta^2 - 1}}{\zeta + 1} \tag{18.16.7}$$

This is the velocity in the z plane, but we leave ζ as a parametric variable, since we cannot invert 18.16.3. For a given ζ, 18.16.3 gives the z location and 18.16.7 gives the corresponding velocity. Streamlines and potential lines can be found in a similar manner using 18.16.4 (where U_1 is eliminated through 18.16.6) and 18.16.3. Note that streamlines in the ζ plane transforms into streamlines in the z plane. These relations constitute a detailed solution for the flow.

It will be illustrative to compute the drag force on the face of the step. Equations for the drag coefficient were developed previously in 18.4.13, 18.4.14, and 18.4.18. They reduce to

$$C_D = \int_0^h C_p \frac{dy}{h} = \int_0^h \left[1 - \frac{q^2}{U^2}\right] \frac{dy}{h} \tag{18.16.8}$$

Along the surface between C and B, $dz = i\,dy$, and since we want to use ζ as a parameter, we rearrange 18.16.8 as

$$C_D = \int_{\zeta=1}^{\zeta=-1} \left[1 - \left(\frac{q}{U}\right)^2\right] \frac{1}{ih} \frac{dz}{d\zeta} d\zeta \tag{18.16.9}$$

The velocity $q^2 = W\overline{W}$ and $dz/d\zeta$ both contain factors that are multivalued and need to be interpreted carefully.

The multivalued quantity in 18.16.2 and 18.16.5 is

$$\sqrt{\zeta^2 - 1} = (\zeta - 1)^{1/2}(\zeta + 1)^{1/2}$$

In order to find the proper form, let us set $\zeta - 1 = r_1 e^{i\theta_1}$ and $\zeta + 1 = r_2 e^{i\theta_2}$ as shown in Fig. 18-29. Hence

$$\sqrt{\zeta^2 - 1} = +\sqrt{r_1 r_2}\, e^{i(\theta_1 + \theta_2)/2} \tag{18.16.10}$$

where we have indicated the positive square root of the real product $r_1 r_2$. The integral 18.16.9 is to be evaluated on the real axis $\zeta = \xi$ from $+1$ to -1. For ζ in this region we have $\theta_1 = \pi$, $\theta_2 = 0$, and so 18.16.10 becomes

$$\sqrt{\zeta^2 - 1} = i\sqrt{r_1 r_2} = i\sqrt{1 - \xi^2}$$

Inserting this formula in 18.16.7 for the complex velocity we find

$$\left(\frac{q}{U}\right)^2 = \frac{W\overline{W}}{U^2} = \left(-\frac{i\sqrt{1-\xi^2}}{\xi+1}\right)\left(+\frac{i\sqrt{1-\xi^2}}{\xi+1}\right) \qquad 18.16.11$$

and in 18.16.1 for the transform we have

$$\frac{dz}{d\zeta} = \frac{h}{i\pi}\frac{\xi+1}{\sqrt{1-\xi^2}}$$

Therefore the integral 18.16.9 becomes

$$C_D = \int_{\xi=1}^{\xi=-1}\left[1 - \frac{1-\xi^2}{(1+\xi)^2}\right]\left[\frac{\xi+1}{i\sqrt{1-\xi^2}}\right]\frac{d\xi}{i\pi}$$

$$= \frac{1}{\pi}\int_{-1}^{+1}\left[\frac{\xi+1}{\sqrt{1-\xi^2}} - \sqrt{\frac{1-\xi}{1+\xi}}\right]d\xi$$

A table of integrals will give the final results as

$$C_D = -\left[\frac{1}{\pi}\sqrt{1-\xi^2}\right]_{-1}^{+1} + \left[\frac{2}{\pi}\sin^{-1}\xi\right]_0^1$$

$$-\left[\frac{1}{\pi}\sqrt{1-\xi^2}\right]_{-1}^{+1} - \left[\frac{2}{\pi}\sin^{-1}\sqrt{\frac{\xi-1}{2}}\right]_{-1}^{+1} = 0$$

The drag coefficient is zero.

The net pressure force on the face of a sharp square face is the same as that due to hydrostatic pressure: the flow causes no drag. This is the same result we reached previously for a smooth-shaped plane nose section. In ideal flow, the existence of the sharp corner on the face does not cause any additional drag. At the stagnation point $C_p = 1$ and a high pressure region occurs. Moving toward the edge, the pressure drops below p_∞ ($C_p < 0$) and continues toward minus infinity as the flow reaches the corner at $y = h$. The impact-pressure effect in the center and the suction effect at the corner exactly cancel each other to give zero drag.

Sharp convex corners have streamlines that must turn rapidly as the flow goes around the corner. To effect the turning, the pressure on the outside is high. The corner itself is a singularity with a discontinuous change in the direction of the streamline, infinite velocity, and a corresponding negative infinity in the pressure. Needless to say, the real flow over a sharp front does not have this character. The flow does not turn the corner smoothly, but

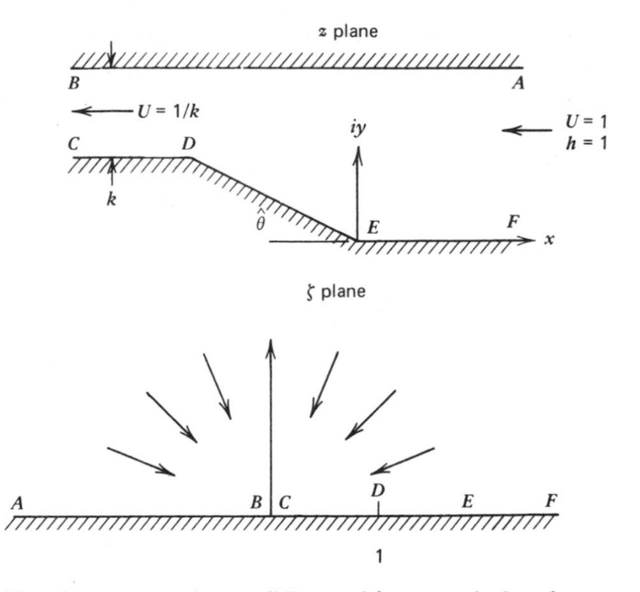

Figure 18-30 Flow in a contraction or diffuser with area ratio k and convergence angle $\hat{\theta}$. The ζ-plane flow is flow into a sink.

separates. Somewhere downstream, the flow reattaches, but the damage has been done, so to speak. When the flow separates, the velocity and the pressure at the corner are finite. Now, the pressure is high over all of the front, and a finite frontal drag occurs. For slender rectangular shapes the typical value of $C_D = 1$ is 60% due to base drag and 40% due to frontal drag.

Since we know the reason for the frontal drag on a square object, we can understand that avoiding or reducing the separated regions should decrease the drag. Rounding the corners aids the flow in turning back to the free-stream direction. When the turning is done smoothly, a low pressure exists near the surface to provide the normal pressure gradient necessary to curve the particle paths. The suction created in this region counteracts the high pressures near the center and reduces the drag.

*18.17 DIFFUSER OR CONTRACTION FLOW

We consider a diffuser that connects two passages with parallel walls, as shown in Fig. 18-30. The figure shows only one diverging wall; however, we can, with equal validity, consider this as one half of a symmetric diffuser with two diverging walls. The width of the small channel will be taken as k, and that of the large channel by h. Let the flow velocity in the large channel be U. Considering the continuity equation, we must have a uniform velocity in the small channel of hU/k. In addition to h/k, a second geometrical parameter is needed to fix the diffuser geometry. We take the angle $\hat{\theta}$. Diffuser angles must be small in actual practice in order to avoid separation. (Angles of 7° or less are required for that purpose, so in many instances diffusers are in fact operated with some flow separation present.) Our solution will only be valid for situations where there is no separation and the boundary layers are thin. The answer may also represent the flow into a contraction by simply changing the sign of the velocity. Practical contraction sections have much larger angles, as flow separation is not such a critical problem in this case.

The Schwarz–Christoffel transformation maps the flow of Fig. 18-30a onto the upper half plane as shown in Fig. 18-30b. We choose to map B–C to the origin, D to $\xi = 1$, and A–F to $\zeta \to \infty$. The ζ image of E is called e and cannot be specified independently. Thus we have the following requirements:

Point	z	Angle θ	Exponent $(\theta/\pi) - 1$	ζ
A–F	$+\infty$	0	Not needed	∞
B–C	$-\infty$	0	Not needed	0
D	$(h-k)(-\cot\theta + i)$	$\pi + \hat{\theta}$	$1 + \hat{\theta}/\pi$	1
E	0	$\pi - \hat{\theta}$	$1 - \hat{\theta}/\pi$	$e = (h/k)^{\pi/\hat{\theta}}$

Applying this information in the Schwarz–Christoffel equation 18.15.1, we have

$$\frac{dz}{d\zeta} = K\zeta^{-1}(\zeta - 1)^{\hat{\theta}/\pi}(\zeta - e)^{-\hat{\theta}/\pi} \qquad 18.17.1$$

The integration of 18.17.1 can be done in closed form if we take $\hat{\theta}/\pi$ as a rational fraction. Therefore we let

$$\frac{\hat{\theta}}{\pi} = \frac{m}{2n} \qquad 18.17.2$$

for m and n integers. Any angle can be approximated as closely as one desires by 18.17.2.

Before we actually integrate 18.17.1, it is useful to give the velocity potential and determine the constants K and e in 18.17.1. The flow in the ζ plane is the flow from a source at the origin. Since the volume flow in the large channel is Uh, the strength of the source in the ζ plane should be twice that amount. The complex potential is therefore

$$F = \frac{Uh}{\pi} \ln \zeta \qquad 18.17.3$$

The corresponding velocity potential in the z plane is

$$W(z) = \frac{dF}{d\zeta}\frac{d\zeta}{dz} = \frac{hU}{K\pi}\left(\frac{\zeta - e}{\zeta - 1}\right)^{m/2n} \qquad 18.17.4$$

Now as $\zeta \to \infty$, $z \to A\text{–}F$, where the velocity $W = U$. For this to be true, the constant K in 18.17.4 must be

$$K = \frac{h}{\pi} \qquad 18.17.5$$

At the other end of the channel, the point $B\text{–}C$, the velocity is $W = hU/k$. Since BC maps to $\zeta = 0$, we substitute $\zeta = 0$ into 18.17.4 to obtain

$$e = \left(\frac{h}{k}\right)^{2n/m} \qquad 18.17.6$$

With these constants the velocity becomes

$$W(z) = U\left[\frac{\zeta - (h/k)^{2n/m}}{\zeta - 1}\right]^{m/2n} \qquad 18.17.7$$

We leave this expression with $\zeta = \zeta(z)$ as a parameter.

Now, we return to the question of integrating 18.17.1. Integrals of the form 18.17.1, with rational exponents 18.17.2, can be separated into partial fractions if we make a variable change by defining s according to

$$s = \left(\frac{\zeta - e}{\zeta - 1}\right)^{1/2n} \qquad 18.17.8$$

Solving this for ζ yields

$$\zeta = \frac{e - s^{2n}}{1 - s^{2n}} \qquad 18.17.9$$

Substitution of 18.17.9, 18.17.5, and 18.17.2 into 18.17.1 changes the integral into

$$dz = 2n\frac{h}{\pi}\left\{\frac{s^{2n-m-1}}{1 - s^{2n}}\,ds - \frac{s^{2n-m-1}}{e - s^{2n}}\,ds\right\} \qquad 18.17.10$$

To simplify the notation, we define

$$I_{mn}(s) = 2n\int \frac{s^{2n-m-1}}{1 - s^{2n}}\,ds \qquad 18.17.11$$

Equation 18.17.10 may now be cast into the form

$$z = \frac{h}{\pi}\left\{I_{mn}(s) - \frac{k}{h}I_{mn}\left(\left(\frac{k}{h}\right)^{1/m}s\right) - \left[1 - \frac{k}{h}\right]I_{mn}(0)\right\} \qquad 18.17.12$$

The integrals I_{mn} in 18.17.11 can be evaluated exactly (Gradshteyn and Ryzhik 1965):

$$I_{mn}(s) = (-1)^{2n-m+1}\ln(1+s) - \ln(1-s)$$

$$-\sum_{j=1}^{n-1} \cos\frac{j(2n-m)\pi}{n} \ln\left[1 - 2s\cos\frac{j\pi}{n} + s^2\right]$$

$$+2\sum_{j=1}^{n-1} \sin\frac{j(2n-m)\pi}{n} \arctan\left[\frac{s-\cos(j\pi/n)}{\sin(j\pi/n)}\right] \qquad 18.17.13$$

The transformation function $z = f(\zeta)$ is given by the combination of 18.17.8, 18.17.12, and 18.17.13. The streamlines in the z plane are easily found, as they are radial lines through the origin in the ζ plane. Similarly the potential lines are circular arcs in the ζ plane.

The velocity potential in the z plane is expressed by 18.17.4. A more useful form is to employ s as a parameter through 18.17.8. In terms of s the velocity potential is

$$W = u - iv = Us^m \qquad 18.17.14$$

Computation of the velocity and its position is accomplished by using ζ as the independent variable and $s(\zeta)$ as an intermediary parameter.

A plane where the points represent the complex velocity $W = u - iv$ is called a *hodograph* plane. A vector from the origin to a certain point is a mirror image of the velocity vector: the direction is $-\theta$ instead of $+\theta$. Figure 18-31 shows the hodograph plane for the flow into a contraction. Lines drawn in this figure represents the velocities that occur on a given streamline. All streamlines start from (1, 0), the uniform upstream flow. Flow along the centerline goes from (1, 0) to $(1/k, 0)$ while the angle remains zero. An interior streamline makes a looping path between these same two points. The loop degenerates into a series of straight lines for the streamline that follows the walls.

Figure 18-32 displays the velocity for a typical contraction as a function of the distance along the streamline. All streamlines begin at $U = 1$ in the wide section. As the concave corner is approached, the velocity decreases and must become zero at the corner itself. The pressure rises according to Bernoulli's equation, with stagnation pressure existing at the corner point. Recall that curved streamlines in inviscid flow mean that a pressure gradient must exist across the streamlines. The high pressure in the corner supplies the pressure gradient to initially turn the streamlines into the contraction. This same effect is seen to a lesser extent on the first streamline in from the wall. The velocity on this streamline first decreases as the corner is approached and then increases as it heads into the contraction. A practical problem sometimes occurs in contractions, as boundary-layer separation can occur in the mild

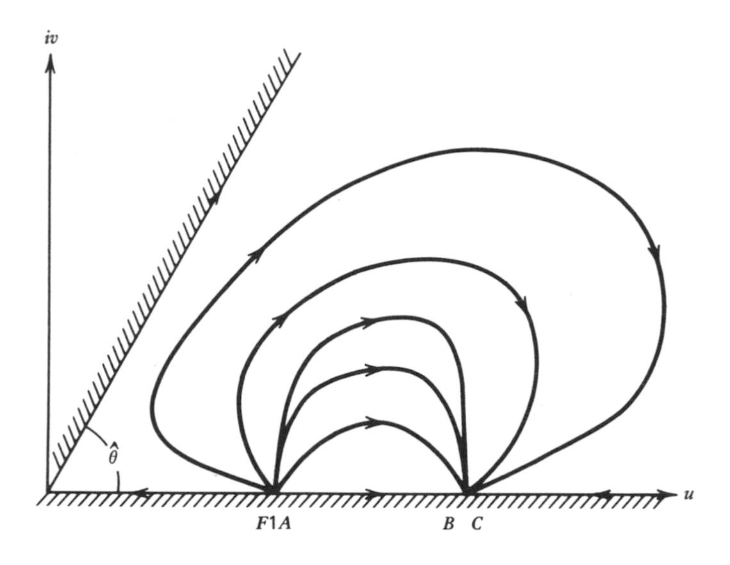

Figure 18-31 Hodograph plane $W = u - iv$ for flow in a contraction. The path of streamlines in this plane is depicted.

adverse pressure gradient at the first corner. It is good design practice to make this initial curvature small to avoid separation.

As we continue to follow the streamline along the contraction wall toward the convex corner, the velocity rises. At the corner itself, it becomes infinite. The corresponding negative infinity in the pressure is needed so that the streamlines can curve around the corner where the radius of curvature is zero. Streamlines that come near this corner have an overshoot in velocity and then, as they proceed into the small section, approach the final velocity from above. Here again is a region of adverse pressure and the possibility of boundary-layer separation. Once more the practical solution is to make this corner gently rounded to reduce the adverse pressure gradient tending to separate the flow. Most contractions in use, even the so-called bell-mouth entrance, produce a nonuniform velocity profile at the end of the geometric entrance. The velocity near the wall tends to be too high and the pressure somewhat low. This is a remnant of the curvature of the streamlines as they pack themselves into the straight section.

Note that not all the streamlines have a low velocity near the concave corner and an overshoot in velocity near the concave corner. Streamlines near the center of the flow display a monotonic increase in velocity as the flow enters the contraction. As a matter of fact, one may prove that the 50% streamline is the demarcation between streamlines with the two kinds of behavior. This result is valid for all contraction ratios and for all angles. The importance of this fact is that the 50% streamline can be used in the design of a smooth wall contraction that has a monotonic velocity change on all streamlines.

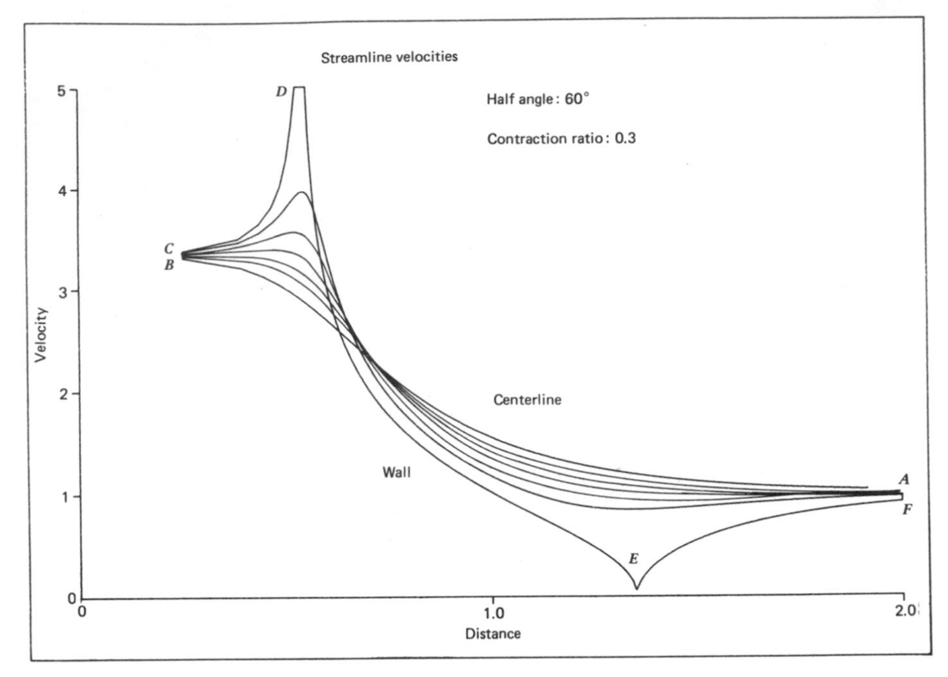

Figure 18-32 Velocities on several streamlines as a function of distance along the streamlines; $\hat{\theta} = 60°$, $k = 0.3$. Graph from L. N. Goenka (1982).

*18.18 GRAVITY WAVES IN LIQUIDS

Ideal-flow theory may be used to describe waves in a liquid where there is a free surface. If a free surface, where the pressure is constant, is displaced from its equilibrium position, gravity causes a higher pressure under the crest than under the troughs. The resulting flow is unsteady, irrotational, and incompressible.

For purposes of analysis we assume that the surface shape is a traveling sine wave of amplitude A, wavelength λ, and phase speed c in the x direction. The y position of the surface is given by

$$y_s \equiv \eta = A \sin\left(\frac{2\pi x}{\lambda} - \frac{2\pi ct}{\lambda}\right) \qquad 18.18.1$$

The wavenumber is defined as $k = 2\pi/\lambda$, while the frequency at a fixed position is $\omega = kc$. Figure 18-33 shows the wave where $y = 0$ is the equilibrium position of the free surface and the bottom is at $y = -h$.

In light of 18.18.1, we introduce nondimensional variables for x and t as

$$X = kx = \frac{2\pi x}{\lambda}$$

$$T = \omega t$$

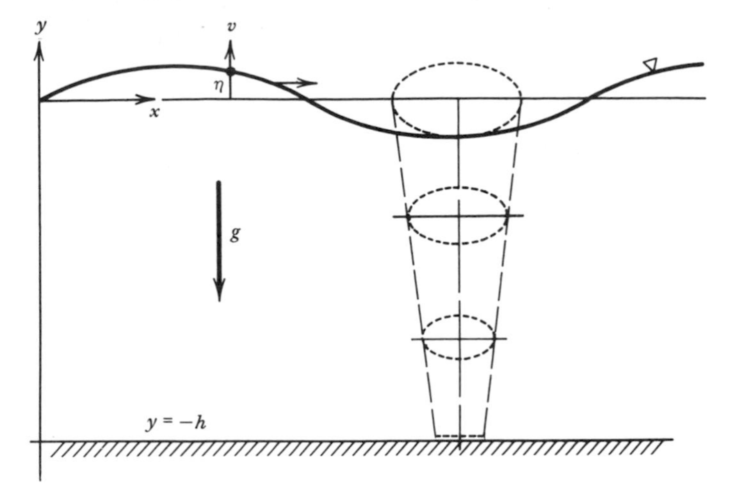

Figure 18-33 Gravity wave in a liquid. Dotted lines show particle paths at various depths.

Hence,the liquid surface is

$$\frac{\eta}{A} = \sin(X - T) \tag{18.18.2}$$

Furthermore, since ideal flow has a potential that obeys

$$\nabla^2\phi = 0 \tag{18.18.3}$$

we should use the same nondimensional scale for y as for x, that is, $Y = ky$. Velocities are nondimensionalized by estimating that the amplitude of the motion A times the frequency indicates the particle velocity. Thus, we let

$$U = \frac{u}{A\omega} \qquad V = \frac{v}{A\omega} \tag{18.18.4}$$

The nondimensional potential

$$\Phi = \frac{\phi}{Ac} \tag{18.18.5}$$

will render $v_i = \partial_i\phi$ consistent with the previous nondimensional forms.

The most complicated part of this problem is caused by the moving free surface. Let r_2 be the y position of a material particle on the surface. The

Lagrangian description of r_2 uses the original particle position x^0 and the Lagrangian time $\hat{t}$:

$$r_2 = r_2(x^0, \hat{t}) = \eta(x, t)$$

In Eulerian variables x, t the vertical particle position is the equation of the free surface. Next, we recall that the vertical velocity of a particle is given by

$$v = \frac{\partial r_2}{\partial \hat{t}} = \frac{D\eta}{Dt} = \frac{\partial \eta}{\partial t} + u\frac{\partial \eta}{\partial x} \qquad \text{at} \quad y = \eta \qquad 18.18.6$$

A fluid particle on the surface must remain on the surface. Equation 18.18.6 is a kinematic boundary condition for the flow. It introduces a nonlinearity into the problem even when the wave shape is assumed. We can make further progress by assuming that the waves have a small amplitude compared to their wavelengths: $Ak \to 0$. This results in the linearized theory of gravity waves. Writing 18.18.6 in nondimensional terms gives

$$V = \frac{\partial(\eta/A)}{\partial T} + kAU\frac{\partial(\eta/A)}{\partial x} \qquad \text{at} \quad y = \frac{\eta}{A}kA$$

Next, we note that $V = \partial\Phi/\partial Y$ and expand the surface value in a Taylor series about $Y = 0$. This yields

$$V = \left.\frac{\partial \Phi}{\partial Y}\right|_0 + \left.\frac{\partial^2 \Phi}{\partial Y}\right|_0 \frac{\eta}{A}kA + \cdots$$

Hence, in the limit $kA \to 0$ the two relations above yield the boundary condition

$$\left.\frac{\partial \Phi}{\partial Y}\right|_0 = \frac{\partial(\eta/A)}{\partial T}$$

$$= -\cos(X - T) \qquad 18.18.7$$

This equation, together with the restriction that no flow crosses the bottom,

$$\frac{\partial \Phi}{\partial Y} = V = 0 \qquad \text{at} \quad Y = -H \qquad 18.18.8$$

constitute the kinematic condition for the solution. Separation of variables applied to $\nabla^2\Phi$ and the boundary conditions above yields the solution as the potential

$$\Phi = \frac{\cosh(Y + H)}{\sinh H}\cos(X - T) \qquad 18.18.9$$

The corresponding velocities are

$$U = -\frac{\cosh(Y+H)}{\sinh H}\cos(X-T)$$
$$V = \frac{\sinh(Y+H)}{\sinh H}\cos(X-T) \qquad 18.18.10$$

The solution above has an arbitrary wavelength and an arbitrary phase speed.

The free surface of the wave has a constant pressure. This information enters the problem through the Bernoulli equation, the only dynamic restriction in the problem. At the surface,

$$\frac{\partial \phi}{\partial t} + \frac{p}{\rho} + \tfrac{1}{2}(u^2 + v^2) + g\eta = C(t)$$

The function $C(t)$ is equivalent to the arbitrary constant in the velocity potential (i.e., $\bar{\phi} = \phi + \int C\,dt$) and may be set equal to zero. The nondimensional form of the Bernoulli equation is

$$\frac{\partial \Phi}{\partial T} + P + \frac{kA}{2}(U^2 + V^2) + \frac{g}{kc^2}\left(\frac{\eta}{A}\right) = 0$$

where $P = p/(\rho c^2 kA)$. For small amplitude ($kA \to 0$), the velocity term may be neglected. Furthermore, differentiation with respect to time eliminates the constant pressure. This yields

$$\frac{\partial^2 \Phi}{\partial T^2} + \frac{g}{kc^2}\frac{\partial(\eta/A)}{\partial T} = 0 \qquad 18.18.11$$

The derivative of Φ is to be evaluated at the surface; however, by expanding from $Y = 0$ in a Taylor series and noting that $kA \to 0$ (the same steps that were used to arrive at 18.18.7), we can use the value at $Y = 0$. Equation 18.18.11 supplies an eigenvalue relation that determines the wave speed c:

$$\frac{kc^2}{g} = \tanh kh \qquad 18.18.12$$

The speed of propagation of a gravity wave is fixed by the wavenumber and the fluid depth. This is the central result.

For special cases we may simplify 18.18.12. In water that is shallow compared to the wavelength ($kh \to 0$), $\tanh kh \approx kh$ and we find

$$c^2 = gh \qquad 18.18.13$$

All waves have a speed that depends upon the depth, but not on the wavelength.

The second special case is when the liquid is deep compared to the wavelength ($kh \to \infty$). Now, $\tanh kh \approx 1$, so we obtain as the simplified form of 18.18.12 the relation

$$c^2 = \frac{g}{k} \tag{18.18.14}$$

Waves have a speed that depends upon their wavelength. Consider that a wave in deep water with an arbitrary shape is composed of several Fourier components. Each component, according to 18.18.14, has a different phase speed. Thus, the shape of the wave will continually change. For this reason the speed–wavenumber equation above is called the *dispersion relation*.

An important physical concept called the *group velocity* can be illustrated by water waves. Envision a disturbance that generates a train of waves composed of several wavelengths. After a while the waves sort themselves out according to their differing phase speeds, and packets of waves of nearly the same wavenumber are traveling together. For illustrative purposes assume two waves of equal amplitude and wavenumbers k and $k + \Delta k$:

$$\eta_1 = A \sin(kx + \omega t)$$

$$\eta_2 = A \sin[(k + \Delta k)x + (\omega + \Delta\omega)t]$$

The surface for these waves is

$$\begin{aligned} \eta &= \eta_1 + \eta_2 \\ &= 2A \cos\left[\tfrac{1}{2}\Delta k\, x - \tfrac{1}{2}\Delta\omega\, t\right] \sin\left[\left(h + \frac{\Delta k}{2}\right)x + \left(\omega + \frac{\Delta\omega}{2}\right)t\right] \end{aligned}$$

The sine part of this expression is a wave that has nearly the basic wavelength k and phase speed $c = \omega/k$. The cosine part is a much longer modulation, corresponding to a group or packet of the sine waves. This envelope moves with a *group velocity* defined by

$$c_g = \frac{\Delta\omega}{\Delta k} = \frac{d\omega}{dk}$$

or, since $\omega = kc$

$$c_g = c - k\frac{dc}{dk} \tag{18.18.15}$$

Evaluating this for water waves using 18.19.14, we find

$$c_g = \tfrac{1}{2}c \tag{18.18.16}$$

The wave packet moves with only one-half the dominant phase speed. Physically this is important because the energy of the group is transported at this velocity. Individual waves move within the pocket at their own speed. Individual components leave the packet at the front while others enter at the rear.

The group-velocity phenomena only occurs in a physical situation where waves of different wavelengths move with different speeds. A dispersion relation $c(k)$ must exist, or else 18.19.15 shows that $c_g = c$. Shallow-water waves, for example, travel without changing shape.

We have spent some time discussing the wave shape and how it moves. Let us now consider the motion of the fluid particles within the flow. The particle position r_i can be expressed in both Lagrangian and Eulerian variables:

$$r_i = r_i(x_i^0, \hat{t}) = r_i(x_i, t)$$

By definition, the fluid velocities are

$$v_i = \frac{\partial r_i}{\partial \hat{t}} = \frac{Dr_i}{Dt} = \frac{\partial r_i}{\partial t} + u\frac{\partial r_i}{\partial x} + v\frac{\partial r_i}{\partial y}$$

Converting to nondimensional form, we have

$$\frac{\partial(r_i/A)}{\partial \hat{T}} = \frac{\partial(r_i/A)}{\partial T} + kA\left[U\frac{\partial(r_i/A)}{\partial X} + V\frac{\partial(r_i/A)}{\partial Y}\right]$$

From the equation above we see that the Lagrangian and Eulerian time derivatives are equivalent for $kA \to 0$. Therefore, we may find the particle paths by integrating 18.18.10 with respect to time. This gives

$$R_1 = \frac{r_i}{A} = -\frac{\cosh(Y+H)}{\sinh H}\cos(X - \hat{T})$$

$$R_2 = \frac{r_2}{A} = -\frac{\sinh(Y+H)}{\sinh H}\sin(X - \hat{T})$$

Here X, Y denote the average particle position and R_1, R_2 the displacement from that position. One may verify that the paths are elliptical motions obeying

$$\left(\frac{R_1}{a}\right)^2 + \left(\frac{R_2}{b}\right)^2 = 1$$

where

$$a = \frac{\cosh(Y+H)}{\sinh H}, \qquad b = \frac{\sinh(Y+H)}{\sinh H}$$

Sketches of the motion are given in Figure 18-33. Note that at the bottom, the amplitude of the motion is $1/\sinh H$. For an infinitely deep fluid this goes to zero, and furthermore the particle paths near the surface become circles.

18.19 CONCLUSION

Ideal flows in internal passages have a unique answer when the velocity component normal to the boundary is specified on the walls and across an inlet and an outflow surface. On the other hand, streaming motions over a closed body do not have a unique solution until a circulation constant Γ is specified. The circulation constant is a global effect of the vorticity distribution in the boundary layer. The flow over airfoil shapes is made unique by the Kutta condition requiring a smooth flow over the trailing edge.

PROBLEMS

18.1 (A) Show that the sum of two velocity fields, potential fields, or stream function fields for an incompressible, irrotational flow is again an incompressible and irrotational flow.

18.2 (A) For the stagnation point flow $F = Uz^2$, find the streamlines, potential lines, and the equations for the velocity components u, v.

18.3 (A) Do Problem 18.2 using cylindrical coordinates and produce ψ, ϕ, v_r, and v_z.

18.4 (A) Verify that the streamlines from a doublet are given by 18.5.3. Find the velocity components v_r and v_θ for this flow.

18.5 (B) A cylinder is moving through a infinite fluid with a velocity $U = a + bt$. What is the pressure at the stagnation point? Next, consider a cylinder that is stationary while the flow stream maintained at p_∞ increases as $U = a + bt$. What is the stagnation pressure for this case?

18.6 (A) Find the streamline equations for a line source and a line vortex both located at the origin. Determine the pressure as a function of distance and compare the diffuser effect with that of a spherical point source flow.

18.7 (A) A line source of strength m is parallel to a wall at a distance h. Find the pressure distribution on the wall where p_0 is the pressure at the stagnation point.

18.8 (C) The flow in a flat slot of width h is nearly uniform at U_0, however, at x_0 it has a small deviation $U(y)$ $[0 \leq y \leq h]$ from its previous history. How rapidly in x will the deviation die out? What types of deviations are first and last to die out?

18.9 (B) Find the flow field potential F and velocity W for a streaming motion U over a source located at $z = -1$ and a sink of equal strength at $z = +1$.

18.10 (A) Consider an elliptic cylinder of length five times the thickness. Find the complex potential and complex velocity for streaming flow without circulation past this object.

18.11 (B) Determine the pressure distribution over the surface in 18.10 as a function of distance s from the stagnation point.

18.12 (C) Show that the exact shape of a Joukowski airfoil with zero thickness is a circular arc.

18.13 (A) Sketch the streamline patterns you would expect for ideal flow over a circular arc at angle of attack α and various values of circulation Γ.

18.14 (B) Find the pressure distribution on a Joukowski airfoil with 3% camber and 10% thickness operating at 4° angle of attack.

18.15 (B) Express the result of Problem 18.14 in terms of the surface distance from the stagnation point.

18.16 (A) What is the angle of attack for zero lift for the airfoil in Problem 18.14?

18.17 (B) Determine the complex velocity for streaming flow past a wall that has a thin vertical plate of height h projecting from the wall.

18.18 (A) A step in a wall is of height h. Find the complex velocity for flow into a sink located at the concave corner.

18.19 (A) Find the complex potential for flow into a sink located at the convex corner in Problem 18.18.

18.20 (A) Find the pressure distribution on the walls of a 45° contraction with an area ratio of 3 : 1.

18.21 (B) Consider a step contraction of area ratio 4 : 1. Find the equation for the 50% streamline and plot on a true scale graph. Compute the pressure along the streamline to verify that it changes monotonically.

18.22 (B) A bicylindrical (co-axial) coordinate system $\zeta = \xi + i\eta$ is related to rectangular coordinates $z = x + iy$ by

$$z = i\cot\left(\tfrac{1}{2}\zeta\right)$$

In the upper half of the z plane, a curve of constant ξ is a circular arc with the center somewhere on the y axis and passing through the points $(-1, 0)$ and $(1, 0)$. For instance, $\xi = n\pi/2$ is half of a circle (cutting through $0, 1$) when $n = 1$. If $1 < n < 2$, the arc cuts the y axis at a value less than one, and if $n = 2$, that is $\xi = \pi$, the arc is a straight line between $(-1, 0)$ and $(1, 0)$. On the other hand, curves of constant η are circles in the right half plane for $\eta > 0$ (in the left half plane for $\eta < 0$). All centers are located on the x axis. All circles in the right half plane enclose the point $(1, 0)$ and those in the left half planes enclose $(-1, 0)$. As we move along a $\xi =$ constant arc in the upper half plane the point $(-1, 0)$ is $\eta = -\infty$, increasing to $\eta = 0$ where the arc cuts the y axis and proceeding to $\eta \to \infty$ at $(+1, 0)$.

Prove that the transformation above is equivalent to

$$x = \frac{\sinh \eta}{\cosh \eta - \cos \xi} \qquad y = \frac{\sin \xi}{\cosh \eta - \cos \xi}$$

Plot the circular arc with $\xi_0 = \frac{4}{3}\pi/2$. On this arc locate the points $n = -\infty$, $-\pi$, 0, π, and $+\infty$.

A biconvex airfoil or strut shape consists of two circular arcs back to back. Consider that the symmetric streaming motion in the upper half plane from right to left is given by the potential

$$F = U\frac{2}{n} i \cot\left(\frac{\zeta}{n}\right)$$

Show that the velocity is given by

$$\begin{aligned}(u^2 + v^2)^{1/2} &= \frac{4U}{n^2}\left(\frac{\sin \frac{1}{2}\zeta \sin \frac{1}{2}\bar{\zeta}}{\sin(\zeta/n)\sin(\bar{\zeta}/n)}\right) \\ &= \frac{4U}{n^2}\left(\frac{\cosh \eta - \cos \xi}{\cosh(2\eta/n) - \cos(2\xi/n)}\right).\end{aligned}$$

18.23 (B) Consider the circular arc strut of 18.22. Determine how the thickness ratio $2t/l$ and the nose half angle depend upon the parameter n.

18.24 (C) Find the pressure distribution on the surface of a circular arc strut with $n = \frac{5}{3}$. Give the pressure as a function of the distance from the nose measured along the surface.

19 Axisymmetric and Three-Dimensional Ideal Flows

The mathematical approach to solving axisymmetric and three-dimensional ideal flows is somewhat different from that used in plane flows. No longer can we use complex-variable theory and its powerful techniques. In spite of the change in the mathematical approach, the physical events and trends in three-dimensional flow are much like their plane-flow counterparts. In this chapter, for example, we shall find that the drag on a nonlifting three-dimensional body is zero, just as it is for plane two-dimensional bodies. Nevertheless, there are some differences between plane flows and three-dimensional flows. For example, we shall find that a finite wing or other three-dimensional body that produces lift will also have a drag force. This is a three-dimensional effect, since in the previous chapter we discovered that the flow over lifting airfoils produced no drag.

19.1 GENERAL EQUATIONS AND CHARACTERISTICS OF THREE-DIMENSIONAL IDEAL FLOWS

Recall from Chapter 17 that any irrotational flow will allow a description in terms of the velocity potential,

$$\mathbf{v} = \nabla\phi \qquad 19.1.1$$

Since the flows we are considering are incompressible, the velocity potential obeys the Laplace equation

$$\nabla \cdot \mathbf{v} = \nabla^2\phi = 0 \qquad 19.1.2$$

The elliptic nature of 19.1.2 implies that all parts of the flow are in communication with each other. Changing a boundary position on one part of the field has an effect at all other points in the flow.

The boundary conditions at a solid wall require that the component normal to the wall be the same as the wall velocity. This is expressed as

$$\mathbf{n} \cdot \mathbf{w}_{\text{wall}} = \mathbf{n} \cdot \mathbf{v} = \mathbf{n} \cdot \nabla\phi \quad \text{at walls} \qquad 19.1.3$$

If the wall does not move, $0 = \mathbf{n} \cdot \nabla\phi$. The mathematical problem for ϕ is completed by specifying the velocity at some arbitrarily chosen positions: the inflow and outflow boundaries for an internal flow, or $r \to \infty$ for an external flow.

By inspecting the governing equations above, we see that they are linear and may be nondimensionalized by a single length scale L and a single velocity scale U. The nondimensional potential ϕ/UL and velocities v_i/U are both functions of x_i/L and the geometry of the boundaries. Thus, this is a geometry-dominated problem. Neither the flow pressure level nor the fluid density has any effect on the velocity or upon the flow pattern. As we noted previously, the absence of time in the differential equation means that the flow is completely determined by the instantaneous position and velocities of the walls and by the fluid velocities at prescribed boundaries. The past history of the flow is of no consequence.

The pressure field that is compatible with the velocity field is determined from the Bernoulli equation for ideal flow. It is

$$\frac{\partial \phi}{\partial t} + \frac{p}{\rho} + gh + \tfrac{1}{2}q^2 = C(t) \qquad 19.1.4$$

If desired, the integration function $C(t)$ may be incorporated into a redefined potential $\hat{\phi} = \phi + \int C\,dt$, since ϕ and $\hat{\phi}$ both produce the same velocity field. From 19.1.4 we see that the proper nondimensional pressure is $p/\rho U^2$; the actual pressure field will increase directly as the density of the fluid and directly as the kinetic energy of the flow, $\frac{1}{2}U^2$.

The discussion above is applicable to general three-dimensional flows. If the flow is axisymmetric, there are only two nonzero velocity components, which depend on only two distance variables. In these cases we may use the stream-function theory of Chapter 12. The stream function $\psi(r, \theta)$ is defined as the only nonzero component of the vector potential $\mathbf{B}$: $B_\theta = \psi/r$ in cylindrical coordinates or $B_\varphi = \psi/(r \sin\theta)$ in spherical coordinates. Recall that the velocities are given by

$$\mathbf{v} = \nabla \times \mathbf{B}$$

Furthermore, when the vorticity is zero we may find $\psi(r, \theta)$ from the relation

$$0 = \nabla \cdot \nabla \mathbf{B}$$

In many problems we shall find that spherical coordinates r, θ are the most useful. For further reference recall from Appendix D the working equations relating the stream function to the velocity components:

$$v_r = \frac{1}{r^2 \sin\theta}\frac{\partial \psi}{\partial \theta}, \qquad v_\theta = \frac{-1}{r \sin\theta}\frac{\partial \psi}{\partial r} \qquad 19.1.5$$

After some simplification, the relation for $\nabla^2 \mathbf{B} = 0$ in spherical coordinates reduces to

$$\frac{\partial^2\psi}{\partial r^2} + \frac{\sin\theta}{r^2}\frac{\partial}{\partial\theta}\left(\frac{1}{\sin\theta}\frac{\partial\psi}{\partial\theta}\right) = 0 \qquad 19.1.6$$

This equation is not the Laplace equation $\nabla^2\phi = 0$ in spherical coordinates, which instead, $\phi = \phi(r, \theta)$ reduces to

$$\frac{\partial}{\partial r}\left(r^2\frac{\partial\phi}{\partial r}\right) + \frac{1}{\sin\theta}\frac{\partial}{\partial\theta}\left(\sin\theta\frac{\partial\phi}{\partial\theta}\right) = 0 \qquad 19.1.7$$

The velocity components in terms of $\phi(r, \theta)$ are

$$v_r = \frac{\partial\phi}{\partial r}, \qquad v_\theta = \frac{1}{r}\frac{\partial\phi}{\partial\theta} \qquad 19.1.8$$

In axisymmetric flow we have our choice of working with $\psi(r, \theta)$ or $\phi(r, \theta)$. For general three-dimensional flows ψ cannot be used, so ϕ is the only choice.

19.2 SWIRLING FLOW TURNED INTO AN ANNULUS

In this section we consider a flow such as that shown in Fig. 19-1, where a stream flows radially toward an axis of symmetry. Upon approaching the axis, the stream turns and flows along the axis in an annular region. An arrangement such as this is used as the inlet section to hydraulic turbines in hydroelectric power plants. We assume that the flow is uniform at the entrance $r = R$ with an inward velocity $v_r = V_r$ and a swirling component $v_\theta = V_\theta$. The swirling velocity component is produced by vanes located far from the axis. In addition, we assume that the flow was placed in the initial condition solely by pressure and gravity forces, so that the vorticity is zero. Hence, because of Helmholtz's theorems the vorticity remains zero.

As an ideal flow, the velocity must obey the continuity condition $\nabla \cdot \mathbf{v} = 0$ and the irrotationality condition $\boldsymbol{\omega} = \nabla \times \mathbf{v} = 0$. The continuity equation in cylindrical coordinates is

$$\frac{1}{r}\frac{\partial}{\partial r}(rv_r) + \frac{1}{r}\frac{\partial v_\theta}{\partial\theta} + \frac{\partial v_z}{\partial z} = 0 \qquad 19.2.1$$

In the entrance section, the flow has no v_z component and v_θ depends only on r; hence 19.2.1 is satisfied by

$$v_r = \frac{R}{r}V_r \qquad \text{(entrance region)} \qquad 19.2.2$$

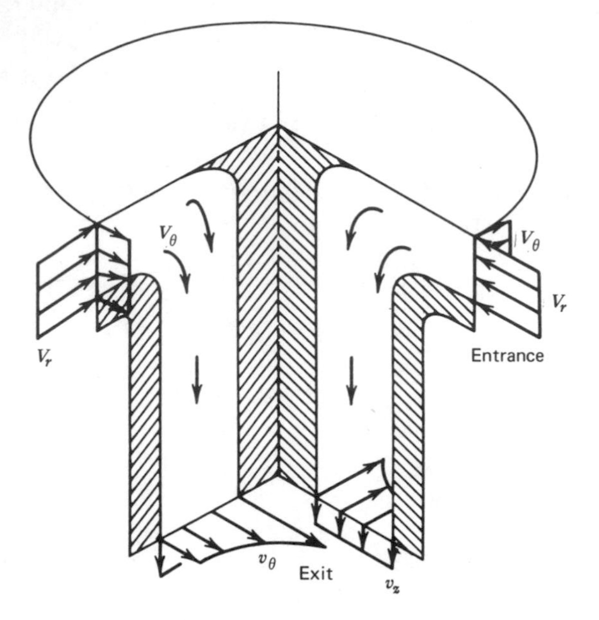

Figure 19-1 Swirling flow into an annulus.

The vorticity condition is

$$\omega_r = 0 = \frac{1}{r}\frac{\partial v_z}{\partial \theta} - \frac{\partial v_\theta}{\partial z}$$

$$\omega_\theta = 0 = \frac{\partial v_r}{\partial z} - \frac{\partial v_z}{\partial r} \qquad 19.2.3$$

$$\omega_z = 0 = \frac{1}{r}\frac{\partial}{\partial r}(rv_\theta) - \frac{1}{r}\frac{\partial v_r}{\partial \theta}$$

Considering again only the flow in the inflow section, we find from the last equation of 19.2.3 that

$$v_\theta = \frac{R}{r}V_\theta \qquad 19.2.4$$

The fact that the vorticity is zero means that the angular momentum rv_θ is conserved as the flow moves toward the axis.

We shall not find out the details of the velocity profiles in the region where the flow is turned to the z direction. However, by noting that none of the velocity components can depend upon θ, we see from the first equation in 19.2.3 that $v_\theta = f(r)$ and thus is independent of z. The last equation of 19.2.3 then shows that 19.2.4 is good for the whole flow field from entrance to exit. If the center section is absent, the flow continues to $r = 0$ and 19.2.4 predicts that v_θ becomes infinite. This, of course, does not happen, as the vorticity of the boundary layer on the back wall is swept into the core region of the resulting vortex. The cone region is then not an ideal flow.

The only remaining step is to assume that the straight walls of the exit annulus force the flow into streamlines that lie on cylindrical surfaces $r =$ const. Hence, $v_r = 0$ in this region, and the continuity equation 19.2.1 is satisfied by $v_z = f(r)$. The second vorticity equation, $\omega_\theta = 0$ in 19.2.3, shows that v_z in fact cannot depend upon r; so we must have a constant value of v_z at the exit. The areas of the inlet and outlet relate this constant to the inlet flow velocity V_r.

This problem illustrates how kinematic conditions determine the velocity profiles in ideal flows. By way of summary the velocities are:

at the entrance,

$$v_z = 0, \qquad v_r = \frac{R}{r} V_r, \qquad v_\theta = \frac{R}{r} V_\theta$$

in the exit annulus,

$$v_z = \frac{A_1}{A_2} V_r, \qquad v_r = 0, \qquad v_\theta = \frac{R}{r} V_\theta$$

The angular momentum RV_θ would be converted into a torque if the flow continued through a hydraulic machine.

19.3 FLOW OVER A WEIR

A weir is a device used in open-channel flows to measure the volume flow rate. A broad-crested weir is shown in Fig. 19-2, where a tranquil flow of depth z_0 occurs upstream of the weir. The surface is depressed an amount h_c as the flow goes over the weir. The weir is broad enough that a uniform flow is established across the top and h_c is easily measured. The water finally spills into the downstream flow as shown in the picture. If the downstream conditions prevent this pattern by backing up the water so that the free surface merely dips slightly, the weir is said to be drowned.

In the neighborhood of the crest we denote the depth of the fluid by d and the position of the free surface by h, and we assume the flow changes slowly in the streamwise direction so that it is reasonable to regard the velocity q as constant across the flow. The global continuity equation requires that

$$d = \frac{Q}{q} \tag{19.3.1}$$

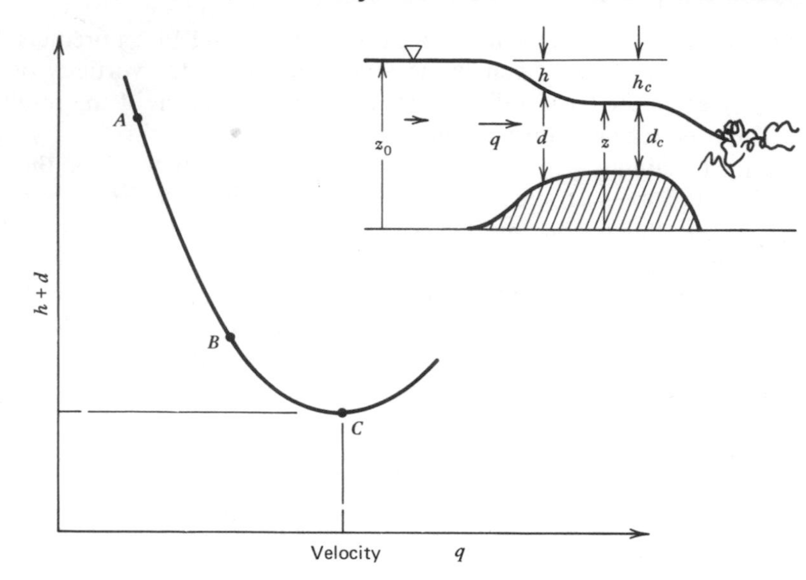

Figure 19-2 Critical flow over a broad-crested weir.

where Q is the total flow per unit width. The Bernoulli equation for any streamline is

$$\frac{p}{\rho} + \tfrac{1}{2}q^2 + gz = C \qquad 19.3.2$$

In particular, the surface streamline has a constant pressure and originates at $z = z_0$, where the velocity is negligible. For this streamline

$$\tfrac{1}{2}q^2 = g(z_0 - z) = gh \qquad 19.3.3$$

Next, we solve 19.3.3 for h and combine with 19.3.1 to give the depth of the bottom measured from reference level of the free surface far upstream:

$$h + d = \frac{Q}{q} + \frac{q^2}{2g} \qquad 19.3.4$$

The behavior of 19.3.4 is shown in Fig. 19-2, where $h + d$ is plotted as a function of q for fixed Q and g. The tranquil flow upstream starts at a low value of q and high $h + d$, as at point A on the figure. As the flow approaches the crest of the weir, q increases. If the lowest value of $h + d$ is at a point such as B, the weir is drowned. In this case the velocity decreases downstream, with the flow state returning toward A. We can deduce little else about this situation, as the exact position of point B depends upon the details of the downstream channel and the weir.

The case of most interest is when the flow reaches the critical point C and the velocity q continues to increase as the flow goes over the weir. The crest of the weir is the minimum of $h + d$, and from 19.3.4 we find the velocity of the water by locating the minimum point. Setting the derivative equal to zero:

$$0 = \frac{d(h+d)}{dq} = -Qq^{-2} + \frac{q}{g}$$

$$q_c = (gQ)^{1/3} \qquad 19.3.5$$

Inserting 19.3.5 into 19.3.3 gives h_c for the crest as

$$h_c = \frac{1}{2}\left(\frac{Q^2}{g}\right)^{1/3} \qquad 19.3.6$$

From 19.3.1 the water depth at the crest is

$$d_c = \frac{Q}{q_c} = Q(gQ)^{-1/3}$$

Finally, we solve for the flow rate Q:

$$Q = g^{1/2} d_c^{3/2} \qquad 19.3.7$$

The flow rate can be calculated from a single measurement of the fluid depth at the crest of the weir. In actual practice a coefficient is inserted in 19.3.6 to account for three-dimensional effects and, in certain situations, viscous effects.

An interesting and important side result of this analysis is that point C marks the place in the flow where the Froude number $\text{Fr} = q^2/(gd_c)$ is equal to unity. When $\text{Fr} = 1$ the velocity of propagation of shallow water waves, c, is exactly equal to the flow velocity (see 18.18.13). Any small disturbance downstream of C that causes a wave is not felt upstream, because the wave cannot make its way beyond the crest of the weir. This divides a *subcritical* flow with $q < c$ from a *supercritical* flow $q > c$.

19.4 POINT SOURCE

The equations that govern the stream function, the velocity potential, and the velocity itself are all linear, and therefore a flow field may be constructed by superposition of elementary flows. One of the most useful elementary solutions is the point source.

Consider a spherical coordinate system, and assume that the flow is purely radial $v_r(r)$. Since $v_\theta = 0$, we see from 19.1.5 that ψ is a function of θ but not a function of r. Hence, the equation that governs ψ, 19.1.6, reduces to

$$\frac{d}{d\theta}\left(\frac{1}{\sin\theta}\frac{d\psi}{d\theta}\right) = 0 \qquad 19.4.1$$

Integrating 19.4.1 gives

$$\psi = -C_1 \cos\theta + C_2$$

Any radial line $\theta = \text{const}$ gives a certain value of ψ and is a streamline in the r, θ plane. When we choose the reference axis $\theta = 0$ to be $\psi = 0$, we find $C_1 = C_2$. This choice of $\psi = 0$ also means that $\psi = 2\pi Q$, where Q is the volume flow between the $\theta = 0$ axis and the streamsurface ψ. When $\theta = \pi$, $\psi = 2C_1$ and all of the source flow lies between this streamline and the $\theta = 0$ axis. We denote the strength of the source (the volume outflow) by m. Then 19.4.1 becomes

$$\psi = \frac{m}{4\pi}[1 - \cos\theta] \qquad 19.4.2$$

The velocity of the fluid is found from 19.4.2:

$$v_r = \frac{1}{r^2 \sin\theta}\frac{\partial\psi}{\partial\theta} = \frac{m}{4\pi r^2} \qquad 19.4.3$$

From 19.4.3 we see that the velocity decays as r^{-2} for a point source. This contrasts with a decay rate of r^{-1} for a line source.

It is also easy to find the velocity potential for a source. The fact that $v_\theta = 0$ coupled with 19.1.8 means that $\phi = \phi(r)$. Equation 19.1.7 shows that

$$r^2 \frac{\partial\phi}{\partial r} = C$$

Integrating this equation and evaluating the constants yields

$$\phi = -\frac{m}{4\pi r} \qquad 19.4.4$$

The correctness of this expression is easily checked by verifying that $v_r = d\phi/dr$.

In plane flow we had two first-order singularities, the line source and the line vortex. In three-dimensional flow we have a point source, but there is no such thing as a point vortex.

19.5 RANKINE NOSE SHAPE

Superposition of a uniform stream and a source located at the origin produces the flow over a smooth blunt-nosed body. In this interpretation the streamline separating the source flow and the uniform stream is the surface of a body that extends to infinity. This particular nose shape has many of the flow characteristics of any smooth blunt body.

The stream function is

$$\psi = \tfrac{1}{2}Ur^2 \sin^2\theta + \frac{m}{4\pi}[1 - \cos\theta] + C \qquad 19.5.1$$

The first term represents a uniform stream flowing from left to right, as the student can verify. The second term is the source flow discussed in Section 19.4. An arbitrary constant has been added in order that we might make $\psi = 0$ on the stream surface separating the source fluid from the free-stream fluid. To do this we first compute the radial velocity,

$$v_r = \frac{1}{r^2 \sin\theta}\frac{\partial\psi}{\partial\theta} = U\cos\theta + \frac{m}{4\pi r^2} \qquad 19.5.2$$

Now, let us locate the point on the $\theta = \pi$ axis where the uniform stream velocity just cancels the source velocity. Setting $v_r = 0$ in 19.5.2, we find that the point $\theta = \pi$, $r = (m/4\pi U)^{1/2}$ is the stagnation point for the nose. The constant C is found from 19.5.1 by setting $\psi = 0$ at this point. The result is $C = -2m/4$. Thus 19.5.1 becomes

$$\psi = \tfrac{1}{2}Ur^2 \sin^2\theta - \frac{m}{4\pi}[1 + \cos\theta] \qquad 19.5.3$$

For a given streamline $2\pi\psi$ is the volume flow between that streamline and the nose-shaped surface.

An equation $R = R(\theta)$ describing the nose surface is obtained by substituting $\psi = 0$ in 19.5.3:

$$R = \left[\frac{m}{2\pi U}\frac{1 + \cos\theta}{\sin^2\theta}\right]^{1/2} \qquad 19.5.4$$

At $\theta = 0$, $R \to \infty$, indicating that the body is semiinfinite. Another form of the surface equation can be obtained by letting $Y(\theta)$ be the height of the surface from the reference axis. Since $y = R\sin\theta$, we see from 19.5.4 that

$$Y = \left[\frac{m}{2\pi U}(1 + \cos\theta)\right]^{1/2} \qquad 19.5.5$$

In this form we can readily see that the body has a finite radius as $\theta \to 0$,

$$h = Y(\theta \to 0) = \sqrt{\frac{m}{\pi U}} \qquad 19.5.6$$

Equation 19.5.6 relates the characteristic body radius h to the source constant and the free stream velocity U.

The θ velocity in the flow is computed from 19.5.3 and 19.1.5. It is

$$v_\theta = \frac{-1}{r\sin\theta}\frac{\partial\psi}{\partial r} = \frac{-1}{r\sin\theta}[Ur\sin^2\theta] = -U\sin\theta \qquad 19.5.7$$

Note that v_θ comes only from the free-stream flow; the source has $v_\theta = 0$.

Figure 19-3 shows a plot of the surface speed and the surface pressure where z is the coordinate along the reference axis measured from the nose. The trends for flow over the axisymmetric nose are the same as those we found previously for a plane nose. The maximum velocity in the axisymmetric case is $1.15U$, compared to $1.26U$ for the plane case. Corresponding pressure coefficients are $C_p = -0.333$ and -0.59 at the minimum points. Pressure curves such as Fig. 19-3 are useful in locating the static parts on pitot-static tubes.

The equation that describes the speed of the fluid on the body surface is found to be

$$\frac{q^2}{U^2} = \left(\frac{v_r}{U}\right)^2 + \left(\frac{v_\theta}{U}\right)^2 = 1 + 2\cos\theta\sin^2\frac{\theta}{2} + \sin^4\frac{\theta}{2} \qquad 19.5.8$$

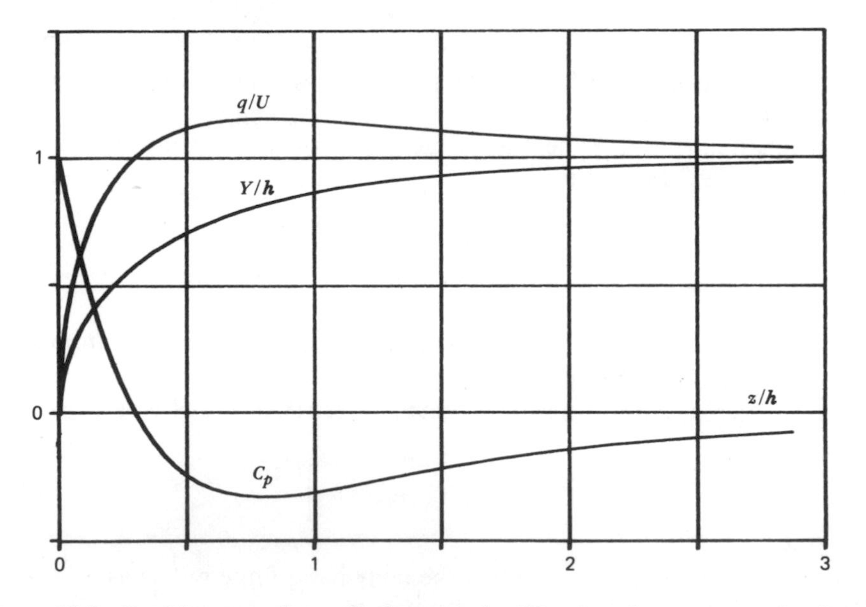

Figure 19-3 Rankine nose shape: Surface speed q/U and surface pressure C_p, plotted together with the body contour Y/h as functions of the distance from the nose divided by the maximum radius h.

The parameter θ is related to the body coordinates by

$$\frac{Y}{h} = \cos\frac{\theta}{2} \tag{19.5.9}$$

$$\frac{Z}{h} = \frac{1}{2}\frac{\cos\theta}{\sin(\theta/2)} \tag{19.5.10}$$

One can easily find the surface pressure coefficient by rearranging Bernoulli's equation into the relation

$$C_p \equiv \frac{p - p_\infty}{\frac{1}{2}\rho U^2} = 1 - \left(\frac{q}{U}\right)^2 \tag{19.5.11}$$

It is possible to integrate the surface pressure equation and show that the net drag force on the body is $p_\infty A$; the same force which would exist if the half body were in a static fluid. Hence, we say that there is no net drag on the nose. There is an exact balance between the drag on the high-pressure region and the suction force on the shoulders. This result is quite general and applies to all nose shapes.

19.6 EXPERIMENTS ON THE DRAG OF SLENDER SHAPES

The drag force on an object is frequently broken out into several parts. The first separation has a rigorous physical basis in that friction drag is easily distinguished, at least in theory, from the drag due to the pressure forces. In the present and preceding chapters we have ignored friction effects, so all of our results have actually referred to the pressure drag. Further division of the pressure drag is not so clear. When a three-dimensional body produces lift, the resulting pressure distribution also gives a drag component. Changing the attitude of a body to increase the lift force also increases the drag. This drag

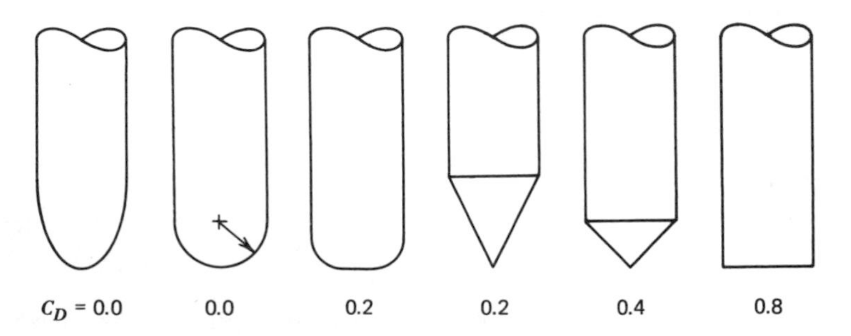

Figure 19-4 Drag coefficient of various-shaped noses. Adapted from Hoerner (1965). For the first two shapes $C_D < .05$.

component is known as the *induced drag* or *drag due to lift*. We shall discuss the origin of this drag in Section 19.10. The pressure drag that remains after the induced drag is subtracted is called *form drag*, because the shape of the body is the primary factor in determining it. Form drag also increases slightly with the angle of attack. From a practical standpoint this increase is small, and especially in the case of wings, it is common to use the term induced drag to mean all drag increases that come from increasing the angle of attack above that for which zero lift occurs. The combination of friction drag and form drag is called *profile drag*. At zero lift the drag on a body is all profile drag.

Ideal-flow theory predicts that the form drag on a nose of any shape will be zero as long as the flow is attached. Flow separation at the corner of the body means the streamlines do not turn as sharply as the surface does. The low-pressure suction region is modified or completely lost, leaving the high-pressure region to cause a drag. The form drag for a blunt nose is shown in Fig. 19-4 to be about $C_D = 0.80$. This drag is isolated by subtracting out the base drag and friction drag of a smooth nose. The approximate form drag of several other nose shapes is also given in Fig. 19-4. The round nose actually has some drag, but it is less than 0.05. As progressively sharper corners and more side separation is encountered, the drag coefficient increases, as one would expect. What is perhaps surprising is how little rounding of the corners

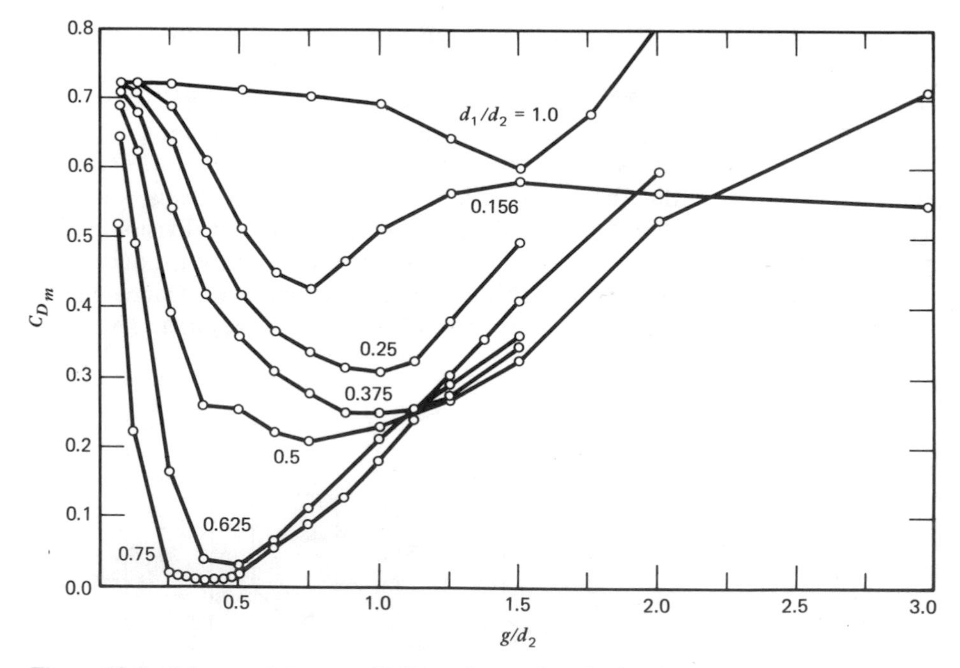

Figure 19-5 Measured drag coefficient of a tandem body. From Roshko and Koenig (1984).

is needed to decrease the drag from 0.8 for a square forebody to 0.2 as for the third shape to the left.

Let us now turn to another aspect of blunt-body drag: that of interference effects. It is common knowledge that a racing car traveling very close behind another car experiences significantly less drag. A similar effect can be used advantageously in a tandem body consisting of a disk in front of the bluff main body. Figure 19-5 is a plot of the drag coefficients measured by Roshko and Koenig (1984). When the ratio of the front-disk diameter d_1 to the main-body diameter d_2 is about 0.75, the tandem body (i.e., the disk and the main body combined) has a minimum drag coefficient of about 0.02. This occurs when the space between the bodies, g, is from 0.25 to $0.5d_2$. Photographs of the flow (Fig. 19-6) show that the shear layer formed as the flow leaves the forward disk turns and smoothly attaches to the main body. In contrast Fig. 19-7 shows the large separated regions at the corners of a body with a sharp face. When the disk is placed too far in front of the main body, the wake does not attach smoothly but sets up an oscillation where the cavity flow becomes unsteady. The drag increases in this situation.

Figure 19-6 Photograph of flow over tandem body at the configuration for lowest drag; $d_1/d_2 = 0.75$ and $g/d_2 = 0.375$. From Roshko and Koenig (1984).

Figure 19-7 Same as Fig. 19-6, except $g/d_2 = 0.125$. The flow separates at the shoulder, and the drag increases.

19.7 FLOW FROM A DOUBLET

The source–sink doublet in three-dimensional flow is depicted in Fig. 19-8. The source is located a distance ϵ to the left of the origin and the sink an equal distance to the right. At an arbitrary field point P, the stream function is the sum of the stream function 19.4.2 for the individual source and sink (in addition to spherical coordinates we use z as the position of P along the reference axis):

$$\psi = \frac{m}{4\pi}[\cos\theta_2 - \cos\theta_1] = \frac{m}{4\pi}\left[\frac{z-\epsilon}{r_2} - \frac{z+\epsilon}{r_1}\right]$$

$$= \frac{mz(r_1 - r_2)}{4\pi r_1 r_2} - \frac{m\epsilon(r_1 + r_2)}{4\pi r_1 r_2} \qquad 19.7.1$$

In order to extract the stream function for a doublet, we must let $\epsilon \to 0$, $m \to \infty$ in such a way that the product $\epsilon m = \mu$ is a constant. When this limit is taken, the radii $r_1, r_2 \to r$. The first term in 19.7.1 is indeterminate,

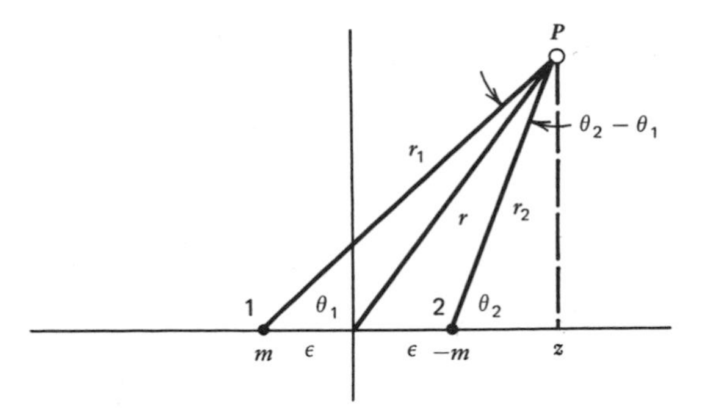

Figure 19-8 Derivation of three-dimensional doublet.

since $m(r_1 - r_2) = \infty \times 0$. We must carefully rearrange this term so that it contains μ.

Consider the law of sines applied to Fig. 19-8. This produces

$$\frac{r_1}{\sin\theta_2} = \frac{r_2}{\sin\theta_1} = \frac{2\epsilon}{\sin(\theta_2 - \theta_1)}$$

Hence, the quantity $r_1 - r_2$ may be expressed as

$$r_1 - r_2 = \frac{2\epsilon[\sin\theta_2 - \sin\theta_1]}{\sin(\theta_2 - \theta_1)}$$

$$= \frac{2\epsilon 2\cos\left[\frac{1}{2}(\theta_2 + \theta_1)\right]\sin\left[\frac{1}{2}(\theta_2 - \theta_1)\right]}{2\sin\left[\frac{1}{2}(\theta_2 - \theta_1)\right]\cos\left[\frac{1}{2}(\theta_2 - \theta_1)\right]}$$

In the second relation the fact that $\sin\theta = 2\sin(\theta/2)\cos(\theta/2)$ has been used. Simplifying the expression above and multiplying by m gives

$$m(r_1 - r_2) = \frac{2\epsilon m\cos(\theta_2/2 + \theta_1/2)}{\cos(\theta_2/2 - \theta_1/2)} \qquad 19.7.2$$

When the limit $\epsilon \to 0$, $m \to \infty$ is applied to 19.7.2, this term becomes $+2\mu\cos\theta$. Hence, 19.7.1 reduces to

$$\psi = \frac{\mu z\cos\theta}{2\pi r^2} - \frac{\mu}{4\pi r} = \frac{\mu}{4\pi r}(\cos^2\theta - 1)$$

$$= -\frac{\mu}{4\pi r}\sin^2\theta \qquad 19.7.3$$

This is the stream function of a doublet oriented with the sink on the positive z

axis. The corresponding expression for the velocity potential turns out to be

$$\phi = \frac{\mu}{4\pi r^2} \cos\theta \qquad 19.7.4$$

Unlike the plane case, where a doublet could be viewed as the limit of either a source–sink or two counterrotating line vortices, the three-dimensional doublet cannot be considered as the limit of a ring vortex as the diameter decreases to zero.

19.8 FLOW OVER A SPHERE

The ideal flow over a sphere is given by the superposition of a uniform stream and a doublet. Summing these contributions and including an arbitrary constant C yields

$$\psi = \frac{Ur^2}{2} \sin^2\theta - \frac{\mu}{4\pi r} \sin^2\theta + C \qquad 19.8.1$$

The velocity component in the radial direction is

$$v_r = \frac{1}{r^2 \sin\theta} \frac{\partial \psi}{\partial \theta} = 2\cos\theta \left[\frac{U}{2} - \frac{\mu}{4\pi r^3} \right] \qquad 19.8.2$$

Now, on the negative axis $\theta = \pi$, the flow is purely radial and the uniform stream exactly cancels the doublet when $v_r = 0$. This occurs at the position r_0 given by

$$r_0 = \left(\frac{\mu}{2\pi U} \right)^{1/3} \qquad 19.8.3$$

Inserting this for μ in 19.8.1 and finding C by taking $\psi = 0$ at $r = r_0$, $\theta = \pi$, leads to the result

$$\psi = \tfrac{1}{2} r_0^2 U \sin^2\theta \left[\left(\frac{r}{r_0} \right)^2 - \left(\frac{r}{r_0} \right)^{-1} \right] \qquad 19.8.4$$

From 19.8.4 we can verify that for arbitrary θ, $r = r_0$ is a spherical stream surface where $\psi = 0$. For future reference note that the corresponding velocity potential is

$$\phi = r_0 U \cos\theta \left[\frac{r}{r_0} + \frac{1}{2} \left(\frac{r}{r_0} \right)^{-2} \right]$$

The velocity components for this flow are

$$v_r = U\cos\theta\left[1 - \left(\frac{r_0}{r}\right)^3\right] \tag{19.8.5}$$

and

$$v_\theta = -\frac{U}{2}\sin\theta\left[2 + \left(\frac{r_0}{r}\right)^3\right]$$

As one might expect, the velocities have fore-and-aft symmetry and no evidence of a wake. On the surface of the sphere the velocity is

$$q = v_\theta = -\tfrac{3}{2}U\sin\theta \tag{19.8.6}$$

The maximum occurs at the equator, where $q = -3U/2$. Recall that for a circular cylinder the maximum was slightly higher, $q = -2U$. We might interpret this effect in terms of inserting a body into a uniform stream. A body that is finite in all dimensions allows the flow to go around in three dimensions, whereas a cylinder forces the flow to squeeze by in only two dimensions. Hence, the velocity at the shoulder of a sphere is less than the velocity at the shoulder of a cylinder.

Velocities 19.8.5 differ from a uniform stream by a term that dies out as r^{-3}. Similar equations for a cylinder (18.6.3) reveal that the effect of a two dimensional body dies out as r^{-2}.

The surface pressures are found by evaluating Bernoulli's equation on the surface of the sphere, using 19.8.6 for the velocity. In terms of the pressure coefficient, the result is

$$C_p = 1 - \tfrac{9}{4}\sin^2\theta \tag{19.8.7}$$

At the equator, the minimum pressure is $C_p = -\frac{5}{4}$. This is not nearly so low as the pressure predicted for ideal flow over a cylinder, $C_p = -3$. Nevertheless, the real flow over a solid body cannot penetrate into the high-pressure region at the rear of the sphere, and the flow separates. The drag coefficient for a sphere follows the same trends as experienced with circular cylinders (see Section 15.6). One difference in the details of the flow is that a dominant Strouhal number and regular vortex shedding are not usually observed on spheres. In some instances a spiral shedding of the vortices is observed, but usually the wake flow is irregular and without a dominant pattern.

There are at least two situations where the potential-flow solution is in reasonable agreement with actual flows. When a sphere is started impulsively from rest (or equivalently, when the fluid is started impulsively around a stationary sphere), the initial motion is the irrotational flow given above. Another case in which the solution is valid is when a sphere undergoes small oscillations back and forth along a line. In this instance, if the frequency is high, the Stokes layer is thin, leaving the main part of the flow irrotational.

Another possible application of the ideal-flow solution is to a spherical drop in an infinite fluid. The motion of the outside fluid can engender a circulatory motion within the drop. Recall that in Section 13.6 a complete solution of the Navier–Stokes equations known as Hill's spherical vortex was discussed. It is an interesting fact that the surface velocity of Hill's vortex and the surface velocity of the ideal flow over a sphere can be exactly matched. If U is the free-stream velocity far away from the sphere, r_0 the radius of the sphere, and A the constant in the vorticity distribution $\omega_\theta = Ar_0$ for Hill's vortex, then when

$$U = \tfrac{2}{15} r_0^2 A$$

the surface velocity of the solutions will match. Patching these two solutions together gives a result that satisfies the Navier–Stokes equations everywhere and has continuous velocities. Its only defect is that the shear stress and pressure at the interface of the sphere are discontinuous. Although this is an interesting result, it does not appear to have any application. High-Reynolds-number situations, such as large raindrops, seem to have separation regions similar to those of solid particles. On the other hand, many problems involving particles and droplets are also at low Reynolds numbers. Hill's vortex solution can also be matched to a low-Reynolds-number solution (Section 21.5).

19.9 D'ALEMBERT'S PARADOX

We can prove that the ideal flow drag on any closed nonlifting body is zero. Consider a body moving in an infinite fluid as shown in Fig. 19-9. A control region is chosen whose boundaries are very far away (in the still fluid) and around the body, where the fluid has the velocity $w_i = (U(t), 0, 0)$ in the x direction. Recall that the kinetic-energy equation (5.16.2) for an arbitrary region is

$$\frac{d}{dt}\int \tfrac{1}{2}\rho v_i v_i \, dV = -\int \left[\rho n_i (v_i - w_i)\tfrac{1}{2} v_k v_k + n_i v_i p\right] dS \qquad 19.9.1$$

Because we are dealing with an incompressible and inviscid flow, the compression-work and viscous-dissipation integrals in 5.16.2 are zero. All terms on the right-hand side of 19.9.1 are zero at the far boundary, where the fluid is still. Moreover, since flow does not cross the surface of the body, $n_i(v_i - w_i) = 0$ in the convection term. This relation also means that $n_i v_i = n_i w_i = n_x w_x = n_x U(t)$

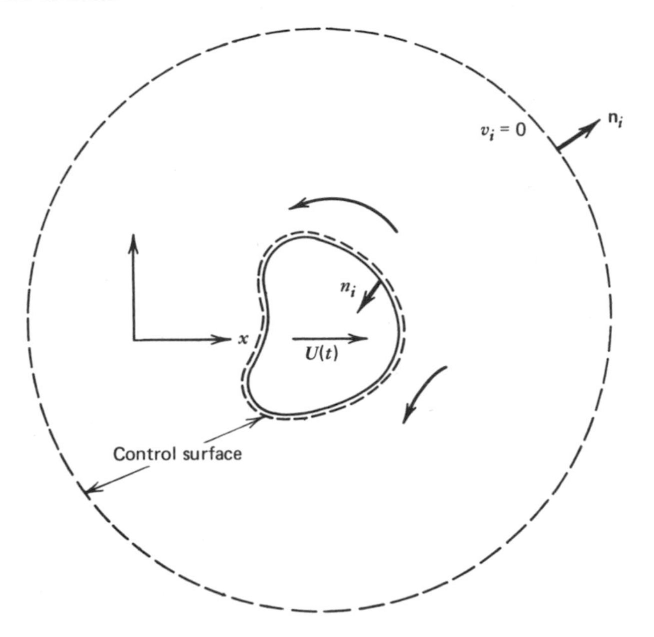

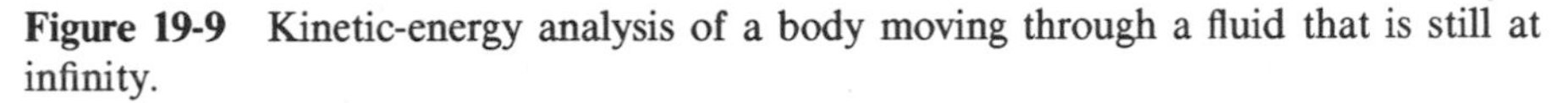

Figure 19-9 Kinetic-energy analysis of a body moving through a fluid that is still at infinity.

on the body. Thus 19.9.1 becomes

$$\frac{d}{dt}\int \tfrac{1}{2}\rho v_i v_i \, dV = -U\int_{S_b} n_x p \, dS$$

$$= UF_D \qquad 19.9.2$$

The x component of the surface pressure is defined to be the drag. The result above states that the work done by moving the body at a speed U is UF_D (there being no motion and hence no work done in the direction of the lift force), and that this work must appear within an inviscid, incompressible flow as an increase in the kinetic energy of the fluid.

Let us first discuss the special case when the motion U is steady. Furthermore, assume the flow is an ideal, potential flow, without singularities, everywhere outside of the body (a nonlifting body). A potential flow of this type may be represented by sources, sinks, and doublets distributed on and within the body. Such a flow is steady with respect to a coordinate system on the body ($\hat{v}_i$). The velocity in ground-based coordinates is $v_i = \hat{v}_i - U_i$, and $v_i v_i = \hat{v}_i \hat{v}_i - 2\hat{v}_i U + U^2$. Therefore, since both $\hat{v}_i$ and U are constant, the integral of $\frac{1}{2}v_i v_i$, the total energy of the fluid motion, is constant. This leads to the conclusion from 19.9.2 that the drag is zero:

$$0 = F_D \qquad 19.9.3$$

The fact that ideal flow theory predicts $F_D = 0$ while any real flow has a drag force is known as D'Alembert's paradox.

Let us consider several flow situations where the assumptions necessary for $F_D = 0$ are not met. First, picture a bluff body moving through the fluid creating a wake region (usually turbulent). Within the wake the potential-flow assumption is invalid. A momentum analysis of this flow, in a manner exactly similar to the cylinder analysis of Section 15.6, would show that the wake has a defect of convected momentum that exactly accounts for the drag. The energy analysis of 19.9.2 yields another point of view. It shows that the drag is proportional to the rate of increase of energy resulting from the lengthening of the wake. A rifle shot into still air generates a wake where the fluid is moving with respect to the ground in a direction toward the bullet. The work done by the bullet moving through the air increases the energy of the fluid in the wake. Subsequently this energy is converted, through turbulence and viscous dissipation, into thermal energy.

For another example, consider a ship in a calm sea. At time zero the ship gets underway and shortly attains a steady speed. We neglect the viscous wake, as discussed above, to study the wake of waves generated by the ship. The waves are irrotational potential motions; however, their energy is never returned. The wave system continues to grow, in principle, throughout the motion. Hence, 19.9.2 indicates that there is a wave drag equal to the rate that energy is carried out in the ever expanding wave wake. A similar wave drag occurs in supersonic flow, where bodies generate shock waves that continually expand away from the body. In either case the waves are generated through the action of pressure at the surface of the body. The existence of the wave drag means that the integral of the x component of these pressures is nonzero.

The last example is the most subtle way by which a drag force is produced. It is called induced drag and is illustrated by the flow over a lifting body such as a wing. Recall that a two-dimensional airfoil has no drag whatsoever. The case of a finite wing is much different, as we shall see in the next section.

*19.10 INDUCED DRAG

In the first approximation, the flow over any section of a wing is just like the flow over an infinite airfoil of the same cross section. This idea, while sufficiently accurate for the lift, needs to be refined if we are to find the drag of the wing, since the drag is much smaller than the lift. The fact that strong vortices are formed at the tip of wings, and more importantly associating them with a drag force, is a contribution from the aeronautical pioneer Lanchester (1906).

Figure 19-10 shows the flow pattern over a wing of span L. Consider that the lift force produced at each chord position must change as we move along the span toward the tip. Near the tip the lift drops very rapidly to zero. Now, the amount of lift at any section is related to the local circulation $\Gamma(z)$

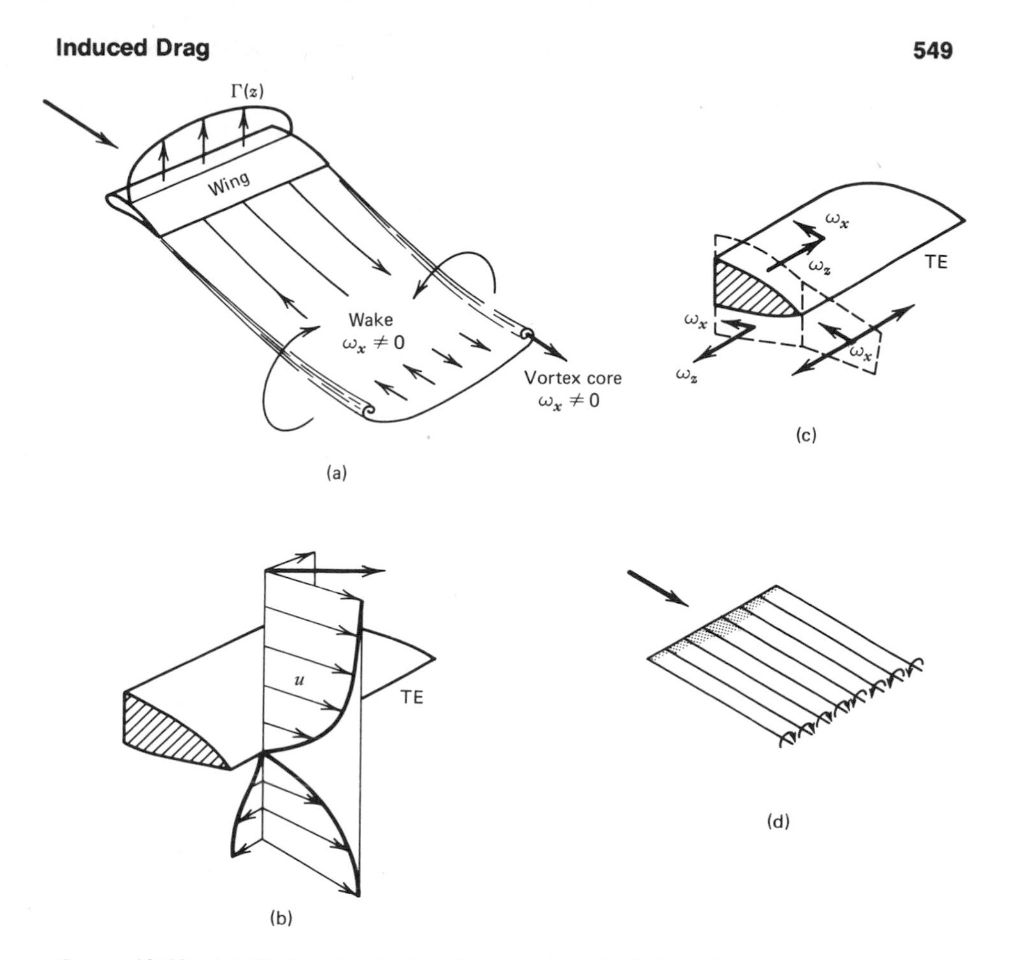

Figure 19-10 (a) Wake of a finite wing, composed of shear layer and vortex cores containing longitudinal vorticity. (b) Close-up of the trailing-edge boundary layers. (c) Skewed flow on top and bottom gives net vorticity in flow direction. (d) Wing and wake modeled as a sequence of line vortices.

by generalizing the Kutta–Joukowski law 18.8.1

$$\frac{dF_L}{dz} = \rho U \Gamma(z)$$

$$F_L = \rho U \int \Gamma(z)\, dz \qquad 19.10.1$$

In turn, the local circulation is the integral of the vorticity in the boundary layer at that section. As $\Gamma(z)$ decreases, each section must have less vorticity. With regard to its effect in ideal flow, we may represent the vorticity in the upper and lower boundary layers by distributed line vortices, where each line represents a given strength of vorticity. (Beginning at the stagnation point, a line vortex is drawn for each area over which $\int \omega \, dA$ has some fixed small value.) As we move outboard the number of these lines on the wing must decrease to correspond to the decreasing lift.

The fact that the lift decreases along the wing has a very slight, but very important, effect on the inviscid flow over the wing. The flow on the lower surface leaves the trailing edge with an outboard velocity component, while that from the upper surface has a slight inboard velocity component. Figure 19-10b depicts the vortex lines as they leave the wing and continue into the shear layer formed by the merging boundary layers. The largest component of vorticity in the boundary layers, the z-component, cancels out when the boundary layers merge into the shear layer (viscous diffusion subsequently destroys this part of the shear layer). However, the vorticity component in the flow direction, arising from the skewed profiles, remains the same for a long distance behind the wing.

Near the tip of the wing these events are accentuated, the flow becomes three-dimensional, and the x-direction line vortices become concentrated into the wing-tip vortex. The core of the vortex is a finite area of dense vorticity. The overall effect of the tip vortex is to take fluid from the underside of the wake and swirl it around to the upper side. This accounts for the inviscid flow toward the tip on the underside and away from the tip on the upperside.

Figure 19-11 is the plan view of a delta wing. In this case the "tip vortices" form at the apex of the delta.

With these events in mind, let us return to the question of D'Alembert's paradox and the inviscid pressure drag on a wing. The tip vortices and the

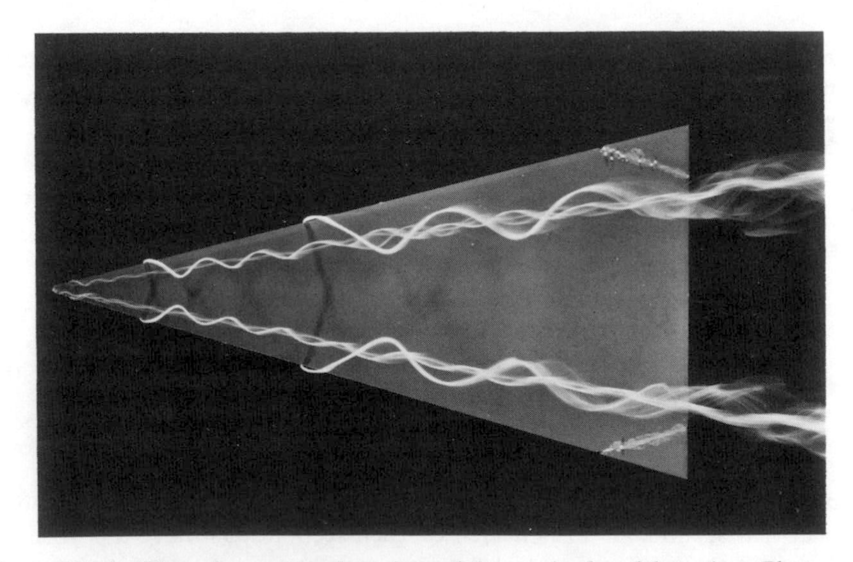

Figure 19-11 Plan view of vortices formed by a slender delta wing. Photograph courtesy of H. Werlé, ONERA, France (see Werlé, 1963).

wake area singular surface in the potential flow. As we cross the shear layer, the inviscid velocity changes direction (the magnitude is the same because the pressure is constant across the shear layer and both streams have the same Bernoulli constant). A potential solution for the flow requires a distribution of vortex lines along the wake (or doublets) as well as sources, sinks, and doublets (or vortices) on the wing surface. Now, when we envision a wing moving in an infinite still medium, the ideal-flow problem continues to change as time goes on. The wake grows longer and longer, containing more and more energy. Thus, the wing leaves a permanent mark in the fluid, just as a bullet does. The wake of a wing is a swirling ideal flow generated by trailing vortex lines imbedded in a shear layer of zero thickness.

Assume that after a certain distance behind the wing, the wake is developed to such a stage that the flow no longer depends upon x and the vortex lines lie in the x direction. Let this position be $x_1 = x_0 + Ut$. (This model is approximate because the vortex wake propels itself in the downward direction with a very slight velocity.) With these assumptions, the integral of the energy in 19.9.2 is broken into three parts: $-\infty$ to x_0, x_0 to x_1, and x_1 to $+\infty$ (Fig. 19-12). The energy in the end pieces is constant, so the only increase in energy is between x_0 and x_1. Hence,

$$UF_D = \frac{d}{dt}\int_{x_0}^{x_1=x_0+Ut}\iint_{-\infty}^{\infty} \tfrac{1}{2}v_iv_i \, dy\, dz\, dx$$

Because the flow at x_1 is fully established, only the upper limit $x_1 = x_0 + Ut$ is a function of time. Hence we apply Leibnitz's formula to the x integration and obtain

$$F_D = \iint_{-\infty}^{\infty} \tfrac{1}{2}v_iv_i \, dy\, dz \qquad \text{at } x_1 \tag{19.10.2}$$

The drag force is equal to the kinetic energy integrated across a plane at a position where the wake flow is fully established.

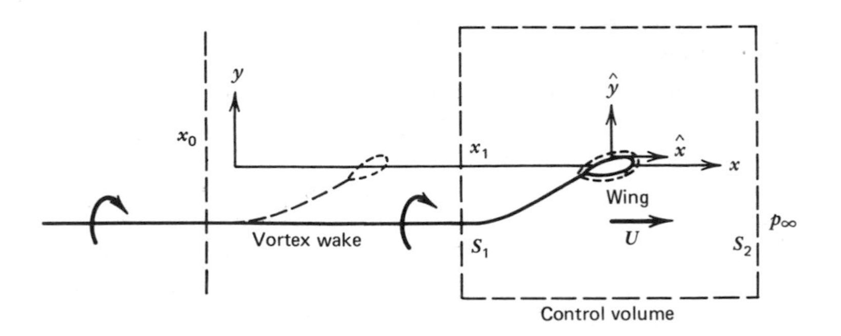

Figure 19-12 Induced drag from increasing length of the vortex wake. The wing, the plane x_1, and the control volume move through the fluid with velocity U.

It is informative to apply a momentum analysis to this problem also. We choose a region that moves along with the wing as shown in Fig. 19-12. Let $\hat{v}_i$ denote the velocity in the wing-fixed system, and retain v_i for the ground fixed system. The flow on the inflow boundary S_2 is uniform: $v_x = U$ and $p = p_\infty$. The outflow is where the vortex lines, which are straight and aligned with the x direction, penetrate the control volume. The Biot–Savart law shows that a system of straight parallel vortex lines can only induce velocities in the plane perpendicular to the lines. Hence at the outflow boundary we have $v_x = 0$ in the ground-based system, and since $v_x = \hat{v}_x - U$, we find that $\hat{v}_x = U$. The x-direction flow into and out of the control volume is uniform. Moreover, this means that the net convection of momentum into the region is zero. The integral x-momentum equation 5.14.1 then reduces to a balance of pressure forces,

$$0 = -\int_{S_1} n_x p\, dS - \int_{S_b} n_x p\, dS - \int_{S_2} n_x p\, dS$$

$$0 = \int_{S_1} n_x (p_\infty - p)\, dS - F_D \qquad 19.10.3$$

The pressure in the plane that cuts across the wake is slightly lower than the free-stream pressure. This creates the induced drag.

We can also connect 19.10.3 to the energy-equation result 19.10.2. In wing-fixed coordinates the Bernoulli equation between points on the inlet and outlet flow planes is

$$p_\infty + \tfrac{1}{2}\rho U^2 = p + \tfrac{1}{2}\rho \hat{v}_i \hat{v}_i$$

Using $\hat{v}_x = v_x + U$, $\hat{v}_y = v_y$, and $\hat{v}_z = v_z$, we have

$$p_\infty - p = \tfrac{1}{2}\rho [v_i v_i - 2 v_x U]$$

Noting that $v_x = 0$ on S_1 and substituting this into 19.10.3 yields the previous result 19.10.2. Thus, the momentum analysis and the energy-equation analysis give equivalent results. The low pressures on S_1 result from the swirling vortex flow giving the fluid an extra high velocity.

The induced drag on a lifting body is the result of the vortex system in the wake and the tip vortices. In turn the wake exists because the circulation (and hence the lift) varies along the wing span and must drop to zero at the tips. Books on aerodynamics (Bertin and Smith 1979, Karamchiti 1966, Kuethe and Chow 1976, Landau and Lifshitz 1959, or Thwaites 1960) solve the potential-flow problem for the wake and show exactly how the induced drag depends upon the distribution of lift along the wing. The result is of the form

$$F_D = C\rho \Gamma_0^2 \qquad 19.10.4$$

where C is a constant that depends on the shape of the lift distribution over the wing, and Γ_0 is the average circulation. The remarkable aspect of 19.10.4 is that increasing the length of a wing does not increase F_D, but it does increase the lift, since

$$F_L = \rho U \Gamma_0 L \qquad 19.10.5$$

These facts may be illustrated in another way. Assume an airplane has a certain weight, requiring for level flight at speed U an equal lift force. We can rearrange 19.10.4 and 19.10.5 to eliminate Γ_0 and find

$$\frac{F_D}{F_L^2} = \frac{C}{\rho U^2 L^2} \qquad 19.10.6$$

This shows that the longest wing will have the smallest induced drag for a given lift. For this reason sailplanes have long wings. Another point of interest is that an elliptical distribution of lift along the wing produces the smallest value of the constant C.

*19.11 LIFTING-LINE THEORY

The task of predicting the lift and induced drag of a wing or blade of given shape is important in engineering design. Modern computer programs to solve this problem are quite complex and are the subject of advanced aerodynamics textbooks and research papers. The purpose of this section is to give a general discussion of the nature of the problem and indicate where assumptions and approximations are necessary to solve the problem. At the end of this discussion we shall outline Prandtl's lifting-line theory, the first successful theory of a finite wing.

Consider the nonseparated potential flow over a given wing. Recall that even in two dimensions this problem is not unique until the Kutta condition is imposed. In three dimensions, viscosity not only determines the Kutta condition, but also determines the exact details of the tip-vortex formation. Most potential-flow computer methods assume the Kutta condition and fix the tip vortex at the trailing edge of the wing tip. Actually the vortex always springs from a position near midchord and slightly inboard. An exact calculation of the flow in this region involves both viscous and inviscid phenomena and hence is very difficult.

With the assumption of where and how the wake leaves the wing, the potential problem is uniquely determined. The problem consists of finding a potential flow with velocities tangent to the wing surface and obeying the Kutta condition. The strength of the vortex lines in the wake and the subsequent location of the wake in the downstream region must be calculated. The potential-flow boundary conditions at the wake are: (a) no flow across the wake, and (b) no change in pressure across the wake. These two conditions at the unknown wake position are equivalent to a single condition at a known location such as at a solid body.

Not only is Prandtl's lifting-line theory of historical interest, but various aspects of it are incorporated in many of the more modern methods. To find the influence of the wake on the flow over the wing, Prandtl modeled the wake as a straight, flat sequence of line vortices coming back from another vortex line, the lifting line, which represents the wing (Fig. 19-10c).

As each vortex line peels off to form the wake, the strength $\Gamma(z)$ of the lifting line decreases. Recall from the previous section that the lift distribution is proportional to the circulation distribution. The set of vortices representing the wake induces a velocity at each point of the flow. If the geometry and strength of the wake are known, it is possible to find the downwash velocity v_{dn} caused by the wake at the position of the lifting line. The lifting line induces no velocities on itself. Of course, the wake cannot really cause a flow through the wing. To remove this absurdity we suppose that the wing moves down at a velocity v_{dn}. This effectively reduces the angle of attack of the wing.

Let the true geometric angle of attack be $\alpha = V_\infty/U_\infty$. The effective angle of attack α_e is the true angle minus the angle induced by the wake,

$$\alpha_e = \alpha - \alpha_i \approx \frac{V_\infty - v_{\text{dn}}}{U_\infty}$$

The induced angle α_i, and hence α_e, is not necessarily the same across the span of the wing. At each location the lift is computed according to two-dimensional airfoil theory using α_e as the angle of attack. The lift vector from this flow is decomposed into components with respect to the true free-stream direction α. Hence, there is a lift component and a drag component. The drag calculated in this manner is the induced drag.

The overall effect of the wake is to induce a downwash that effectively reduces the angle of attack and causes the lift vector to turn and generate a drag component along the stream direction.

*19.12 ADDED MASS

Let us consider once more a body moving through a still fluid with a velocity $U_i(t)$. Equation 19.9.2 revealed that the work U_iF_i (we now let the velocity and the force be vectors) is equal to the rate of change of the kinetic energy of the entire flow:

$$U_iF_i = \frac{d}{dt}\int \tfrac{1}{2}\rho v_i v_i \, dV \qquad 19.12.1$$

For example, a sphere moving at $U(t_1)$ is accelerated by an external force to a higher speed $U(t_2)$. The flow field at the second speed has more kinetic energy; hence by 19.12.1 an additional drag force F_D was necessary to generate this motion. Even an ideal flow, where $F_D = 0$ before and after the acceleration, has a drag force during the acceleration phase. Previously it has been noted that ideal flows are dominated by kinematics and that the instantaneous velocity field is determined by the shape of the body and its instantaneous velocity, a quasisteady situation. The pressure field, on the other hand, is not quasisteady, as it is governed by Bernoulli's equation, which contains the unsteady term $\partial\phi/\partial t$ (19.1.4). When a body is accelerated, the surface pressures must increase to supply the forces necessary to accelerate the fluid.

The added drag force may be found from 19.12.1, which relates the work and the rate of change of the fluid energy. A useful way to evaluate 19.12.1 is to use Green's theorem,

$$\int [\nabla\phi \cdot \nabla\phi + \phi\nabla^2\phi]\, dV = \int \phi\frac{\partial\phi}{\partial n}\, dS$$

In potential flow $v = \nabla\phi$ and $\nabla^2\phi = 0$, so the equation above may be used to transform the volume integral in 19.12.1 into a surface integral. The result is

$$U_i F_i = \tfrac{1}{2}\rho\frac{d}{dt}\int_{S_2} \phi\frac{\partial\phi}{\partial n}\, dS = -\tfrac{1}{2}\rho\frac{d}{dt}\int_{S_b} \phi\mathbf{n}\cdot\mathbf{v}\, dS \qquad 19.12.2$$

The integral of $\phi\, \partial\phi/\partial n$ over the surface at infinity is zero because $\partial\phi/\partial n = -\mathbf{v}\cdot\mathbf{n}|_\infty$ is zero. The sign of the second integral of 19.12.2 has been changed so that $\mathbf{n}$ is outward from the body.

Before we compute the force in 19.12.2, it is good to list the transformation equations between the body-fixed system $\hat{x}_i$ and the ground system x_i:

$$U_i(t) = \text{translational velocity of the body}$$

$$x_i^0(t) = \text{location of a reference body point}$$

$$= \int U_i(t)\, dt \qquad 9.12.3$$

$$x_i = \hat{x}_i + x_i^0$$

$$v_i = \hat{v}_i + U_i$$

The velocity potentials are defined by $\hat{\mathbf{v}} = \hat{\nabla}\hat{\phi}$ and $\mathbf{v} = \nabla\phi$. In order for $\nabla\phi = \hat{\nabla}\phi + \mathbf{U}$ to hold, we must have

$$\phi = \hat{\phi} + \hat{x}_i U_i \qquad 19.12.4$$

[Actually $\phi = \hat{\phi} + \hat{x}_i U_i + C(t)$ is acceptable, but the choice $C = 0$ is best for the Bernoulli equation.] The pressure $\hat{p}$ computed from incompressible flow equations in the moving reference frame are pseudopressures (10.7.6). They are related to the true pressure p by

$$p = \hat{p} - \rho \hat{x}_i \frac{dU_i}{dt} \qquad 19.12.5$$

The irrotationality and incompressibility conditions are the same in either coordinate system; however, the Bernoulli equations are

$$\frac{\partial \phi}{\partial t} + \frac{p}{\rho} + \tfrac{1}{2} v_i v_i = \frac{p_\infty}{\rho} \qquad 19.12.6$$

and

$$\frac{\partial \hat{\phi}}{\partial \hat{t}} + \frac{\hat{p}}{\rho} + \tfrac{1}{2} \hat{v}_i \hat{v}_i = \frac{p_\infty}{\rho} + \tfrac{1}{2} U_i U_i \qquad 19.12.7$$

One may verify that 19.12.6 and 19.12.7 are compatible by using the relations above.

As a useful example of the added-mass concept we consider a sphere. A sphere moving through a still fluid has the velocity potential

$$\phi = \tfrac{1}{2} r_0^3 \frac{U_i\left(x_i - x_i^0\right)}{|x_i - x_i^0|^3} \qquad 19.12.8$$

Let us restrict U_i to motion along a certain line and take a spherical coordinate system with the center at x_i^0. In order to evaluate 19.12.2, note that at the surface of the body $|x_i - x_i^0| = \hat{r}_0$, $(x_i - x_i^0)U_i = \hat{r}_0 U \cos\theta$, $\mathbf{v} \cdot \mathbf{n} = U \cos\theta$, and $dS = 2\pi \hat{r}_0 \sin\theta \, d\theta$. Hence, 19.2.2 becomes

$$UF = -\tfrac{1}{2}\rho \frac{d}{dt} \int_0^\pi \pi U^2 \hat{r}_0^3 \cos^2\hat{\theta} \sin\hat{\theta} \, d\hat{\theta} \qquad 19.12.9$$

Simplifying, we arrive at

$$F = -\tfrac{2}{3}\pi r_0^3 \rho \frac{dU}{dt} \qquad 19.12.10$$

This is the drag force on a sphere undergoing a linear acceleration dU/dt in ideal flow. The form of this equation allows us to interpret this effect as adding a certain mass of the fluid to the true mass of the sphere. In this case the added mass is one-half the volume of the sphere times the fluid density. In other words, we may neglect the force required to accelerate the fluid if we give the sphere a virtual mass of $\left(\rho_0 + \tfrac{1}{2}\rho\right)V$.

Let us consider the ideal flow about an arbitrary body. In body-fixed coordinates the potential will have a form where the space and time functions are separable, that is,

$$\hat{\phi} = \Phi(\hat{x}_i)U(\hat{t})$$

This, together with the fact that $\mathbf{v} \cdot \mathbf{n} = \mathbf{n} \cdot \mathbf{U}$ on the body surface, means that the integral in 19.12.2 will always break down into U times a surface integral that is independent of time. Hence, for all potential flows, the drag required to accelerate the fluid may always be expressed in terms of the mass of a certain volume of fluid added to the mass of the object. The size of the volume of fluid depends on the shape of the object and the direction of motion. For example, the added mass of a circular disk moving normal to its plane is $8\rho_f r_0^3/3$. Since $\frac{8}{3} \approx \pi$, this is roughly the mass of a fluid cylinder r_0 in radius and r_0 in height.

A body of a general shape has a different added mass for linear acceleration in each coordinate direction, and another added mass for angular acceleration about each axis. A more general development (Batchelor 1967, Landau and Lifshitz 1959, and Yih 1969) leads to an added-mass tensor.

For a specific application of the theory, consider the initial acceleration of a sphere of density ρ_0 in a fluid of density ρ_f. A bubble in a liquid or a helium balloon in air are specific examples. Let M_0 be the mass of the sphere, and M_f the mass of an equal volume of fluid. The motion of the ball is subjected to forces from its weight, it buoyancy, and its added mass. The equation of motion is

$$M_0 \frac{dU}{dt} = -M_0 g + M_f g - \tfrac{1}{2} M_f \frac{dU}{dt} \qquad 19.12.11$$

Rearranging this yields

$$\frac{dU}{dt} = \frac{M_f - M_0}{M_0 + \frac{1}{2}M_f} g = \frac{\rho_f - \rho_0}{\rho_0 + \frac{1}{2}\rho_f} g \qquad 19.12.12$$

In the case of a bubble $\rho_0 \ll \rho_f$, and the acceleration is $2g$ upwards, the buoyancy force exactly supplies the force required to accelerate the liquid. The opposite extreme, a heavy ball in a light fluid, gives an acceleration of $-g$ of course.

As a second example, consider a sphere in a fluid that is oscillating with $v_x(\infty) = A \sin \Omega t$. A situation of this type is a bubble in a tank of vibrating liquid, or a light particle in a sound field where the wavelength is long compared to the particle diameter. We neglect weight and buoyancy to concentrate on the motion of the particle induced by the oscillating fluid. In the ground-fixed system x_i, the sphere velocity is $U(t)$ and the pressure is p. The x-direction force on the sphere is

$$M_0 \frac{dU}{dt} = F = -\int_{S_b} n_x p \, dS \qquad 19.12.13$$

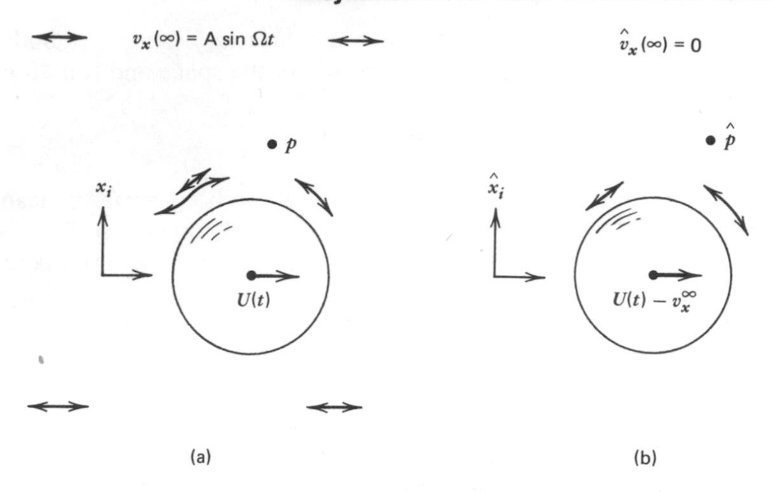

Figure 19-13 Bubble in an oscillating liquid: (a) ground-based system, (b) imaginary flow with a coordinate system chosen so the fluid at infinity is still.

In order to evaluate 19.12.13 we introduce a coordinate system in which the fluid at infinity is at rest. Let an $\hat{x}_i$ system be defined so that (Fig. 19-13)

$$\hat{v}_x = v_x - v_x^\infty \qquad 19.12.14$$

In the $\hat{x}_i$ system, the sphere moves as $U - v_x^\infty$. Now, imagine a potential flow in the $\hat{x}_i$ system. The velocities will be related to the real flow by 19.12.14, and the pressures by 19.12.5 modified for this situation to

$$\hat{p} = p + \rho \hat{x}_x \frac{dv_x^\infty}{d\hat{t}} \qquad 19.12.15$$

The surface pressures in the imaginary flow must integrate to give the added-mass effect. Hence,

$$\tfrac{1}{2} M_f \frac{d(U - v_x^\infty)}{dt} = \int_{S^+} n_x \hat{p}\, dS \qquad 19.12.16$$

The final differential equation is obtained by multiplying 19.12.15 by n_x and integrating over S_b, combining with 19.12.16, and inserting the result into 19.12.13. These steps produce

$$M_0 \frac{dU}{dt} = -\tfrac{1}{2} M_f \frac{d}{dt}(U - v_x^\infty) + M_f \frac{dv_x^\infty}{dt} \qquad 19.12.17$$

In the ground system x_i, the last term is sometimes interpreted as a buoyancy force caused by an acceleration dv_x^∞/dt of the fluid at infinity.

Equation 19.12.17 may be rearranged and solved to produce

$$U = \frac{3\rho_f}{\rho_f + 2\rho_0} v_x^\infty \qquad 19.12.18$$

This result is actually valid for an arbitrary time function $v_x^\infty(t)$. For the limiting case of a heavy particle in a light fluid we have $U = 0$; for a particle and fluid of the same density, $U = v_x^\infty$, and for a light particle in a liquid, $U = 3v_x^\infty$. A bubble in a liquid will oscillate with three times the amplitude of the liquid oscillation. The reason for this is seen as follows: a certain pressure gradient is needed in the liquid for acceleration. This same gradient acts on the bubble, which has almost no mass of its own, but does have an added mass. Since the added mass of the bubble is only one-half that of the fluid in an equivalent volume, the bubble responds with greater acceleration.

Added-mass effects are usually associated with liquids, because the density is large. In air the effect is usually less important, but upon occasion it may be significant. This is especially true if the forces are small, as in a loudspeaker, or the object is very large, such as a parachute. Added mass has been included in this section because the easily worked examples are all ideal-flow problems. The principle that accelerating a body requires extra surface pressure to accelerate the fluid in the neighborhood of the body applies to any flow, however. What distinguishes ideal flow is that the extra force is simply related to the acceleration and the geometry, so that it may be related to the mass of a certain volume of fluid added to the mass of the object. Only in this special case is the result so simple.

19.13 CONCLUSION

Three-dimensional ideal flows have many of the same characteristics that plane two-dimensional flows have. One significant difference is in the model of thin wakes and shear layers. In plane flow a shear layer is idealized as a vortex sheet where the tangential velocity jumps and the pressure is constant. On the other hand, a plane two dimensional airfoil wake, because the net vorticity is zero, has no jump in tangential velocity, and hence the inviscid streamlines have exactly the same properties on either side of the wake. Three-dimensional wakes are quite different in that the pressure and magnitude of the velocity can be the same on both sides while the direction of the velocity vectors changes. These wakes are like a plane wake in one direction and a plane shear layer in the other direction. Ideal-flow models of three-dimensional wakes must contain a surface with a longitudinal vortex sheet. Disregarding this singular surface leads to D'Alembert's paradox: the drag is predicted to be zero.

The induced drag of a lifting body comes from growth in the length of a three dimensional wake. The work done by the drag force increases the energy of the flow by making the wake longer. The ultimate destruction of this energy is, of course, a viscous effect.

The Kutta condition was used in plane flows to select the proper circulation constant for the flow and to select a unique flow pattern. The same difficulty exists in three-dimensional flows. We must specify where the wake and the tip vortices leave a wing in order to have a unique solution.

PROBLEMS

19.1 (B) An ideal flow exists above a small hole in a flat wall. The flow consists of a line vortex with the axis going through the hole and perpendicular to the wall plus a sink flow into the hole. Find the velocity and pressure on the wall as a function of the distance and the strength parameters of the flow.

19.2 (B) Compute the surface speed and pressure coefficient on a Rankine nose as a function of distance along the surface.

19.3 (A) A smooth nose shape with an open separation region at the shoulder experiences a drag force. According to 19.9.2 there must be an increase in the energy of the flow. How does this occur?

19.4 (C) Consider a geometrically smooth nose shape where the flow separates from the shoulders and a turbulent region extends back along the body. How does 19.9.2 apply to determine the pressure drag on the nose? Next, consider a case with a thin closed separation bubble near the shoulder. What is the pressure drag for this situation?

19.5 (A) A flat plate is moved according to $U = At$ normal to its plane. What is the initial drag?

19.6 (B) A bubble of radius r_0 is in a tank of water. The take is moved according to $U = At$. What is the motion of the bubble?

19.7 (C) A hydrogen filled balloon 0.5 m in diameter is held by a 1 m string in a bus. The bus goes around a 90° turn with a 500 ft radius at 40 mi/hr. What is the motion of the balloon?

20 Boundary Layers

Boundary layers are thin regions in the flow where viscous forces are important. Although the name "boundary layer" originally referred to the layer of fluid next to the wall, we may also apply the term to a jet or a thin shear layer between two streams of different velocities. The essential ideas are that the layer is thin in the direction across the streamlines and that viscous stresses are important only within the layer.

In Chapter 17, where the physical motivation and derivation of the boundary-layer equations was given, we found that boundary layers are a high-Reynolds-number phenomenon. They constitute a correction to the main inviscid flow, which does not meet the no-slip condition at the wall. The boundary layer and the inviscid flow are coupled together through a boundary condition. The inviscid-flow velocity along the wall at any x wall position, $u_{\text{invis}}(x)$, is equal to the boundary-layer velocity $u_e = u(x, y \to \infty)$ as the distance y from the surface, measured in boundary-layer coordinates, becomes infinite. The geometry of the wall does not directly enter into the boundary-layer analysis. It does, of course, affect the inviscid flow that determines u_{invis} and thus, through $u_e(x)$, the boundary layer itself.

20.1 BLASIUS FLOW OVER A FLAT PLATE

The simplest boundary layer one can imagine forms when a thin flat plate is placed in a uniform stream of velocity u_0 so that it is perfectly aligned with the streamlines. Such a plate does not cause any disturbance in the inviscid flow; hence the inviscid velocity at all points of the surface has the constant value u_0. We take a coordinate system with the origin at the leading edge of the plate as shown in Fig. 20-1. It turns out that the length of the plate is not important, and for the present we assume that the plate extends to infinity.

The boundary conditions for the boundary-layer equations require that we match the inviscid stream,

$$u(x \geq 0, y \to \infty) = u_e(x) = u_0 \qquad 20.1.1$$

and that the no-slip condition is met,

$$u(x > 0, y = 0) = 0 \qquad 20.1.2$$

$$v(x > 0, y = 0) = 0 \qquad 20.1.3$$

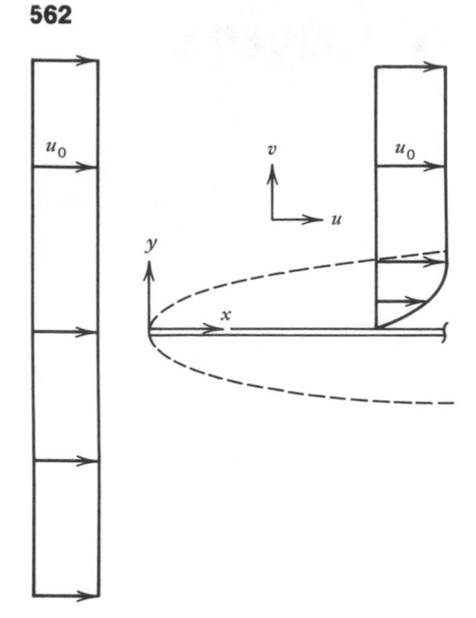

Figure 20-1 Blasius flow: a flat plate in a uniform stream.

In addition, we need to specify an initial condition. For a sharp leading edge the appropriate assumption is

$$u(x = 0, y > 0) = u_0 \qquad 20.1.4$$

The leading edge itself is a mathematical discontinuity in u; if we approach $x = 0$, $y = 0$ along the surface of the plate, we find $u = 0$, but if we approach from the free stream (i.e., $x = 0$, $y \to 0$), we find $u = u_0$.

The boundary-layer equations when u_e is a constant are

$$\frac{\partial u}{\partial x} + \frac{\partial v}{\partial y} = 0 \qquad 20.1.5$$

$$u\frac{\partial u}{\partial x} + v\frac{\partial u}{\partial y} = \nu\frac{\partial^2 u}{\partial y^2} \qquad 20.1.6$$

The problem consisting of 20.1.1 through 20.1.6 was solved first by H. Blasius (1908), while he was a student of Prandtl's.

Blasius's problem is a nonlinear partial differential equation that does not have a simple-closed form answer. The successful approach is to look for a similarity solution. The fact that neither x nor y has a natural measuring scale in the boundary data leads us to suspect that the solution $u(x, y)$ does not depend upon x and y separately but only upon some combination of them. Simple dimensional analysis of the function $u(x, y; u_0, \nu)$ does not lead us to a similarity variable, so we try another approach using a physical argument. Assume for a moment that we want to solve the problem only out to a certain distance L on the plate. Then the variable $x^* = x/L$ will be the proper nondimensional variable for this direction. As we well know, the y direction is dominated by a diffusion process, so, according to Rayleigh's argument for viscous diffusion, we expect the nondimensional y distance to be scaled by the diffusion depth at $x = L$, that is,

$$y^* = \frac{y}{\delta} = \frac{y}{\sqrt{\nu L/u_0}}$$

Since L could be any position and was artificially introduced into the problem, we seek a combination of x^* and y^* that will eliminate L. Hence, we define the similarity variable as

$$\eta \equiv \frac{y^*}{\sqrt{x^*}} = \frac{y}{\sqrt{\nu x/u_0}} \qquad 20.1.7$$

The natural scale for y is the diffusion length at a distance x down the plate.

The streamwise velocity u is nondimensionalized by u_0:

$$u^* \equiv \frac{u}{u_0} \qquad 20.1.8$$

Observe that this definition makes all the boundary conditions on u (20.1.1, 20.1.2, and 20.1.4) into pure numbers. It is customary in boundary-layer work to denote the nondimensional stream function by f. For the problem at hand, the proper nondimensional form is

$$f(\eta) \equiv \frac{\psi}{\sqrt{\nu x u_0}} \qquad 20.1.9$$

With 20.1.7 and 20.1.8 it can be verified that the stream-function relation $u = \partial\psi/\partial y$ becomes

$$u^* = \frac{df}{d\eta} = f'(\eta) \qquad 20.1.10$$

The continuity equation will be satisfied exactly if we relate v to the stream function. Since

$$v = -\frac{\partial \psi}{\partial x}$$

We insert 20.1.9 for ψ and use chain rules along with 20.1.7 to show that

$$v^* \equiv \frac{v}{\sqrt{\nu u_0/x}} = \tfrac{1}{2}[\eta f' - f] \qquad 20.1.11$$

The equation above may be considered as the natural way to nondimensionalize v.

When the relations 20.1.10 and 20.1.11 are inserted into the momentum equation 20.1.6, we find that an ordinary differential equation results.

$$f''' + \tfrac{1}{2}ff'' = 0 \qquad 20.1.12$$

A successful similarity variable not only reduces the partial differential equation; it also must make the boundary conditions collapse in an appropriate way. In terms of $f(\eta)$, the boundary conditions are evaluated using 20.1.10 and 20.1.11. They are

$$\begin{aligned} u(x, y = 0) = 0 \quad &\Rightarrow \quad f'(\eta = 0) = 0 \\ v(x, y = 0) = 0 \quad &\Rightarrow \quad f(\eta = 0) = 0 \\ u(x, y \to \infty) = u_e \quad &\Rightarrow \quad f'(\eta \to \infty) = 0 \\ u(x = 0, y) = u_e \quad &\Rightarrow \quad f'(\eta \to \infty) = 0 \end{aligned} \qquad 20.1.13$$

The last two conditions collapse to give the same boundary condition in terms of $f(\eta)$. We still have consistency between the differential equation to be solved and the number of boundary conditions that are applied.

Blasius's problem is a nonlinear, two-point boundary-value problem and may be solved by standard techniques such as the RK-34 routine of Appendix E. Figure 20-2 gives the results of a computation using RK-34 and a value of $f''(0) = 0.33206$.

As a matter of interest, the thickness of the boundary layer, taken as the point where the u velocity becomes $0.99u_0$, occurs at $\eta = 4.9$. Hence at $y = \delta_{99}$, the definition 20.1.7 yields

$$\frac{\delta_{99}}{x} = 4.9\sqrt{\frac{\nu}{xu_0}} = 4.9\,\mathrm{Re}_x^{-1/2} \qquad 20.1.14$$

$$\mathrm{Re}_x \equiv \frac{u_0 x}{\nu} \qquad 20.1.15$$

Typical values of δ_{99} are: for a stream of water at 1 m/sec with x at 1 meter from the leading edge, $\mathrm{Re}_x \approx 10^6$ and $\delta_{99} = 0.5$ cm; for air under the same conditions, $\mathrm{Re}_x \approx 6.7 \times 10^4$ and $\delta_{99} = 1.9$ cm. Even for these modest velocities, the boundary layers are very thin. The formula above is valid only for $\mathrm{Re} < 3 \times 10^6$, because the flow becomes unstable for higher Reynolds numbers and transition to a turbulent boundary layer occurs.

Another result of practical importance is the wall friction. The friction coefficient is found from $f''(0)$. It is

$$C_f \equiv \frac{\tau_0}{\frac{1}{2}\rho u_0^2} = 2f''(0)\,\mathrm{Re}_x^{-1/2} = \frac{0.664}{\mathrm{Re}_x^{1/2}} \qquad 20.1.16$$

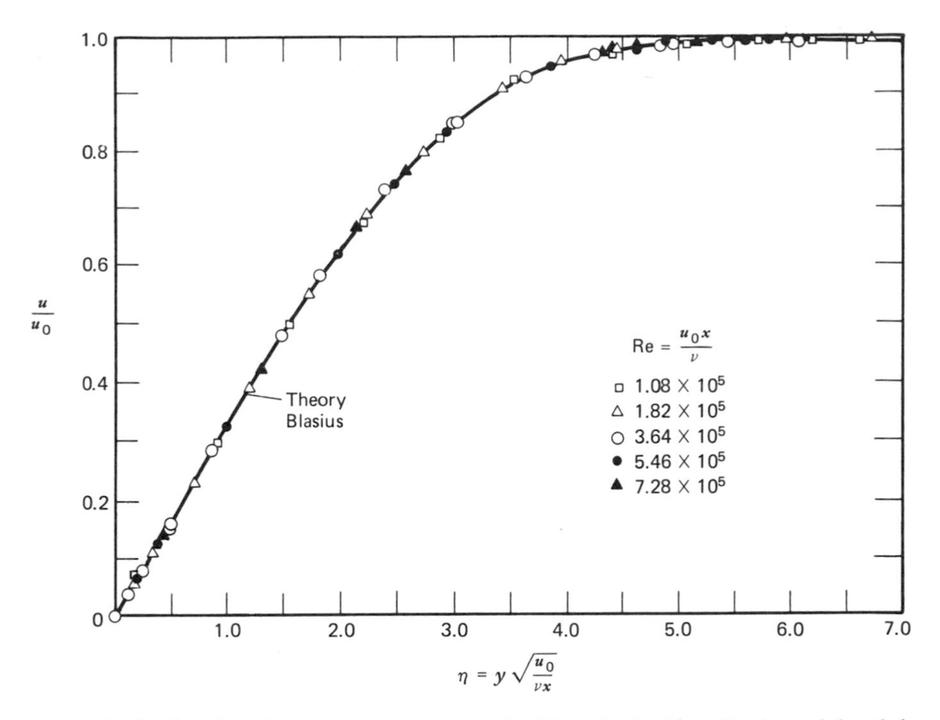

Figure 20-2 Results of experiments compared with analysis. The effective origin of the plate is found from data at two different positions. The graph adapted from Schlichting (1950). Nikuradse (1942) made the original measurements.

Note that C_f compares the wall stress with the dynamic pressure of the flow. For a practical Reynolds number of 10^4 the shear stress is 0.0066 times the dynamic pressure. As a general rule, wall shear stresses in fluid flows are roughly 1% of the dynamic pressure (recall that the dynamic pressure indicates the range over which the normal pressure force varies on a body). This statement holds true for turbulent boundary layers as well. This is the reason that friction forces may be ignored in many practical cases. Only for large surfaces or at low Reynolds numbers does the accumulated effect of wall shear stress compare with the pressure forces.

The drag force on a plate of length L may be found by integrating the friction coefficient 20.1.16 from $x = 0$ to $x = L$. The result is

$$C_D = \frac{F_b}{\frac{1}{2}\rho u_0^2 L} = \frac{1}{L}\int_0^L C_f(x)\,dx$$

$$= 1.338\,\text{Re}_L^{-1/2} \qquad 20.1.17$$

Although the skin friction is infinite at $x = 0$, this singularity is $\sim x^{-1/2}$ and integrates to give a finite force on the plate.

The vertical velocity v within the boundary layer is very small. To compare v with u_0 we rearrange 20.1.11 into

$$\frac{v}{u_0} = \tfrac{1}{2}\,\mathrm{Re}_x^{-1/2}[\eta f' - f] \qquad 20.1.18$$

As we move out in the boundary layer ($\eta \to \infty$), the factor in brackets takes on a constant value, so that

$$\eta \to \infty, \qquad \frac{v}{u_0} \sim 0.861\,\mathrm{Re}_x^{-1/2}$$

At first it might appear unusual that v does not go to zero as $y \to \infty$. This will be explained in Section 20.7.

The vorticity in the boundary layer is dominated by the velocity profile $u(y)$. From the definition of vorticity we find for $\omega_z = \omega$

$$\omega = -\frac{\partial u}{\partial y} + \frac{\partial v}{\partial x} \qquad 20.1.19$$

Converting this expression into boundary-layer variables where ω is scaled by u_0/δ yields

$$\omega^* \equiv \frac{\omega}{u_0}\sqrt{\frac{\nu x}{u_0}} = -\frac{df'}{d\eta} + \frac{1}{\mathrm{Re}_x}[f - \eta f' - \eta^2 f'']$$

$$\omega^* = f''(\eta) + O[\mathrm{Re}^{-1}] \qquad 20.1.20$$

The total amount of vorticity at any x position of the boundary layer is found by integrating 20.1.20. The result is

$$\int_0^\infty \omega^*\,d\eta = -f'(\infty) + f'(0) = -1$$

The constant value means that no new vorticity is entering the flow. This agrees with our ideas that a vorticity flux enters the flow along a wall only when a pressure gradient exists along the wall.

Boundary-layer theory fails at a sharp leading edge. There is some small region in the neighborhood of $x = 0$ where the distance down the plate is about the same as the distance over which upstream diffusion of viscous effects can occur. In this region the assumptions made to derive the boundary-layer equations are invalid. Fortunately, for engineering purposes this region is very small and does not need to be taken into account in computing the drag. However, one effect that does remain in the flow is an apparent shift in the origin of the boundary layer. The geometric origin of the plate and the apparent origin of the boundary layer do not exactly coincide.

It is usually said that the plate has a sharp leading edge. This is an idealization, of course. It might be better to imagine a plate of finite thickness h and a well-rounded leading edge. From our study of the inviscid flow over a rounded nose shape in Chapter 18 we know that the flow would stagnate to zero velocity at the front, accelerate to a high velocity around the shoulder, and finally approach u again as we moved a few h downstream from the nose. The corresponding pressures on the surface are the stagnation-point value, a value below the free-stream value, and finally a gradual buildup back to the free-stream value (cf. Section 18.4). Through Chapter 13 we have come to expect that a vorticity flux will be produced at a surface wherever there is a pressure gradient along the surface. Hence, the nose region is a strong source of vorticity; the top has vorticity of one sign and the bottom has vorticity of the opposite sign. As we let h become small, the nose of the plate becomes a doublet of vorticity flux. Since the downstream portion of the plate has zero pressure gradient, the vorticity within the layer originates entirely at the leading edge.

20.2 DISPLACEMENT THICKNESS

The thickness δ of a boundary layer, defined as the point where the u velocity becomes equal to 99% of $u_e(x)$, is useful in gauging the influence of viscous diffusion. Another thickness of importance is the *displacement thickness* δ^*. Consider the boundary layer shown in Fig. 20-3a. Let the streamline ψ_0 be located at y_0 in the inviscid flow, where the velocity is $u_e(x)$. Next, we imagine a completely inviscid flow, Fig. 20-3b, where the velocity u_e extends downward but that the flow ends at $y = \delta^*$, the position where the boundary layer displaces the ideal flow so that the same mass flow is obtained. Mathematically δ^* is defined (for compressible flow the density is included, with ρ_e denoting the free-stream value) by

$$\int_0^{y_0} \rho u \, dy = \int_{\delta^*}^{y_0} \rho_e u_e \, dy \qquad 20.2.1$$

Although the physical meaning of the equation above is clear, this is not a convenient form. Note that since $u_e(x)$ does not depend on y, the following holds:

$$\rho_e u_e \delta^* = \int_0^{\delta^*} \rho_e u_e \, dy$$

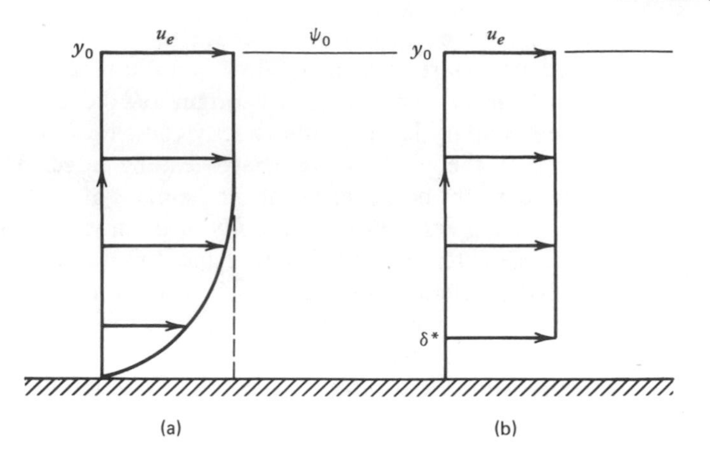

Figure 20-3 Displacement thickness. The inviscid flow above the boundary layer shown in (a) would reach to the position δ^* if it were continued toward the wall until the same flow rate was achieved.

Adding this equation to 20.2.1 and rearranging produces a definition for δ^*,

$$\delta^* = \int_0^{y_0} \left(1 - \frac{\rho u}{\rho_e u_e}\right) dy \qquad 20.2.2$$

Of course, for incompressible flow the density cancels out. The upper limit is replaced by $y_0 \to \infty$, with the understanding that u/u_e is the velocity profile that occurs in the boundary layer.

The displacement thickness tells us the effect of the boundary layer upon the inviscid flow. As a first approximation, the inviscid flow slips past a wall coinciding with the actual wall. The displacement thickness indicates how much equivalent inviscid fluid has been taken into the boundary layer. Since the boundary layer pushes the inviscid flow outward, a second approximation, if we were to require one, would be to imagine that the inviscid flow occurs over a body that has the displacement thickness added to each wall position. A completely inviscid flow between $y = \delta^*$ and $\psi(y_0)$ has the same mass flow as the actual flow has between $y = 0$ and the same location.

The exact nature of the velocity profile determines the displacement thickness, as the definition 20.2.2 shows. For the Blasius profile of the previous section, the displacement thickness is

$$\delta^* = 1.72\sqrt{\frac{\nu x}{u_0}} \qquad 20.2.3$$

Thus, δ^* is roughly $\frac{1}{3}$ of the 99% thickness δ.

20.3 KÁRMÁN MOMENTUM INTEGRAL

In addition to the displacement thickness, boundary layers have another thickness called the *momentum thickness*. It is defined as

$$\Theta \equiv \int_0^{\infty} \frac{u}{u_e}\left(1 - \frac{u}{u_e}\right) dy \qquad 20.3.1$$

and occurs in von Kármán's integral of the boundary-layer equations.

Let us proceed to derive the momentum integral by integrating the boundary-layer momentum equation 17.4.25 from $y = 0$ to $y = h$. The position $y = h$ is outside the boundary layer, where $u \approx u_e$ and where all derivatives $\partial u/\partial y$, $\partial^2 u/\partial y^2, \ldots$ are zero. The equation is

$$\int_0^h \left(u\frac{\partial u}{\partial x} + v\frac{\partial u}{\partial y}\right) dy = \int_0^h u_e \frac{du_e}{dx} dy + \int_0^h \frac{\mu}{\rho}\frac{\partial^2 u}{\partial y^2} dy$$

Integrating the shear stress term and rearranging produces

$$\int_0^h \left(u\frac{\partial u}{\partial x} - u_e\frac{du_e}{dx} + v\frac{\partial u}{\partial y}\right) dy = \frac{\mu}{\rho}\frac{\partial u}{\partial y}\bigg|_0^y = -\frac{\tau_0}{\rho} \qquad 20.3.2$$

The wall shear stress is denoted as τ_0. Now for a solid wall where $v(y = 0) = 0$, we may express v at any position y by integrating the continuity equation:

$$v = \int_0^y \frac{\partial v}{\partial y} dy' = -\int_0^y \frac{\partial u}{\partial x} dy'$$

The third term in 20.3.2, if we insert the above expression for v, becomes

$$\int_0^h v\frac{\partial u}{\partial y} dy = \int_0^h \left(-\int_0^y \frac{\partial u}{\partial x} dy'\right)\frac{\partial u}{\partial y} dy$$

This integral is of the form $\int w\, dz = wz - \int z\, dw$, where

$$w = \int_0^y \frac{\partial u}{\partial x} dy' \quad \Rightarrow \quad dw = -\frac{\partial u}{\partial x} dy$$

and

$$dz = \frac{\partial u}{\partial y} dy \quad \Rightarrow \quad z = u$$

Hence, integration by parts gives

$$\int_0^h v\frac{\partial u}{\partial y} dy = -u_e\int_0^h \frac{\partial u}{\partial x} dy + \int_0^h u\frac{\partial u}{\partial x} dy$$

With this, 20.3.2 becomes

$$\int_0^h \left(u\frac{\partial u}{\partial x} - u_e\frac{du_e}{dx} - u_e\frac{\partial u}{\partial x} + u\frac{\partial u}{\partial x} \right) dy = -\frac{\tau_0}{\rho}$$

Next, we add and subtract $u(du_e/dx)$ and rearrange the terms into the form

$$-\int_0^h \left[u\left(\frac{du_e}{dx} - \frac{\partial u}{\partial x} \right) + (u_e - u)\frac{\partial u}{\partial x} \right] dy - \int_0^h (u_e - u)\frac{du_e}{dx}\, dy = -\frac{\tau_0}{\rho}$$

or equivalently

$$\frac{d}{dx}\left\{ u_e^2 \int_0^h \frac{u}{u_e}\left(1 - \frac{u}{u_e} \right) dy \right\} + u_e\frac{du_e}{dx}\int_0^h \left(1 - \frac{u}{u_e} \right) dy = \frac{\tau_0}{\rho}$$

As $h \to 0$ the integrals are by definition the displacement and momentum thicknesses; thus

$$\frac{d}{dx}(u_e^2\Theta) + u_e\frac{du_e}{dx}\delta^* = \frac{\tau_0}{\rho} \qquad 20.3.3$$

This is the integral momentum equation originally given by von Kármán (1921) using physical arguments, and subsequently derived mathematically by Pohlhausen (1921).

The momentum thickness Θ, when multiplied by ρu_e, gives an indication of the momentum of the boundary-layer flow. As a typical example of momentum thickness, one can compute Θ from the definition 20.3.1 using the Blasius flat-plate profile. The answer is

$$\Theta = 0.664\sqrt{\frac{\nu x}{u_0}}$$

Very roughly, Θ is one-third of δ^*, which in turn is roughly one-third of δ. These ratios depend upon the shape of the velocity profiles, of course. Only in special cases, such as the Blasius problem, are they constant.

20.4 KÁRMÁN–POHLHAUSEN APPROXIMATE METHOD

In the early days of boundary-layer theory the electronic computer was not available and the integral method of solving boundary-layer problems became highly developed. The Kármán–Pohlhausen method was one of the first techniques used; a comprehensive survey of various methods and comparison of their accuracy is given in Rosenhead (1964). Many complicated problems, such as turbulent flows, heat-transfer problems, and combustion problems, are still solved using methods of the integral type. Because of these applications in other areas, it is useful to study how the method is applied to laminar flows.

An essential part of an integral method is an assumption for the form of the velocity profile. Pohlhausen assumed a fourth-order polynomial. For our example we take the cubic

$$u^* \equiv \frac{u}{u_e} = a + b\eta + c\eta^2 + d\eta^3 \tag{20.4.1}$$

where

$$\eta = \frac{y}{\delta(x)} \tag{20.4.2}$$

The profile must meet several boundary conditions. At the wall: the no-slip condition, the momentum equation, and the derivative of the momentum equation with respect to y require that

$$u = 0, \qquad -u_e\frac{du_e}{dx} = \nu\left.\frac{\partial^2 u}{\partial y^2}\right|_0, \qquad \left.\frac{\partial^3 u}{\partial y^3}\right|_0 = 0, \ldots \tag{20.4.3}$$

A smooth approach of u to u_e is enforced at the finite position $y = \delta$. This gives

$$u = u_e, \qquad \left.\frac{\partial u}{\partial y}\right|_\delta = 0, \qquad \left.\frac{\partial^2 u}{\partial y^2}\right|_\delta = 0, \ldots \tag{20.4.4}$$

For a specific boundary layer where $u_e(x)$ is given, the conditions 20.4.3 and 20.4.4 together with the momentum integral 20.3.3 allow one to find the coefficients a, b, c, and d in the assumed profile as functions of x. As profiles with more coefficients are introduced, more smoothness is required in the approach of u to u_e at $y = \delta$; that is, the higher-order derivatives $\partial^n u/\partial y^n|_\delta$ are required to be zero.

As a relatively simple example let us solve the Blasius problem where $u_e = u_0$, a constant. Applying the first two boundary conditions of 20.4.3 and the first two of 20.4.4 shows that the profile is

$$u^* = \tfrac{3}{2}\eta - \tfrac{1}{2}\eta^3 \tag{20.4.5}$$

Inserting 20.4.5 into the definition of the displacement thickness yields

$$\delta^* = \delta\int_0^1 (1 - u^*)\, d\eta = \tfrac{3}{8}\delta \tag{20.4.6}$$

The same process applied to the definition of the momentum thickness produces

$$\Theta = \delta\int_0^1 u^*(1 - u^*)\, d\eta = \tfrac{117}{840}\delta \tag{20.4.7}$$

We may also use 20.4.5 to find the wall shear stress,

$$\frac{\tau_0}{\rho} = \nu \left.\frac{\partial u}{\partial y}\right|_0 = \nu \frac{3}{2} \frac{u_0}{\delta} \qquad 20.4.8$$

Now we are in a position to substitute for all terms in the von Kármán momentum integral. Equation 20.3.3 for $u_e = u_0$ becomes

$$u_e^2 \frac{d\Theta}{dx} = \frac{\tau_0}{\rho}$$

$$u_0^2 \frac{117}{840} \frac{d\delta}{dx} = \frac{3}{2} \frac{\nu u_0}{\delta}$$

Integration gives

$$\delta = \sqrt{\frac{840}{39}} \sqrt{\frac{\nu x}{u_0}} = 4.64 \sqrt{\frac{\nu x}{u_0}} \qquad 20.4.9$$

With $\delta(x)$ known, all items in the profile 20.4.5 have been determined. The coefficient in 20.4.9 compares well with the exact value of 4.9 in 20.1.14.

In this problem with $u_e = u_0$, the shape of the profile was constant; only the thickness δ changed with x. In the general case where $u_e(x)$ is given by a pressure gradient, the profile will contain a shape parameter, which also changes with distance along the flow.

20.5 FALKNER–SKAN SIMILARITY SOLUTIONS

In the stagnation-point flow (Hiemenz flow) or the streaming flow over a flat plate (Blasius flow), we have seen how similarity methods combine two independent variables into one. Falkner and Skan (1931) investigated the boundary-layer equations to see what specific types of external flows $u_e(x)$ would allow similarity solutions.

We start the derivation of the Falkner–Skan solutions by assuming that a scaling function $u_e(x)$ makes the velocity profile similar:

$$\frac{u(x, y)}{u_e(x)} = f'(\eta) \qquad 20.5.1$$

where $\delta(x)$ is a scaling function for y; that is,

$$\eta = \frac{y}{\delta(x)} \qquad 20.5.2$$

At this stage $u_e(x)$ and $\delta(x)$ are undetermined functions. We analyze the boundary-layer equations to determine $u_e(x)$ and $\delta(x)$. Now, at any x location the stream function is

$$\psi = \int_0^y \frac{\partial \psi}{\partial y} dy = \delta u_e \int_0^\eta \frac{u}{u_e} d\left(\frac{y}{\delta}\right) = u_e \delta \int_0^\eta f' \, d\eta$$

Since $\psi = 0$ at $y = 0$ for all x, we have $f(0) = 0$ and

$$\frac{\psi}{u_e \delta} = f(\eta) \tag{20.5.3}$$

An expression for the vertical velocity v is found by using its relation to the stream function and applying chain rules to 20.5.3. The result is

$$v = -\frac{\partial \psi}{\partial x} = -f \cdot [u_e \delta]' + \eta u_e f' \delta' \tag{20.5.4}$$

The primes on u_e and δ stand for differentiation with respect to x. When we employ the stream function to compute v, we have implicitly satisfied the continuity equation.

The momentum equation for a steady boundary-layer flow is

$$u \frac{\partial u}{\partial x} + v \frac{\partial u}{\partial y} = u_e u_e' + \nu \frac{\partial^2 u}{\partial y^2}$$

Performing the indicated operations on 20.5.1 and using 20.5.4 gives the momentum equation as

$$u_e f' \left[u_e' f' - \frac{\eta}{\delta} u_e \delta' f'' \right] + \left[\eta u_e \delta' f' - f \cdot (\delta u_e)' \right] \frac{u_e}{\delta} f'' = u_e u_e' - \frac{\nu u_e}{\delta^2} f''' \tag{20.5.5}$$

This equation may be organized into the following form:

$$f''' + \alpha f f'' + \beta (1 - f'^2) = 0 \tag{20.5.6}$$

where, for convenience, the coefficients α and β are defined by

$$\alpha = \frac{\delta}{\nu} \frac{d}{dx} (u_e \delta) \tag{20.5.7}$$

$$\beta = \frac{\delta^2}{\nu} \frac{du_e}{dx} \tag{20.5.8}$$

Equation 20.5.6 will be an ordinary differential equation for $f(\eta)$ only if α and β do not depend upon x but are constant. Thus, the equations above furnish relations that determine those $u_e(x)$ and $\delta(x)$ functions that result in self-similar boundary layers. The solutions of the coupled equations 20.5.7 and 20.5.8 is made easy if we recognize that

$$2\alpha - \beta = \frac{1}{\nu}\frac{d}{dx}(\delta^2 u_e)$$

Hence

$$(2\alpha - \beta)(x - x_0) = \frac{1}{\nu}\delta^2 u_e \tag{20.5.9}$$

The free constants in 20.5.9 are chosen as follows. We may set $x_0 = 0$; then $x = 0$ is the place where either $\delta = 0$ or $u_e = 0$. Solving 20.5.9 yields

$$\delta = \sqrt{\frac{(2\alpha - \beta)\nu x}{u_e}} \tag{20.5.10}$$

The common situation is for u_e and x both to be positive; however, in certain situations x and u_e have opposite signs. We choose $2\alpha - \beta = 1$ or -1 according as u_e and x have the same or opposite signs. Note that any value of $2\alpha - \beta$ is acceptable. If y/δ is a similarity variable, then $y/C\delta$ is also a similarity variable for any constant C. Thus, 20.5.10 becomes

$$\delta = \sqrt{\pm\frac{\nu x}{u_e}} \tag{20.5.11}$$

[This is the original definition of δ. Several later papers use $\alpha = 1$, so that their δ is $(2 - \beta)^{1/2}$ times 20.5.11.]

The external velocity $u_e(x)$ is found by inserting 20.5.10 into 20.5.8. This gives

$$\beta = \pm\frac{x}{u_e}\frac{du_e}{dx} \tag{20.5.12}$$

Integrating produces

$$u_e = u_0\left(\frac{x}{L}\right)^m \tag{20.5.13}$$

where the arbitrary constants u_0 and L have the same signs as u and x. The exponent m is

$$m = \begin{Bmatrix} \beta \\ -\beta \end{Bmatrix} \quad \text{if} \quad u_e \text{ and } x \text{ have} \begin{cases} \text{the same sign} \\ \text{opposite signs} \end{cases}$$

Self-similar boundary layers occur when the external velocity is the simple power law 20.5.13. The similarity variable for these flows is

$$\eta = \frac{y}{\delta} = \frac{y}{\sqrt{\pm\dfrac{\nu x}{u_e}}} = \frac{y}{\sqrt{\pm\dfrac{\nu L}{u_0}\left(\dfrac{x}{L}\right)^{m+1}}} \qquad 20.5.14$$

The equation that governs the stream function is 20.5.6 with α and β eliminated in favor of m. When u_e and x have the same sign, this equation is

$$f''' + \tfrac{1}{2}(m+1)ff'' + m(1 - f'^2) = 0 \qquad 20.5.15$$

This is known as the Falkner–Skan equation. The two arbitrary constants α and β in 20.5.6 have been reduced to one constant m by fixing the scale for the function $\delta(x)$.

For flow over a solid wall we require $u = 0$, $v = 0$ at the wall and $u = u_e$ at $\eta \to \infty$. Equations 20.5.1 and 20.5.3 show that these conditions imply that

$$\begin{aligned} f(0) &= 0 \\ f'(0) &= 0 \\ f'(\infty) &= 1 \end{aligned} \qquad 20.5.16$$

We refrain from specifying an initial velocity profile at $x = x_1$, as this would overdetermine the problem. Whatever profile comes out of the solution is the initial profile required for similarity. [If $x_1 = 0$, this fact is hidden by the presence of a singular point in the $\eta(x, y)$ transformation.] The Falkner–Skan equation 20.5.14 is a nonlinear differential equation, so we cannot easily know what values of m will give a solution, and if that solution, once obtained, is unique.

Several typical velocity profiles for different values of m are given in Fig. 20-4. As long as $m > 0$ the solutions are known to exist, and they are also unique. As far as the boundary layer is concerned, the details of how the inviscid flow produces $u_e(x)$ are inconsequential. Nevertheless, it is useful to identify several simple ideal flows that lead to Falkner–Skan boundary layers. For $m > 0$ they are:

- $m = 0$: Blasius flow over a flat plate with a sharp leading edge. Also the local flow at any cusp leading edge.
- $0 < m < 1$: Flow over a wedge with half angle $\theta_{1/2} = m\pi/(m+1)$ with $0 < \theta_{1/2} < \pi/2$.
- $m = 1$: Hiemenz flow toward a plane stagnation point.
- $1 < m < 2$: Flow into a corner with $\theta_{1/2} > \pi/2$. A flow of this type may be difficult to produce experimentally.
- $2 < m$: No corresponding simple ideal flow.

In all of the above cases there is one unique boundary layer profile.

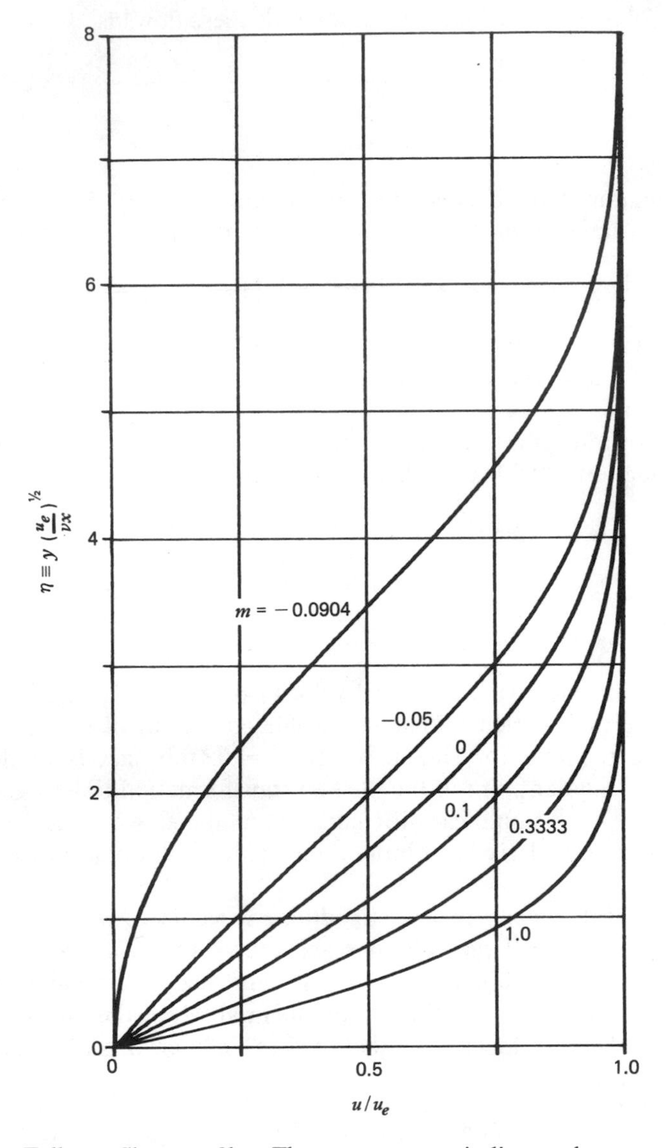

Figure 20-4 Falkner–Skan profiles. The parameter m indicates the external velocity variation through $u_e = u_0 x^m$.

Many people have contributed to the classification and computation of Falkner–Skan flows. Rosenhead (1964) contains a good summary by C. W. Jones and E. J. Watson. From these works some of the complicated behavior at negative values of m can be pieced together. When $-0.0904 < m < 0.0$, there are an infinite number of solutions for each value of m (Fig. 20-4). However, not all of these solutions are physically acceptable. One of the main arguments in establishing boundary-layer theory is that the viscous effects are confined to a thin region near the wall. In light of this fact people have proposed that a boundary layer should approach the free stream exponentially:

$$1 - u^* = 1 - f' \sim Ae^{-B\eta} \qquad \text{as} \quad \eta \to \infty$$

If this condition is applied there are only two known acceptable solutions for each m in the range $-0.0904 < m < 0$. One of the solutions has $u > 0$ for all η; while the other has the interesting characteristic that there is backflow for a small region near the wall [Stewartson's (1954) reverse-flow profiles]. When m is exactly equal to -0.0904, only one solution exists. This profile has zero shear stress at the wall and therefore is on the verge of separating for all x.

For $-1 < m < -0.0904$, all solutions for a given m tend to oscillate about $f' = 1$ as η becomes infinite. At each value of m, one of these solutions has just one region where the velocity $f' > 1$, and then $f \to 1$ exponentially. Because a laminar boundary layer with these supervelocities may be difficult to produce experimentally, some workers reject these solutions as physically impossible.

The case $m = -1$ with u_e and x having opposite signs, $u = -a_0/x$ $(u_0 = -a_0)$, represents a solid wall in the flow field of an ideal line sink. When two walls are present the problem represents the flow into a wedge. The differential equation in this case has an exact closed-form solution. On the other hand, the equivalent problem with the sign changed so the flow comes from a source ($m = -1$ and $u_e = +u_0/x$) has no solution. This means that boundary-layer theory does not produce a similarity solution for flows in a flat-wall diffuser. These flows require a nonsimilar solution.

Most of the complicated behavior in Falkner–Skan solutions happens when m is between -1 and 0. When $m < -1$ we again find a unique solution. All solutions in the range $m < -1$ have the flow going from large x toward $x = 0$. Hence, the flows are strongly accelerated with $u = -a_0(x/L)^m$, $m < -1$.

20.6 ARBITRARY TWO-DIMENSIONAL LAYERS: CRANK–NICOLSON DIFFERENCE METHOD

An arbitrary inviscid flow over a wall provides us with $u_e(x)$ as an external velocity. When $u_e(x)$ does not follow a Falkner–Skan variation, we have a nonsimilar boundary layer where $u(x, y)$ cannot be reduced to a function of a single variable η. With the aid of computers, two-dimensional boundary layers may be calculated with relative ease using finite-difference techniques.

Of the many methods that have been proposed, two methods are the most popular: the Crank–Nicolson method, and the method developed by Keller and Cebeci (1971). Since the latter method is given a detailed account in the book of Cebeci and Bradshaw (1977) and in the review article of Keller (1980), we shall study the former, for which Blottner (1970, 1975) is the standard review paper. A simplified program for incompressible laminar boundary layers is given in Appendix G.

To start, consider the equations and boundary conditions that constitute a well-posed boundary layer problem on the domain $x[0, L]$, $y[0, \infty]$. We are to find u and v that satisfy the equations

$$\frac{\partial u}{\partial x} + \frac{\partial v}{\partial y} = 0 \qquad 20.6.1$$

$$u\frac{\partial u}{\partial x} + v\frac{\partial u}{\partial y} = u_e\frac{du_e}{dx} + \nu\frac{\partial^2 u}{\partial y^2} \qquad 20.6.2$$

No slip is allowed on the wall; hence

$$u(x, y = 0) = 0, \qquad v(x, y = 0) = 0 \qquad 20.6.3$$

At infinity the boundary layer matches a given inviscid flow $u_e(x)$:

$$u(x, y \to \infty) = u_e(x) \qquad 20.6.4$$

The function u_e has a characteristic velocity scale u_0. The final condition is to specify the initial profile

$$u(x = 0, y) = u_i(y) \qquad 20.6.5$$

Equations 20.6.1 through 20.6.5 give the mathematical problem to be solved.

Since the boundary layer grows as we proceed along the surface, it is an advantage to adjust the scale of the y axis so the boundary layer has a nearly constant thickness. This is done by introducing Falkner–Skan-like variables defined according to

$$\xi \equiv \frac{x}{L}, \qquad \eta = y\left[\frac{\nu x}{u_e(x)}\right]^{-1/2} = \frac{y}{L}\left[\frac{L}{x}\frac{u_e}{u_0}\frac{u_0 L}{\nu}\right]^{1/2} \qquad 20.6.6$$

Notice that L and u_0 do not really occur in η, but have been inserted to display the physical variables with the inviscid scales L and u_0. Velocities for the boundary layer are defined with an unusual notation where the prime refers to partial differentiation with respect to η. The nondimensional velocities are

$$f'(\xi, \eta) \equiv \frac{\partial f}{\partial \eta} \equiv \frac{u}{u_e} \qquad 20.6.7$$

$$V(\xi, \eta) \equiv \frac{v}{u_e}\left[\frac{u_e x}{\nu}\right]^{1/2} \qquad 20.6.8$$

Although $f(\xi, \eta)$ is the stream function, the method does not solve for f, but keeps the two dependent variables defined above.

In terms of the new variables, the boundary conditions corresponding to 20.6.3 and 20.6.5 are

$$f'(\xi, 0) = 0, \qquad V(\xi, 0) = 0 \tag{20.6.9}$$

and

$$f'(0, \eta) = f_i'(\eta) \tag{20.6.10}$$

The given inviscid-velocity turns out to enter the computations in two ways. The boundary condition 20.6.4 is

$$f'(\xi, \eta \to \infty) = 1 \tag{20.6.11}$$

On the other hand, within the differential equation itself $u_e(x)$ enters in the form

$$\beta(\xi) \equiv \frac{x/L}{u_e/u_0} \frac{d(u_e/u_0)}{d(x/L)} = \frac{x}{u_e} \frac{du_e}{dx} \tag{20.6.12}$$

One boundary layer differs from another because they have different free-stream variations $\beta(\xi)$ and/or different initial velocity profiles $f_i'(\eta)$.

Transforming the differential equations 20.6.1 and 20.6.2 yields the continuity equation,

$$\xi \frac{\partial f'}{\partial \xi} + \beta f' + \frac{\eta}{2}(\beta - 1)\frac{\partial f'}{\partial \eta} + \frac{\partial V}{\partial \eta} = 0 \tag{20.6.13}$$

and the momentum equation,

$$f'\xi \frac{\partial f'}{\partial \xi} + \overline{V}\frac{\partial f'}{\partial \eta} = [1 - f'^2]\beta + \frac{\partial^2 f'}{\partial \eta^2} \tag{20.6.14}$$

A new variable $\overline{V}$ has been introduced in the equation above for convenience. It is defined by

$$\overline{V} \equiv V + \tfrac{1}{2}\eta f' \cdot (\beta - 1) \tag{20.6.15}$$

The mathematical problem to be solved is now 20.6.9 through 20.6.15.

A finite-difference grid is placed over the computation domain as depicted in Fig. 20-5, where grid points are denoted by m, n. The wall $\eta = 0$ is the grid point $n = 1$ and $\eta = \infty$ is taken to be the finite position $\eta = 8$ where $n = 81$. Thus, the mesh spacing is $\Delta\eta = 0.1$ and the formula for η is

$$\eta = 0.1(n - 1)$$

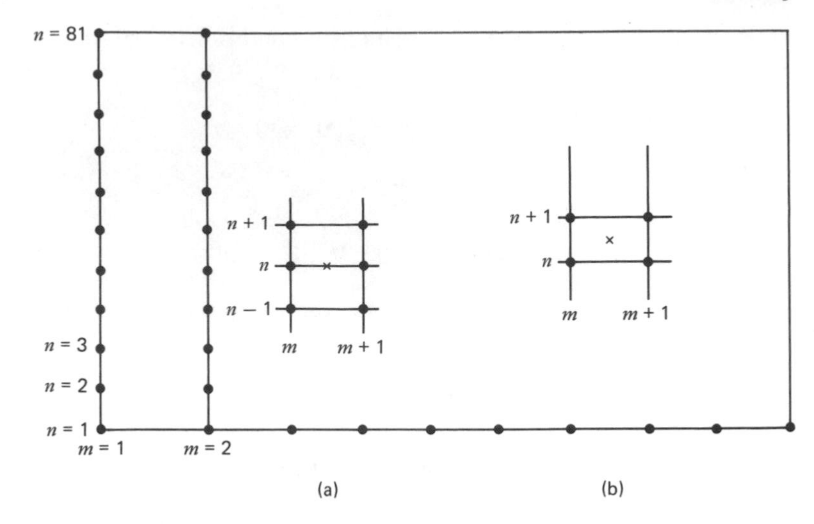

Figure 20-5 Grid and computational molecules for the Crank–Nicolson boundary-layer program: (a) momentum-equation molecule, (b) continuity-equation molecule.

In the ξ direction the grid points are denoted as $m = 1, 2, \ldots$ with an increment $\Delta\xi$, which we may specify arbitrarily.

The boundary-layer problem is parabolic, so the solution marches forward in ξ from a known initial profile $f_i'(\eta)$ at $m = 1$. When the solution at $m = 2$ is found, this acts as an initial condition for $m = 3$. This process continues for as long as the boundary data $\beta(\xi)$ are specified.

The finite-difference equations are written for the computational molecule shown in Fig. 20-5(a). The center of the molecule is within the grid at $m + \frac{1}{2}, n$. We are essentially writing the difference equations about this point. The first term in 20.6.14 is approximated using a centered-difference formula as follows:

$$\xi f' \frac{\partial f'}{\partial \xi} \approx \frac{\xi_{m+\frac{1}{2}} f'_{mn} [f'_{m+1,n} - f'_{mn}]}{\Delta\xi} \qquad 20.6.16$$

Recall that we know all values on line m and seek values for $m + 1$. Equation 20.6.16 has been linearized by evaluating the f' coefficient at m, n, where it is known, instead of at $m + \frac{1}{2}, n$. The next convective term in 20.6.14 is likewise linearized by evaluating $\overline{V}$ at m, n, where it is known. In addition, for this term we find $\partial f'/\partial\eta$ by averaging the centered-difference formula at $m + 1$ and at m. The result is

$$\overline{V} \frac{\partial f'}{\partial \eta} \approx \frac{1}{2} \overline{V}_{mn} \left[\frac{f'_{m+1,n+1} - f'_{m+1,n-1}}{2\Delta\eta} + \frac{f'_{m,n+1} - f'_{m,n-1}}{2\Delta\eta} \right] \qquad 20.6.17$$

The idea of averaging $\partial f/\partial \eta$ at m and $m+1$ to represent $\partial f/\partial \eta$ at $m+\frac{1}{2}$ is the essential characteristic of the Crank–Nicolson method. The same process is applied to the second-derivative term to get

$$\frac{\partial^2 f'}{\partial \eta^2} \approx \frac{1}{2}\left[\frac{f'_{m+1,\,n+1} - 2f'_{m+1,\,n} + f'_{m+1,\,n-1}}{(\Delta\eta)^2} + \frac{f_{m,\,n+1} - 2f_{mn} + f_{m,\,n-1}}{(\Delta\eta)^2}\right] \qquad 20.6.18$$

The last remaining term in 20.6.14 is also linearized as follows:

$$\beta[1 - f'^2] = \beta_{m+\frac{1}{2}}[1 - f'_{m+1,\,n} f'_{mn}] \qquad 20.6.19$$

All of the finite-difference relations above for the momentum equation are linear in the f' unknowns at level $m+1$. Equally important is the fact that the momentum equation is now decoupled from the continuity equation. The finite-difference form of 20.6.14 contains only known values of $\overline{V}_{mn}$. This allows us to solve 20.6.14 for f' and then use the continuity equation 20.6.13 to find $\overline{V}$ for known values of f'.

The final finite-difference momentum equation is obtained by substituting 20.6.16 through 20.6.19 into 20.6.14. The resulting equation contains unknowns $f'_{m+1,\,n+1}$, $f'_{m+1,\,n}$, and $f'_{m+1,\,n-1}$. It has the form

$$A_{mn} f'_{m+1,\,n+1} + B_{mn} f'_{m+1,\,n} + C_{mn} f'_{m+1,\,n-1} = D_{mn} \qquad 20.6.20$$

where the coefficients are

$$A_{mn} = \frac{\overline{V}_{mn}}{4\,\Delta\eta} - \frac{1}{2(\Delta\eta)^2}$$

$$B_{mn} = \xi_{m+\frac{1}{2}} f'_{mn} \frac{1}{\Delta\xi} + \beta_{m+\frac{1}{2}} f'_{mn} + \frac{1}{(\Delta\eta)^2}$$

$$C_{mn} = -\left[\frac{\overline{V}_{mn}}{4\,\Delta\eta} + \frac{1}{2(\Delta\eta)^2}\right] \qquad 20.6.21$$

$$D_{mn} = \beta_{m+\frac{1}{2}} + \xi_{m+\frac{1}{2}} f'^2 \frac{1}{\Delta\xi} - \overline{V}\frac{f'_{m,\,n+1} - f'_{m,\,n-1}}{4\,\Delta\eta} + \frac{f'_{m,\,n+1} - 2f'_{mn} + f_{m,\,n-1}}{2(\Delta\eta)^2}$$

Since all f''s in 20.6.20 are at the $m+1$ level as unknowns and all f''s in

20.6.21 are known at the m level, it is permissible to drop the m and $m + 1$ notation. In this form 20.6.20 is

$$A_n f'_{n+1} + B_n f'_n + C_n f'_{n-1} = D_n \tag{20.6.22}$$

The next step is to write out 20.6.22 for grid points $n = 2, 3, \ldots, 80$ across the boundary layer. This produces 79 equations in 81 unknowns as follows:

$$\begin{array}{lllllll} C_2 f'_1 + B_2 f'_2 + A_2 f'_3 & & & & & & = D_2 \\ & C_3 f'_2 + B_3 f'_3 + A_3 f'_4 & & & & & = D_3 \\ & & C_4 f'_3 + B_4 f'_4 + A_4 f'_5 & & & & = D_4 \\ & & & \ddots & & & \vdots \\ & & & & C_{79} f'_{78} + B_{79} f'_{79} + A_{79} f'_{80} & & = D_{79} \\ & & & & & C_{80} f'_{79} + B_{80} f'_{80} + A_{80} f'_{81} & = D_{80} \end{array} \tag{20.6.23}$$

Boundary conditions on f' supply the fact that $f'_1 = 0$ and $f'_{81} = 1$, thus eliminating two unknowns. The first and last equations in 20.6.23 become

$$\begin{gathered} B_2 f'_2 + A_2 f'_3 = D_2 \\ C_{80} f'_{79} + B_{80} f'_{80} = D_{80} - A_{80} \end{gathered} \tag{20.6.24}$$

The equations 20.6.23 are now a tridiagonal system of linear equations for f'_2 through f'_{80}. The Crank–Nicolson method is implicit because we solve for all the f''s at one time. We cannot isolate a single equation for f'_n as a function of known quantities. Implicit methods are known to be very stable and allow large step sizes in the ξ direction with good accuracy.

The solution of 20.6.23 is not very difficult, primarily because the system is tridiagonal. The boundary-layer program in Appendix G solves 20.6.23 by the Thomas algorithm. In order to understand the program completely, we briefly review the Thomas method. For a tridiagonal system of unknowns f'_n it is true that f'_n and f'_{n+1} are related by a linear equation

$$f'_n = E_n f'_{n+1} + F_n \tag{20.6.25}$$

or, with a change of subscripts,

$$f'_{n-1} = E_{n-1} f'_n + F_{n-1} \tag{20.6.26}$$

In these equations the coefficients E_n and F_n are numbers that depend on the A, B, C coefficients of 20.6.23. They are found as follows: Substitute 20.6.26 into 20.6.22 to obtain

$$A_n f'_{n+1} + B_n f'_n + C_n \left(E_{n-1} f'_n + F_{n-1} \right) = D_n \tag{20.6.27}$$

Rearrange this result into

$$f'_n = -\frac{A_n}{B_n + C_n E_{n-1}} f'_{n+1} + \frac{D_n - C_n F_{n-1}}{B_n + C_n E_{n-1}}$$

Comparing this with 20.6.25 shows that

$$E_n = -\frac{A_n}{B_n + C_n E_{n-1}} \tag{20.6.28}$$

$$F_n = \frac{D_n - C_n F_{n-1}}{B_n + C_n E_{n-1}} \tag{20.6.29}$$

These recursive relations allow us to calculate E_n and F_n from known values of E_{n-1} and F_{n-1}. Starting values E_1 and F_1 are chosen by writing 20.6.25 with $n = 1$, that is,

$$f'_1 = E_1 f'_2 + F_1 \tag{20.6.30}$$

In a general boundary layer f'_2 takes on different values, while the boundary condition requires f'_1 to be zero. The only way for 20.6.30 to be valid for arbitrary f'_2 is if $E_1 = 0$ and $F_1 = 0$. With these starting values, all E_n and F_n are calculated using 20.6.28 and 20.6.29.

The solution of the momentum equation for the $m = 2$ line can now be completed. Beginning with $f'_{81} = 1$, 20.6.26 will produce f'_{80}. Continued application of this equation gives f'_n at all positions across the boundary layer.

The continuity equation 20.6.13 is employed to find the vertical velocity V. Let the computational molecule consist of four points as shown in Fig. 20-5(b). The center of the points, $m + \frac{1}{2}$, $n - \frac{1}{2}$, is considered as the expansion point for the finite-difference approximations. Employing centered differences and averaging as needed for the Crank–Nicolson method, we approximate the terms as follows:

$$\frac{\partial V}{\partial \eta} = \frac{1}{2}\left\{\frac{V_{m+1,n} - V_{m+1,n-1}}{\Delta\eta} + \frac{V_{m,n} - V_{m,n-1}}{\Delta\eta}\right\}$$

$$\tfrac{1}{2}\eta(\beta - 1)\frac{\partial f'}{\partial \eta} = \tfrac{1}{4}\eta_{n-\frac{1}{2}}\left(\beta_{m+\frac{1}{2}} - 1\right)\left\{\frac{f'_{m+1,n} - f'_{m+1,n-1}}{\Delta\eta} + \frac{f_{m,n} - f_{m,n-1}}{\Delta\eta}\right\}$$

$$\beta f' = \tfrac{1}{4}\beta_{m+\frac{1}{2}}\left\{f'_{m+1,n} + f'_{m+1,n-1} + f'_{m,n} + f'_{m,n-1}\right\} \tag{20.6.31}$$

$$\xi\frac{\partial f'}{\partial \xi} = \tfrac{1}{2}\xi_{m+\frac{1}{2}}\left\{\frac{f'_{m+1,n} - f_{m,n}}{\Delta\xi} + \frac{f'_{m+1,n-1} - f'_{m,n-1}}{\Delta\xi}\right\}$$

Everything is known in the equations above except $V_{m+1,n}$; hence we may substitute 20.6.31 into 20.6.13 and solve explicitly for the unknown velocity. The result is

$$V_{m+1,n} = V_{m+1,n-1} + V_{m,n-1} - V_{m,n} + 2\,\Delta\eta\left\{A_n^c f'_{m+1,n} + B_n^c f'_{m+1,n-1} + C_n^c f'_{m,n} + D_n^c f'_{m,n-1}\right\} \qquad 20.6.32$$

where

$$\begin{aligned} A_n^c &= -\tfrac{1}{4}\beta_{m+\frac{1}{2}} - \frac{1}{2\,\Delta\xi}\xi_{m-1/2} - \frac{1}{4\,\Delta\eta}\eta_{n+1/2}(\beta_{m+1/2} - 1) \\ B_n^c &= - \quad \dots \quad - \quad \dots \quad + \quad \dots \\ C_n^c &= - \quad \dots \quad + \quad \dots \quad - \quad \dots \\ D_n^c &= - \quad \dots \quad + \quad \dots \quad + \quad \dots \end{aligned} \qquad 20.6.33$$

Stepping across the layer from $n = 1$, where the wall condition $V_1 = 0$ is employed, to the outer edge, $n = 81$, gives all V values on line $m + 1$. The variable $\overline{V}$ is related to V by our algebraic equation 20.6.15. The finite-difference form of this equation is

$$\overline{V}_{m+1,n} = V_{m+1,n} + \tfrac{1}{2}\eta_n f'_{m+1,n}(\beta_{m+1} - 1) \qquad 20.6.34$$

This equation is evaluated for all n.

In principle the calculation has progressed one step from $m = 1$ to $m + 1 = 2$. The step from $m = 2$ to $m = 3$ is exactly the same as the first step so we simply rename the data, $f'_{2,n} \to f'_{1,n}$ and $\overline{V}_{2,n} \to \overline{V}_{1,n}$, and repeat the calculation procedure using the new data as initial conditions. This process is continued for the entire ξ length of the boundary layer.

In describing the computing method, the initial conditions were glossed over by assuming $f_i'(\eta)$ and $V_i(\eta)$ are given functions. Now we return to this question. Assume the initial position in the boundary layer is $\xi = 0$ (It is usually, but not always, permitted to set $x_i = 0$ at the beginning of the calculation). With $\xi = 0$, the continuity equation 20.6.13 becomes

$$\beta_i f_i' + \frac{\eta}{2}(\beta_i - 1)\frac{\partial f_i'}{\partial \eta} + \frac{\partial V}{\partial \eta} = 0$$

Multiplying by $d\eta$, integrating from 0 to η, and noting that $f_i(0) = f_i'(0) = V(0) = 0$ produces

$$V_i = -\beta_i f_i - \frac{\beta_i - 1}{2}(\eta f_i' + f_i) \qquad 20.6.35$$

This equation relates V_i to the initial profile f_i. Recall that we specify initial conditions on $u = f'$, from which f may be found by integration if necessary, but no initial conditions should be given for V. The boundary-layer equations themselves, in particular 20.6.35, give a restraining equation between initial values of u and v.

Next consider the momentum equation 20.6.14, evaluated at $\xi = 0$ with 20.6.15 inserted for $\overline{V}$:

$$\left[V_i + \tfrac{1}{2}\eta f_i'(\beta_i - 1)\right] f_i'' = \left[1 - f_i'^2\right]\beta_i + f_i'''$$

Introducing 20.6.35 reduces the equation above to an ordinary differential equation for f_i,

$$f_i''' + \tfrac{1}{2}(\beta_i + 1) f_i f_i'' + \beta_i\left(1 - f_i'^2\right) = 0 \qquad 20.6.36$$

Equation 20.6.36 is recognized as the Falkner–Skan equation 20.5.15, where $\beta_i = m$. [Recall that m is the power in the equation $u_e = u_0(x/L)^m$.]

If we choose $f_i(\eta)$ as one of the similarity profiles from the Falkner–Skan family, and the corresponding f_i' and V_i using 20.6.35, we have a set of initial conditions that obey the differential equations that govern the problem (though this is not actually required). The Falkner–Skan profiles supply most of the initial conditions we might need:

$\beta_i = 1,$	plane stagnation point
$1 < \beta_i < 0,$	wedge of half angle $\theta_{1/2} = \beta\pi/(\beta + 1)$
$\beta_i = 0,$	flat plate with sharp leading edge

If the body is blunt, such as the rounded nose of an airfoil, the stagnation-point solution is appropriate for starting the calculation. If the body has a pointed front, the solution corresponding to the proper wedge angle is used. If the body has a sharp or cusp leading edge, the solution for $\beta_i = 0$ is used. This is true irrespective of the pressure gradient at the leading edge. For example, consider a converging channel that causes the flow to accelerate, and suppose we insert a flat plate in the middle of this channel. The proper initial condition for this flow is $\beta_i = 0$ even though $dp/dx \neq 0$ at the leading edge.

A flow chart for the boundary-layer program of Appendix G is shown in Fig. 20-6. The program contains only the boundary-layer calculation where the initial conditions f_i and f_i' are to be read in at the beginning. As the program is given, the external velocity $u_e(x/L)/u_0$ is to be read in a parametric form $u_e(m)/L$, $x(m)/L$ at a finite number of points. The only way in which this information enters the boundary layer equations is through the function $\beta(m)$ defined by 20.6.12. The program evaluates 20.6.12 with a forward-difference equation. Since the finite-difference boundary-layer equations require both β_m

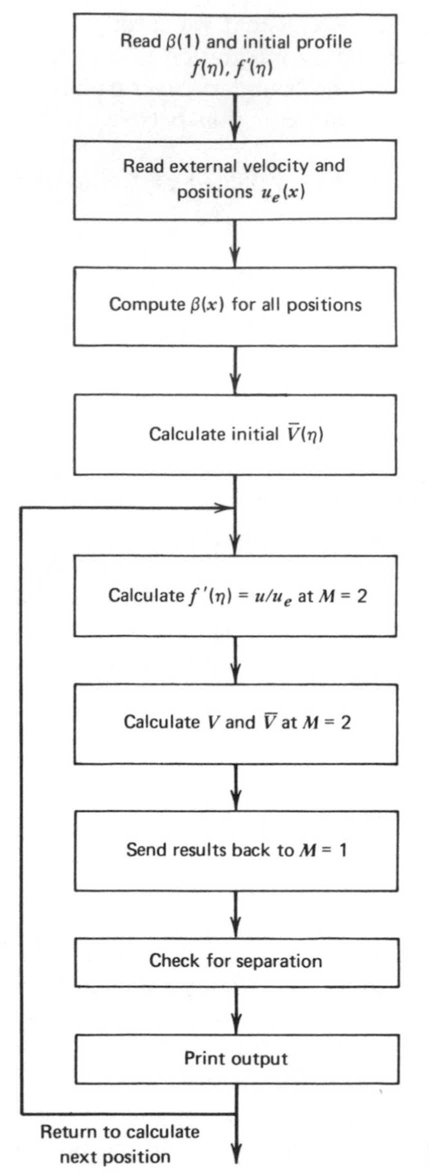

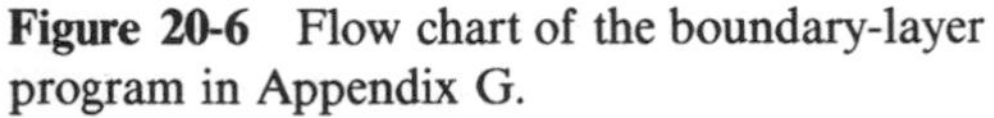

Figure 20-6 Flow chart of the boundary-layer program in Appendix G.

and $\beta_{m+1/2}$, we make the approximation $\beta_m = \beta_{m+1/2}$. If a second-order accurate method is used to determine β_m, then the distinction between β_m and $\beta_{m+1/2}$ would be more appropriate.

The program first reads $u_e(x)$ and the initial data $f_i(\eta)$ and $f_i'(\eta)$, and calculates $\beta(x)$. Then it starts a DO loop to advance the boundary-layer profile $u/u_e = f'(\xi, \eta)$ from ξ_m to ξ_{m+1}. When this calculation is complete, the program uses the new f' values in a calculation of $V(\xi, \eta)$ and $\overline{V}(\xi, \eta)$. A check is made after each step to see if the flow has separated. If separation occurs, the calculation stops at once. If no separation occurs, the calculation continues until the last position $x = L$ is reached.

*20.7 VERTICAL VELOCITY

The vertical velocity that results from a boundary-layer calculation needs to be interpreted carefully. Recall that during the derivation of the boundary-layer equations in Section 17.4, the first guess was that v/u_0 would approach zero as $\text{Re} = u_0 L/\nu$ becomes infinite. Intuitively this is reasonable. As we stand outside the boundary layer and watch the layer become thinner and thinner because $\text{Re} \to \infty$, the vertical velocity vanishes. Within the boundary layer the picture is quite different. The estimate $v/u_0 \to 0$ is not good enough on the boundary-layer scale. Small vertical currents convect momentum into regions where the u-velocity is much different. This effect is represented by the term $v\,\partial u/\partial y$ in the momentum equation. In deriving the boundary-layer equation, since $\partial u/\partial y \to \infty$ and $v \to 0$ as $\text{Re} \to \infty$, we were required to make a sharper estimate of v in order to find out if the indeterminate form $0 \cdot \infty$ was zero or finite. The result was that the effect caused by $v\,\partial u/\partial y$ is important within the boundary layer. Here the correct nondimensional vertical velocity is (17.4.5)

$$v^* \equiv \frac{v}{u_0}\left(\frac{L}{\delta}\right) = \frac{v}{u_0}\,\text{Re}^{1/2} \qquad 20.7.1$$

The nondimensional v velocity used in the Crank–Nicolson program was essentially of this same form (20.6.8):

$$V \equiv \frac{v}{u_e}\left[\frac{u_e x}{\nu}\right]^{1/2} = \frac{v}{u_0}\left(\frac{u_0 L}{\nu}\right)^{1/2}\left[\frac{x/L}{u_e/u_0}\right]^{1/2}$$

The factor in the square brackets results from the transformation to ξ, η variables.

Consider the inviscid flow near the wall. Figure 20-7 shows an outside view of a flow, where the inviscid surface velocity $u_e(x)$ is expressed in boundary-layer coordinates x, y. Next to the wall the inviscid-flow continuity equation (in dimensional form) is written in boundary-layer coordinates as

$$\frac{du_e}{dx} + \frac{\partial v}{\partial y} = 0$$

Integrating this equation a small distance in the y direction and considering the integrand to be constant yields

$$v_{\text{inviscid region}} = \int_0^y \frac{\partial v}{\partial y} dy = -\frac{du_e}{dx}\int_0^y dy$$

$$\sim -\frac{du_e}{dx} y \quad \text{as} \quad y \to 0 \qquad 20.7.2$$

We may interpret this as the inviscid vertical velocity (expressed in boundary-layer coordinates, however) that would exist if the boundary had zero thickness. At this point we know three things about the inviscid velocities near a wall: the velocity along the wall is $u_e(x)$, the v velocity into the wall is zero, and v grows linearly as we leave the wall in accord with 20.7.2.

Next, turn attention to the boundary layer. A more exact calculation of the vertical velocity can be made by integrating through the boundary layer. Consider again the expression for v:

$$v = \int_0^y \frac{\partial v}{\partial y} dy = -\int_0^y \frac{\partial u}{\partial x} dy = \frac{d}{dx}\int_0^y (u_e - u)\, dy - \frac{du_e}{dx} y \qquad 20.7.3$$

When y is taken in the outer portion of the boundary layer, the integral becomes the displacement thickness by 20.2.2. Hence, a boundary-layer analysis gives

$$v(y \to \infty) = -\frac{du_e}{dx} y + \frac{d}{dx}(u_e \delta^*) \qquad 20.7.4$$

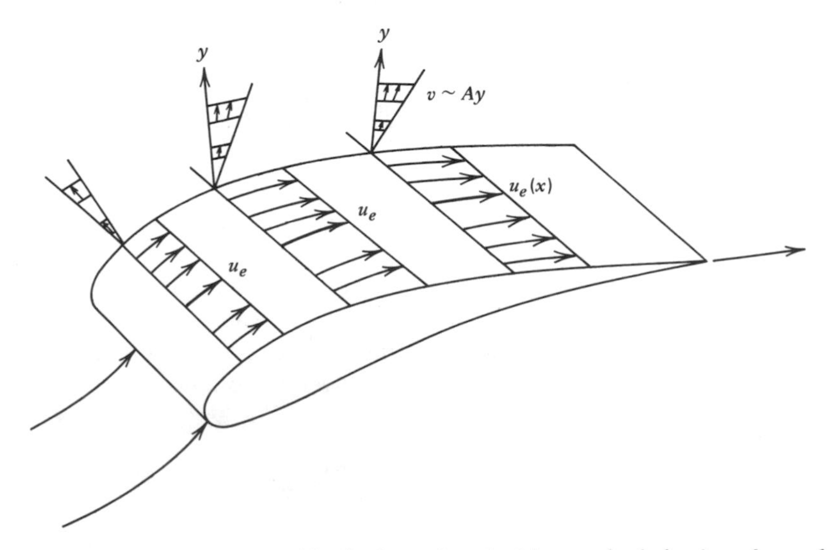

Figure 20-7 A view from outside the boundary looking at the behavior of u and v in the inviscid flow at the top of the layer.

This equation offers another interpretation of the influence of the displacement thickness on the inviscid flow (Lighthill 1958). The first term represents the inviscid velocity as found in 20.7.2. The second term is the influence of the boundary-layer profile on the v velocity in the outer regions. This is often viewed as a correction effect that the boundary layer imposes upon the inviscid flow:

$$v_{\text{B.L.inv.corr.}} = \frac{d}{dx}(u_e \delta^*) \qquad 20.7.5$$

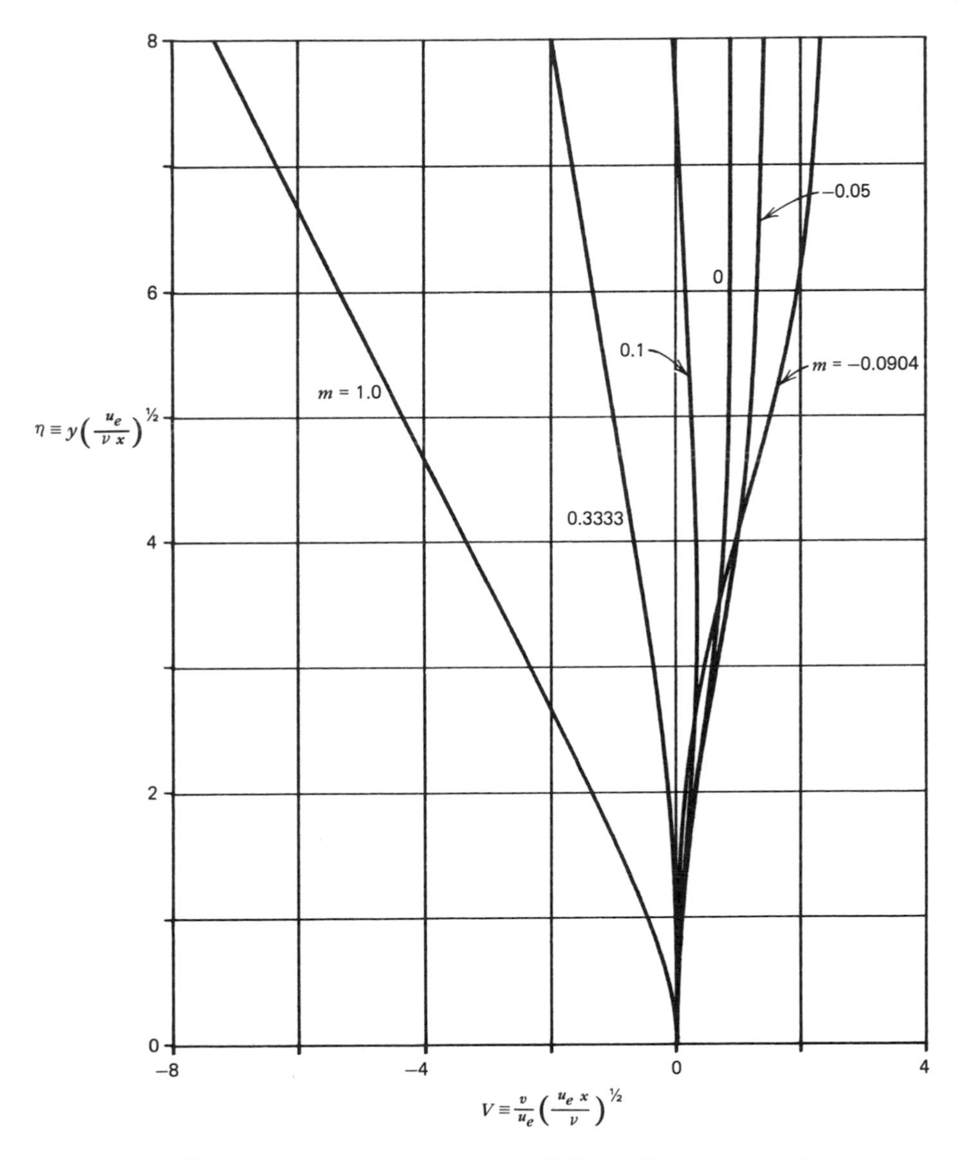

Figure 20-8 Vertical-velocity profiles for Falkner–Skan boundary layers.

In an inviscid analysis the first-order condition is that $v = 0$ on the wall. The condition 20.7.5 is the second-order condition.

Figure 20-8 gives the vertical-velocity results $v(\eta)$ for several Falkner–Skan boundary-layer flows. For $\eta \to \infty$, v approaches the appropriate nondimensional form of 20.7.4, that is

$$V(\infty) = \frac{v}{u_e}\left(\frac{u_e x}{\nu}\right)^{1/2} = -\beta\eta + \frac{x}{u_e\delta}\frac{d}{dx}(u_e\delta^*)$$

The second term is the v intercept (here δ is 20.5.11). We can expect $V(\infty)$ to become infinite for any value of $\beta \neq 0$.

Boundary layers satisfy the no-slip condition at the wall and match the inviscid flow as $y \to \infty$. In this match u approaches $u_e(x)$ and v approaches a curve with the proper slope $(-du_e/dx)$ for v in the inviscid region near the wall.

20.8 JOUKOWSKI-AIRFOIL BOUNDARY LAYER

The boundary layer on the upper surface of a Joukowski airfoil is a typical example of a nonsimilar boundary layer that can be computed using the method of Section 20.7. Figure 20-9a gives the external velocity $u_e(s)/u_0$ and the associated function $\beta(s)$. To emphasize that the distance is measured along the surface, here, and also in the next section, we use s instead of x. The boundary layer begins with a stagnation point, $\beta_i = 1.0$ (the surface location and flow speed for the airfoil change very rapidly in this region and must be calculated exactly without using the linearizing approximations that were employed in Section 18.11). A short distance away from the stagnation point, the flow accelerates even more rapidly, as indicated by the fact that β is larger than one. β then begins a sharp fall, crossing zero when u_e/u_0 reaches its maximum value of 1.49. For this airfoil, the maximum u_e is close to the leading edge but is fairly broad. From here on, the flow gradually decelerates in an adverse pressure gradient.

Velocity profiles of $f' = u/u_e$ are shown as functions of $\eta = y[u_e/s\nu]^{1/2}$ in Fig. 20-10. The initial profile is the Hiemenz stagnation-point result ($\beta_i = m = 1$). The circle on the profile at $\eta = 2.36$ denotes the boundary-layer thickness δ_{99}. Due to the rapid acceleration of the flow away from the stagnation point, the boundary layer becomes thinner and the profile flattens. For example, as $s/L = 0.0055$ the thickness has dropped to $\delta_{99} = 2.24$. Subsequently the velocity u_e reaches a peak, deceleration begins, and the boundary layer thickens.

In some cases where the u_e/u_0 curve has a sharp peak, the boundary layer separates in the nose region. This would probably occur on this airfoil if the angle of attack were higher. If flow separation does occur, the boundary layer will usually become turbulent and immediately reattach to the surface (of

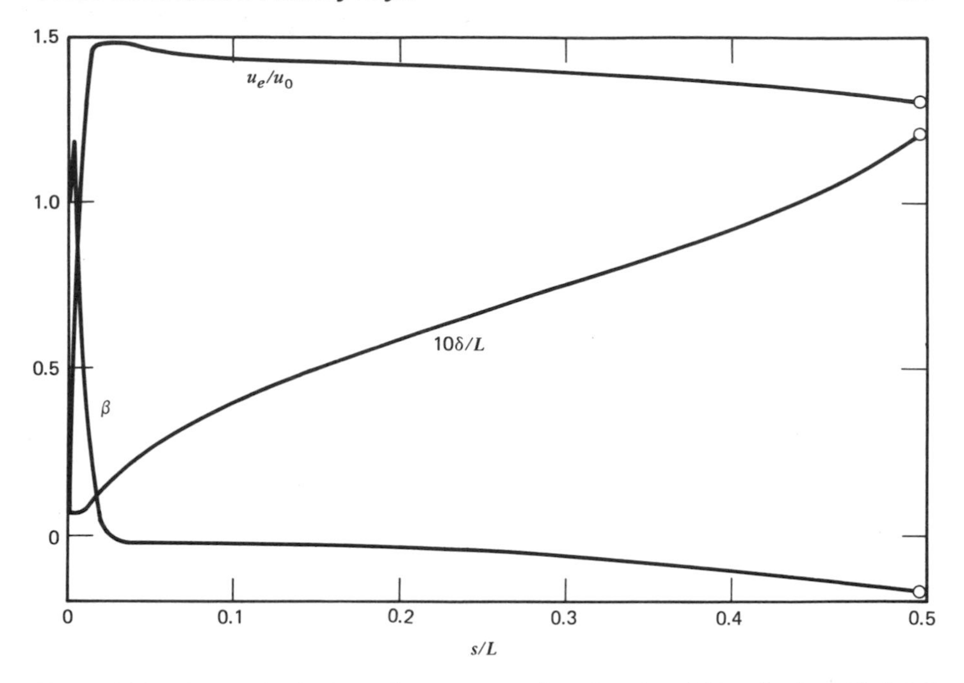

Figure 20-9 External velocity and pressure-gradient function β for a Joukowski airfoil of $t/L = 0.09$ and $h/L = 0.05$ at 3° angle of attack. The circle marks separation.

course, at very high angles of attack the reattachment cannot be maintained and the airfoil stalls completely). A separation bubble that becomes turbulent and reattaches brings outside fluid into intimate contact with the surface. A turbine blade for a jet engine would have extremely high heat transfer at the attachment point.

Over the remainder of the upper surface, there is a mild adverse pressure gradient, which causes the boundary layer to thicken. Since the pressure gradient acts uniformly through the layer, it slows all the particles within the boundary layer with equal effectiveness. The particles near the wall have very little momentum; thus the adverse pressure does not have to act very long before these particles are slowed to zero velocity and turn around to move upstream. At this point the boundary-layer calculation is stopped. It is common to say that the boundary layer has separated. If separation does not occur, we can assume that our boundary-layer calculation is reasonably good. If separation is predicted, then the analysis needs to be modified. A boundary layer can predict its own demise, but the location of the separation is not usually given correctly.

The separated region is no longer a thin region satisfying the boundary-layer assumptions. Moreover, the effective body that the inviscid flow sees includes in some way the separated-flow region. This modifies the u_e velocity on the surface and hence affects the boundary layer. Boundary-layer calculations

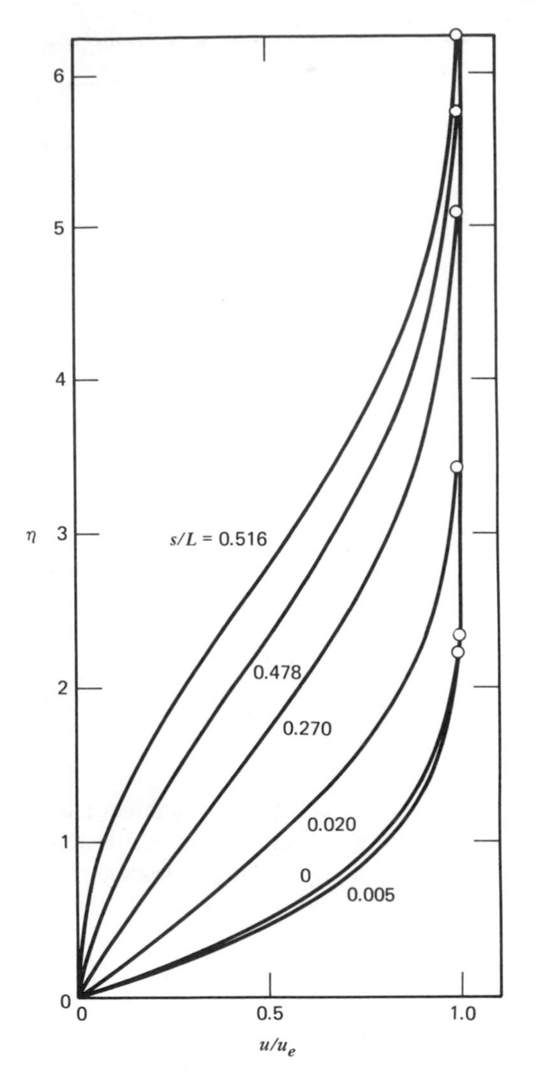

Figure 20-10 Velocity profiles for the Joukowski airfoil in boundary layer coordinates. The circle marks δ_{99}.

made with measured pressure distributions, rather than the ideal distributions, can come fairly close to predicting the location of the separation point.

Continuity considerations determine the V velocity. In an accelerating layer the velocity is negative as the streamlines move closer to the surface. In a decelerating layer V is positive, as the particles must move away from the wall so that the same mass flow between the wall and a given streamline is maintained. According to the arguments of Section 20.7, V becomes unbounded with a slope related to du_e/ds.

It is important to note that the results so far have been presented in boundary-layer variables. We have not specified whether the airfoil is 2 inches or 2 meters long. Moreover, we have not specified the free-stream velocity u_0 or the kinematic viscosity ν of the fluid. All the results so far, including the prediction of separation, are independent of L, u_0, and ν, and hence of the Reynolds number $u_0 L/\nu$. Let us express the boundary-layer results in terms of the inviscid scales u_0 and L. The relations are

$$\frac{s}{L} = \xi, \qquad \frac{y}{L} = \eta\left[\frac{\xi u_0}{\mathrm{Re}\, u_e(\xi)}\right]^{1/2}, \qquad \frac{u}{u_0} = \frac{u_e(\xi)}{u_0} f'(\xi, \eta)$$

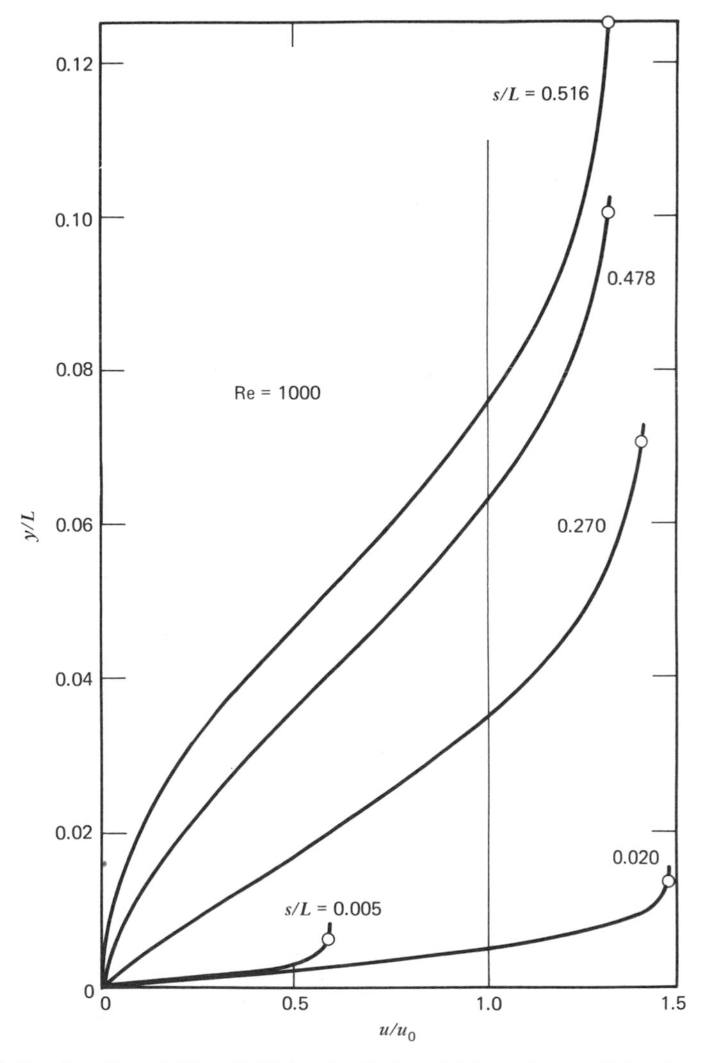

Figure 20-11 Profiles of Fig. 20-10 in physical variables u/u_0, y/L for Re = 1000.

where $\text{Re} = u_0 L/\nu$. Figure 20-11 shows the velocity profiles and the boundary-layer thickness δ_{99}/L plotted for a value of Re = 1000, which from a practical standpoint is a very low number. The stagnation point thickness $\delta_{99}/L = 0.0075$ increases by a factor of 10 to $\delta_{99}/L = 0.122$ at the separation point. The thickness of the entire layer will decrease as $1/\sqrt{\text{Re}}$ as Re becomes larger; for $\text{Re} = 10^6$ the layer would be approximately $\frac{1}{30}$ as thick. Boundary-layer theory is a high-Reynolds-number theory that yields profiles that are independent of Re when expressed in boundary-layer variables. The actual physical variables are determined once the flow Reynolds number is given.

20.9 BOUNDARY LAYER ON A BRIDGE PILING

The inviscid flow over a section consisting of a circular arc was given in Problem 18.22. This shape might represent a bridge piling or a streamlined strut. If the object is thin enough, it is know as a biconvex airfoil. Figure 20-12 gives the surface velocity u_e/u_0 for the special case where the half angle at the nose is 30°. This corresponds to a thickness ratio of $t/l = 0.267$ ($n = \frac{5}{3}$ in Problem 18.24).

The inviscid flow in the neighborhood of the front edge must locally have the same character as the flow over a semiinfinite wedge with the same angle. Thus, for the initial boundary-layer profile we use the Falkner–Skan profile corresponding to a 30° wedge half angle. From the formula $\theta_{1/2} = m\pi/(m+1)$ we find that $m = \beta_i = 0.2$. The surface velocity on the wedge starts with a stagnation point and then increases as $u_e \propto s^{0.2}$. The boundary layer starts with no thickness whatever. This is seen by substitution of 20.5.10 into 20.5.8:

$$\delta = \left(\frac{\nu s}{u_e}\right)^{1/2} = \left(\frac{\nu}{u_0 L}\right)\left(\frac{s}{L}\right)^{(1-m)/2}$$

For $0 \le m < 1$—that is, all wedge angles from zero up to but not including the 90° stagnation profile—the initial boundary-layer thickness is zero. This is regarded as a failure of boundary-layer theory, as there must be some small region near the leading edge where vorticity diffuses forward against the flow. The ξ, η boundary-layer coordinates hide this singularity. We can view the initial profile $f_i(\eta)$ as the profile we would find if we approached the leading edge from positive values of ξ.

As shown in Fig. 20-12, the surface velocity rises rapidly at first and then slowly increases to a maximum at the midchord position. Thereafter, u_e/u_0 will fall again in a mirror image of the behavior on the front half. In contrast to the Joukowski airfoil, the bridge piling has a long acceleration region followed by a gentle deceleration. The function $\beta(\xi)$ begins at 0.2, as discussed above, and rises gently to a maximum of 0.243 at $\xi = 0.130$. The β function becomes zero at the midchord point (as it must when u_e/u_0 reaches a

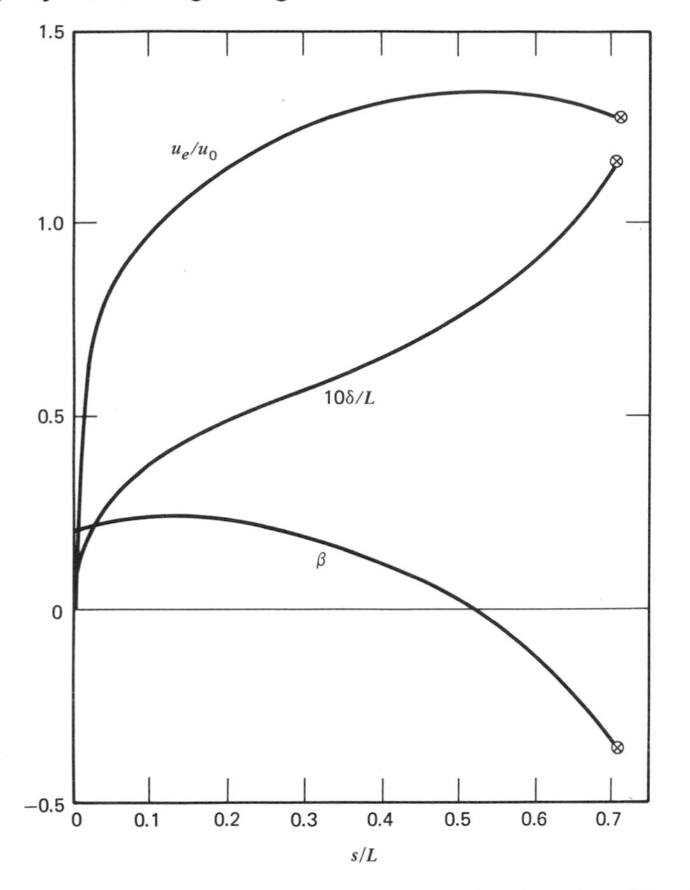

Figure 20-12 External velocity, pressure-gradient function β, and resulting boundary-layer thickness for flow over a biconvex strut with 30° half angle at the nose. The circle marks separation.

maximum) and then becomes negative when the flow decelerates. Although the adverse pressure gradient is very mild, the flow separates at $\xi = 0.708$. (Velocity profiles are shown in Fig. 20-13). At this point u_e/u_0 is 1.28, only a very slight decrease from the maximum value 1.34.

The physical thickness δ_{99}/L is also shown in Fig. 20-12. The layer begins with $\delta_{99} = 0$ and grows throughout its length. Even on the forward portion, $\xi < 0.1$, the thinning of the layer by acceleration cannot overcome the thickening influence of viscosity. Take note that near separation, a rapid increase in δ_{99} occurs.

As shown by this example, laminar boundary layers separate quickly once they encounter an adverse pressure gradient. To counteract this tendency, engineers sometimes design airfoils or diffusers so that the boundary layer becomes turbulent before the adverse pressure region is reached. Turbulent

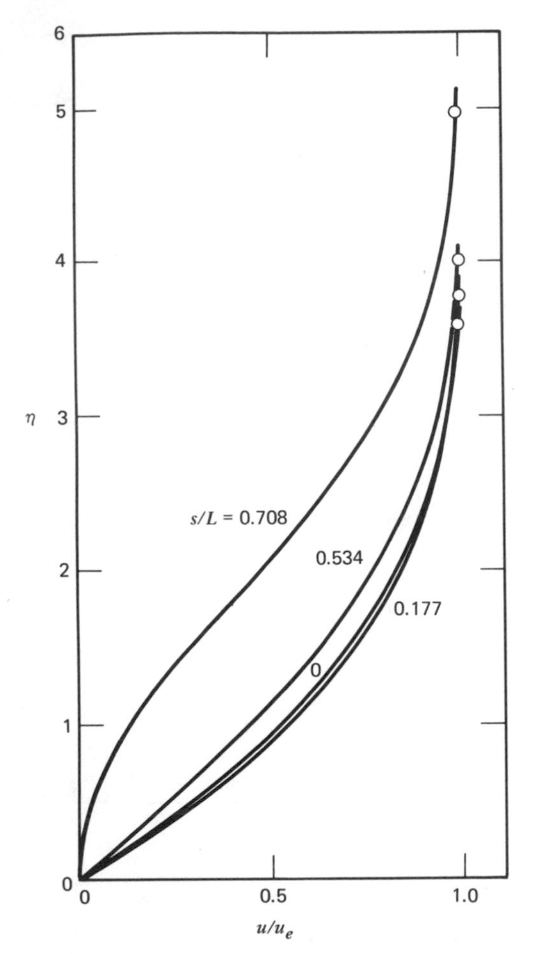

Figure 20-13 Velocity profiles in the boundary layer on a biconvex strut.

layers do not separate as readily, because the turbulence continually mixes high- and low-momentum fluid within the layer. This prevents the pressure gradient from slowing a large chunk of fluid as needed for separation. Turbulent boundary layers do not always prevent separation; however, if it occurs, the position of turbulent separation is further downstream than for a laminar layer.

20.10 BOUNDARY-LAYER SEPARATION

The pressure force imposed on the boundary layer by the inviscid flow acts with the same magnitude on all particles within the layer. Those particles next to the wall have the least momentum and are the first to be stopped and turned around to form a region of reverse flow. In steady laminar flow, separation is identified as the place where the shear stress is zero ($\partial u/\partial y = 0$). We cannot make a general rule about when and where separation will occur. If an adverse pressure gradient of $dp/dx > 0$ acts long enough, separation is likely. No more definite statement can be made, since we know that some Falkner–Skan boundary layers with very mild adverse gradients ($-0.0906 < \beta < 0$) do not separate.

The inviscid flow $u_e(x)$ and the resulting boundary-layer profiles are independent of the Reynolds number. Hence, so is the prediction of separation. (An exception to this rule occurs when a boundary layer changes from laminar to turbulent flow, as in the case of a sphere or cylinder.) Even though the separation may lead to a large wake, which in turn modifies the inviscid flow, the position and occurrence of separation is fairly insensitive to Reynolds number. A boundary-layer calculation based upon measured dp/dx accounts for the effect of the wake upon the inviscid flow and satisfactorily predicts the proper location for separation. Flow in the large wake region does not satisfy the boundary-layer assumptions, and we do not carry the boundary-layer calculation past the separation point.

The flow in a large separated region is three-dimensional even if the boundary layer up to separation is roughly two-dimensional. This is evident in the surface streamlines made visible by putting oil on the surface of the wing shown in Fig. 20-14.

We class separated flows into two types: those with large wakes as discussed above, and those that are thin and reattach to the wall rather soon, forming a separation bubble. Separation bubbles can be calculated using the boundary-layer equations (a review article is Williams 1977). The methods used are called *inverse methods*, because they do not use a specified pressure gradient as the boundary condition.

When a pressure gradient is specified to calculate a boundary layer, the problem breaks down at the separation point. Goldstein (1948) solved the

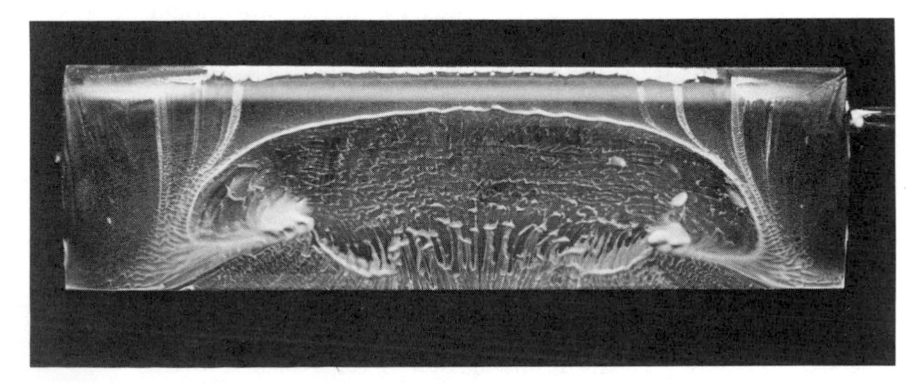

Figure 20-14 Plan view of the trailing-edge stall pattern on a Clark Y-14 airfoil. The pattern is made visible by the oil-flow technique. Flow is from top to bottom. Photograph courtesy of A. Winkelmann, University of Maryland.

boundary-layer equations at a separation point and found a singular solution where the vertical velocity and displacement thickness become infinite while the shear stress goes to zero in such a way that $d\tau/dx$ is infinite. Later it was found that the singular behavior is not an inherent property of the boundary-layer equations, but is associated with the specified pressure gradient. Catherall and Mangler (1966) modified the boundary-layer problem by specifying the displacement thickness and allowing the appropriate $u_e(x)$, and hence dp/dx, to be found as part of the solution. This problem can be integrated through the separation point without any difficulty. Another inverse method, which is easier to use, is to specify the wall shear instead of the displacement thickness. The requirement for a regular solution through $\tau_0 = 0$ is that the pressure gradient be modified so that $d\beta/dx > 0$ at the separation point.

Note that even in thin separation bubbles there is an interaction between the inviscid flow and the boundary layer. Specifying the wall shear or the displacement thickness is a trick to take account of the interaction. We do not really know τ_0 or δ^* beforehand. Iteration between the inviscid flow and the inverse boundary-layer calculation is required to find two compatible solutions.

A minor difficulty in continuing the boundary-layer calculation into the separated region is caused by the reverse flow. Recall that the timelike term in the boundary-layer momentum equation is $u\,\partial u/\partial x$. The marching direction for this parabolic equation is determined by the u coefficient of this term. Information travels downstream but not upstream. In the separated region, a very thin portion near the wall has negative u, so the proper marching direction is reversed. In formulating inverse methods special account of this effect is taken by upwind-differencing $u\,\partial u/\partial x$ or, alternately even ignoring the term by setting $u\,\partial u/\partial x$ equal to zero (the convection effect is extremely small anyway).

In addition to numerical calculations of separation bubbles, there exists an asymptotic theory called Stewartson's triple-deck theory. The interested reader should see Stewartson (1954) for an introduction to these ideas.

In closing this section we should note an important fact about the velocity profiles as they approach separation. All boundary-layer profiles have a point of inflection whenever the pressure gradient is adverse. This may be deduced as follows. Consider the geometry of a velocity profile, where u is everywhere positive and at the wall $\partial u/\partial y$ is positive. Furthermore, the momentum equation evaluated at the wall shows that

$$\frac{1}{\rho}\frac{dp}{dx} = \nu\,\frac{\partial}{\partial y}\left(\frac{\partial u}{\partial y}\right)\bigg|_{\text{wall}}$$

A positive pressure gradient means that $\partial u/\partial y$ increases as we leave the wall. If $\tau \approx \partial u/\partial y$ increases, it must have a maximum within the flow, since $\tau \approx 0$ at infinity. A maximum in τ implies an inflection point in the velocity profile at the same position.

20.11 AXISYMMETRIC BOUNDARY LAYERS

This section deals with boundary layers that are symmetric about an axis and have no swirl. Figure 20-15 shows the coordinate geometry of an external flow over a cylindrical shape and an internal flow of a confined fluid. In both cases we assume that the boundary-layer thickness is small compared to the radius of curvature in both the longitudinal and the lateral directions. Mangler (1945) derived the boundary-layer equations for this situation. They differ from those for the plane case only by the appearance in the continuity equation of r_0, the distance from the axis to the surface:

$$\frac{\partial}{\partial x}(r_0 u) + r_0 \frac{\partial v}{\partial y} = 0$$

$$\frac{\partial p}{\partial y} = 0 \qquad 20.11.1$$

$$u\frac{\partial u}{\partial x} + v\frac{\partial u}{\partial y} = u_e \frac{du_e}{dx} + \nu \frac{\partial^2 u}{\partial y^2}$$

Mangler also gave a mathematical transformation that sends the axisymmetric problem into an equivalent plane problem. Consider the transformation

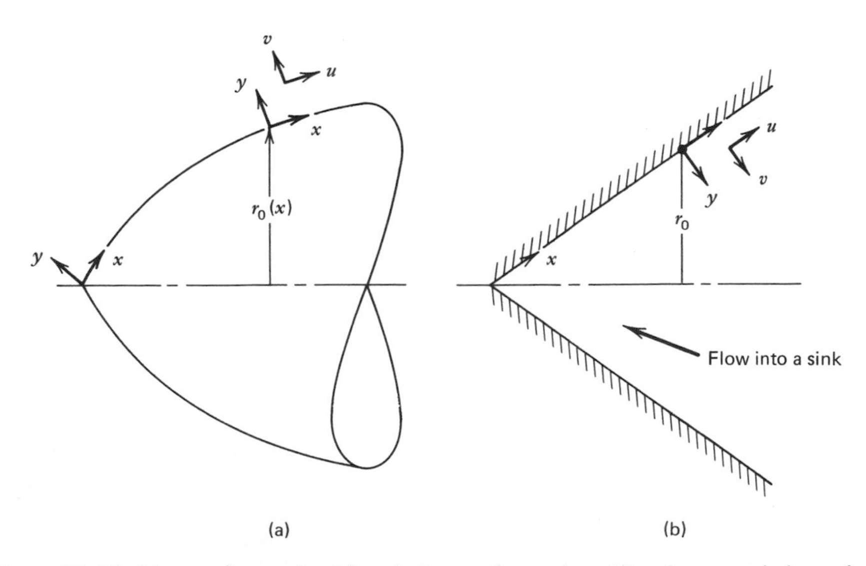

Figure 20-15 Nomenclature for Mangler's transformation: (a) axisymmetric boundary layer, (b) flow into a conical passage.

defined by

$$\hat{x} = \int_0^x \left(\frac{r_0}{L}\right)^2 dx, \qquad \hat{y} = \frac{r_0}{L} y$$

$$\hat{u} = u, \qquad \hat{v} = \frac{L}{r_0}\left[v + \frac{y}{r_0} u \frac{dr_0}{dx}\right] \tag{20.11.2}$$

where L is an arbitrary length scale. Substitution of the transformation 10.11.2 [note that $\partial\hat{x}/\partial x = (r_0/L)^2$ and $\partial\hat{y}/\partial x = (y/L)\, dr_0/dx$] into 20.11.1 produces the boundary-layer equations for a plane flow, namely

$$\frac{\partial \hat{u}}{\partial \hat{x}} + \frac{\partial \hat{v}}{\partial \hat{y}} = 0$$

$$\hat{u}\frac{\partial \hat{u}}{\partial \hat{x}} + \hat{v}\frac{\partial \hat{u}}{\partial \hat{y}} = \hat{u}_e \frac{d\hat{u}_e}{d\hat{x}} + \nu \frac{\partial^2 \hat{u}}{\partial \hat{y}^2} \tag{20.11.3}$$

One may verify that the boundary conditions also transform appropriately. A solution for an axisymmetric boundary layer may be found by considering an equivalent plane boundary layer defined by Mangler's transformation. Notice that when r_0 is constant, the transformation is trivial. The boundary layer on the outside or inside of a cylinder is the same as a plane layer as long as the boundary-layer thickness is much smaller than the radius.

We consider as an example the axisymmetric stagnation point on a blunt body. The inviscid velocity near the stagnation point has the form

$$\frac{u_e}{u_0} = \frac{x}{L} \tag{20.11.4}$$

L is a characteristic dimension of the body, and $u_0 = \alpha u_\infty$ is related to the free-stream velocity u_∞ by a factor α that depends upon the shape of the body. For this case $r_0 = x$, and 20.11.2 becomes

$$\frac{\hat{x}}{L} = \int_0^{x/L} \left(\frac{x}{L}\right)^2 d\left(\frac{x}{L}\right) = \frac{1}{3}\left(\frac{x}{L}\right)^3$$

$$\frac{\hat{y}}{L} = \frac{r_0}{L}\frac{y}{L} = \frac{x}{L}\frac{y}{L} \tag{20.11.5}$$

These equations define the point $\hat{x}$, $\hat{y}$ in the plane flow which is equivalent to the point x, y in the axisymmetric flow. Next, consider how the external velocity 20.11.4 transforms. Using $\hat{u}_e = u_e$, we obtain

$$\frac{\hat{u}_e}{u_0} = \frac{x}{L} = \left(3\frac{\hat{x}}{L}\right)^{1/3} \tag{20.11.6}$$

The equivalent plane flow has the external flow $u_e \approx \hat{x}^{1/3}$, where the reference velocity constant is unchanged but the length constant is $L/3 = \hat{L}$. The solution to a boundary layer obeying 20.11.5 is the Falkner–Skan flow for $m = \frac{1}{3}$ ($\theta_{1/2} = 45°$ wedge). Assuming this solution is known, the u velocity at the point x, y in the axisymmetric flow would be

$$\frac{u(x, y)}{u_0} = f'\left(\frac{\hat{y}}{L} \to \frac{xy}{L^2}, \frac{\hat{x}}{L} \to \frac{1}{3}\left(\frac{x}{L}\right)^3\right)$$

A corresponding formula for v is found from 20.11.2.

The flow toward an axisymmetric stagnation point is a special case of the streaming flow over a cone of a given angle. Since the external flow over a cone obeys $u_e \propto x^n$ and the surface position is $r_0 = x \sin\theta_{1/2}$, Mangler's transformation produces $\hat{u}_e \propto \hat{x}^{n/3}$. Again, this is one of the profiles from the Falkner–Skan group. Unfortunately there is no simple mathematical relation between the cone angle θ and the exponent n. Whitehead and Canetti (1950) give a graphical presentation of the relation.

For a second example consider the flow on the inside of a cone of angle θ (Fig. 20-15b). Fluid is drawn in through the orifice at the apex, and we assume that the Reynolds number is high enough so that the flow in a region somewhat removed from the orifice is a potential flow. The potential flow is modeled as the flow into a point sink. Along the cone walls

$$\frac{u_e}{-a_0} = \left(\frac{x}{L}\right)^{-2} \qquad 20.11.7$$

where a_0 and L are positive constants. The flow u_e is negative, since it is against the direction of increasing x. This formula is valid irrespectve of the angle, $0 < \theta < \pi$, since the flow into a sink is independent of θ.

Mangler's transformation is computed using the fact that $r_0 = x\sin\theta$. From 20.11.2 we find

$$\frac{\hat{x}}{L} = \frac{1}{3}\sin^2\theta\left(\frac{x}{L}\right)^3, \qquad \frac{\hat{y}}{L} = \sin\theta\,\frac{x}{L}\frac{y}{L} \qquad 20.11.8$$

The external velocity is obtained using the fact that $\hat{u} = u$ and $\hat{u}_e = u_e$. So from 20.11.6 we get

$$\frac{\hat{u}_e}{-u_0} = \left(\frac{3}{\sin^2\theta}\frac{\hat{x}}{L}\right)^{-2/3} \qquad 20.11.9$$

This is also a Falkner–Skan flow, with velocity scale $\hat{u}_0 = -a_0$, length scale $\hat{L} = \frac{1}{3}L\sin^2\theta$, and exponent $m = -\frac{2}{3}$. [This is a case where the signs of u_e and x are different, so the Falkner–Skan solution would proceed with $\beta = -m = \frac{2}{3}$ (20.5.12), and since $2\alpha - \beta = -1$, $\alpha = -\frac{1}{6}$.] The nondimensional velocity profiles $\hat{f}'(\hat{\eta})$ are the same for all values of θ; however, the $\hat{x}$, $\hat{y}$ position, and hence the $\hat{\eta}$ position, depend upon θ through 20.11.8.

The corresponding problem where flow comes from a source and moves radially along the cone surface would result in 20.11.9 without the minus sign. Since in this case $\hat{x}$ and $\hat{u}$ have the same sign, we now have $\beta = m = -\frac{2}{3}$. A Falkner–Skan solution does not exist for this case. This means that a conical diffuser does not have a similarity boundary-layer solution. In the actual case, the initial boundary layer at the entrance and the finite length determine the flow.

20.12 JETS

Figure 20-16 shows a jet of width h issuing into the same ambient fluid. The major assumption is that the Reynolds number $u_0 h/\nu$ is large, causing a long thin jet to which the boundary-layer approximation may be made. We consider the plane two-dimensional case; however, the method of analysis applies equally well to a round jet.

The flow in and near the mouth of the jet depends on the details of the flow before the fluid actually exits from the orifice. Somewhat downstream from the orifice, all jets decay in the same manner regardless of the original jet profile. Our discussion will apply only to this downstream region.

It is true that there is a contradiction in the idea of a high-Reynolds-number laminar jet. Actual jet profiles are very unstable and lead to a turbulent jet in a

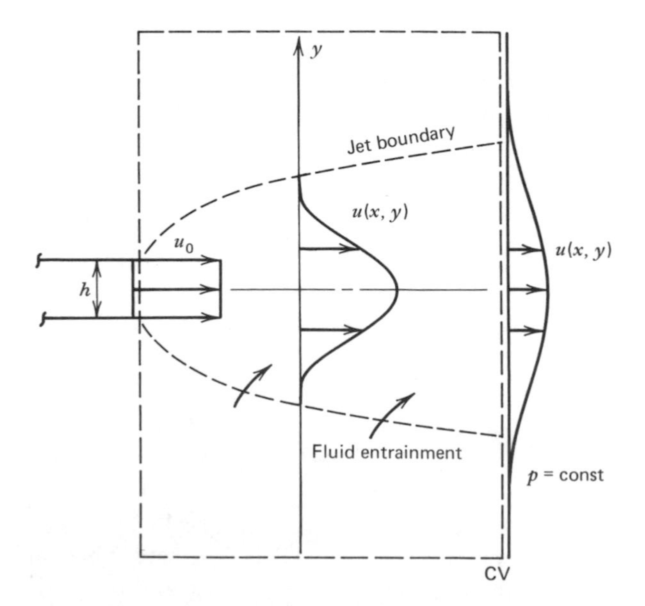

Figure 20-16 Plane laminar jet into an infinite medium.

very short distance. This problem is nevertheless useful on at least two accounts: it supplies the laminar profile that a stability analysis shows is unstable, and it can be adapted with only a slight change to apply to the actual turbulent jet. The problem also has an interesting history. Schlichting (1933) did the original solution. At a much later date it was shown to be the high-Reynolds-number limit of an exact solution of the complete Navier–Stokes equation (Squire 1951).

To begin, consider a control region that cuts across the exit plane of the jet and again at an arbitrary station downstream. We assume that the pressure at the two planes is the same, an assumption compatible with boundary-layer analysis. The momentum equation shows that the momentum carried across any station must be a constant,

$$M = \rho u_0^2 h = \rho \int_{-\infty}^{\infty} u^2(y)\, dy \qquad 20.12.1$$

Since the jet entrains ambient fluid, the flow rate across successive cross sections is not the same, and M is the only property of the flow at the orifice that is carried downstream. We take M as a given number and require that the jet velocity profile satisfy 20.12.1.

The boundary-layer equations for a constant-pressure jet may be expressed in terms of the stream function. They are

$$\psi_y \psi_{xy} - \psi_x \psi_{yy} = \nu \psi_{yyy} \qquad 20.12.2$$

where

$$u = \psi_y, \qquad v = -\psi_x \qquad 20.12.3$$

We seek a similarity solution where Ax^p is the scaling of the stream function and Bx^q is the scaling of the y distance:

$$\psi = Ax^p f(\eta) \qquad 20.12.4$$

$$\eta = \frac{y}{Bx^q} \qquad 20.12.5$$

The powers p and q will be found as part of the analysis, while A and B are chosen so that f and η are dimensionless. The boundary conditions require a symmetric flow about the centerline ($\psi = 0$) and $u \sim f'$ vanishes far away from the centerline; thus

$$f(\eta = 0) = 0, \qquad f''(\eta = 0) = 0, \qquad f'(\eta \to \infty) = 0 \qquad 20.12.6$$

Substituting 20.12.4 into 20.12.2 yields

$$\frac{AB}{\nu} x^{p+q-1}\left[(p - q)f'^2 - pff''\right] = f''' \qquad 20.12.7$$

For similarity the factor in front of the brackets must be independent of x. Hence, the condition $p + q = 1$ must be met. A second condition is found by substituting 20.12.4 into 20.12.1; this produces

$$M = \rho A^2 B^{-1} x^{2p-q} \int_{-\infty}^{\infty} f'^2 \, d\eta \qquad 20.12.8$$

M will be constant for all x only if $2p = q$. We satisfy these conditions if $p = \frac{1}{3}$ and $q = \frac{2}{3}$.

Equation 20.12.7 now becomes

$$f''' + \frac{AB}{3\nu}\left[f'^2 + ff''\right] = 0 \qquad 20.12.9$$

The solution to this equation is

$$f = \tanh \eta \qquad 20.12.10$$

The corresponding velocity profiles are

$$\frac{u}{u_{\max}} = f' = \operatorname{sech}^2 \eta \qquad 20.12.11$$

The simple formulas 20.12.10 and 20.12.11 are obtained when

$$A = \left(\frac{9\nu M}{2\rho}\right)^{1/3}, \qquad B = \left(\frac{48\nu^2\rho}{M}\right)^{1/3} \qquad 20.12.12$$

It is informative to write the final profile in dimensional variables:

$$u = \left(\frac{3M^2}{32\rho^2\nu}\right)^{1/3} x^{-1/3} \operatorname{sech}^2\left[y\left(\frac{M}{48\rho\nu^2}\right)^{1/3} x^{-2/3}\right] \qquad 20.12.13$$

We can now see that the maximum velocity is the coefficient of the sech function and that it decays like $x^{-1/3}$. Furthermore, if we define the jet thickness as a point where u is a certain fraction of $u_{\max}$, the locus of such points is $yx^{-2/3}$ = const. Hence, the jet width grows as $x^{2/3}$.

The viscous forces at the edge of the jet accelerate ambient fluid and entrain it into the jet. From 20.12.13 the flow rate may be calculated as

$$Q = \int_{-\infty}^{\infty} u \, dy = \left(\frac{36M\nu}{\rho}\right)^{1/3} x^{1/3} \qquad 20.12.14$$

Q grows like $x^{1/3}$, a very slow growth.

The origin is a singularity in the solution. At $x = 0$ the maximum velocity is infinite while the thickness and flow rate are zero. This breakdown in the validity of the boundary-layer analysis is the same type of difficulty we previously encountered in the external flow over a wedge. In a practical application of the formulas above we should replace x by $x - x_0$, where x_0 is an unknown effective origin.

The modifications necessary to adapt the analysis to an axisymmetric jet or to turbulent jets can be found in the books listed on the supplemental reading list.

20.13 FAR WAKE OF NONLIFTING BODIES

The vorticity generated as the flow goes over a body is swept into the wake. There is just as much positive vorticity as negative vorticity, in the sense that the integral of ω over any cross section of the wake is zero. As this vorticity is carried downstream, the positive vorticity in the lower half plane diffuses toward the centerline, where it meets with negative vorticity diffusing from the upper half plane. Furthermore, at the outer edges of the wake, the vorticity diffuses outward to spread the wake. This effect is more than compensated for in the core of the wake, where the vorticity is being diminished. The net result is a decrease of the integral of $\frac{1}{2}\boldsymbol{\omega}\cdot\boldsymbol{\omega}$ over a cross section.

The results we shall derive in this section were first given by Tollmien (1931); for a comprehensive survey of many other aspects of wakes the reader should consult Berger (1971). Let us propose that the boundary-layer equations govern the wake flow because the Reynolds number of the body that creates the wake is large. Outside the boundary layer the velocity is u_0 (Fig. 20-17). We assume that similarity exists in terms of the velocity defect,

$$u \equiv u_0 - u_w \qquad 20.13.1$$

where u_w is the velocity profile across the wake. In terms of u the boundary-layer equation becomes

$$(u_0 - u)\frac{\partial u}{\partial x} + v\frac{\partial u}{\partial y} = \nu\frac{\partial^2 u}{\partial y^2} \qquad 20.13.2$$

In manner similar to the jet analysis, we introduce a stream function $u = \partial\psi/\partial y$, $v = -\partial\psi/\partial x$, and assume similarity of the form

$$\psi \equiv Ax^p f(\eta)$$
$$\eta \equiv \frac{y}{Bx^q} \qquad 20.13.3$$

The constants A, B, p, and q will be chosen in the course of the analysis.

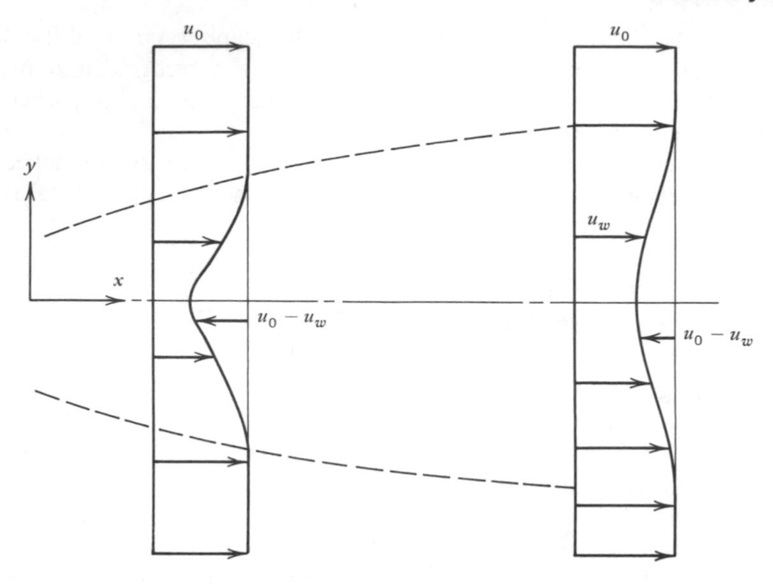

Figure 20-17 Laminar wake of a nonlifting body.

Substituting 20.13.3 into 20.13.2 produces terms in the following proportions:

$$u_0 \frac{\partial u}{\partial x} \propto u_0 B^2 x^{2q-1}[(p-q)f' - qf'']$$

$$-u\frac{\partial u}{\partial x} + v\frac{\partial u}{\partial y} \propto AB^2 x^{p+q-1}[(q-p)f'^2 + 2q\eta f'f'' - pff''] \quad 20.13.4$$

$$\nu \frac{\partial^2 u}{\partial y^2} \propto \nu f'''$$

In addition to these boundary-layer equations, the wake must also satisfy an integral constraint.

Recall from Section 15.6 that the drag force F_D is related to the constant momentum thickness of the wake as follows:

$$F_D = \rho u_0^2 \Theta = \rho u_0 \int \left[\frac{u_w}{u_0} - \left(\frac{u_w}{u_0} \right)^2 \right] dy$$

Introducing 20.13.1 and 20.13.3 puts this expression in the form

$$F_D = \rho A x^p \int \left[u_0 - \frac{A}{B} x^{p-q} f' \right] f' \, d\eta \qquad 20.13.5$$

The task of choosing p and q so that all terms in 20.13.4 and 20.13.5 are independent of x cannot be accomplished. The problem does not allow similarity.

The first term in 20.13.4 is independent of x if $q = \frac{1}{2}$. Likewise, the first term in the drag expression is independent of x if $p = 0$. If we assume these values, then the remaining terms in both equations contain $x^{-1/2}$. Hence, we propose to place an additional restriction on the problem by allowing x to become infinite. For this reason the solution is called a *far-wake* solution. Formally we are solving the problem as an asymptotic expansion where the x coordinate is the expansion parameter.

The equations governing the far wake are

$$f''' + \frac{u_0 B^2}{2\nu}[\eta f'' + f'] = 0 \qquad 20.13.6$$

$$F_D = \rho A u_0 \int f' \, d\eta = \rho u_0 Q \qquad 20.13.7$$

The relation $F_D = \rho u_0 Q$ may be verified by noting the dimensional form of 20.13.5. The differential equation 20.13.6 has the solution $f' = \exp(-\eta^2)$ when the coefficient in front of the bracket is two. Thus, we choose

$$B = \left(\frac{4\nu}{u_0}\right)^{1/2} \qquad 20.13.8$$

Inserting $f' = \exp(-\eta^2)$ into 20.13.7 reveals that A is given by

$$A = \frac{F_D}{\rho u_0 \pi} \qquad 20.13.9$$

The final expressions are

$$\eta = y\sqrt{\frac{u_0}{4\nu x}} \qquad 20.13.10$$

$$\psi = \frac{F_D}{2\rho u_0} \operatorname{erf} \eta \qquad 20.13.11$$

$$u_0 - u_w = \left(\frac{F_D}{4\pi\rho^2 u_0 \nu x}\right)^{1/2} \exp(\eta^2) \qquad 20.13.12$$

As the wake decays, the velocity defect decreases as $x^{-1/2}$ while the thickness of the wake increases parabolically, $y_\delta \sim x^{1/2}$.

This analysis may also be extended to apply to axisymmetric and to turbulent wakes. Hence, many of the conclusions are relevant to these cases as well as to plane laminar wakes.

The drag force on a body plays a very important role in determining the wake. From previous work we found that the momentum thickness of the wake is a constant given by $\Theta = F_D/\rho u_0^2$. This same expression (20.13.5) in the far wake reduces to $Q = F_D/\rho u_0 = u_0\Theta$ (20.13.7), where Q is the defect-velocity volume flow. Q represents the flow one would observe in a quiescent fluid after a moving object, such as an automobile, had passed by. From the point of view of a fixed body in an infinite stream, the integral which gives Q is actually the displacement thickness of the wake $Q = u_0\delta^*$. Hence $\delta^* = \Theta =$ const in the far wake of any nonlifting body.

The wake of a lifting body differs significantly from the results above. The circulation necessary to produce the lift requires that vortices leave the body. This additional vorticity in the streamwise direction changes the picture so that the analysis given above is no longer valid.

Recall that we originally proposed to consider a body at a high Reynolds number and apply boundary-layer theory to the wake. However, the resulting problem was not simple enough, so we further assumed that the distance x from the body was very large (the far wake). There is a surprise ending to this story. The answer obtained is valid for all Reynolds numbers greater than zero. The Oseen equations of slow viscous flow—low Re—give exactly the same wake solution as that found above. All bodies produce a parabolic wake for $\text{Re} > 0$. The actual influence of the body Reynolds number is such that we do not need to go very far downstream when Re is high.

20.14 SHEAR LAYERS

Consider two streams that are accelerated to different velocities while being separated by a thin flat plate as shown in Fig. 20-18. At $x = 0$ the plate ends and the two streams merge. The boundary layers that build up on the plate before the streams merge are ignored, and we assume that the initial profiles are $u = u_1$ in the upper fluid $y > 0$ and $u = u_2$ in the lower fluid $y < 0$. The constant u_1 will always be greater than zero; however, u_2 can be set equal to zero. This case corresponds to the flow off the lip of a large cavity or the shear layer at the lip of a jet issuing into a large room.

Lessen (1949) gave the solution for the case when the two fluids are the same, and Lock (1951) extended the analysis to apply to fluids with different densities and viscosities.

As the two streams merge at the lip of the plate, we assume that the pressure of both streams has the same value, so that the streamline leaves the plate horizontally. The subsequent position of the dividing streamline depends on the inviscid-flow field. For the case of completely uniform inviscid flows u_1 and u_2, this streamline will be straight at $y = 0$. Since pressures do not change across boundary layers, the subsequent growth of the layers does not modify the original inviscid position of the dividing streamline.

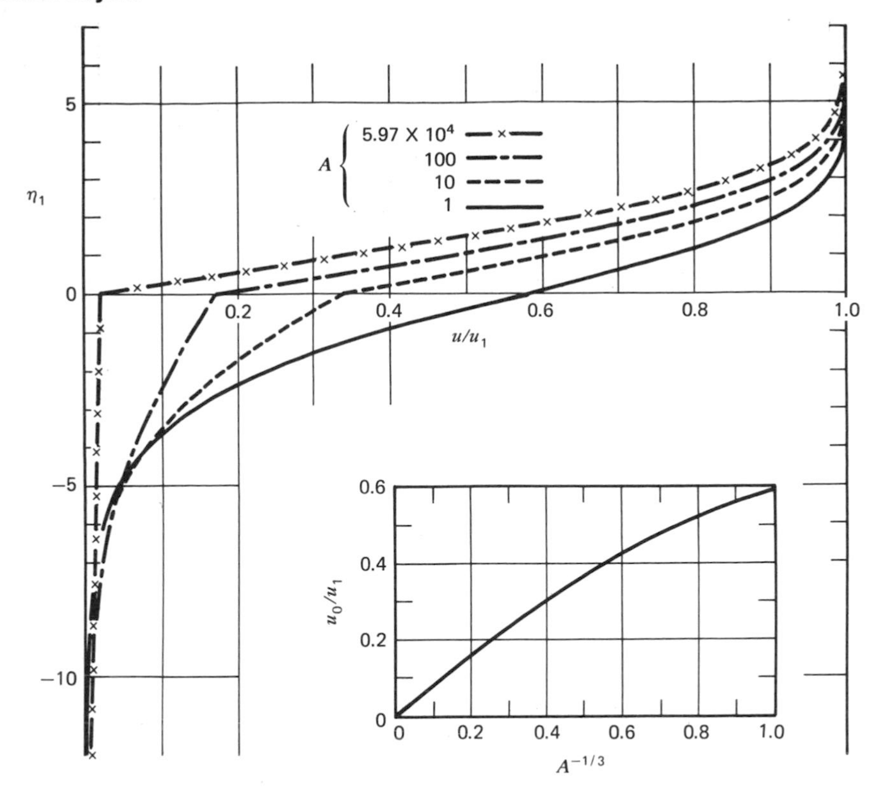

Figure 20-18 Velocity profiles between shear layers. $A = \rho_1\mu_1/\rho_2\mu_2$. From Lock (1951). Inset shows centerline velocity as a function of A.

With the absence of a pressure gradient, the boundary-layer equation is

$$u\frac{\partial u}{\partial x} + v\frac{\partial u}{\partial y} = \nu\frac{\partial^2 u}{\partial y^2} \qquad 20.14.1$$

The Blasius similarity variable is introduced:

$$\eta_\alpha = \left(\frac{u_1}{\nu_\alpha x}\right)^{1/2} y \qquad 20.14.2$$

In this section we use a subscript α to indicate a variable that may be used in either the lower fluid ($\alpha = 2$) or the upper fluid ($\alpha = 1$). The variable η_1 is used in the upper fluid ($y > 0$), while η_2 is used in the lower fluid ($y < 0$). The reference velocity for both fluids is u_2, since u_1 may take on the value of zero.

When the stream function

$$\psi_\alpha = (\nu_\alpha u_1 x)^{1/2} f_\alpha(\eta_\alpha) \qquad 20.14.3$$

is introduced, the velocities are

$$u_\alpha = \frac{u_\alpha}{u_1} = f'_\alpha \tag{20.14.4}$$

$$v_\alpha = \frac{1}{2}\left(\frac{u_1 \nu_\alpha}{x}\right)^{1/2}[\eta_\alpha f'_\alpha - f_\alpha] \tag{20.14.5}$$

and the differential equation 20.14.1 transforms into the Blasius equation,

$$2f'''_\alpha + f_\alpha f''_\alpha = 0 \tag{20.14.6}$$

The boundary conditions on the far stream show that

$$\begin{aligned} f_1(\infty) &= 1 \\ f_2(-\infty) &= \frac{u_2}{u_1} \end{aligned} \tag{20.14.7}$$

The interface between the fluids is by choice the streamline $\psi = 0$. In addition, the viscous interface conditions requiring continuity of the velocity and the shear stress imply that

$$f'_1(0) = f'_2(0) \tag{20.14.8}$$

and

$$f''_1(0) = \left(\frac{\rho_2 \mu_2}{\rho_1 \mu_1}\right)^{1/2} f''_2(0) \tag{20.14.9}$$

In many previous viscous-diffusion problems $\nu = \mu/\rho$ has occurred as the only important physical fluid property. In this problem the shear-stress condition $\mu_1\, \partial u_1/\partial y|_0 = \mu_2\, \partial u_2/\partial y|_0$ introduces the viscosity by itself.

The velocity profiles shown in Fig. 20-18 are from Lock (1951). These results were computed numerically. The case $\rho_2\mu_2/\rho_1\mu_1 = 5.97 \times 10^4$ represents air flowing over water with $u_2 = 0$. The figure also contains a graph of the interface velocity u_0 as a function of the parameter $\rho_2\mu_2/\rho_1\mu_1$ for $u_2 = 0$. When the same fluid is in both layers, the interface velocity is $u_0/u_1 = 0.58$.

*20.15 ASYMPTOTIC THEORY AND MATCHING

In simplified situations, boundary-layer theory may be organized as asymptotic theory in the mathematical sense. One of the reasons for doing this is to bring the matching conditions between the boundary layer and the inviscid flow into sharper focus. The other benefit is that asymptotic theory can produce corrections to compensate for things that are neglected in the first calculation. For example, the displacement thickness of the boundary layer requires a modification of the inviscid flow. The modification appears as the second term in an asymptotic expansion, while the basic flow, the inviscid flow for zero boundary-layer thickness, is the first term. The boundary layer itself receives several corrections in its second-order term. The fact that the body surface has longitudinal and (possibly) transverse curvature was ignored in the first boundary-layer calculation, but will cause an effect in the second term. Likewise, any weak vorticity in the inviscid flow causes a second-order effect. The last second-order effect is the new external velocity in the inviscid flow because of the displacement thickness. In other words, the boundary layer modifies the inviscid flow, which in turn produces a new surface speed for the boundary layer.

The different second-order effects were first investigated by several different researchers. Subsequently, Van Dyke (1962) and Maslen (1963) gave a general development of the theory. We shall outline the approach of Van Dyke.

Consider a smooth semiinfinite surface as shown in Fig. 20-19. As a simplification assume that the inviscid flow is free of vorticity. Within the flow an orthogonal coordinate system ξ, η with velocity components u, v, is established. The body surface coincides with a constant-η line; we define s as the distance along the surface and n as the distance perpendicular to the surface. All quantities are nondimensionalized by the free-stream velocity u_0, the nose radius L, and the density ρ.

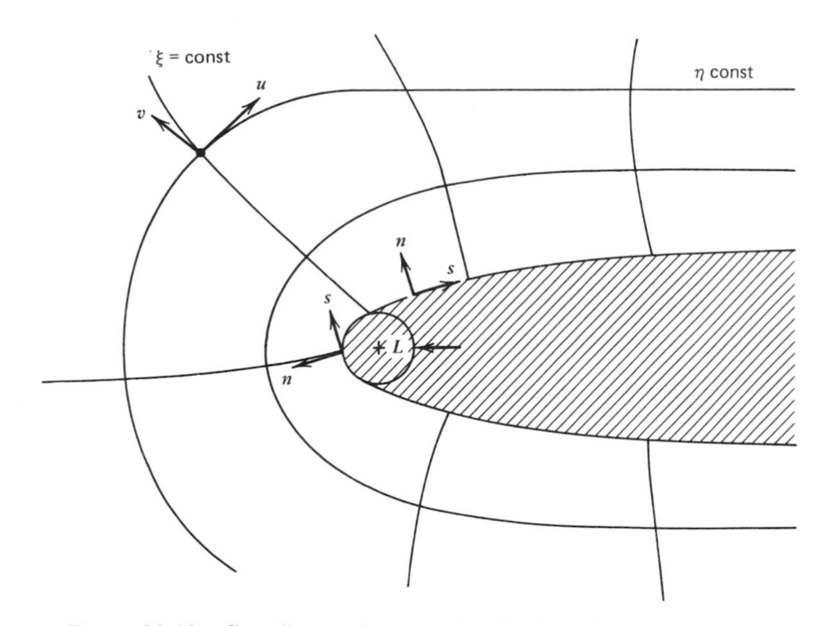

Figure 20-19 Coordinates for second-order boundary-layer matching.

We know that the flow is governed by the Navier–Stokes equations, where $\mathrm{Re} = u_0 L/\nu$ is a parameter. Assume the following asymptotic expansions are valid for $\mathrm{Re} \to \infty$:

$$\begin{aligned} u(\xi, \eta; \mathrm{Re}) &\sim U_1(\xi, \eta) + \mathrm{Re}^{-1/2} U_2(\xi, \eta) \\ v(\xi, \eta; \mathrm{Re}) &\sim V_1(\xi, \eta) + \mathrm{Re}^{-1/2} V_2(\xi, \eta) \\ p(\xi, \eta; \mathrm{Re}) &\sim P_1(\xi, \eta) + \mathrm{Re}^{-1/2} P_2(\xi, \eta) \\ \psi(\xi, \eta; \mathrm{Re}) &\sim \Psi_1(\xi, \eta) + \mathrm{Re}^{-1/2} \Psi_2(\xi, \eta) \end{aligned} \qquad 20.15.1$$

Substituting these relations into the Navier–Stokes equation and allowing $\mathrm{Re} \to \infty$ yields the Euler equations for U_1, V_1, and P_1 as the first-order inviscid flow. The second-order variables U_2, V_2, and P_2 are also governed by inviscid equations (with a linearized form of the momentum equation). Since the continuity equation $\nabla \cdot \mathbf{v} = 0$ and the irrotational conditions $\nabla \times \mathbf{v} = 0$ are linear, the second-order flow is also a potential flow.

The variables U_1, V_1 satisfy the upstream boundary condition, so in second order $U_2 = V_2 = 0$ will be appropriate conditions. On the body, the normal velocity $V_1(s, 0) = 0$ is specified, and the slip velocity $U_1(s, 0)$ is free to be determined by the solution. The second-order velocity V_2 will normally be required to complete the second-order potential flow. This velocity can only be found by matching with the boundary-layer solution. Hence, we must proceed to solve for the first-order boundary layer before we can, in principle, solve the second-order inviscid flow.

The first-order inviscid flow is a valid approximation to the Navier–Stokes problem everywhere in the flow except on the line representing the body surface. The no-slip condition required for the Navier–Stokes problem is not satisfied, because $U_1(s, 0)$ has finite values. The solution is singular on the boundary of the domain, the body surface, where a local coordinate system s, n is erected and the nondimensional inner variable is N defined:

$$N \equiv \mathrm{Re}^{1/2} \cdot n \qquad 20.15.2$$

This variable magnifies the region near the wall, since a fixed value of N moves closer to the wall (n decreases) as $\mathrm{Re} \to \infty$. The square-root scaling of Re is required because this is how the size of the singular region changes as $\mathrm{Re} \to \infty$.

The boundary-layer or inner expansions are

$$\begin{aligned} u(s, n; \mathrm{Re}) &\sim u_1(s, N) + \mathrm{Re}^{-1/2} u_2(s, N) \\ v(s, n; \mathrm{Re}) &\sim \mathrm{Re}^{-1/2} v_1(s, N) + \mathrm{Re}^{-1} v_2(s, N) \\ p(s, n; \mathrm{Re}) &\sim p_1(s, N) + \mathrm{Re}^{-1/2} p_2(s, N) \\ \psi(s, n; \mathrm{Re}) &\sim \mathrm{Re}^{-1/2} \psi_1(s, N) + \mathrm{Re}^{-1} \psi_2(s, N) \end{aligned} \qquad 20.15.3$$

Note that the first gauge function for v and ψ is $\mathrm{Re}^{-1/2}$. If we had written the expansion as

$$v \sim v_0 + \mathrm{Re}^{-1/2} v_1 + \mathrm{Re}^{-1} v_2$$

we would have found that v_0 is zero. An alternative way to write the equations of 20.15.3 is

$$v\,\mathrm{Re}^{1/2} \sim v_1 + \mathrm{Re}^{-1/2} v_2$$

This form shows that the dependent variable v needs to be rescaled in the boundary layer; as the layer becomes thinner, the vertical currents are correspondingly smaller. The proper nondimensional v in the boundary layer is $v\,\mathrm{Re}^{1/2}$.

Substituting 20.15.3 into the Navier–Stokes equations and letting $\mathrm{Re} \to \infty$ produces the boundary-layer equations. Solutions to these equations satisfy the no-slip condition at the wall, but need an additional boundary condition as $N \to \infty$. This condition is supplied by matching between the inviscid solution and the boundary layer.

Recall that Van Dyke's matching rule is

m-term inner expansion of *p*-term outer expansion

= *p*-term outer expansion of *m*-term inner expansion

Applying this rule with $m = 1, p = 1$ to the u expansion of 20.15.1 and 20.15.2 gives the familiar condition

$$u_1(s, N \to \infty) = U_1(\xi = s, \eta = 0) \qquad 20.15.4$$

This is the missing boundary condition on u_1, since the inviscid problem for U_1 may be found directly.

Next let us consider an $m = 2, p = 2$ match. The operations on the inviscid expansion are

2-term outer:

$$u \sim U_1(s, n) + U_2(s, n)\,\mathrm{Re}^{-1/2}$$

rewritten inner variable:

$$\sim U_1(s, N\,\mathrm{Re}^{-1/2}) + U_2(s, N\,\mathrm{Re}^{-1/2})\,\mathrm{Re}^{-1/2}$$

expanded ($R \to \infty$, N fixed) by a Taylor series:

$$\sim U_1(s,0) + U_{1n}(s,0) N\,\mathrm{Re}^{-1/2} + \cdots$$

$$+ U_2(s,0)\,\mathrm{Re}^{-1/2} + U_{2n}(s,0) N\,\mathrm{Re}^{-1} + \cdots$$

truncated at two terms in Re:

$$\sim U_1(s,0) + [U_2(s,0) + NU_{1n}(s,0)]\,\mathrm{Re}^{-1/2} \qquad 20.15.5$$

The operations on the boundary-layer expansion are
2-term inner:

$$u \sim u_1(s, N) + u_2(s, N)\,\mathrm{Re}^{-1/2}$$

rewritten in n:

$$\sim u_1(s, n\,\mathrm{Re}^{1/2}) + u_2(s, n\,\mathrm{Re}^{1/2})\,\mathrm{Re}^{-1/2}$$

The expression above must be expanded for $\mathrm{Re} \to \infty$, n fixed. To do this step we need to assume asymptotic expansions for u_1 and u_2 as $N \to \infty$. Assume $u_1(s, n) \sim u_1(s, \infty) + o[N^{-1}]$ as $N \to \infty$, and a similar expression for $u_2(s, N)$. Actually this is a generous assumption, as we know that boundary layers decay exponentially to the free stream. Here we only assume that the next term is of an order smaller than N^{-1}. The next step is

expanded ($\mathrm{Re} \to \infty$, n fixed)

$$u \sim u_1(s, \infty) + o[n^{-1}\,\mathrm{Re}^{-1/2}]$$

$$+ \cdots + u_2(s, \infty)\,\mathrm{Re}^{-1/2} + o[n^{-1}\,\mathrm{Re}^{-1}] + \cdots$$

Truncated at two terms in Re:

$$\sim u_1(s, \infty) + u_2(s, \infty)\,\mathrm{Re}^{-1/2} \qquad \text{as} \quad n \to \infty$$

Before matching, we must express the equation above and 20.15.5 in the same variable, n or N. We choose the inner variable N (intermediate expansions, for which the matching theoretically should be done, usually have the same form as the inner expansion). Hence

$$u \sim u_1(s, \infty) + u_2(s, \infty)\,\mathrm{Re}^{-1/2}$$

Matching this with 20.15.5 and noting 20.15.4 shows that

$$u_2(s, \infty) = U_2(s,0) + NU_{1n}(s,0)$$

A better form of this equation is found by inserting the fact that $U_{1n}(s,0) = -KU_1(s,0)$ where K is the wall curvature [differentiate the Bernoulli equation $P + \frac{1}{2}V_1^2 + \frac{1}{2}U_1^2 = C$ with respect to n, evaluate at the wall, and combine with the normal momentum equation (17.3.8) $P_{1n} = -KU_1^2$]. This produces

$$u_2(s, \infty) = U_2(s,0) - KNU_1(s,0) \qquad \text{as} \quad N \to \infty \qquad 20.15.6$$

The second-order boundary-layer velocity u_2 matches the second-order inviscid velocity U_2 at the wall plus a correction due to longitudinal wall curvature.

Now we turn to consider the vertical velocity. A match with $m = 1, p = 1$ simply gives the inviscid boundary condition,

$$V_1(s,0) = 0 \qquad 20.15.7$$

If we differentiate the series for V with respect to n before a match with $m = 1$, $p = 1$ is performed, we get

$$v_{1N}(s,\infty) = V_{1n}(s,0) \qquad 20.15.8$$

The boundary-layer v_1 does not match the inviscid V_1; however, the slopes match. Furthermore, if we take a two-term outer and a one-term inner expansion, we find that matching yields

$$V_2(s,0) = \lim_{N\to\infty} \left[v_1(s,N) - N v_{1N}(s,N) \right]$$

Another form of this equation uses 20.15.8 and the fact that the continuity equation on the wall is $V_{1n}(s,0) = -U_{1s}(s,0)$. The result is

$$V_2(s,0) = \lim_{N\to\infty} \left[v_1(s,N) + N U_{1s}(s,0) \right] \qquad 20.15.9$$

This is the missing boundary condition for the second-order inviscid velocity in terms of the first-order boundary layer. To see that this is the displacement effect, compare it with 20.7.4 and 20.7.5.

To summarize, the outer flow expansion is valid over the plane area of the flow field with the exception of a certain line. Boundary-layer theory is an expansion that extends the solution from the line. The boundary-layer expansion itself may be singular at certain points on the line: the leading edge, trailing edge, or points where the wall slope is discontinuous. Additional expansions are required to make the solution valid in the neighborhood of these point singularities.

*20.16 ENTRANCE FLOW INTO A CASCADE

We shall study a slightly different way in which inviscid flows fail to be uniformly valid. Consider the entrance flow problem for which a complete solution was given in Section 15.4. The computer solution given there required an increasingly long computation domain as Re became large. In fact, the size of the hydrodynamic entrance region increases linearly with Re. For high Re this may be broken into several parts. Van Dyke (1970) is also responsible for the organization of this problem.

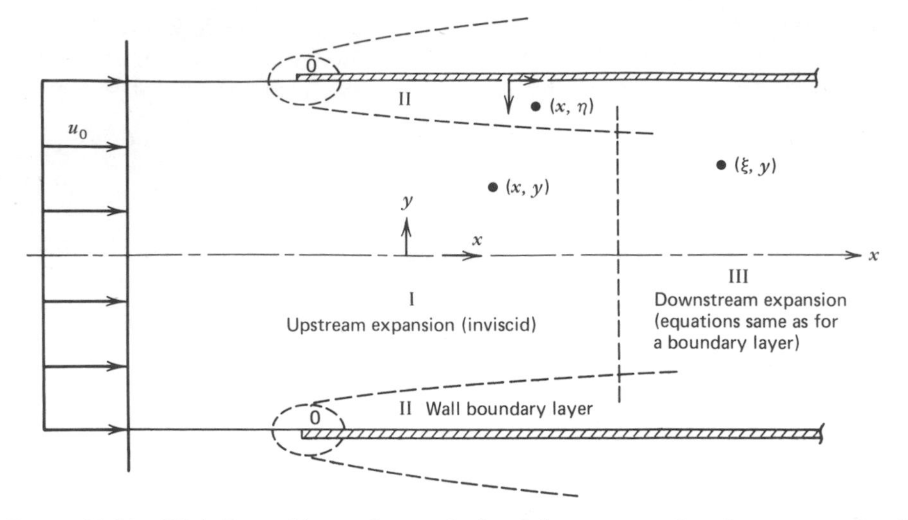

Figure 20-20 High-Reynolds-number analysis of the entrance flow into a cascade of plates.

Figure 20-20 is a sketch of the geometry and velocity profiles. The uniform stream u_0 at $x = -\infty$ feels the presence of the plate and begins slowing down before the leading edges are reached. At the same time a slight acceleration of the centerline fluid is observed (due to continuity, of course). Within the cascade, the profiles gradually change until the parabolic profile of uniform flow is reached as $x \to \infty$.

The Re $\to \infty$ (for fixed x) limit of the Navier–Stokes equations where u_0, ρ, and h are the scales, leads to the Euler equations. Any fixed point x, y is in inviscid flow when Re becomes high enough. The solution is a constant velocity $U_1 = u/u_0 = 1$, $V_1 = 0$. This answer is valid in the region marked I in Fig. 20-20, that is, upstream and into a core area within the cascade. This solution fails to meet the no-slip condition along both walls (the classic boundary-layer failure), and it also fails to satisfy the parabolic profile prescribed as the downstream condition for the complete Navier–Stokes problem.

Regions II on both walls are boundary layers on a flat plate with a uniform stream. Hence, the Blasius solution gives the correction to the inviscid flow in these regions. For the sake of completeness, note that the Blasius boundary layer fails at the leading edge, the region marked 0. Several solutions for this region have been given, as typified by Davis (1972).

Region III is a new type of failure of the inviscid flow. For internal flow, the region $x \to \infty$ is described incorrectly by the solution $U_1 = u_0$. The nonuniform region is not a thin region near a wall, but quite a large region within the cascade. It is true that the boundary layers of regions II would grow and ultimately fill the slot, but this is not the failure mechanism of region III. Region III begins long before the boundary layers gain any significant thick-

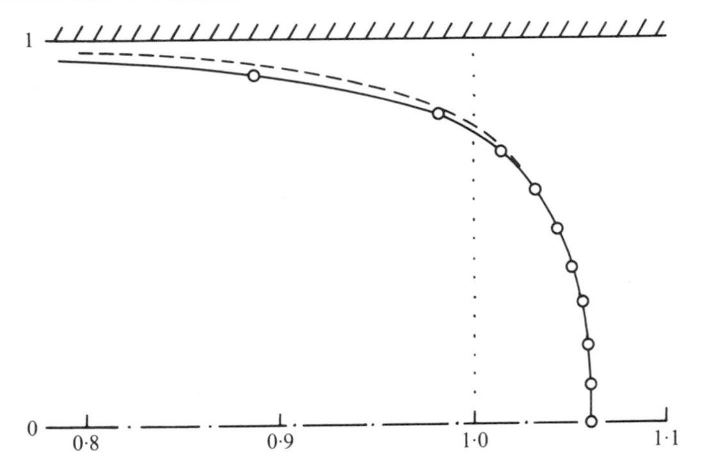

Figure 20-21 Velocity profiles at the entrance $x = 0$ for Re = 75, van Dyke (1970). Dashed line is the two term upstream expansion, solid line is composite expansion with Blasius boundary layer, and circles are the Navier-Stokes numerical solution of Wang and Longwell (1964).

ness. Region III is due to the failure of $U_1 = u_0$ to meet the parabolic profile. The position where region III begins is $x \sim 0(\text{Re})$.

A solution for region III is formulated by defining a stretched x variable

$$\xi = \frac{x}{\text{Re}} \qquad 20.16.1$$

Now as $\text{Re} \to \infty$, a $\xi = \text{const}$ position moves downstream proportionally. Since $\partial\psi/\partial x = -v$, we have $\partial\psi/\partial\xi = -\text{Re}\cdot v$, implying that the region-III scaling for v should be changed by defining $v^{\text{III}} = v\,\text{Re}$. All other variable retain the nondimensional forms of the original Navier–Stokes equations. The momentum equations for region III are

$$\begin{aligned} uu_\xi + v^{\text{III}}u_y &= -p_\xi + \text{Re}^{-2}u_{\xi\xi} + u_{yy} \\ \text{Re}^{-2}uv_\xi^{\text{III}} + \text{Re}^{-2}v^{\text{III}}v_y^{\text{III}} &= -p_y + \text{Re}^{-1}v_{\xi\xi} + \text{Re}^{-1}v_{yy} \end{aligned} \qquad 20.16.2$$

The ordering of terms in these equations is different from that in the usual boundary layer, although the limit $\text{Re} \to \infty$ gives the boundary-layer equations.

The boundary conditions for 20.16.2 are the no-slip condition on the walls and matching with the first-order inviscid flow:

$$u(\xi = 0, y) = u_0 \qquad 20.16.3$$

Equations 20.16.2 automatically produce the parabolic profile as $\xi \to \infty$. The problem above is termed the downstream expansion, and the region-I solution is called the upstream expansion.

There is a question whether we should call the downstream expansion (20.16.2 and 20.16.3) a boundary layer or not. The differential equations for the problem are the boundary-layer equations, but the boundary conditions and method of matching are quite unlike those of Section 20.15. The shear layers in this problem are thin to start out, but by the end of region III they cover the tube. The idea used in the original solution of the entrance-flow problem by Schlichting (1934) was that two boundary layers on the walls grew and simultaneously interacted with an inviscid core. This must be considered as an approximate solution for the downstream expansion in the light of Van Dyke's formulation.

Van Dyke went on to calculate the second-order inviscid flow, the upstream expansion. This flow accounts for the displacement thickness of the side-wall boundary layers. Figure 20-21 gives the velocity profile at $x = 0$ calculated for $\mathrm{Re} = 75$ with the second-order inviscid expansion and compared with a complete Navier–Stokes calculation. Because of the good agreement, we can assert that the physical event that modifies the flow ahead of the plates is inviscid. The flow is essentially the inviscid flow into a channel modified for the displacement thickness of the boundary layers. Although our complete solution in Section 15.5 at the same Re would be slightly more accurate, we would not be able to distinguish inviscid effects from viscous diffusion effects as in the $\mathrm{Re} \to \infty$ theory.

*20.17 THREE-DIMENSIONAL BOUNDARY LAYERS

The idea that a flow has thin regions where viscous forces are important is, of course, not restricted to plane flows. There are several unique physical events that may occur in three-dimensional boundary layers. We shall discuss several of these before we introduce the three-dimensional boundary-layer equations.

The first important effect is lateral convergence or divergence of the flow. An axisymmetric stagnation point illustrates this; the streamlines diverge when they move away from the stagnation point. By the continuity equation we expect that divergence will cause thinning of the boundary layer. Conversely, convergence should cause thickening. An axisymmetric stagnation point has a thickness about 80% of its two-dimensional counterpart. Indeed, in axisymmetric flow without swirl, streamline convergence is the only three-dimensional effect. It is intimately tied to the increase or decrease in the surface radius $r_0(x)$. Recall that r_0 only occurs in the continuity equation 20.11.1, and that this is the only way in which the axisymmetric boundary-layer equations (without swirl) differ from the two-dimensional equations.

A second, and perhaps the most important, three-dimensional effect is the secondary flow caused by transverse pressure gradients. In plane flows the longitudinal pressure gradient causes the flow either to speed up or to slow down. Another degree of freedom exists in three-dimensional layers. Now, transverse pressure gradients are allowed to change the direction of the free stream by pushing sideways on the particles. It is still true that the pressure is constant across a boundary layer, so the full force of the transverse pressure is also applied to the low-velocity particles deep within the boundary layer. This causes the low-velocity particles to have a more tightly curved path than the high-velocity particles at the edge of the layer. As an example, consider the situation where a fluid is in solid-body rotation about an axis that is perpendicular to a wall. Far from the wall, the circular streamlines of the rotation can only exist if the pressure increases in the radial direction so that the centrifugal force is balanced. This balance is upset near the wall, where the viscous forces slow the particles and thereby reduce the centrifugal force. The undiminished pressure force causes the fluid near the wall to flow inward in the boundary layer. The picture is completed by an outward flow along the axis to satisfy the conservation of mass.

Another example of this same effect is found in the cyclone separator as shown in Fig. 20-22. The fluid is injected tangentially in the large-diameter portion of the chamber. The momentum of the fluid is used to establish a pressure gradient from the core to the outer wall. The mass flow through the machine is relatively small and is really just the secondary flow. The slanted

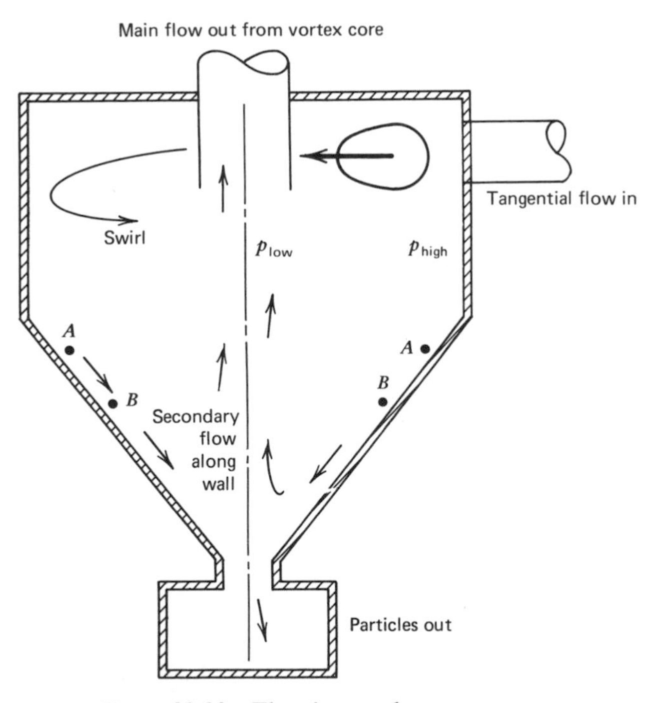

Figure 20-22 Flow in a cyclone separator.

wall is where the boundary-layer events take place. Because the point A is farther from the rotation axis than B, it has a higher pressure. Hence, when viscous forces slow down the particles at the wall, the pressure gradient pushes them toward the apex. This boundary-layer fluid, which has lost most of its swirling velocity, then turns and flows through the core and out the top through the *vortex finder*. The vortex finder is placed inside the chamber to inhibit the same process on the upper wall (it is sometimes an advantage if the flow in the top region will separate).

The chamber at the bottom has little or no flow. It is used to collect the particles that are to be separated from the fluid. As the particles enter the cyclone with the main stream, they are pushed toward the walls, since they are too dense to be turned by the pressure gradient. Once concentrated in this region, they are swept toward the apex by the secondary boundary-layer flow. Now, as the flow turns to ascend through the core (and here a shallow angle to the core helps), the inertia of the particles separates them from the flow and they collect in the lower chamber.

To summarize: the transverse pressure gradient in a boundary layer causes streamlines within the layer to curve more tightly than those of the free stream. Thereby a secondary crossflow develops.

Another important aspect of three-dimensional boundary layers is caused by the secondary flow itself. We might term this a spreading transverse influence. Consider all the particles in the boundary layer at a certain wall location at a certain time. As time goes on, these particles are carried downstream. Particles in the external flow follow certain streamlines to new positions, while particles within the layer, because of the secondary flow, follow streamlines to different transverse positions from those of the external flow. As the flow goes along, particles that were originally neighbors tend to separate. The next section shall illustrate this effect more fully, and we defer discussion until then.

We now discuss another aspect in which three-dimensional boundary layers differ from two-dimensional ones. Recall that the wall geometry plays no direct role in two-dimensional layers. The equations and boundary conditions were free of any wall curvature or coordinate-scale effects. The situation in three-dimensional layers is slightly different. Consider an orthogonal net x, z on the surface of the body with y everywhere perpendicular to the wall. A point in the boundary layer is given (for sufficiently smooth surfaces and thin boundary layers) by the vector position

$$\mathbf{R} = \mathbf{r}(x, z) + y\mathbf{n}(x, z) \qquad 20.17.1$$

The scale factors of this net are

$$h_x \equiv \left|\frac{d\mathbf{r}}{dx}\right|, \qquad h_y = 1, \qquad h_z \equiv \left|\frac{d\mathbf{r}}{dz}\right| \qquad 20.17.2$$

A complete derivation of the three-dimensional boundary-layer equations may be found in Rosenhead (1964) or Moore (1964) (see also Dwyer 1982). For steady flows the equations are

$$\frac{u}{h_x}\frac{\partial u}{\partial x} + \frac{w}{h_z}\frac{\partial u}{\partial z} + v\frac{\partial u}{\partial y} + \frac{uw}{h_x h_y}\frac{\partial h_x}{\partial z} - \frac{w^2}{h_x h_z}\frac{\partial h_z}{\partial x} = -\frac{1}{\rho h_x}\frac{\partial p}{\partial x} + \nu\frac{\partial^2 u}{\partial y^2} \qquad 20.17.3$$

$$\frac{u}{h_x}\frac{\partial w}{\partial x} + \frac{w}{h_z}\frac{\partial w}{\partial z} + v\frac{\partial w}{\partial y} - \frac{u^2}{h_x h_z}\frac{\partial h_x}{\partial z} + \frac{uw}{h_x h_y}\frac{\partial h_z}{\partial x} = -\frac{1}{\rho h_z}\frac{\partial p}{\partial z} + \nu\frac{\partial^2 w}{\partial y^2} \qquad 20.17.4$$

$$0 = \frac{\partial p}{\partial y} \qquad 20.17.5$$

$$\frac{1}{h_x h_y}\left\{\frac{\partial}{\partial x}(h_z u) + \frac{\partial}{\partial z}(h_x w)\right\} + \frac{\partial v}{\partial y} = 0 \qquad 20.17.6$$

The equations above differ from their two-dimensional counterparts by some extra convection terms (a result of the fact that the flow may now have a w velocity) and the explicit appearance of the coordinate scale factors $h_x(x, z)$ and $h_z(x, z)$.

Three-dimensional boundary layers have three unknown velocities as functions of three coordinates—a considerably more complicated situation than the two-dimensional case. Moreover, surface-curvature effects now enter the problem through h_x and h_z. It is not possible to pin down the exact way in which curvature influences the boundary layer, because we still have a great deal of flexibility in placing an orthogonal net on the surface. For the most general surface $h_x(x, z)$ and $h_z(x, z)$ will exist and exert some influence in the convection terms. If a surface is developable, that is, if it can in principle be constructed from a flat sheet without stretching, then it is possible to choose a coordinate system where $h_x = h_z = 1$. Hence, it is only for surfaces more complicated than developable surfaces that the curvature actually affects the boundary-layer equations.

Boundary conditions for the equations are the no-slip condition and as $y \to \infty$ a match of the velocities u and w with the corresponding inviscid components u_e and w_e evaluated at the wall. The pressure in the layer is the inviscid pressure. Compatibility conditions between u_e, w_e, and the pressure are found by assuming that $u \to u_e$ and $w \to w_e$ exponentially fast in y. This is equivalent to striking all derivatives with respect to y in 20.17.3 and 20.17.4

and replacing u and w with u_e and w_e. The resulting compatibility equations, unlike the two-dimensional case, do not lead to Bernoulli's equation for the free stream. The reason is that 20.17.3 through 20.17.6 allow an arbitrary distribution of vertically directed vorticity ω_y at $y \to \infty$. Several interesting boundary solutions do have $\omega_y \neq 0$ in the free stream. If the inviscid flow does have $\omega_y = 0$, this extra information may be used in conjunction with the compatibility conditions to derive Bernoulli's relation.

The three-dimensional boundary-layer equations are so complicated that numerical computers must be used in all but the most simplest cases. Because of the need to take account of the surface geometry and the spreading transverse influence, these programs are themselves very complicated.

*20.18 BOUNDARY LAYER WITH A CONSTANT TRANSVERSE PRESSURE GRADIENT

In this example we consider a flat plate with a leading edge at $x = 0$. The flow approaches the leading edge at an angle θ_0 (Fig. 20-23). For $x > 0$ the boundary conditions are

$$\begin{aligned} u &\to u_0 = \text{const} \\ w &\to w_0 = u_0(a + bx) \end{aligned} \qquad 20.18.1$$

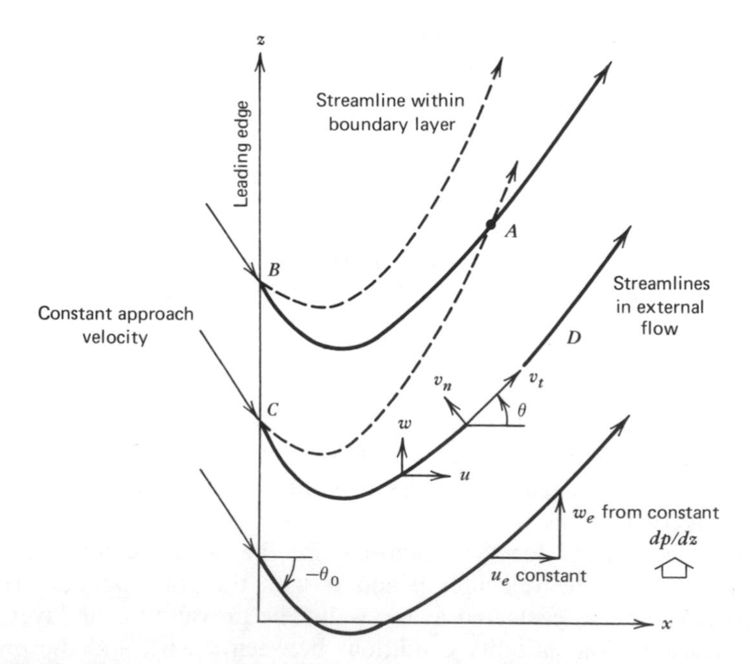

Figure 20-23 Plan view of the boundary layer on a plate with a constant transverse pressure gradient.

(This flow has a constant external vorticity ω_y.) The compatibility conditions show that this corresponds to pressure gradients

$$\frac{\partial p}{\partial x} = 0, \qquad -\frac{1}{\rho}\frac{\partial p}{\partial z} = u_0 \frac{\partial w_0}{\partial x} = u_0^2 b \tag{20.18.2}$$

Hence, the constant b indicates the magnitude of the transverse pressure gradient. If $b = 0$, the external flow transverses the plate at a constant angle

$$\theta = \theta_0 = \arctan \frac{w_0}{u_0} = \arctan a \tag{20.18.3}$$

The constant a is the sweep angle of the plate with respect to the free stream. The special case $b = 0$ is of interest, as it shows the effect of sweep on a Blasius boundary layer.

Two researchers, Sowerby (1954) and Loos (1955), independently solved this problem. We give only the highlights; the student who is interested in the details should consult to original works.

The inviscid-flow streamlines $z_0(x)$ are easily found from 20.18.1:

$$\frac{dz_0}{dx} = \frac{w_0}{u_0} = a + bx$$

$$z_0 = ax + \tfrac{1}{2}bx^2 + c \tag{20.18.4}$$

where c is a constant that gives z_0 at the leading edge. The streamlines are a system of translated parabolas as depicted in Fig. 20-24.

The boundary-layer equations 20.17.3–20.17.6 for this problem become

$$u\frac{\partial u}{\partial x} + v\frac{\partial u}{\partial y} = \nu\frac{\partial^2 u}{\partial y^2} \tag{20.18.5}$$

$$u\frac{\partial w}{\partial x} + v\frac{\partial w}{\partial y} = bu_0^2 + \nu\frac{\partial^2 w}{\partial y^2} \tag{20.18.6}$$

$$\frac{\partial u}{\partial x} + \frac{\partial v}{\partial y} = 0 \tag{20.18.7}$$

The x-momentum and continuity equations do not contain w, and in fact, they are identical to the Blasius problem. The solution for the x direction follows immediately as

$$\frac{u}{u_0} = f'(\eta) \tag{20.18.8}$$

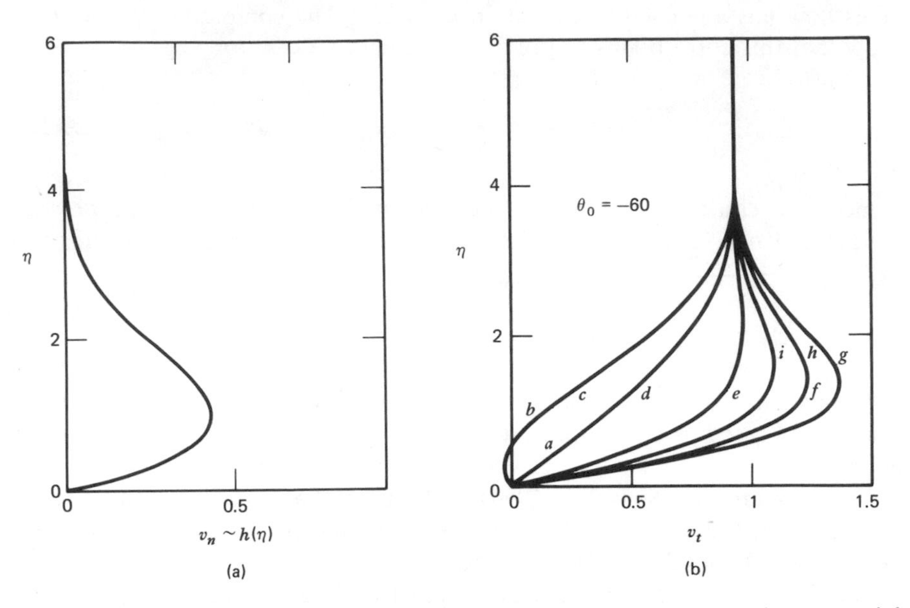

Figure 20-24 Typical velocity profiles: (a) transverse profiles $h(\eta)$, (b) tangential profiles for various flow angles: a, $\theta = -60°$; b, $\theta = -40°$; c, $\theta = -20°$ (same curve as b); d, $\theta = 0°$ (same as a); e, $\theta = 20°$; f, $\theta = 40°$; g, $\theta = 60°$; h, $\theta = 80°$ (same as f); i, $\theta = 90°$.

where

$$\eta \equiv y\sqrt{\frac{u_0}{\nu x}} \tag{20.18.9}$$

and $f(\eta)$ is the Blasius function obeying

$$ff'' + 2f''' = 0 \tag{20.18.10}$$

The fact that the problem separates out so that u and v may be solved without regard for w is known as the *sweep-independence principle*. It applies to any yawed cylindrical shape, no matter what the cross section (flat plate, circle, ellipse, airfoil). In all of these flows the boundary layer for flow in the direction normal to the leading edge may be calculated as a two-dimensional boundary layer. Subsequently this flow is used to calculate the cross-flow component parallel to the leading edge. As a practical example there is a strong spanwise outflow near the trailing edge of a swept wing.

The transverse flow due to the constant pressure force is a little more complicated. The assumed form of the velocity is

$$\frac{w}{w_0} = \frac{w_0(x)}{u_0}f'(\eta) + bxh(\eta) \tag{20.18.11}$$

Note that the first term is a Blasius component rescaled to match $w_0(x)$ (20.18.1). Substitution of 20.18.11 into the different equations and considerable algebra will yield the following equation for $h(\eta)$:

$$h'' + \tfrac{1}{2}fh' - f'h + 1 - (f')^2 = 0 \qquad 20.18.12$$

Appropriate boundary conditions are $h(0) = h(\infty) = 0$. Loos realized that the equation for h was the same equation solved and tabulated previously by Mager and Hansen (1952).

An informative way to look at the results is to refer the velocities to the inviscid-flow streamline. Let v_∞ be the magnitude of the inviscid velocity at any point, and v_t and v_n the velocity components along and normal to the inviscid streamline. It turns out that if one uses the local inviscid streamline angle θ as a variable, the velocity in the tangent direction is

$$\frac{v_t}{v_\infty} = \tfrac{1}{2}\sin 2\theta\,(\tan\theta - \tan\theta_0)h(\eta) + f'(\eta) \qquad 20.18.13$$

and the velocity in the normal direction is

$$\frac{v_n}{v_\infty} = \tfrac{1}{2}\cos^2\theta\,(\tan\theta - \tan\theta_0)h(\eta) \qquad 20.18.14$$

where f' is the Blasius function and h is the solution of 20.18.12.

In discussing the results let us first consider a uniform stream with a flat plate at a sweep angle θ_0. This is the case $b = 0$ and $\theta = \theta_0$. Equations 20.18.13 and 20.18.14 show that the boundary layer is simply a Blasius profile with no secondary flow at all. [A general statement may be made about the absence of secondary flows. If the external inviscid streamlines coincide with geodesic curves of the surface, then the boundary-layer streamlines have the same direction as the main stream and no secondary flow exists (Squire 1965).]

A typical example with a transverse pressure gradient is shown in Fig. 20-24. Here tangential velocity profiles are presented as a function of θ for the case $\theta_0 = -60°$. The normal velocity profiles are not given because 20.18.14 shows that they are simply a multiplicative factor times $h(\eta)$. The first profile of Fig. 20-24b at $\theta = -60°$ is the Blasius profile. As we proceed along the flow to $\theta = -40°$ and $\theta = -20°$, the profiles show negative values of v_t near the wall. This is not separation or even a region of backflow; the normal velocity at the corresponding positions is large enough to keep the flow moving toward positive values of x.

Moving farther along in the flow, that is, to higher values of θ, we see that v_t develops an overshoot. The maximum velocity is within the boundary layer. Another view is given by the polar diagram of Fig. 20-25. This very interesting phenomena occurs frequently in three-dimensional boundary layers. To explain it we must consider the history of particles in the upper part of the boundary layer. Near the leading edge, viscosity retards the particles into the Blasius profile while pressure begins to accelerate the flow in the transverse direction to develop the $h(\eta)$ profile. Particles in the upper part of this profile (from the inflection point outward, to be precise) are experiencing a net viscous force in the same direction as the pressure force. This aiding viscous force dies out as $y \to \infty$, where particles receive only the pressure force. Hence, there is a thin region at the top of the boundary layer where the particles experience a pressure force and a net viscous force in the same direction. These particles have more transverse acceleration than the external stream, which experiences only the pressure force. Farther along in the flow, the stream turns to higher θ, and the extra velocity from the viscous acceleration appears as an overshoot in the v_t profiles. This is another example of viscous acceleration increasing the Bernoulli constant.

The surface streamlines may be computed exactly and they are also parabolas. The equation is

$$z_s(x) = ax + \tfrac{1}{2}bx^2\left(1 + \frac{h'(0)}{f''(0)}\right) + C_0 \qquad 20.18.15$$

The curvature of the surface streamlines is greater than that of the external flow by a factor $1 + h'(0)/f''(0) \approx 1\tfrac{1}{3}$. Note that this particular flow never separates.

One of the most important things to be learned from this problem concerns the spreading lateral influence mentioned in Section 20.17. Wang (1971) pointed out that particle paths are subcharacteristics of the three-dimensional boundary-layer equations. In a sense, a particle carries its initial conditions and history wherever it goes within the boundary layer. Return to Fig. 20-23,

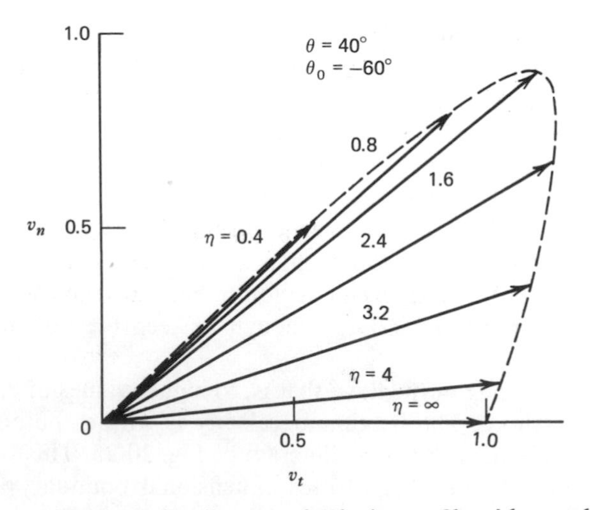

Figure 20-25 Polar diagram of velocity profile with overshoot.

where a surface streamline *CA* and two external flow streamlines *CD* and *BA* are shown. All the particles in the boundary layer at point *A* have come from the region *ABC*. Changing the initial conditions or boundary conditions anywhere in the region *ABC* has an effect on the velocity profile at *A*. From another standpoint, consider the boundary layer at point *C*. Particles that pass through *C* ultimately influence the flow within the region *ACD*. This is called the region of influence of point *C*. As we proceed in the flow direction, the region of influence spreads in the lateral direction. Any numerical calculation scheme for boundary layers needs to take account of the spreading lateral influence.

*20.19 HOWARTH'S STAGNATION POINT

The plane or axisymmetric stagnation point we studied previously is a very special case. In order to introduce a more general way in which a flow meets a body, consider a uniform stream that approaches a circular cylinder that has a sinusoidal radius variation as shown in Fig. 20-26. At each maximum and minimum of the radius (points *A*, *B*, and *C*) there is a stagnation point. From *C* to *B* and from *A* to *B* there is an attachment line dividing the flow that goes on either side of the cylinder. The arrows on the surface streamlines indicate the direction of the ideal inviscid flow a short distance directly above the surface.

To analyze the boundary layer in the neighborhood of one of these stagnation points, we take a coordinate system with x in the upward direction, y normal to the surface, and z in the longitudinal direction. The inviscid surface velocity in the neighborhood of one of these points, and for zero

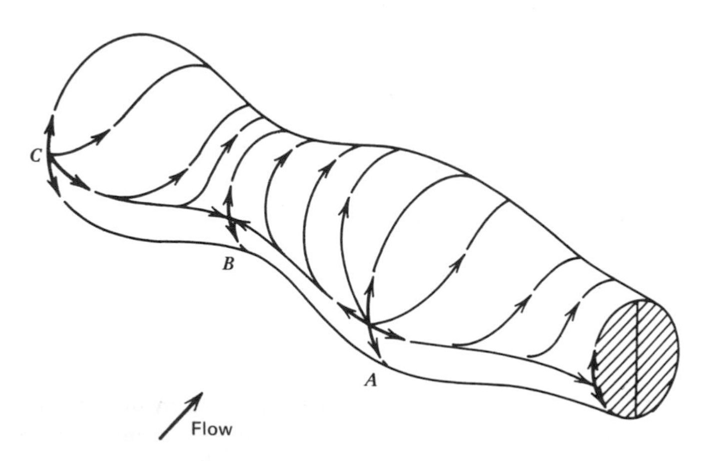

Figure 20-26 Ideal flow over a wavy cylinder. Surface streamlines emanate from nodal attachment points *A* and *C*.

free-stream vorticity, was shown by Howarth (1951) to be

$$u_e = ax, \qquad w_e = bz \tag{20.19.1}$$

where a and b are constants that depend upon the free-stream velocity and the size and shape of the body. Note that $b = 0$ is the plane-flow case, while $b = a$ is the axisymmetric case. There is no loss of generality in requiring that $|a| > |b|$. Streamlines in the external flow are given by the equation

$$x_0 = cz^{1/\alpha} \tag{20.19.2}$$

where

$$\alpha \equiv b/a \tag{20.19.3}$$

and c is a constant that gives a specific streamline. The cases $0 < \alpha \leq 1$ display nodal attachment (stagnation) points, $\alpha = 0$ is the plane-flow case (stagnation line), and the cases $-1 < \alpha < 0$ display saddle points of attachment where b is negative and a positive. Typical patterns of streamlines are shown in Fig. 20-27.

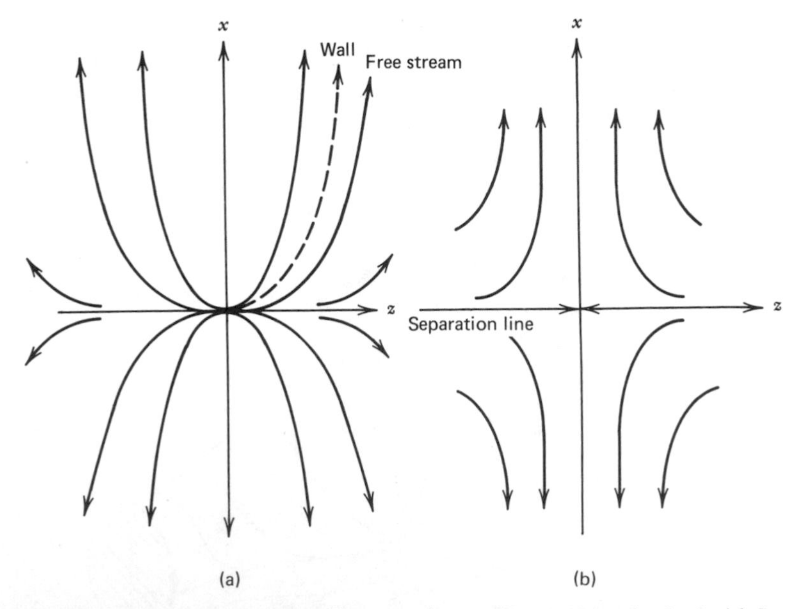

Figure 20-27 Plan view of streamlines on the surface and in the inviscid flow for a typical Howarth stagnation point: (a) nodal attachment point as points A and C of Fig. 20-26, (b) saddle attachment point as point B of Fig. 20-26.

The boundary-layer equations for this flow are

$$u\frac{\partial u}{\partial x} + v\frac{\partial u}{\partial y} + w\frac{\partial u}{\partial z} = a^2 x + \nu\frac{\partial^2 u}{\partial y^2} \tag{20.19.4}$$

$$u\frac{\partial w}{\partial x} + v\frac{\partial w}{\partial y} + w\frac{\partial w}{\partial z} = b^2 z + \nu\frac{\partial^2 w}{\partial y^2} \tag{20.19.5}$$

$$\frac{\partial u}{\partial x} + \frac{\partial v}{\partial y} + \frac{\partial w}{\partial z} = 0 \tag{20.19.6}$$

Howarth's stagnation-point analysis is presented in variables similar to those used in the plane-flow case. They are

$$\eta \equiv \left(\frac{a}{\nu}\right)^{1/2} y \tag{20.19.7}$$

$$\frac{u}{u_e} = \frac{u}{ax} = f'(\eta) \tag{20.19.8}$$

$$\frac{w}{w_e} = \frac{w}{bz} = g'(\eta) \tag{20.19.9}$$

These assumptions lead to a coupled set of equations with α as a parameter:

$$(f')^2 - (f + \alpha g)f'' = 1 + f''' \tag{20.19.10}$$

$$(g')^2 - \left(g + \frac{1}{\alpha}f\right)g'' = 1 + \frac{1}{\alpha}g''' \tag{20.19.11}$$

and with boundary conditions

$$f = g = f' = g' = 0 \qquad \text{at} \quad \eta = 0$$

$$f' = g' = 1 \qquad \text{at} \quad \eta \to \infty$$

Howarth gave numerical solutions for $\alpha = 0$, 0.25, 0.5, 0.75, and 1. Davey (1961) realized that the saddle points of attachment were missing and produced the results for negative α.

The pressure gradient in the x direction is (by assumption) always greater than that in the z direction. Hence within the boundary layer particles curve more rapidly toward the x direction than do those in the external flow. For example, wall streamlines are given by

$$x_s = cz^{f''(0)/[g''(0)\alpha]} \tag{20.19.12}$$

$f''(0)/g''(0)$ depends on α but is always greater than one. Thus, the exponent in 20.19.12 is greater than the exponent of 20.19.2.

We have solved for the flow in the neighborhood of the points A, B, and C in Fig. 20-26. Note how important these flows, and the separation lines from C to B and from A to B, are to determining the boundary layer over the entire cylinder. Mathematically all the inviscid surface flow from C to B and over the top of the cylinder originates at the stagnation point C (or within ϵ of it, to be exact). Likewise, all particles from $y = 0$ to $y = \delta$ in the stagnation point at C (or within ε of the point C) are carried out towards B and over the top of the cylinder in an ever thinning sheet near the surface of the boundary layer, the upper portions of the boundary layer being entrained from the external flow. Hence, by the effect of lateral spreading, the flow near the nodal point C has an influence, very much diminished at remote locations of course, on the entire boundary layer from C to B.

The points A, B, and C are called *singular points* of the surface streamlines because the shear stress (and vorticity) is zero at these points. Point B is a *saddle point*, because only two distinct streamlines pass through it. All other surface streamlines that approach B turn away. Points A and C are termed *nodal points*, because all streamlines in the neighborhood pass through these points. In fact, all streamlines except one come in to the point along a tangent line. A nodal point where all streamlines come together without a common tangent is a *focal node* (or focus).

The axisymmetric stagnation point is a focal node. Furthermore, the attachment point for an axisymmetric swirling flow such as von Kármán's problem, where the streamlines are spiral paths, is a focus of the spiral-node type. These three categories—saddles, nodes, and focal nodes—are all the possible patterns of surface streamlines in the neighborhood of singular points. The same classification is also valid when the flow direction is reversed and the singular points are points where the flow separates; that is, one has saddles, nodes, and focal points of separation.

*20.20 SEPARATION

Fluid flows frequently take on three-dimensional patterns and separate from the walls. The slightest transverse pressure force sends the wall streamlines on highly curved paths, modifies the effective shape of the wall, and makes the entire flow three-dimensional. Separation under these conditions is so complicated that little quantitative information, either experimental or computational, is known. Most information that exists has been obtained from flow-visualization experiments. For this reason, this section will mainly deal with the qualitative aspects of separation and the classification of different streamline patterns.

Consider, as a simple example, the idealized separation bubble shown in Fig. 20.28. The flow approaches a saddle separation point at A, goes over the

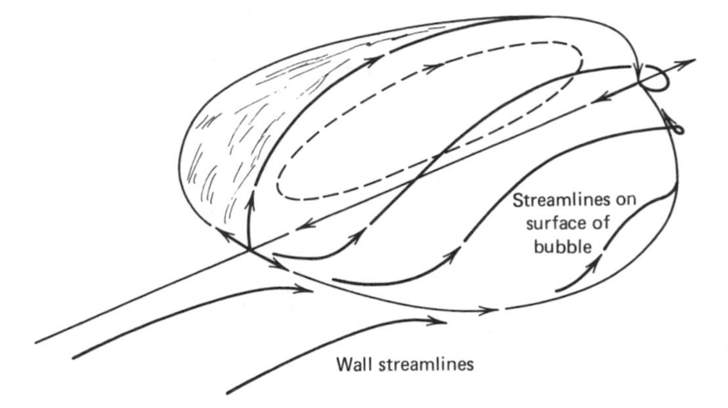

Figure 20-28 Sketch of an idealized closed separation bubble.

bubble, and then reattaches at a saddle attachment point. The fluid within the bubble stays within the bubble, and the fluid outside has its origin far upstream. The best and most general definition we can give for three-dimensional separation is that the two regions have fluid coming from widely separated places of origin. There is no general local mathematical criterion for three-dimensional separation, such as we have for two-dimensional separation.

We may classify separation by the types of singular points that occur on the surface, the location of certain attachment and separation lines, and the pattern of arrangement of these elements. In Fig. 20-28 there are two saddle points. The line on the surface that separates the two regions of fluid is called the *separation line*. The shear stress on the separation line is zero only at the saddle points on the ends of the line. Everywhere else the stress is nonzero. (The basis for this topological description is the postulate that the surface streamlines and vortex lines are continuous vector fields. Some people prefer to speak of shear-stress trajectories, rather than surface streamlines, because the magnitude of the velocity on the wall is zero, whereas the shear stress is not. This is permissible in a Newtonian fluid, where the stress lines and the limiting streamlines coincide.) The discussion of singular points may be extended to any chosen plane of the flow by finding the trajectories of the projections of the velocity vectors onto that plane, that is, solutions to $d\mathbf{r} \times (\mathbf{v} \text{ projection}) = 0$. In special cases such as planes of symmetry, the projected lines are streamlines, but this is not generally so.

Separation lines are lines on the surface where the surface streamlines tend to accumulate. Consider for example a Couette flow system with Taylor vortices as shown in Fig. 20-29. Surface streamlines cross the cell and then approach the separation line very rapidly. For visual purposes, such lines are often drawn as merging with the separation line, although this is only possible at a point where the shear stress is zero (a continuous vector field allows merging only when the vector is zero). A close-up detail is shown in Fig. 20-29,

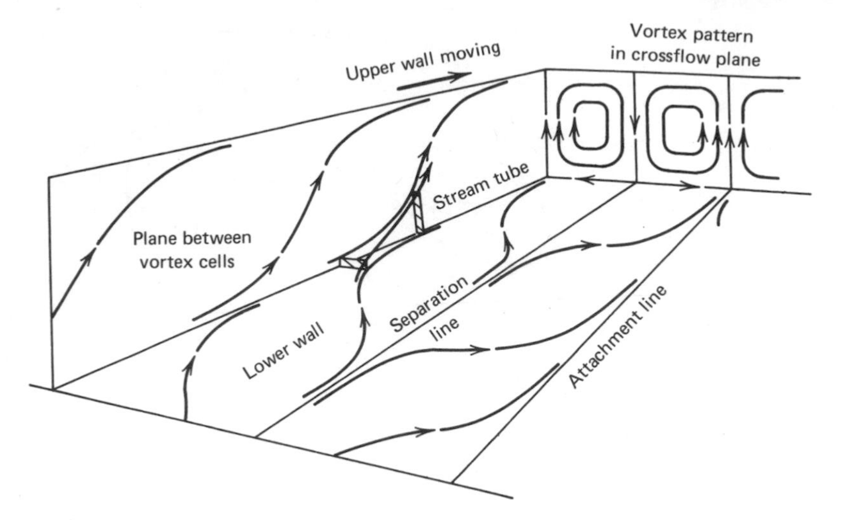

Figure 20-29 Streamlines in a Taylor vortex cell of a Couette flow. The wall and the surfaces separating vortex cells intersect in lines of separation and lines of attachment.

where the geometry of a streamtube is sketched. As the fluid separates from the wall rapidly, the distance between the surface streamline and the separation line decreases exponentially. Flow-visualization techniques cannot resolve such minute distances, and in the records from such experiments the streamlines appear to merge with the separation line. On the other side of a cell the reverse of these events take place. The line is now a line of attachment and appears to be a source of surface streamlines. In this flow the separation surface from one wall becomes an attachment surface on the other wall. This is typical of closed separation regions.

In more complex patterns of separation, several nodal and saddle points occur in combination with separation–attachments lines. Nodal focus points also play a role whenever a vortex flow springs from a wall. For example, the vortex that forms at the tip of a wing begins not at the trailing edge, but at a point near midspan on the upper surface. A focus point must occur somewhere in this vicinity as the origin of the streamline in the core of the vortex. The exact nature of this flow has not been determined.

Separated regions may also exist within the flow itself, free of the walls. The vortices generated by the delta wing offer good examples of this situation. Any vortex, once it is formed is susceptible to *vortex breakdown*, a transition from a small high-velocity vortex into a larger lower-velocity vortex. There are three major types, along with some minor variations (see Fig. 20-30). The type known as spiral breakdown has a singular point where the streamlines split

---

Figure 20-30 Three typical types of vortex breakdown: (a) double helix, (b) axisymmetric, and (c) spiral. Photographs courtesy of T. Sarpkaya, Naval Postgraduate School.

(a)

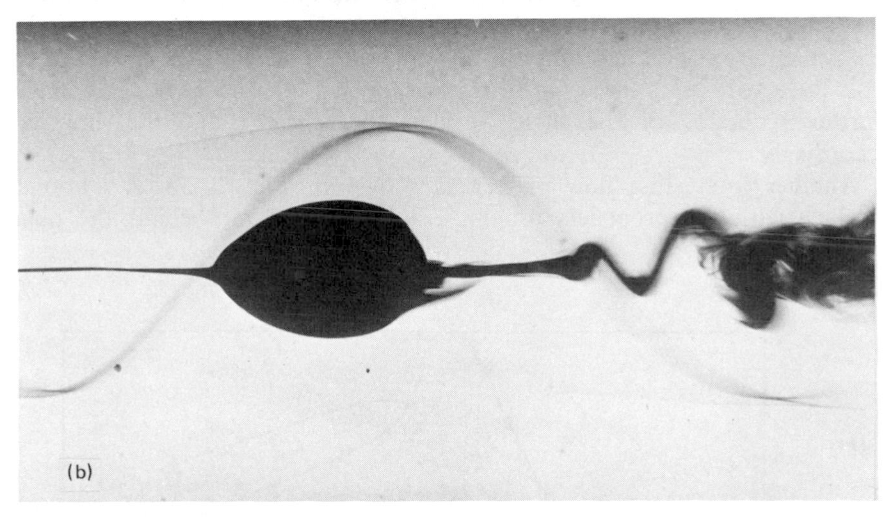
(b)

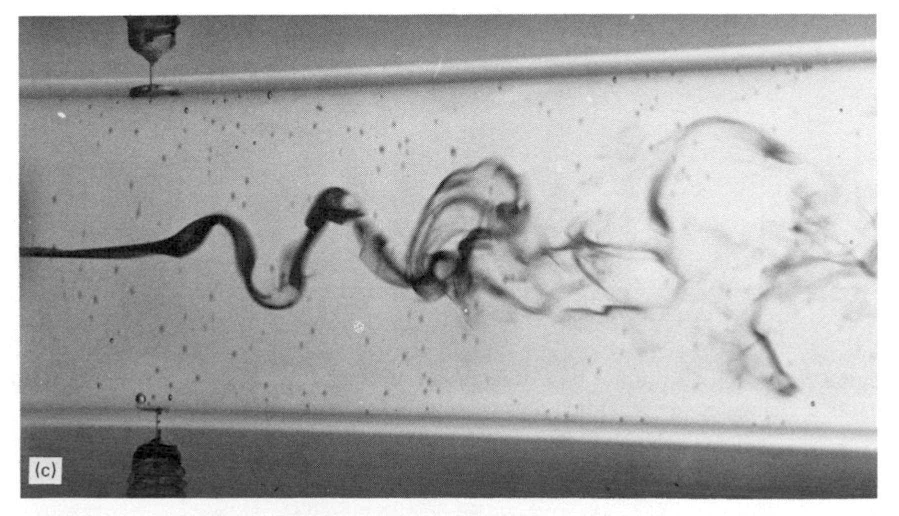
(c)

and fan out into a double helix. Other types have a free separation bubble containing a slow-moving fluid (the bubble flow is sometimes unsteady). The reader interested in vortex bursting can find more information in articles by Sarpkaya (1971), Leibovich (1978), and Berger (1982).

The core of a vortex is a low-pressure region, and when it is near a surface it naturally decreases the pressures on the surface. For example, the lift and drag models for a delta wing include the strength and location of the vortex cores. Another illustration of this influence of vortices is shown in Fig. 20-31. The unexpectedly high drag of some "hatchback" automobiles motivated the tests shown in the figure. A rounded nose shape was terminated with a flat base at various angles. In principle, all of the pressure drag is the result of the beveled base. When the base is square or at a modest angle, the separated region is disorganized with a lot of turbulence. The pressure is somewhat lower than ambient and results in a significant drag. As the bevel angle is reduced there is a large, unexpected jump in the drag. Flow visualization reveals that this is caused by a change in the pattern of flow separation. Two strong vortices are now attached at the top of the bevel and stay close to the surface as they follow down the base region and trail off into the wake. The low pressure associated with these vortices is the cause of the unexpected high drag for these angles. Further reduction of the angle weakens the vortices, and the drag falls accordingly.

Another interesting flow pattern is shown in Fig. 20-32. A stationary airplane with the propellers running sends a jet of air behind the plane.

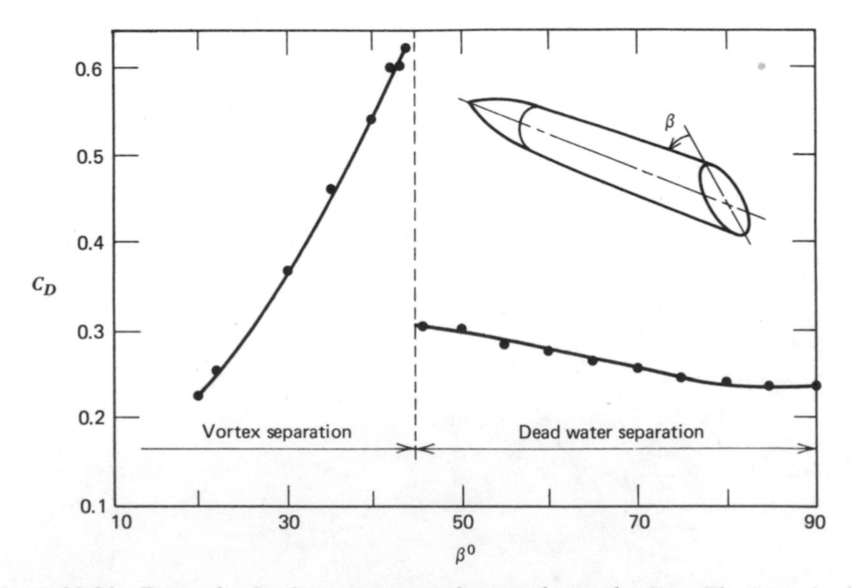

Figure 20-31 Drag of a flat base area at various angles to the flow. The jump in drag marks a change in the separated-flow pattern. Adapted from T. Morel (1978).

Figure 20-32 Vortex generated by flow into airplane propeller. Core of the vortex is made visible by ground moisture from a recent rainstorm. Photograph courtesy of J. Cornish III, Lockheed-Georgia Co.

Continuity requires a return flow. As the flow is gathered together, a boundary layer develops along the ground, and as clearly seen in the figure, a vortex ascends from a stagnation point on the ground. The intense velocities generated can carry debris from the ground into the propeller.

20.21 CONCLUSION

Boundary-layer theory is one of those inventions that allow a giant step in understanding to be taken. Prandtl originally preferred the term transition layer, but Blasius (1908) employed the term boundary layer, and it became popular among researchers as the subject expanded. The idea of a boundary layer completed the theory of attached flows at high Reynolds numbers and placed potential flow in its proper perspective. Tani (1977) gives a historical sketch of the subject.

In a more general context boundary-layer theory is historically the first example of a singular perturbation. A field is governed by two competing events with different characteristic length scales. As the ratio of these lengths becomes large, one phenomenon dominates everywhere except in a transition region. The characteristics of the transition region can only be found by rescaling the region to retain the proper physical events.

PROBLEMS

20.1 (A) Solve the Blasius boundary layer problem using $f''(0) = 0.33206$. Solve again to find the value of $f''(0)$ to seven significant figures.

20.2 (A) Verify the displacement thickness formula for a Blasius boundary layer.

20.3 (A) Solve the Blasius boundary layer using the von Kármán momentum integral with $u = u_0 \sin(\pi y/2\delta)$ as the assumed profile.

20.4 (B) Use the K-P 4 method to solve for the Howarth problem where $u_e = u_0(1 - a(L/x))$.

20.5 (C) The Blasius equation is unchanged if $f(\eta) \to af_0(a\eta)$ for any constant a. That is, f and f_0 both obey the Blasius equation. If f obeys the Blasius B.C., what B.C. are equivalent for f_0? If f_0 is found using $f_0(0) = f_0'(0) = 0$ and $f_0''(0) = 1$, how can the constant a be found so that $f(\eta)$ is the flat plate result? The solution has now been obtained without shooting to the far boundary condition?

20.6 (C) Derive the boundary layer form of the energy equation given in 10.8.8.

20.7 (B) Construct a computer program to solve the Falkner–Skan equation over the range $\eta[0, 8]$ and print out f and f' every 0.1 increment in η. Solve the case $\beta = 0.2$.

20.8 (A) Find the Falkner–Skan equation when u_e and x have opposite signs.

20.9 (B) Find the exact closed form solution to the Falkner–Skan flow along a wall into a line sink.

20.10 (B) Prove that a sharp leading edge in a pressure gradient flow has $\beta_i = 0$ just the same as the Blasius flow.

20.11 (B) Modify the computer program in Appendix G to give a second order accurate table of β_m and $\beta_{m+1/2}$ for given data $u_e(x)$.

20.12 (B) Consider the boundary layer under a line source of strength m located a fixed distance h above the wall. Locate the region of adverse pressure gradient. Repeat this problem where the line source moves by self-induction.

20.13 (B) Compute the boundary layer on an ellipse as given in Problem 18.11.

20.14 (B) Compute the boundary layer on a Joukowski airfoil as found in Problem 18.15.

20.15 (B) Compute the boundary layer on the circular arc strut of Problem 18.22.

20.16 (B) Compute the boundary layer on a flat plate aligned with the flow from a line source. The leading edge of the plate is a distance r_0 from the source which has a strength Q.

20.17 (B) Was an implicit assumption made that restricts the program of Appendix G from computing a boundary layer where u_e is negative and going from large $+x$ toward zero?

20.18 (C) Derive the boundary layer equations for a liquid film falling down a vertical flat plate from an arbitrary initial state. State the appropriate boundary conditions.

20.19 (C) Derive the proper three-dimensional boundary layer forms for the equations $\boldsymbol{\omega} = \nabla \times \mathbf{v}$, $\nabla \cdot \boldsymbol{\omega} = 0$, and the three vorticity equations 13.3.5.

20.20 (B) Find the boundary layer for flow over a Rankine nose and determine if laminar separation will occur.

20.21 (C) Consider a boundary layer where $f_i'(\eta)$ is given but is not one of the Falkner–Skan profiles. Let the boundary layer begin at $\xi_0 \neq 0$, that is $f_i'(\eta) = f'(\xi_0, \eta)$. Derive an integral equation for $V_i(\eta)$. Program an iteration scheme to find $V_i(\eta)$. Solve the constant-pressure boundary layer with the initial profile $u/U_0 = 2\sin(\pi y/2d)$ for $0 < y < d$ and $u/U_0 = \frac{3}{2} + \frac{1}{2}\cos[\pi(y-d)/2d]$ for $d < y < 3d$. Will the Blasius profile result far downstream?

21 Low-Reynolds-Number Flows

Flows where the Reynolds number is very low form another class of incompressible flows having common physical events. For a domain characterized by the length L and a motion characterized by the velocity U, we can imagine that the Reynolds number $\mathrm{Re} = LU\rho/\mu$ is small either because the fluid is very viscous ($\mu \to \infty$), or because the inertia, or density, is very small ($\rho \to 0$). From a physical point of view we may think of $\mathrm{Re} \to 0$ as either the flow of a massless fluid or the flow of a highly viscous fluid. These flows are frequently called "creeping" flows.

In practical cases the quantity that actually becomes small, compared to human sizes of course, is the length or the velocity. A small particle of dirt or a droplet of liquid settling out of the air does so very slowly, at a low Reynolds number. The size range of engineering interest extends all the way down to aerosols, where the continuum assumption itself must be modified. As another example, the width of the gap in a oil-lubricated bearing is typically very small: 0.001 inch or less. The flow of groundwater, oil, or natural gas through porous rock formations furnishes yet another example of a low-Reynolds-number situation. When considering the flow through small passages or around tiny objects, we focus our interest on a small region in space. The physical events are those of a viscous, massless fluid.

The flows mentioned above are either confined flows, such as the bearing, or unconfined flows, such as the particles in an infinite fluid. Low-Reynolds-number flows can also occur as parts of larger flow fields. Consider, for example, the leading edge of a flat plate aligned with the free stream. As the flow separates to go on either side of the plate, the velocity on the stagnation streamline must become zero, and viscous diffusion can therefore extend a slight distance in front of the edge. The size of this region is ν/U, very small in most instances. In a small neighborhood near the leading edge, the flow is viscous and a local Reynolds number $x_i U/\nu$ is less than one.

The term low-Reynolds-number is commonly used several different meanings. Frequently it merely designates flows where Re is lower than typical values. Here, and throughout this chapter, we use the term to mean flows where $\mathrm{Re} \to 0$.

21.1 GENERAL RELATIONS FOR Re → 0; STOKES EQUATIONS

Consider a flow field with characteristic length L and velocity U. The proper nondimensional distance $x_i^* = x_i/L$ and velocities $v_i^* = v_i/U$ contain these scales. In previous work for moderate and high Reynolds numbers, the pressure was scaled with the dynamic pressure, $\frac{1}{2}\rho U^2$, implying that inertia and pressure effects are the same size. However, this scaling is inappropriate for Re → 0 because inertia effects, including the dynamic pressure, are becoming very small. In a very viscous flow the pressure force must become large to balance the viscous stresses. The appropriate nondimensional pressure is

$$p^{**} = \frac{p - p_0}{\mu U/L} \tag{21.1.1}$$

With these variables, the nondimensional momentum equation becomes

$$\text{Re}\frac{Dv^*}{Dt^*} = -\nabla^* p^{**} + \nabla^{*2} v^* \tag{21.1.2}$$

For Re → 0 this simplifies to

$$\nabla^* p^{**} = \nabla^{*2} v^* \tag{21.1.3}$$

The net forces on a fluid particle must add to zero (p^{**} may include the body force as in Section 10.5). Flows governed by 21.1.3 are termed *Stokes flows*. Stokes (1851) was the first to propose this simplification, in his paper concerning the motion of a pendulum.

Several other forms for 21.1.3 are (in dimensional form)

$$\begin{aligned} 0 &= \nabla \cdot \mathbf{T} \\ \nabla p &= \mu \nabla^2 \mathbf{v} \\ \nabla p &= \nabla \cdot \boldsymbol{\tau} \\ \nabla p &= \mu \nabla \times \boldsymbol{\omega} \end{aligned} \tag{21.1.4}$$

The characterization of Stokes flows as the flow of a massless fluid, in contrast to its characterization as the flow of a very viscous fluid, is emphasized by noting that setting $\rho = 0$ in the dimensional equations produces the proper simplified forms for Re → 0. Furthermore, all results and conclusions arrived at for Stokes flows are independent of the fluid density. For instance, a particle of ash from a volcano settles at the same velocity at an altitude of 18,000 meters that it does at sea level, in spite of the fact that the air density changes by a factor of 10.

Taking the divergence of 21.1.3 and using the continuity equation shows that the pressure is governed by

$$\nabla^{*2} p^{**} = 0 \tag{21.1.5}$$

Thus, whenever boundary conditions appropriate to this equation may be prescribed in a problem (and some lubrication problems are of this type), we may solve for the pressure independently of the velocity. With the pressure known, 21.1.3 furnishes a mathematical problem for the velocity field.

Taking the curl of 21.1.3 yields a differential equation for the velocity alone:

$$\begin{aligned} 0 &= \nabla^{*2}(\nabla^* \times \mathbf{v}) \\ &= \nabla^{*2}\boldsymbol{\omega} \end{aligned} \tag{21.1.6}$$

This also happens to be the low-Reynolds-number form for the vorticity equation. The velocity and vorticity fields are completely determined by viscous diffusion. The inertia-like effect of convection of vorticity is absent, as is the kinematic effect of stretching the vortex lines. Low-Reynolds-number flows are so slow that these effects are of a smaller magnitude than the viscous diffusion of vorticity.

In 21.1.4 the dynamic viscosity μ appears as a proportionality constant between the pressure field and the velocity field or the vorticity field. The magnitude of the pressure increases directly as the viscosity increases (this, of course, is the reason for the scaling 21.1.1). As an application of this fact, note that in lubrication problems a change in the viscosity by a factor of 2, other things being equal, increases the load-carrying ability of the presure forces by a factor of 2.

As a general feature, all properties of a Stokes flow are governed by linear equations; 21.1.5 for p, 21.1.6 for v_i or ω_i, and a linear equation for τ_{ij} arising from the viscous law for τ_{ij} and the linearity of v_i. The linear property may be used to great advantage in adding flow fields to produce new flows. We saw previously in Chapters 18 and 19 how two or more ideal flows could be added because the potential ϕ, the stream function ψ, and the velocity v_i were all governed by linear equations. Pressures in ideal flows are not additive, because the Bernoulli equation is quadratic in v_i. On the other hand, in Stokes flows, pressures and viscous stresses are also governed by linear equations, and superposition is therefore allowed. Since forces are simply the integration of stresses and pressures, the superposition of forces is also allowed in Stokes flow.

Time does not appear explicitly in the equations governing Stokes flows. Thus, these flows are quasisteady. Any time-dependent motion of a massless fluid arising from unsteady boundary conditions is quasistatic. The validity of this conclusion rests on the assumption that the time scale in the substantial derivative term of 21.1.2 is L/U. The unsteady motion of a boundary condi-

Figure 21-1 Viscous flow over a block shows symmetry at Re = 0.02. From Taneda (1979).

tion introduces another independent time scale into the problem. If we use Ω to characterize the imposed unsteadiness, then when $\Omega L/U$ is of order one, our conclusion that the flow is quasisteady is valid. However, if $\Omega L/U$ is very large (a high-frequency case), then the unsteady term $\rho\ \partial u/\partial t$ must be retained in 21.1.3. Such flows have local inertia effects but no convective inertia effects. Hence, whether a time-dependent flow at a low Reynolds number requires a momentum-storage term or not depends on the frequency parameter of the imposed motion.

The linearity of Stokes flows leads to many nice mathematical results. For example, it is possible to show that the viscous flow with the least energy dissipation, for given geometry and boundary conditions, is the Stokes solution (this result is due to Helmholtz). Another theorem states that the solution of 21.1.6 for prescribed geometry and boundary conditions on the velocity is mathematically unique.

Perhaps one of the most useful mathematical properties of Stokes flows is a direct result of the linearity. Consider what happens if we reverse the velocity ($v_i \rightarrow -v_i$) of a certain flow problem. All equations and boundary conditions are still satisfied. The stresses change direction ($\tau_{ij} \rightarrow -\tau_{ij}$), and the pressure changes sign [$p - p_0 \rightarrow -(p - p_0)$; the reference p_0 may be adjusted also]. Thus Stokes flows are reversible, in the sense that the reverse flow is also a Stokes flow. These facts allow one to argue that a velocity pattern about a symmetrical object or in a symmetrical flow channel must also be symmetrical.

The corresponding pressure distribution is antisymmetric. An application of this principle to symmetrical objects in an infinite fluid shows that these objects have no wakes. The downstream flow has the same streamline pattern and velocity magnitudes as the upstream pattern. Viscous diffusion of the vorticity proceeds upstream and downstream with equal effectiveness. Figure 21-1 shows the symmetrical flow over a block.

Lubrication theory and confined flows in general lead to well-structured solutions to the low-Reynolds-number equations given above. For flows where the domain is infinite, we shall find that the situation is quite different. Stokes flows on an unbounded domain are not uniformly valid from a mathematical standpoint. The difficulty is analogous to the one we discovered for high-Reynolds-number flows: the ideal flow was incorrect at the surface of the body, where a boundary layer analysis was needed to complete the theory. Stokes flows on an infinite domain turn out to be singular at infinity. A theory that includes Stokes flows and gives the correct behavior at infinity is constructed using what are called the Oseen equations. We shall discuss these equations later in the chapter.

21.2 GLOBAL EQUATIONS FOR STOKES FLOW

The integral momentum and kinetic-energy equations for Stokes flow take on very simple forms. The integral momentum equation may be derived by the appropriate simplification of 5.14.1 or by integrating 21.1.4 over a volume of interest and applying the theorem of Gauss. The result by either method is

$$0 = \int - n_i p \, dS + \int n_j \tau_{ji} \, dS \qquad 21.2.1$$

The fact that there is always a local balance between pressure and viscous forces translates into a statement that the pressure and viscous forces on any finite region are also in balance.

Let us apply 21.2.1 to a body moving in an infinite fluid. For a sketch see Fig. 21-2. The integration region consists of the fluid outside the body up to a remote boundary that we call S_∞. Since the drag force is defined as

$$F_i \equiv \int_{S_b} (n_i p - n_j \tau_{ji}) \, dS \qquad 21.2.2$$

the global force balance 21.2.1 becomes

$$F_i = -\int_{S_\infty} (n_i p - n_j \tau_{ji}) \, dS \qquad 21.2.3$$

The surface forces on the remote boundary must die out very slowly. If we

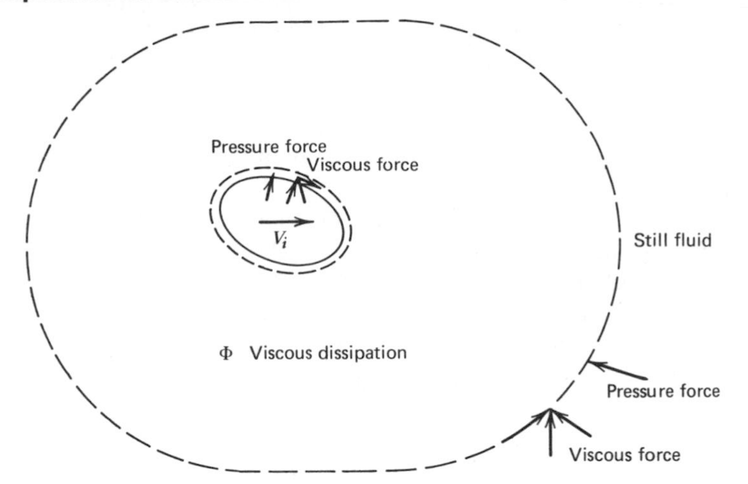

Figure 21-2 Control volume for flow over an object moving in an infinite medium.

move S_∞ to even more remote positions, the surface forces decrease but S_∞ increases in such a way that a finite force is maintained. This is an example of one of the major characteristics of low-Reynolds-number flows: the influence of a body on the flow extends very far in all directions.

In flows at any finite nonzero Reynolds number we are able to relate the drag force to events in the wake, a specific region downstream of the body. Outside the wake the velocity and pressure are essentially the free-stream values. The drag is associated with a defect in the momentum convected in the wake, or with decreased surface forces in the wake. Stokes flows on the other hand have no wakes, so the drag force is transmitted to remote locations in all directions as a surface force. A word of caution is necessary with regard to this result. Stokes flows themselves are not valid at infinite distances from the bodies. The Oseen theory, which corrects the Stokes flow, does display a wake behavior.

Next, we turn our attention to the work done by surface forces. Consider the following mathematical identity for the rate that work is done per unit volume by surface forces:

$$-\partial_i(pv_i) + \partial_j(\tau_{ji}v_i) = -v_i\partial_i p + v_i\partial_j\tau_{ji} + \tau_{ji}\partial_i v_j$$

The total surface force work is equal to the work to accelerate the fluid plus the viscous dissipation (Section 5.10). From 21.1.4 we see that the first two terms on the right hand side are always balanced in Stokes flow. Integrating this equation over a finite region and using Stokes' theorem on the left hand side yields the following:

$$\int (n_i v_i p - n_i \tau_{ij} v_j)\, dS = \int \Phi \, dV \qquad 21.2.4$$

where

$$\Phi \equiv \tau_{ij} \partial_j v_i$$

All the work done by surface forces at the boundary goes to produce viscous dissipation within the region.

Note how the work–dissipation result 21.2.4 applies to the flow field about a body moving in an infinite fluid. Again we take a region that surrounds and encloses the fluid out to a remote surface S_∞. The fluid velocity at S_∞ is approaching zero, so the surface integral in 21.2.4 becomes zero there. On the surface of the body the velocity is a uniform value V_i, so that the integral in 21.2.4 over the body becomes $V_i F_i$. Thus, 21.2.4 is rewritten as

$$F_i V_i = -\int \Phi \, dV \qquad 21.2.5$$

The left-hand side is negative because the drag force and the velocity are in opposite directions. Equation 21.2.5 offers an alternative method for computing the drag force in Stokes flows.

21.3 STOKES FLOW OVER A SPHERE

Consider a spherical coordinate system r, θ, φ with the origin at the center of a fixed sphere of radius r_0. Because the flow has symmetry, we can employ the stream function. The low-Reynolds-number form of the stream-function equation is most simply found by setting $\rho = 0$ ($\nu = \mu/\rho \to \infty$) in the complete equation from Appendix D. (Consider U, L, μ chosen as dimensional scales and set equal to one. Then $\rho = \rho UL/\mu = \text{Re}$.) The resulting equation is

$$\left[\frac{\partial^2}{\partial r^2} + \frac{\sin\theta}{r^2} \frac{\partial}{\partial \theta} \left(\frac{1}{\sin\theta} \frac{\partial}{\partial \theta} \right) \right]^2 \psi = 0 \qquad 21.3.1$$

The expression within the brackets is sometimes called the E^2 operator; 21.3.1 is $E^4 \psi = 0$.

Ideal flows satisfy the equation $E^2 \psi = 0$, and thus we might at first think that they are also low-Reynolds-number solutions. They are solutions to the governing equations. For example, ideal flows satisfy the creeping-flow equation $\nabla p = \nabla \cdot \tau$, since $\nabla \cdot \tau = 0$ (The compatible pressure field would be $p = \text{const.}$) However, the Stokes problem includes no-slip boundary conditions, which ideal flows do not satisfy.

The velocity components are related to the stream function by

$$v_r = \frac{1}{r^2 \sin\theta} \frac{\partial \psi}{\partial \theta}, \qquad v_\theta = -\frac{1}{r \sin\theta} \frac{\partial \psi}{\partial r} \qquad 21.3.2$$

From these relations we can see that the no-slip condition at the surface of the sphere is satisfied by

$$\psi(r = r_0) = 0, \qquad \left.\frac{\partial \psi}{\partial r}\right|_{r=r_0} = 0 \tag{21.3.3}$$

At infinity the flow is to approach a uniform stream from right to left as shown in Fig. 21-3:

$$v_r \sim -U\cos\theta, \quad v_\theta \sim U\sin\theta \qquad \text{as} \quad r \to \infty$$

These relations are satisfied by

$$\psi \sim -\frac{r^2}{2} U \sin^2\theta \qquad \text{as} \quad r \to \infty \tag{21.3.4}$$

This is the stream function for a uniform stream. The mathematical problem consisting of 21.3.1 and boundary conditions 21.3.3 and 21.3.4 can be solved by the method of separation of variables. It may be verified that the solution to the problem is

$$\frac{\psi}{r_0^2 U} = -\frac{1}{2}\left(\frac{r}{r_0}\right)^2 \sin^2\theta \left[\frac{1}{2}\left(\frac{r_0}{r}\right)^3 - \frac{3}{2}\left(\frac{r_0}{r}\right) + 1\right] \tag{21.3.5}$$

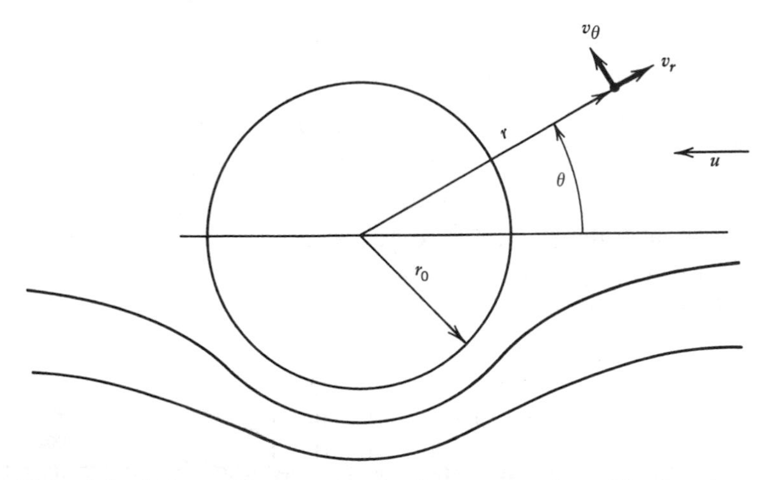

Figure 21-3 Spherical coordinate system for flow over a sphere. The free stream comes from the right.

The corresponding velocity components are

$$\frac{v_r}{u} = -\frac{1}{2}\cos\theta\left[\left(\frac{r_0}{r}\right)^3 - 3\left(\frac{r_0}{r}\right) + 2\right]$$
$$\frac{v_\theta}{U} = \frac{1}{4}\sin\theta\left[\left(\frac{r_0}{r}\right)^3 + 3\left(\frac{r_0}{r}\right) - 4\right] \qquad 21.3.6$$

These velocity components give a symmetric pattern without a wake.

A straightforward calculation shows that the vorticity is

$$\frac{\omega}{U/r_0} = -\frac{3}{2}\left(\frac{r_0}{r}\right)^2 \sin\theta \qquad 21.3.7$$

The vorticity is zero on the reference axis of the coordinate system, rises to a maximum at the shoulder of the sphere, and then drops to zero again at $\theta = \pi$.

It is interesting to compute the surface forces, which according to the general remarks of Section 21.1 dominate the flow. The pressure may be computed by integrating $dp = \nabla p \cdot d\mathbf{x}$, where $\nabla p = \nabla^2 \mathbf{v}$ is used with the known velocity components. The result is

$$p = p_\infty + \frac{3}{2}\frac{\mu U}{r_0}\left(\frac{r_0}{r}\right)^2 \cos\theta \qquad 21.3.8$$

The pressure in the flow approaching the sphere, $0 \le \theta \le \pi/2$, rises higher than p_∞, since $\cos\theta$ is positive in this quadrant. On the downstream side of the sphere, pressures are always lower than p_∞. This antisymmetrical behavior is typical of Stokes flow. Note that the pressure gradient is never adverse.

The pressure reaches a maximum at the forward stagnation point and a minimum at the rear stagnation point. The values are

$$p_0 - p_\infty = \pm\frac{3}{2}\frac{\mu U}{r_0} \qquad 21.3.9$$

This equation illustrates the fact that the magnitude of the pressure in a given velocity field increases directly with the viscosity of the fluid. For comparison we may cast 21.3.9 into the form of the pressure coefficient

$$\frac{p_0 - p_\infty}{\frac{1}{2}\rho U^2} = \frac{6}{\text{Re}} \qquad 21.3.10$$

Although this nondimensional form is only appropriate for moderate or high Reynolds numbers, it reveals that low-Re effects cause the forward stagnation pressure on a sphere to become much larger than ideal-flow value.

The viscous stresses in the flow are given by

$$\tau_{r\theta} = \frac{3}{2}\frac{\mu U}{r_0}\sin\theta\left(\frac{r_0}{r}\right)^4$$

$$\tau_{r\varphi} = \tau_{\varphi\theta} = 0 \qquad 21.3.11$$

$$\tau_{rr} = -2\tau_{\varphi\varphi} = -2\tau_{\theta\theta} = \frac{3}{2}\frac{\mu U}{r_0}\cos\theta\left[\left(\frac{r_0}{r}\right)^2 - \left(\frac{r_0}{r}\right)^4\right]$$

It is apparent from these equations that the shear stresses, the normal viscous stresses, and the pressure are all about the same magnitude. In Stokes flows, normal viscous stresses play an important role in determining the flow patterns. Because we are dealing with a Newtonian fluid, the viscous stress vector at the surface of a body must lie on the surface. Therefore at r_0 all the normal viscous stresses in 21.3.11 vanish.

Now we come to the major result of engineering importance for low-Re flow over a sphere, the drag force. Integration of the pressure and viscous forces over the surface of the sphere yields

$$F = 6\pi\mu r_0 U \qquad 21.3.12$$

This equation is called Stokes's law. It is known from experiments to be very good for Re < 0.5. It is accurate to about 10% at Re = 1. For higher Re the accuracy is lost very rapidly, but Stokes's law is still a lower bound. Theoretical considerations show that any flow where the convective terms are important must have a higher drag than 21.3.12. The proper drag law for Stokes flow over any body is $F/\mu r_0 U = \text{const}$. If we insist on comparing with high-Reynolds-number theories and cast 21.3.12 in the usual nondimensional form, we find

$$C_D = \frac{F}{\frac{1}{2}\rho U^2 \pi r_0^2} = \frac{24}{\text{Re}} \qquad 21.3.13$$

The inverse dependence on Re is required because the density does not actually influence the drag.

*21.4 NONUNIFORMITY OF STOKES FLOWS ON INFINITE DOMAINS

Stokes flows on an infinite domain are not uniformly valid. Oseen (1910) was the first to point this out. The breakdown of the Stokes analysis is a little different from the difficulties we encountered in ideal flows. In the Re → ∞ case it was not possible to satisfy the boundary conditions at the wall. It was thus obvious that the ideal solution was singular. The situation in Stokes flow is different in that the breakdown occurs because our order estimates for the terms in the differential equation are wrong. All the boundary conditions for the problem have been satisfied. Oseen's method of showing this breakdown was to compare the viscous terms, which determine the flow by balancing the pressure force, with the inertia terms, which were neglected. In making this comparison we compute each term using the Stokes-flow results for a sphere.

The largest stress term in the θ-momentum equation is

$$\frac{1}{r\sin\theta}\frac{\partial}{\partial\theta}(\tau_{\theta\theta}\sin\theta) \sim \frac{\mu U}{r_0^2}\left(\frac{r_0}{r}\right)^3 \quad \text{as} \quad r \to \infty$$

The largest convective term is

$$\rho v_r \frac{\partial v_r}{\partial r} \sim \frac{\rho U^2}{r_0}\left(\frac{r_0}{r}\right)^2 \quad \text{as} \quad r \to \infty$$

The ratio of these terms is

$$\frac{\text{convective term}}{\text{viscous term}} \sim \left(\frac{\rho U r_0}{\mu}\right)\left(\frac{r}{r_0}\right) \quad \text{as} \quad r \to \infty \qquad 21.4.1$$

From the relation above we see that the Stokes limit Re $\to 0$ is not uniformly valid for all r/r_0. The convective terms are smaller than the stress terms for Re $\to 0$ and any fixed r/r_0. On the other hand, for any small value of Re, it is possible to go to remote values of r/r_0 where the convective terms are indeed important. This is the hallmark of a nonuniform limit. Stokes flows are singular at infinity.

In light of the discussion above it might at first appear that it is not a well-founded problem to consider the flow over a particle in an infinite medium as Re $\to 0$. Do the walls or boundaries, which must exist in any real flow, always invalidate the problem? To answer this question let us consider a particle 1 mm in diameter that is held in a wind tunnel having a 3-m cross section with a slow but steady flow of 1 mm/sec. For a typical gas, the Reynolds number of this flow is 0.1–0.2. The region of validity of Stokes solution may be estimated from 21.4.1. When Re = 0.1, the convective terms will be of the same order as the stress terms at a position r_c/a of the order of 10: $r_c \sim 1$ cm. Hence, Stokes flow has become invalid at a position relatively near the particle. The walls of the wind tunnel are still remote, and we would be correct in assuming that they have no influence on the flow near the sphere.

Next consider what happens if we keep the same flow conditons and lower the Reynolds number. This could be done by using lighter and lighter molecules, so that $\rho \to 0$. As Re $\to 0$ the radius r_c/a where Stokes flow becomes invalid becomes larger and larger. Ultimately r_c/a is comparable to the distance to the wind-tunnel wall. Now the problem is no longer simply the flow over a small particle in an infinite medium as Re $\to 0$, but it is the flow over a small particle in a wind tunnel as Re $\to 0$.

The conclusion is that the problem for flow over a particle in an infinite medium as Re $\rightarrow 0$ is perfectly good for some range of Reynolds numbers near but not equal to zero. The Stokes flow equations give the answer near the body, but they are not good as $r \rightarrow \infty$. We must have another set of equations and another solution for regions that are far from the body as defined by 21.4.1. The Oseen equations, valid in this region, will be discussed in Section 21.14.

If Stokes's equations are not valid for $r \rightarrow \infty$, how is it that the solutions are able to satisfy the far-stream boundary condition $v_i = U_i$? The answer to the question is that it was just good luck. There is no mathematical or physical reason why Stokes flows should satisfy a boundary condition at $r \rightarrow \infty$. In fact, if we took Stokes's answer as a first approximation, we would find that the second approximation would not satisfy any far boundary condition (this is known as Whithead's paradox). Furthermore, if we attempted to solve Stokes's equations for any two-dimensional object—for example, a circular cylinder in an infinite stream—we would find that the free-stream boundary condition could not be met. This is known as the Stokes's paradox. It is resolved by computing an Oseen flow that is valid far away from the two-dimensional body and in principle matches the Stokes flow near the body.

*21.5 STOKES FLOW OVER FLUID SPHERES

A liquid droplet in an inviscid liquid and a gas bubble in a liquid are common situations of practical importance. Frequently the problem is a steady state where the weight force, the buoyancy force, and the drag force are all in balance. The solution proceeds under the assumption that the droplet is spherical.

Consider a fluid sphere at the origin of a spherical coordinate system as shown in Fig. 21-3. Equation 21.3.1 for the stream functions governs ψ for the outer flow and also ψ^i for the flow on the inside of the droplet. The boundary condition of a uniform stream at infinity (21.3.4) is retained. The condition of no flow across the surface of the droplet gives

$$\psi(r = r_0) = \psi^i(r = r_0) = 0 \qquad 21.5.1$$

The fluid at the interface r_0 may have a tangential velocity v_θ, but we require that both sides have the same value. From 21.3.2 this leads to the condition on the stream function that

$$\left.\frac{\partial \psi}{\partial r}\right|_{r_0} = \left.\frac{\partial \psi^i}{\partial r}\right|_{r_0} \qquad 21.5.2$$

The shear stress is also required to match on either side; thus

$$\mu \frac{\partial}{\partial r}\left(\frac{1}{r}\frac{\partial \psi}{\partial r}\right)_{r_0} = \mu^i \frac{\partial}{\partial r}\left(\frac{1}{r}\frac{\partial \psi^i}{\partial r}\right)_{r_0} \qquad 21.5.3$$

The boundary conditions demanding continuity of v_θ and $\tau_{r\theta}$ across the interface replace the usual condition that the surface velocity of a solid is known. The shear-stress expression above introduces the viscosity ratio

$$\sigma \equiv \frac{\mu}{\mu_i} \tag{21.5.4}$$

as a parameter in the problem. Gas bubbles in a liquid are modeled by $\sigma \to \infty$, while the results for rigid spheres are retrieved by $\sigma \to 0$.

The problem for Stokes flow over a fluid bubble was first solved independently by Rybczynski (1911) and by Hadamard (1911). Just as in the case of the rigid sphere, the technique of separation of variables leads to a solution. The stream function for the outer flow is

$$\psi = -\frac{1}{2}Ur^2\sin^2\theta\left[1 - \left(\frac{r_0}{r}\right)\frac{3+2\sigma}{2(1+\sigma)} + \left(\frac{r_0}{r}\right)^3\frac{1}{1+\sigma}\right] \tag{21.5.5a}$$

while within the sphere

$$\psi^i = \frac{1}{4}Ur^2\sin^2\theta\,\frac{\sigma}{1+\sigma}\left[1 - \left(\frac{r}{r_0}\right)^2\right] \tag{21.5.5b}$$

The solution within the sphere turns out to be Hill's spherical vortex, which was given in Chapter 12. Hill's vortex is a solution to complete unapproximated Navier–Stokes equations. At low Reynolds numbers, the pressure corresponding to Hill's solution may be calculated from $\nabla p = \mu\nabla^2\mathbf{v}$. Since we neglect the inertia terms in this equation, the resulting pressures are valid only when $\text{Re} \to 0$. The vorticity and velocity distributions are in principle valid for all Reynolds numbers.

The fluid velocity on either side of the interface is found using 21.3.2. It is

$$v_\theta(r = r_0) = v_\theta^i(r = r_0) = \frac{1}{2}U\sin\theta\,\frac{\sigma}{1+\sigma} \tag{21.5.6}$$

A calculation of the shear stress on both sides of the interface reveals that

$$\tau_{r\theta}(r = r_0) = \tau_{r\theta}^i(r = r_0) = \frac{3}{2}\frac{\mu U}{r_0}\sin\theta\,\frac{1}{1+\sigma} \tag{21.5.7}$$

Equation 21.5.7 shows that the special case of a gas bubble in a liquid, $\sigma \to \infty$, has no shear at the interface.

Next, let us investigate the normal forces, both pressure and viscous, that act at the interface. In formulating the problem it was not necessary to specify any conditions on the normal stresses. Indeed, to do so would overdetermine the answer. On the outside of the sphere, a pressure reference p_0 is taken at the equator, while p_0^i is a similar reference for the inside motion. These reference values may be functions of time, as would be the case for a sphere moving in a liquid where the hydrostatic pressure gradient is significant. The pressure and normal viscous stress on the outside are

$$p = p_0 + \rho g r_0 \cos\theta + \frac{\mu U}{r_0}\cos\theta\,\frac{3 + 2\sigma}{2(1+\sigma)} \qquad 21.5.8$$

$$\tau_{rr} = -2\frac{\mu U}{r_0}\cos\theta\,\frac{\sigma}{1+\sigma} \qquad 21.5.9$$

It is apparent that p and τ_{rr} are of comparable size, so that the normal viscous force contributes strongly to the total drag force on the sphere. In fact, for a gas bubble ($\sigma \to \infty$), the contribution to the drag from the normal viscous force is twice that of the pressure. The other extreme case is the rigid particle. Equation 21.5.8 shows that $\tau_{rr} = 0$ when $\sigma \to 0$. This is, of course, just an example of the fact that Newtonian fluids cannot have normal viscous stresses at a solid wall.

The total drag force on the outside of the spherical interface is found by integrating the surface forces. The resulting formula is called the Hadamard–Rybczynski drag law. It is

$$F_D = 6\pi\mu r_0 U \frac{1 + \frac{2}{3}\sigma}{1+\sigma} \qquad 21.5.10$$

If $\sigma = 0$, Stokes's drag law is obtained for rigid spheres, while for the gas bubble limit $\sigma \to \infty$ the coefficient of 6π becomes 4π.

A frequent application of the Hadamard–Rybczywski formula is to the case where a fluid sphere of density ρ^i rises or falls through a still fluid of density p. At equilibrium the weight, drag, and bouyancy forces balance. The resulting expression can be solved to yield the terminal velocity equation,

$$U_t = \frac{2}{9}\frac{g r_0^2}{\mu}(\rho^i - \rho)\left[\frac{1+\sigma}{1+\frac{2}{3}\sigma}\right] \qquad 21.5.11$$

This equation offers an experimental method to verify the drag formula 21.5.10. Overall, the experiments confirm the theory as long as the fluid sphere is reasonably large; a considerable amount of practical information is summarized in Clift, Grace, and Weber (1978). The situation changes for very small spheres; they show little internal circulation and have a drag force approaching that of rigid particles. The cause of this deviation is thought to be an accumulation of surface-active molecules in the interface. Such molecules contaminate most liquids and are very difficult to purge. As the drop or bubble moves through the fluid, the contaminant molecules collect on the interface. This gives the interface a viscous characteristic of its own. Small spheres, because of their relatively small surface area, are more susceptible to this effect.

In the calculation above we have dealt exclusively with the pressures and stresses on the outside of the interface. The normal stresses that act on the inside surface of the interface are calculated from Hill's vortex solution. They turn out to be

$$p^i = p_0^i + \rho^i g r_0 \cos\theta - 5\frac{\mu^i U}{r_0}\cos\theta\left[\frac{\sigma}{1+\sigma}\right] \qquad 21.5.12$$

$$\tau_{rr}^i = -2\frac{\mu^i U}{r_0}\cos\theta\left[\frac{\sigma}{1+\sigma}\right] \qquad 21.5.13$$

In order to compare the normal forces on either side of the interface we let $F_n \equiv p - \tau_{rr}$. The change in F_n is found from 21.5.8, 21.5.9, 21.5.12, and 21.5.13. It is

$$F_n - F_n^i = p_0 - p_0^i + (\rho - \rho^i) r_0 \cos\theta + \frac{\mu U}{r_0}\cos\theta\left[\frac{9+6\sigma}{2+2\sigma}\right] \qquad 21.5.14$$

According to the elementary theory for an interface, surface tension should account for this jump. Hence,

$$F_n - F_n^i = \frac{\gamma}{2r_0} \qquad 21.5.15$$

From 21.5.14 we see that when ρ, ρ^i, and U take on arbitrary values, the surface tension must be a function of θ in order to maintain a spherical shape. We would not expect this to be the normal situation.

When a bubble or drop is falling or rising through an infinite fluid at its terminal velocity, a very striking result is obtained. Substitution of 21.5.11 into 21.5.14 shows that the imbalance of hydrostatic pressure across the interface is exactly balanced by the normal stress. The result is a constant jump in the normal force across the interface,

$$F_n - F_n^i = p_0 - p_0^i = \frac{\gamma}{2r_0} \qquad 21.5.16$$

A bubble or droplet at its terminal velocity can have a constant surface tension and has no tendency to distort from its spherical shape. Experiments confirm that bubble and drop retain a spherical shape for all low-Re flows.

*21.6 STOKES FLOW OVER PARTICLES OF ARBITRARY SHAPE

The influence of the geometry of a particle on the creeping-flow drag force is of interest for practical reasons. In many engineering or natural situations, fine particles are suspended in air or water. They may be spherical, angular, rodlike, or platelike.

In Stokes flow the drag of an object is fairly insensitive to the exact shape. Sharp corners and sharp edges are not as important as the surface area upon which the pressure and viscous forces act. A theorem due to Hill and Power (1956) states that the Stokes drag of an object must be larger than the drag of any inscribed figure but smaller than that of any circumscribed figure. A sphere circumscribed around the object has a larger drag than the object, though it has no sharp corners; while a sphere inscribed within the object must have a smaller drag than the object.

The general motion of a settling particle can be quite complicated. Odd-shaped particles tumble and spin as they move in erratic paths. Since the drag force is different when the flow approaches the particle in different orientations, the velocity and direction of motion are constantly changing. Particles with somewhat regular but asymmetric shapes can make a graceful spiral as they settle through a fluid.

As examples of falling particles we consider spheroids (ellipsoids of revolution) as depicted in Fig. 21-4. The reference axis, about which the generating line is rotated to form the spheroid, makes an angle φ with respect to the gravity vector. Let a be the maximum radius and $2b$ be the length of the spheroid. If $e \equiv b/a > 1$, the figure is a *prolate* spheroid; if $e < 1$, it is an *oblate* spheroid. The value $e = 0$ indicates a disk of radius a, $e = 1$ a sphere of the same radius, and $e \to \infty$ a needle of length $2b$.

The fact that the equations governing pressure and stress are linear is used to great advantage in this problem. It means that we may consider the flow as the superposition of a flow aligned with the reference axis with magnitude $U \cos \theta$ and a flow normal to the axis of magnitude $U \sin \theta$. The angle θ is the direction of motion of the particle with respect to the reference axis; conse-

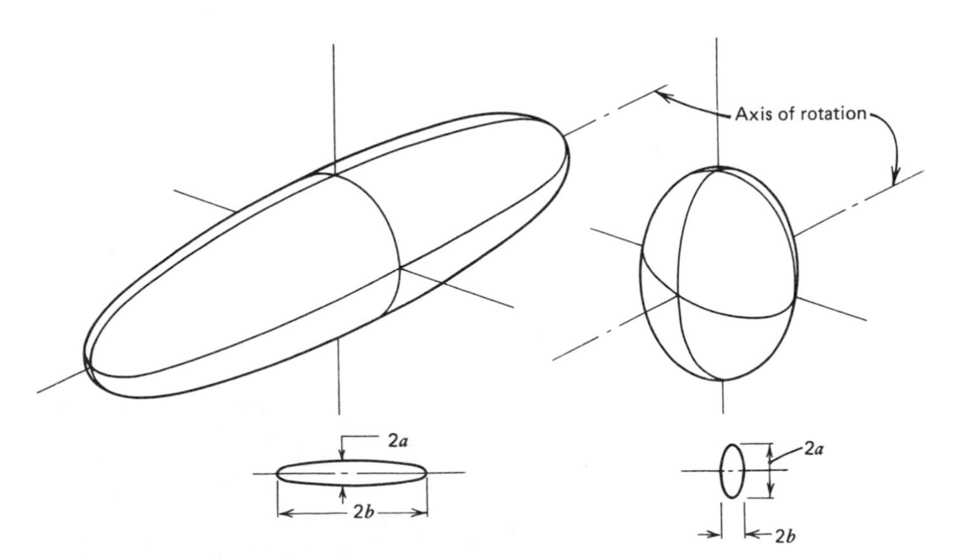

Figure 21-4 Ellipsoids of revolution. The maximum radius is a; the length is 2b.

quently $\varphi - \theta$ gives the direction of motion with respect to gravity. The force on the particle is decomposed into F_1 aligned with the reference axis and F_2 normal to the reference axis:

$$F_1 = C_1 \mu U \cos\theta \qquad F_2 = C_2 \mu U \sin\theta \tag{21.6.1}$$

The constants C_1 and C_2 depend only on the particle geometry. Exact relations for the constants may be found by analysis. (Indeed, Stokes flow is one of the oldest areas in analytical fluid dynamics: this problem was first solved by Oberbeck in 1876.) Useful approximations are (see Clift, Grace, and Weber, 1978)

$$C_1 = 6\pi a\left(\frac{4+e}{5}\right) \qquad C_2 = 6\pi a\left(\frac{3+2e}{5}\right) \tag{21.6.2}$$

These relations can be applied in the range $0 \le e \le 5$ with less than 10% error. At $e = 1$ they are exact.

Needlelike objects ($e \to \infty$) are described by the following approximate formula:

$$C_1 = \frac{2\pi b}{\ln 2e - 0.5} \qquad C_2 = \frac{4\pi b}{\ln 2e + 0.5} \tag{21.6.3}$$

The drag of a very thin needle falling normal to its axis is twice the drag of the same needle falling along its axis.

As a specific example, consider a thin needle falling so that $\theta = 30°$. Since $C_2 = 2C_1$, the ratio of the normal force to the axial force is found from 21.6.1 to be

$$\frac{F_2}{F_1} = 2\frac{\sqrt{3}/2}{1/2} = 2\sqrt{3}$$

At equilibrium the net force is balanced by the weight and thus must be in the vertical direction. This means that the inclination of the particle is

$$\varphi = \arctan\frac{F_2}{F_1} = 73.9°$$

The trajectory of the particle is a line at the angle $\varphi - \theta = 43.9°$. In this case the position of the effective force and the centroid of the particle coincide. Thus, the particle falls steadily without rotating.

*21.7 INTERFERENCE EFFECTS

Viscous diffusion causes the influence of a particle to be felt at large distances. To see this, recall the radial velocity relation 21.3.6 for flow over a sphere:

$$v_r = -\frac{1}{2} U \cos\theta \left[2 - 3\frac{r_0}{r} + \left(\frac{r_0}{r}\right)^3 \right] \qquad 21.7.1$$

The three terms within the brackets show the behavior as we go away from the particle. The constant 2 produces the undisturbed uniform stream, while the other two terms show the remote influence of the particle. Creeping-flow effects decay at the slow rate of r_0/r. In contrast, in ideal flow the velocity decays as $(r_0/r)^3$ [in 21.7.1, $(r_0/r)^3$ represents an ideal flow doublet, while r_0/r is termed a Stokeslet].

As the first example we consider two spherical particles a distance l apart. The flow field around each particle alters the flow at the location of the other. The net effect can be summarized by noting a change in the drag coefficient of the particle. Let λ denote the ratio of the drag of the particle to the drag of a sphere in an unbounded fluid,

$$F = \lambda \, 6\pi\mu r_0 U \qquad 21.7.2$$

Experimental and theoretical information on $\lambda(l/2r_0)$ is given in Fig. 21-5 for two cases: spheres moving parallel and perpendicular to the line of centers.

Due to the reversibility of Stokes flows, the flow pattern must be symmetric and there is no tendency for the particles to move toward or away from each other. Mutual interaction causes the spheres to rotate as they fall. As shown in the figure, the drag of a sphere in the presence of another sphere is reduced slightly.

Strong interference effects occur when a particle is in a closed vessel or near a wall. Approximate formulas for λ show that the wall has a retarding effect (increases the drag). For example, a plane wall a distance l from the particle gives

$$\lambda = \begin{cases} 1 + \dfrac{9}{8}\dfrac{r_0}{l} & \text{for motion perpendicular to the wall} \\[2ex] 1 + \dfrac{9}{16}\dfrac{r_0}{l} & \text{for motion parallel to the wall} \end{cases}$$

In the latter case there is no tendency for the particle to move toward or away

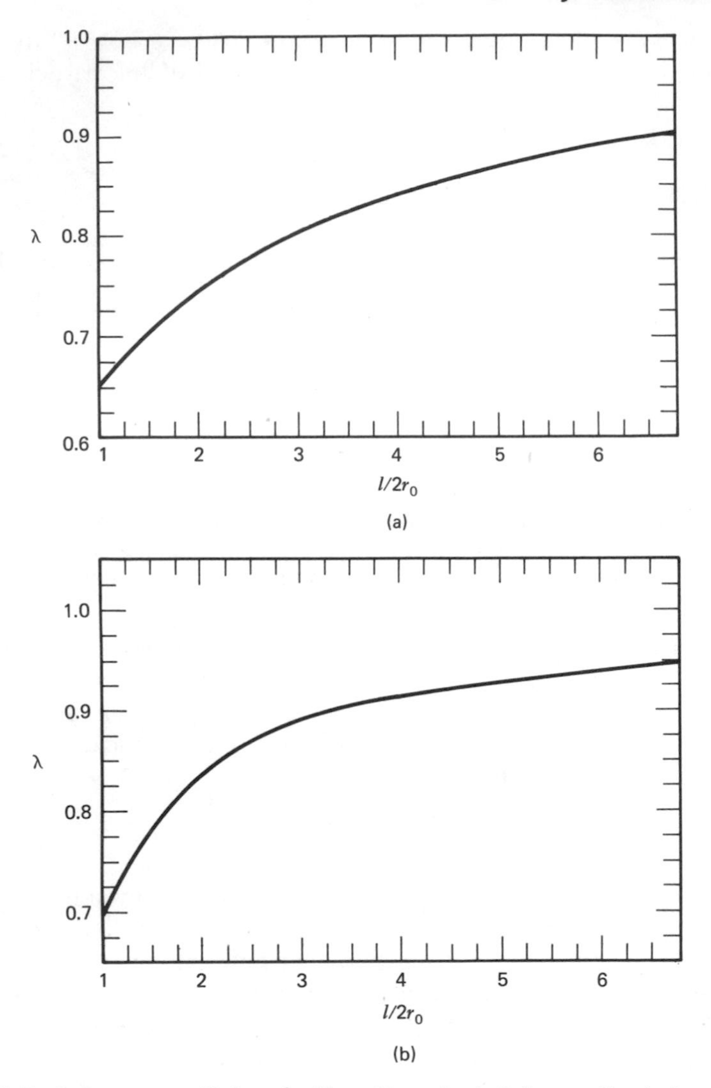

Figure 21-5 Influence coefficient λ (drag/drag in infinite medium) for a sphere influenced by another sphere a distance l away: motion (a) along and (b) perpendicular to the line of centers. Adapted from Happel and Brenner (1965).

from the wall, but the particle does experience a torque, which induces a rotation.

Swarms of bubbles or suspensions of particles usually experience an increase in drag. While a pair of particles in an infinite fluid may fall faster than a single particle, a suspension of particles falls slower. This effect is called *hindered settling*. As a consequence in a fluid from which particles are settling one frequently observes a sharp demarcation separating clear fluid and particle-laden fluid. Particles in the floc settle slower than isolated particles, so that any particle that is left behind has ample opportunity to catch up. Brownian motion causes diffusion of the particles, which tends to smear out the interface. Competition between these effects determines the sharpness of the concentration interface.

21.8 FLOW IN TUBES AND CHANNELS WITH VARYING CROSS SECTIONS

Flow in porous media, and to a certain extent lubrication problems, have a common element in that the flow goes through very small passages where the cross section changes slowly. The flow is driven by a pressure gradient and retarded by strong viscous forces. As an example, we consider a two-dimensional slot as shown in Fig. 21-6.

Assume a length of channel L across which the pressure drop Δp is imposed. Allow the width of the channel to vary slowly but in an arbitrary manner $h(x)$. The wall slope dh/dx has a typical value H', which must be small, while the slot has a typical width h_0. We estimate that the x-direction length over which a significant change in the cross section occurs is h_0/H'.

Let us proceed to estimate the order of magnitude of the flow-field quantities. Since the flow is assumed to be driven by the pressure gradient, the velocity scale for the x direction is similar to that for flow in a plane slot, that is,

$$u_s = \frac{\Delta p\, h_0^2}{\mu L} \tag{21.8.1}$$

The transverse velocity v has a much smaller scale. Denote the v velocity scale by v_s, and estimate the terms in the continuity equation as follows:

$$\frac{\partial u}{\partial x} + \frac{\partial v}{\partial y} = 0$$

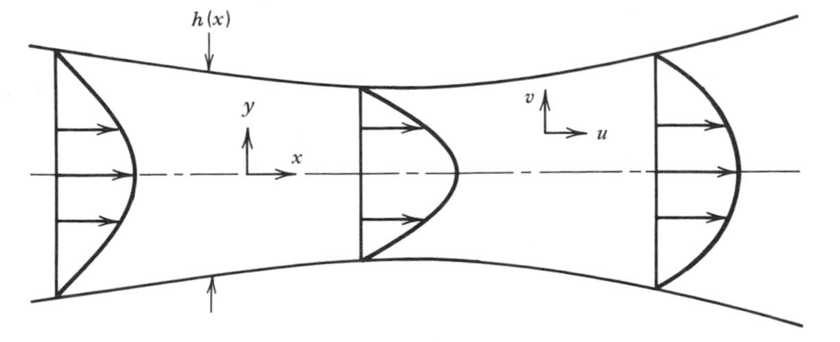

Figure 21-6 Flow in a slot with a slowly changing area is locally a Poiseuille flow.

Since the area changes occur on the length scale h_0/H', we estimate the terms as

$$\frac{u_s}{h_0/H'} + \frac{v_s}{h_0} = 0$$

By requiring that the continuity equation have both terms the same size, we find that the vertical velocity scale is

$$v_s = H'u_s \qquad 21.8.2$$

where u_s is given by 21.8.1. From another viewpoint 21.8.2 is simply a statement that the slope of a streamline is about the same as the wall slope.

Next, consider the transverse momentum equation. From this equation we can estimate the size of the transverse pressure gradient. The equation is

$$\rho u \frac{\partial v}{\partial x} + \rho v \frac{\partial v}{\partial y} = -\frac{\partial p}{\partial y} + \mu \frac{\partial^2 v}{\partial x^2} + \mu \frac{\partial^2 v}{\partial y^2}$$

The terms have the following sizes:

$$\frac{\rho u_s^2 H'}{h_0/H'} + \frac{\rho u_s^2 H'^2}{h_0} = \frac{\partial p}{\partial y} + \frac{\mu u_s H'}{h_0^2/H'^2} + \frac{\mu u_s H'}{h_0^2}$$

The key assumption in our analysis is that $H' \to 0$. If this is true, then the largest term above is the viscous term $\mu\, \partial^2 v/\partial y^2$. At most, the pressure gradient will be of this size. Hence, by using 21.8.1 we have

$$\frac{\partial p}{\partial y} \sim \frac{\mu u_s H'}{h_0^2} = H' \frac{\Delta p}{L} \sim H' \frac{\partial p}{\partial x} \qquad 21.8.3$$

The y pressure gradient will be smaller than the x pressure gradient by a factor H'; therefore we take p as a function of x alone.

The x-momentum equation is considered next. The equation is

$$\rho u \frac{\partial u}{\partial x} + \rho v \frac{\partial u}{\partial y} = -\frac{dp}{dx} + \mu \frac{\partial^2 u}{\partial x^2} + \mu \frac{\partial^2 u}{\partial y^2}$$

Estimates for the size of each term are

$$\frac{\rho u_s^2}{h_0/H'} + \frac{\rho u_s^2 H'}{h_0} = \frac{\Delta p}{L} + \frac{\mu u_s}{h_0^2/H'^2} + \frac{\mu u_s}{h_0^2}$$

Reorganizing into nondimensional groups yields

$$\left(\frac{\rho u_s h_0}{\mu}\right)(H' + H') = \frac{\Delta p\, h_0^2}{\mu L u_s} + H'^2 + 1$$

Thus as $H' \to 0$, the dominant terms are

$$0 = -\frac{dp}{dx} + \mu\frac{\partial^2 u}{\partial y^2} \tag{21.8.4}$$

At this point, in order to simplify the algebra, we assume the wall is symmetrical about the centerline. Then the solution is

$$u = -\frac{h^2}{8\mu}\frac{dp}{dx}\left[1 - \left(\frac{2y}{h}\right)^2\right] \tag{21.8.5}$$

This is Poiseuille flow determined by the local slot width and local pressure gradient.

Global continuity requires that the flow rate be constant. Integration of 21.8.5 yields

$$Q = -\frac{h^3}{12\mu}\frac{dp}{dx} \tag{21.8.6}$$

Using this we can write 21.8.5 in terms of Q and h and eliminate dp/dx. The overall pressure drop over the distance L is

$$p_2 - p_1 = \int_0^L \frac{dp}{dx}\,dx = -12\mu Q \int_0^L h^{-3}\,dx$$

or

$$Q = \frac{p_1 - p_2}{12\mu}\left(\int_0^L h^{-3}\,dx\right)^{-1} \tag{21.8.7}$$

The relations 21.8.7, 21.8.6, and 21.8.5 constitute the solution for viscous flow in a narrow channel. It is interesting to note that the approximations require $H' \to 0$ and $\mathrm{Re} \cdot H' \to 0$ but do not require $\mathrm{Re} \to 0$. We anticipate that the solution might be valid for finite Re if H' were smooth enough. In fact, for a perfectly flat wall, the solution is valid in principle for all Re.

21.9 REYNOLDS EQUATIONS FOR LUBRICATION THEORY

The dominant events in lubrication problems involve pressure, viscosity, and moving walls at a slight incline angle. A typical bearing has a gap width of 0.001 in. or less, and the convergence between the walls may be as small as 1/5000. Unlike the previous section, where the pressure gradient is imposed to cause the flow, the pressure gradients in a bearing are generated by two events. First, the moving wall on one side sweeps fluid into a narrowing passage through the action of viscous shear forces. The local velocity profile from this effect is the Couette profile $u = Uy/h$. A local flow rate due to the Couette motion, $Q_c = \frac{1}{2}Uh$, would be large where h is large and small where h is small. This, of course, cannot happen, because continuity demands that the overall flow rate be constant. Hence, the flow sets up a pressure gradient to supply a Poiseuille component that redistributes the fluid and maintains a constant flow rate.

The simplified equations appropriate for lubrication theory are known as Reynolds equations. Several ideas from the previous section on channel flow are used to derive them. Figure 21-7 shows a channel of height $h(x)$. For simplicity let the upper wall have only a vertical velocity V, which may be a function of time, and the lower wall a steady horizontal velocity U. Although the lower wall only moves in the x direction, we allow for a z-direction flow that may be set up by the pressure gradient. The following scales are appropriate:

$$\begin{aligned} x, z &\sim L \\ y &\sim h_0 \\ u, w &\sim U \\ v &\sim \frac{Uh_0}{L} \\ p &\sim \frac{\mu UL}{h_0^2} \end{aligned} \qquad 21.9.1$$

In contrast with the previous section, where a given pressure drop determined

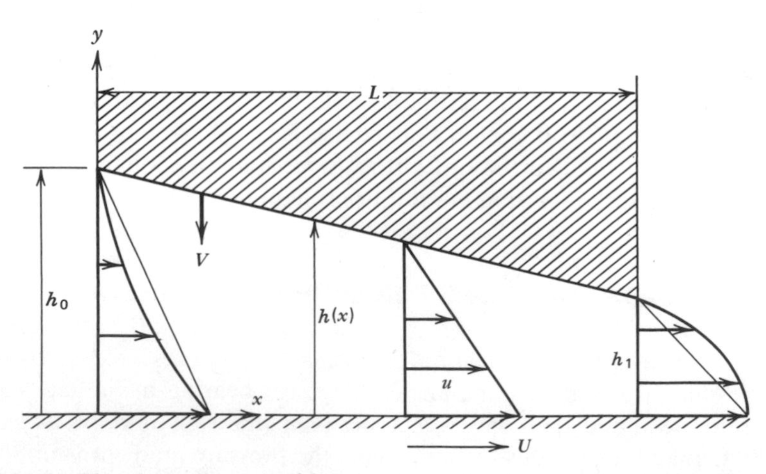

Figure 21-7 Flow in a slipper-pad bearing is locally the sum of a Couette flow and a Poiseuille flow.

the velocity scale, here we have a characteristic velocity U that induces the pressure scale $\mu UL/h_0^2$.

If the scales given in 21.9.1 are used to estimate the terms in the y-momentum equation and the limit $h_0/L \to 0$ is taken, one finds that $\partial p/\partial y = 0$. Thus, the pressure is only a function of x and z.

A similar process applied to the x- and z-momentum equations yields

$$0 = -\frac{\partial p}{\partial x} + \mu \frac{\partial^2 u}{\partial y^2}$$
$$0 = -\frac{\partial p}{\partial z} + \mu \frac{\partial^2 w}{\partial y^2} \qquad 21.9.2$$

Partial integration over y and application of boundary conditions

$$u(y=0) = U, \qquad u(y=h) = 0$$
$$w(y=0) = 0, \qquad w(y=h) = 0$$

produces the velocity profiles

$$u = \frac{1}{2\mu}\frac{\partial p}{\partial x}(y^2 - yh) + \left(1 - \frac{y}{h}\right)U \qquad 21.9.3$$

$$w = \frac{1}{2\mu}\frac{\partial p}{\partial z}(y^2 - yh) \qquad 21.9.4$$

In all lubrication problems the local profiles are a combination of a Couette flow and a Poiseuille flow given by the local gap and pressure gradient. The only remaining question, but the crucial one with regard to supporting a load with the bearing, is to determine the pressure distribution.

The Reynolds equation for the pressure is derived by integrating the continuity equation over the y direction to find

$$\int_0^h \frac{\partial u}{\partial x}\,dy + \int_0^h \frac{\partial w}{\partial z}\,dy = -\int_0^h \frac{\partial v}{\partial y}\,dy = -V \qquad 21.9.5$$

This equation is changed by noting that $-V = \partial h/\partial t$ and that Leibnitz's theorem requires

$$\frac{\partial}{\partial x}\int_0^h u\,dy = \int_0^h \frac{\partial u}{\partial x}\,dy + U\frac{\partial h}{\partial x}$$

Substituting these relations into 21.9.5 and employing velocity profiles 21.9.3 and 21.9.4 allows the integrals to be evaluated. The result is

$$\frac{1}{\mu}\left[\frac{\partial}{\partial x}\left(h^3\frac{\partial p}{\partial x}\right) + \frac{\partial}{\partial z}\left(h^3\frac{\partial p}{\partial z}\right)\right] = 6U\frac{\partial h}{\partial x} - 12\frac{\partial h}{\partial t} \qquad 21.9.6$$

This is the Reynolds equation for lubrication in a channel $h(x, t)$ with the lower wall moving at velocity U. It is the basic equation in lubrication theory. The pressure distribution may be found by knowing the geometry and motion of the walls. Once the pressure is known, the proportion of Poiseuille and Couette components in the velocity profiles 21.9.3 is fixed.

21.10 SLIPPER-PAD BEARING

The principles of lubrication are demonstrated nicely by the slipper pad. Figure 21-7 shows a flat pad inclined at an angle α. The pad length is L, the entrance gap is h_0, and the exit gap is h_1. Let the lower wall move with a velocity U. In accord with 21.9.1 we define nondimensional variables as

$$p^* = \frac{p - p_0}{\mu UL/h_0^2}$$
$$x^* = \frac{x}{L}, \qquad h^* = \frac{h}{h_0} \qquad 21.10.1$$

The wall location is given by

$$h^* = 1 - Ax^*, \qquad \text{where} \quad A = \frac{\alpha L}{h_0} = \frac{h_0 - h_1}{h_0} \qquad 21.10.2$$

For an infinitely wide bearing, the flow is assumed to be one-dimensional. This allows 21.9.5 to be integrated once to yield

$$h^{*3}\frac{dp^*}{dx^*} = 6(h^* - h_m^*)$$

In this equation the constant of integration has been denoted as h_m^*, the gap at the location where the pressure gradient is zero. This particular point is where the pressure is a maximum and the velocity profile is the linear Couette profile without any Poiseuille component. This being the case, the flow rate through the bearing is simply

$$Q = \frac{Uh_m}{2}, \qquad \text{or} \quad Q^* = \frac{2Q}{Uh_0} = h_m^* \qquad 21.10.3$$

A second integration of the pressure equation, together with boundary conditions that $p^* = 0$ at $h^* = 1$ and at $h^* = 1 - A$, yields

$$p^* = 6A^{-1}(h^{*-1} - 1) - 3A^{-1}h_m^*(h^{*-2} - 1) \qquad 21.10.4$$

with

$$h_m^* = 2\frac{1-A}{2-A} \tag{21.10.5}$$

We now insert 21.10.5 and 21.10.2 into 21.10.4 to obtain the pressure as a function of x^*:

$$p^* = \frac{6x^*}{1-Ax^*}\left[1 - \frac{1-A}{2-A}\frac{2-Ax^*}{1-Ax^*}\right] \tag{21.10.6}$$

The pressure curves contain the slope parameter A. We can show that the maximum pressure is

$$p_m^* = \frac{3A}{2(2-A)(1-A)} \tag{21.10.7}$$

From 21.10.5 and 21.10.2 this occurs at

$$x_m^* = \frac{1}{2-A} \tag{21.10.8}$$

To interpret these results, consider the case $A = 0$, implying the upper and lower walls are parallel. This flow is simply a Couette flow with a constant pressure along the pad. If a small convergence is added, so that $A > 0$, then a maximum pressure $p_m^* \approx 3A/4$ occurs at $x_m^* \approx \frac{1}{2}$. For the first half of the pad, the pressure rises, indicating a Poiseuille component opposing the flow into the bearing. The Poiseuille component changes sign for $x^* > x_m^*$ and causes an increase in the flow as the gap narrows. We can roughly describe the situation by saying that the fluid dragged into the converging channel by the viscous shear forces piles up to create a high pressure near the center $x^* = \frac{1}{2}$. The pressure gradient between the center and either end induces a Poiseuille flow toward both ends of the bearing. The Poiseuille component subtracts from the flow rate for the first half of the bearing length and then adds for the other half.

The magnitudes of pressure generated by this mechanism are truly remarkable. In dimensional terms

$$p_m - p_0 = \frac{\mu U_0 L}{h_0^2}\left[\tfrac{3}{4}A + O(A^2)\right]$$

This shows that h_0 plays the strongest role fixing the pressure.

Recall that the Stokes flow equations are reversible. If we imagine that the direction of motion of the wall is reversed, then the flow is dragged into a diverging channel and all velocities are reversed. More importantly, the theory predicts that the pressure curve will change sign. Unless the reference pressure at both ends of the wedge is very high, we predict negative pressures within the bearing. Cavitation actually prevents the pressures from taking on large negative values.

Suppose that two wedges are placed end to end so that an hourglass shape is produced. The flow converges through the first wedge and then diverges through the second wedge. This behavior is much like a journal bearing in that the eccentricity between the shaft and the housing gives the same pattern of area changes. In principle we have just solved this problem. The flow into the converging section reaches a pressure maximum and then returns to $p^* = 0$ at the position of the minimum gap. The flow in the diverging section is a reflection of that in the converging channel, and the pressure p^* is the negative of that at corresponding points in the entrance channel. It is, in fact, typical for journal bearings to operate with cavitation due to this effect. The region of cavitation vapor does not completely block the channel, and the lubricating fluid is carried over the top to flow into the converging section again.

*21.11 SQUEEZE FILM LUBRICATION; VISCOUS ADHESION

There is an intriguing physical similarity between the "wringing" together of smooth surfaces and the operation of a crankshaft bearing in an automobile. In the latter case, the power stroke of the piston causes a normal motion $V(t)$ between the crankshaft and the bearing (this effect is actually dominant over the hydrodynamic journal-bearing effect.) The case of attempting to separate smooth surfaces by pulling in the normal direction is just the opposite, as we are trying to generate a normal motion $V(t)$ that increases the gap height. Suppose two highly polished flat surfaces with a liquid coating are brought together. If the fluid gap is small, it is impossible to pull them apart in the normal direction, though sliding motion is quite easy.

Recall that when the Reynolds equations were derived in Section 21.9, a normal motion of the upper surface $V(t)$ was allowed. Consider a bearing of length L with a uniform but time-dependent gap $h(t)$. The Reynolds equation 21.9.5 becomes

$$\frac{1}{\mu}\frac{d}{dx}\left(h^3\frac{dp}{dx}\right) = -12\frac{dh}{dt}$$

Let the origin be halfway from either end of the bearing pad. Integration of the equation above gives

$$\frac{dp}{dx} = -\frac{12\mu}{h^3}\frac{dh}{dt}x$$

The Poiseuille component is zero at $x = 0$, where there is no flow at all, and is a maximum at the ends ($x = \pm L/2$). All of the flow into or out of the gap must cross the ends.

The pressure distribution is found by further integration:

$$p - p_0 = -\frac{6\mu L^2}{h^3}\frac{dh}{dt}\left[\left(\frac{2x}{L}\right)^2 - 1\right]$$

Note that the pressure varies as h^{-3}. Smoothly polished materials allow h to become very small, and the pressures are consequently very large.

*21.12 LOCAL FLOW SOLUTIONS

Many general flows have local regions where the length scale of the flow, and hence the Reynolds number, is small. The leading edge of a sharp plate, a salient edge of a body, and a very small orifice in a thin plate are typical examples. Frequently we can find local solutions to the Stokes equations that represent the flow in these small regions. Often these solutions are not unique and depend in some way on the flow further away. For instance, Stokes flow may separate at any angle from a smooth wall. The actual angle depends on events outside the Stokes flow region. (Note also that Stokes flow does not couple directly to boundary-layer theory. There are regions between the local Stokes region and the boundary layer where more complex behavior occurs.) In this section we shall review several aspects of local Stokes flows.

First consider flows with axial symmetry. Stokes flow in a conical channel of an arbitrary angle is one such closed-form solution. The special case where the cone angle is π represents the viscous flow into a sink bounded by a flat wall. In this case the solution is (in spherical coordinates)

$$\psi = \frac{Q}{2\pi}(1 - \cos^3\theta)$$

$$v_r = \frac{3Q}{2\pi r^2}\cos^2\theta$$

The velocity drops off as r^{-2} (a continuity effect that is the same in ideal flow) and dies out like $\cos^2\theta$ as one approaches the walls at $\theta = \pm\pi/2$.

Another closed-form solution of this type can be found where the walls are a hyperbola rotated about an axis of symmetry. A limiting case in this geometry represents a hole of diameter d in a thin wall. The flow rate of seepage through the hole due to a pressure difference Δp is found to be in direct proportion to the pressure drop:

$$Q = \frac{\Delta p\, d^3}{24\mu}$$

Experiments show that this expression is valid for $\mathrm{Re}_d < 6.4$ (see Happel and Brenner, 1965).

Plane two-dimensional Stokes flow gives some interesting insights into the behavior of separation and attachment streamlines. Consider the flow (v_r, v_θ) within the wedge $-\beta \leq \theta \leq +\beta$ in polar coordinates. Solutions of the biharmonic equation for the stream function are of the form

$$\psi = r^{n+1}\{A\cos(n+1)\theta + B\sin(n+1)\theta$$
$$+C\cos(n-1)\theta + D\sin(n-1)\theta\}$$

Here A, B, C, and D are constants and n takes on any value whatever (for $n = 0$, the solution has a slightly different form). Results are summarized by Michael and O'Neil (1977) and by Hasimoto and Sano (1980). Because the equations are linear, several solutions for different n may be added. Two of the constants in the expression above are determined by the no-slip condition on the walls $\theta = \pm\beta$. The other constants remain to be fixed by global considerations about the flow far away from the wedge.

Several special cases occur. First we consider $\beta = \pi/2$, that is, a smooth wall. Investigating the equation above for this case, Dean (1950) showed that separation (or attachment) can occur at any angle without restriction. The second special case is a cusped leading or trailing edge, which is represented by a wedge of $\beta = \pm\pi$. In this case too, the separation can leave the corner at any angle without restriction.

The general case of arbitrary β shows that symmetrical and unsymmetrical flow patterns occur. If $2\beta < 146.3°$ (159.1° for symmetrical flow), a very strange thing happens. Solutions of the form above are not possible (Dean and Montagnon, 1949): a simple streaming motion into and out of the corner cannot exist. When the wedge becomes this small a sequence of counterrotating vortices form as shown in Fig. 21-8. These are known as Moffatt (1964) vortices.

When β is above the critical angle, the dominant-mode solutions show no separation for the symmetrical solution and separation along the bisector for the unsymmetrical solution. Hence, it is possible for a trailing edge of a finite angle to separate along the bisector or, on the other hand, for the flow to actually turn and flow smoothly around the corner. In the latter case separation occurs a little way away from the edge, as separation from a smooth wall. Another example of this is at the corner of a cavity in a wall, $\beta = 3\pi/2$. Numerical calculations confirm that the flow goes around the corner and into the cavity before separating from the wall.

The characteristics of the local solutions discussed above—in particular, the fact that a streaming motion cannot exist in a corner with a wedge angle less than $\beta = 146.3°$—leads to some unusual patterns of separation regions and recirculation eddies. For a typical example, consider the situation of a circular cylinder held at a fixed distance from a flat wall. If the gap between the cylinder and the wall is large, then a streaming motion goes smoothly around the cylinder as one would expect. As the gap is made progressively narrower a series of flow patterns with closed eddies occurs. The first change is that a separation bubble occurs on the wall in front of the cylinder. The flow goes over this attached eddy and then through the gap between the cylinder and the wall. The symmetry requirement for Stokes flow demands that another eddy be attached to the wall on the downstream side of the cylinder (Fig. 21-9). As the gap height is reduced further, the wall eddies become larger and two additional eddies form on the cylinder (Davis and O'Neil, (1977)). The flow now goes in a sinuous path as it travels over the wall eddy, then down around the eddy attached to the cylinder, and finally through the gap to repeat the pattern on the downstream side. The condition when the cylinder touches the wall produces, in theory, the infinite sequence of Moffatt vortices in the channel between the cylinder and the wall.

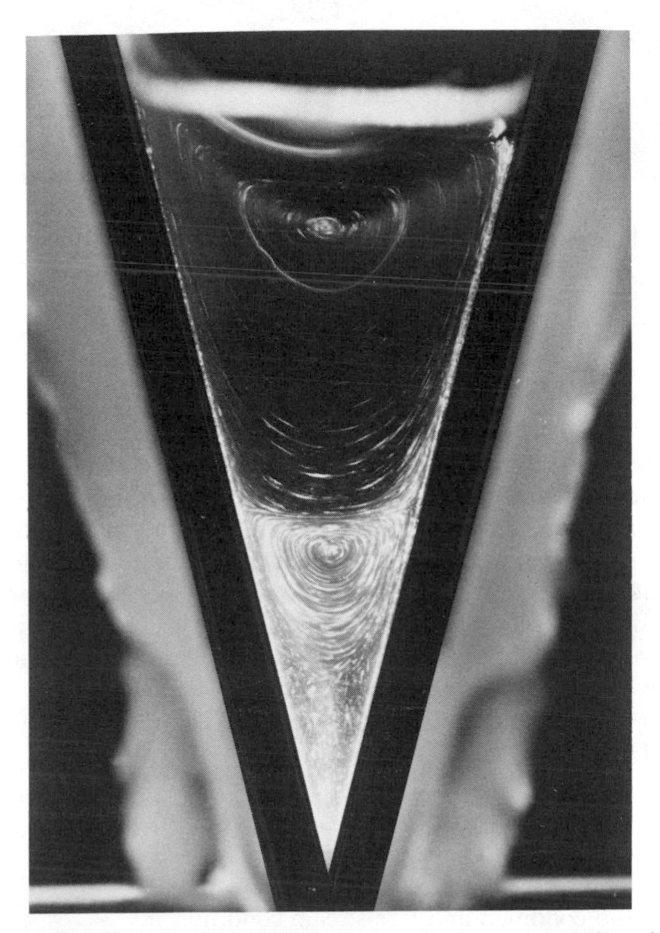

Figure 21-8 Moffatt vortices formed by rolling a cylinder over a V-shaped notch. The Reynolds number is 0.17. A 90-minute exposure was required for this photograph. From Taneda (1979).

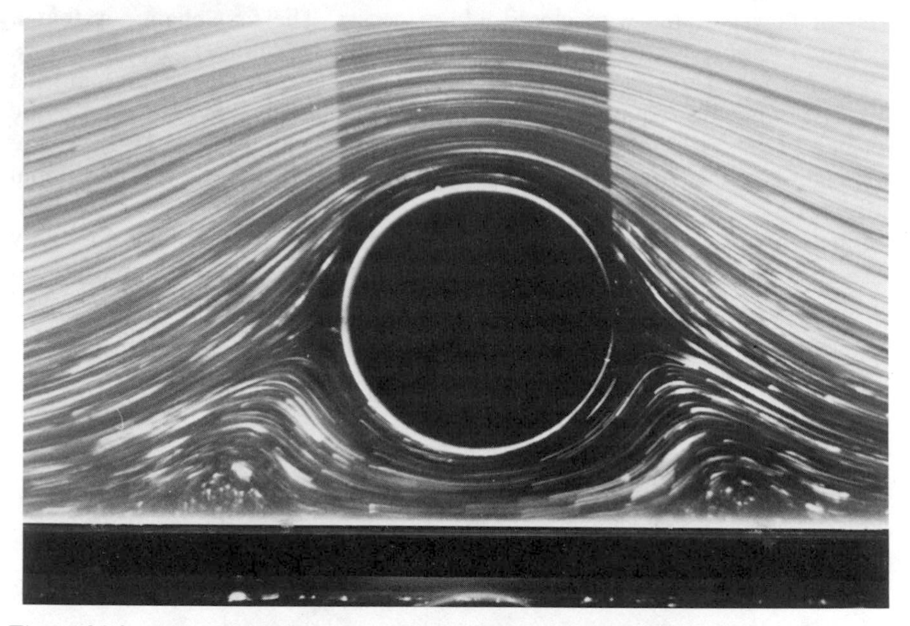

Figure 21-9 Re = 0.011 shear flow over a cylinder near a wall. Vortices form on the upstream and downstream sides. Photograph from Taneda (1979).

*21.13 STOKES FLOW NEAR A CIRCULAR CYLINDER

It is impossible to solve Stokes problem for the streaming flow over any two-dimensional object in an infinite stream. This fact is known as the Stokes paradox. In Section 21.4 it was shown that the Stokes equations are only valid near an object and cannot be expected to yield valid results in the far stream. In three dimensions, that is, for bodies that are finite in all directions, this difficulty is not as important, because it is mathematically possible to have the Stokes solution satisfy the free-stream boundary condition. In contrast, plane two-dimensional flows around semiinfinite bodies cannot be made to merge smoothly with a uniform stream. These solutions are valid only near the body and not far away.

Consider the Stokes flow past a cylinder. The velocity obeys the linear equation 21.1.3

$$0 = -\nabla^* p^{**} + \nabla^2 \mathbf{v}^*$$

with the no-slip condition and free-stream condition

$$v^*(r^* = 1) = 0, \qquad v^*(r^* \to \infty) = 1$$

In terms of the stream function, the mathematical problem is

$$\nabla^4\psi \equiv \left[\frac{\partial^2}{\partial r^2} + \frac{1}{r}\frac{\partial}{\partial r} + \frac{1}{r^2}\frac{\partial^2}{\partial \theta^2}\right]^2 \psi = 0 \qquad 21.13.1$$

$$\psi(1) = 0, \qquad \frac{\partial \psi}{\partial r}(1) = 0 \qquad 21.13.2$$

$$\psi = Ur_0 r \sin\theta \qquad \text{as} \quad r \to \infty \qquad 21.13.3$$

The difficulty in the far field is revealed if we seek a solution by separation of variables. The solution that satisfies 21.13.1 and 12.13.2 and the $\sin\theta$ part of 21.13.3 is

$$\psi = CUr_0^2\left[\frac{r_0}{r} - \frac{r}{r_0} + 2\frac{r}{r_0}\ln\frac{r}{r_0}\right]\sin\theta \qquad 21.13.4$$

It is impossible to match to the uniform stream as $r \to \infty$ because of the $r \ln r$ term. The solution diverges too rapidly. This is the origin of the Stokes paradox. All external plane-flow solutions to the Stokes equations have this same characteristic. When people say they have a plane-flow solution for the streaming motion past a certain body or group of bodies, they actually mean that they have the near-field solution. Far from the body, the solution will grow like $r \ln r$, and hence it will not match a uniform stream.

Recall that in the case of a sphere 21.4.1 showed that the ratio of convective terms, which are neglected in Stokes approximation but nevertheless may be estimated using the Stokes flow answer, to the viscous terms was

$$\frac{\text{convective term}}{\text{viscous term}} \sim \text{Re} \cdot \frac{r}{r_0} \qquad \text{as} \quad r \to \infty$$

Thus, for any $\text{Re} > 0$ a far position r/r_0 may be found where the conditions for the Stokes approximation are violated. In plane flow past a cylinder a similar calculation shows that

$$\frac{\text{convective term}}{\text{viscous term}} \sim \text{Re} \cdot \frac{r}{r_0}\ln\frac{r}{r_0} \qquad \text{as} \quad r \to \infty$$

From this we see that Stokes approximation becomes invalid slightly closer to the body in plane flow than it does in three-dimensional flow.

*21.14 OSEEN'S EQUATIONS

Oseen (1910) discovered the nonuniform character of Stokes solutions and proceeded to give a modified approximation that is uniformly valid. It is interesting that Oseen did not derive his equations as a mathematically rational approximation to the Navier–Stokes equations. His method is an ad hoc, intuitive construction for the purpose of correcting the deficiencies of the Stokes solutions. It gives improved results, but by its very nature it cannot be used as a basis for further refinements. Much later, Proudman and Pearson (1957) and Kaplun (1957) produced an organized theory of matched asymptotic expansions for low-Reynolds-number streaming flows. This theory is very elaborate, as the matching process is very delicate, especially in the plane-flow case. The simplicity of Oseen's equations offers a striking example of how ideas constructed with intuition and inductive reasoning can be as valuable as those produced by rational deduction. In fact the former usually precede the latter.

The failure of the Stokes equations occurs in the far field, where the convective term is just as large as the viscous term. Oseen reasoned that he could replace $(\mathbf{v} \cdot \nabla)\mathbf{v}$ by the linear approximation $U\,\partial v_i/\partial x$, where u is the free-stream velocity. Thus, he proposed the momentum equation

$$\rho U \frac{\partial v_i}{\partial x} = -\partial_i p + \nu \partial_j \partial_j v_i \qquad 21.14.1$$

This equation, or its equivalent form in terms of the stream function, gives answers that are uniformly valid for the whole flow field as Re $\to$ 0. Near the body the viscous terms are dominant and the linearized convective term in 21.14.1 makes very little contribution to the flow. In fact, the Oseen approximation $U\,\partial v_i/\partial x$ and the Stokes approximation of zero are errors of the same order in comparison with the true $(\mathbf{v} \cdot \nabla)\mathbf{v}$ term; therefore both equations are acceptable in this region. Far away from the body the velocity differs only slightly from U, so Oseen's linearized convection term is a valid approximation. Thus, Oseen achieved a remarkable improvement over the Stokes equations.

We may verify the statements above by looking at Oseen's solution for flow over a sphere. The stream function in spherical coordinates for this flow is

$$\frac{\psi}{Ur_0^2} = \frac{1}{4}\left[2\left(\frac{r}{r_0}\right)^2 + \left(\frac{r_0}{r}\right)\right]\sin^2\theta$$

$$-\frac{3}{2\,\mathrm{Re}}(1+\cos\theta)\left[1 - \exp\left(-\frac{\mathrm{Re}}{2}\frac{r}{r_0}(1-\cos\theta)\right)\right] \qquad 21.14.2$$

(Note that the nondimensional form of Oseen's equations contains the Reynolds number.) Near the surface of the sphere r/r_0 is small, so the exponential term in the last bracket may be expanded in a series. The result shows that 21.14.2 is identical to Stokes solution to order one:

$$\frac{\psi}{Ur_0^2} = \frac{1}{4}\left[2\left(\frac{r}{r_0}\right)^2 - 3\left(\frac{r}{r_0}\right) + \frac{r_0}{r}\right]\sin^2\theta + O[\mathrm{Re}]$$

Equation 21.14.2 is not the true Oseen solution for flow over a sphere in that it does not satisfy the no-slip condition at the wall exactly but only to $O[\mathrm{Re}]$. Nevertheless, we should consider this adequate, as the Oseen equations themselves are only valid to this order near the sphere.

As a matter of practical interest, the Oseen theory produces the drag law

$$F_D = 6\pi\mu U r_0 \left(1 + \tfrac{3}{8}\,\mathrm{Re}\right)$$

The matched asymptotic theory mentioned above shows that the next term is $\frac{9}{40}\,\mathrm{Re}\ln\mathrm{Re}$.

21.15 CONCLUSION

Low-Reynolds-number flows are flows where the inertia of the fluid is not important. In confined flows this idea leads to a uniformly valid theory. Unconfined flows, on the other hand, are subject to some complications. The theory of streaming flow over a body as $\mathrm{Re} \to 0$ is a singular problem with the nonuniform region occurring at infinity. Fortunately, from an engineering standpoint this complication is not important for three-dimensional objects. It is possible to find the flow near the object even though the answer is in error in the far stream. Two-dimensional objects give more difficulty. The singular behavior at infinity does not allow a solution of the Stokes equations to match with a uniform stream. The velocity becomes infinite as $r \ln r$ when $r \to \infty$ (Stokes paradox). If we want to know how the streaming velocity U affects the Stokes flow near a two-dimensional body, then the singular region at infinity must be dealt with by a matched asymptotic expansion.

An ingenious ad hoc method of dealing with streaming flows at low Reynolds number is given by Oseen's equations, consisting of the Stokes equations plus a linearized convective term $U\,\partial v_i/\partial x$. Using these equations gives an approximation that is valid near the body and also in the free stream.

PROBLEMS

21.1 (A) What is the ratio of p^{**} to p^{*}?

21.2 (A) Prove that the vorticity flux $\mathbf{n}\cdot\nabla\boldsymbol{\omega}$ through any closed surface is zero when $\mathrm{Re} \to 0$.

21.3 (A) Verify that 21.3.5 is a solution to the low Re equations and that it produces the velocity components and the vorticity relations given in the text.

21.4 (A) What is the terminal velocity in air of a water droplet 0.01 mm in diameter.

21.5 (A) Compare the E^2 and ∇^2 operators.

21.6 (A) A spheroid with $4b = a$ has its axis at 50° to the horizontal. What is its angle of decent.

21.7 (A) Compute $\mathbf{n} \cdot \nabla \boldsymbol{\omega}$ at the surface of a sphere. Find its integral over the surface.

21.8 (B) Find the pressure drop vs flow rate relation for a slot with width given by $h = h_0 + A \sin(2\pi x/L)$.

21.9 (A) Verify that 21.13.4 satisfies 21.13.1.

21.10 (B) Find the pressure distribution for a bearing with a constant gap h_0 for $0 \leq x < L/2$ followed by a smaller gap h_1 over the last half of the bearing.

21.11 (A) Find the load carrying capacity and the drag force for problem 21.10.

21.12 (A) Find the drag force on the plane slipper pad bearing.

22 Introduction to Stability

It is the perverse nature of fluid flows to become unstable. To exist, a flow pattern must not only be a solution to the Navier–Stokes equations, but it must also be stable. Any real flow contains slight deviations in the boundaries, irregularities in the incoming stream, or any of many other possible imperfections that cause the velocity and pressure to depart slightly from the nominal steady-state values. Situations where the departure is damped out and the flow returns to its steady values are stable. Typically, flows at low Reynolds numbers, where the damping effect of viscosity is strong, are stable. Most ideal inviscid flows, especially those that are free of vorticity, are also stable.

Several things may happen to unstable flows. The most prominent is turbulence. In Reynolds's famous experiments on pipe flow, he noted that the transition from a stable laminar flow to a turbulent flow depended upon exceeding a critical value of the Reynolds number. Pipe flows, wall boundary layers, jets, and shear layers are examples of flows that become turbulent. The second type of behavior of an unstable flow was characterized by Taylor's famous experiments on Couette flow. When a cylindrical Couette flow becomes unstable, it makes a transition to a new pattern containing Taylor vortices. The new pattern itself may at some point become unstable and give way to yet another pattern. Ultimately, at high enough Reynolds number, turbulence develops.

Hydrodynamic stability theory deals solely with predicting if a given flow pattern is stable or not. The approach is to consider a given steady flow V_i, which satisfies the governing equations. A perturbation of some type is added. The velocity is then

$$v_i(x_i, t) = V_i(x_i) + v_i'(x_i, t)$$

In this chapter we review only the linear theory where v_i' is taken to be much smaller than V_i. (We could put a small amplitude parameter ϵ in front of v_i' in the equation above, but since we only work to first order in ϵ, it is easy to keep track of the proper order of terms without this parameter. This follows the customary notation employed in this field.) When the equation $v_i = V_i + v_i'$ is substituted into the governing equations, they become a system to determine v_i' with a known basic flow V_i. In generating these equations, products of terms containing v_i' are discarded because they are of order ϵ^2. Thus, the small-amplitude assumption results in a linear system of equations for v_i'. The term "linear stability theory" is frequently employed to indicate this approach. A theory of this nature can only mark the beginnings of any instability, as the growth of the disturbance soon invalidates the linearity assumption. The most we can expect of linear stability theory is that it will tell us what types of disturbances will grow and the critical values of the Reynolds number, or equivalent parameter, at which this will happen.

Stability theory is mathematically a complex subject. It has seen slow development because of the great difficulty in solving problems. In many instances heuristic and imaginative methods have been employed to solve for stability boundaries.

22.1 NORMAL MODES AS PERTURBATIONS

Since the equations for a small disturbance are linear, it is natural to draw on our experience with simple equations and propose that a disturbance can be decomposed into normal modes of various wavelengths. Although they are almost always used, it is only in a few special flows that the normal modes can be shown to form a complete set. Consider a basic flow in the x direction that is parallel and depends only on the y coordinate, that is, $V_i = (V_x(y), 0, 0)$. A normal-mode disturbance is a traveling wave with an amplitude that depends on y. For this flow it is assumed to be the real part of

$$v_i' = \hat{v}_i(y) e^{i(\alpha x + \beta z - \alpha c t)} \tag{22.1.1}$$

Here $\hat{v}(y)$ is a complex amplitude function, α (real) is the wavenumber in the x direction, β (real) is the wavenumber in the z direction, and c is a complex wave speed. The total wavenumber has the magnitude

$$k = (\alpha^2 + \beta^2)^{1/2} \tag{22.1.2}$$

[The wavevector $\mathbf{k} = (\alpha, 0, \beta)$.]

It is useful to separate the complex wave speed into parts as

$$c = c_R + ic_I \tag{22.1.3}$$

With this notation 22.1.1 becomes

$$v_i' = \hat{v}_i(y) e^{i(\alpha x + \beta z - \alpha c_R t)} e^{\alpha c_I t} \tag{22.1.4}$$

The physical importance of c is evident in this form. At a fixed point in the flow the disturbance mode oscillates with a frequency $\omega = \alpha c_R$ as waves with wavelength α, β pass by. The phase velocity of the mode is in the direction α, β with magnitude

$$c_\varphi = \frac{\alpha c_R}{k} = \frac{\omega}{k} \tag{22.1.5}$$

A disturbance in the flow direction ($\beta = 0$, so that $k = \alpha$) has the phase velocity c_R.

For the stability question, the most important quantity is c_I (assuming $\alpha > 0$). The growth or decay of 22.1.4 occurs as follows:

$c_I < 0$	Flow is stable
$c_I > 0$	Flow is unstable
$c_I = 0$	Flow is neutrally stable

A point where $c_I = 0$ but a small change in a flow parameter, say the Reynolds number, moves it into a region where $c_I > 0$ is called *marginally stable*. Such points are important because their locus marks the boundary between stable and unstable conditions.

22.2 KELVIN–HELMHOLTZ INVISCID SHEAR-LAYER INSTABILITY

When the basic flow and the perturbation are governed by the Euler equations, the stability problem is called (for obvious reasons) an inviscid stability problem, and it is thought of as characterized by the limit $\mathrm{Re} \to \infty$. The Kelvin–Helmholtz instability of an inviscid shear layer is an important example of this type of flow. Figure 22-1 shows a shear layer and the resulting

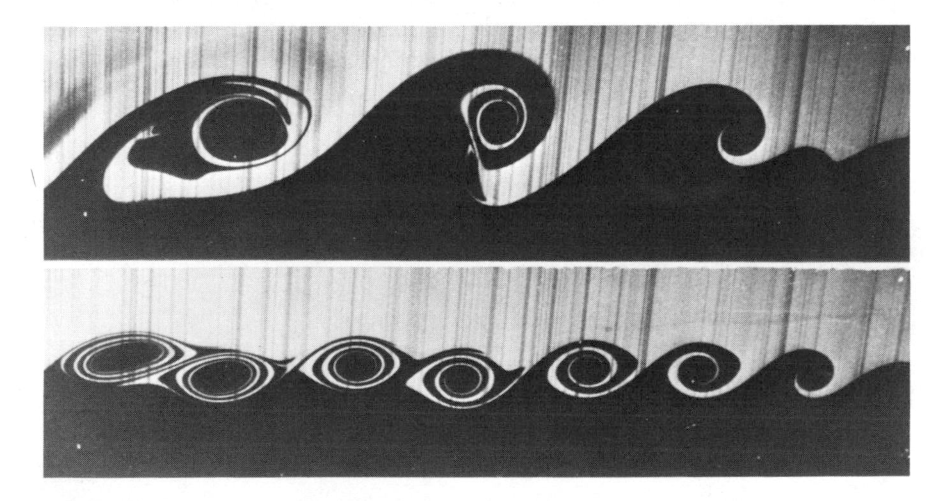

Figure 22-1 Kelvin-Helmholtz instability of a shear layer. Upper water stream which contains a fluorescent dye. moves faster than lower stream. A perturbation is introduced to initiate the growth in a regular pattern. Frequency is halved in lower picture. Photograph courtesy of F. A. Roberts, P. E. Dimotakis, and A. Roshko, California Institute of Technology.

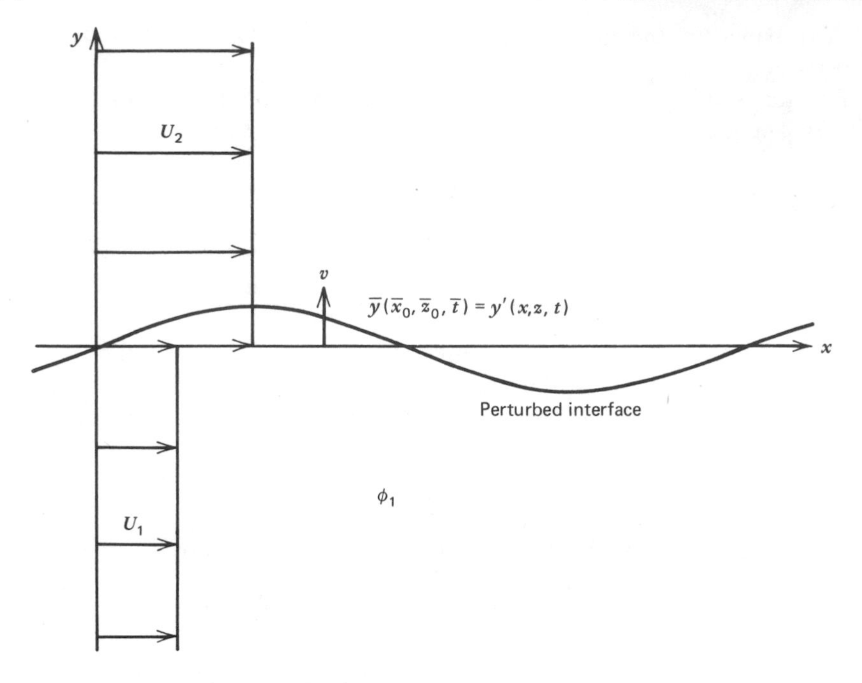

Figure 22-2 Shear layer nomenclature.

growth of the instability. In this particular experiment a splitter plate separates the flows, so that the age of the perturbation increases in the downstream direction. The mathematical problem is posed in a slightly different form where two uniform streams in the x direction slip past each other with the velocity discontinuity in the $y = 0$ plane, as shown on Fig. 22-2.

In inviscid-flow theory, the slip surface constitutes a vortex sheet of uniform density. Let us consider the equations that govern the flow including any perturbation. The flow above the sheet has a velocity potential ϕ_2 and that below the sheet ϕ_1. Incompressible, irrotational flows satisfy

$$\nabla^2\phi_1 = 0 \quad \text{and} \quad \nabla^2\phi_2 = 0 \tag{22.2.1}$$

with boundary conditions

$$\begin{aligned} \nabla\phi_1 &= U_1 \quad \text{as} \quad y \to -\infty \\ \nabla\phi_2 &= U_2 \quad \text{as} \quad y \to +\infty \end{aligned} \tag{22.2.2}$$

These conditions require that the perturbations die out far from the interface.

The interface is displaced slightly by the perturbed flow. Let a material particle on the interface have the vertical position $y = \bar{y}(\bar{x}_0, \bar{z}_0, \bar{t})$, where $\bar{y}$

stands for the Lagrangian function of the initial particle position $\bar{x}_0$, $\bar{z}_0$ and the Lagrangian time $\bar{t}$. We can also express this variable as a Eulerian function of x, z, t:

$$y = \bar{y}(\bar{x}_0, \bar{z}_0, \bar{t}) = y'(x, z, t)$$

The vertical velocity of a particle at the interface is

$$v = \frac{\partial \bar{y}}{\partial \bar{t}} = \frac{Dy'}{Dt}$$

For particles on the upper side, where $v = \partial \phi_2 / \partial y$, this is

$$\frac{\partial \phi_2}{\partial y} = \frac{\partial y'}{\partial t} + u_2 \frac{\partial y'}{\partial x} + w_2 \frac{\partial y'}{\partial z} \quad \text{at} \quad y = y' \qquad 22.2.3$$

On the lower side of the interface, the corresponding particle may have a different vertical velocity because the sliding velocities along the interface are different. For the lower side we have

$$\frac{\partial \phi_1}{\partial y} = \frac{\partial y'}{\partial t} + u_1 \frac{\partial y'}{\partial x} + w_1 \frac{\partial y'}{\partial z} \quad \text{at} \quad y = y' \qquad 22.2.4$$

Continuity of the interface is assured because y' is the same function in both of the expressions above.

Dynamics enters the problem through the unsteady Bernoulli equation,

$$\frac{\partial \phi}{\partial t} + \tfrac{1}{2}(\nabla \phi)^2 + \frac{p}{\rho} = C(t)$$

Our last boundary condition requires that the pressure be continuous across the interface. For $y = y'$ we have

$$\frac{\partial \phi_1}{\partial t} + \tfrac{1}{2}(\nabla \phi_1)^2 - C_1 = \frac{\partial \phi_2}{\partial t} + \tfrac{1}{2}(\nabla \phi_2)^2 - C_2 \qquad 22.2.5$$

Equations 22.2.1 through 22.2.5 govern the inviscid motion of a slip interface and the flows on either side.

The basic flow U_1, U_2 satisfies the problem with $y' = 0$, and the Bernoulli relation then reduces to

$$C_1 - \tfrac{1}{2}U_1^2 = C_2 - \tfrac{1}{2}U_2^2 \qquad 22.2.6$$

Perturbations from the basic flow are denoted by a prime. Thus, the potentials are

$$\phi_1 = U_1 x + \phi_1'$$

$$\phi_2 = U_2 x + \phi_2'$$

When these relations are substituted into 22.2.1 we find that

$$\nabla^2\phi_1' = \nabla^2\phi_2' = 0 \qquad 22.2.7$$

While 22.2.2 shows that the perturbations die out at infinity:

$$\begin{aligned} \nabla\phi_1' &= 0 \quad \text{as} \quad y \to -\infty \\ \nabla\phi_2' &= 0 \quad \text{as} \quad y \to +\infty \end{aligned} \qquad 22.2.8$$

The surface conditions 22.2.3 and 22.2.4 are transferred back to the basic surface $y = 0$ (see Section 16.3) and linearized by dropping products of primed quantities. This results in

$$\left.\frac{\partial \phi_1'}{\partial y}\right|_0 = \frac{\partial y'}{\partial t} + U_1\frac{\partial y'}{\partial x} \quad \text{and} \quad \left.\frac{\partial \phi_2'}{\partial y}\right|_0 = \frac{\partial y'}{\partial t} + U_2\frac{\partial y'}{\partial x} \qquad 22.2.9$$

In a similar manner the perturbation quantities are introduced into the Bernoulli equation 22.2.5, the relation is expanded about the basic surface $y = 0$, and the steady flow relation 22.2.6 is subtracted. The final equation is

$$\frac{\partial \phi_1'}{\partial t} + U_1\frac{\partial \phi_1'}{\partial x} = \frac{\partial \phi_2'}{\partial t} + U_2\frac{\partial \phi_2'}{\partial x} \quad \text{at} \quad y = 0 \qquad 22.2.10$$

The mathematical problem for y', ϕ_1', and ϕ_2' consists of 22.2.7 through 22.2.10.

The flow has been divided into a steady basic flow and a time-dependent perturbation. Next, we assume the disturbance can be represented by a composition of normal modes of the form

$$\begin{Bmatrix} y' \\ \phi_1' \\ \phi_2' \end{Bmatrix} = \begin{Bmatrix} \hat{y} \\ \hat{\phi}_1(y) \\ \hat{\phi}_2(y) \end{Bmatrix} e^{i(\alpha x + \beta z - \alpha c t)} \qquad 22.2.11$$

As discussed in Section 22.1, $c = c_R + ic_I$ is a complex wave speed. Note that $\hat{y}$ is a constant that gives the original amplitude of the interface displacement. It keys the size of all perturbed quantities. If $c_I = 0$, this displacement is unstable and grows exponentially in time.

Substituting 22.2.11 into 22.2.7 with boundary conditions 22.2.8 shows that the amplitude functions are

$$\begin{aligned} \hat{\phi}_1(y) &= A_1 e^{-ky} \\ \hat{\phi}_2(y) &= A_2 e^{-ky} \end{aligned} \qquad 22.2.12$$

where $k = (\alpha^2 + \beta^2)^{1/2}$. Substituting 22.2.11 and 22.2.12 into the interface conditions 22.2.9 yields

$$A_1 = i\alpha\frac{\hat{y}}{k}(U_1 - c)$$
$$A_2 = -i\alpha\frac{\hat{y}}{k}(U_2 - c) \qquad 22.2.13$$

The final relation is obtained by inserting 22.2.13, 22.2.12, and 22.2.11 into the Bernoulli relation 22.2.10. This gives

$$(U_1 - c)^2 = -(U_2 - c)^2$$

The final step is to solve for the complex wave speed:

$$c = \tfrac{1}{2}(U_1 + U_2) \pm i\tfrac{1}{2}|U_2 - U_1|$$
$$= c_R + ic_I \qquad 22.2.14$$

A flow with $c_I > 0$ is unstable. Hence, waves of all wavenumbers α, β are unstable if we take the plus sign in 22.2.14 (the possibility of stable decay for the minus sign exists; however, we are more interested in situations that allow instability). All shear layers $U_1 \neq U_2$ are inviscidly unstable to disturbances of all wavelengths.

Consider a disturbance of a certain wavelength $k = (\alpha^2 + \beta^2)^{1/2}$ and an arbitrary orientation. Since the growth rate is $\exp(\alpha c_I t)$, the wave that is oriented in the flow direction $k = \alpha$ will grow the fastest. The plane speed of this wave is

$$c_\varphi = \frac{\alpha c_R}{k} = c_R = \tfrac{1}{2}(U_1 + U_2)$$

The disturbance travels at the average speed of the main flows.

Kelvin–Helmboltz instability is extremely common, as many flows are essentially thin free shear layers (e.g. the initial region of a jet). Viscous effects in a shear layer, which spread the velocity profile over a finite thickness, stabilize perturbations with wavelength comparable to the shear-layer thickness. Waves much longer are still governed by the analysis we have just completed.

22.3 STABILITY PROBLEM FOR NEARLY PARALLEL VISCOUS FLOWS

When viscous effects are also included, the stability equations become much more complicated. A degree of simplicity can be restored if we consider a flow with two components $U(x, y)$, $V(x, y)$. Furthermore, it is assumed the flow is nearly parallel: $V \ll U$ and $\partial U/\partial x \ll \partial U/\partial y$. Velocity and pressure perturbations v' and p' are added to the main flow, so that

$$v = (U + u', V + v', w')$$
$$p = P + p' \qquad 22.3.1$$

Both the steady main flow U, V, P and the flow with the perturbation satisfy the Navier–Stokes equations. The variables in 22.3.1 are nondimensional. Typical scales U_∞, L, ρ will be specified for each particular problem.

The x component of the momentum equation for the flow is

$$\frac{\partial u'}{\partial t} + (U + u')\frac{\partial}{\partial x}(U + u') + (V + v')\frac{\partial}{\partial y}(U + u')$$
$$= -\frac{\partial}{\partial x}(P + p') + \frac{1}{\text{Re}}\nabla^2(U + u') \qquad 22.3.2$$

The same equation for the main flow is

$$U\frac{\partial U}{\partial x} + V\frac{\partial U}{\partial y} = -\frac{\partial P}{\partial x} + \frac{1}{\text{Re}}\nabla^2 U \qquad 22.3.3$$

Equation 22.3.2 is expanded and linearized by dropping products of u', v', w', V, and $\partial U/\partial x$. When 22.3.3 is subtracted, we arrive at an x-momentum equation for the perturbation:

$$\frac{\partial u'}{\partial t} + U\frac{\partial u'}{\partial x} + v'\frac{\partial U}{\partial y} = -\frac{\partial p'}{\partial x} + \frac{1}{\text{Re}}\nabla^2 u' \qquad 22.3.4$$

By similar steps the y- and z-direction momentum equations are found:

$$\frac{\partial v'}{\partial t} + U\frac{\partial v'}{\partial x} = -\frac{\partial p'}{\partial y} + \frac{1}{\text{Re}}\nabla^2 v' \qquad 22.3.5$$

and

$$\frac{\partial w'}{\partial t} + U\frac{\partial w'}{\partial x} = -\frac{\partial p'}{\partial z} + \frac{1}{\text{Re}}\nabla^2 w' \qquad 22.3.6$$

Applying the same process to the continuity equation yields

$$\nabla \cdot \mathbf{v}' = 0 \qquad 22.3.7$$

The above relations 22.3.4 through 22.3.7 are a linear system of equations for $\mathbf{v}'$, p' when a specified main flow $U(y)$ is given. The dependence of the main flow in the x direction has been suppressed by the "nearly parallel" assumption. Essentially we are treating any chosen location x as if the profile at that station continued upstream and downstream without changing.

If the flow is confined between two walls, the no-slip condition requires that the perturbation $\mathbf{v}'$ vanish at both walls. If the flow extends to infinity, we require that the perturbation velocity vanish there also. No boundary conditions on the pressure are required.

Next, we note that the problem is linear and propose that an arbitrary disturbance may be decomposed into normal modes described by

$$\begin{aligned} v_i' &= \hat{v}_i(y)\, e^{i(\alpha x + \beta z - \alpha c t)} \\ p' &= \hat{p}(y)\, e^{i(\alpha x + \beta z - \alpha c t)} \end{aligned} \qquad 22.3.8$$

For a chosen wavenumber α, β (both real) and specified Reynolds number Re, the substitution of 22.3.8 into 22.3.4 through 22.3.7 produces an eigenvalue problem where certain solutions $\hat{v}_i(y)$, $\hat{p}(y)$, the eigenfunctions, occur for specific values of c, the eigenvalue. Performing the substitution just described produces the following set of equations:

$$\begin{aligned} i\alpha(U - c)\hat{u} + \hat{v}\frac{dU}{dy} &= -i\alpha\hat{p} + \frac{1}{\text{Re}}\left[\frac{d^2\hat{u}}{dy^2} - (\alpha^2 + \beta^2)\hat{u}\right] \\ i\alpha(U - c)\hat{v} &= -\frac{d\hat{p}}{dy} + \frac{1}{\text{Re}}\left[\frac{d^2\hat{v}}{dy^2} - (\alpha^2 + \beta^2)\hat{v}\right] \\ i\alpha(U - c)\hat{w} &= -i\beta\hat{p} + \frac{1}{\text{Re}}\left[\frac{d^2\hat{w}}{dy^2} - (\alpha^2 + \beta^2)\hat{w}\right] \\ i\alpha\hat{u} + i\beta\hat{w} + \frac{d\hat{v}}{dy} &= 0 \end{aligned} \qquad 22.3.9$$

These equations govern the viscous stability of a normal mode in any nearly parallel flow.

A remarkable simplification of the stability problem was revealed by Squire (1933). He showed that for any unstable three-dimensional disturbance, there is a corresponding two-dimensional disturbance ($\hat{w} = 0$) that is more unstable. This allows us to seek the stability boundary of the flow with a two-dimensional disturbance $\hat{u}$, $\hat{v}$ and be assured that this is sufficient to find the lowest limit of stability. To prove Squire's theorem, consider the following transformation of variables:

$$\begin{aligned} \alpha^* &= (\alpha^2 + \beta^2)^{1/2}, & c^* &= c \\ \alpha^* u^* &= \alpha\hat{u} + \beta\hat{w}, & v^* &= \hat{v} \\ \frac{p^*}{\alpha^*} &= \frac{\hat{p}}{\alpha}, & \alpha^*\,\text{Re}^* &= \alpha\,\text{Re} \end{aligned} \qquad 22.3.10$$

These relations are used to transform the set 22.3.9. The first and third equations in 22.3.9 are added together, while the second and fourth are simply transformed. The result is the three equations

$$i\alpha^*(U - c^*)u^* + v^*\frac{dU}{dy} = -i\alpha^* p^* + \frac{1}{\text{Re}^*}\left[\frac{d^2u^*}{dy^2} - \alpha^{*2}u^*\right]$$

$$i\alpha^*(U - c^*)v^* = -\frac{dp^*}{dy} + \frac{1}{\text{Re}^*}\left[\frac{d^2u^*}{dy^2} - \alpha^* v^*\right] \qquad 22.3.11$$

$$i\alpha^* u^* + \frac{dv^*}{dy} = 0$$

These equations are identical with 22.3.9 when $\beta = 0$ and $\hat{w} = 0$ are taken in the former set. Through Squire's transformation 22.3.10, any solution for a two-dimensional disturbance u^*, v^* of wavenumber α^* and wave speed c^* may be used to describe an equivalent three-dimensional disturbance of wavenumber α, β, and wave speed c.

Recall from 22.1.4 that the growth rate of a disturbance is $e^{\alpha c_I t}$. Thus, for any three-dimensional disturbance of wavenumber α, β and wave speed $c = c_R + ic_I$, the equivalent two-dimensional disturbance has a larger x-direction wavenumber, since $\alpha^* = (\alpha^2 + \beta^2)^{1/2}$, and is more unstable in the sense that $\alpha^* c_I^* > \alpha c_I$ (note that $c_I^* = c_I$ in Squire's transformation). Moreover, the Reynolds number of the equivalent flow is lower because $\text{Re}^* = \text{Re} \cdot \alpha/\alpha^*$. In seeking the marginal-stability curve of a flow, we are usually interested in the smallest Reynolds number for which any disturbance is unstable. Squire's transformation shows that if this value is found for a two-dimensional disturbance, we have, in fact, determined the smallest value for both two- and three-dimensional disturbances. As a matter of convention, in further work we shall use the notation of 22.3.9 in the two-dimensional problem 22.3.11. This is equivalent to setting $\beta = \hat{w} = 0$ in 22.3.9.

22.4 ORR–SOMMERFELD EQUATION

Incompressible, two-dimensional flows may be formulated in terms of a stream function. Since the perturbations satisfy $\nabla \cdot \mathbf{v}' = 0$, it is permissible to introduce a perturbation stream function ψ' defined by

$$u' = \frac{\partial \psi'}{\partial y}, \qquad v' = -\frac{\partial \psi'}{\partial x} \qquad 22.4.1$$

The normal-mode assumption for the stream function is

$$\psi' = \phi(y)e^{i(\alpha x + \beta z - ct)} \qquad 22.4.2$$

where $\phi(y)$ is the complex amplitude function for the stream function. (This is the universally accepted symbol. The student should carefully note that this ϕ has no connection with the velocity potential.) Substituting 22.4.2 into 22.4.1 and using 22.1.4 gives

$$\hat{u} = \frac{d\phi}{dy}, \qquad \hat{v} = -i\alpha\phi \tag{22.4.3}$$

These relations reduce the continuity equation in 22.3.11 to an identity.

A single equation for $\phi(y)$ is found by substituting 22.4.3 into the first equation of 22.3.11, differentiating with respect to y so that $d\hat{p}/dy$ occurs, and then eliminating $d\hat{p}/dy$ in the second equation of 22.3.11. The result may be written as

$$(U - c)\left[\frac{d^2\phi}{dy^2} - \alpha^2\phi\right] - \phi\frac{d^2U}{dy^2} = \frac{1}{i\alpha\,\mathrm{Re}}\left[\frac{d^2}{dy^2} - \alpha^2\right]^2\phi \tag{22.4.4}$$

This equation is the cornerstone of hydrodynamic stability theory. It was derived first by Orr (1907) and Sommerfeld (1908). Although the equation is linear, it is notoriously difficult to solve. Consider, for example, the Blasius boundary-layer profile. The first successful solutions of the Orr–Sommerfeld equation for this flow were by Tollmien (1929) and Schlichting (1933), over twenty years after the equation was discovered. These first attempts were approximations. A significantly different approach was taken by Lin (1945), and finally, what might be called an exact solution was found by Jordinson (1970), over a half a century after the beginning.

The Orr–Sommerfeld equation, with no-slip boundary conditions $\phi = d\phi/dy = 0$ at two locations in the flow, is to be solved for a given velocity profile $U(y)$, Reynolds number Re, and wavenumber α. The equation not only determines the eigenfunction $\phi(y)$, but also the complex wave speed $c = c_R + ic_I$ as the eigenvalue. A bounded flow has, for fixed Re and α, a discrete set of eigenvalues $c_1, c_2, c_3, \ldots$. A boundary-layer flow that is unbounded has a discrete set and an additional continuous spectrum. Consider the eigenvalues as a function of α and Re (for a given flow):

$$\begin{aligned} c_R &= c_R(\alpha, \mathrm{Re}) \\ c_I &= c_I(\alpha, \mathrm{Re}) \end{aligned} \tag{22.4.5}$$

Recall that the perturbation growth is $e^{\alpha c_I t}$ and that the flow is unstable for $c_I > 0$. If at least one neutral stability mode $c_I = 0$ exists, then setting $c_I = 0$ in 22.4.5 gives a curve of neutral stability $c_I(\alpha, \mathrm{Re}) = 0$. If, in addition, it is possible to show that c_I changes sign as one crosses the neutral curve, then this is also the curve of marginal stability. It separates a stable and an unstable region in α, Re space.

22.5 INVISCID STABILITY OF NEARLY PARALLEL FLOWS

When Re $\to \infty$ in the Orr–Sommerfeld equation, we have a simplified form known as Rayleigh's equation:

$$\frac{d^2\phi}{dy^2} - \left[\alpha^2 + \frac{1}{U - c}\frac{d^2U}{dy^2}\right]\phi = 0 \qquad 22.5.1$$

Since an inviscid flow slips past a wall, only the boundary condition $\phi = 0$ is enforced at two places in the flow. It is particularly important to realize that Rayleigh's equation describes a perturbation that behaves in an inviscid manner. The main flow profile $U(y)$ may be a viscous-dominated flow. Thus, it is perfectly reasonable to talk about the inviscid stability of the Blasius boundary-layer profile. This is, in fact, the behavior of the marginal stability curve $c_I(\alpha, \text{Re}) = 0$ as Re $\to \infty$.

Rayleigh proved two characteristics of inviscid stability that are of general utility in predicting stability characteristics. The first is known as *Rayleigh's point-of-inflection theorem*. It may be stated as follows:

> A necessary (but not sufficient) condition for inviscid instability is that the basic profile $U(y)$ has a point of inflection.

This is a very powerful result, as we may conclude that any profile without a point of inflection is stable as Re $\to \infty$. Plane Poiseuille flow, plane Couette flow, the Blasius boundary layer, and all boundary layers with positive pressure gradients are therefore stable to inviscid disturbances.

A slightly stronger result is known as *Fjørtoft's theorem*. It may be stated as follows: If y_0 is the position of a point of inflection [$d^2U/dy^2 = 0$ in the basic profile $U(y)$ and $U_0 = U(y_0)$], then a necessary (but not sufficient) condition for inviscid instability is that $(d^2U/dy^2)(U - U_0) < 0$ somewhere in the flow. Figure 22-3 illustrates this theorem. Both profiles a and b have points of inflection, and hence by Rayleigh's theorem they are possibly unstable. The stronger theorem of Fjørtoft shows the first case, Fig. 22-3a, is also stable because $(d^2U/dy^2)\ (U - U_0) > 0$. The second case is possibly unstable by both criteria.

Rayleigh's second result concerns the neutral stability mode with $c_I = 0$. For this mode $c = c_R$ and the denominator $U - c$ in 22.5.1 is real. Rayleigh proved that the phase speed c_R must lie between the maximum and minimum values of the profile $U(y)$. Thus, $U - c = 0$ at some point within the flow. This position is called the *critical layer*, because the inviscid equation 22.5.1 is singular at this location. The wave propagation velocity and the flow velocity are matched at the critical layer.

When we regard Rayleigh's equation 22.5.1 as the asymptotic form of the Orr–Sommerfeld equation 22.4.4 as Re $\to \infty$, we note that the inviscid eigen-

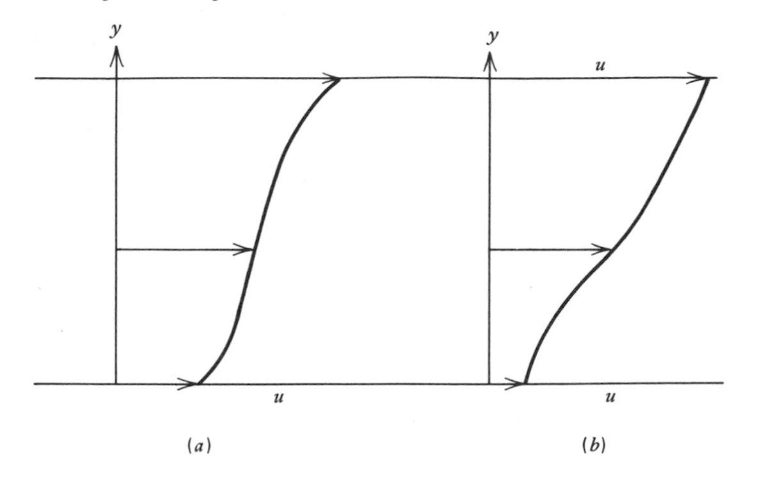

Figure 22-3 Inviscid instability of shear flows. Both cases are possibly unstable by Rayleigh's theorem but only (b) is possibly unstable by Fjørtoft's theorem.

functions $\phi(y)$ cannot be uniformly valid. Two boundary conditions ($d\phi/dy = 0$) have been dropped. Since 22.5.1 is of second order while 22.4.4 is of fourth order, viscous effects on the perturbation must occur at both walls (or at the wall and the free stream in an unbounded flow). Another nonuniform region occurs at the critical layer. Here $U - c = 0$ causes singular behavior in the Rayleigh equation, while the Orr–Sommerfeld equation has no difficulty at this point. The critical layer, where the wave speed of the inviscid perturbation matches the flow speed, is another place where viscous effects on the perturbation are important. Lin (1945) showed that when a profile has a point of inflection, the critical layer is located there.

22.6 VISCOUS STABILITY OF NEARLY PARALLEL FLOWS

Inertia and pressure forces must be in balance in an inviscid flow. An unstable velocity perturbation, say an increase in speed, through the Bernoulli equation must produce a pressure gradient, which continues to accelerate the particle. Intuitively, the addition of viscosity to this picture would be expected to be stabilizing. Indeed, a direct retarding force is one effect of viscosity. For any basic profile, one can usually find complete stability if the flow Reynolds number is low enough.

A second, and somewhat unexpected, effect of viscosity is destabilizing. A slight amount of viscosity can destabilize an otherwise stable profile. The reason has to do with the diffusion of the net shear stress. Viscous diffusion introduces a time lag. We have already seen an example of this mechanism in our study of the Stokes problem in Section 11.2. Recall the case where a free stream oscillates parallel to and above a fixed wall. The maximum amplitude was not in the free stream but at an intermediate distance within the Stokes layer. This effect was explained by noting that viscous stresses generated at the wall, take some time to diffuse to the free stream. After a time lag of one-half cycle, the pressure force has changed sign, and it now combines with the viscous force to accelerate the particles to a large amplitude. Viscous instability is this same diffusional time lag operating in a traveling-wave perturbation. The viscous stress generated by the perturbation diffuses across the flow in the y direction. After a certain time it is in the proper location and has the proper phase to add to the pressure mechanisms of the traveling wave and generate instability.

The classic example of viscous instability is the Blasius boundary layer. Figure 22-4 shows a plot of the marginal-stability curve in the α, Re plane. This curve divides the plane into stable and unstable regions. The flow is stable to disturbances of all wavenumbers at low Re. This is the stabilizing effect of viscous damping. Likewise, as Re $\rightarrow \infty$ by Rayleigh's point-of-inflection theorem the flow is completely stable for all α. Within the loop of the marginal-stability curve is a range of wavenumbers that are unstable. These unstable waves are called Tollmien–Schlichting waves. The point on the marginal curve with the lowest Reynolds number is called the *critical point*. The Blasius boundary layer has a critical Reynolds number of Re $= U\delta^*/\nu = 520$. The boundary layer is completely stable until this Reynolds number is reached. At

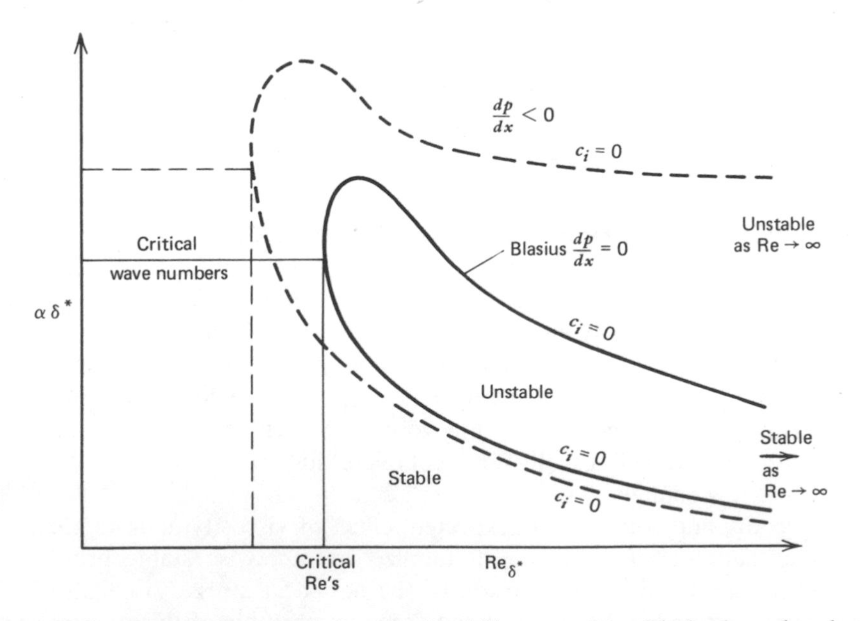

Figure 22-4 General shape of the marginal stability curve for a Blasius boundary layer and for layers with adverse pressure gradients.

this point the layer is unstable to a Tollmien–Schlichting wave with $\alpha\delta^* = 0.30$. The wavelength of this unstable mode is $L = 2\pi/\alpha = 2\pi\delta^*/0.3 \approx 18\,\delta^* \approx 6\delta$. Thus, the first unstable wave has a very long wavelength. The boundary layer is always stable to short wavelengths. As the boundary layer grows, Re increases and the band of unstable wavelength becomes larger. Of course, the loop closes and stability returns as $\text{Re} \to \infty$. The boundary layer never actually reaches this state, because the end result of the previous instability is a turbulent flow, which regenerates itself.

Boundary layers in an adverse pressure gradient must have a point of inflection. By Fjørtoft's theorem these profiles may have inviscid instability. Indeed, it turns out that they do. The upper branch of the marginal-stability curve (see Fig. 22-4) now has a finite limit α_∞ as $\text{Re} \to \infty$. Tollmien–Schlichting waves with $0 < \alpha < \alpha_\infty$ (long wavelengths) are unstable to inviscid mechanisms. Perhaps more important from a practical standpoint is the fact that the critical Reynolds number is now lower and the range of unstable wavenumbers is larger. Adverse pressure gradients rapidly promote the transition to turbulence.

As a final example in this section, we consider the marginal-stability curve for a shear layer. Figure 22-5 was computed by Betchov and Szewczyk (1963) for a shear-layer profile $U = \tanh(y/L)$. This is actually a model of a shear layer making a transition from U_1 to U_2 by the profile $U = (U_2 - U_1)\tanh(y/L) + U_1$. Only at $\text{Re} = 0$ is the profile completely stable for all

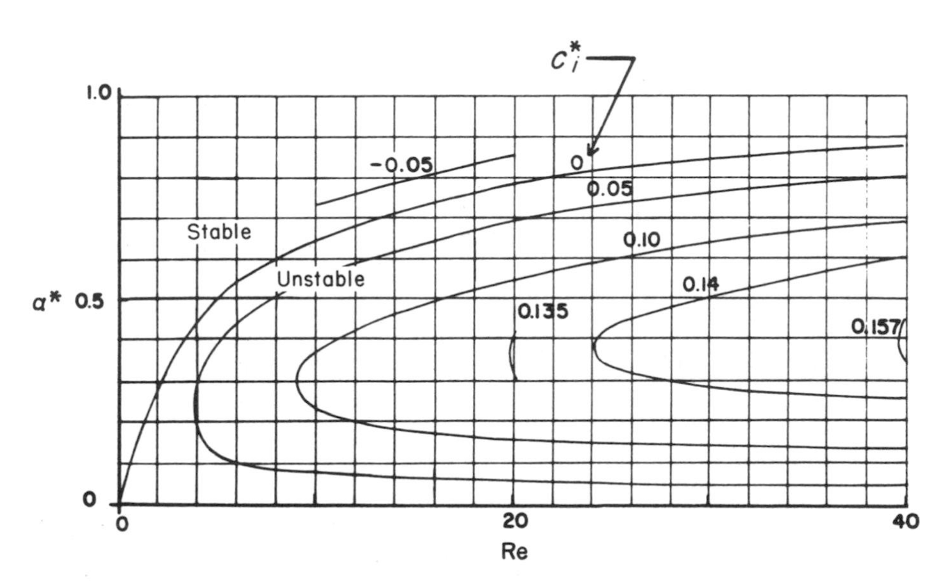

Figure 22-5 Curves of constant amplification $c_i^* = c_i\alpha L/U_0$ in the wave number $\alpha^* = \alpha L$ vs. Re plane for a shear layer $u = U_0\tanh(y/L)$. Marginal stability is given by $c_i^* = 0$. Viscosity makes the flow completely stable only at $\text{Re} = 0$. From Betchov and Szewczyk (1963). Photograph courtesy of A. Szewczyk, Notre Dame University.

Table 22-1

Flow	$\frac{U(y)}{U_0}$	$U_0 L/\nu$ $= \mathrm{Re}_c$	$\alpha_c L$ $= \alpha^*$	0–α^*_∞ inviscid	Remarks
Shear layer	$\tanh(y/L)$	0	0	0–1.0	Kelvin–Helmholtz, $\mathrm{Re} \to \infty$
Jet	$\mathrm{sech}^2(y/L)$	4	0.2	0–2.0	Even mode
Falkner–Skan separating profile	$\beta = -0.199$	64	1.24	0–0.8	$L = \delta^*$
Blasius	$\beta = 0$	520	0.30	0	$L = \delta^*$
Stagnation	$\beta = 1$	14,000	—	0	$L = \delta^*$
Flow into a sink	$\beta = \infty$	21,700	0.17	0	$L = \delta^*$
Poiseuille (plane)	$1 - (y/L)^2$	5,780	1.02	0	L = half width
Couette (plane)	y/L	∞	—	0	Stable

wavenumbers. At $\mathrm{Re} \to \infty$ the flow is unstable to long wavelengths ($0 \leq \alpha L \leq 1$). This is essentially the Kelvin–Helmholtz instability. The only new aspect is that the actual profile of the shear layer causes wavelengths shorter than the shear-layer thickness L to be stable. Between these extremes of Reynolds number the marginal-stability curve has a smooth monotonic variation. Viscosity has a purely stabilizing effect in this flow.

Table 22-1 gives stability characteristics of several common flows. The first three flows have points of inflection in the profiles. They are unstable as $\mathrm{Re} \to \infty$ for the range of wavenumbers denoted as 0–$\alpha^*_\infty L$. The critical Reynolds numbers are very low for these flows. Furthermore, the stability curve is somewhat insensitive to the shape of the main velocity profile $U(y)$. All the remaining flows are stable as $\mathrm{Re} \to \infty$; they do not have a point of inflection. In these flows the viscous instability mechanism operates. They have a larger critical Reynolds number and show more sensitivity to the form of the basic flow profile. Note in the last entry that plane Couette flow is linearly stable.

22.7 EXPERIMENTAL VERIFICATION OF STABILITY THEORY

The usual historical sequence in fluid mechanics is that phenomena are observed, documented experimentally, and subsequently explained by analysis and theory. Linear stability theory is an interesting reversal of this sequence. The prediction of the existence of Tollmien–Schlichting waves and of the neutral-stability curve lacked verification for many years. Finally, experiments on the Blasius boundary layer confirmed the theory.

The Blasius boundary layer is known to become turbulent at $\mathrm{Re}^*_\delta \approx 3000$. Because an unstable wave grows very slowly at first, the critical Reynolds number for stability is much lower than this value. Moreover,

Tollmien–Schlichting waves are only the first stage in a natural transition. The waves slowly change form, become three-dimensional, and nonlinear processes determine the final transition. Depending on the details of the flow situation, there are several different mechanisms and events that can lead ultimately to turbulence. The idealized transition beginning with the linear instability of a Tollmien–Schlichting wave (T-S wave) is usually called a natural transition. In most instances the imperfections in the experimental arrangement introduce large disturbances that trigger the transition process.

A sequence of experiments conducted at the National Bureau of Standards ultimately produced measurements of the neutral-stability curve. To do this they constructed a special wind tunnel, which had a very low turbulence level in the free stream. The rms velocity fluctuation averaged in three directions was less than 0.03% of the free-stream velocity. Figure 22-6 gives the stability

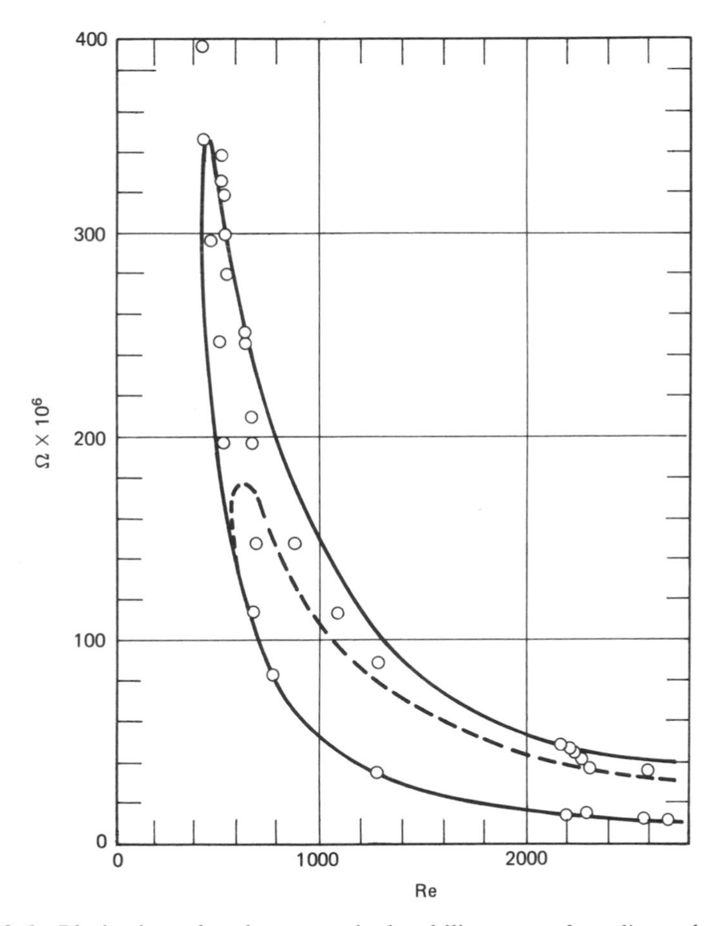

Figure 22-6 Blasius boundary layer marginal stability curves from linear theory using approximate methods. Experimental points by Schubauer and Skramstad (1947), solid line by Shen (1954), dashed line by Schlichting (1933b). Re $= U\delta^*/\nu$ and $\Omega = \alpha c_r/\text{Re}$.

curve determined by vibrating a ribbon in the boundary layer and observing the stability of the disturbance. The frequency of the disturbance is plotted, as this is experimentally easier to determine than the wavenumber. Also shown in the figure are the stability curves of Schlichting (1933) and some improved calculations by Shen (1954). The agreement between theory and experiment at this time was close. With the development of computers, more exact methods of solving the Orr–Sommerfeld equations became possible. Exact solutions revealed that the previous curves are slightly wrong; $\mathrm{Re}_c = 520$ in the exact calculations, compared to an experimental value of 450.

Attempts to resolve the discrepancy have centered on accounting for the growth of the boundary layer. Several nonparallel stability calculations, involving quite different approximations and approaches, have been done. All nonparallel theories give a correction in the proper direction; however, the amount of the correction varies widely. The subject remains controversial. Figure 22-7 shows the work of Saric and Nayfeh (1977).

The flow pattern in a T-S wave is sketched in Fig. 22-8. Flow visualization of these waves requires extreme care to avoid extraneous disturbances. A view

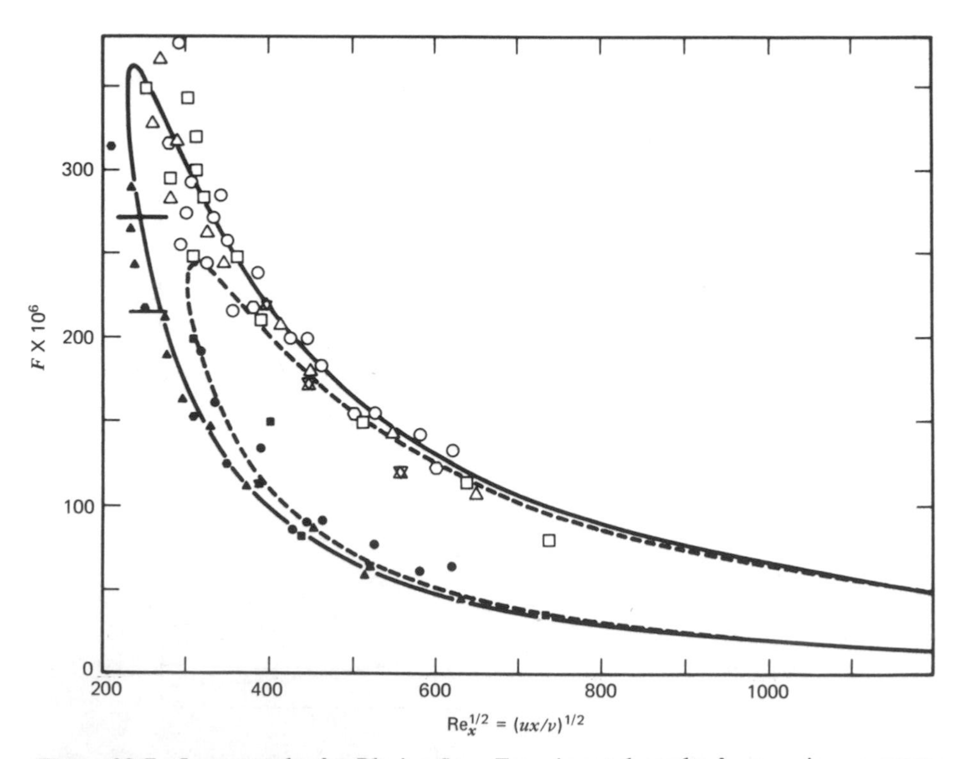

Figure 22-7 Later results for Blasius flow. Experimental results from various sources. Theory; – – – linear, — nonparallel theory of Saric and Nayfeh (1977). This graph plotted using $\mathrm{Re}_x^{1/2}$ instead of Re_{δ^*} courtesy of W. Saric, VPI and State University. $F = 2\pi f \nu / U^2$.

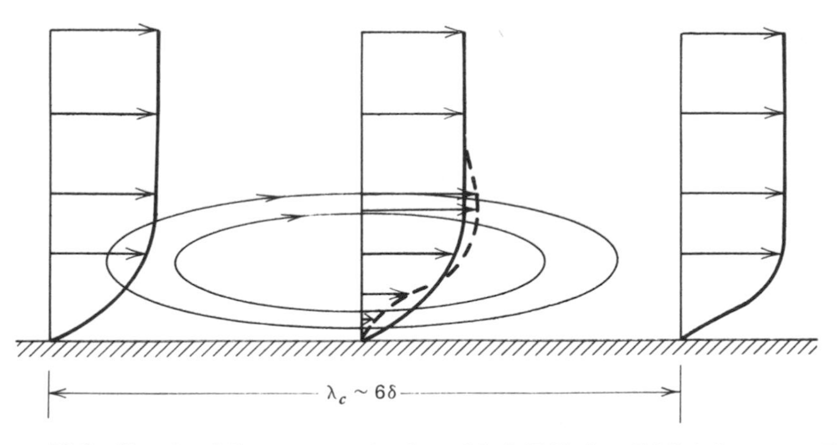

Figure 22-8 Sketch of flow pattern in the critical Tollmien–Schlichting wave. Perturbation streamlines are given in a coordinate system moving with the wave.

Figure 22-9 Plan view of smoke visualization of T-S waves in a Blasius boundary layer. Photograph courtesy of A. Thomas, Lockheed-Georgia Co.

looking down on the flow over a flat plate in Fig. 22-9 shows a pattern of T-S waves. This figure also shows the formation of Λ shaped structures (see also Fig. 23-4). These are nonlinear three-dimensional events.

Poiseuille flow in a round pipe has been a particularly challenging flow for stability theory to resolve. Engineers regard the transition Reynolds number as being around 2000. Reynolds himself measured critical values up to 13,000. Very careful experiments have maintained laminar flow to above 50,000. Stability theory has so far failed to explain the origin of the transition in Poiseuille flow. All calculations of the linear stability of two-dimensional disturbances show that the flow is stable. Investigations into axisymmetric disturbances and finite-amplitude disturbances have likewise failed to show instability. Other sources of the instability, such as slow rotation or instability of the inlet flow, have also been proposed. In many practical engineering cases it is known that the boundary layers of the entrance flow become turbulent before the Poiseuille profile has developed. Turbulent flow within the pipe is then self-sustaining.

22.8 INVISCID INSTABILITY OF FLOWS WITH CURVED STREAMLINES

Streamlines of the main flow usually curve because a pressure gradient exists in the direction across the streamlines. This pressure gradient supplies the centripetal force to turn the particle trajectories. A particle disturbed in a curved flow can disrupt the balance between the pressure gradient and the centrifugal effect to such an extent that the flow becomes unstable. This instability mechanism is inviscid.

The main result on this topic is *Rayleigh's circulation criterion*. It gives a necessary and sufficient condition for the stability to axisymmetric disturbances of a velocity profile with circular streamlines. The fact that the streamlines are circles is not too restrictive, as we might imagine that a flow with a local radius of curvature R is somewhat like a circular flow with the same radius of curvature. Consider an axisymmetric flow in cylindrical coordinates. The θ-momentum equation for a flow where $\partial(\ )/\partial\theta = 0$ is

$$\frac{\partial v_\theta}{\partial t} + v_r \frac{\partial v_\theta}{\partial r} + \frac{v_r v_\theta}{r} + v_z \frac{\partial v_\theta}{\partial z} = 0 \qquad 22.8.1$$

Multiplying this equation by r and rearranging shows that the *reduced circulation* $\gamma \equiv r v_\theta$ for a particle is constant, that is,

$$\frac{D\gamma}{Dt} = 0 \qquad 22.8.2$$

Any axisymmetric disturbance must occur in such a way as to maintain $\gamma = \text{const}$ for each material element.

Let us assume that a ring of fluid at location r_1, z_1 is disturbed so that it is interchanged with a ring of equal volume at r_2, z_2. The kinetic energy of the elements (per unit volume) is $\text{KE} = \frac{1}{2}\rho v_\theta^2 = \frac{1}{2}\rho\gamma^2/r^2$. In the unperturbed state the energy of both elements is

$$\text{KE}_A = \frac{\rho}{2}\left[\left(\frac{\gamma_1}{r_1}\right)^2 + \left(\frac{\gamma_2}{r_2}\right)^2\right]$$

Since the perturbation must occur at constant γ, the energy of the perturbed state is

$$\text{KE}_B = \frac{\rho}{2}\left[\left(\frac{\gamma_1}{r_2}\right)^2 + \left(\frac{\gamma_2}{r_1}\right)^2\right]$$

Next, we calculate the change in energy of the flow:

$$\text{KE}_B - \text{KE}_A = \tfrac{1}{2}\rho\left(\gamma_2^2 - \gamma_1^2\right)\left(r_1^{-2} - r_2^{-2}\right) \qquad 22.8.3$$

If the energy in state B is larger than that of state A, the perturbation requires a finite amount of energy. We would not expect this to be available internally, and thus this situation is stable. Conversely, a disturbance that leads to a lower total energy liberates energy, which can be used to make the disturbance grow; thus the flow is unstable. Without loss of generality we assume $r_2 > r_1$. Hence, for a stable flow, where 22.8.3 is positive, we must have

$$\gamma_2^2 > \gamma_1^2 \qquad \text{for} \quad r_2 > r_1 \qquad 22.8.4$$

The square of the reduced circulation increases outward in a stable swirling flow. In terms of the velocity or angular velocity profiles, 22.8.4 is

$$\begin{aligned} \left(r_2 v_{\theta 2}\right)^2 &> \left(r_1 v_{\theta 1}\right)^2 \\ \left(r_2^2 \Omega_2\right)^2 &> \left(r_1^2 \Omega_1\right)^2 \end{aligned} \qquad 22.8.5$$

Rayleigh's criterion is a useful method of estimating inviscid instability, even though it tells nothing about stability of the flow with respect to three-dimensional disturbances.

The rotary flow between coaxial cylinders offers an example to which Rayleigh's criterion may be applied. We consider the general situation where the inner cylinder and the outer cylinder may both be rotated: R_1, Ω_1 for the inside and R_2, Ω_2 for the outside. The laminar viscous flow solution may be expressed in terms of the angular velocity as

$$\frac{\Omega(r)}{\Omega_1} = A + B\left(\frac{R_1}{r}\right)^2$$

where

$$A = \frac{\dfrac{\Omega_2}{\Omega_1} - \left(\dfrac{R_1}{R_2}\right)^2}{1 - \left(\dfrac{R_1}{R_2}\right)^2}$$

$$B = \frac{1 - \dfrac{\Omega_2}{\Omega_1}}{1 - \left(\dfrac{R_1}{R_2}\right)^2}$$

It is customary to consider that the outer cylinder has positive rotation while the inner cylinder can have either sign. Consider first the case when both cylinders rotate in the same direction. The quantity $r^2\Omega$ will always increase with r as long as the outer speed satisfies

$$\Omega_2 \geq \left(\frac{r_1}{r_2}\right)^2 \Omega_1 \qquad 22.8.6$$

Equation 22.8.6 is called the Rayleigh line. If the inner cylinder is fixed and the outer is rotated, the flow is stable. If the outer cylinder is fixed and the inner is rotated, Rayleigh's criterion indicates instability. Viscosity does, however, stabilize this situation until a certain speed is reached. More will be said on this in the next section.

Situations where the cylinders rotate in opposite directions are always unstable to inviscid disturbances. In this case, where Ω_2 is positive and Ω_1 is negative, Ω^2 decreases in a region near the inner cylinder until the radius where $\Omega = 0$ is reached. From this point outward Ω increases. The flow is unstable in the inner region $\Omega < 0$, but stable in the outer region.

22.9 TAYLOR INSTABILITY OF COUETTE FLOW

Viscosity plays only its stabilizing role in Couette flows. A chart of the stability characteristics is given as Fig. 22-10. The viscous stability of these flows was first determined by Taylor (1921, 1923) both experimentally and theoretically. The theoretical problem is quite difficult, and most work is done using a thin-gap assumption. This assumption takes centrifugal effects out of the main flow but retains them partially in the disturbance equations. The problem, simplified for axisymmetric disturbances, contains a parameter called the Taylor number. Several definitions are in use. A typical one is

$$\mathrm{Ta} \equiv \frac{R_1(R_2 - R_1)^3(\Omega_1^2 - \Omega_2^2)}{\nu^2} \qquad 22.9.1$$

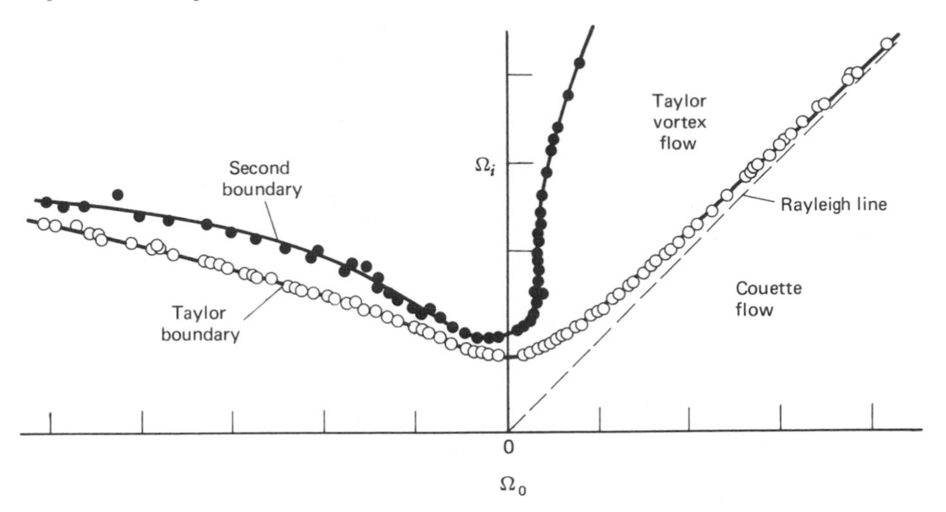

Figure 22-10 Stability chart for Taylor vortex behavior. From Coles (1956).

Essentially Ta represents the centrifugal effect divided by the viscous effect. Upon crossing Taylor's first stability boundary, one encounters a second stable laminar flow pattern with toroidal vortices. For inner rotation only, this boundary is Ta = 1708. These vortices were mentioned previously in Section 9.3. The new flow pattern is an example of the principle of "exchange of stabilities". Taylor vortices and Couette flow are both stable laminar flow patterns.

Taylor vortices themselves become unstable at higher rotation rates, where they give way to wavy Taylor vortices as shown in Fig. 22-11. The single second transition line in Fig. 22-10 is far from the complete story. Many states of different mode numbers can be attained in the wavy patterns. Moreover, when the outer cylinder is allowed to rotate, one can achieve extremely complex flows such as the braided pattern of Fig. 22-12. For the highest rotation rates turbulent flow finally ensues. Figure 22-13 shows turbulent Taylor vortices. (If the outer cylinder is rotated fast enough, a turbulent flow without vortices can be attained.)

Centrifugal effects in combination with viscous damping can lead to some striking flow patterns. Two other flow situations with curved streamlines are worthy of a brief description. One is the flow in a curved pipe or channel, and the other is the flow of a boundary layer over a concave wall.

Consider a curved channel where the radius of curvature is much larger than the channel width. The flow is driven by a pressure gradient, and when the radius of curvature is large, the Poiseuille velocity profile is established. Rayleigh's criterion shows that the flow on the inside half of the channel is stable, while that on the outer half is unstable. This situation is very much like the Taylor problem in that viscosity stabilizes a profile that is inviscidly unstable. The parameter that measures the curvature effect compared to

Figure 22-11 Wavy Taylor vortices. Photograph courtesy of L. Koschmieder (1979), University of Texas.

viscous effects is called the Dean number. Exceeding a critical value of the Dean number results in an instability and the development of a second stable laminar flow pattern with a toroidal vortex, or in some cases several vortices, superimposed on the main flow. This is the origin of the secondary flow in curved pipes and channel bends.

A somewhat similar flow situation exists when a boundary layer flows over a concave wall. The case of interest is when the curvature is in the longitudinal direction as shown in Fig. 22-14. If the boundary layer is thin compared to the radius of curvature R, a condition required for our standard boundary-layer analysis, then the pressure is constant through the layer and centrifugal stability is not important. On the other hand, if δ is a reasonable fraction of the radius of curvature, the first effect on the boundary-layer equations is simply to

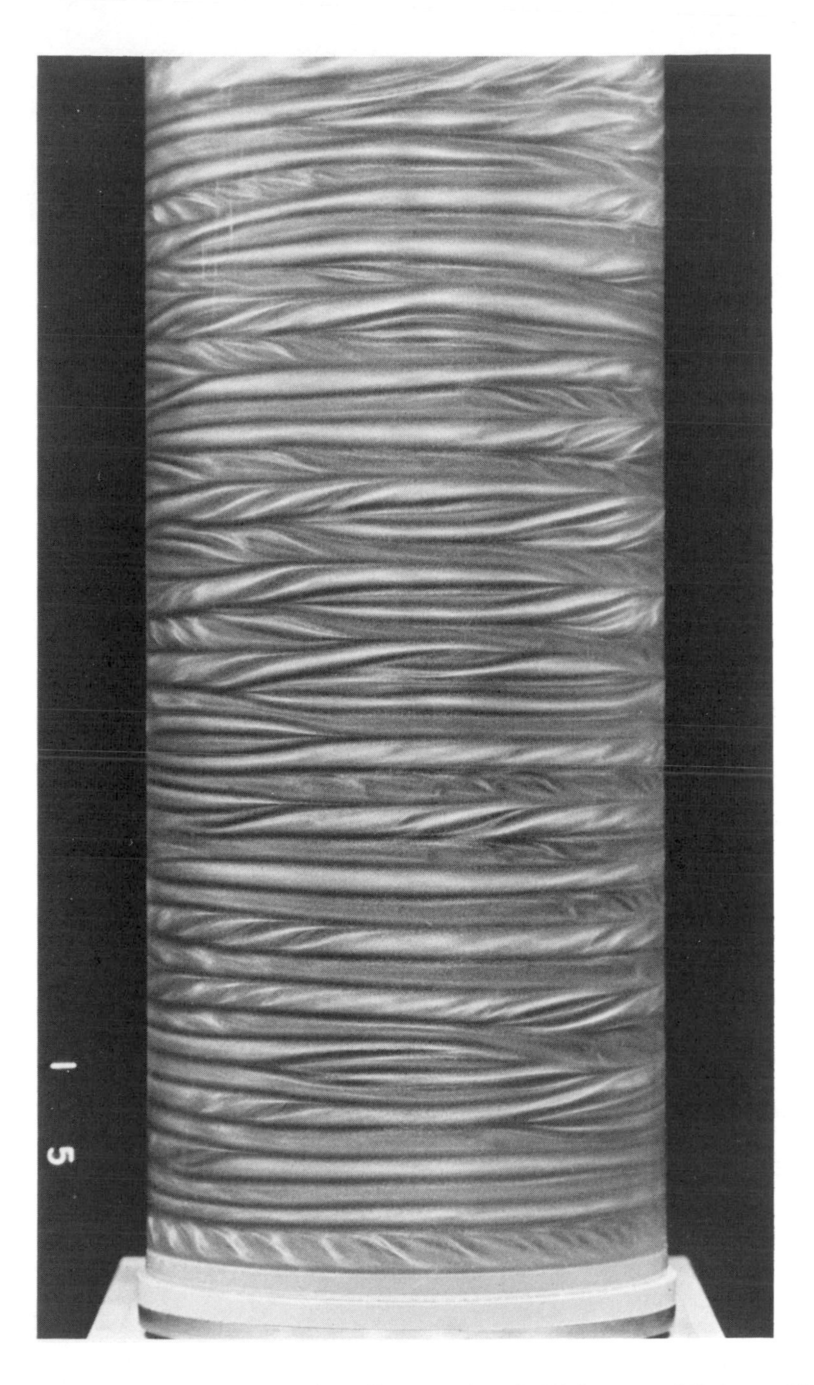

Figure 22-12 Braided Taylor vortices. From Andereck, Dickman and Swinney (1983).

Figure 22-13 Turbulent Taylor vortices. (Photograph courtesy of L. H. Zhang and H. L. Swinney.) University of Texas.

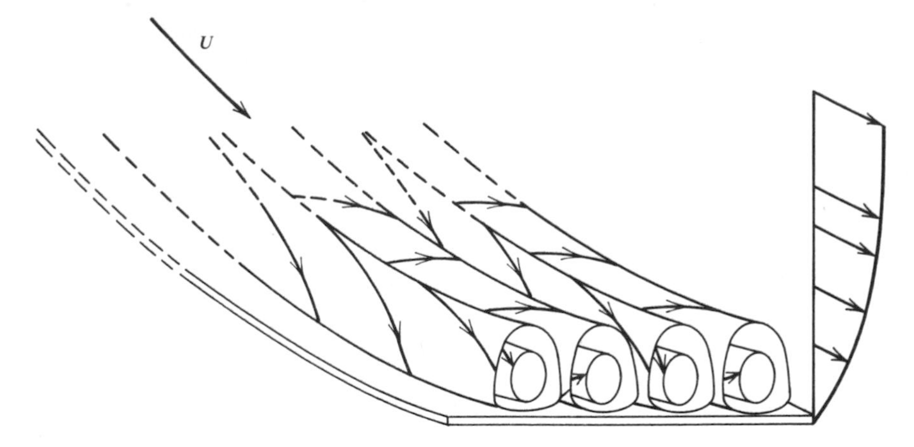

Figure 22-14 Schematic of Görtler vortices in a boundary layer on a concave wall.

modify the transverse momentum equation from $\partial p/\partial y = 0$ to $\partial p/\partial y = -\rho u^2/R_0$. The possibility of centrifugal instability now exists in this boundary layer. Once again we have a flow that is inviscidly unstable according to Rayleigh's criterion, and viscosity plays a stabilizing role. In this flow the parameter comparing the centrifugal effect with the viscous effect is called the Gortler number. It is

$$G = \frac{u_e \Theta}{\nu}\left(\frac{\Theta}{R_0}\right)^{1/2}$$

Here u_e is the external velocity and Θ is the momentum thickness. Gortler vortices (the secondary flow pattern shown in Fig. 22-14) occur when G exceeds about 0.3. This number is not very sensitive to the shape of the velocity profile; $G_c = 0.311$ for the Blasius profile, and $G_c = 0.278$ for a straight-line profile between u_e and the wall value zero.

22.10 STABILITY OF REGIONS OF CONCENTRATED VORTICITY

We have seen many examples where an inviscid flow is determined by a distribution of vorticity in a small isolated region. The velocity field in these flows can in principle be found by the Biot–Savart law,

$$v = \frac{1}{4\pi}\int \frac{\boldsymbol{\omega} \times d\mathbf{r}}{|\mathbf{r} - \mathbf{r}_0|^3}$$

From this point of view the Helmholtz instability of a vortex sheet is caused by warping the sheet so that the integral is in a new location, with a rearrangement of the vorticity within the sheet. The sheet is no longer of uniform vorticity, but has a lumpy distribution. Both of these effects cause a new velocity field through the equation above. In particular, a certain point on the sheet has a new velocity induced by the remainder of the sheet. If this self-induction mechanism causes further growth, then the vorticity distribution is unstable.

Since a vortex sheet is unstable, what about a line vortex? A small perturbation of a vortex line is neutrally stable. There are two modes. In one mode a displacement travels down the vortex in the form of a helix, while in the second mode the vortex remains in a plane as a spinning sine wave. Neutral stability means that there is no tendency for these perturbations to either grow or decay. Because we are dealing with small perturbations, the problem is linear and disturbances of different wavelengths behave independently. The problem is actually nonlinear for finite disturbance. In this case different wavelengths interact and unstable growth occurs.

Consider next a plane two-dimensional flow that contains several line vortices. These vortices move with the local fluid velocity in conformity with Helmholtz's laws. We have previously calculated (Section 13.9) how two counterrotating vortices are self-propelled. This situation is neutrally stable in the sense that a small disturbance simply produces a new configuration, which propels itself in a new direction or with a new speed. There is no tendency for the perturbation to grow or die. Two corotating vortices form a similar situation except that the vortices rotate around each other.

Several line vortices leads to an unstable configuration in the following sense. A perturbation of the location or strength of a vortex alters the effect that it produces on the motion of all the others. The subsequent motion goes through a different history than if the perturbation were absent. A famous exception to this statement is the Kármán vortex street. von Kármán found that an infinite array of counterrotating vortices will be neutrally stable only if the spacing between the vortices and the rows is $b/h = 3.56$. In any other configuration the vortex street is unstable.

Another inviscid line-vortex instability is called the Crow (1970) instability. It was first studied in connection with the trailing vortex wake of airplanes. Figure 22-15 shows a vortex trail made by an airplane at very high altitude in smooth stable air. Water vapor, the result of combustion in the jet engines, has condensed and migrated to mark the core of the vortex from each wing tip. Crow considered that the wake was like two infinite line vortices separated by a given distance b. His analysis showed that the vortices were most unstable to a symmetric oscillation in the vortex location with wavelength 8.6b. The pictures clearly show the steady growth of a long-wavelength disturbance. The growth continues until the vortices touch and pinch off to form vortex rings. It is necessary to have a very smooth atmosphere to observe this sequence of events. Contrast Fig. 22-15 with Fig. 1-2, where atmospheric turbulence has formed kinks of various wavelengths in each vortex. Self-induction and interaction of these several wavelengths will soon scramble these vortices before the Crow mode has a chance to develop.

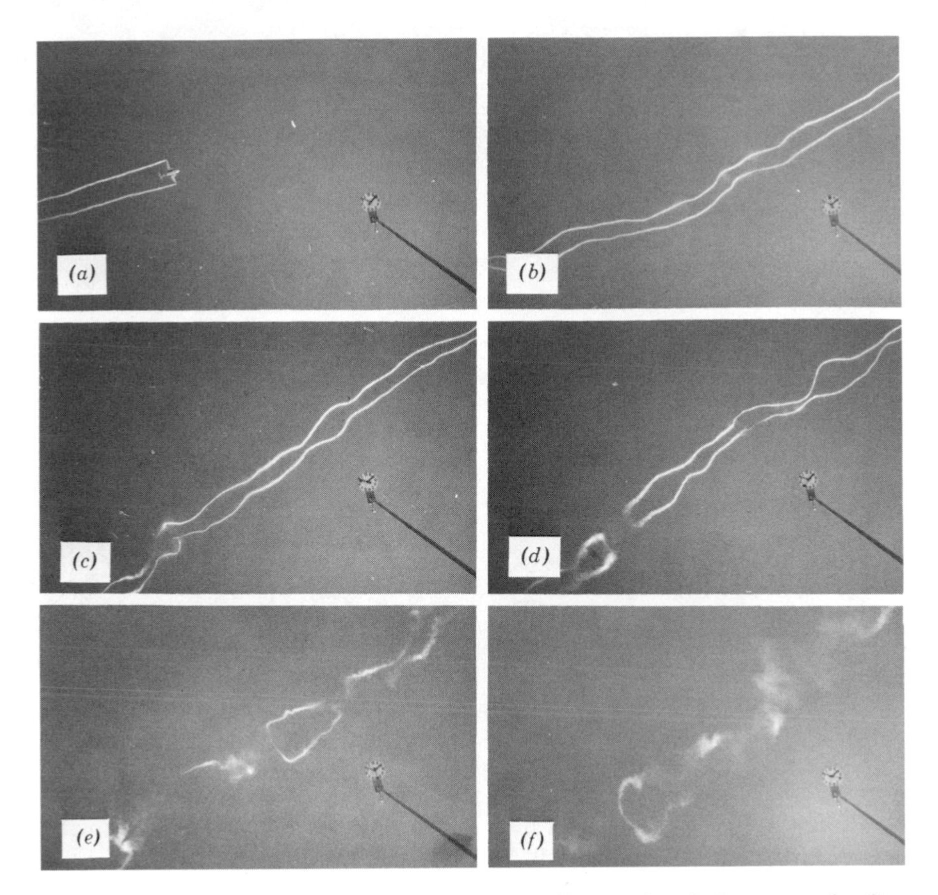

Figure 22-15 The inviscid instability of a pair of line vortices is known as the Crow instability. The time lapse between each picture is 10 sec. Photograph courtesy of I. Tombach (1974) AeroVironment Inc.

22.11 SOME OTHER INSTABILITIES

Brief mention of a few additional instabilities in fluids will be made. Another instability named after Taylor (1950), also called the Rayleigh–Taylor instability, occurs in the acceleration of two fluid layers with different densities. For example, in the earth's gravitational field the heavier fluid must be on the bottom for stability: removing the cover from a jar of water held upside down would lead to a disaster. Taylor proved that a density discontinuity in a fluid was unstable to any acceleration from the light fluid into the heavy fluid and stable for acceleration in the opposite direction.

A capillary instability is illustrated in Fig. 22-16. This is essentially an inviscid instability where a cylinder of liquid from a jet is so thin that surface tension is important. The surface-tension law is $p = p_\infty + \sigma(R_1^{-1} + R_2^{-1})$,

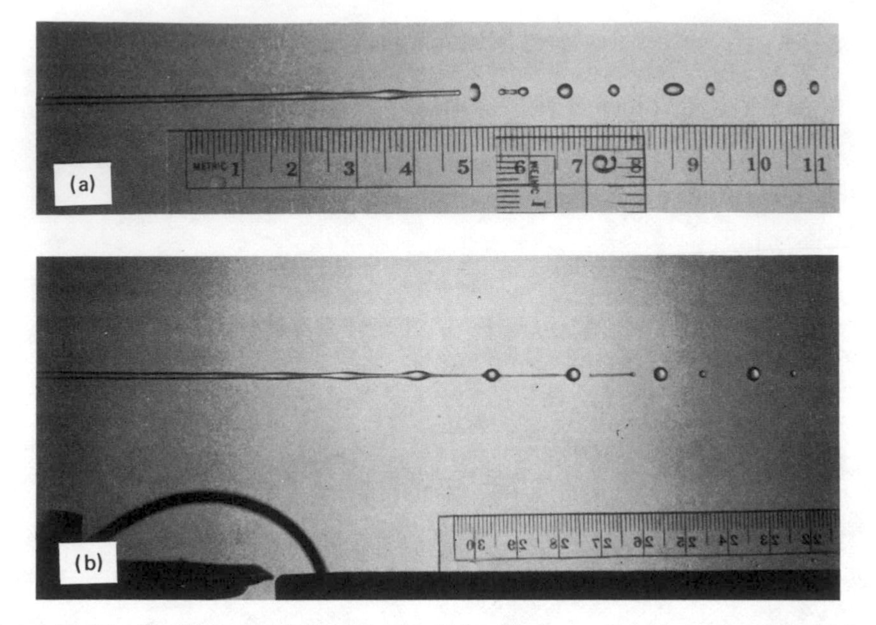

Figure 22-16 Capillary instability of a liquid jet. Photograph (a) is water while (b) is a more viscous glycerine–water solution. From Donnelley and Glaberson (1966).

where R_1 and R_2 are the radii of curvature of the surface. When the surface is deformed, the pressure distribution within the liquid changes. The primary effect is that the fiber has a higher internal pressure where the radius is smaller. The resulting pressure gradient then drives the fluid toward the regions of larger radii. The jet is unstable to all axisymmetric disturbances with wavelengths greater than the circumference of the jet. The wavelength that grows the fastest is $\lambda = 9.0r$. Note in the figure that the final breakup of the jet into droplets produces two distinct sizes. Small satellite drops are formed between the large drops.

Line vortices with a finite core where viscosity produces a certain profile undergo an instability known as vortex breakdown. Before the breakdown, the vortex has a small core with high swirling velocities. After breakdown the core is much larger, the swirling velocity is lower, and frequently some turbulence exists. Because the flow outside the core is inviscid, the total circulation, and hence the integral of the vorticity, remains constant. Some view vortex breakdown as analogous to a hydraulic jump; a transition between two alternative flow states (the second state has less energy, and hence the excess is dissipated by viscosity or turbulence). Recall that Fig. 20-30 showed three different types of vortex breakdown as classified by Sarpkaya (1971). More details on these phenomena may be found in Faler and Leibovich (1978). The final example is a vortex ring. A circular line vortex is stable to small perturbations in its shape. Smoke rings often last for a considerable time. However, if the ring has a large core it is subject to a viscous instability frequently called the Widnall instability. Pictures of this phenomenon are to be found in Van Dyke (1982).

22.12 CONCLUSION

The initial instability of several flows has been satisfactorily explained by linear stability theory. Subsequent experiments have confirmed the validity of the approach.

It is interesting to note that potential flows of a homogeneous fluid are stable. Instabilities arise when regions of the flow carry vorticity. The exact distribution of vorticity is not as important as its existence in many instances —for example, in the Kelvin–Helmholtz instability of a shear layer or the Crow instability of two line vortices. In these cases the amplification of disturbances occurs through inviscid processes; nevertheless, the existence of the vortical region is an essential element.

Viscous instability is a counterexample to the expectation that viscosity is a stabilizing force. Viscous instability is a time-delay phenomenon. In several instances we have observed viscous forces accelerating a fluid to a higher velocity than would be achieved by the pressure forces acting alone. In these cases, and in the case of viscous instability, a net shear stress generated at one time diffuses slowly through the flow. At some later time it is in the proper position to add to a net pressure force. The result is an overshoot in the velocity profile or a growth in the instability wave. Viscous instability is the only theoretical reason a Blasius boundary layer is unstable.

23 Introduction to Turbulent Flows

Instability in a fluid flow will result in another stable laminar flow pattern, an example of the exchange of stabilities, or in a turbulent flow. Turbulent flows contain self-sustaining velocity fluctuations in addition to the main flow. Most flows in industry and in nature are at high enough Reynolds number so that they are turbulent.

Consider for a moment the turbulent flow in a pipe or over a flat plate. As we view positions further and further downstream, we can reasonably expect that the state of the flow will become independent of the inlet or leading-edge conditions. The particular events that occurred at the inlet and during the transition process will be forgotten. The final local flow state will have universal characteristics that are the result of the local turbulent processes in the immediate neighborhood of the point in question. This type of reasoning—that is, seeking universal characteristics—is behind all turbulence research. Unless these universal characteristics exist, there is no hope of bringing order out of the chaos. We only have experimental assurance that our proposition of a single asymptotic turbulence state actually exists.

Turbulence as a field of study has many facets. Some come to the field needing only to know the wall shear stress and the mean velocity profiles. Others wish to have a knowledge of mass diffusion and concentration statistics. Many wish to know the details of turbulent motions and how they interact. At the most abstract level, turbulence is viewed as an example of the mathematical phenomena called strange attractors. An introduction cannot hope to cover such a large field. Our purpose will be to treat several topics, but only in a descriptive way. The student should gain a vocabulary and the proper orientation to proceed to more thorough treatments.

23.1 TYPES OF TURBULENT FLOWS

Turbulent flows are not all alike. The universal characteristics of a turbulent jet and those of flow in a pipe display certain differences. We can roughly classify turbulent flows into three groups; gridlike flows, wall shear layers, and free shear layers (mixing layers).

Grid turbulence is a special type of turbulence that violates the definition because it is not self-sustaining. To generate this flow a grid of (say) circular cylinders is placed perpendicular to a uniform stream. The vortices generated by the cylinders interact, and, after a certain distance, a homogeneous, isotropic field of turbulence is achieved. The turbulence exists without preference for direction, and it decays so slowly that the variations in the flow direction are not important to the decay process. Many experiments have been carried out on grid turbulence to see how decay occurs in this idealized situation.

Wall shear layers are the second class. The presence of a wall has a dominant effect on the processes that produce turbulence. For example, the turbulent characteristics of flow in a pipe are determined by the wall. The core region and the presence of the wall on the other side have only a minor influence. Boundary layers and all internal flows are also in this category.

Free shear layers include not only the typical mixing layer between two fluids moving at different speeds, but also all sorts of jets and wakes. A transition region near the origin of these flows, precedes the turbulent region. Downstream, the extent of the turbulent region always grows. It is thought that these flows develop universal characteristics at distances far from the origin. Flows that develop a state that depends only on the local flow quantities—for example, the local value of the mean velocity and the local jet thickness—are said to be *self-preserving* or self-similar. Different turbulence characteristics may become self-preserving at different stages. The mean velocity profile in a jet becomes self-preserving about 8 diameters downstream. On the other hand, the turbulence properties—that is, the statistics of the fluctuations—are still not self-preserving after a distance of 40 diameters.

The postulate of a self-preserving state after some transition process is forgotten must be experimentally documented for each flow. Research indicates that turbulent flows contain remnants of their initial conditions much longer than one would expect.

23.2 CHARACTERISTICS OF TURBULENT FLOWS

It is probably not wise to make a rigid definition of a turbulent flow. On the other hand, the flows we call turbulent do have certain properties in common. For example, waves on the surface of water should not be classified as turbulence. There are too many mechanisms that generate waves, and the active interaction of the air complicates this flow. It does not possess the mechanisms and characteristics of the types of turbulent flows mentioned in the previous section. In this section we shall describe the major qualities of turbulent flows, with the reservation that some of these characteristics are not found in every turbulent flow.

Turbulent flows have irregular fluctuations of velocity in all three directions. The intensity of the fluctuations is variable, but is customarily 10% or less of the mean velocity. A time history of the velocity at a point looks like a random signal. Nevertheless, there is structure to the fluctuations, so it is not absolutely accurate to say that the fluctuations are random (mathematicians have a definition of the term "random variable", which turbulent irregularities do not meet).

The irregularities in the velocity field have certain spatial structures known as *eddies*. This is a vague term that may be applied to any spatial flow pattern that persists for a short time. An eddy may be like a vortex, an imbedded jet, a mushroom shape, or any other recognizable form. Large eddies are quite evident in the picture of the boundary layer in Fig. 23-1. Eddies are not isolated; small eddies exist inside larger eddies, and even smaller eddies exist inside the small eddies. Such small-scale motions are also visible in Figure 23-1. One of the main characteristics of turbulence is a continuous distribution of eddy sizes. A flow where the irregularities are limited to a few separated frequency bands does not qualify as a turbulent flow.

The turbulence in a flow is self-sustaining. Processes, which are not well defined or understood, occur that generate more turbulence and maintain the irregular motion. Once a flow becomes unstable and turbulence develops, it does not simply die out and repeat the process as a limit-cycle oscillation. Turbulence, once initiated, continues and perpetuates itself without diminishing. Thus, the transition mechanisms, the original instability, do not necessarily play a role in sustaining turbulence.

A gradient in the mean velocity profile is another characteristic. This *mean shear* must exist for the turbulence to be self-sustaining. We shall verify this fact later. In shear layers, boundary layers, jets, and wakes, the region where turbulence exists coincides with the region of mean shear. The reason that grid turbulence decays is that it has no mean shear.

In confined flows, turbulence may grow to cover the entire flow, but in all other cases the turbulent region has a limited extent. This is not to say that some fluid is always turbulent and other fluid is always nonturbulent. Another characteristic of turbulent flows is that they *entrain* nonturbulent fluid, so that the extent of the turbulent region grows. Consider the jet as an example. The fluid that composes the jet continues to increase by entrainment as the jet extends further from the origin.

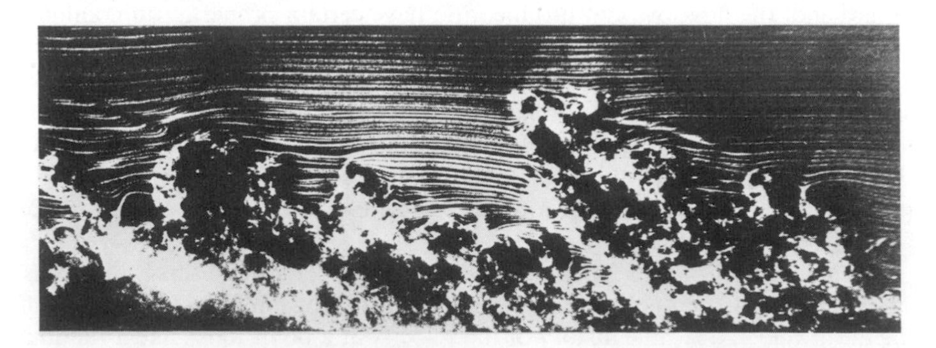

Figure 23-1 Smoke wire flow visualization of a turbulent boundary layer at Re = 3500. Flow is from right to left. Photograph courtesy of T. Corke, Y. Guezennec, and H. Nagib, Illinois Institute of Technology.

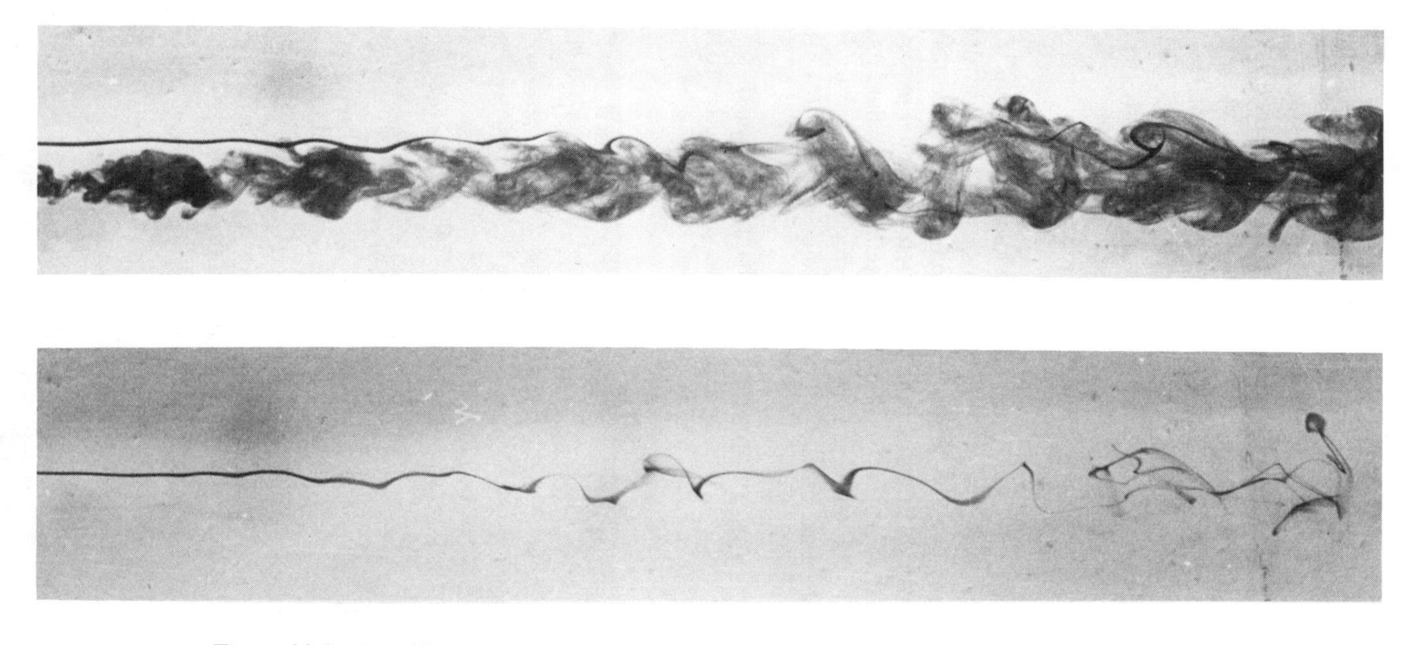

Figure 23-2 Jet of liquid in a flowing stream. Photograph (a) shows an external dye streak and a dyed jet; (b) jet without dye. From Tritton (1977).

The rigorous way to decide whether fluid is turbulent or nonturbulent is based on vorticity. By definition turbulent fluid has vorticity, nonturbulent fluid does not. Figure 23-2 shows a jet of dye submerged in a uniform laminar flow of lower velocity (this spreads out the events for better visualization). A filament of fluid outside the jet is also marked. The figure shows that the outside filament oscillates slowly for a while and then is entrained into the jet. Once within the jet, the streak is subjected to straining motions from eddies of all sizes. The irregular motions when the streak is outside the jet are nonturbulent, nonvortical potential fluctuations. The interface between the turbulent and nonturbulent fluid is very sharp. It is easy to see in Fig. 23-2 that the unsteady potential motions have only large-scale motions, while the turbulent motions have a continuous range of eddy sizes as stated. Figure 23-3 shows the wake of a bullet and the sharp interface between the turbulent and nonturbulent fluid.

Turbulent flows are diffusive. Just as random molecular motions in a gas are responsible for viscous diffusion, thermal diffusion, and mass diffusion, a turbulent eddy can transport fluid from a region of low momentum and deposit it in a region of high momentum. Although the actual process is more complicated than that, it is clear that turbulence tends to mix fluid and thereby has a diffusive effect. The term *eddy diffusion* is frequently used to distinguish this effect from molecular diffusion. Eddy diffusion can be 10 or 100 times stronger than molecular diffusion.

All turbulent flows involve processes that change the length scale of the eddies. Once again, not much is actually known about these processes, but

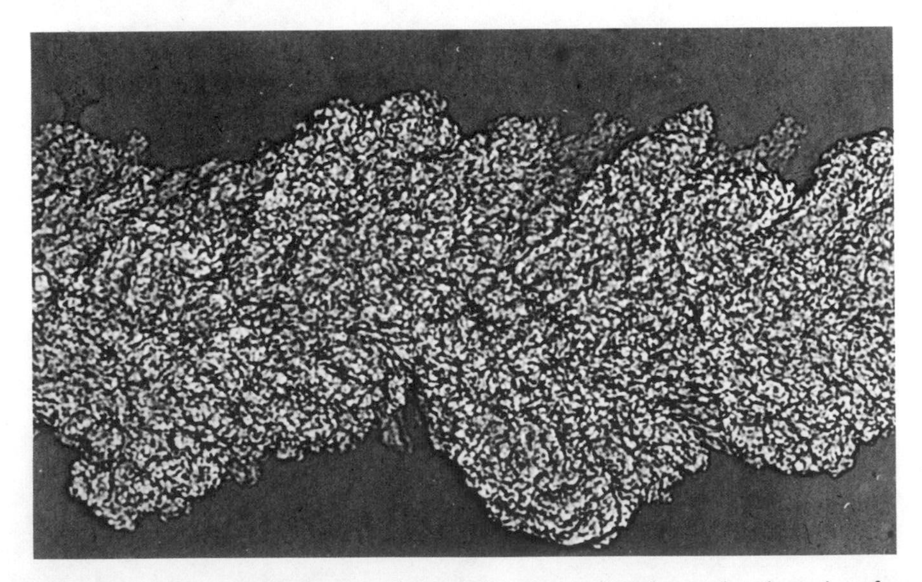

Figure 23-3 This classic picture of the wake of a bullet shows the sharp interface between turbulent and nonturbulent fluid. Photograph supplied by S. Corrsin from an experiment conducted by Ballistic Research Laboratory, Aberdeen Proving Ground.

there is no doubt that they are a major characteristic of turbulent flows. These processes act in both directions. A modest-size irregularity or eddy grows and becomes a large eddy. The largest eddies in a flow are about as large as the thickness of the turbulent region. The size of the largest eddies in a boundary layer is about δ, and that of the largest eddies in a jet is about equal to the local jet diameter.

Processes also occur that reduce the eddy size. Turbulent eddies are continually formed with smaller and smaller length scales. There is also a limit to this process. When the spatial extent of an eddy becomes very small, viscous forces, because of the steep velocity gradient, become very important. They tend to destroy the smallest eddies and hence viscosity puts a lower limit on the eddy size.

The last major characteristic of turbulent flows is that they are *dissipative*. Any flow with viscosity has viscous dissipation, but turbulent flows have much more of it because the small-scale eddies have sharp velocity gradients. The energy dissipated in the small eddies dominates that dissipated in the largest eddies and in the mean flow. Since the small eddies dissipate energy and tend to destroy themselves, the scale-changing process that produces smaller eddies is a necessary element of self-sustaining turbulence.

23.3 TRANSITION

Transition to a fully developed turbulent state may occur in several ways. In pipe flow, turbulent elements called *puffs* form when the pipe Reynolds number is in the range $2000 < \text{Re} < 2700$. Puffs originate in the region where the laminar profile is fully developed. They are started by a large disturbance

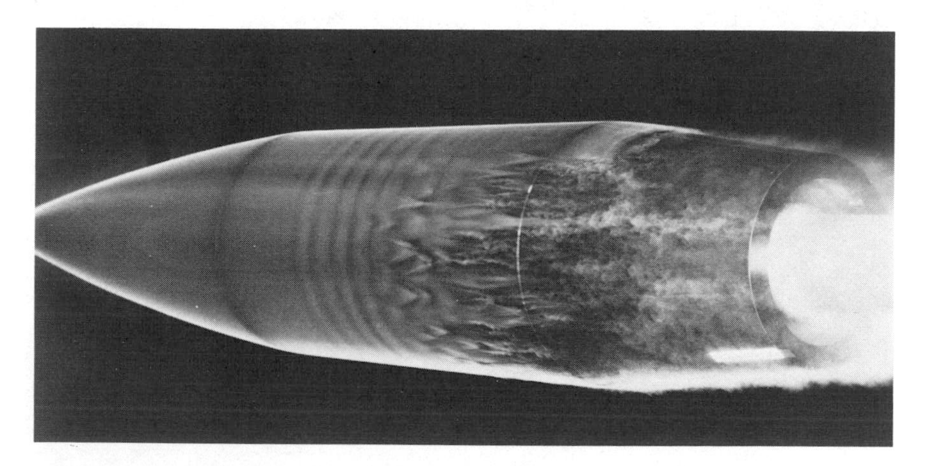

Figure 23-4 Smoke in the flow over a cylindrical body shows transition. T-S waves form Λ or hairpin shapes that break down. Photograph courtesy of T. J. Mueller and R. C. Nelson, University of Notre Dame.

at the inlet, according to Wygnanski and Champagne (1973). When Re > 3200 (the region from 2700 to 3200 being uncertain), another turbulent structure called a *slug* is the main transition mechanism. Slugs form in the hydrodynamic entrance region, where the flow still consists of a core flow and a boundary layer. A slug immediately fills the pipe cross section and continues to grow at the ends as it is swept downstream. By either process, the flow far downstream becomes completely turbulent.

Turbulence does not develop in boundary layers by a single well-defined series of steps. Free-stream turbulence, wall roughness, the presence of trip wires, and even acoustic noise may initiate turbulence in a boundary layer. Natural transition is the idealized process that begins with the Tollmien–Schlichting instability discussed in the previous chapter. As the Tollmien–Schlichting waves grow, they develop a three-dimensional instability. Frequently flow-visualization pictures reveal Λ-shaped *hairpin eddies* as in Fig. 23-4 (and also Fig. 22-9). This figure shows how the hairpin eddies break up and coalesce into a continuous range of eddy sizes. After this, the boundary

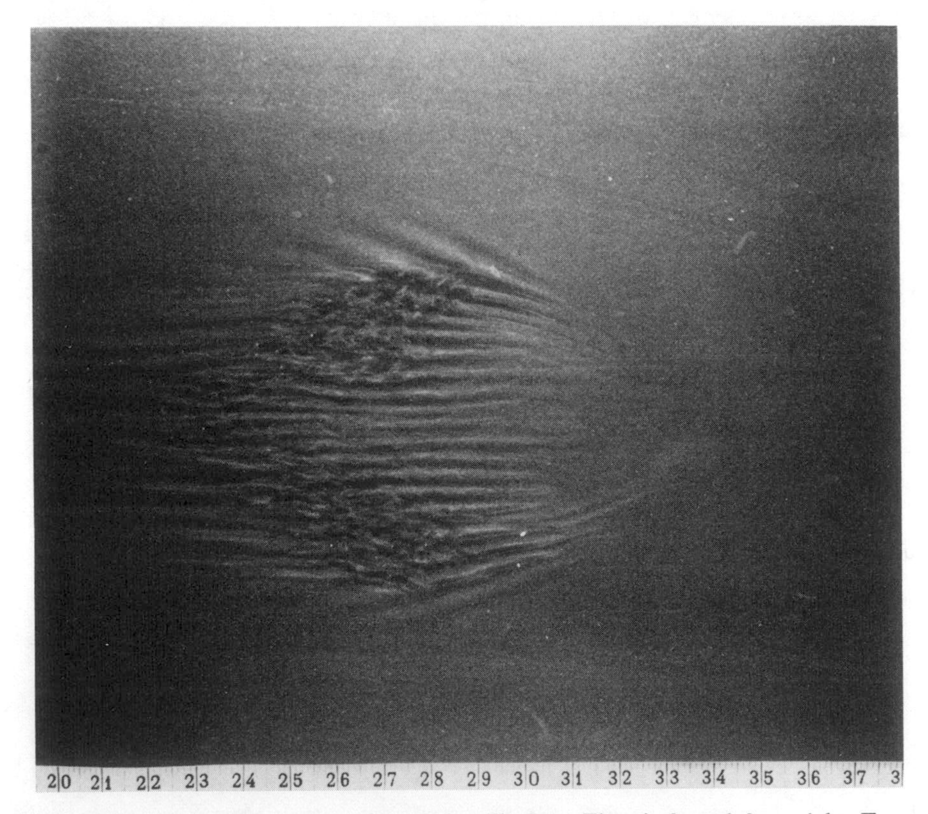

Figure 23-5 Turbulent spot in plane Poiseuille flow. Flow is from left to right. From Carlson, Widnall, and Peeters (1982).

layer looks turbulent, although it takes some additional time to form the finer statistical structure.

There is a very interesting transition structure called a *turbulent spot*. In a boundary layer or other flow with a viscous wall region, a sharp disturbance can generate a turbulent region that expands and is carried downstream. Figure 23-5 shows such a spot. It is shaped like an arrowhead which points in the same direction as the flow. The front of the spot moves at about $0.9U$, while the back moves at only $0.5U$. The spot expands laterally so that it makes an angle of 22.5° with the point of origin. Note the wavelike structures at the sides of the spot. After the spot passes over a given point, the flow is again laminar. Turbulent spots grow in vortical fluid and are self-sustaining. Currently it is not clear what processes within the spot are characteristic of fully developed turbulence and what processes are unique to the spot.

23.4 REYNOLDS DECOMPOSITION

Computers do not have enough storage to solve the Navier–Stokes equations for all the detailed unsteady velocity profiles of a turbulent flow. (However with the aid of models of turbulent processes they can compute the large-scale turbulent events.) Since we must give up hope of an exact computation of turbulence, we lower our sights and seek to find only the time-averaged properties. The basic engineering approach today is the same as Reynolds used when he decomposed the instantaneous velocity into a mean part and a fluctuation (turbulence literature usually uses u_i rather than v_i for the velocity):

$$\begin{aligned} \tilde{u}_i(x_i, t) &= U_i(x_i) + u_i(x_i, t) \\ \tilde{p}(x_i, t) &= P(x_i) + p(x_i, t) \end{aligned} \qquad 23.4.1$$

In the final equations, capital letters will refer to mean quantities and lowercase letters will refer to turbulent fluctuations. The mean velocity is defined as the time average for a period T that is long enough to get an accurate average:

$$\begin{aligned} U_i(x_i) &\equiv \frac{1}{T}\int_0^T \tilde{u}_i(x_i, t)\, dt \\ &= \overline{\tilde{u}_i} \end{aligned} \qquad 23.4.2$$

An overbar denotes the time average. We assume that differentiation and time averaging are commuting mathematical operations. For example,

$$\overline{\partial_i \tilde{u}_j} = \partial_i \overline{\tilde{u}_j} = \partial_i U_i$$

The time average of a fluctuation is, of course, zero: $\overline{u_i} = 0$.

A *turbulence intensity* can be defined for each velocity component as the root mean square referenced to a characteristic mean flow velocity U_0. It is given by

$$I_x \equiv \frac{\left(\overline{u_1 u_1}\right)^{1/2}}{U_0}$$

The overall turbulence intensity is defined as

$$I \equiv \frac{\left(\tfrac{1}{3}\overline{u_i u_i}\right)^{1/2}}{U_0} \qquad 23.4.3$$

If the turbulence is isotropic, then $\overline{u_1 u_1} = \overline{u_2 u_2} = \overline{u_3 u_3}$ and the turbulence intensity I is equal to the component intensities.

Another quantity of interest, especially in engineering computation methods, is the *turbulent kinetic energy*. It is

$$K = \tfrac{1}{2}\overline{u_i u_i} \qquad 23.4.4$$

(Lowercase k will be reserved to represent wavenumbers.)

Consider the continuity equation $\partial_i \tilde{u}_i = 0$. If we time-average this equation, we find that the mean velocities also obey the same equation,

$$\partial_i U_i = 0 \qquad 23.4.5$$

Next, we substitute 23.4.1 into $\partial_i \tilde{u}_i = 0$ and subtract 23.4.5 to arrive at the fact that the fluctuations themselves are incompressible,

$$\partial_i u_i = 0 \qquad 23.4.6$$

Both the mean flow and the fluctuations satisfy the continuity equation separately.

Let us turn to the momentum equation next. It is

$$\rho \partial_0 \tilde{u}_i + \rho \partial_j(\tilde{u}_j \tilde{u}_i) = -\partial_i \tilde{p} + \mu \partial_j \partial_j \tilde{u}_i$$

Inserting 23.4.1 and time-averaging leads to

$$\rho \partial_j (U_j U_i) + \rho \partial_j \left(\overline{u_j u_i}\right) = -\partial_i P + \mu \partial_j \partial_j U_i \qquad 23.4.7$$

This momentum equation governs the time-averaged properties of the flow. It contains a new effect, $-\rho\overline{u_j u_i}$, called the *Reynolds stresses*. Through this term the details of the turbulence make their imprint upon the mean velocity profile U_i.

As an illustration, consider the Reynolds stress $-\rho\overline{u_2 u_1}$ in a mean flow $U_1(x_2)$. If a slight excess velocity u_1 has a tendency to occur at the same time a positive transverse velocity u_2 occurs, then extra u_1 momentum is being transported across the flow. This is the diffusion characteristic of turbulent flows. Something in the turbulent processes of the flow causes a correlation, so that the time average $-\overline{u_1 u_2}$ is not zero. Note that Reynolds stresses depend on the position in the flow; for instance, they must be zero at a wall.

When we time-average the Navier–Stokes equations, we lose information about the details of the flow. The effect of the turbulent structure has been distilled into the Reynolds stresses. However, the Reynolds stresses are not known. We have generated new unknowns, and the number of equations is insufficient to solve the problem. This is called the *closure problem*. An ad hoc, special assumption must be made about the nature of the turbulent flow. This assumption is equivalent to determining the Reynolds stresses. All turbulent calculation techniques contain such an assumption at some stage in their formulation.

*23.5 CORRELATIONS OF FLUCTUATIONS

The statistical approach to uncovering the structure of turbulence uses space–time correlations. Let $X(x_i, t)$ and $Y(x_i, t)$ be two fluctuating properties of the turbulent flow. For simplicity assume that X and Y have been normalized by their rms values. Is there a relation between X at the point x_i^A and Y at another point x_i^B in the flow? If a turbulent process occurs that tends to produce a relationship, we can express this fact by means of the integral that defines the *correlation coefficient*,

$$R_{XY}\left(x_i^A, x_i^B\right) \equiv \frac{1}{T}\int_0^T X\left(x_i^A, t\right)Y\left(x_i^B, t\right)\,dt \qquad 23.5.1$$

Another notation is to replace x_i^A, x_i^B by x_i, r_i, where $x_i = x_i^A$ and $r_i = x_i^B - x_i^A$. Equation 23.5.1 shows the average instantaneous correlation of events at two points in space. In many cases the event at point A does not arrive at point B until sometime later. This is especially true at large separations where the event is convected between A and B by the flow velocity. Such events can be extracted if we delay one variable by a time τ. Allowing τ to range over different values will show if a delayed correlation exists. Thus, the more general definition of a space-time correlation is

$$R_{XY}(x_i, r_i, \tau) \equiv \overline{X(x_i, t)Y(x_i + r_i, t + \tau)} \qquad 23.5.2$$

The Reynolds stresses are (proportional to) such correlations, where X and Y are velocity fluctuations and $r_i = 0$, $\tau = 0$.

When X and Y are the same variable, R_{XX} is the *autocorrelation*. Figure 23-6 shows the autocorrelation of the pressure fluctuations on the wall under a turbulent boundary layer. Curves for several fixed distances $r_i = \xi, 0, 0$ are

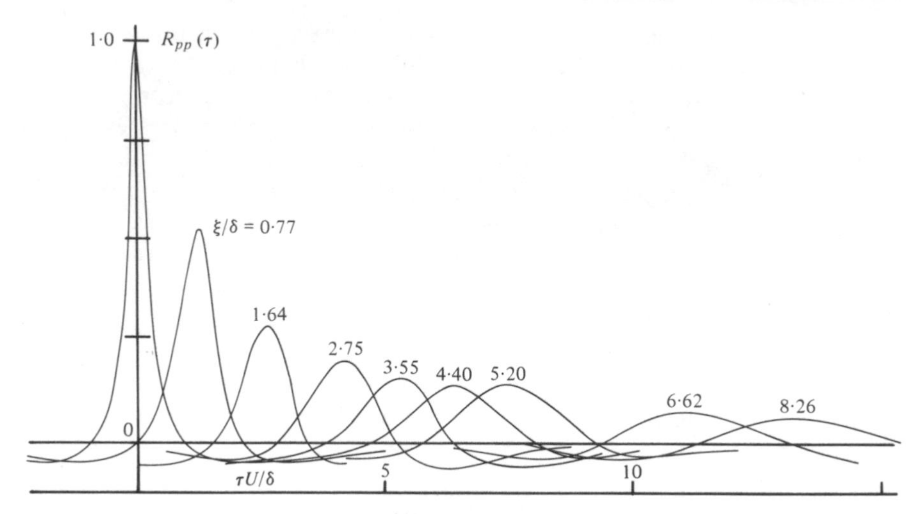

Figure 23-6 Cross-correlation of wall pressure fluctuations under a turbulent boundary layer. The distance between microphones is ξ/δ while the time lag is τ. From Panton et al. (1980).

drawn as a function of the delay time. The autocorrelation at $r_i = 0$, $\tau = 0$ is by definition unity. If we let $\tau = 0$, 23.5.2 indicates, as a function of r_i, the spatial structure of the turbulence. In general, eddies with scales longer than r_i contribute to this correlation, while scales much shorter than r_i do not. As r_i increases to become larger than the largest eddies, the autocorrelation approaches zero. An indication of the largest eddies that cause the fluctuation X is given by defining an *integral length scale*

$$L_{XX_1} \equiv \int R_{XX}(x_i, r_1, \tau = 0)\, dr_1$$

Correlations tend to emphasize large-scale effects and to hide smaller-scale ones.

From a practical standpoint it is difficult to hold $\tau = 0$ and vary the measurement position r_i. The concept of *convection velocity* sometimes overcomes this difficulty. Many turbulent flows have a large mean velocity and much smaller fluctuations. *Taylor's hypothesis* is the assumption that a large convection velocity U_c sweeps a frozen turbulent state past the position of interest. This connects the flow-direction space separation and time by $r_1 = U_c t$. Invoking the convection hypothesis allows us to interpret $R_{XX}(x_i = 0, r_i = 0, \tau)$ as indicating the spatial structure of the turbulence in the flow direction. Referring to Figure 23-6, we use the convective hypothesis to interpret the spatial size of turbulent eddies that cause the largest wall pressure fluctuations, that is, let

$$R_{pp}(x_i = 0, r_i = 0, \tau) \approx R_{pp}(x_1 = U_c\tau, 0, 0)$$

Since the extent of the positive region of R_{pp} (at $r_i = 0$) is about $\tau U/\delta \approx 2$, we surmise that the spatial extent of the large turbulent eddy is probably twice this size:

$$L_E \approx U_c\tau \approx \frac{U_c}{U}\frac{U\tau}{\delta}\delta \approx \frac{U_c}{U}4\delta \approx 3\delta$$

Here the convective velocity has been taken as about 0.8 of the free velocity.

Next, we consider the correlation R_{pp} for large separation r_i. The decay of the correlation with an increase $= \xi$ in r_1, the microphone separation distance, is caused by two effects. First, the pressure-producing eddies change their structure as they traverse the distance r_1. To the extent that this modification is random, it will drive R_{pp} toward zero. Second, the convection process over the distance r_1 may not be uniform. The convection velocity itself is a statistical variable. If perfectly coherent eddies were convected to long distances by a varying convection velocity, we would find no correlation at the remote point. The arrival time would be random. Thus, the decay of R_{pp} with r_1 is a mixture of these two effects.

Cross correlations are formed when X and Y are distinct: $\overline{u_1 u_2}$ for example. Cross correlations have proved useful in determining the general nature of large-scale eddies. If turbulence were a single process with definite scale, this method would be more revealing.

23.6 DISTINCTIVE REGIONS OF A WALL LAYER

The inner region of a turbulent boundary layer and the region near the wall of a pipe or channel have a common turbulent structure. To find this structure we must focus attention on the flow very close to the wall. The idea is that if we are near the wall, say $y < 0.1\delta$ or $y < 0.1r_0$, all turbulent flows are dynamically similar. In making this analysis we are considering flows where changes in conditions in the flow direction are very slow. The flows have a self-preserving nature, so that local parameters, such as the turbulent-layer thickness or the wall shear, may be used to describe the flow state completely. If there is a rapid change of the flow conditions, the turbulent processes and the resulting mean flow take a long time to respond. This disrupts the equilibrium and invalidates the self-preservation assumption.

Figure 23-7a sketches some typical mean velocity profiles $U(y)$ very near the wall. They begin at zero and trail off into an unknown shape as y becomes large. The flow in that region depends on events in the outer portion of the boundary layer or in the core region of the pipe. Consider, from the standpoint of dimensional analysis, what physical parameters should change $U(y)$. First, we expect density, viscosity, and a wall roughness parameter ϵ (a length) to be important. The crucial question is how the outer flow influences the flow near the wall. We are too near the wall for U_e or U_{max} to be used directly. The outer

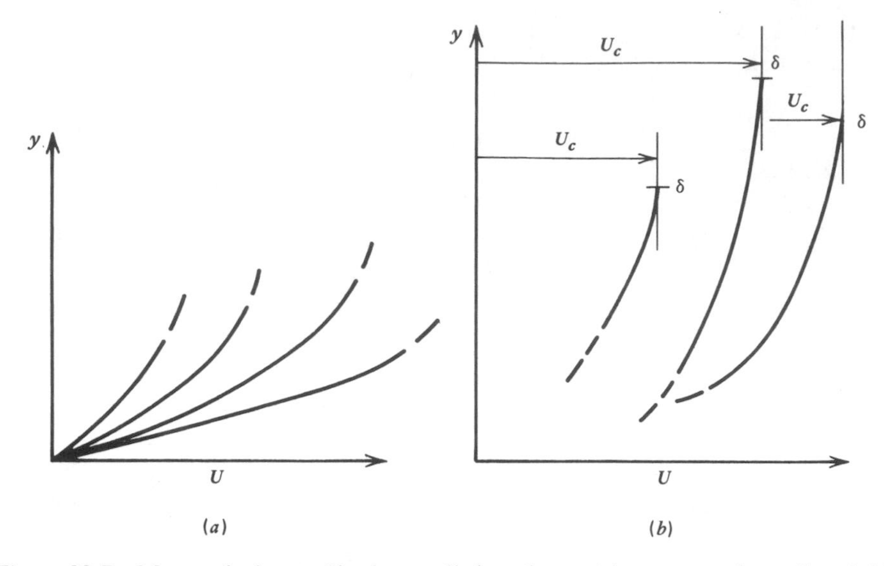

Figure 23-7 Mean velocity profiles in a wall shear layer: (a) very near the wall and (b) in the outer region.

region has a velocity profile of its own, which modifies the effect of U_e. The situation is very much like the stagnation-point flow on a body. The velocity U_∞ far from the body is not the characteristic velocity for the stagnation-point flow. U_∞ influences the stagnation-point flow, but that influence is modified by the flow field around the body before reaching the stagnation point. In the same way U_e or U_{max} influences the wall region, but that influence is modified by the turbulent flow in the outer layer. Outer layers with different properties have different effects on the wall region. What characteristics of the wall profile can we use to indicate the outer-flow effect? For increasing y (in the wall region) neither $U(y)$ nor its slope approaches a constant value that could be used as a characteristic. The answer to the question is the slope of the velocity profile at the wall, $dU/dy|_0$. Equivalently, since $\mu = \nu\rho$ is already in the list of variables, we may take the wall shear τ_0 as the parameter. The proposition is that the outer-flow influence may be represented in some complicated way by τ_0.

With these assumptions, the velocity profile near the wall has the functional form

$$U = U(y, \rho, \nu, \tau_0, \epsilon)$$

Dimensional analysis predicts that this relation will involve three nondimensional variables. Before listing the variables, it is useful to note that only ρ and τ_0 contain the dimension of mass. They are traditionally combined to produce a parameter with the dimensions of a velocity. We define the *friction velocity* as

$$u_* \equiv \sqrt{\frac{\tau_0}{\rho}} \tag{23.6.1}$$

Introducing this variable leads to the nondimensional groups

$$u^+ \equiv \frac{U}{u_*}, \qquad y^+ \equiv \frac{yu_*}{\nu}, \qquad \epsilon^+ = \frac{\epsilon u_*}{\nu}$$

Furthermore, the functional relation for the velocity profile is

$$u^+ = f(y^+, \epsilon^+) \tag{23.6.2}$$

This relation is called the *law of the wall.* It rests on the proposition that turbulent flows near walls are similar. Experiments show that this is indeed true.

Experimentally determined velocity profiles for the law-of-the-wall region were given in Figure 8-5. The region is further divided into three layers: the

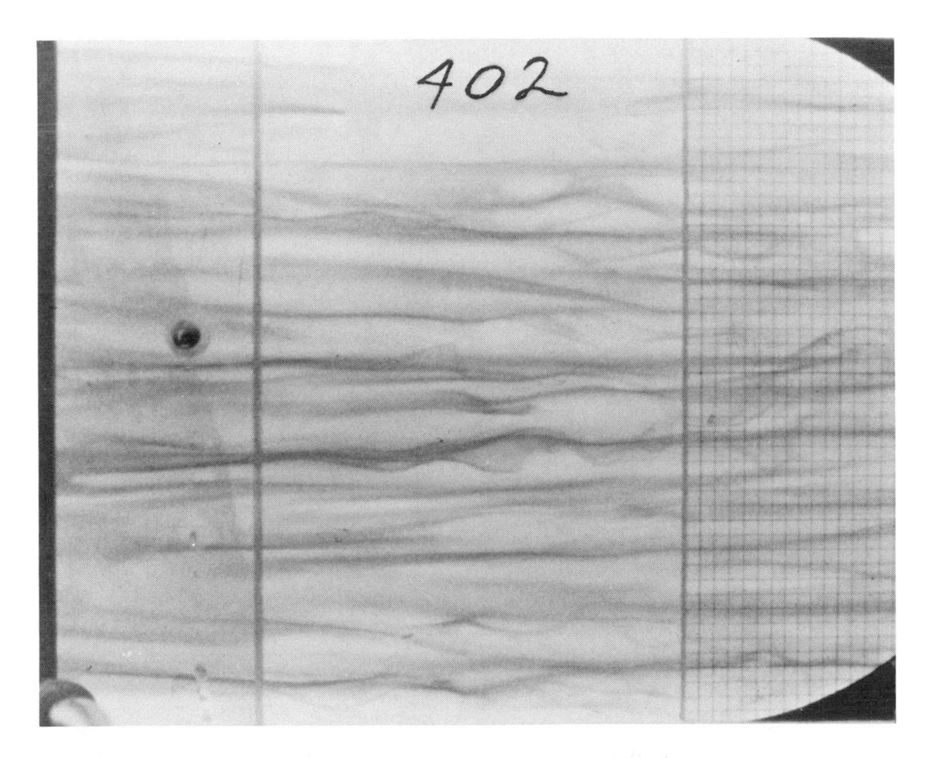

Figure 23-8 Streaks in the sublayer. Photograph courtesy of W. Tiederman, Purdue University. This research is described in Donohue, Tiederman, and Reischman (1972).

viscous sublayer, the buffer layer, and the overlap layer (also known as the fully turbulent region, the inertial sublayer, or simply the log region). The *viscous sublayer* extends from the wall out to about $y^+ = 5$ to 8. In this region the mean velocity profile is linear, as if the region were devoid of turbulence. However, flow visualization, such as that given in Figure 23-8, shows slow moving oscillations known as sublayer *streaks*. These are sets of counter-

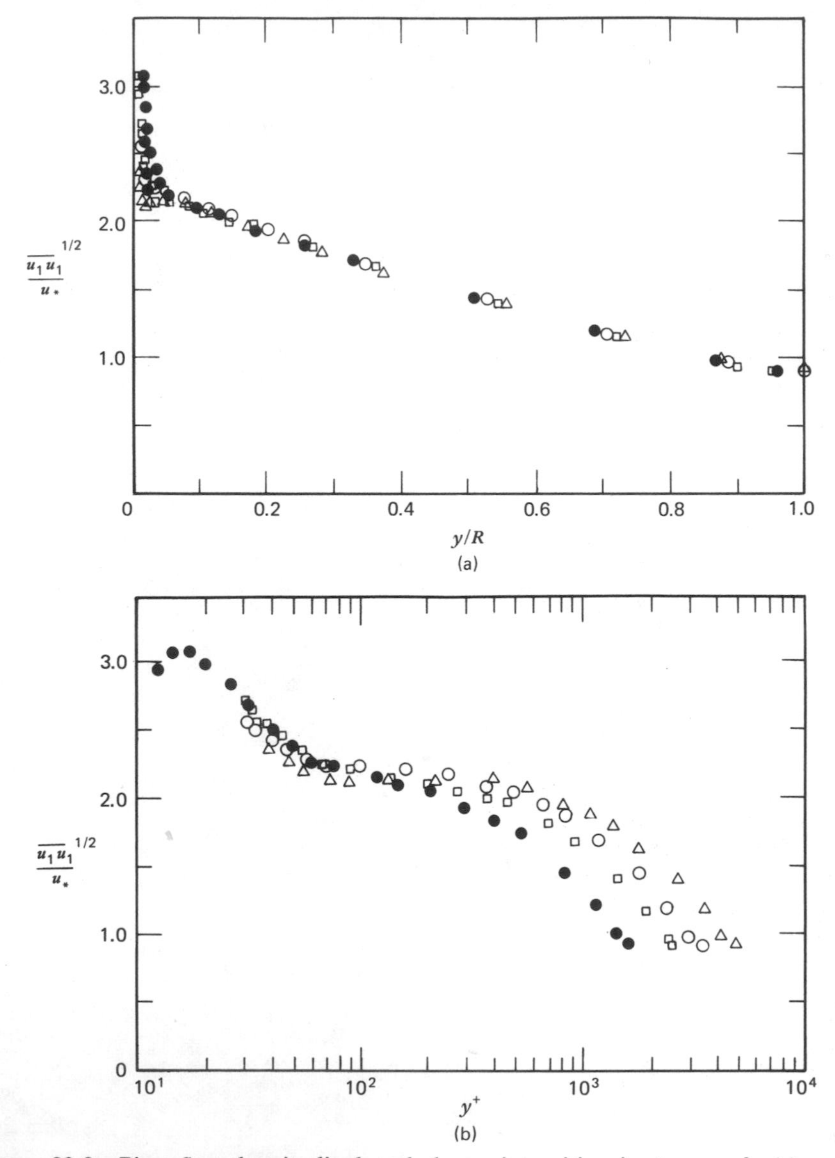

Figure 23-9 Pipe flow longitudinal turbulence intensities in terms of; (a) outer variable y/R, (b) wall variable y^+. The transverse turbulence intensities are given in (c) while the Reynolds stress is displayed in (d) along with the total stress given by – – –. From Perry and Abell (1975).

rotating vortices, which are about $100y^+$ units across one set and $1000y^+$ units long. The region between vortices has a slightly lower velocity and tends to collect the flow-visualization marker to form the streaks.

The *buffer region* from $y^+ = 8$ to $y^+ = 30–50$ is the scene of the most active turbulence production. Figure 23-9 shows that measurements of turbulence intensities peak in this region. Note that the u_1 turbulence intensity is about twice as large as u_2 and u_3 intensities. Although the viscous sublayer and buffer region are physically very thin (in pipe flow at Re = 50,000 the position $y^+ = 50$ corresponds to $y/R_0 = 0.04$), they are extremely important as the major site where wall turbulence "originates".

An important fact about these regions is that the total stress, which consists of the viscous shear stress and the Reynolds stress, is always constant. In fact,

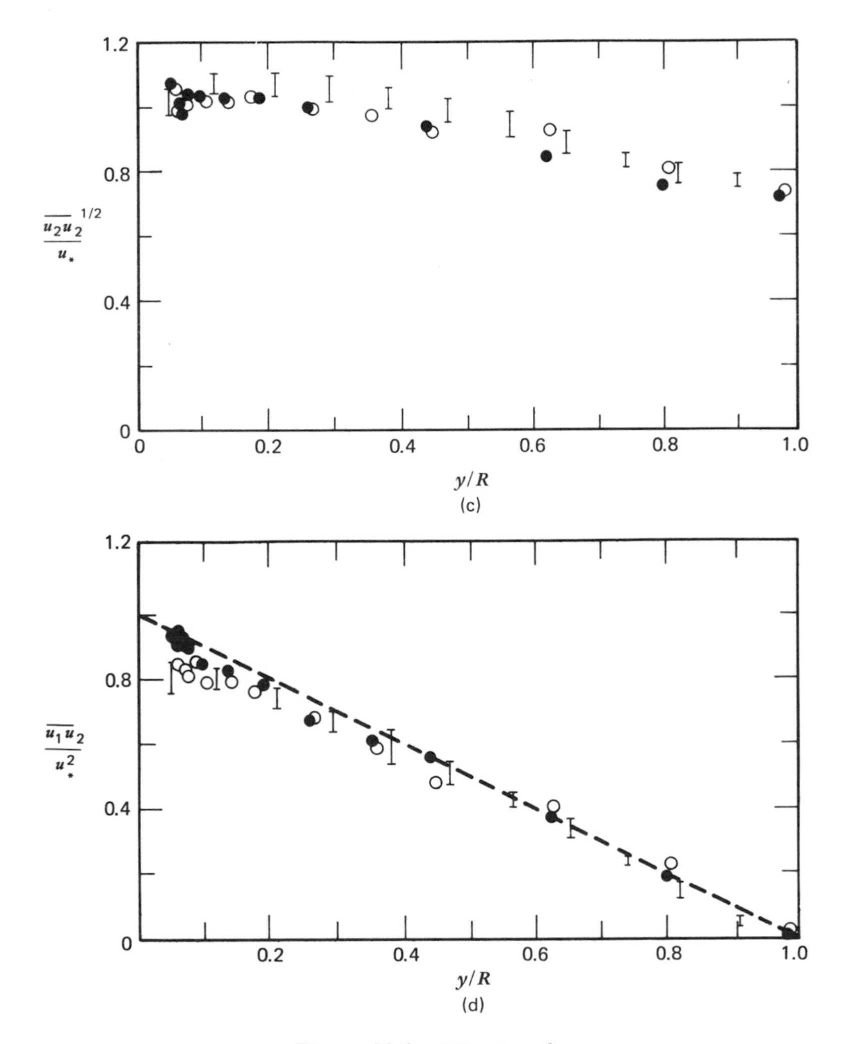

Figure 23-9 (*Continued*)

the term "constant-stress region" is sometimes used to refer to the entire region where the law of the wall is valid. At the wall itself, the stress τ_0 is 100% viscous shear because the fluctuation must vanish. Through the buffer layer τ decreases while the Reynolds stress $-\rho\overline{u_1 u_2}$ increases. At the outer edge of the buffer layer, the stress is essentially all turbulent Reynolds stress and is equal to the wall value τ_0 [substantiating arguments for these ideas are given, for instance, in Tennekes and Lumley (1972)]. The Reynolds stress continues to be constant through the overlap region.

Characteristics of the outer flow will be considered next; then we shall return to consider the overlap region. Figure 23-7b displays typical mean velocity profiles $U(y)$ for this region. The self-preserving or local-similarity assumption means that we can characterize the outer flow by a slowly changing thickness δ (or R_0 in a pipe) and the external velocity U_e (or U_{max}). Experience with boundary-layer theory suggests that the local pressure gradient dp/dx should also be important. To perform a dimensional analysis we must list all the things that influence $U(y)$. The first few are obviously y, ρ, δ, U_e, and dp/dx. Note that the viscosity has been left off this list.

The role that viscosity plays in turbulence at high Reynolds numbers is surprisingly limited. It is true that turbulent fluid always has vorticity—a quantity we associate with viscous action. Nevertheless, many important turbulent processes are effectively the unsteady, inviscid flow of a fluid-carrying vorticity. The one role that viscosity always does play is to stop the creation of smaller and smaller eddies. There comes a point where the corresponding velocity gradients are so steep that viscosity prevents their formation. When the Reynolds number is high, the small-scale eddies are so small that they do not contribute to the Reynolds stress. In this state of affairs $U(y)$ is independent of viscosity. The idea that profiles are independent of ν as $\mathrm{Re} \to \infty$ goes by the unlikely name of *Reynolds-number similarity*. It is a very important and often used concept in turbulence theory. It is because of Reynolds-number similarity that the list of outer-flow variables does not include ν.

The outer-velocity profiles of Figure 23-7b begin at U_e and trail off in an unknown way as they approach the wall region. This effect is caused by the Reynolds stress received from the inner region. Since the total stress is constant in the inner region, we may take τ_0 as the parameter representing the action of the wall layer on the outer flow.

The second thing to notice about Figure 23-7b is that U_e acts only as a reference: the scale for the change in $U(y)$ is determined by the Reynolds stress that the wall layer applies to the inner layer. Thus, the velocity in the outer layer has the functional form

$$U(y) - U_e = f\left(y, \delta, \rho, \frac{dp}{dx}, u_*\right) \tag{23.6.3}$$

Here again $u_* \equiv (\tau_0/\rho)^{1/2}$ and ρ are taken in lieu of τ_0 and ρ.

Dimensional analysis of 23.6.3 yields a functional form known as the *defect law*:

$$\frac{U(y) - U_e}{u_*} = F\left(\frac{y}{\delta}, \frac{\delta}{\rho u_*^2}\frac{dp}{dx}\right) \qquad 23.6.4$$

Velocity profiles in the outer layer are referenced to U_e and scaled by u_*. This reflects the fact that they are dominated by the Reynolds stress common to both regions. An external pressure gradient dp/dx causes an additional force $\delta\, dp/dx$ on the region. This is compared with $\rho u_*^2 = \tau_0$ in the pressure-gradient parameter of 23.6.4. Defect profiles for zero-pressure-gradient layers were given in Figure 8-6. Note that u_* is a scale for both the inner and outer regions.

The law of the wall and the defect law must somehow match together. This matching does not occur at a certain point, as one might first expect, but over a large region in y/δ. On the basis of dimensional analysis, Millikan (see Section 8.9) showed that the velocity profile in the matching region must be

$$u^+ = \frac{1}{k}\ln y^+ + C \qquad 23.6.5$$

when expressed in wall variables, or

$$\frac{U - U_e}{u_*} = \frac{1}{k}\ln\frac{y}{\delta} + B \qquad 23.6.6$$

when expressed in outer variables. The Kármán constant k is empirically found to be $k = 0.4$. The constant C in 23.6.5 depends on the wall roughness but not on the pressure gradient; the constant B in 23.6.6 depends on the pressure gradient but not on the wall roughness. A smooth wall has $C = 5.0$, and zero pressure gradient yields $B = -1.0$.

The *equilibrium layer* or *overlap layer* is where 23.6.2 and 23.6.4 are both valid. It begins at about $y^+ = 30$–50 and extends to about $y/\delta = 0.2$. This upper limit, in wall units, may be $y^+ = 300$ or more, depending upon the Reynolds number. Likewise, the lower limit, in outer variables, decreases with increasing Reynolds number. The relation between wall units and outer units is

$$\frac{y}{\delta} = y^+ \cdot \left(\frac{\nu}{u_*\delta}\right) \qquad 23.6.7$$

As the Reynolds number $u_*\delta/\nu$ increases, the wall layer becomes infinitely thin compared to δ—a boundary layer within a boundary layer. Simultaneously the logarithmic region expands toward the wall. The overlap layer may be counted as part of either the wall region or the outer region.

The last region, which extends from $y/\delta = 0.2$ to $y/\delta = 1$, is known as the *wake region*. Cole (1956) provided a good empirical equation to fit the velocity profile throughout the outer region. It is the logarithmic equation in wall variables plus a function $W(y/\delta)$:

$$\frac{U(y)}{u_*} = \frac{1}{k}\ln\left(\frac{u_* y}{\nu}\right) + C + \frac{1}{k}\Pi\, W\left(\frac{y}{\delta}\right) \qquad 23.6.8$$

where a simplified algebraic form of W is

$$W = 2\sin^2\left(\frac{y}{\delta}\frac{\pi}{2}\right) \qquad 23.6.9$$

$W(y/\delta)$ is called the *wake function*, and Π is a pressure-gradient parameter. In pipe flow $\Pi = \frac{1}{4}$ with $\delta = R$, while in a constant-pressure boundary layer $\Pi = 0.6$.

23.7 REYNOLDS STRESS

Turbulence affects the mean velocity directly through the Reynolds stresses in 23.4.7. Figure 23-9d gives the distribution of $\overline{u_1 u_2}$ through a pipe. It indicates that the Reynolds stress is high in the region of high mean shear. In boundary-layer flows, including wakes and shear layers, or in pipe flows, the gradient-diffusion concept introduced by Boussinesq (1877) is a useful engineering model. It is

$$\overline{u_1 u_2} = \nu_t \frac{dU}{dy} \qquad 23.7.1$$

The *eddy viscosity* ν_t depends on the turbulence in the flow and hence is a function of position. It is definitely not a fluid property.

Many simple flows may be calculated using the mixing length concept introduced by Taylor. Prandtl's (1925) mixing-length hypothesis for momentum relates ν_t to a mixing length l by the equation

$$\nu_t = l^2\left|\frac{dU}{dy}\right| \qquad 23.7.2$$

This equation was originally derived by analogy with the kinetic-theory model for molecular viscosity. The mixing length is much like the mean free path in a gas. Even though the eddies in turbulence are incapable of the independent action that molecules have, the concept works well with fairly simple assumptions for l. Equations 23.7.1 and 23.7.2 solve the closure problem once the behavior of l is assumed. The mixing-length model is the least sophisticated way of predicting turbulent flows.

Since the Reynolds-stress motions obey Reynolds-number similarly, they are inviscid, and l depends only on the position in the flow. In free shear layers, jets, wakes, and the wake portion of boundary layers, the mixing length is a certain fraction of the mean shear width:

$$l = C\delta$$

The eddies producing the Reynolds stress always scale with the transverse width of the turbulent region. In the overlap region of a turbulent boundary layer (where viscosity is also not important) another assumption is required. The argument used here is that the eddies are bounded by the wall, and therefore at a position y, the Reynolds stress motions must scale with the distance from the wall:

$$l = ky$$

The Kármân constant k in this relation will yield the logarithmic-region mean-velocity profile when $l = ky$ is used in 23.7.2 which in turn is inserted in 23.7.1 and integrated with the assumption $\rho\overline{u_1 u_2} = \tau_0$ is constant (hence the name constant stress layer).

It is possible to derive an exact equation that governs the Reynolds stresses $\overline{u_i u_j}$. An equation for u_i is produced by subtracting the time-averaged equation 23.4.7 from the Navier–Stokes equation, where $\tilde{u}_i = U_i + u_i$. This equation is multiplied by u_j. The process is repeated with the roles of i and j reversed to yield a second equation. Adding these equations and time-averaging yields the final result:

$$U_k \partial_k \left(\overline{u_i u_j}\right) = \underbrace{-\partial_k\left(\overline{u_k u_i u_j}\right) - \frac{1}{\rho}\left[\partial_i\left(\overline{u_j p}\right) + \partial_j\left(\overline{u_i p}\right)\right]}_{\text{diffusion terms}}$$

$$\underbrace{-\left(\overline{u_i u_k}\partial_k U_j + \overline{u_j u_k}\partial_k U_i\right) + \frac{1}{\rho}\left(\overline{p\,\partial_j u_i} + \overline{p\,\partial_i u_j}\right)}_{\text{production terms}}$$

$$\underbrace{+\nu\left(\overline{u_j \partial_k \partial_k u_i} + \overline{u_j \partial_k \partial_k u_i}\right)}_{\text{destruction terms}} \qquad 23.7.3$$

The nine equations above can only be solved when closure assumptions are made about the effects on the right-hand side. Our main interest is in interpreting the various effects that change the Reynolds stresses. The first two terms are called diffusion effects because the quantities of interest are differentiated. They are imagined to redistribute turbulence, because their integrated effect (over a region where fluctuations on the boundary vanish) must be zero.

The production of Reynolds stress is due to the terms containing $\partial_k U_l$. Mean shear is an essential part of turbulence. Without it the turbulence decays. Production also occurs through the pressure–strain correlation (note that the pressure diffusion and pressure–strain production actually combine mathematically into only one group: $\overline{u_i \partial_j p} + \overline{u_j \partial_i p}$. This form emphasizes that the actual pressure force that generates or destroys a fluctuation u_i is $\partial_j p$).

The last term in 23.7.3 is the viscous destruction of Reynolds stress. Implicit in this name is the assumption that viscous forces only affect the small eddies and are mainly a retarding influence.

Turbulence calculations where the terms on the right-hand side of 23.7.3 are modeled are called second-order closure models. They are quite complicated in their most general form.

*23.8 KINETIC ENERGY

An equation governing the kinetic energy of the mean flow is derived by multiplying 23.4.7 by U_i. It is

$$\rho U_j \partial_j \left(\tfrac{1}{2} U_i U_i \right) = -U_i \partial_i P + \mu U_i \partial_j \partial_j U_i - U_i \partial_j \left(\rho \overline{u_i u_j} \right) \qquad 23.8.1$$

Turbulence adds or subtracts energy from the mean flow through the gradient of the Reynolds stress. The total overall effect is to decrease the mean energy. In order to show this, consider the identity

$$\partial_j \left(\rho U_i \overline{u_i u_j} \right) = \rho \overline{u_i u_j} \partial_j U_i + U_i \partial_j \left(\rho \overline{u_i u_j} \right)$$

We integrate this over a thin slice of boundary layer or pipe flow where changes in the flow direction are negligible. After applying Gauss's theorem to the left-hand side, the result is

$$\int n_j \rho U_i \overline{u_i u_j} \, dS = \int \rho \overline{u_i u_j} \partial_j U_i \, dV + \int U_i \partial_j \left(\rho \overline{u_i u_j} \right) dV$$

The left-hand side is zero, since the Reynolds stress is zero on the wall or in the free stream and the integrals on the ends of the slice cancel each other. Hence, the overall mean kinetic-energy production by turbulence is

$$-\int U_j \partial_j \left(\rho \overline{u_i u_j} \right) dV = \int \rho \overline{u_i u_j} \partial_j U_i \, dV < 0 \qquad 23.8.2$$

In a typical shear layer $\overline{u_i u_j}$ is always negative and $\partial_j U_i$ is always positive; thus the net effect is negative. On balance, the Reynolds-stress effect tends to decrease the mean kinetic energy in 23.8.1.

Next, we take up the kinetic energy of the turbulent fluctuations. The relation governing the turbulent kinetic energy is the trace of 23.7.3, namely (recall $K \equiv \frac{1}{2}\overline{u_i u_i}$)

$$U_k \partial_k K = -\partial_k \left(\overline{u_k K} \right) - \frac{1}{\rho} \partial_i \left(\overline{u_i p} \right) - \overline{u_i u_k} \partial_k U_i + \nu \overline{u_i \partial_k \partial_k u_i} \qquad 23.8.3$$

The first two terms on the right-hand side of 23.8.3 are diffusion terms, which redistribute K. The main production term in this equation is the Reynolds stress multiplying the mean shear. On the average, the Reynolds stress effect that produces turbulence balances that which destroys mean flow energy. The overall volume-average energy production by Reynolds stresses in 23.8.3 is equal to the destruction of mean-flow kinetic energy by Reynolds stresses in 23.8.1 in light of the equality in 23.8.2. Be aware that there is a certain arbitrariness in the way production and diffusion effects are defined in 23.8.3 (Brodkey et al., 1973). When we time-average these effects, we lose information and cannot strictly apply these interpretations to instantaneous motions.

The viscous-destruction term in 23.8.3 is frequently replaced by the viscous dissipation, that is,

$$\varepsilon \equiv \nu \partial_k u_i \partial_k u_i \approx -\nu \overline{u_i \partial_k \partial_k u_i}$$

This is based on the idea that viscosity affects only the smallest eddies and since the net viscous work for the entire region must be zero, we assume this is also true locally:

$$0 \approx \partial_k\left(\overline{u_i \partial_k u_i}\right) = \overline{u_i \partial_k \partial_k u_i} + \overline{\partial_k u_i \partial_k u_i}$$

The viscous work that retards the formation of the smallest eddies is equal to the work that heats the fluid.

Note that the pressure–velocity correlation is a diffusive effect in 23.8.3. The pressure–strain correlation, which appeared in 23.7.3 as a production term, does not occur in 23.8.3, because $\overline{p \partial_i u_i} = 0$ (since $\partial_i u_i = 0$). Pressure–velocity correlations only transfer turbulent energy that the Reynolds-stress mean-shear term produces. To illustrate this, consider the simple shear flow where $U_1(x_2)$, $U_2 = 0$, $U_3 = 0$, and all turbulence properties are uniform in the x_2 and x_3 directions. The turbulent intensities $\frac{1}{2}\overline{u_1 u_1}$, $\frac{1}{2}\overline{u_2 u_2}$, and $\frac{1}{2}\overline{u_3 u_3}$ are controlled by the appropriate equation in 23.7.3. Hence

$$\begin{aligned}
U_1 \partial_1\left(\tfrac{1}{2}\overline{u_1 u_1}\right) &= 0 \\
&= -\partial_2\left(\overline{u_2 \tfrac{1}{2} u_1 u_1}\right) + \frac{1}{\rho}\overline{p \partial_1 u_1} - \overline{u_1 u_2}\partial_2 U_1 - \tfrac{1}{3}\varepsilon \\
U_1 \partial_1\left(\tfrac{1}{2}\overline{u_2 u_2}\right) &= 0 \\
&= -\partial_2\left(\overline{u_2 \tfrac{1}{2} u_2 u_2}\right) + \frac{1}{\rho}\overline{p \partial_2 u_2} + \frac{1}{\rho}\partial_2 \overline{u_2 p} - \tfrac{1}{3}\varepsilon \qquad 23.8.4 \\
U_1 \partial_1\left(\tfrac{1}{2}\overline{u_3 u_3}\right) &= 0 \\
&= -\partial_2\left(\overline{u_2 \tfrac{1}{2} u_3 u_3}\right) + \frac{1}{\rho}\overline{p \partial_3 u_3} - \tfrac{1}{3}\varepsilon
\end{aligned}$$

Here, the viscous destruction has been assumed to act equally on all components.

Some interesting physics is implied by 23.8.4. On the average turbulence production $\overline{u_1 u_2} \partial_2 U_1$ goes entirely into the $\overline{u_1 u_1}$ equation, while pressure–strain production terms $\overline{p \partial_1 u_1}$, $\overline{p \partial_2 u_2}$, and $\overline{p \partial_3 u_3}$ are in all three equations. Since $\overline{p \partial_i u_i} = 0$, these terms redistribute turbulent energy that was originally generated as $\frac{1}{2}\overline{u_1 u_1}$ energy into turbulent energy in the other directions. Recall that in a wall layer $\frac{1}{2}\overline{u_1 u_1}$ was much higher than the other components in Figure 23-9. Net production occurs only in the u_1 direction, but viscous destruction is roughly equal for all components. Pressure–strain is the major intracomponent transfer mechanism.

*23.9 ENERGY SPECTRUM

The spectrum of longitudinal turbulent energy is given in Fig. 8-7. These data were taken on a turbulent jet, but they are typical of any turbulent flow. The easiest way to view a spectrum is simply as a function that produces the proper energy when it is integrated over all wavenumbers, that is, $F_1(k_1)$ has the property that

$$\overline{u_1 u_1} = \int_0^\infty F_1(k)\, dk, \qquad \text{or} \qquad \frac{d}{dk}\left(\overline{u_1 u_1}\right) = F_1(k_1) \tag{23.9.1}$$

This equation does not really define $F_1(k_1)$ properly, but it gives one of its major properties. F_1 is a measure of the energy between k_1 and $k_1 + dk_1$ in a Fourier decomposition of $u_1(r_1)$. [Actually $F_1(k_1)$ and the autocorrelation $R_{11}(r_1)$ are Fourier transforms of each other.] Most of the energy is at low frequencies, in the larger eddies. The energy decreases at higher wavenumbers because the velocity fluctuations are less severe. Kolmogorov applied dimensional analysis to the energy spectrum to show that it has an overlap region where

$$F_1 \propto k^{-5/3} \tag{23.9.2}$$

This is the special case mentioned in Section 8.9. The range of k_1 for 23.9.2 to apply is called the *inertial subrange*.

The arguments that lead to the inertial subrange are important because they reveal some physical aspects of turbulence. Consider first the low-wavenumber (large-eddy) region of the spectrum where most of the $\overline{u_1 u_1}$ energy exists. These eddies are anisotropic and bear the mark of the way in which they were formed. Jets, wakes, and wall layers have different-shaped spectra in this region. The one common aspect of all these flows is that viscosity does not affect the main energy-carrying eddies (Reynolds-number similarity). The primary instability and scale building processes are inviscid. We let L and u_0 be length and velocity scales that characterize the large eddies. Dimensionally the energy spectrum is

$$F_1 = F(k_1, u_0, L)$$

The essence of Kolmogorov's argument is that inviscid processes also redistribute turbulent energy into smaller and smaller eddies until they are so small that viscous forces restrain their growth. This *energy cascade* is very long when the Reynolds number of the flow is large. The only thing in common between large and small eddies is that the rate at which energy is put into forming the large eddies by the Reynolds stress term must be equal to the viscous dissipation at the smaller scales.

The energy of a large eddy is about u_0^2. As the eddy turns over once, we assume that a certain fraction of its energy is lost into smaller scales. Since the turnover time is L/u_0, the rate at which energy is lost is proportional to

$$\varepsilon = \frac{u_0^3}{L} \tag{23.9.3}$$

In terms of ε, instead of u_0, the large-scale spectrum is

$$F_1 = F(k_1, \varepsilon, L) \tag{23.9.4}$$

The nondimensional form is a function of $k_1 L$,

$$\bar{F} \equiv \frac{F_1}{\varepsilon^{2/3} L^{5/3}} = \bar{F}(k_1 L) \tag{23.9.5}$$

This corresponds to 8.9.5 in the generalized Millikan analysis with ϵ = Re.

Consider two flows with the same u_0 and L but with different viscosities, so that the turbulence Reynolds numbers are different. These flows have the same large-scale structure, since the main turbulence-building processes are inviscid. The difference occurs in the fine-scale structure. When ν is low, the small eddies can become much smaller before the viscous forces retarding their formation is effective. The range of eddy sizes at high Re is wider, and the smallest eddies are smaller.

As Re $\to \infty$ the small eddies are so far down the cascade that the scale L is no longer important. On the other hand, the viscosity is important. For high wavenumbers, the spectrum has the form

$$F_1 = f(k_1, \varepsilon, \nu) \tag{23.9.6}$$

We can nondimensionalize this into

$$\bar{f} \equiv \frac{F_1}{\varepsilon^{1/4} \nu^{5/4}} = \bar{f}(k_1 \eta) \tag{23.9.7}$$

where η, the *Kolmogorov length*, is defined as

$$\eta \equiv \left(\frac{\nu^3}{\varepsilon}\right)^{1/4} \tag{23.9.8}$$

The Kolmogorov length is a length measure for the smallest eddies in the turbulence. Note that the L and η are related by

$$\frac{L}{\eta} = \text{Re}^{3/4} \tag{23.9.9}$$

As Re $\to \infty$ the difference in scale between the large and small eddies becomes very great.

From 23.9.5, 23.9.7, 23.9.8 and 23.9.9 we see that

$$\bar{f} = \text{Re}^{5/4}\bar{F}$$

Comparing this with 8.9.8 (where $Y_1 = Y_2 = 0$), we find that $m = -\frac{5}{4}$, and comparing 23.9.9 with 8.9.7, we find that $n = \frac{3}{4}$. Thus, according to 8.9.20 the spectrum in the overlap region (the inertial subrange) follows the law

$$\bar{F} = A \cdot (k_1 L)^{-5/3}$$

in outer variables, or

$$\bar{f} = B \cdot (k_1 \eta)^{-5/3} \tag{23.9.10}$$

in inner variables. This latter function is, in principle, universally valid for all turbulent flows. Thus, we have derived 23.9.2 by dimensional analysis.

At one time it was thought that the small-scale eddies would be isotropic and also uniformly distributed in space. The experimental evidence is that these eddies occur in patches (a phenomenon called intermittency of dissipating eddies) and that they are not isotropic. Corrections to 23.9.10 for these facts turn out to be negligible.

The Kolmogorov length is the smallest turbulent fluctuation that can withstand the damping effect of viscosity. Since η is typically 0.1–1 mm, there is no question that the continuum hypothesis applies to common turbulent flows.

23.10 EQUILIBRIUM BOUNDARY LAYERS

A turbulent boundary layer in a zero pressure gradient develops into a self-preserving state where it grows very slowly at a nearly linear rate. Typically, $d\delta/dx$ produces an angle of 1° or less. The growth rate depends very slightly on the local Reynolds number $u_*\delta/\nu$. Pressure gradients that accelerate the flow decrease the turbulence production, and if they are strong enough, the flow can relaminarize. Adverse pressure gradients, on the other hand, cause a thickening of the layer and a faster growth rate.

Clauser (1954, 1956) reasoned that it should be possible to produce special pressure gradients that would give self-similar defect-velocity profiles. He called these flows *equilibrium boundary layers*. They are analogous to the Falkner–Skan similarity solutions for laminar boundary layers. When the defect law 23.6.8 is written in outer variables it becomes

$$\frac{U(y) - U_e}{u_*} = \frac{1}{k}\ln\left(\frac{y}{\delta}\right) + \frac{\Pi}{k}[W - 2]$$

Thus, an equilibrium boundary layer is one where Π is constant, so that the wake component is fixed at a certain value. This is the empirical definition of an equilibrium boundary layer. Theoretical analysis shows that these layers have a constant ratio (an equilibrium, if you like) of the pressure force retarding the layer to the Reynolds stress at the bottom of the defect layer (the similarity does not include the viscous region $y^+ < 30$, but the total stress is constant through this region). Since the Reynolds stress is equal to τ_0, a theoretical definition of an equilibrium layer is $\beta = \text{const}$, where

$$\beta \equiv \frac{\delta}{\tau_0}\frac{dp}{dx} \qquad 23.10.1$$

A third definition is also useful. An experimental definition of an equilibrium layer is one where $U_e \approx x^{-a}$ for any constant a. Of course, the three definitions are related, but in a complicated way.

Many computer models of turbulent boundary layers are based on data obtained from equilibrium layers. They can accurately predict boundary layers that slowly change their equilibrium state.

The wall layer $y^+ < 30$ typically is 5% or less of the total width δ. For momentum calculations this region may be neglected. Its major effect is to supply Reynolds stress to the outer layer. The friction velocity u_* characterizes the stress. An implicit shear-stress law for equilibrium boundary layers is obtained by evaluating 23.6.8 at $y = \delta$. The result is

$$\frac{U_e}{u_*} = \frac{1}{k}\ln\left(\frac{u_*\delta}{\nu}\right) + C + \frac{2}{k}\Pi \qquad 23.10.2$$

Turbulent boundary layers are very complicated because the major part of the flow is inviscid and has all the momentum. This region receives Reynolds stress from a thin wall region where viscosity is important. In some unknown way these regions interact to determine the stress u_*. Moreover, as $\text{Re} \to \infty$, $u_*/U_e \to 0$.

23.11 STRUCTURE OF WALL-LAYER TURBULENCE

Detailed studies of turbulence have revealed several distinctive motions of the fluid. Nevertheless, researchers do not claim to have a complete picture. The main object of structure studies is to define a loop or sequence of events responsible for the self-sustaining nature of turbulence. Flow visualization has

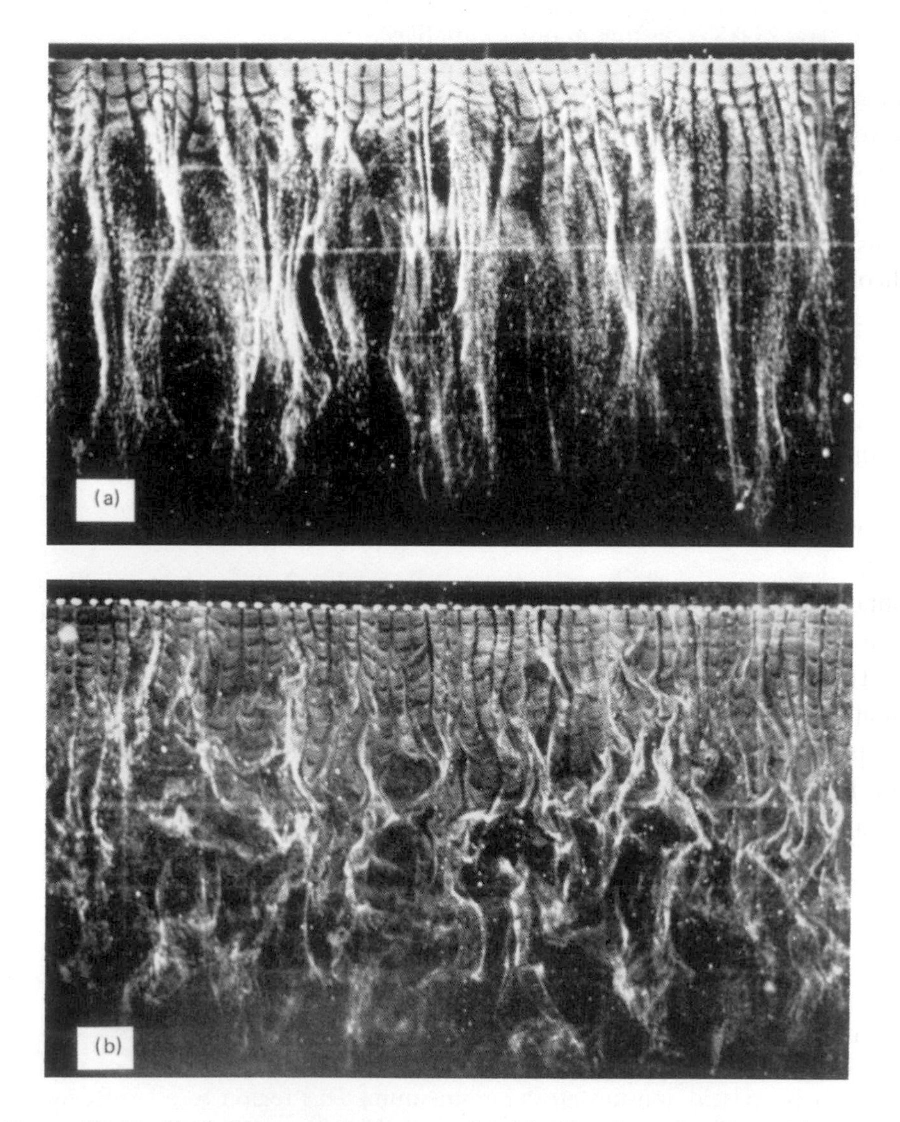

Figure 23-10 Turbulence structure at various heights in a low-Reynolds-number boundary layer is visualized with hydrogen bubbles in water. Flow is from top to bottom; (a) sublayer $y^+ = 2.7$, (b) buffer region $y^+ = 38$,

been used to identify important events, and subsequently special measurement of their characteristics have been made. Explanations of the dynamics of the events are vague at best. Indeed, we have only speculation as to what constitutes a complete loop of events.

Since turbulence has several overall universal characteristics, it is natural to suppose that some universal turbulence events are responsible. The task is first to identify the distinctive events, sometimes termed coherent structures; then to find their relationships to one another; and finally to determine what they do to produce a given universal characteristic of turbulence. For example,

Figure 23-10 (*Continued*) (c) log region $y^+ = 101$, and (d) wake region $y^+ = 407$. From Kline, et al. (1967).

consider the self-sustaining characteristic of turbulence. Without Reynolds stresses, the turbulence loses its self-sustaining property. Any turbulent eddy has $u_1 u_2 \neq 0$; however, it may be an incidental motion that does not make an essential contribution to the time averaged Reynolds stress $\overline{u_1 u_2}$. The terms *active motion* and *inactive motion* are sometimes used to make this distinction. We must not only identify coherent structures, but we must also determine their role. As another illustration, consider the very small eddies being destroyed by viscous forces. These eddies are important to the overall turbulent energy balance, but they play no role in the Reynolds stress and hence no role in determining the mean-velocity profile. They are essential to describing the energy interactions but relatively unimportant to the momentum interactions.

In the present section we shall briefly discuss some of the coherent events that are observed in wall layers. More details are given by Cantwell (1981). A similar discussion for shear layers is in Section 23-12. Figure 23-10 shows a low-speed water channel where hydrogen bubbles are used to mark the flow. The wire that produces the bubbles is parallel to the wall and perpendicular to the flow at a certain height in the layer. The sublayer is shown in Fig. 23-10a. Here, the collection of bubbles into the sublayer streaks is apparent (see also Fig. 23-8). A picture from the buffer region where the turbulence is much more intense is shown in Fig. 23-10b. Researchers have identified several typical events in the inner wall region. The wall streaks, marked with either hydrogen bubbles or with dye, slowly lift up into the buffer region, where they undergo a distinct oscillation. The oscillation ends abruptly as the marking fluid breaks up into a larger region with many smaller-scale eddies. This region has an outward trajectory. The entire process is called *bursting*. It makes major contributions to the Reynolds stress. Other motions in the buffer layer have also been named. The term *sweep* refers to high-speed fluid that comes coherently from the logarithmic region into the buffer region. *Ejection* refers to a structure of slow-moving fluid that abruptly leaves the buffer region toward the logarithmic region. The interactions of these motions and the triggering mechanisms are not clear.

The overlap region (log region) is shown in Fig. 23-10c. Note the larger variety of eddy sizes in this picture. In the final view, Fig. 23-10d, the bubble wire is located in the wake region where both large-scale potential motions and pockets of turbulent motion exist. Recall the similar structure of jets and wakes depicted in Figs. 23-2 and -3. The interface between turbulent and nonturbulent fluid is always very sharp.

In the boundary layer the position $y = \delta$ does not mark the end of the turbulent fluid. Delta is defined by the mean velocity profile. Turbulent fluid may be found out to $\delta > 1.2$, while potential fluctuations exist to over twice this distance. Large-scale motions in the wake region take nonturbulent fluid deep within the boundary layer, say to $y = 0.3\delta$. Vorticity at the turbulent–nonturbulent interface slowly contaminates this fluid, and it too becomes turbulent. This is the *entrainment* process discussed earlier.

23.12 JETS, WAKES, AND FREE SHEAR LAYERS

Flows with regions of free shear unbounded by walls are highly unstable and become turbulent very close to their origin. A low-Reynolds-number jet in water, Fig. 23-11, and a somewhat higher-Reynolds-number jet in air, Fig. 23-12, illustrate many of the properties of turbulent flows. Note in particular the size of the small size eddies compared to the local jet width. These figures are a visual illustration of 23.9.9. Figure 23-13 depicts jet, wake, and free shear flows and their characteristic widths and velocities. Each flow has a characteristic integral that is constant at any station. In the jet, the momentum constant is

$$M = \rho \int U^2 \, dy \qquad 23.12.1$$

while the centerline velocity $U_c(x)$ decays and the half width $\Delta(x)$ increases. In the wake, the momentum defect is constant:

$$D = \rho \int U(U_0 - U) \, dy \qquad 23.12.2$$

while the centerline velocity $U_e(x)$ approaches the free-stream velocity U_0, and $\Delta(x)$ increases. In the free shear layer, the flow rate must be conserved:

$$U_0 \Delta = \int U \, dy \qquad 23.12.3$$

while the width $\Delta(x)$ increases for a constant velocity difference U_0. We can find out quite a lot about these flows by just applying the two major assumptions about turbulent flows—self-preservation (local similarity) and Reynolds-number similarity (inviscid Reynolds stresses). The self-preservation assumption means:

Jet
$$\frac{U(x, y)}{U_c(x)} = f\left(\frac{y}{\Delta(x)}\right) \qquad 23.12.4$$

Wake
$$\frac{U_0 - U(x, y)}{U_c(x)} = f\left(\frac{y}{\Delta(x)}\right) \qquad 23.12.5$$

Shear layer
$$\frac{U(x, y) - U_1}{U_0} = f\left(\frac{y}{\Delta(x)}\right) \qquad 23.12.6$$

In each case we expect these assumptions to be valid only downstream, after the turbulence has had a chance to develop some sort of equilibrium state so

Figure 23-11 Turbulent water jet visualized in a plane on the axis. At Re = 2300 the picture resolution exceeds the Kolmogorov scale in the lower half. Also note the spiral structure. From Dimotakis, Miake-Lye and Papantoniou (1983).

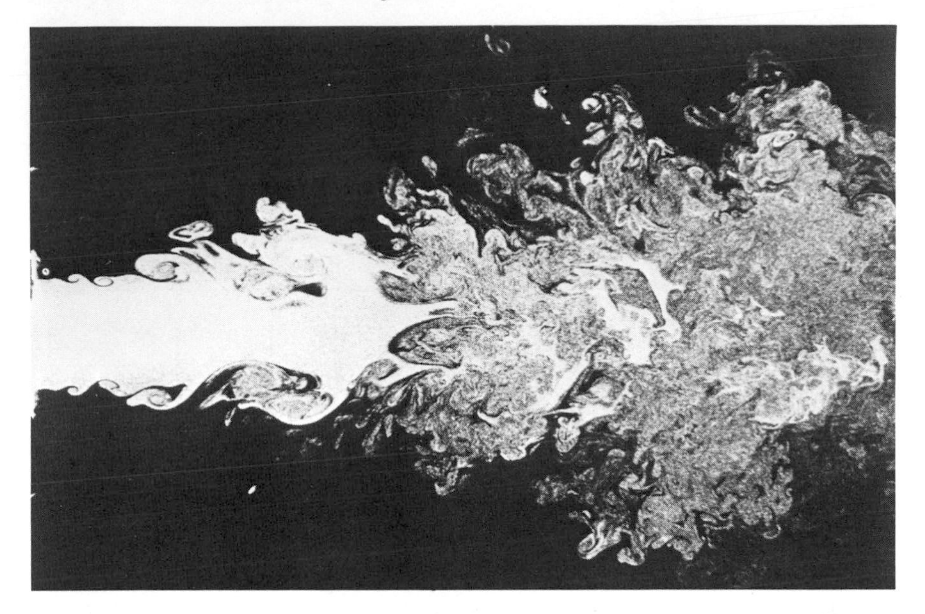

Figure 23-12 Turbulent air jet at a higher Reynolds number (11,000) shows an increase in fine scale structure. Photograph courtesy of J. L. Balint, M. Ayrault, and J. P. Schon, Ecole Centrale de Lyon, France.

that the Reynolds stresses can establish the mean profile. In the case of the wake we must further restrict the answer to the far wake where $U_c \to U_0$.

When these local-similarity assumptions are inserted into the appropriate integral constants, we get a relation between U_c and Δ. For example, inserting 23.12.4 into 23.12.1 yields

$$M = \rho U_c^2 \Delta \int f^2 \, d\left(\frac{y}{\Delta}\right) \qquad 23.12.7$$

Since the integral is a constant, the decay of the centerline velocity U_c and the growth of the jet Δ must be such that $U_c^2 \Delta$ is constant. A similar substitution for the wake shows that $U_c \Delta$ is constant. However in the case of the free shear layer 23.12.3 is satisfied for any growth rate for $\Delta(x)$.

The assumption of Reynolds-number similarity means that as $\text{Re} \to \infty$ the flows become independent of viscosity. Let us use this fact in a dimensional analysis of the centerline velocity decay. Omitting viscosity, we propose that

$$U_c = U_c(M, \rho, x)$$

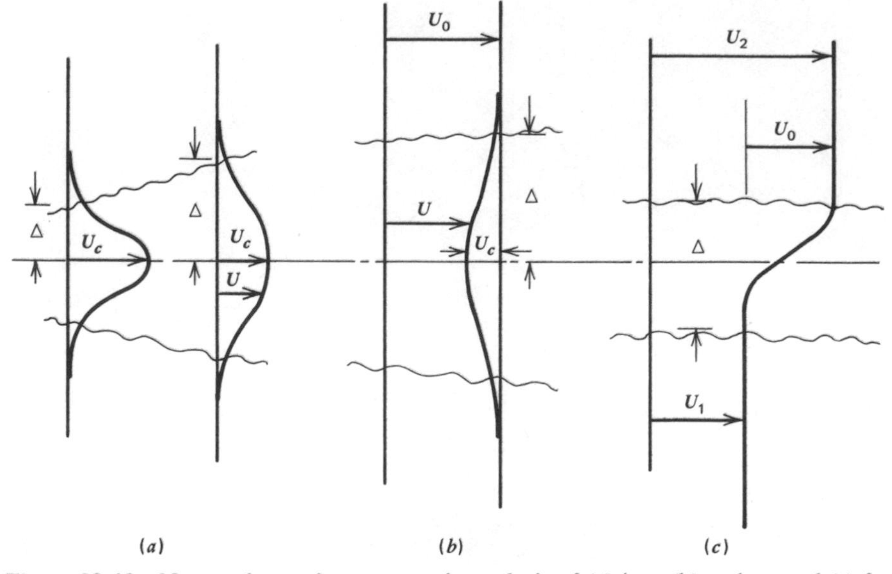

Figure 23-13 Nomenclature for asymptotic analysis of (a) jets, (b) wakes, and (c) free shear layers.

Dimensional analysis shows that one nondimensional parameter is needed:

$$U_c\sqrt{\frac{\rho x}{M}} = \text{const}$$

Thus, we have determined that $U_c \sim x^{-1/2}$. Furthermore, since 23.12.7 demands that $U_c^2\Delta$ be constant, the jet must spread linearly: $\Delta \sim x$. An even more instructive argument is to assume $\Delta = \Delta(M, \rho, x)$ and try a dimensional analysis. It turns out that M is the only variable containing time and therefore must be stricken from the list as an improper choice. Now $\Delta = \Delta(\rho, x)$ is not a proper assumption, because ρ is the only variable that contains mass. Therefore, we must have $\Delta = \Delta(x)$, implying Δ/x is a single constant independent of the fluid density or the jet momentum. All jets grow at the same rate. Measurements show that the half angle of plane jet spreading is roughly 4°. The mean velocity profiles become self-preserving about 5 jet diameters downstream from the mouth. Details of the turbulent structure can take much longer to become self-similar.

Completing the analysis above for the wake shows that $U_c \sim x^{-1/2}$ and $\Delta \sim x^{1/2}$. Plane free shear layers have $U_0 = \text{const}$ and $\Delta \sim x$. All of these results hinge on the assumptions that the profiles are locally similar and that the turbulent motions that determine the Reynolds stresses are essentially inviscid.

As a final topic in this section, we note a few facts about turbulent structure in free shear layers. Figure 23-14 is a schlieren photograph of the shear layer between two different gases. Flow is from left to right with the upper stream faster. A splitter plate separating the streams ends on the left side just out of view. The density difference between the gases aids the photography but does not affect the flow pattern. The dominant features are the large coherent vortex eddies which are very regular and go entirely across the shear layer. This very regular growth and two-dimensional character of the large scale eddies continues downstream as the shear layer grows. Further downstream the eddies are not only larger but fewer in number.

The large eddies grow by two mechanisms: they entrain fluid outside the shear layer into their edges, and they swallow up vortices that are already in the shear layer. This latter process is called *vortex pairing* and is illustrated in Figure 23-15 where the camera moves to follow the same eddies. It is obviously the mechanism by which the number of vortices is reduced. Vortex pairing has also been observed at the beginning of turbulent jets. However, eddies in jet flows become three-dimensional rather quickly, and vortex pairing is very difficult to observe in the downstream portion of the jet.

Figure 23-14 Growth of a free shear layer between streams with different speeds. The lower portion of the photo is a side view while a plan view is reflected in a mirror in the upper portion. Large coherent eddies continue to grow for all downstream distances. From Brown and Roshko (1974).

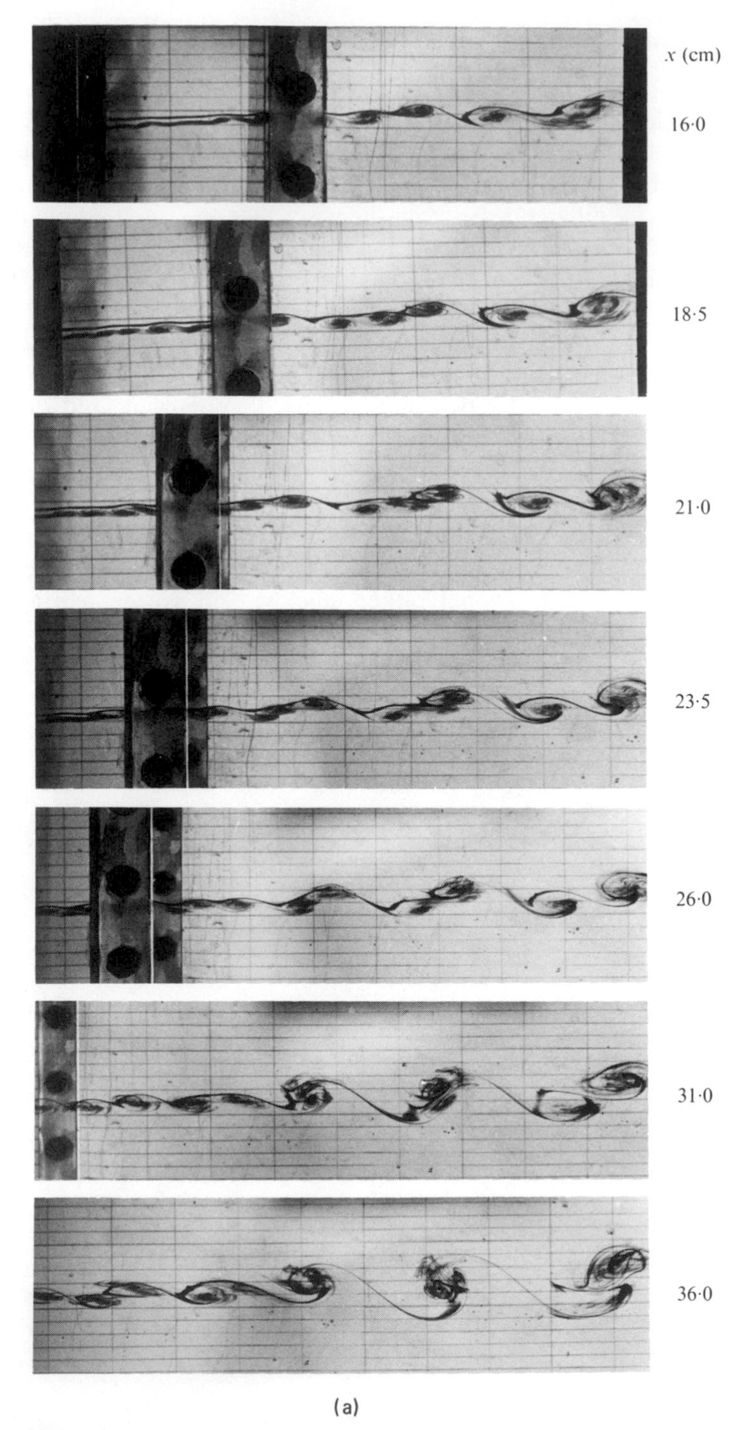

(a)

Figure 23-15 Vortex pairing shown in step by step sequence. From Winant and Browand (1974).

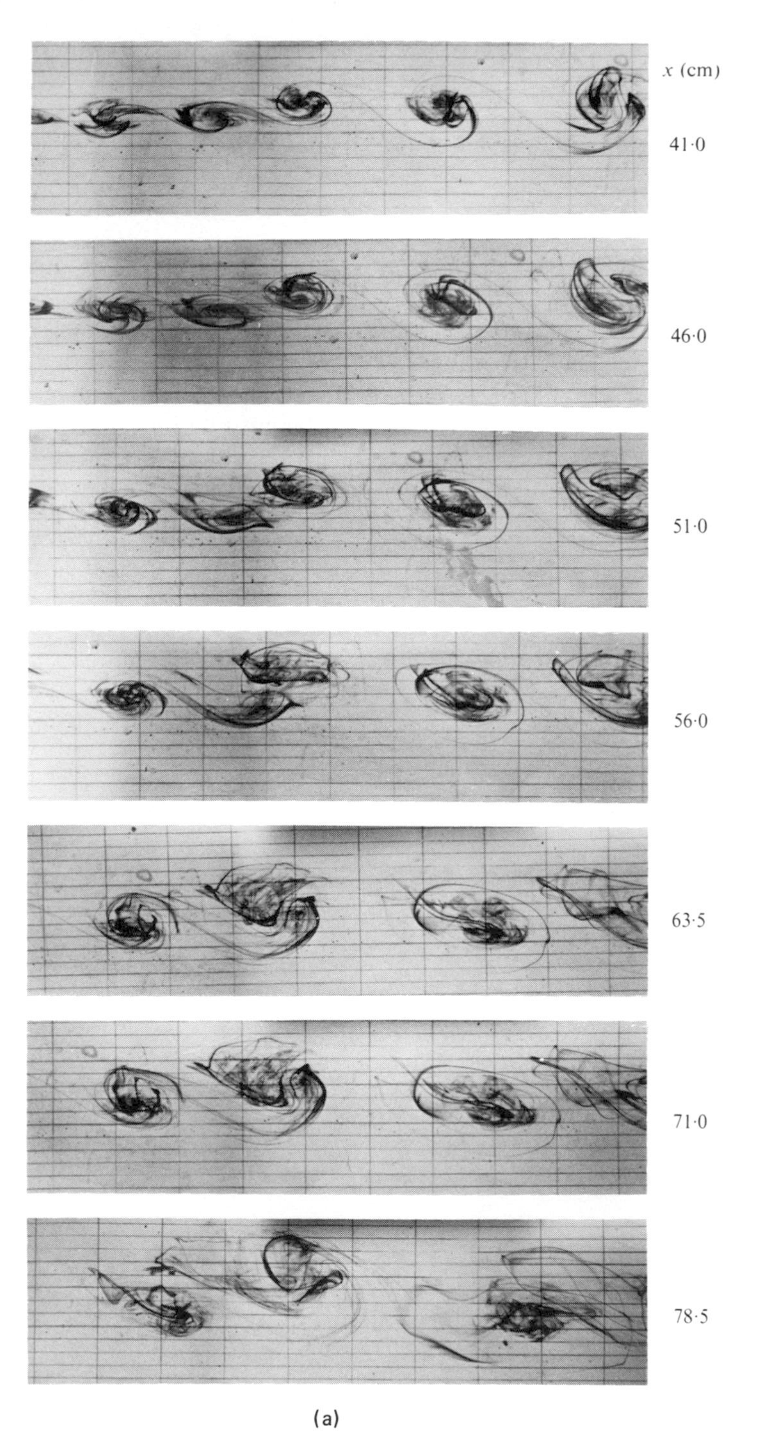

(a)

Figure 23-15 (*Continued*)

Small scale eddies are also evident in these last two figures. In particular note that the secondary instability produces longitudinal striations in the plan view of Fig. 23-14.

23.13 CONCLUSION

Turbulence is a unique phenomenon in the behavior of fields. It is characterized by irregular fluctuations that are self-sustaining. Vorticity exists in the flow before turbulence develops, and a mean vorticity is necessary to sustain it. On the other hand, the large turbulent motions appear to grow by inertia and pressure effects without regard for the viscosity of the fluid. Viscous forces play a subordinate role in limiting the size of the smallest eddies. Two flows at different Reynolds numbers have the same large-scale features, but the range of eddy sizes is larger when the viscosity is lower.

A Properties of Fluid

	Temperature	Pressure	Density	Kinematic viscosity
Air	15°C	101.3 kPa	1.225 kg/m^3	14.5×10^{-6} m^2/sec
	59°F	2116 psf	0.07648 lb_m/ft^3	156×10^{-6} ft^2/sec
Water	15°C	1497 Pa[a]	999.2 kg/m^3	1.138×10^{-6} m^2/sec
	59°F	31.27 psf	62.38 lb_m/ft^3	12.25×10^{-6} ft^2/sec

	Temperature	Specific gravity	Kinematic viscosity
Ethyl alcohol	15°C	0.79	1.70×10^{-6} m^2/sec
Gasoline	15°C	0.68–0.74	$0.46–0.88 \times 10^{-6}$ m^2/sec
Oil (SAE 30)	15°C	0.88–0.94	5.0×10^{-4} m^2/sec
	99°C		1.0×10^{-4} m^2/sec
Glycerine 100%	20°C	1.26	6.5×10^{-4} m^2/sec
50% H_2O	20°C	1.13	5.3×10^{-4} m^2/sec
Mercury	15°C	13.6	12×10^{-4} m^2/sec

Conversion Factors and Constants

Standard gravity acceleration 9.807 m/sec^2

32.17 ft/sec^2

Length 1 m = 3.281 ft

Mass 1 kg = 2.205 lb_m

[a] Vapor.

B Differential Operations in Cylindrical and Spherical Coordinates

Table B-1 Differential Operations in Cylindrical Coordinates (r, θ, z)

$$(\nabla \cdot \mathbf{v}) = \frac{1}{r}\frac{\partial}{\partial r}(rv_r) + \frac{1}{r}\frac{\partial v_\theta}{\partial \theta} + \frac{\partial v_z}{\partial z}$$

$$(\nabla^2 s) = \frac{1}{r}\frac{\partial}{\partial r}\left(r\frac{\partial s}{\partial r}\right) + \frac{1}{r^2}\frac{\partial^2 s}{\partial \theta^2} + \frac{\partial^2 s}{\partial z^2}$$

$$\begin{aligned}(\boldsymbol{\tau}:\nabla \mathbf{v}) = {} & \tau_{rr}\left(\frac{\partial v_r}{\partial r}\right) + \tau_{r\theta}\left(\frac{1}{r}\frac{\partial v_r}{\partial \theta} - \frac{v_\theta}{r}\right) + \tau_{rz}\left(\frac{\partial v_r}{\partial z}\right) \\ & + \tau_{\theta r}\left(\frac{\partial v_\theta}{\partial r}\right) + \tau_{\theta\theta}\left(\frac{1}{r}\frac{\partial v_\theta}{\partial \theta} + \frac{v_r}{r}\right) + \tau_{\theta z}\left(\frac{\partial v_\theta}{\partial z}\right) \\ & + \tau_{zr}\left(\frac{\partial v_z}{\partial r}\right) + \tau_{z\theta}\left(\frac{1}{r}\frac{\partial v_z}{\partial \theta}\right) + \tau_{zz}\left(\frac{\partial v_z}{\partial z}\right)\end{aligned}$$

$$[\nabla s]_r = \frac{\partial s}{\partial r} \qquad [\nabla \times \mathbf{v}]_r = \frac{1}{r}\frac{\partial v_z}{\partial \theta} - \frac{\partial v_\theta}{\partial z}$$

$$[\nabla s]_\theta = \frac{1}{r}\frac{\partial s}{\partial \theta} \qquad [\nabla \times \mathbf{v}]_\theta = \frac{\partial v_r}{\partial z} - \frac{\partial v_z}{\partial r}$$

$$[\nabla s]_z = \frac{\partial s}{\partial z} \qquad [\nabla \times \mathbf{v}]_z = \frac{1}{r}\frac{\partial}{\partial r}(rv_\theta) - \frac{1}{r}\frac{\partial v_r}{\partial \theta}$$

$$[\nabla \cdot \boldsymbol{\tau}]_r = \frac{1}{r}\frac{\partial}{\partial r}(r\tau_{rr}) + \frac{1}{r}\frac{\partial}{\partial \theta}\tau_{\theta r} + \frac{\partial}{\partial z}\tau_{zr} - \frac{\tau_{\theta\theta}}{r}$$

$$[\nabla \cdot \boldsymbol{\tau}]_\theta = \frac{1}{r^2}\frac{\partial}{\partial r}(r^2\tau_{r\theta}) + \frac{1}{r}\frac{\partial}{\partial \theta}\tau_{\theta\theta} + \frac{\partial}{\partial z}\tau_{z\theta} + \frac{\tau_{\theta r} - \tau_{r\theta}}{r}$$

$$[\nabla \cdot \boldsymbol{\tau}]_z = \frac{1}{r}\frac{\partial}{\partial r}(r\tau_{rz}) + \frac{1}{r}\frac{\partial}{\partial \theta}\tau_{\theta z} + \frac{\partial}{\partial z}\tau_{zz}$$

$$[\nabla^2 \mathbf{v}]_r = \frac{\partial}{\partial r}\left(\frac{1}{r}\frac{\partial}{\partial r}(rv_r)\right) + \frac{1}{r^2}\frac{\partial^2 v_r}{\partial \theta^2} + \frac{\partial^2 v_r}{\partial z^2} - \frac{2}{r^2}\frac{\partial v_\theta}{\partial \theta}$$

$$[\nabla^2 \mathbf{v}]_\theta = \frac{\partial}{\partial r}\left(\frac{1}{r}\frac{\partial}{\partial r}(rv_\theta)\right) + \frac{1}{r^2}\frac{\partial^2 v_\theta}{\partial \theta^2} + \frac{\partial^2 v_\theta}{\partial z^2} + \frac{2}{r^2}\frac{\partial v_r}{\partial \theta}$$

$$[\nabla^2 \mathbf{v}]_z = \frac{1}{r}\frac{\partial}{\partial r}\left(r\frac{\partial v_z}{\partial r}\right) + \frac{1}{r^2}\frac{\partial^2 v_z}{\partial \theta^2} + \frac{\partial^2 v_z}{\partial z^2}$$

$$[\mathbf{v} \cdot \nabla \mathbf{w}]_r = v_r\left(\frac{\partial w_r}{\partial r}\right) + v_\theta\left(\frac{1}{r}\frac{\partial w_r}{\partial \theta} - \frac{w_\theta}{r}\right) + v_z\left(\frac{\partial w_r}{\partial z}\right)$$

$$[\mathbf{v} \cdot \nabla \mathbf{w}]_\theta = v_r\left(\frac{\partial w_\theta}{\partial r}\right) + v_\theta\left(\frac{1}{r}\frac{\partial w_\theta}{\partial \theta} + \frac{w_r}{r}\right) + v_z\left(\frac{\partial w_\theta}{\partial z}\right)$$

Table B-1 *(Continued)*

$$[\mathbf{v} \cdot \nabla \mathbf{w}]_z = v_r\left(\frac{\partial w_z}{\partial r}\right) + v_\theta\left(\frac{1}{r}\frac{\partial w_z}{\partial \theta}\right) + v_z\left(\frac{\partial w_z}{\partial z}\right)$$

$$\{\nabla \mathbf{v}\}_{rr} = \frac{\partial v_r}{\partial r}$$

$$\{\nabla \mathbf{v}\}_{r\theta} = \frac{\partial v_\theta}{\partial r}$$

$$\{\nabla \mathbf{v}\}_{rz} = \frac{\partial v_z}{\partial r}$$

$$\{\nabla \mathbf{v}\}_{\theta r} = \frac{1}{r}\frac{\partial v_r}{\partial \theta} - \frac{v_\theta}{r}$$

$$\{\nabla \mathbf{v}\}_{\theta\theta} = \frac{1}{r}\frac{\partial v_\theta}{\partial \theta} + \frac{v_r}{r}$$

$$\{\nabla \mathbf{v}\}_{\theta z} = \frac{1}{r}\frac{\partial v_z}{\partial \theta}$$

$$\{\nabla \mathbf{v}\}_{zr} = \frac{\partial v_r}{\partial z}$$

$$\{\nabla \mathbf{v}\}_{z\theta} = \frac{\partial v_\theta}{\partial z}$$

$$\{\nabla \mathbf{v}\}_{zz} = \frac{\partial v_z}{\partial z}$$

$$\{\mathbf{v} \cdot \nabla \boldsymbol{\tau}\}_{rr} = (\mathbf{v} \cdot \nabla)\tau_{rr} - \frac{v_\theta}{r}(\tau_{r\theta} + \tau_{\theta r})$$

$$\{\mathbf{v} \cdot \nabla \boldsymbol{\tau}\}_{r\theta} = (\mathbf{v} \cdot \nabla)\tau_{r\theta} + \frac{v_\theta}{r}(\tau_{rr} - \tau_{\theta\theta})$$

$$\{\mathbf{v} \cdot \nabla \boldsymbol{\tau}\}_{rz} = (\mathbf{v} \cdot \nabla)\tau_{rz} - \frac{v_\theta}{r}\tau_{\theta z}$$

$$\{\mathbf{v} \cdot \nabla \boldsymbol{\tau}\}_{\theta r} = (\mathbf{v} \cdot \nabla)\tau_{\theta r} + \frac{v_\theta}{r}(\tau_{rr} - \tau_{\theta\theta})$$

$$\{\mathbf{v} \cdot \nabla \boldsymbol{\tau}\}_{\theta\theta} = (\mathbf{v} \cdot \nabla)\tau_{\theta\theta} + \frac{v_\theta}{r}(\tau_{r\theta} + \tau_{\theta r})$$

$$\{\mathbf{v} \cdot \nabla \boldsymbol{\tau}\}_{\theta z} = (\mathbf{v} \cdot \nabla)\tau_{\theta z} + \frac{v_\theta}{r}\tau_{rz}$$

$$\{\mathbf{v} \cdot \nabla \boldsymbol{\tau}\}_{zr} = (\mathbf{v} \cdot \nabla)\tau_{zr} - \frac{v_\theta}{r}\tau_{z\theta}$$

$$\{\mathbf{v} \cdot \nabla \boldsymbol{\tau}\}_{z\theta} = (\mathbf{v} \cdot \nabla)\tau_{z\theta} + \frac{v_\theta}{r}\tau_{zr}$$

$$\{\mathbf{v} \cdot \nabla \boldsymbol{\tau}\}_{zz} = (\mathbf{v} \cdot \nabla)\tau_{zz}$$

where the operator $(\mathbf{v} \cdot \nabla) = v_r\frac{\partial}{\partial r} + \frac{v_\theta}{r}\frac{\partial}{\partial \theta} + v_z\frac{\partial}{\partial z}$

Adapted from Bird, Armstrong, and Hassager (1977) by permission of John Wiley & Sons.

Table B-2 Summary of Differential Operations in Spherical Coordinates (r, θ, φ)

$$(\nabla \cdot \mathbf{v}) = \frac{1}{r^2}\frac{\partial}{\partial r}(r^2 v_r) + \frac{1}{r\sin\theta}\frac{\partial}{\partial\theta}(v_\theta \sin\theta) + \frac{1}{r\sin\theta}\frac{\partial v_\varphi}{\partial\varphi}$$

$$(\nabla^2 s) = \frac{1}{r^2}\frac{\partial}{\partial r}\left(r^2\frac{\partial s}{\partial r}\right) + \frac{1}{r^2\sin\theta}\frac{\partial}{\partial\theta}\left(\sin\theta\frac{\partial s}{\partial\theta}\right) + \frac{1}{r^2\sin^2\theta}\frac{\partial^2 s}{\partial\varphi^2}$$

$$(\boldsymbol{\tau}:\nabla\mathbf{v}) = \tau_{rr}\left(\frac{\partial v_r}{\partial r}\right) + \tau_{r\theta}\left(\frac{1}{r}\frac{\partial v_r}{\partial\theta} - \frac{v_\theta}{r}\right) + \tau_{r\varphi}\left(\frac{1}{r\sin\theta}\frac{\partial v_r}{\partial\varphi} - \frac{v_\varphi}{r}\right)$$

$$+\tau_{\theta r}\left(\frac{\partial v_\theta}{\partial r}\right) + \tau_{\theta\theta}\left(\frac{1}{r}\frac{\partial v_\theta}{\partial\theta} + \frac{v_r}{r}\right) + \tau_{\theta\varphi}\left(\frac{1}{r\sin\theta}\frac{\partial v_\theta}{\partial\varphi} - \frac{v_\varphi}{r}\cot\theta\right)$$

$$+\tau_{\varphi r}\left(\frac{\partial v_\varphi}{\partial r}\right) + \tau_{\varphi\theta}\left(\frac{1}{r}\frac{\partial v_\varphi}{\partial\theta}\right) + \tau_{\varphi\varphi}\left(\frac{1}{r\sin\theta}\frac{\partial v_\varphi}{\partial\varphi} + \frac{v_r}{r} + \frac{v_\theta}{r}\cot\theta\right)$$

$$[\nabla s]_r = \frac{\partial s}{\partial r} \qquad [\nabla\times\mathbf{v}]_r = \frac{1}{r\sin\theta}\frac{\partial}{\partial\theta}(v_\varphi\sin\theta) - \frac{1}{r\sin\theta}\frac{\partial v_\theta}{\partial\varphi}$$

$$[\nabla s]_\theta = \frac{1}{r}\frac{\partial s}{\partial\theta} \qquad [\nabla\times\mathbf{v}]_\theta = \frac{1}{r\sin\theta}\frac{\partial v_r}{\partial\varphi} - \frac{1}{r}\frac{\partial}{\partial r}(r v_\varphi)$$

$$[\nabla s]_\varphi = \frac{1}{r\sin\theta}\frac{\partial s}{\partial\varphi} \qquad [\nabla\times\mathbf{v}]_\varphi = \frac{1}{r}\frac{\partial}{\partial r}(r v_\theta) - \frac{1}{r}\frac{\partial v_r}{\partial\theta}$$

$$[\nabla\cdot\boldsymbol{\tau}]_r = \frac{1}{r^2}\frac{\partial}{\partial r}(r^2\tau_{rr}) + \frac{1}{r\sin\theta}\frac{\partial}{\partial\theta}(\tau_{\theta r}\sin\theta) + \frac{1}{r\sin\theta}\frac{\partial}{\partial\varphi}\tau_{\varphi r} - \frac{\tau_{\theta\theta}+\tau_{\varphi\varphi}}{r}$$

$$[\nabla\cdot\boldsymbol{\tau}]_\theta = \frac{1}{r^3}\frac{\partial}{\partial r}(r^3\tau_{r\theta}) + \frac{1}{r\sin\theta}\frac{\partial}{\partial\theta}(\tau_{\theta\theta}\sin\theta) + \frac{1}{r\sin\theta}\frac{\partial}{\partial\varphi}\tau_{\varphi\theta} + \frac{(\tau_{\theta r}-\tau_{r\theta}) - \tau_{\varphi\varphi}\cot\theta}{r}$$

$$[\nabla\cdot\boldsymbol{\tau}]_\varphi = \frac{1}{r^3}\frac{\partial}{\partial r}(r^3\tau_{r\varphi}) + \frac{1}{r\sin\theta}\frac{\partial}{\partial\theta}(\tau_{\theta\varphi}\sin\theta) + \frac{1}{r\sin\theta}\frac{\partial}{\partial\varphi}\tau_{\varphi\varphi} + \frac{(\tau_{\varphi r}-\tau_{r\varphi}) + \tau_{\varphi\theta}\cot\theta}{r}$$

$$[\nabla^2\mathbf{v}]_r = \frac{\partial}{\partial r}\left(\frac{1}{r^2}\frac{\partial}{\partial r}(r^2 v_r)\right) + \frac{1}{r^2\sin\theta}\frac{\partial}{\partial\theta}\left(\sin\theta\frac{\partial v_r}{\partial\theta}\right) + \frac{1}{r^2\sin^2\theta}\frac{\partial^2 v_r}{\partial\varphi^2}$$

$$-\frac{2}{r^2\sin\theta}\frac{\partial}{\partial\theta}(v_\theta\sin\theta) - \frac{2}{r^2\sin\theta}\frac{\partial v_\varphi}{\partial\varphi}$$

$$[\nabla^2\mathbf{v}]_\theta = \frac{1}{r^2}\frac{\partial}{\partial r}\left(r^2\frac{\partial v_\theta}{\partial r}\right) + \frac{1}{r^2}\frac{\partial}{\partial\theta}\left(\frac{1}{\sin\theta}\frac{\partial}{\partial\theta}(v_\theta\sin\theta)\right)$$

$$+\frac{1}{r^2\sin^2\theta}\frac{\partial^2 v_\theta}{\partial\varphi^2} + \frac{2}{r^2}\frac{\partial v_r}{\partial\theta} - \frac{2\cot\theta}{r^2\sin\theta}\frac{\partial v_\varphi}{\partial\varphi}$$

$$[\nabla^2\mathbf{v}]_\varphi = \frac{1}{r^2}\frac{\partial}{\partial r}\left(r^2\frac{\partial v_\varphi}{\partial r}\right) + \frac{1}{r^2}\frac{\partial}{\partial\theta}\left(\frac{1}{\sin\theta}\frac{\partial}{\partial\theta}(v_\varphi\sin\theta)\right)$$

$$+\frac{1}{r^2\sin^2\theta}\frac{\partial^2 v_\varphi}{\partial\varphi^2} + \frac{2}{r^2\sin\theta}\frac{\partial v_r}{\partial\varphi} + \frac{2\cot\theta}{r^2\sin\theta}\frac{\partial v_\theta}{\partial\varphi}$$

$$[\mathbf{v}\cdot\nabla\mathbf{w}]_r = v_r\left(\frac{\partial w_r}{\partial r}\right) + v_\theta\left(\frac{1}{r}\frac{\partial w_r}{\partial\theta} - \frac{w_\theta}{r}\right) + v_\varphi\left(\frac{1}{r\sin\theta}\frac{\partial w_r}{\partial\varphi} - \frac{w_\varphi}{r}\right)$$

$$[\mathbf{v}\cdot\nabla\mathbf{w}]_\theta = v_r\left(\frac{\partial w_\theta}{\partial r}\right) + v_\theta\left(\frac{1}{r}\frac{\partial w_\theta}{\partial\theta} + \frac{w_r}{r}\right) + v_\varphi\left(\frac{1}{r\sin\theta}\frac{\partial w_\theta}{\partial\varphi} - \frac{w_\varphi}{r}\cot\theta\right)$$

Table B-2 (*Continued*)

$$[\mathbf{v}\cdot\nabla\mathbf{w}]_{\varphi} = v_r\left(\frac{\partial w_\varphi}{\partial r}\right) + v_\theta\left(\frac{1}{r}\frac{\partial w_\varphi}{\partial \theta}\right) + v_\varphi\left(\frac{1}{r\sin\theta}\frac{\partial w_\varphi}{\partial \varphi} + \frac{w_r}{r} + \frac{w_\theta}{r}\cot\theta\right)$$

$$\{\nabla\mathbf{v}\}_{rr} = \frac{\partial v_r}{\partial r}$$

$$\{\nabla\mathbf{v}\}_{r\theta} = \frac{\partial v_\theta}{\partial r}$$

$$\{\nabla\mathbf{v}\}_{r\varphi} = \frac{\partial v_\varphi}{\partial r}$$

$$\{\nabla\mathbf{v}\}_{\theta r} = \frac{1}{r}\frac{\partial v_r}{\partial \theta} - \frac{v_\theta}{r}$$

$$\{\nabla\mathbf{v}\}_{\theta\theta} = \frac{1}{r}\frac{\partial v_\theta}{\partial \theta} + \frac{v_r}{r}$$

$$\{\nabla\mathbf{v}\}_{\theta\varphi} = \frac{1}{r}\frac{\partial v_\varphi}{\partial \theta}$$

$$\{\nabla\mathbf{v}\}_{\varphi r} = \frac{1}{r\sin\theta}\frac{\partial v_r}{\partial \varphi} - \frac{v_\varphi}{r}$$

$$\{\nabla\mathbf{v}\}_{\varphi\theta} = \frac{1}{r\sin\theta}\frac{\partial v_\theta}{\partial \varphi} - \frac{v_\varphi}{r}\cot\theta$$

$$\{\nabla\mathbf{v}\}_{\varphi\varphi} = \frac{1}{r\sin\theta}\frac{\partial v_\varphi}{\partial \varphi} + \frac{v_r}{r} + \frac{v_\theta}{r}\cot\theta$$

$$\{\mathbf{v}\cdot\nabla\boldsymbol{\tau}\}_{rr} = (\mathbf{v}\cdot\nabla)\tau_{rr} - \left(\frac{v_\theta}{r}\right)(\tau_{r\theta} + \tau_{\theta r}) - \left(\frac{v_\varphi}{r}\right)(\tau_{r\varphi} + \tau_{\varphi r})$$

$$\{\mathbf{v}\cdot\nabla\boldsymbol{\tau}\}_{r\theta} = (\mathbf{v}\cdot\nabla)\tau_{r\theta} + \left(\frac{v_\theta}{r}\right)(\tau_{rr} - \tau_{\theta\theta}) - \left(\frac{v_\varphi}{r}\right)(\tau_{\varphi\theta} + \tau_{r\varphi}\cot\theta)$$

$$\{\mathbf{v}\cdot\nabla\boldsymbol{\tau}\}_{r\varphi} = (\mathbf{v}\cdot\nabla)\tau_{r\varphi} - \left(\frac{v_\theta}{r}\right)\tau_{\theta\varphi} + \left(\frac{v_\varphi}{r}\right)[(\tau_{rr} - \tau_{\varphi\varphi}) + \tau_{r\theta}\cot\theta]$$

$$\{\mathbf{v}\cdot\nabla\boldsymbol{\tau}\}_{\theta r} = (\mathbf{v}\cdot\nabla)\tau_{\theta r} + \left(\frac{v_\theta}{r}\right)(\tau_{rr} - \tau_{\theta\theta}) - \left(\frac{v_\varphi}{r}\right)(\tau_{\theta\varphi} + \tau_{\varphi r}\cot\theta)$$

$$\{\mathbf{v}\cdot\nabla\boldsymbol{\tau}\}_{\theta\theta} = (\mathbf{v}\cdot\nabla)\tau_{\theta\theta} + \left(\frac{v_\theta}{r}\right)(\tau_{r\theta} + \tau_{\theta r}) - \left(\frac{v_\varphi}{r}\right)(\tau_{\theta\varphi} + \tau_{\varphi\theta})\cot\theta$$

$$\{\mathbf{v}\cdot\nabla\boldsymbol{\tau}\}_{\theta\varphi} = (\mathbf{v}\cdot\nabla)\tau_{\theta\varphi} + \left(\frac{v_\theta}{r}\right)\tau_{r\varphi} + \left(\frac{v_\varphi}{r}\right)[\tau_{\theta r} + (\tau_{\theta\theta} - \tau_{\varphi\varphi})\cot\theta]$$

$$\{\mathbf{v}\cdot\nabla\boldsymbol{\tau}\}_{\varphi r} = (\mathbf{v}\cdot\nabla)\tau_{\varphi r} - \left(\frac{v_\theta}{r}\right)\tau_{\varphi\theta} + \left(\frac{v_\varphi}{r}\right)[(\tau_{rr} - \tau_{\varphi\varphi}) + \tau_{\theta r}\cot\theta]$$

$$\{\mathbf{v}\cdot\nabla\boldsymbol{\tau}\}_{\varphi\theta} = (\mathbf{v}\cdot\nabla)\tau_{\varphi\theta} + \left(\frac{v_\theta}{r}\right)\tau_{\varphi r} + \left(\frac{v_\varphi}{r}\right)[\tau_{r\theta} + (\tau_{\theta\theta} - \tau_{\varphi\varphi})\cot\theta]$$

$$\{\mathbf{v}\cdot\nabla\boldsymbol{\tau}\}_{\varphi\varphi} = (\mathbf{v}\cdot\nabla)\tau_{\varphi\varphi} + \left(\frac{v_\theta}{r}\right)(\tau_{r\varphi} + \tau_{\varphi r}) + \left(\frac{v_\varphi}{r}\right)(\tau_{\theta\varphi} + \tau_{\varphi\theta})\cot\theta$$

where the operator $(\mathbf{v}\cdot\nabla) = v_r\frac{\partial}{\partial r} + \frac{v_\theta}{r}\frac{\partial}{\partial\theta} + \frac{v_\varphi}{r\sin\theta}\frac{\partial}{\partial\varphi}$

Adapted from Bird, Armstrong, and Hassager (1977) by permission of John Wiley & Sons.

C Basic Equations in Rectangular, Cylindrical, and Spherical Coordinates

Table C-1 The Equation of Continuity

Rectangular Coordinates (x, y, z):

$$\frac{\partial \rho}{\partial t} + \frac{\partial}{\partial x}(\rho v_x) + \frac{\partial}{\partial y}(\rho v_y) + \frac{\partial}{\partial z}(\rho v_z) = 0$$

Cylindrical Coordinates (r, θ, z):

$$\frac{\partial \rho}{\partial t} + \frac{1}{r}\frac{\partial}{\partial r}(\rho r v_r) + \frac{1}{r}\frac{\partial}{\partial \theta}(\rho v_\theta) + \frac{\partial}{\partial z}(\rho v_z) = 0$$

Spherical Coordinates (r, θ, φ):

$$\frac{\partial \rho}{\partial t} + \frac{1}{r^2}\frac{\partial}{\partial r}(\rho r^2 v_r) + \frac{1}{r\sin\theta}\frac{\partial}{\partial \theta}(\rho v_\theta \sin\theta) + \frac{1}{r\sin\theta}\frac{\partial}{\partial \varphi}(\rho v_\varphi) = 0$$

Adapted from Bird, Stewart, and Lightfoot (1960) by permission of John Wiley & Sons.

Table C-2 Components of the Rate-of-Strain Tensor $S_{ij} = \partial_{(i}v_{j)} = \frac{1}{2}\partial_i v_j + \frac{1}{2}\partial_j v_i$

Rectangular Coordinates (x, y, z):

$$S_{xx} = \frac{\partial v_x}{\partial x} \qquad S_{yx} = S_{xy} = \frac{1}{2}\left[\frac{\partial v_y}{\partial x} + \frac{\partial v_x}{\partial y}\right]$$

$$S_{yy} = \frac{\partial v_y}{\partial y} \qquad S_{zy} = S_{yz} = \frac{1}{2}\left[\frac{\partial v_z}{\partial y} + \frac{\partial v_y}{\partial z}\right]$$

$$S_{zz} = \frac{\partial v_z}{\partial z} \qquad S_{xz} = S_{zx} = \frac{1}{2}\left[\frac{\partial v_x}{\partial z} + \frac{\partial v_z}{\partial x}\right]$$

Cylindrical Coordinates (r, θ, z):

$$S_{rr} = \frac{\partial v_r}{\partial r} \qquad S_{\theta r} = S_{r\theta} = \frac{1}{2}\left[r\frac{\partial}{\partial r}\left(\frac{v_\theta}{r}\right) + \frac{1}{r}\frac{\partial v_r}{\partial \theta}\right]$$

$$S_{\theta\theta} = \frac{1}{r}\frac{\partial v_\theta}{\partial \theta} + \frac{v_r}{r} \qquad S_{z\theta} = S_{\theta z} = \frac{1}{2}\left[\frac{1}{r}\frac{\partial v_z}{\partial \theta} + \frac{\partial v_\theta}{\partial z}\right]$$

$$S_{zz} = \frac{\partial v_z}{\partial z} \qquad S_{rz} = S_{zr} = \frac{1}{2}\left[\frac{\partial v_r}{\partial z} + \frac{\partial v_z}{\partial r}\right]$$

Spherical Coordinates (r, θ, φ):

$$S_{rr} = \frac{\partial v_r}{\partial r} \qquad S_{\theta r} = S_{r\theta} = \frac{1}{2}\left[r\frac{\partial}{\partial r}\left(\frac{v_\theta}{r}\right) + \frac{1}{r}\frac{\partial v_r}{\partial \theta}\right]$$

$$S_{\theta\theta} = \frac{1}{r}\frac{\partial v_\theta}{\partial \theta} + \frac{v_r}{r} \qquad S_{\varphi\theta} = S_{\theta\varphi} = \frac{1}{2}\left[\frac{\sin\theta}{r}\frac{\partial}{\partial \theta}\left(\frac{v_\varphi}{\sin\theta}\right) + \frac{1}{r\sin\theta}\frac{\partial v_\theta}{\partial \varphi}\right]$$

$$S_{\varphi\varphi} = \frac{1}{r\sin\theta}\frac{\partial v_\varphi}{\partial \varphi} + \frac{v_r}{r} + \frac{v_\theta \cot\theta}{r} \qquad S_{r\varphi} = S_{\varphi r} = \frac{1}{2}\left[\frac{1}{r\sin\theta}\frac{\partial v_r}{\partial \varphi} + r\frac{\partial}{\partial r}\left(\frac{v_\varphi}{r}\right)\right]$$

Adapted from Bird, Armstrong, and Hassager (1977) by permission of John Wiley & Sons.

Table C-3 Components of the Stress Tensor for Newtonian Fluids

Rectangular Coordinates (x, y, z)	Cylindrical Coordinates (r, θ, z)	Spherical Coordinates (r, θ, φ)
$\tau_{xx} = \mu\left[2\frac{\partial v_x}{\partial x} - \frac{2}{3}(\nabla \cdot \mathbf{v})\right]$	$\tau_{rr} = \mu\left[2\frac{\partial v_r}{\partial r} - \frac{2}{3}(\nabla \cdot \mathbf{v})\right]$	$\tau_{rr} = \mu\left[2\frac{\partial v_r}{\partial r} - \frac{2}{3}(\nabla \cdot \mathbf{v})\right]$
$\tau_{yy} = \mu\left[2\frac{\partial v_y}{\partial y} - \frac{2}{3}(\nabla \cdot \mathbf{v})\right]$	$\tau_{\theta\theta} = \mu\left[2\left(\frac{1}{r}\frac{\partial v_\theta}{\partial \theta} + \frac{v_r}{r}\right) - \frac{2}{3}(\nabla \cdot \mathbf{v})\right]$	$\tau_{\theta\theta} = \mu\left[2\left(\frac{1}{r}\frac{\partial v_\theta}{\partial \theta} + \frac{v_r}{r}\right) - \frac{2}{3}(\nabla \cdot \mathbf{v})\right]$
$\tau_{zz} = \mu\left[2\frac{\partial v_z}{\partial z} - \frac{2}{3}(\nabla \cdot \mathbf{v})\right]$	$\tau_{zz} = \mu\left[2\frac{\partial v_z}{\partial z} - \frac{2}{3}(\nabla \cdot \mathbf{v})\right]$	$\tau_{\varphi\varphi} = \mu\left[2\left(\frac{1}{r\sin\theta}\frac{\partial v_\varphi}{\partial \varphi} + \frac{v_r}{r} + \frac{v_\theta \cot\theta}{r}\right) - \frac{2}{3}(\nabla \cdot \mathbf{v})\right]$
$\tau_{xy} = \tau_{yx} = \mu\left[\frac{\partial v_x}{\partial y} + \frac{\partial v_y}{\partial x}\right]$	$\tau_{r\theta} = \tau_{\theta r} = \mu\left[r\frac{\partial}{\partial r}\left(\frac{v_\theta}{r}\right) + \frac{1}{r}\frac{\partial v_r}{\partial \theta}\right]$	$\tau_{r\theta} = \tau_{\theta r} = \mu\left[r\frac{\partial}{\partial r}\left(\frac{v_\theta}{r}\right) + \frac{1}{r}\frac{\partial v_r}{\partial \theta}\right]$
$\tau_{yz} = \tau_{zy} = \mu\left[\frac{\partial v_y}{\partial z} + \frac{\partial v_z}{\partial y}\right]$	$\tau_{\theta z} = \tau_{z\theta} = \mu\left[\frac{\partial v_\theta}{\partial z} + \frac{1}{r}\frac{\partial v_z}{\partial \theta}\right]$	$\tau_{\theta\varphi} = \tau_{\varphi\theta} = \mu\left[\frac{\sin\theta}{r}\frac{\partial}{\partial \theta}\left(\frac{v_\varphi}{\sin\theta}\right) + \frac{1}{r\sin\theta}\frac{\partial v_\theta}{\partial \varphi}\right]$
$\tau_{zx} = \tau_{xz} = \mu\left[\frac{\partial v_z}{\partial x} + \frac{\partial v_z}{\partial z}\right]$	$\tau_{zr} = \tau_{rz} = \mu\left[\frac{\partial v_z}{\partial r} + \frac{\partial v_r}{\partial z}\right]$	$\tau_{\varphi r} = \tau_{r\varphi} = \mu\left[\frac{1}{r\sin\theta}\frac{\partial v_r}{\partial \varphi} + r\frac{\partial}{\partial r}\left(\frac{v_\varphi}{r}\right)\right]$
$(\nabla \cdot \mathbf{v}) = \frac{\partial v_x}{\partial x} + \frac{\partial v_y}{\partial y} + \frac{\partial v_z}{\partial z}$	$(\nabla \cdot \mathbf{v}) = \frac{1}{r}\frac{\partial}{\partial r}(rv_r) + \frac{1}{r}\frac{\partial v_\theta}{\partial \theta} + \frac{\partial v_z}{\partial z}$	$(\nabla \cdot \mathbf{v}) = \frac{1}{r^2}\frac{\partial}{\partial r}(r^2 v_r) + \frac{1}{r\sin\theta}\frac{\partial}{\partial \theta}(v_\theta \sin\theta) + \frac{1}{r\sin\theta}\frac{\partial v_\varphi}{\partial \varphi}$

Table C-4 Momentum Equations in Terms of τ[a]

Rectangular Coordinates (x, y, z):

$$\rho\left(\frac{\partial v_x}{\partial t} + v_x\frac{\partial v_x}{\partial x} + v_y\frac{\partial v_x}{\partial y} + v_z\frac{\partial v_x}{\partial z}\right) = \left[\frac{\partial}{\partial x}\tau_{xx} + \frac{\partial}{\partial y}\tau_{yx} + \frac{\partial}{\partial z}\tau_{zx}\right] - \frac{\partial p}{\partial x} + \rho g_x$$

$$\rho\left(\frac{\partial v_y}{\partial t} + v_x\frac{\partial v_y}{\partial x} + v_y\frac{\partial v_y}{\partial y} + v_z\frac{\partial v_y}{\partial z}\right) = \left[\frac{\partial}{\partial x}\tau_{xy} + \frac{\partial}{\partial y}\tau_{yy} + \frac{\partial}{\partial z}\tau_{zy}\right] - \frac{\partial p}{\partial y} + \rho g_y$$

$$\rho\left(\frac{\partial v_z}{\partial t} + v_x\frac{\partial v_z}{\partial x} + v_y\frac{\partial v_z}{\partial y} + v_z\frac{\partial v_z}{\partial z}\right) = \left[\frac{\partial}{\partial x}\tau_{xz} + \frac{\partial}{\partial y}\tau_{yz} + \frac{\partial}{\partial z}\tau_{zz}\right] - \frac{\partial p}{\partial z} + \rho g_z$$

Cylindrical Coordinates (r, θ, z):

$$\rho\left(\frac{\partial v_r}{\partial t} + v_r\frac{\partial v_r}{\partial r} + \frac{v_\theta}{r}\frac{\partial v_r}{\partial \theta} - \frac{v_\theta^2}{r} + v_z\frac{\partial v_r}{\partial z}\right) = \left[\frac{1}{r}\frac{\partial}{\partial r}(r\tau_{rr}) + \frac{1}{r}\frac{\partial}{\partial \theta}\tau_{\theta r} + \frac{\partial}{\partial z}\tau_{zr} - \frac{\tau_{\theta\theta}}{r}\right] - \frac{\partial p}{\partial r} + \rho g_r$$

$$\rho\left(\frac{\partial v_\theta}{\partial t} + v_r\frac{\partial v_\theta}{\partial r} + \frac{v_\theta}{r}\frac{\partial v_\theta}{\partial \theta} + \frac{v_r v_\theta}{r} + v_z\frac{\partial v_\theta}{\partial z}\right) = \left[\frac{1}{r^2}\frac{\partial}{\partial r}(r^2\tau_{r\theta}) + \frac{1}{r}\frac{\partial}{\partial \theta}\tau_{\theta\theta} + \frac{\partial}{\partial z}\tau_{z\theta} + \frac{\tau_{\theta r} - \tau_{r\theta}}{r}\right] - \frac{1}{r}\frac{\partial p}{\partial \theta} + \rho g_\theta$$

$$\rho\left(\frac{\partial v_z}{\partial t} + v_r\frac{\partial v_z}{\partial r} + \frac{v_\theta}{r}\frac{\partial v_z}{\partial \theta} + v_z\frac{\partial v_z}{\partial z}\right) = \left[\frac{1}{r}\frac{\partial}{\partial r}(r\tau_{rz}) + \frac{1}{r}\frac{\partial}{\partial \theta}\tau_{\theta z} + \frac{\partial}{\partial z}\tau_{zz}\right] - \frac{\partial p}{\partial z} + \rho g_z$$

Spherical Coordinates (r, θ, φ):

$$\rho\left(\frac{\partial v_r}{\partial t} + v_r\frac{\partial v_r}{\partial r} + \frac{v_\theta}{r}\frac{\partial v_r}{\partial \theta} + \frac{v_\varphi}{r\sin\theta}\frac{\partial v_r}{\partial \varphi} - \frac{v_\theta^2 + v_\varphi^2}{r}\right) = \left[\frac{1}{r^2}\frac{\partial}{\partial r}(r^2\tau_{rr}) + \frac{1}{r\sin\theta}\frac{\partial}{\partial \theta}(\tau_{\theta r}\sin\theta)\right] + \left[\frac{1}{r\sin\theta}\frac{\partial}{\partial \varphi}\tau_{\varphi r} - \frac{\tau_{\theta\theta} + \tau_{\varphi\varphi}}{r}\right] - \frac{\partial p}{\partial r} + \rho g_r$$

$$\rho\left(\frac{\partial v_\theta}{\partial t} + v_r\frac{\partial v_\theta}{\partial r} + \frac{v_\theta}{r}\frac{\partial v_\theta}{\partial \theta} + \frac{v_\varphi}{r\sin\theta}\frac{\partial v_\theta}{\partial \varphi} + \frac{v_r v_\theta}{r} - \frac{v_\varphi^2\cot\theta}{r}\right) = \left[\frac{1}{r^3}\frac{\partial}{\partial r}(r^3\tau_{r\theta}) + \frac{1}{r\sin\theta}\frac{\partial}{\partial \theta}(\tau_{\theta\theta}\sin\theta)\right]$$
$$+ \left[\frac{1}{r\sin\theta}\frac{\partial}{\partial \varphi}\tau_{\varphi\theta} + \frac{(\tau_{\theta r} - \tau_{r\theta}) - \tau_{\varphi\varphi}\cot\theta}{r}\right] - \frac{1}{r}\frac{\partial p}{\partial \theta} + \rho g_\theta$$

$$\rho\left(\frac{\partial v_\varphi}{\partial t} + v_r\frac{\partial v_\varphi}{\partial r} + \frac{v_\theta}{r}\frac{\partial v_\varphi}{\partial \theta} + \frac{v_\varphi}{r\sin\theta}\frac{\partial v_\varphi}{\partial \varphi} + \frac{v_\varphi v_r}{r} + \frac{v_\theta v_\varphi}{r}\cot\theta\right) = \left[\frac{1}{r^3}\frac{\partial}{\partial r}(r^3\tau_{r\varphi}) + \frac{1}{r\sin\theta}\frac{\partial}{\partial \theta}(\tau_{\theta\varphi}\sin\theta)\right]$$
$$+ \left[\frac{1}{r\sin\theta}\frac{\partial}{\partial \varphi}\tau_{\varphi\varphi} + \frac{(\tau_{\varphi r} - \tau_{r\varphi}) + \tau_{\varphi\theta}\cot\theta}{r}\right] - \frac{1}{r\sin\theta}\frac{\partial p}{\partial \varphi} + \rho g_\varphi$$

[a] For symmetric τ set $\tau_{ij} = \tau_{ji}$.

Table C-5 Momentum Equations for a Newtonian Fluid with Constant Density (ρ) and Constant Viscosity (μ)

Rectangular Coordinates (x, y, z):

$$\rho\left(\frac{\partial v_x}{\partial t} + v_x\frac{\partial v_x}{\partial x} + v_y\frac{\partial v_x}{\partial y} + v_z\frac{\partial v_x}{\partial z}\right) = \mu\left[\frac{\partial^2 v_x}{\partial x^2} + \frac{\partial^2 v_x}{\partial y^2} + \frac{\partial^2 v_x}{\partial z^2}\right] - \frac{\partial p}{\partial x} + \rho g_x$$

$$\rho\left(\frac{\partial v_y}{\partial t} + v_x\frac{\partial v_y}{\partial x} + v_y\frac{\partial v_y}{\partial y} + v_z\frac{\partial v_y}{\partial z}\right) = \mu\left[\frac{\partial^2 v_y}{\partial x^2} + \frac{\partial^2 v_y}{\partial y^2} + \frac{\partial^2 v_y}{\partial z^2}\right] - \frac{\partial p}{\partial y} + \rho g_y$$

$$\rho\left(\frac{\partial v_z}{\partial t} + v_x\frac{\partial v_z}{\partial x} + v_y\frac{\partial v_z}{\partial y} + v_z\frac{\partial v_z}{\partial z}\right) = \mu\left[\frac{\partial^2 v_z}{\partial x^2} + \frac{\partial^2 v_z}{\partial y^2} + \frac{\partial^2 v_z}{\partial z^2}\right] - \frac{\partial p}{\partial z} + \rho g_z$$

Cylindrical Coordinates (r, θ, z):

$$\rho\left(\frac{\partial v_r}{\partial t} + v_r\frac{\partial v_r}{\partial r} + \frac{v_\theta}{r}\frac{\partial v_r}{\partial \theta} - \frac{v_\theta^2}{r} + v_z\frac{\partial v_r}{\partial z}\right) = \mu\left[\frac{\partial}{\partial r}\left(\frac{1}{r}\frac{\partial}{\partial r}(r v_r)\right) + \frac{1}{r^2}\frac{\partial^2 v_r}{\partial \theta^2} + \frac{\partial^2 v_r}{\partial z^2} - \frac{2}{r^2}\frac{\partial v_\theta}{\partial \theta}\right] - \frac{\partial p}{\partial r} + \rho g_r$$

$$\rho\left(\frac{\partial v_\theta}{\partial t} + v_r\frac{\partial v_\theta}{\partial r} + \frac{v_\theta}{r}\frac{\partial v_\theta}{\partial \theta} + \frac{v_r v_\theta}{r} + v_z\frac{\partial v_\theta}{\partial z}\right) = \mu\left[\frac{\partial}{\partial r}\left(\frac{1}{r}\frac{\partial}{\partial r}(r v_\theta)\right) + \frac{1}{r^2}\frac{\partial^2 v_\theta}{\partial \theta^2} + \frac{\partial^2 v_\theta}{\partial z^2} + \frac{2}{r^2}\frac{\partial v_r}{\partial \theta}\right] - \frac{1}{r}\frac{\partial p}{\partial \theta} + \rho g_\theta$$

$$\rho\left(\frac{\partial v_z}{\partial t} + v_r\frac{\partial v_z}{\partial r} + \frac{v_\theta}{r}\frac{\partial v_z}{\partial \theta} + v_z\frac{\partial v_z}{\partial z}\right) = \mu\left[\frac{1}{r}\frac{\partial}{\partial r}\left(r\frac{\partial v_z}{\partial r}\right) + \frac{1}{r^2}\frac{\partial^2 v_z}{\partial \theta^2} + \frac{\partial^2 v_z}{\partial z^2}\right] - \frac{\partial p}{\partial z} + \rho g_z$$

Spherical Coordinates (r, θ, φ):

$$\rho\left(\frac{\partial v_r}{\partial t} + v_r\frac{\partial v_r}{\partial r} + \frac{v_\theta}{r}\frac{\partial v_r}{\partial \theta} + \frac{v_\varphi}{r\sin\theta}\frac{\partial v_r}{\partial \varphi} - \frac{v_\theta^2 + v_\varphi^2}{r}\right)$$

$$= \mu\left[\frac{\partial}{\partial r}\left(\frac{1}{r^2}\frac{\partial}{\partial r}(r^2 v_r)\right) + \frac{1}{r^2\sin\theta}\frac{\partial}{\partial\theta}\left(\sin\theta\frac{\partial v_r}{\partial\theta}\right) + \frac{1}{r^2\sin^2\theta}\frac{\partial^2 v_r}{\partial\varphi^2} - \frac{2}{r^2\sin\theta}\frac{\partial}{\partial\theta}(v_\theta\sin\theta) - \frac{2}{r^2\sin\theta}\frac{\partial v_\varphi}{\partial\varphi}\right] - \frac{\partial p}{\partial r} + \rho g_r$$

$$\rho\left(\frac{\partial v_\theta}{\partial t} + v_r\frac{\partial v_\theta}{\partial r} + \frac{v_\theta}{r}\frac{\partial v_\theta}{\partial \theta} + \frac{v_\varphi}{r\sin\theta}\frac{\partial v_\theta}{\partial \varphi} + \frac{v_r v_\theta}{r} - \frac{v_\varphi^2\cot\theta}{r}\right)$$

$$= \mu\left[\frac{1}{r^2}\frac{\partial}{\partial r}\left(r^2\frac{\partial v_\theta}{\partial r}\right) + \frac{1}{r^2}\frac{\partial}{\partial\theta}\left(\frac{1}{\sin\theta}\frac{\partial}{\partial\theta}(v_\theta\sin\theta)\right) + \frac{1}{r^2\sin^2\theta}\frac{\partial^2 v_\theta}{\partial\varphi^2} + \frac{2}{r^2}\frac{\partial v_r}{\partial\theta} - \frac{2\cot\theta}{r^2\sin\theta}\frac{\partial v_\varphi}{\partial\varphi}\right] - \frac{1}{r}\frac{\partial p}{\partial\theta} + \rho g_\theta$$

$$\rho\left(\frac{\partial v_\varphi}{\partial t} + v_r\frac{\partial v_\varphi}{\partial r} + \frac{v_\theta}{r}\frac{\partial v_\varphi}{\partial \theta} + \frac{v_\varphi}{r\sin\theta}\frac{\partial v_\varphi}{\partial \varphi} + \frac{v_\varphi v_r}{r} + \frac{v_\theta v_\varphi}{r}\cot\theta\right)$$

$$= \mu\left[\frac{1}{r^2}\frac{\partial}{\partial r}\left(r^2\frac{\partial v_\varphi}{\partial r}\right) + \frac{1}{r^2}\frac{\partial}{\partial\theta}\left(\frac{1}{\sin\theta}\frac{\partial}{\partial\theta}(v_\varphi\sin\theta)\right) + \frac{1}{r^2\sin^2\theta}\frac{\partial^2 v_\varphi}{\partial\varphi^2} + \frac{2}{r^2\sin\theta}\frac{\partial v_r}{\partial\varphi} + \frac{2\cot\theta}{r^2\sin\theta}\frac{\partial v_\theta}{\partial\varphi}\right] - \frac{1}{r\sin\theta}\frac{\partial p}{\partial\varphi} + \rho g_\varphi$$

D Streamfunction Relations in Rectangular, Cylindrical, and Spherical Coordinates

Table D-1 Streamfunction for Plane Two-Dimensional Flow: Cartesian Coordinates

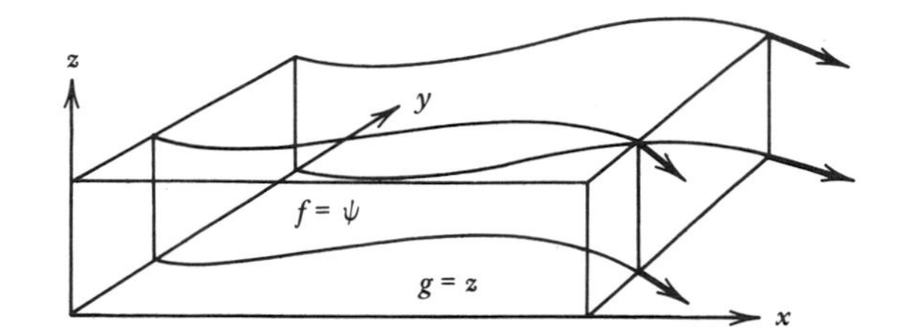

Coordinates: x, y, z

Velocities: $u(x, y), \quad v(x, y), \quad w = 0$

Streamsurfaces: $f = \psi(x, y), \quad g = z, \quad \nabla g = (0, 0, 1)$

Vector potential: $B_x = 0, \quad B_y = 0, \quad B_z = \psi$

$\mathbf{v} = \nabla f \times \nabla g = \nabla \times \mathbf{B}$: $\quad u = \dfrac{\partial \psi}{\partial y}, \quad v = -\dfrac{\partial \psi}{\partial x}$

$-\boldsymbol{\omega} = \nabla^2 \mathbf{B}$: $\quad -\omega_z = \dfrac{\partial^2 \psi}{\partial x^2} + \dfrac{\partial^2 \psi}{\partial y^2} = \nabla^2 \psi$

Vorticity equation: $\quad \dfrac{\partial \omega_z}{\partial t} + \dfrac{\partial \psi}{\partial y} \dfrac{\partial \omega_z}{\partial x} + \dfrac{\nabla \psi}{\partial x} \dfrac{\partial \omega_z}{\partial y} = \nu \nabla^2 \omega_z$

where $\quad -\nabla^2 \omega_z = \psi_{xxxx} + 2\psi_{xxyy} + \psi_{yyyy}$

Flow rate: $\quad Q = (f_2 - f_1)(g_2 - g_1)$

$$g_1 = z_1 = 0, \quad g_2 = z_2 = 1$$

$$Q = \psi_2 - \psi_1$$

Table D-2 Streamfunction for Plane Two-Dimensional Flow: Cylindrical Coordinates

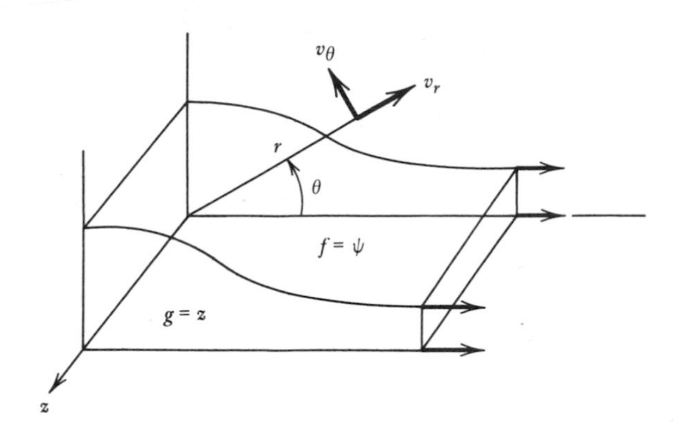

Coordinates: r, θ, z

Velocities: $v_r(r,\theta), \quad v_\theta(r,\theta), \quad v_z = 0$

Streamsurfaces: $f = \psi(r,\theta), \quad g = z, \quad \nabla g = (0,0,1)$

Vector potential: $B_r = 0, \quad B_\theta = 0, \quad B_z = \psi$

$\mathbf{v} = \nabla f \times \nabla g = \nabla \times \mathbf{B}$: $\quad v_r = \dfrac{1}{r}\dfrac{\partial \psi}{\partial \theta}, \quad v_\theta = -\dfrac{\partial \psi}{\partial r}$

$-\boldsymbol{\omega} = \nabla^2 \mathbf{B}$:

$$-\omega_z = \frac{\partial^2 \psi}{\partial r^2} + \frac{1}{r}\frac{\partial \psi}{\partial r} + \frac{1}{r^2}\frac{\partial^2 \psi}{\partial \theta^2}$$

$$= \frac{1}{r}\frac{\partial}{\partial r}\left(r\frac{\partial \psi}{\partial r}\right) + \frac{1}{r^2}\frac{\partial^2 \psi}{\partial \theta^2}$$

Vorticity equation:

$$\frac{\partial \omega_z}{\partial t} + v_r\frac{\partial \omega_z}{\partial r} + v_\theta \frac{1}{r}\frac{\partial \omega_z}{\partial \theta} = \nu \frac{1}{r}\frac{\partial}{\partial r}\left(r\frac{\partial \omega_z}{\partial r}\right) + \nu\frac{1}{r^2}\frac{\partial^2 \omega_z}{\partial \theta^2}$$

Flow rate:

$$Q = (f_2 - f_1)(g_2 - g_1)$$

$$g_1 = z_1 = 0, \quad g_2 = z_2 = 1$$

$$Q = \psi_2 - \psi_1$$

Table D-3 Streamfunction for Axisymmetric Flow: Cylindrical Coordinates

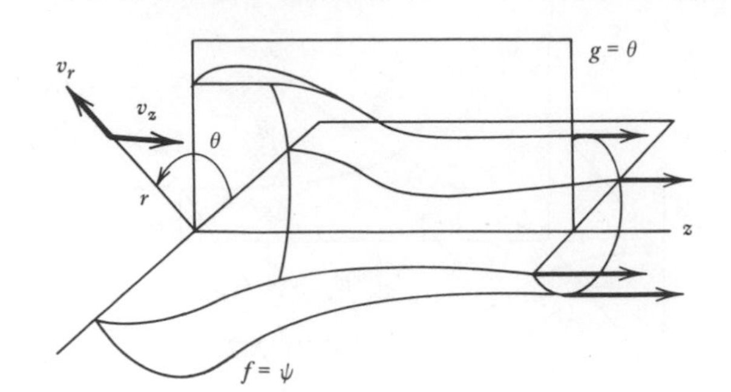

Coordinates: r, θ, z

Velocities: $v_r(r, z), \quad v_\theta = 0, \quad v_z(r, z)$

Streamsurfaces: $f = \psi(r, z), \quad g = \theta, \quad \nabla g = (0, r^{-1}, 0)$

Vector potential: $B_r = 0, \quad B_\theta = \dfrac{\psi}{r}, \quad B_z = 0$

$\mathbf{v} = \nabla f \times \nabla g = \nabla \times \mathbf{B}$: $\quad v_r = -\dfrac{1}{r}\dfrac{\partial \psi}{\partial z}, \quad v_z = \dfrac{1}{r}\dfrac{\partial \psi}{\partial r}$

$-\boldsymbol{\omega} = \nabla^2 \mathbf{B}$: $\quad -\omega_\theta = \dfrac{\partial}{\partial r}\left(\dfrac{1}{r}\dfrac{\partial \psi}{\partial r}\right) + \dfrac{\partial^2}{\partial z^2}\left(\dfrac{\psi}{r}\right)$

Vorticity equation: $\quad \dfrac{\partial \omega_\theta}{\partial t} + v_r \dfrac{\partial \omega_\theta}{\partial r} + v_z \dfrac{\partial \omega_\theta}{\partial z} = \dfrac{\omega_\theta v_r}{r} + \nu \dfrac{\partial}{\partial r}\left(\dfrac{1}{r}\dfrac{\partial}{\partial r}(r\omega_\theta)\right) + \nu \dfrac{\partial^2 \omega_\theta}{\partial z^2}$

Flow rate:

$$Q = (f_2 - f_1)(g_2 - g_1)$$

$$g_1 = \theta_1 = 0, \quad g_2 = \theta_2 = 2\pi$$

$$Q = 2\pi(\psi_2 - \psi_1)$$

$$= 2\pi\psi_2 \quad \text{if} \quad \psi_1 = 0 \text{ is } z \text{ axis}$$

Table D-4 Streamfunction for Axisymmetric Flow: Spherical Coordinates

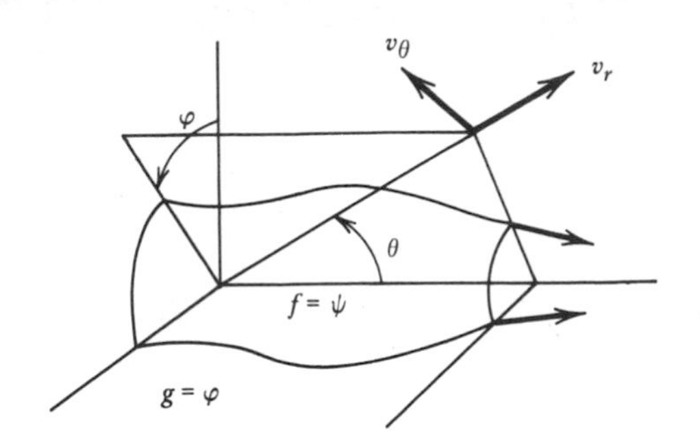

Coordinates: r, θ, φ

Velocities: $v_r(r,\theta), \quad v_\theta(r,\theta), \quad v_\varphi = 0$

Streamsurface: $f = \psi(r,\theta), \quad g = \varphi, \quad \nabla g = \left(0, 0, \dfrac{1}{r\sin\theta}\right)$

Vector potential: $B_r = 0, \quad B_\theta = 0, \quad B_\varphi = \dfrac{\psi}{r\sin\theta}$

$\mathbf{v} = \nabla f \times \nabla g = \nabla \times \mathbf{B}$: $\quad v_r = \dfrac{1}{r^2\sin\theta}\dfrac{\partial\psi}{\partial\theta}, \quad v_\theta = \dfrac{-1}{r\sin\theta}\dfrac{\partial\psi}{\partial r}$

$-\boldsymbol{\omega} = \nabla^2\mathbf{B}$: $\quad -\omega_\varphi = \dfrac{1}{r^2}\dfrac{\partial}{\partial r}\left[r^2\dfrac{\partial}{\partial r}\left(\dfrac{\psi}{r\sin\theta}\right)\right] + \dfrac{1}{r^2}\dfrac{\partial}{\partial\theta}\left[\dfrac{1}{\sin\theta}\dfrac{\partial}{\partial\theta}\left(\dfrac{\psi}{r}\right)\right]$

Vorticity equation:

$$\frac{\partial\omega_\varphi}{\partial t} + v_r\frac{\partial\omega_\varphi}{\partial r} + \frac{v_\theta}{r}\frac{\partial\omega_\varphi}{\partial\theta} = \frac{\omega_\varphi}{r}(v_r + v_\theta\cot\theta)$$

$$+\frac{\nu}{r^2}\frac{\partial}{\partial r}\left(r^2\frac{\partial\omega_\varphi}{\partial r}\right) + \frac{\nu}{r^2}\frac{\partial}{\partial\theta}\left[\frac{1}{\sin\theta}\frac{\partial}{\partial\theta}(\sin\theta\,\omega_\varphi)\right]$$

Flow rate: $Q = (f_2 - f_1)(g_2 - g_1)$

$$g_1 = \varphi_1 = 0, \qquad g_2 = \varphi_2 = 2\pi$$

$$Q = 2\pi(\psi_2 - \psi_1)$$

$$Q = 2\pi\psi_2 \qquad \text{if} \quad \psi_1 = 0 \text{ on } \theta = 0 \text{ axis}$$

E Computer Code for Runge–Kutta Integration

SUBROUTINE RK34: A NUMERICAL SOLUTION OF ORDINARY DIFFERENTIAL EQUATIONS BY W. L. OBERKAMPF

This program in FORTRAN IV solves up to 50 simultaneous, first-order, ordinary differential equations. The system of equations must be an initial value problem of the form:

$$y'_1 = f_1(x, y_1, y_2, \ldots, y_n)$$

$$y'_2 = f_2(x, y_1, y_2, \ldots, y_n)$$

$$\vdots$$

$$y'_n = f_n(x, y_1, y_2, \ldots, y_n)$$

where y_i, $i = 1, 2, \ldots, n$, are the dependent variables, x is the independent variable, and $y' \equiv dy/dx$.

The method used is formula 2 from Fehlberg (1969). The subroutine advances the solution one step each time it is called by the statement:

CALL RK34 (N, X, Y, H, EMAX, HMIN)

Subroutine arguments are:

N number of first-order, differential equations. N must be ≤ 50. (integer)

X independent variable. When the subroutine is called the first time, the initial value should be stored in X. (real)

Y array of dependent variables. When the subroutine is called the first time, the initial conditions should be stored in Y. The dimension of Y in the calling program must be 50. (singly dimensioned, real)

H integration step size. When the subroutine is called the first time, a reasonable estimate of H should be made. The subroutine will adjust the step size to meet the error criterion. If it is desired to integrate with a constant step size, the subroutine may be rigged as follows. Each time just before the subroutine is called, set H equal to the desired step size. This value of H must be small enough to produce truncation error less than EMAX. (real)

EMAX per-step, relative, truncation error criterion. (real)

HMIN minimum integration step size with which the subroutine will advance the solution. (real)

Subroutine RK34 requires the user to supply one additional subroutine subprogram. The subroutine subprogram calculates the derivatives of y_i, that is, $f_i(x, y_1, y_2, \ldots, y_n)$. The subroutine must have three arguments and have the name FUNCT, that is,

SUBROUTINE FUNCT (X, Y, F)

X independent variable. (real)

Y array of dependent variables. The dimension of Y in FUNCT must be 50. (singly dimensioned, real)

F array of derivatives of Y. The dimensions of F in FUNCT must be 50. (singly dimensioned, real)

```
      SUBROUTINE RK34(N,X,Y,H,E,HMIN)
C     PROGRAMMED BY W. L. OBERKAMPF
      DIMENSION Y(50),YTEMP(50),YHAT(50),F(50),
     CA(50),B(50),C(50),D(50)
      DATA C0,C2,C3,CH0,CH2,CH3,CH4,A1,A2,A3,B10,B20,B21,
     CB30,B31,B32,B40,B42,B43/.1612244898,.5998345284,
     C.2389409818,.1557823129,.6205184777,.1681436539,.0555555556,
     C.2857142857,.4666666667,.9210526316,.2857142857,.0855555556,
     C.3811111111,.5574792244,-1.406455023,1.770028430,
     C.1612244898,.5998345284,.2389409818/
    1 XTEMP=X
      DO 2 I=1,N
    2 YTEMP(I)=Y(I)
      CALL FUNCT(XTEMP,YTEMP,F)
      XTEMP=X+A1*H
      DO 3 I=1,N
      A(I)=H*F(I)
    3 YTEMP(I)=Y(I)+B10*A(I)
      CALL FUNCT(XTEMP,YTEMP,F)
      XTEMP=X+A2*H
      DO 4 I=1,N
      B(I)=H*F(I)
    4 YTEMP(I)=Y(I)+B20*A(I)+B21*B(I)
      CALL FUNCT(XTEMP,YTEMP,F)
      XTEMP=X+A3*H
      DO 5 I=1,N
      C(I)=H*F(I)
    5 YTEMP(I)=Y(I)+B30*A(I)+B31*B(I)+B32*C(I)
      CALL FUNCT(XTEMP,YTEMP,F)
      XTEMP=X+H
      DO 6 I=1,N
      D(I)=H*F(I)
    6 YTEMP(I)=Y(I)+B40*A(I)+B42*C(I)+B43*D(I)
      CALL FUNCT(XTEMP,YTEMP,F)
      DO 7 I=1,N
      YTEMP(I)=Y(I)+C0*A(I)+C2*C(I)+C3*D(I)
      YHAT(I)=Y(I)+CH0*A(I)+CH2*C(I)+CH3*D(I)+CH4*H*F(I)
```

```
  7 A(I)=ABS(YHAT(I)-YTEMP(I))
    DO 8 I=1,N
    C(I)=ABS(YHAT(I))
    IF(C(I).LE.A(I)*1.E-3) C(I)=1.
    B(I)=A(I)/C(I)
    IF(B(I).GT.E) GO TO 11
  8 CONTINUE
  9 X=X+H
    ETEMP=E/16.
    IFLAG=0
    DO 10 I=1,N
    Y(I)=YHAT(I)
 10 IF(B(I).GT.ETEMP) IFLAG=1
    IF(IFLAG.EQ.1) RETURN
    H=H+H
    RETURN
 11 IF(H.GT.HMIN) GO TO 12
    PRINT 100,E,HMIN
    GO TO 9
 12 H=H/2.
    GO TO 1
100 FORMAT (/10X51HTHE PER-STEP RELATIVE TRUNCATION ERROR CRITERION OF
   CE8.1,51H COULD NOT BE SATISFIED WITH A MINIMUM STEP SIZE OF,
   CE8.1//)
    END
```

F Computer Code for Entrance Flow into a Cascade

```
C               COMPUTER CODE FOR ENTRANCE FLOW INTO A CASCADE
C
      DIMENSION U(202,21),V(202,21),VORT(202,21),PSI(202,21)
      DIMENSION VORTN1(202,21)
      DIMENSION CP(2,102),SUM(102),DYVOR(102),AVEDY(102)
      DIMENSION CPP(2,102)
    2 FORMAT(I3,10X,4F15.6)
    1 FORMAT(11F10.4)
C
C      SET INPUT DATA
C
      II=50 $ JJ=10 $ AL=20.    $ RE=10. $ EPSI=1.E-04 $ EVORT=1.E-04
C
C      COMPUTE DX , BETA , AND F
C
      AIIP=FLOAT(II)+.5
      DX=  AL/AIIP
      BETA=AL*FLOAT(JJ)/AIIP
      BETA2=BETA**2
      PI=3.14159286
      BE2P1=BETA2+1.
      DT=.5/((1.+BETA)/DX+BE2P1*2./RE/DX**2)
      E=(COS(PI/FLOAT(2*II+1))+BETA2*COS(PI/FLOAT(JJ)))/BE2P1
      ETA=E**2
      F=2.*(1-SQRT(1.-ETA))/ETA
C
C       SET INITIAL CONDITIONS AND BOUNDARY VALUES
C
      III=2*II+1
      DO 20 I=1,III
      DO 10 J=1,JJ+1
      U(I,J)=1.
      V(I,J)=0.
      VORT(I,J)=0.
   10 PSI(I,J)=(FLOAT(J)-1.)/FLOAT(JJ)
   20 CONTINUE
      DO 25 I=II+2,III
      U(I,JJ+1)=0.
      V(I,JJ+1)=0.
      PSI(I,JJ+1)=1.
   25 VORT(I,JJ+1)=3.
      DO 27 J=1,JJ+1
      Y=(FLOAT(J)-1.)/FLOAT(JJ)
      V(III+1,J)=0.
      U(III+1,J)=1.5*(1.-Y**2)
      PSI(III+1,J)=1.5*Y-.5*Y**3
   27 VORT(III+1,J)=3.*Y
C
C       SOLVE FOR VORTICITY AT INTERIOR POINTS
C
   30 ISW2=0
      DO 38 I=2,III
      DO 37 J=2,JJ
      DELSQ=VORT(I+1,J)+VORT(I-1,J)+BETA2*VORT(I,J+1)+BETA2*VORT(I,J-
     11)-2.*BE2P1*VORT(I,J)
```

```
      IF(U(I,J).GT.0.) GO TO 31
      CONU=U(I+1,J)*VORT(I+1,J)-U(I,J)*VORT(I,J)
      GO TO 32
   31 CONU=U(I,J)*VORT(I,J)-U(I-1,J)*VORT(I-1,J)
   32 IF(V(I,J).GT.0) GO TO 35
      CONV=V(I,J+1)*VORT(I,J+1)-V(I,J)*VORT(I,J)
      GO TO 36
   35 CONV=V(I,J)*VORT(I,J)-V(I,J-1)*VORT(I,J-1)
   36 DVORT=DT/DX*(-(CONU+BETA*CONV)+2./RE/DX*DELSQ)
      VORTN1(I,J)=VORT(I,J)+DVORT
   37 IF(DVORT.GT.EVORT) ISW2=1
   38 CONTINUE
C
C       UPDATE VORTICITY MATRIX
C
      DO 40 I=2,III
      DO 39 J=2,JJ
   39 VORT(I,J)=VORTN1(I,J)
   40 CONTINUE
C
C       SOLVE FOR STREAM FUNCTION
C
   41 ISW1=0
      DO 50 I=2,III
      DO 49 J=2,JJ
      DSTR=PSI(I+1,J)+PSI(I-1,J)+BETA2*PSI(I,J+1)+BETA2*PSI(I,J-1)-2.*
     1BE2P1*PSI(I,J)+VORT(I,J)*DX**2
      PSI(I,J)=PSI(I,J)+F/2./BE2P1*DSTR
   49 IF(DSTR.GT.EPSI) ISW1=1
   50 CONTINUE
      IF(ISW1.GT.0) GO TO 41
C
C      CALCULATE U AND V VELOCITIES AT INTERIOR POINTS
C
      DO 60 I=2,III
      DO 59,J=2,JJ
      U(I,J)=(PSI(I,J+1)-PSI(I,J-1))*BETA/2./DX
   59 V(I,J)=(PSI(I-1,J)-PSI(I+1,J))/2./DX
   60 CONTINUE
C
C      CALCULATE CENTERLINE & STAG STREAMLINE VALUES OF U
C
      DO 75 I=2,III
   75 U(I,1)=PSI(I,2)*BETA/DX
      DO 80 I=2,II+1
   80 U(I,JJ+1)=(1.-PSI(I,JJ))*BETA/DX
C
C      CALCULATE VORT ON THE WALLS
C
      DO 90 I=II+2,III
      VORTEMP=VORT(I,JJ+1)
      VORT(I,JJ+1)=(1.-PSI(I,JJ))*2*BETA2/DX**2
      DVORT=VORT(I,JJ+1)-VORTEMP
   90 IF(DVORT.GT.EVORT) ISW2=1
      IF(ISW2.GT.0) GO TO 30

C
C      PRINT OUTPUT
C
      WRITE(6,1) ((PSI(I,J),J=1,11,1),I=1,103,1)
      WRITE(6,1) ((U(I,J),J=1,11,1),I=1,103,1)
      WRITE(6,1) ((V(I,J),J=1,11,1),I=1,103,1)
      WRITE(6,1) ((VORT(I,J),J=1,11,1),I=1,103,1)
      END
```

G Computer Code for Boundary-Layer Analysis

```
C
C*********** PROGRAM TO SOLVE LAMINAR BOUNDARY LAYER PROBLEMS. *********
C  ADAPTED FROM A PROGRAM WRITTEN BY H.A.DWYER,E.D.DOSS,& A.L.GOLDMAN
C                UNIVERSITY OF CALIFORNIA-DAVIS
C
      DIMENSION A(81),B(81),C(81),D(81),E(81),F1(81),F(122),FP(81,2),
     2V(81,2),VBAR(81),UE(500),ALENG(500),BETA(500),BETAM(500)
      REAL NU
      INTEGER SEP,DELPRNT
C
C *********************** FLOW VARIABLES ******************************
C *********** FP(N,M)     VELOCITY RATIO U/UE  *************************
C **************** N = POSITION ACROSS BOUNDARY LAYER  ****************
C ******M = 1, UPSTREAM VALUES.  M = 2, DOWNSTREAM VALUES. *************
C *****   MM IS THE NUMBER OF STREAMWISE POSITIONS TO BE CALCULATED ***
C ** DN = STEP SIZE IN Y DIRECTION, DX = STEP SIZE IN X DIRECTION ******
C
C       READ IN THE INITIAL PROFILE
C
      READ(5,99) BETAM(1)
      READ(5,100)(F(N),FP(N,1),N = 1,81)
      DN = .1
C
C       READ IN THE EXTERNAL VELOCITY UE(M) AT THE X-POSITIONS ALENG(M)

C         COMPUTE BETAM(M)=BETA AT M
C
      READ(3,101) MM,DELPRNT
      READ(3,100) (UE(M),ALENG(M),M=1,MM)
      MPRINT=DELPRNT
      BETAM(2)=ALOG(UE(3)/UE(2))/ALOG(ALENG(3)/ALENG(2))
      DO 20 M=3,MM-1
   20 BETAM(M)=ALOG(UE(M+1)/UE(M-1))/ALOG(ALENG(M+1)/ALENG(M-1))
C
C     EXTRAPOLATE TO GET BETAM(MM)
C
      DX1=ALENG(MM-1)-ALENG(MM-2)
      DX2=ALENG(MM)-ALENG(MM-1)
      B1=BETAM(MM-1)-BETAM(MM-2)
      BETAM(MM)=BETAM(MM-1)+B1*DX2/DX1
C
C          COMPUTE BETA(M)=BETA AT M+1/2
C
      DO 25 M=1,MM-1
   25 BETA(M)=BETAM(M)
C
C  CALCULATION OF VBAR, THE FIRST STEP IN THE B. L. CALCULATIONS.
C
      DO 3000 N=1,81
      V(N,1)=-BETAM(1)*F(N)-0.5*(BETAM(1)-1.)*(FLOAT(N-1)*DN*FP(N,1)
     1-F(N))
      VBAR(N)=V(N,1)+0.5*(BETAM(1)-1.)*FLOAT(N-1)*DN*FP(N,1)
 3000 CONTINUE
      I=0
      WRITE(6,350)I,ALENG(1),UE(1),BETAM(1)
```

```
      WRITE(6,550)(N,FP(N,1),V(N,1),N=1,81)
C
C           BEGIN BOUNDARY LAYER CALCULATIONS.
C
      DO 8000 M = 1,MM-1
      X=ALENG(M+1)
      DXI=X-ALENG(M)
      XI=X-0.5*DXI
C
C     CALCULATION OF A,B,C,D,E,F,AND NEW VALUES OF VELOCITY
C
      DO 3500 N = 2,80
      A(N) = 1./(2.*DN*DN) - VBAR(N)/(4.*DN)
      B(N) = -(XI*FP(N,1) /DXI + BETA(M)*FP(N,1) + 1./(DN**2))
      C(N) = 1./(2.*DN*DN) + VBAR(N)/(4.*DN)
      D(N) = -(FP(N+1,1) - 2.*FP(N,1) + FP(N-1,1))/(2.*DN*DN) + VBAR(N)*
     1(FP(N+1,1) - FP(N-1,1))/(4.*DN) - BETA(M) -XI*(FP(N,1)**2)/DXI
 3500 CONTINUE
      E(1) = 0.0
      F1(1) = 0.0
      DO 4000 N = 2,80
      E(N) = -A(N)/( B(N) + C(N)*E(N - 1) )
      F1(N) = (D(N) - C(N)*F1(N - 1) )/( B(N) + C(N)*E(N - 1) )
 4000 CONTINUE
      FP(81,2) = 1.0
      DO 4500 N = 1,80
      NDWN = 82 - N
      FP(NDWN-1,2) = E(NDWN-1)*FP(NDWN,2) + F1(NDWN-1)
 4500 CONTINUE
      SEP = 0
      IF( FP(2,2) .LT. 0.0 ) SEP = 1
C
C      CONTINUITY EQUATION........NEW VALUES OF VBAR.
C
      V(1,1) = 0.0
      V(1,2) = 0.0
      DO 5000 N = 2,81
      AC =- BETA(M)/4.-XI/(2.*DXI)-(FLOAT(N-1)-.5)*(BETA(M)-1.)/4.
      BC =-BETA(M)/4.-XI/(2.*DXI) +(FLOAT(N-1)-.5)*(BETA(M)-1.)/4.
      CC =-BETA(M)/4.+XI/(2.*DXI) -(FLOAT(N-1)-.5)*(BETA(M)-1.)/4.
      DC =-BETA(M)/4.+XI/(2.*DXI) +(FLOAT(N-1)-.5)*(BETA(M)-1.)/4.
      V(N,2) = V(N-1,2) + V(N-1,1) - V(N,1) + (AC*FP(N,2) + BC*FP(N-1,2)
     1 + CC*FP(N,1) + DC*FP(N-1,1))*2.*DN
 5000 CONTINUE
      DO 5500 N = 2,81
      VBAR(N) = V(N,2)+FP(N,2)*FLOAT(N-1)*DN*(BETAM(M+1)-1.)/2.
C
C    SEND M=2 RESULTS BACK TO M=1
C
      FP(N,1) = FP(N,2)
      V(N,1) = V(N,2)
 5500 CONTINUE
C
C      THE END OF THE PROGRAM.  OUTPUT STATEMENTS.
C
```

```
      WRITE(6,350) M+1,ALENG(M+1),UE(M+1),BETAM(M+1)
      IF (SEP.EQ.1) GO TO 7500
      IF(M.LT.MPRINT) GO TO 8000
      MPRINT=MPRINT+DELPRNT
 7500 WRITE(6,550)(N,FP(N,1),V(N,1),N=1,81)
      IF (SEP.EQ.1) GO TO 11
 8000 CONTINUE
      WRITE(6,700)
      GO TO 9000
   11 WRITE(6,701)
   99 FORMAT(F20.10)
  100 FORMAT(2F12.9)
  101 FORMAT(2I10)
  350 FORMAT( 5X,'M=',I4,5X,'X/L=',F6.4,5X,'UE/UO=',F6.4,5X,'BETAM=',F
     18.4,5X)
  550 FORMAT(4(8X,I3,2X,F9.5,2X,F9.5))
  700 FORMAT(//41H *** THE END ***     THATS ALL SHE WROTE.)
  701 FORMAT(//'FLOW SEPARATED SO CALCULATION WAS STOPPED')
 9000 CONTINUE
      END
```

Supplemental Reading List

Chapter 1	Feynman, R. P., R. B. Leighton and M. Sands (1963)
	Van Dyke, M. D. (1982)
Chapter 2	Callen, H. B. (1960)
Chapter 3	Prager, W. (1961)
	Bird, R. B., W. E. Stewart, and E. W. Lightfoot (1960) Appendix A
	Jeffreys, H. (1963)
	Aris, R. (1962)
Chapter 4	Prager, W. (1961)
	Goldstein, S. (1960)
Chapter 5	Goldstein, S. (1960)
	Batchelor, G. K. (1967)
	Serrin, J. (1959)
	Bird, R. B., W. E. Stewart, and E. W. Lightfoot (1960)
Chapter 6	Those listed under Chapter 5 plus:
	Bird, R. B., R. C. Armstrong, and O. Hassager (1977)
Chapter 7	None
Chapter 8	Sedov, L. I. (1959)
	Bridgeman, P. W. (1922)
	Langhaar, H. L. (1951)
	Zierip, J. (1971)
Chapter 9	None
Chapter 10	Lagerstrom, P. A. (1964)
Chapter 11	Schlichting, H. (1950)
	Rosenhead, L. (1963)
	White, F. M. (1974)
Chapter 12	Batchelor, G. K. (1967)
	Yih, C-S (1969)
Chapter 13	Batchelor, G. K. (1967)
	Rosenhead, L. (1963)
Chapter 14	Batchelor, G. K. (1967)
	Aris, R. (1962)
Chapter 15	Roache, P. J. (1972)
	Tritton, D. J. (1977)
Chapter 16	Cole, J. D. and J. Kevorkian (1981)
	Bender, C. M. and S. A. Orszag (1978)
	Nayfeh, A. H. (1973)
	Van Dyke, M. D. (1964)

Chapter 17 Rosenhead, L. (1963)
Schlichting, H. (1950)
Chapter 18 Churchill, R. V., et al (1974)
Batchelor, G. K. (1967)
Currie, I. (1974)
Lamb, H. (1932)
Milne-Thomson, L. M. (1960)
Prandtl, L. and O. G. Tietjiens (1934)
Chapter 19 Karamcheti, K. (1966)
Schlichting, H. and E. Truckenbrodt (1979)
Chapter 20 Rosenhead, L. (1963)
Schlichting, H. (1978)
White, F. M. (1974)
Chapter 21 Moore, F. K. (1964)
Rosenhead, L. (1963)
Happel, J. and H. Brenner (1965)
Chapter 22 Betchov, R. and W. O. Criminale (1967)
Drazin, P. G. and W. H. Reid (1981)
Chapter 23 Hinze, J. O. (1975)
Tennekes, H. and J. L. Lumley (1972)
Townsend, A. A. (1976)

References

Abramowitz, M. and I. A. Stegun (1964). *Handbook of Mathematical Function*, U.S. Government Printing Office, Washington, D.C.

Ames, W. F. (1977). *Numerical Methods in Partial Differential Equations*, Academic, New York.

Andereck, C. D., R. Dickman, and H. L. Swinney (1983). *Phys. Fluids*, **26**, p. 1395.

Aris, R. (1962). *Vector Tensors, and the Basic Equations of Fluid Mechanics*, Prentice-Hall, Englewood Cliffs, N.J.

Arndt, R. E. A. (1981). In *Annual Review of Fluid Mechanics*, Vol. 13, Annual Reviews Inc., Palo Alto, Calif.

Batchelor, G. K. (1967). *An Introduction to Fluid Dynamics*, Cambridge University Press, London.

Bender, C. M. and S. A. Orszag (1978). *Advanced Mathematical Methods for Scientists and Engineers*, McGraw-Hill, New York.

Berger, S. A. (1971). *Laminar Wakes*, Elsevier, New York.

Berger, S. A. (1983). In *Annual Review of Fluid Mechanics*, Vol. 15, Annual Reviews Inc., Palo Alto, Calif.

Berker, R. (1963). In *Handbuch der Physik*, Vol. VIII, pt. 2, Springer, Berlin.

Bertin, J. J. and M. L. Smith (1979). *Aerodynamics*, Prentice-Hall, Englewood Cliffs, N.J.

Betchov, R. and A. B. Szewczyk (1963). *Phys. Fluids*, **6**, p. 1391.

Betchov, R. and W. O. Criminale (1967). *Stability of Parallel Flows*, Academic, New York.

Bird, R. B., R. C. Armstrong, and O. Hassager (1977). *Dynamics of Polymeric Liquids*, Vol. 1, Wiley, New York.

Bird, R. B., W. E. Stewart, and E. N. Lightfoot (1960). *Transport Phenomena*, Wiley, New York.

Blasius, H. (1908). *Z. Angew. Math. Phys.*, **56**, p. 1 (translation, NACA TM 1256).

Blottner, F. G. (1970). *AIAA J.*, **8**, p. 193.

Blottner, F. G. (1975). NATO-AGARD Lecture Series No. 73, Chapter 3.

Blumen, G. W. and J. D. Cole (1969). *J. Math. Mech.*, **18**, p. 1025.

Bödewadt, U. T. (1940). *Z. Angew. Math. Mech.*, **20**, p. 241.

Boussinesq, J. (1877). *Mem. Pres. Acad. Sci., Paris*, **23**, p. 46.

Brand, L. (1957). *Arch. Rat. Mech. Anal.*, **1**, p. 35.

Bridgman, P. W. (1922). *Dimensional Analysis*, Yale University Press, New Haven, Conn.

Brodkey, R. S., S. G. Nychas, J. L. Taraba, and J. M. Wallace (1973). *Phys. Fluids*, **16**, p. 2010.

Brown, G. L. and A. Roshko (1974). *J. Fluid Mech.* **64**, p. 775.

Buckingham, E. (1914). *Phys. Rev., Ser. 2*, **4**, p. 345.

Burkhalter, J. E. and E. L. Koschmieder (1973). *J. Fluid Mech.*, **58**, p. 547.

Callen, H. B. (1960). *Thermodynamics*, Wiley, New York.

Cantwell, B. J. (1981) In *Annual Review of Fluid Mechanics*, Vol. 13, Annual Reviews Inc., Palo Alto, Calif.

Carrier, G. F., M. Krook, and C. E. Pearson (1966). *Functions of a Complex Variable; Theory and Technique*, McGraw-Hill, New York.

Carslaw, H. S. and J. C. Jaeger (1947). *Conduction of Heat in Solids*, Clarendon Press, Oxford.

Carlson, D. R., S. E. Widnall, and M. F. Peeters (1982). *J. Fluid Mech.*, **121**, p. 487.

Catherall, D. and K. W. Mangler (1966). *J. Fluid Mech.*, **26**, p. 163.

Cebeci, T. and P. Bradshaw (1977). *Momentum Transfer in Boundary Layers*, McGraw-Hill, New York.

Churchill, R. V., J. W. Brown, and R. F. Verhey (1974). *Complex Variables and Applications*, McGraw-Hill, New York.

Champagne, F. H. (1978). *J. Fluid Mech.*, **86**, p. 67.

Cimbala, J., H. Nagib, and A. Roshko (1981). *Bulletin Am. Phy. Soc.* **26**, p. 1257.

Clauser, F. H. (1954). *J. Aero. Sci.*, **21**, p. 91.

Clauser, F. H. (1956). *Adv. Appl. Mech.*, **4**, p. 1.

Clift, R., J. R. Grace, and M. E. Weber (1978). Academic, New York.

Cole, J. D. and J. Kevorkian (1981). *Perturbation Methods in Applied Mathematics*, Springer, New York.

Coles, D. E. (1956). *J. Fluid Mech.*, **1**, p. 191.

Coles, D. E. (1965). *J. Fluid Mech.*, **21**, p. 385.

Crow, S. D. (1970). *AIAA J.*, **8**, p. 2172.

Currie, I. G. (1974). *Fundamental Mechanisms of Fluids*, McGraw-Hill, New York.

Davey, A. (1961). *J. Fluid Mech.*, **10**, p. 593.

Davis, R. T. (1972). *J. Fluid Mech.*, **51**, p. 417.

Davis, A. M. J. and M. E. O'Neill (1977). *J. Fluid Mech.*, **81**, p. 551.

Dean, W. R. (1950). *Proc. Camb. Phil. Soc.*, **46**, p. 293.

Dean, W. R. and P. E. Montagnon (1949). *Proc. Cambridge Phil. Soc.*, **45**, p. 389.

Delany, N. K. and N. E. Sorenson (1953). NACA Tech. Note 3038.

Dimotakis, P. E., R. C. Miake-Lye, and D. A. Papantoniou (1983). *Phy. Fluids*, **26**, p. 3185.

Donnelley, R. J. and W. Glaberson (1966). *Proc. Roy. Soc. London Ser. A*, **290**, p. 547.

Donohue, G. L., W. G. Tiederman, and M. M. Reischman, *J. Fluid Mech.*, **56**, p. 559.

Drazin, P. G. and W. H. Reid (1981). *Hydrodynamic Stability*, Cambridge University Press, Cambridge.

Dwyer, H. A. (1982). In *Annual Review of Fluid Mechanics*, Annual Reviews Inc., Palo Alto, Calif.

Emmons, H. W. (1949). *Proceedings of the First Symposium on Applied Mathematics*, American Mathematics Society, Providence.

Faler, J. H. and S. Leibovich (1978). *J. Fluid Mech.*, **86**, p. 313.

Falkner, V. M. and S. W. Skan (1931). *Phil. Mag.*, **12**, p. 865.

Fehlberg, E. (1969). NASA TR R-315.

Feynman, R. P., R. B. Leighton, and M. Sands (1963). *Lectures on Physics*, Addison-Wesley, Reading, Mass.

Finn, R. K. (1953). *J. Appl. Phys.*, **24**, p. 771.

Fornberg, B. (1980). *J. Fluid Mech.*, **98**, p. 819.

Fourier, J. B. J. (1822). *Théorie Analytic de Chaleur*, Paris, translation published by Dover.

Glauret, M. B. (1957). *Proc. Roy. Soc. London Ser. A* **242**, p. 108.

Goenka, L. N. (1982). M. S. Thesis, University of Texas.

Goldstein, S. (1948). *Q. J. Mech. Appl. Math.*, **1**, p. 43.

Goldstein, S. (1960). *Lectures in Fluid Mechanics*, Interscience, New York.

Goldstein, S. (ed.) (1965). *Modern Developments in Fluid Dynamics*, Dover, New York.

Gradshteyn, I. S. and I. W. Ryzhik (1965). *Tables of Integrals, Series and Products*, Academic Press, New York.

Guggenheim, E. A. (1949). *Thermodynamics*, North-Holland, Amsterdam; Interscience, New York.

Hadamard, J. S. (1911). *Comp. Rend. Acad. Sci.*, **152**, p. 1735.

Hagen, G. (1839). *Poggendorff's Ann. Phys. Chem.* **46**(2), p. 423.

Happel, J. and H. Brenner (1965). *Low Reynolds Number Hydrodynamics*, Prentice-Hall, New York.

Hasimoto, H. and O. Sano (1980). In *Annual Reviews of Fluid Mechanics*, Vol. 12, Annual Reviews Inc., Palo Alto, Calif.

Helmholtz, H. (1867). *Phil Mag.*, **33**, (4), p. 485 (translated from *Crelle's J.* (1858), **55**).

Hiemenz, K. (1911). *Dingler's Polytech. J.*, **326**, p. 311.

Hill, M. J. M. (1894). *Phil. Trans. Roy. Soc. London Ser. A*, **185**, p. 213.

Hill, R. and G. Power (1956). *Q. J. Mech. Appl. Math.*, **9**, p. 313.

Hinze, J. O. (1975). *Turbulence*, McGraw-Hill, New York.

Hoerner, S. F. (1964). *Fluid-Dynamic Drag*, Hoerner Fluid Dynamics, Brick Town, N.J.

Homann, F. (1936). *Forsch. Arb. Ing.-Wes.*, **7**, p. 1.

Howarth, L. (1951). *Phil Mag.*, **42**(7), p. 1433.

Huntley, H. E. (1953). *Dimensional Analysis*, McDonald, London.

Illingworth, C. R. (1950). *Proc. Camb. Phil. Soc.*, **46**, p. 469.

Jeffreys, H. (1963). *Cartesian Tensors*, Dover, New York.

Jordinson, R. (1970). *J. Fluid Mech.*, **43**, p. 801.

Kaplun, S. (1957). *J. Math. Mech.*, **6**, p. 595.

Karamcheti, K. (1966). *Principles of Ideal-Fluid Aerodynamics*, Wiley, New York. (Now published by Krieger Publishing, Malabar, FL.)

Karman, T. Von (1921). *Z. Angew. Math. Mech.*, **1**, p. 233.

Keller, H. B. (1978). In *Annual Review of Fluid Mechanics*, Vol. 10, Annual Reviews Inc., Palo Alto, Calif.

Keller, H. B. and T. Cebeci (1971). *Lecture Notes in Physics*, Vol. 8, Proceedings of the Second International Conference on Numerical Methods in Fluid Dynamics, p. 92.

Knapp, R. T., J. W. Daily, and F. G. Hammitt (1970). *Cavitation*, McGraw-Hill, New York.

Kolmogorov, A. N. (1941a). *Comp. Rend. Acad. Sci. U.R.S.S.*, **30**, p. 301.

Kolmogorov, A. N. (1941b). *Comp. Rend. Acad. Sci. U.R.S.S.*, **32**, p. 16.

Koschmieder, E. L. (1979). *J. Fluid Mech.*, **93**, p. 515.

Kuethe, A. M. and C. Y. Chow (1976). *Foundations of Aerodynamics*, Wiley, New York.

Lagerstrom, P. A. (1964). In Vol. IV, *Princeton University Series of High Speed Aerodynamics and Jet Propulsion*, Princeton University Press (F. K. Moore, Ed.).

Lamb, H. (1932). *Hydrodynamics*, Cambridge University Press, London (Dover, 1945).

Lanchester, L. (1907). *Aerodynamics*, Constable, London.

Landau, L. D. and E. M. Lifshitz (1959). *Fluid Mechanics*, Pergamon Press, New York.

Langhaar, H. L. (1951). *Dimensional Analysis and the Theory of Models*, Wiley, New York.

Leibovich, S. (1978). In *Annual Review of Fluid Mechanics*, Vol. 10, Annual Reviews Inc., Palo Alto, Calif.

Lessen, M. (1949). NACA Rept. 979.

Lin, C. C. (1945). *Q. Appl. Math*, **3**, pp. 117, 218, 277.

Lighthill, M. J. (1958). *J. Fluid Mech.*, **4**, p. 383.

Lighthill, M. J. (1963). In *Laminar Boundary Layers*, L. Rosenhead (Ed.), Clarendon Press, Oxford.

Lock, R. C. (1951). *Q. J. Mech. Appl. Math.*, **4**, p. 42.

Loos, H. G. (1955). *J. Aeronaut. Sci.*, **22**, p. 35.

Macagno, E. O. (1971). *J. Franklin Institute*, **292**, p. 391.

Mager, A. and A. G. Hansen (1952). NACA TN. No. 2658.

Mangler, W. (1945). Ber. Aerodyn. Versuchsanst. Goett. Rept. 45-A-17.

Maslen, S. H. (1963). *AIAA J.*, **1**, p. 33.

Maxwell, J. C. (1871). *Proc. London Math. Soc.*, **3**, p. 224.

Michael, D. H. and M. E. O'Neill (1977). *J. Fluid Mech.*, **80**, p. 785.

Millikan, C. B. (1938). Proceedings of the 5th International Conference on Applied Mechanics, Cambridge, Mass., p. 386.

Milne-Thomson, L. M. (1960). *Theoretical Hydrodynamics*, Macmillan, New York.

Minnaert, M. (1933). *Phil. Mag.*, **16**, p. 235.

Moffatt, H. K. (1964). *J. Fluid Mech.*, **18**, p. 1.

Moore, F. K. (Ed.) (1964). *Theory of Laminar Flows*, Princeton University Press, Princeton, N. J.

Morel, T. (1978). *Soc. Automotive Eng.*, Paper 780267.

Mueller, T. H. (1980). AIAA 13th Fluid and Plasmadynamics Conference, Proceedings.

Nayfeh, A. H. (1973). *Perturbation Methods*, Wiley, New York.

Nikuradse, J. (1942). *Laminare Reibungsschichten an der längsangeströmten Platte*, Monograph, Zentrale f. wiss. Berichtswesen, Berlin.

Oberbeck, H. A. (1876). *Crelles' J.*, **81**, p. 62.

Obert, E. F. (1960). *Concepts of Thermodynamics*, McGraw-Hill, New York.

Orr, W. M. F. (1907). *Proc. R. Irish Acad.*, **27**, p. 9.

Oseen, C. W. (1910). *Ark. Mat. Astr. Fys.*, **6**, No. 29.

Panton, R. L. (1968). *J. Fluid Mech.*, **31**, p. 819.

Panton, R. L., A. L. Goldman, R. L. Lowery, and M. M. Reischman, (1980). *J. Fluid. Mech.*, **97**, p. 299.

Phillips, H. B. (1933). *Vector Analysis*, Wiley, New York.

Perry, A. E. and C. J. Abell (1975). *J. Fluid Mech.*, **67**, p. 257.

Plesset, M. S. and A. Prosperetti, (1977). In *Annual Reviews of Fluid Dynamics*, Vol. 9, p. 145, Annual Reviews Inc., Palo Alto, Calif.

Pohlhausen, K. (1921). *Z. Angew. Math. Mech.*, **1**, p. 252.

Poiseuille, J. L. M. (1840). *Comp. Rend.*, **11**, pp. 961 and 1041; **12**, p. 112.

Potter, M. C., and J. F. Foss, (1975). *Fluid Mechanics*, Ronald Press, New York.

Prager, W. (1961). *Introduction to Mechanics of Continua*, Ginn, Boston.

Prandtl, L. (1904). *Verhandlunger IIIrd.* International Mathematiker Kongresser, Heidelberg, p. 484 (translation as NASA Memo 452).

Prandtl, L. (1925). *Z. Angew. Math. Mech.*, **5**, p. 136.

Prandtl, L. and O. G. Tietjens (1934). *Applied Hydro- and Aeromechanics*, McGraw-Hill, New York. (Dover edition, 1957).

Proudman, I. and J. R. A. Pearson (1957). *J. Fluid Mech.*, **2**, p. 237

Rayleigh, Lord (1879). *Proc. Royal Soc.*, **29**, p. 71. (see also (1915)).

Rayleigh, Lord (1911). *Phil. Mag.*, **21**, p. 397.

Rayleigh, Lord (1915). *Nature*, **95**, p. 66.

Rayleigh, Lord (1917). *Phil. Mag.*, **34**(6), p. 94.

Riabouchinsky, D. (1911). *L'Aérophile*, p. 407.

Riabouchinsky, D. (1915). *Nature*, **95**, p. 2387.

Richardson, E. G. and E. Tyler (1929). *Proc. Phys. Soc. London*, **42**, p. 1.

Richardson, S. M. and A. R. H. Cornish (1977). *J. Fluid Mech.*, **82**, p. 309.

Roache, P. J. (1972). *Numerical Fluid Dynamics*, Hermosa Press, Albuquerque, N. Mex.

Roberts, F. A., P. E. Dimotakis, and A. Roshko (1984). *J. Fluid Mech.* (to appear).

Rogers, M. H. and G. N. Lance (1960). *J. Fluid Mech.*, **7**, p. 617.

Rosenhead, L. (Ed.) (1963). *Laminar Boundary Layers*, Oxford University Press, London.

Roshko, A. (1961). *J. Fluid Mech.*, **10**, p. 345.

Roshko, A. and K. Koenig (1984). *J. Fluid Mech.* (to appear).

Rouse, H. and S. Ince (1957). *History of Hydraulics*, Institute of Hydraulic Research, University of Iowa, Ames.

Rybczynski, W. (1911). *Bull. Int. Acad. Sci. Cracov*, **1911A**, p. 40.

Saric, W. S. and A. H. Nayfeh (1977). NATO-AGARD Conference Proc. No. 224, Laminar-Turbulent Transition.

Sarpkaya, T. (1971). *J. Fluid Mech.*, **45**, p. 545.

Schlichting, H. (1932). *Phys. Z.*, **33**, p. 327.

Schlichting, H. (1933). *Z. Angew. Math. Mech.*, **13**, p. 260.

Schlichting, H. (1934). *Z. Angew. Math. Mech.*, **14**, p. 368.

Schlichting, H. (1950). *Boundary Layer Theory*, McGraw-Hill, New York (7th ed. 1979).

Schlichting, H. and E. Truckenbrodt (1979). *Aerodynamics of the Airplane*, McGraw-Hill, New York.

Schubauer, G. B. and H. K. Skramstad (1947). *J. Aero. Sci.*, **14**, p. 69.

Scriven, L. E. (1960). *Chem. Eng. Sci.*, **12**, p. 98.

Sedov, L. I. (1959). *Similarity and Dimensional Methods in Mechanics*, M. Holt (Ed.), Academic Press, New York.

Serrin, J. (1959). In *Handbuch der Physik*, VIII/1, p. 252 (in English).

Sexl, T. (1930). *Z. Phys.*, **61**, p. 349.

Shen, S. F. (1954). *J. Aeronaut. Sci.*, **21**, p. 62.

Sherman, F. S. (1955). NACA Tech. Note 3298.

Sommerfeld, A. (1908). *Atti 4th Congr. Int. Math. Rome*, **3**, p. 116.

Sommerfeld, A. (1956). *Thermodynamics and Statistical Mechanics*, Academic Press, New York.

Sowerby, L. (1954). Aeronautical Research Council London, Rept. 16832.

Squire, H. B. (1933). *Proc. Roy. Soc. London Ser. A*, **142**, p. 621.

Squire, H. B. (1951). *Q. J. Mech.*, **4**, p. 321.

Squire, H. B. (1965). *Aeronaut. Q.*, **16**, p. 302.

Stewartson, K. (1954). *Proc. Camb. Phil. Soc.*, **50**, p. 454.

Stokes, G. G. (1845). *Trans. Camb. Phil. Soc.*, **8**, p. 287.

Stokes, G. G. (1851). *Trans. Camb. Phil. Soc.*, **9**, pt. II, p. 8.

Sudarshaw, E. and N. Mukundew (1974). *Classical Dynamics*, Wiley, New York.

Taneda, S. (1956). *J. Phys. Soc. Japan*, **11**, p. 302.

Taneda, S. (1979). *J. Phys. Soc. Japan*, **46**, p. 1935.

Tani, I. (1977). In *Annual Review of Fluid Mechanics*, Vol. 9, Annual Reviews Inc., Palo Alto, Calif.

Taylor, G. I. (1921). *Proc. Roy. Soc. London Ser. A*, **100**, p. 114.

Taylor, G. I. (1923). *Proc. Roy. Soc. London Ser. A*, **104**, p. 213.

Taylor, G. I. (1950). *Proc. Roy. Soc. London Ser. A*, **201**, p. 192.

Tennekes, H. and J. L. Lumley, (1972). *Introduction to Turbulence*, MIT Press, Cambridge.

Thom, A. (1933). *Proc. Roy. Soc. London Ser. A* **141**, p. 651.

Thoman, D. C. and A. B. Szewczyk (1968). *Phys. Fluids*, **12**, SII, p. 76.

Thorpe, J. F. (1962). *Am. J. Phys.* **30**, p. 637.

Thwaites, B. (Ed.) (1960). *Incompressible Aerodynamics*, Oxford University Press, London.

Tollmien, W. (1929). *Nachr. Ges. Wiss. Goett*, p. 21 (translation as NACA Tech. Memo 609).

Tollmien, W. (1931). *Handbuch der Experimental-ischen Physik*, Vol. IV, pt. 1, Leipzig.

Tombach, I. (1974). *Sixth Conference on Aerospace and Aeronautical Meterology*, American Meterological Society.

Townsend, A. A. (1976). *The Structure of Turbulent Shear Flow*, Cambridge University Press, London.

Tritton, D. J. (1959). *J. Fluid Mech.*, **6**, 547.

Tritton, D. J. (1977). *Physical Fluid Dynamics*, Van Nostrand Reinhold, New York.

Truesdell, C. A. (1954). *Kinematics of Vorticity*, University of Indiana Press, Bloomington.

Truesdell, C. A. (1968). *Essays in the History of Mechanics*, Springer, Berlin.

Van Driest, E. (1946). *J. Appl. Mech. ASME*, **13**, p. A-34.

Van Dyke, M. (1962). *J. Fluid Mech.*, **14**, p. 161.

Van Dyke, M. (1964). *Perturbation Methods in Fluid Mechanics*, Academic, New York. (Now published by Parabolic Press, Stanford, Calif.)

Van Dyke, M. (1970). *J. Fluid Mech.*, **44**, p. 813.

Van Dyke, M. (1982). *An Album of Fluid Motion*, Parabolic Press, Stanford, Calif.

Vashy, A. (1892). *Annales télégraphique*, **19**, p. 25.

Wang, K. C. (1971). *J. Fluid Mech.*, **48**, p. 397.

Wang, Y. L. and P. A. Longwell (1964). *AIChE J.*, **10**, p. 323.

Werlé, H. (1963). *la Houille Blanche*, **28**, p. 330.

Werlé, H. (1980). *Resh. Aérosp.*, **5**, p. 35.

White, F. M. (1974). *Viscous Fluid Flow*, McGraw-Hill, New York.

Whitehead, L. G. and G. S. Canetti (1950). *Phil. Mag.*, **41**(5), p. 988.

Wieselsberger, C. (1921). *Phys. Z.*, **22**, p. 321.

Williams, W. (1892). *Phil. Mag.*, **34**(5), p. 234.

Williams, J. C. (1977). In *Annual Review of Fluid Mechanics*, Vol. 9, Annual Reviews Inc., Palo Alto, Calif.

Winant, C. D. and F. K. Browand (1974). *J. Fluid Mech.*, **78**, p. 237.

Wygnanski, I. and F. H. Champagne (1973). *J. Fluid Mech.*, **59**, p. 281.

Yih, C. S. (1969). *Fluid Dynamics*, McGraw-Hill, New York. (Now published by West River Press, Ann Arbor, Mich.)

Zierip, J. (1971). Similarity Laws and Modeling, Gas dynamic Series, Vol. 2, Marcel Dekker.

Index